T0180724

Communications in Computer and Information Science 1591

More information about this series at https://link.springer.com/bookseries/7899

Valentina E. Balas · G. R. Sinha ·
Basant Agarwal · Tarun Kumar Sharma ·
Pankaj Dadheech · Mehul Mahrishi (Eds.)

Emerging Technologies in Computer Engineering

Cognitive Computing and Intelligent IoT

5th International Conference, ICETCE 2022
Jaipur, India, February 4–5, 2022
Revised Selected Papers

Springer

Editors
Valentina E. Balas
Aurel Vlaicu University of Arad
Arad, Romania

G. R. Sinha
Myanmar Institute of Information Technology
Mandalay, Myanmar

Basant Agarwal
Indian Institute of Information Technology Kota
Jaipur, India

Tarun Kumar Sharma
Shobhit University
Gangoh, India

Pankaj Dadheech
SKIT
Jaipur, India

Mehul Mahrishi
SKIT
Jaipur, India

ISSN 1865-0929 ISSN 1865-0937 (electronic)
Communications in Computer and Information Science
ISBN 978-3-031-07011-2 ISBN 978-3-031-07012-9 (eBook)
https://doi.org/10.1007/978-3-031-07012-9

This Springer imprint is published by the registered company Springer Nature Switzerland AG
The registered company address is: Gewerbestrasse 11, 6330 Cham, Switzerland

Preface

The 5th International Conference on Emerging Technologies in Computer Engineering: Cognitive Computing and Intelligent IoT (ICETCE 2022) was held virtually during February 4–5, 2022, at the Centre of Excellence (CoE) for "Internet of Things", Swami Keshvanand Institute of Technology, Management & Gramothan, Jaipur, Rajasthan, India. The conference focused on the recent advances in the domain of artificial and cognitive intelligence through image processing, machine/deep learning libraries, frameworks, and the adoption of intelligent devices, sensors, and actuators in the real world.

There were five keynotes covering the different areas of the conference. Seeram Ramakrishna, Vice President (Research Strategy) and Professor, Faculty of Engineering, National University of Singapore, discussed "BioTech Befriending Circular Economy". Sumit Srivastava, Professor, Department of Information Technology, Manipal University Jaipur and Senior Member, IEEE Delhi Section, presented a "Machine Learning Model for Road Accidents". Arun K. Somani, Associate Dean for Research, College of Engineering, Iowa State University, discussed "Hardware Perspective of Neural Architecture Search". He explained the neural network processing hardware, image and weight reuse without layer fusion, and various case studies like ASIC. Pratibha Sharma, Technical Test Lead, Infosys Ltd., started with a brief introduction to IoT; she then talked about different waves of technologies like IoT, AI, and robotics and explained the three As of IoT, namely 'Aware, Autonomous, and Actionable'. S. C. Jain, Professor, Department of Computer Science and Engineering, Rajasthan Technical University introduced the basics of bank transactions, cybercrimes, and security measures. He also talked about blockchain for emerging business applications and discussed Blockchain as a Service (BaaS).

The conference received 235 submissions, out of which 60 papers were finally accepted after rigorous reviews. The online presentations, fruitful discussions, and exchanges contributed to the conference's success. Papers and participants from various countries made the conference genuinely international in scope. The diverse range of presenters included academicians, young scientists, research scholars, postdocs, and students who brought new perspectives to their fields.

Through this platform, the editors would like to express their sincere appreciation and thanks to our publication partner Springer, the authors for their contributions to this publication, and all the reviewers for their constructive comments on the papers.

We would also like to extend our thanks to our industry collaborators: IBM, Infosys Technologies Ltd., and Natural Group.

May 2022

Valentina E. Balas
G. R. Sinha
Basant Agarwal
Tarun Kumar Sharma
Pankaj Dadheech
Mehul Mahrishi

Organization

General Chair

Seeram Ramakrishna	National University of Singapore, Singapore

Conference Chair

Arun K. Somani	Iowa State University, USA

Organizing Chairs

Anil Chaudhary	SKIT, India
C. M. Choudhary	SKIT, India

Technical Program Chairs

Valentina E. Balas	"Aurel Vlaicu" University of Arad, Romania
G. R. Sinha	IIIT Bangalore, India
Basant Agarwal	IIIT Kota, India
Tarun Kumar Sharma	Shobhit University, India
Pankaj Dadheech	SKIT, India
Mehul Mahrishi	SKIT, India

Program Committee

Sandeep Sancheti	Marwadi University, India
R. K. Joshi	IIT Bombay, India
Mitesh Khapara	IIT Madras, India
R. S. Shekhawat	Manipal University, India
Y. K. Vijay	VGU, India
Mahesh Chandra Govil	NIT Sikkim, India
Neeraj Kumar	Chitkara University Institute of Engineering and Technology, India
Manoj Singh Gaur	IIT Jammu, India
Virendra Singh	IIT Bombay, India
Vijay Laxmi	MNIT, India
Neeta Nain	MNIT, India
Namita Mittal	MNIT, India

Emmanuel Shubhakar Pilli	MNIT, India
Mushtaq Ahmed	MNIT, India
Yogesh Meena	MNIT, India
S. C. Jain	RTU Kota, India
Lakshmi Prayaga	University of West Florida, USA
Ehsan O. Sheybani	Virginia State University, USA
Sanjay Kaul	Fitchburg State University, USA
Manish Pokharel	Kathmandu University, Nepal
Nistha Keshwani	Central University of Rajasthan, India
Shashank Gupta	BITS Pilani, India
Rajiv Ratan	IIIT Delhi, India
Vinod P. Nair	University of Padua, Italy
Adrian Will	National Technological University, Tucumán, Argentina
Chhagan Lal	University of Padua, Italy
Subrata Mukhopadhyay	IEEE Delhi Section, India
K. Subramanian	IEEE Delhi Section, India
A. Murali M. Rao	IEEE Delhi Section, India
Nisheeth Joshi	Banasthali Vidhytapith, India
Maadvi Sinha	Birla Institute of Technology, India
M. N. Hoda	IEEE Delhi Section, India
Daman Dev Sood	IEEE Delhi Section, India
Soujanya Poria	Nanyang Technological University, Singapore
Vaibhav Katewa	University of California, Riverside, USA
Sugam Sharma	Iowa State University, USA
Xiao-Zhi Gao	Lappeenranta University of Technology, Finland
Deepak Garg	Bennett University, India
Pranav Dass	Galgotias University, India
Linesh Raja	Manipal University, India
Vijender Singh	Manipal University, India
Gai-Ge Wang	Jiangsu Normal University, China
Piyush Maheshwari	Amity University, Dubai, UAE
Wenyi Zhang	University of Michigan, USA
Janos Arpad Kosa	Kecskemet College, Hungary
Dongxiao He	Tianjin University, China
Thoudam Doren Singh	IIIT Manipur, India
Dharm Singh Jat	Namibia University of Science and Technology, Namibia
Vishal Goyal	Punjabi University, India
Dushyant Singh	MNNIT Allahabad, India
Amit Kumar Gupta	DRDO, Hyderabad
Pallavi Kaliyar	University of Padua, Italy

Sumit Srivastava	IEEE Delhi Section and Manipal University, India
Ripudaman Magon	Natural Group, India
Sumit Srivastava	Pratham Software Pvt. Ltd., India
Nitin Purohit	Wollo University, Ethiopia
Ankush Vasthistha	National University of Sigapore, Singapore
Dinesh Goyal	PIET, India
Reena Dadhich	University of Kota, India
Arvind K. Sharma	University of Kota, India
Ramesh C. Poonia	Christ University, India
Vijendra Singh	Manipal University, India
Mahesh Pareek	ONGC, New Delhi, India
Kamaljeet Kaur	Infosys Ltd., India
Rupesh Jain	Wipro Technologies Ltd., India
Gaurav Singhal	Bennett University, India
Lokesh Sharma	Galgotia University, India
Ankit Vidhyarthi	Bennett University, India
Vikas Tripathi	Graphic Era University, India
Prakash Choudhary	NIT Manipur, India
Smita Naval	IIIT Kota, India
Subhash Panwar	Government Engineering College, Bikaner, India
Richa Jain	Banasthali University, India
Pankaj Jain	JECRC University, India
Gaurav Meena	Central University of Rajasthan, India
Vishnu Kumar Prajapati	Governmentt Polytechnic College, Sheopur, India
Rajbir Kaur	LNM Institute of Information Technology, India
Poonam Gera	LNM Institute of Information Technology, India
Madan Mohan Agarwal	Birla Institute of Technology, India
Ajay Khunteta	Poornima College of Engineering, India
Subhash Gupta	Birla Institute of Technology
Devendra Gupta	Governmentt Polytechnic College, Tonk, India
Baldev Singh	Vivekanand Institute of Technology, India
Rahul Dubey	Madhav Institute of Technology and Science, India
Smita Agarwal	Global Institute of Technology, India
Sonal Jain	JK Lakshmipat University, India
Manmohan Singh	Chameli Devi Group of Institutions, India
Shrawan Ram	MBM Engineering College, India

Contents

Cognitive Computing

Internet of Things (IoT)

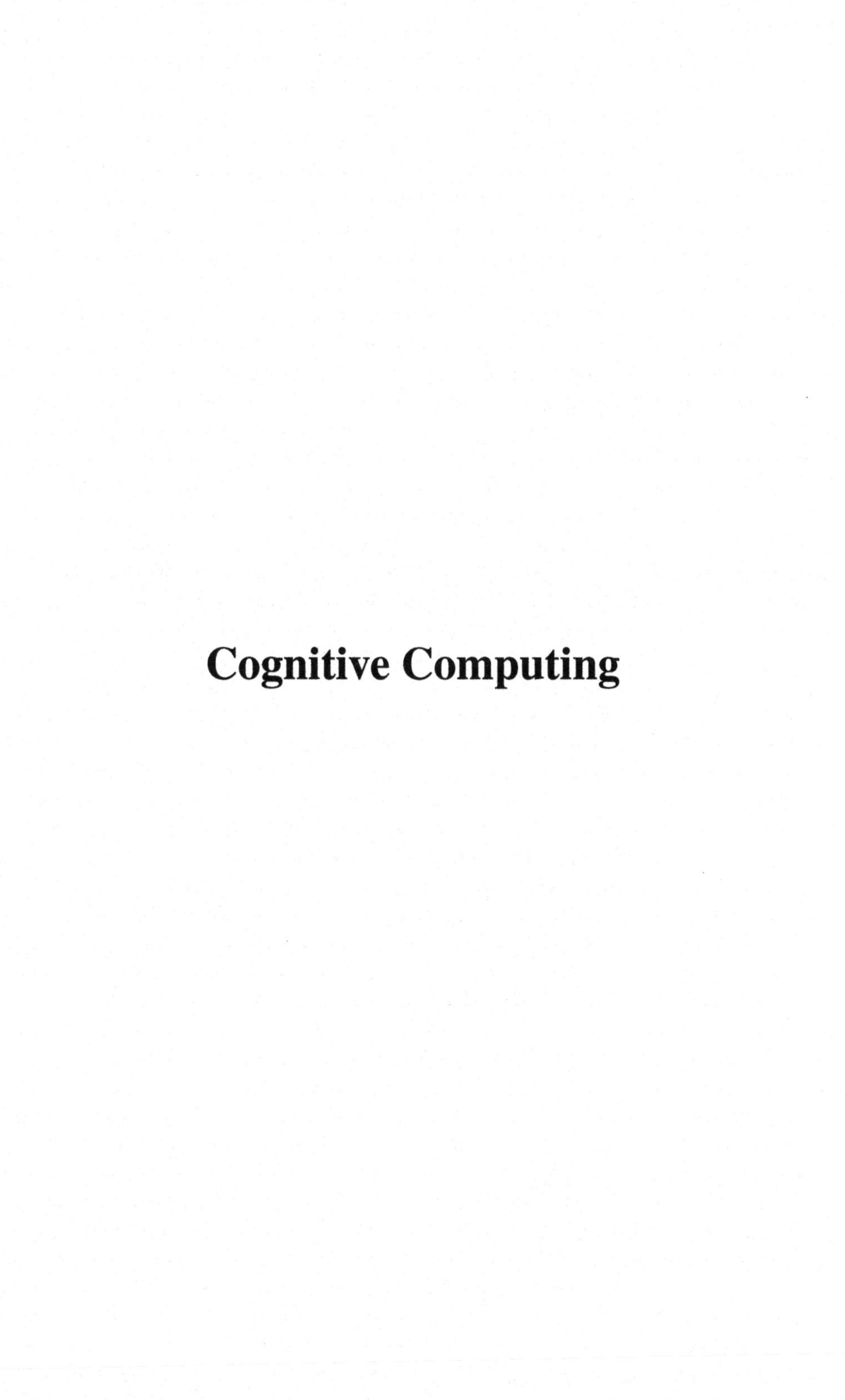

Cognitive Computing

Game-Based Learning System for Improvising Student's Learning Effectively: A Survey

E. S. Monish[1], Ankit Sharma[1], Basant Agarwal[1(✉)], and Sonal Jain[2]

[1] Department of Computer Science and Engineering, Indian Institute of Information Technology Kota (IIIT Kota), MNIT Campus, Jaipur, India
basant.cse@iiitkota.ac.in

[2] Department of Computer Science and Engineering, JK Lakshmipat University, Jaipur, India

Abstract. Game-based learning approaches and 21st-century skills have been gaining a lot of attention from researchers. Given there are numerous researches and papers that support the effect of games on learning, a growing number of researchers are determined to implement educational games to develop 21st-century skills in students. This review shows how Game-Based learning techniques impact 21st Century skills. 22 recent papers have been analyzed and categorized according to learning outcome, age group, game design elements incorporated, and the type of game (Game genre) implemented. The range of game genres and game design elements as well as learning theories used in these studies are discussed. The impact of implementing machine learning strategies and techniques in educational games has also been discussed. This study contributes to the ongoing research on the use of gaming features for the development of innovative methodologies in teaching and learning. This study aims to shed light on the factors and characteristics to be considered when implementing a game-based learning system. This study provides valuable insight to future researchers and game designers with issues and problems related to educational game design and implementation.

1 Introduction

Gaming is a very popular leisure activity in teenagers and adults. The time and money spent on video games have spiked exponentially in the past 2 years due to lockdown in almost every country. Using this trend to our advantage and incorporating learning methods with games will have a huge positive impact, improving students' performance and broadening their knowledge. Even After years of research and experiments we still don't have an effective learning method that would captivate students and improve the learning methodology. This is why there is a need to implement new and efficient methods that merge with the modern style and capabilities and incorporate Human-Computer interactions.

We know that students' inclination towards video games and computer games is the exact opposite of that they have towards schools and classrooms. Yet this very inclination is all that is required for the students to learn and practice their curriculum regularly. This proclivity towards games causes students to be more inclined, competitive, cooperative, and consistently seeking information and solutions to solve problems. The gaming

V. E. Balas et al. (Eds.): ICETCE 2022, CCIS 1591, pp. 3–18, 2022.
https://doi.org/10.1007/978-3-031-07012-9_1

industry has reached the 30-billion-dollar mark in market value; therefore, it would be reasonable to try to incorporate learning methodologies with games and interactive modules, this is what this study demonstrates.

How is game-based learning methodology productive for students? What captivates students and adolescents to games is not the violence, gaming experience, or the story, but the learning they impart and the sense of involvement. Computers and games facilitate students with the learning experience and opportunities. What students gain from playing games might take years to attain in traditional classrooms and teachings. Students learn to implement strategies to overcome and solve problems spontaneously, they learn to gather and process information from multiple sources and make a conclusive decision, and also understand complex structures by experimentation. And most importantly they learn to collaborate and cooperate with other members/students, improving their social skills and building up self-confidence.

2 Game-Based Learning

Games meet the requirements and needs of the students and can adapt and change quickly to changes in trends which makes them the most popular computer interaction. Games provide a new mode of interaction virtually for the students to explore and learn. There are multiple benefits of game-based learning systems: they captivate and interests students/youngsters, they provide an appropriate environment that keeps the students focused on the task given. Learning is more effective and fruitful when it is not forced upon. Game-based learning methodologies make the student the center of learning, which makes learning simpler and more efficient. Gee (2003) [1] argues that "the actual significance of a good game is that they facilitate learners to recreate themselves in new, virtual environments and help achieve entertainment and deep learning simultaneously."

Traditional teaching methods have failed to attract youngsters, and are not able to satisfy the requirements of the information society. And due to the convenient availability of network and interactivity which in turn results in improving time and location flexibility, interactive learning/game-based learning has become the popular trend for learning and educational purposes. The issue with traditional learning methodology is that they are often abstruse with many theories, proofs, case studies, and problems, but they lack an important feature required for efficient learning: practical exercise and involvement. The game-based learning methodology provides abundant features such as Outcomes, feedback, win states, competition, challenges, tasks, goals, problem solving, stories, and so on (Felix & Johnson, 1993; Prensky, 2001) [2], which increase the motivation to learn.

2.1 Outcomes and Impact of Game-Based Learning

Past research indicates that the most commonly observed learning outcome of game-based learning was improved knowledge (Connolly et al., 2012 [3]). Some other learning outcomes that were observed were problem-solving skills and critical thinking and "affective and motivational outcomes" (Connolly et al., 2012 [3]).

This paper categorizes the recent research based on the skillset discussed and mentioned in the recent research. The general 21st-century skillset is defined as Critical

Thinking, creativity, collaboration, and communication (Binkley et al., 2012) [4]. Critical Thinking comprises skills like reasoning, problem-solving, computational thinking, and decision making (Binkley et al., 2012) [4]. Creativity includes uniqueness, innovative thinking, originality, inventiveness, and the ability to perceive failures as an opportunity to improve (Binkley et al., 2012) [4].

2.2 Elements of Game-Based Learning

Lepper (1985) [5] suggested that the degree of proclivity that students have towards learning through game-based learning methods is determined by the characteristics of the game. While developing a computer game, developers must take into consideration the features and characteristics of the game. There are many different views of different researchers, on essential features required for a game-based learning system. The most important feature to take into consideration is the diversity of challenges and learning resources included in the game. They motivate learners and keep them captivated to learn and practice consistently. The instructions and educational content integrated into the game must be structured efficiently in order to increase difficulty and challenges. A well-defined structure motivates a student and keeps them interested to solve further challenges and complete what they started.

Research shows that game-based learning methods can motivate students to learn by including elements such as Fantasy, rules/goals, sensory stimuli, challenges, mystery, collaboration, competition, and control.

3 Method

3.1 Research Questions

Despite the growing popularity of game-based learning techniques evidence and reviews of ongoing research are required to evaluate the potential of game-based learning. This paper aims to determine the effect of Game-Based learning techniques on the attainment of learning outcomes aligned with 21st-century skill-sets. This paper also aims to find the effect of Game-based learning systems on respected age groups. The elements of game design and the game genre of recent research have been reviewed and categorized.

3.2 Searching Method

The keywords used to collect research papers were ("Game-Based Learning") or ("Education Games") or ("Digital Learning") or ("Design-based Learning") or ("Games for learning") or ("Games for teaching") of ("digital games") or ("21^{st}century learning outcomes"). All the papers from this keyword search were analyzed and categorized. To be included in this review, papers had to (a) depict the effect of Game-based learning on learning outcomes (21st-century skill set) (b) must include the age group or the target audience the paper aims to demonstrate. (c) dated from 1 January 2021. Out of these papers, 22 papers investigated the impact of Game-Based learning on 21st-century skills. These 22 papers were analyzed and reviewed.

3.3 Categorization of Outcomes

Outcomes of the selected researches were categorized as the 21st-century skillset: critical thinking, creativity, communication, and collaboration.

3.4 Categorization of Ages

The particular age group the papers aim to experiment with or analyze was determined. Further, these researches were categorized into Elementary Schools, High Schools, Higher Education (Undergraduate or Graduate), Adults, and Undisclosed. Where Adults refer to individuals who are no longer a student in educational institutions or organizations.

3.5 Categorization of Gaming Elements

Further, the research papers were categorized into the types of elements of game-based learning methods they incorporated. The elements of game-based models used and demonstrated were analyzed and categorized.

3.6 Categorization of Game Genres

The type of Gaming genres incorporated with the GBL models was analyzed to find a relation or trend between educational concepts incorporated and the type of gaming genre. The most common gaming genres used in GBL are Puzzle/Quiz, Simulation, Sandbox, Strategy, Adventure, Role-Playing, and Virtual Reality.

4 Results

4.1 Learning Outcomes

As depicted in Fig. 1, Out of these twenty-two papers around (71%) of these papers reported the influence of games on critical-thinking skills, whereas (23%) of the papers focused on collaboration and only 4 papers considered collaboration as a learning outcome, and only 5 papers considered communication as a learning outcome.

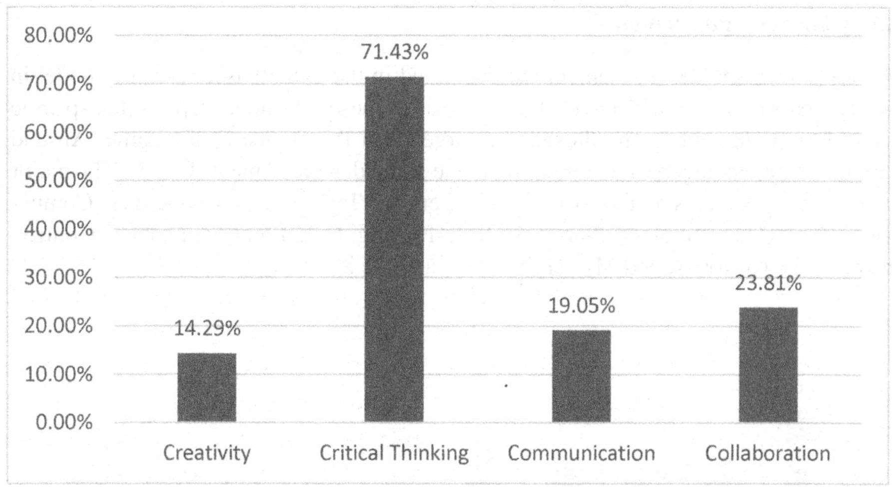

Fig. 1. 21st century skills targeted as learning outcomes in recent game-based learning studies

4.2 Age Groups

Participants' age in the 22 papers was analyzed ranged from elementary school to adults (shown in Fig. 2). The majority of the papers focused on High School students followed by Elementary Schools and Higher Studies. Only 1 paper focused on the effect of Game-Based Learning techniques on Adults and 1 paper was undisclosed.

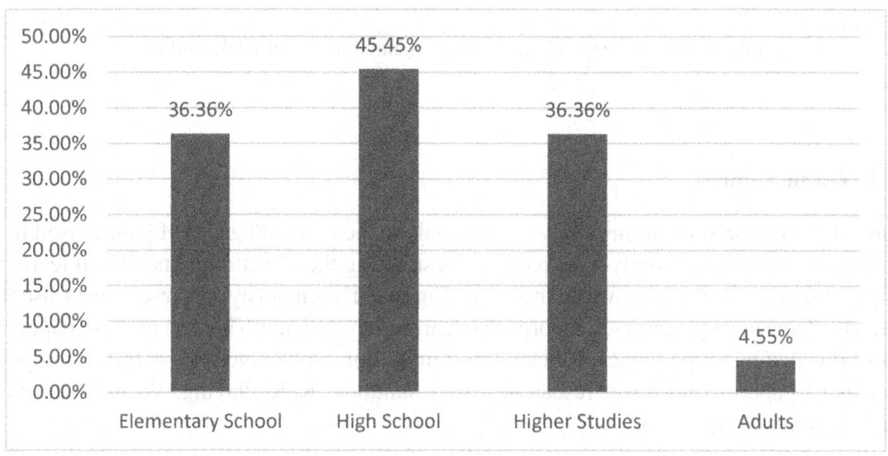

Fig. 2. Age group of individuals targeted in recent game-based learning studies

4.3 Game Design Elements

The types of game design elements incorporated in the game-based learning model in these 22 papers were analyzed. Further analysis of these 22 studies depicts that specific game design elements were chosen and targeted in the choice of the game. Around 85% of the papers incorporated more than one game design element (Fig. 3). The major game design elements analyzed in these papers are Tasks/Goals, Challenges, Competition, Exploration, Strategy, Narrative, Role-Playing, Collaboration, Fantasy, Control, Interactivity, Curiosity, and Mystery as seen in Table 2.

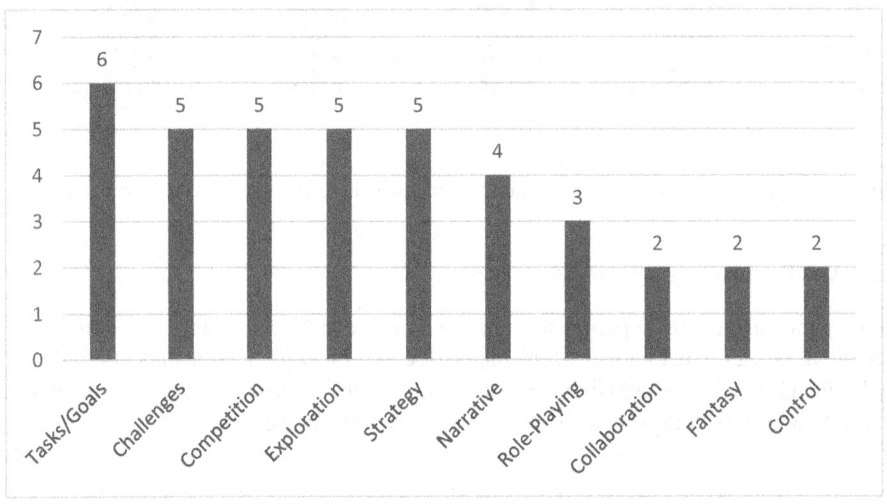

Fig. 3. Game design elements implemented in recent GBL studies

4.4 Game Genres

One of the major aims of this study was to analyze the type and genre of games used in different researches. Twenty-Two papers investigating the effect of Game-Based learning techniques on learning were analyzed (Fig. 4). The majority of these papers used Puzzles/Quizzes type games to incorporate learning to students. Thirteen of these papers used Puzzle/Quiz type games, followed by Simulation type games (4 papers), Strategy games (3 Papers), and fewer researches used Sandbox, Role Playing, Virtual Reality, and Adventure Type games.

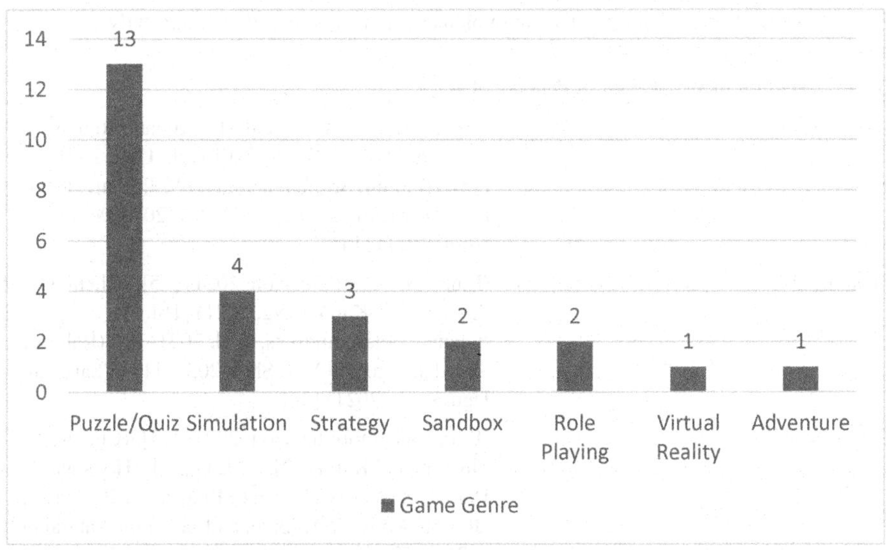

Fig. 4. Type of games (Genre) implemented in recent GBL studies

Table 1. Age categories

Age Category	Count of Papers	Study
Elementary School	8	(Kangas, Marjana) [6], (Squire, Burt DeVane et.. al) [7], (Giannakoulas, Andreas, et al. 2021) [8], (Hooshyar, Danial, et al.) [9], (Vidergor, Hava E, 2021) [10], (Rahimi, Seyedahmad, and V. J. Shut, 2021) [11], (Araiza-Alba, Paola, et al.) [12], (Mårell-Olsson, Eva., 2021) [13]
High School	10	(Nguyen Thi Thanh Huyen et al.) [26], (Squire, Burt DeVane et,. al) [7], (Giannakoulas, Andreas, et al. 2021) [8], (Hijriyah, Umi, 2021) [25], (Januszkiewicz, Barbara, 2021) [18], (Gürsoy, Gülden, 2021) [19], (Rahimi, Seyedahmad, and V. J. Shut, 2021) [11], (Saricam, Ugur et al., 2021) [20], (Mårell-Olsson, Eva., 2021) [13], (Demir, Ümit, 2021) [14]
Higher Studies	8	(Liunokas, Yanpitherszon, 2021) [15], (Cheung, Siu Yin, and Kai Yin Ng, 2021) [16], (Mazeyanti Mohd Ariffin et al. 2021) [17], (Januszkiewicz, Barbara, 2021) [18], (Kurniati, Eka, et al., 2021) [21], (Joseph, Mickael Antoine, and Jansirani Natarajan, 2021) [22], (Rahimi, Seyedahmad, and V. J. Shut, 2021) [11], (Chang, Wei-Lun, and Yu-chu Yeh, 2021) [23]
Adults	1	(Rahimi, Seyedahmad, and V. J. Shut, 2021) [11]
Not Defined	1	(Clark, Douglas, et al., 2021) [24]

Table 2. Game design element implemented in game-based learning study.

Game Design Elements	Count of Papers	Study
Tasks/ Goals	6	(Squire, Burt DeVane et al.) [7], (Giannakoulas, Andreas, et al. 2021) [8], (Hijriyah, Umi, 2021) [25], (Rahimi, Seyedahmad, and V. J. Shut, 2021) [11], (Saricam, Ugur et al., 2021) [20], (Demir, Ümit, 2021) [14]
Challenges	5	(Liunokas, Yanpitherszon, 2021) [15], (Cheung, Siu Yin, and Kai Yin Ng, 2021) [16], (Giannakoulas, Andreas, et al. 2021) [8], (Rahimi, Seyedahmad, and V. J. Shut, 2021) [11], (Saricam, Ugur et al., 2021) [20]
Competition	5	(Liunokas, Yanpitherszon, 2021) [15], (Cheung, Siu Yin, and Kai Yin Ng, 2021) [16], (Hooshyar, Danial, et al.) [9], (Kurniati, Eka, et al., 2021) [21], (Joseph, Mickael Antoine, and Jansirani Natarajan, 2021) [22]
Exploration	5	(Squire, Burt DeVane et al.) [7], (Rahimi, Seyedahmad, and V. J. Shut, 2021) [11], (Januszkiewicz, Barbara, 2021) [18], (Hijriyah, Umi, 2021) [25], (Saricam, Ugur et al., 2021) [20]
Strategy	5	(Squire, Burt DeVane et al.) [7], (Cheung, Siu Yin, and Kai Yin Ng, 2021) [16],], (Giannakoulas, Andreas, et al. 2021) [8], (Hooshyar, Danial, et al.) [9], (Joseph, Mickael Antoine, and Jansirani Natarajan, 2021) [22]
Narrative	4	(Kangas, Marjana) [6], (Mazeyanti Mohd Ariffin et al. 2021) [17], (Januszkiewicz, Barbara, 2021) [18], (Gürsoy, Gülden, 2021) [19]
Role-Playing	3	(Giannakoulas, Andreas, et al. 2021) [8], (Gürsoy, Gülden, 2021) [19], (Mårell-Olsson, Eva., 2021) [13]
Collaboration	2	(Nguyen Thi Thanh Huyen et al.) [26], (Vidergor, Hava E, 2021) [10]
Fantasy	2	(Kangas, Marjana) [6], (Mårell-Olsson, Eva., 2021) [13]
Control	2	(Clark, Douglas, et a, 2021) [24], (Rahimi, Seyedahmad, and V. J. Shut, 2021) [11],
Interactivity	2	(Clark, Douglas, et al., 2021) [24], (Rahimi, Seyedahmad, and V. J. Shut, 2021) [11],
Curiosity/Mystery	1	(Vidergor, Hava E, 2021) [10], (Rahimi, Seyedahmad, and V. J. Shut, 2021) [11]

5 Discussion

The current review aims to analyze recent research and depict the effect of game-based learning techniques on learning outcomes and 21st-century skills. Our initial search resulted in multiple papers dated from 1 January 2021. In the end, 22 papers met our inclusion criteria and were selected for this particular review. These researches had diverse effects on learning outcomes and 21st-century skills. The majority of the studies depicted that game-based learning techniques had a positive effect on the critical thinking ability of the student. Around 70% of these papers showed that GBL had a positive effect on the critical thinking ability of a student, followed by collaboration. From these 22 papers, 5 papers revealed that GBL had a positive impact on collaboration. Only 4 of these papers showed results about communication and very few papers for creativity.

With regard to the audience targeted in these researches/the age-groups of the audience, most of these researches were focused on High School students, followed by Elementary Schools and Higher Studies students. Surprisingly only 1 study focused on adults, making it difficult to analyze the effect of GBL on learning outcomes and 21st-century skills of adults. And 1 paper did not disclose the age group of its audience or participants.

A further review of these 22 papers investigating the game design elements used or incorporated in these games showed that tasks/goals were a common and frequent element of game design. 13 of these papers included tasks and goals in the game to motivate the student to complete these tasks and to be persistent in learning. Challenges, Competition, Exploration, and Strategy were the next most frequently targeted game design elements. Competition motivates students to think quickly and to come up with solutions to overcome tasks and problems quickly. Competition is an important feature that ensures extensive and consistent learning. Competitions between groups of individuals improve teamwork, cooperation and develop social skills. Competitions allow students to express their knowledge and skills and transform learning into achievements. These achievements give learners a sense of achievement motivating them to work harder and practice further to achieve more. This thirst for achievement keeps students consistent in practicing and learning concepts. Competition boosts individuals' self-esteem and helps them to develop new skill sets and creativity.

Challenges make learning interesting and appealing. Learners put in persistent efforts to keep up with the challenges and to solve them. The sense of achievement gained by solving challenges motivates individuals to continue learning and practicing. Malone and Lepper (1987) [27] have stated that students require an adequate number of challenges with varying difficulty levels. Failure to incorporate challenges with varying difficulty levels demotivates learners and might appear unappealing to them. A structured implementation of challenges with progressive difficulty and various goals is essential to motivate learners and to fascinate them. Exploration refers to the freedom given to the student within the virtual gaming environment. Games that allow students to explore and roam freely motivate them to learn more and explore more giving making them curious and excited. Strategical games allow students to think quickly and efficiently to come up with effective solutions to overcome tasks and challenges.

Collaboration is one of the most important game design elements implemented in modern games and game-based learning methods. Students work with each other forming groups to understand and process the concepts provided. They collaborate and share their thoughts, ideas, and opinions to come up with a solution or an efficient method to understand a concept. This constructive technique instructs students to collaborate and implement effective learning environments and methodologies. Learners share their prior skillsets and knowledge with their peers and class. Although collaboration is a key feature in game-based learning, only very few recent researchers have implemented this feature in their game design.

Other game design elements like Role-Play, Fantasy, Interactivity, and Curiosity/Mystery were not frequently observed in these researches, whereas some researches do incorporate these game design elements. Educational games can implement all these game design elements to provide opportunities for students to develop 21st-century skills and to motivate learning.

Another major aim of this review was to analyze the type of game genre used/ preferred in these papers. It is important for the developers to study the changing demand and trends to ensure maximum participation of students and individuals. Determining the perfect gaming genre is the most crucial part of developing a game. Developers and analysts need to study changing market trends and demands to decide on a gaming genre to develop. The game genre decides how the educational content must be incorporated and integrated into the game. It decides the structure and the feature of the game. Hence selecting the genre must be done after proper analysis and research.

A review of recent papers shows that majority of these papers used Puzzle/Quiz type games to incorporate learning in games. 13 out of 22 papers implemented Puzzles/Quiz type games followed by simulation-type games. Puzzles and Quizzes are commonly used as they are easy to develop and maintain and are cheaper to implement. Puzzles and Quiz-type games help students learn educational concepts easily and quickly as compared to other gaming genres. These quiz-type games are direct and straightforward whereas other game genres are indirect and time taking. Puzzles/Quizzes improve the critical thinking ability of a student hence having a direct effect on the learning outcome.

Secondly, recent researches use simulation-type games to incorporate learning into games. Simulation games replicate or simulate real-world features and aspects of real-life into a virtual environment. These games allow users to create, destroy and modify real-world entities within a virtual environment making learning fun and effective. 4 out of 22 papers preferred Simulation type games. Whereas 3 out of the 22 papers preferred Strategical games. Strategical games improve the logical thinking ability of a student. Strategy games require strategic thinking to succeed and proceed further into the game, rather than using brute force or power methods. Strategical games improve motivate students to create and come up with creative and effective strategies to solve problems and solutions. These games motivate students to think creatively.

Other common types of games preferred in these papers are Sandbox, Virtual Reality, Adventure, and Roleplaying. Hence deciding the type of game/ Game genre is really important while designing an educational game. Each game genre focuses on a particular type of learning outcome. Analyzing and deciding on a proper gaming genre is important to implement an effective game-based learning model.

6 Machine Learning and Game-Based Learning

Education and Learning are shifting towards online platforms exponentially. Online courses, classes, and other educational content are becoming popular and are being used widely by students and teachers. The rapid development and increasing popularity of Artificial Intelligence and Machine Learning algorithms and Applications of AI have attracted the attention of educational researchers and practitioners (Luckin & Cukurova, 2019) [28]. (Garcia et al., 2007 [29]; Tsai et al., 2020 [30]) Researchers contemplate AI as a medium for providing precision education. Machine Learning and AI techniques like Neural Networks, Deep Learning, Reinforcement learning, etc.. can be used to support learning and teaching, facilitating learners with an efficient and effective medium for learning. The unusually large data storage and effective computing have caused researchers to incorporate AI with learning and teaching. Game-Based Learning systems can incorporate AI and ML algorithms and techniques to improve the learning experience and to ensure effective learning. Artificial Intelligence in Games can have numerous advantages like Predictive Grading, Student Review, AI Chatbots, Recommender Systems, and Intelligent Gaming Environments.

(Dan Carpenter et al., 2020) [31] Used Natural Language Processing (NLP) Techniques to assess the depth of student reflection responses. To analyze the assessment of student reflection, a game-based learning environment CRYSTAL ISLAND was used to collect data and responses from students. In CRYSTAL ISLAND, students adopt the character of a scientist or a science detective who investigates a recent outbreak among a group of scientists. Students solve the challenge or tasks by submitting a diagnosis explaining possible causes, type of pathogen, transmission source, and recommended treatment plan or prevention plan. This research used NLP techniques to analyze student reports and evaluate the depth of students' understandings and reflections.

(Andrew Emerson, Elizabeth B. Cloude, et al., 2020) [32] used Facial Expression Analysis and Eye Tracking to improve the multimodal data streams for learning analysis. Learning analytics depicted from multimodal data streams for learning analytics while capturing students' interaction with the game-based learning environment helps the developers to get a deeper understanding of the GBL system. Gameplay features, Eye Tracking, and facial expressions captured from the learners were used to predict the posttest performance and interest after interacting with the game-based learning environment CRYSTAL ISLAND.

(Jennifer Sabourin, Lucy Shores, et al., 2012) [33] used machine learning models to predict a student's SRL (Self-Regulatory Learning) abilities early into their interaction with the game-based learning environment CRYSTAL ISLAND. Self-Regulatory Learning (SRL) is "the process by which students activate and sustain cognitions, behaviors, and effects that are systemically directed toward the attainment of goals". Self-Regulated learning behaviors like goal setting and monitoring have proved to be essential for a student's success in a vast variety of online learning environments.

Teachers and educators find difficulty in monitoring the progress of students based on the knowledge gained by them. They can only depict the progress of an individual by analyzing their exam scores or marks. This may not be the most effective method to evaluate a students' progress. Marks/Exam scores may not be a reliable feature to evaluate a students', since exams might not be able to depict the maximum potential of a student due to factors like fear, time, and competition. But by using AI and Machine Learning algorithms we can measure the progress of an individual by monitoring the performance of the student over a period of time. The computational efficiency and the ability to store large data allow AI and ML techniques to analyze students' performance and progress. Students' performance in every module/chapter can be reviewed and analyzed separately by running analyzing and predicting each learning module separately. Further, we can use AI techniques to implement a recommender system that recommends the particular field or module where the students can improve their performance. This process of analyzing and predicting students' performance is a rigorous and time taking process for the teachers, whereas the machine learning algorithms could compute these simultaneously and effectively. Most importantly using AI and ML in game-based learning environments for evaluating a students' performance and progress is a spontaneous process. We can get the evaluated review of the student immediately after the completion of the prebuild learning modules.

Further AI and ML techniques can be used to analyze learners' ability to think, and to learn patterns to come up with more effective and challenging learning modules for the learners. This continuous learning and implementation of learning modules ensure progressive and effective learning of the student. The modules can be made more challenging and difficult if the student shows epidemic progress. And the modules can be made easier if the progress remains stagnant or fails to improve. This way using ML models in game-based learning systems can ensure effective and progressive learning.

Students' behavior and emotions can also be analyzed by using ML and Deep Learning techniques to get a better understanding of students' behavior towards certain modules and concepts. This can further be used to understand how the student feels about that particular module or concept and further improve his progress by recommending multiple modules of that particular concept.

7 Limitations and Further Research

This current review has a few limitations. First, the effect of Game-based learning has not been measured quantitatively. Hence the extent of the impact on the learning outcomes cannot be depicted from this review. Only qualitative studies were selected and analyzed. Secondly, this review focused only on papers that were published since 1 January 2021. This paper will provide future researchers and practitioners with great insights into recent research concerning the impact of game-based learning on learning outcomes and its effect on 21st-century skills.

The learning outcomes analyzed from these 22 papers depicted that majority of these papers observed a positive effect on the critical thinking ability of students and learners but very few of these papers focused on developing communication, collaboration, and creativity. While improving critical thinking ability is crucial to any game-based learning model, improving Creativity, Communication, and Collaboration skills are also essential to engage students and motivate learning. Further research is required to implement GBL methods to improve Creativity, Communication, and collaboration skills in an individual. Further research to develop games that focus on developing collaboration, creativity, and communication is essential to ensure efficient and motivated learning.

The majority of these recent papers have focused on developing learning outcomes for Elementary school, High school, and Higher Studies students. Further research to motivate learning in adults is required. Further research to implement games intended to improve the learning outcomes of adults is required. Learning must not be limited to particular age groups. Every individual of every age group must be given the resources and opportunity to develop skills and learn.

The majority of these papers implemented Puzzle/Quiz type games to develop or improve learning outcomes in students. Further researchers can try to implement games of different genres like Simulation, Strategy, Adventure, and Role-play to attract learners and motivate learning. Implementing games of different genres can attract learners from all age groups. But these researches show why that particular gaming genre is chosen for that particular age group. Hence the relation between gaming genre and different age groups is essential while implementing a game-based learning model. Further research to depict the relation between gaming genres and age groups is required to implement an effective game-based learning model to motivate learning.

There is scope for further research in this field as there are many more factors and characteristics to take into consideration before implementing a game-based learning system. Even though we aim to achieve independent, self-motivated learning there is a requirement for the presence of a teacher or trained individual to debrief the learners and guide them throughout the learning process. There is a need for further research to examine the game model described in this paper. It is critical to evaluate how much effect these learning models have on the learning efficiency of students. Research by Parker and Lepper (1992) [34] showed that incorporating learning content with fantasy contexts and animations enhances and improvises the learning experience. Students found learning materials and concepts expressed through fantasy contexts, stories, and animated visualizations interesting. Further research on the factors determining the type, features, and context of the game is required before implementing them. Surveys and experiments must be conducted to determine the trend and demand of gaming genres within a given locality, age group, or time period is strongly required.

There is an essential need to implement and design new methods to incorporate learning materials in a structured format within the game effectively. The game must contain a smooth transition from one topic to another for the learner to understand it. There is a regular need for improving the features of the game to meet the change in trends and changing demands of learners. The game must be flexible so that developers can modify the game as per user feedback and outcomes. If we succeed in implementing a game considering all the above-stated factors and elements, we may develop a solution for effective teaching and learning, helping millions of students and teachers all around the world, it would be a breakthrough in the modern educational system and teaching methodologies.

Competition is an essential feature for efficient learning, it helps bring out the best in a student. Research shows that competition motivates an individual to work harder, to put in more effort and time, to research and study further, and to develop confidence. When games incorporate competition learning becomes fun and exciting as learners get to compete and challenge with peers and individuals from their class, school, and from all around the world. Competition is an important feature that ensures extensive and consistent learning. Competitions between groups of individuals improve teamwork, cooperation and develop social skills. Competitions allow students to express their knowledge and skills and transform learning into achievements. These achievements give learners a sense of achievement motivating them to work harder and practice further to achieve more. This thirst for achievement keeps students consistent in practicing and learning concepts. Competition boosts individuals' self-esteem and helps them to develop new skill sets and creativity.

Game-based learning methods must incorporate collaborative and competitive learning to get fruitful results. Collaborative and competitive learning are two key factors required to motivate and engage learners to be persistent in learning and practicing. Integrating these two features into a game will make learning efficient and also will engage students in continuous and persistent learning. Game-based learning methods can incorporate both collaborative and competitive learning to improve motivation and increase the efficiency of learning. Individuals from every age group are inclined towards competitive and collaborative games. Research shows that there has been an exponential increase in the number of individuals playing competitive and multiplayer games.

8 Conclusion

The paper reviews the impact of Game-Based Learning on the development of 21st-century skills. Learners develop cognitive skills, collaborative learning, leadership skills, and develop confidence while learning. Universities and schools are looking for innovative and effective teaching methods to incorporate efficient learning to students. This paper indicates that further research and study about the potential of using a Game-Based learning approach to develop 21st-century skills is required and can be proven fruitful even though only a few studies targeted communication, collaboration, and creativity as learning outcomes.

References

1. Gee, J.P.: What video games have to teach us about learning and literacy. Comp. Entertain. (CIE) **1**(1), 20 (2003)
2. de Felix, J.W., Johnson, R.T.: Learning from video games. Comp. Sch. **9**(2–3), 119–134 (1993)
3. Connolly, T.M., et al.: A systematic literature review of empirical evidence on computer games and serious games. Comp. Edu. **59**(2), 661–686 (2012)
4. Binkley, M., et al.: Defining twenty-first century skills. In: Assessment and teaching of 21st century skills. Springer, Dordrecht, pp. 17–66 (2012)
5. Lepper, M.R., Chabay, R.W.: Intrinsic motivation and instruction: conflicting views on the role of motivational processes in computer-based education. Edu. Psychol. **20**(4), 217–230 (1985)
6. Kangas, M.: Creative and playful learning: learning through game co-creation and games in a playful learning environment. Thinking skills and Creativity **5**(1), 1–15 (2010)
7. Squire, K.D., Ben, D., Shree, D.: Designing centers of expertise for academic learning through video games. Theory into practice **47**(3), 240–251 (2008)

8. Giannakoulas, A., et al.: A Proposal for an educational game platform for teaching programming to primary school students. In: International Conference on Technology and Innovation in Learning, Teaching and Education. Springer, Cham (2020)
9. Hooshyar, D., et al.: An adaptive educational computer game: effects on students' knowledge and learning attitude in computational thinking. Comp. Hum. Behav. **114**, 106575 (2021)
10. Vidergor, H.E.: Effects of digital escape room on gameful experience, collaboration, and motivation of elementary school students. Comp. Edu. **166**, 104156 (2021)
11. Rahimi, S., Shute, V.J.: The effects of video games on creativity: a systematic review. Handbook of lifespan development of creativity **37** (2021)
12. Araiza-Alba, P., et al.: Immersive virtual reality as a tool to learn problem-solving skills. Comp. Edu. **64**, 104121 (2021)
13. Mårell-Olsson, E.: Using gamification as an online teaching strategy to develop students' 21st-century skills. IxD&A: Interaction Design and Architecture (s) **47**, 69–93 (2021)
14. Demir, Ü.: The effect of unplugged coding education for special education students on problem-solving skills. Int. J. Comp. Sci. Edu. Sch. **4**(3), 3–30 (2021)
15. Liunokas, Y.: The efficacy of using cup stacking game in teaching speaking to indonesian english as foreign language (EFL) students. IDEAS: J. Eng. Lang. Teach. Lear, Ling. Lite. **9**(2), 521–528 (2021)
16. Cheung, S.Y., Ng, K.Y.: Application of the educational game to enhance student learning. Frontiers in Education **6**(Frontiers) (2021)
17. Ariffin, M.M., Aszemi, N.M., Mazlan, M.S.: CodeToProtect©: C++ programming language video game for teaching higher education learners. J. Physics: Conf. Seri. **1874**(1) (2021). IOP Publishing
18. Januszkiewicz, B.: Looking for the traces of polish heritage on the map of ukraine. linguistic educational game. Łódzkie Studia Etnograficzne **60**, 297–302 (2021)
19. Gürsoy, G.: Digital storytelling: developing 21st century skills in science education. Euro. J. Edu. Res. **10**(1), 97–113 (2021)
20. Saricam, U., Yildirim, M.: The effects of digital game-based STEM activities on students' interests in STEM fields and scientific creativity: minecraft case. Int. J. Technol. Edu. Sci. **5**(2), 166–192 (2021)
21. Kurniati, E., et al.: STAD-jeopardy games: A strategy to improve communication and collaboration skills' mathematics pre-service teachers. In: AIP Conference Proceedings **2330**(1) (2021). AIP Publishing LLC
22. Joseph, M.A., Natarajan, J.: Muscle anatomy competition: games created by nursing students. J. Nurs. Edu. **60**(4), 243–244 (2021)
23. Chang, W.-L., Yeh, Y.-C.: A blended design of game-based learning for motivation, knowledge sharing and critical thinking enhancement. Technol. Pedag. Edu. 1–15 (2021)
24. Clark, D., et al.: Rethinking science learning through digital games and simulations: genres, examples, and evidence. Learning science: Computer games, simulations, and education workshop sponsored by the National Academy of Sciences, Washington, DC (2009)
25. Hijriyah, U.: Developing monopoly educational game application on XI grade high school student's about cell teaching material. In: 7th International Conference on Research, Implementation
26. Nguyen Thi Thanh Huyen, K.T.T.N.: Learning vocabulary through games: the effectiveness of learning vocabulary through games. Asian Efl Journal (2002)
27. Wolfe, J., Crookall, D.: Developing a scientific knowledge of simulation/gaming. Simulation & Gaming **29**(1), 7–19 (1998). Malone and Lepper (1987)
28. Luckin, R., Cukurova, M.: Designing educational technologies in the age of AI: a learning sciences-driven approach. Br. J. Edu. Technol. **50**(6), 2824–2838 (2019)
29. García, P., et al.: Evaluating bayesian networks' precision for detecting students' learning styles. Computers & Education **49**(3), 794–808 (2007)

30. Tsai, Y.-L., Tsai, C.-C.: A meta-analysis of research on digital game-based science learning. J Comput Assist Learn. **36**, 280–294 (2020). https://doi.org/10.1111/jcal.12430
31. Carpenter, D., et al.: Automated analysis of middle school students' written reflections during game-based learning. In: International Conference on Artificial Intelligence in Education. Springer, Cham (2020)
32. Emerson, A., et al.: Multimodal learning analytics for game-based learning. British J. Edu. Technol. **51**(5), 1505–1526 (2020)
33. Sabourin, J., et al.: Predicting student self-regulation strategies in game-based learning environments. In: International Conference on Intelligent Tutoring Systems. Springer, Berlin, Heidelberg (2012)
34. Parker, L., Lepper, M.: Effects of fantasy contexts on children's learning and motivation: making learning more fun. J. Pers. Soc. Psychol. **62**, 625–633 (1992). https://doi.org/10.1037/0022-3514.62.4.625

Vector Learning: Digit Recognition by Learning the Abstract Idea of Curves

Divyanshu Sharma(✉), A. J. Singh, and Diwakar Sharma

Department of Computer Science, Himachal Pradesh University, Shimla, India
divyanshusharma1337@gmail.com

Abstract. A method has been developed that solves digit recognition problem through the abstract idea of lines and curves. The human brain can easily classify digits because it learns from the formation of the digit, it learns from its geometry. Human brain will view the digits as combination of lines and curves and not just mere arrangement of pixel values. For the neuron network to exhibit human level expertise, it must also learn from the formation of digits. Traditional techniques learn from pixel values of the image and hence lack the sense of the geometry of the number. With the help the of this developed method the program begins to have a gist of the geometry and the shape of the number. It develops the idea that a combination of some lines and curves forms the digit. The curves can be easily represented by tangent vectors. The program uses this set of vectors instead of pixel values for training. Results show that the program, which is learning through the abstract idea of curves classifies digits far better than the program learning from mere pixel values and is even able to recognize hugely deformed digits. It also becomes capable of recognizing digits invariant to the size of the number.

Keywords: Neural network · Digit recognition · Curves · Vectors

1 Introduction

The brain of human can easily classify objects and patterns. For artificial intelligence to be able to recognize patterns just like human beings, many models and techniques have been developed. The neural network has achieved great success in solving pattern recognition problems. Neural network works using many artificial neurons that behave similarly to human neuron cells. W. Pitts and W.S. McCulloch developed the building block for the neural network in 1943 [1]. They created a model which behaved similarly to human neurons. They achieved this by developing a propositional logic centered on the "all-or-none" characteristic of the neuron of a human. In 1958, Rosenblatt created perceptron [2] which is known to be the first artificially created neural network. Perceptron was build for image recognition. The perceptron consisted of an array containing 400 photocells which were connected randomly to neurons. Weights were updated during learning using the electric motors and were encoded using potentiometers. The next advancement was the model with large amount of layers which was made by Alexey Ivakhnenko and Valentin Grigor'evich Lapa in 1967 [3]. The next landmark came when Dreyfus was

V. E. Balas et al. (Eds.): ICETCE 2022, CCIS 1591, pp. 19–29, 2022.
https://doi.org/10.1007/978-3-031-07012-9_2

successfully able to implement the backpropagation algorithm [4] in 1973 [5]. Using it, Dreyfus was able to adjust the weights of the controllers from the error gradients. Deep learning became an even stronger tool for pattern recognition when LeCun [6] developed a spatial method for handwritten digit recognition and this gave rise to Convolutional Neural Networks [7].

Since then many techniques developed and are still being developed for solving handwritten digit classification problems [8–22].

All these techniques have received remarkable success in solving pattern recognition. However, these techniques require a lot of time to train from a huge amount of data. Even after training the model for so long, it may fail when changing even a single pixel value. This is known as single pixel attack [23]. Neural networks can be easily fooled using such attacks [24–29]. The reason of such a failure is because it does not know the concept or idea for pattern recognition. In case of digit recognition, it lacks the concept that a digit is formed using lines and curves. It only knows the pixel values of the image and hence it lacks the human-level expertise to recognize digits.

2 Inserting the Abstract Idea of Curves

As humans, we know a digit is formed when a series of lines or curves are placed one after another in a certain way to form a certain digit. Similarly, to grasp the notion of the formation of a digit, the program must have an abstract idea of curves and that these curves when put together in a certain order, form a geometric shape of a certain digit.

One of the many ways to introduce the idea of curves is by using vectors. A vector is a mathematical quantity having direction as well as magnitude. A curve can be divided into numerous small vectors. This set of vectors when put together in order, forms the geometric shape of a digit. This introduces a sense of curve into the program. It knows which direction to move to form a certain shape of a certain digit.

This set of vectors can be used as an input to the neural network. Instead of training with pixel values of the image, the program now trains with the digit's shape. This makes the program classify the digits more accurately, as now it knows how to form them.

As the training proceeds, the program starts to make sense that which shape is of which digit. It forms a generalized idea that what shape should be made using the curves (or vectors) to make a certain digit.

3 Formation of Vectors

Any digit can be represented by continuous small line segments. These line segments along with their direction (the angle of their rotation from the x-axis) form a set of vectors. Each set of vectors corresponds to the formation of a certain digit. These vectors when put together in order starts forming the curves or lines that are necessary to form a digit as shown in Fig. 1.

Hence by learning through vectors, the program will get some sense of curves that when put together forms the geometric shape of a digit.

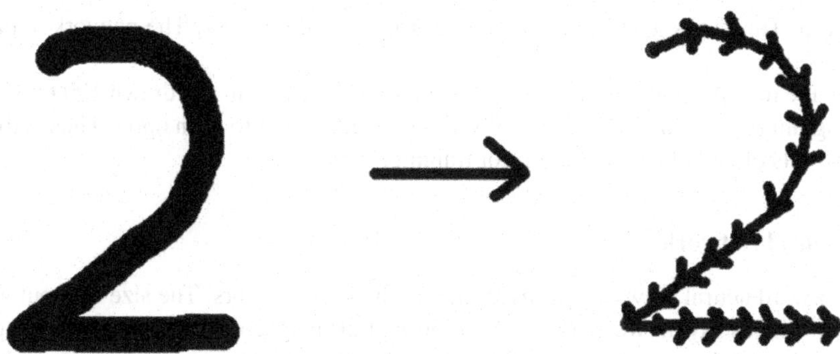

Fig. 1. Transforming image to set of vectors.

For the sake of simplicity, each digit is divided into 50 line segments. The vector can then be calculated by subtracting the vectors corresponding to the two endpoints of the particular line segment as shown in Eq. (1) and (2).

$$V_y = y_{i+1} - y_i \tag{1}$$

$$V_x = x_{i+1} - x_i \tag{2}$$

Using the above vector, the angle this vector with the x-axis can be calculated using Eq. (3) as follows:

$$\theta = \tan^{-1} \frac{V_y}{V_x} \tag{3}$$

This angle represents the vector in radial form. The radial form of vector representation gives us this angle represents the vector in radial form. The radial form of vector representation gives us the advantage of using only one variable instead of two for normalized vectors. Now instead of using components of a vector in the x-direction (Vx) and y-direction (Vy), the angle of the direction (ɵ) can be used. This set of angles are supplied as input to the deep neural network.

4 Model

4.1 Creating the Dataset

The model cannot be trained with the help of hand-drawn images as it provides only pixel values. A new program has to be made that converts the hand-drawn digit by the user into a set of vectors. With the help of OpenGL [30] and C++, a program that converts the digit drawn by the user into a corresponding set of vectors (in radial form) is developed. A window is provided by the program, in which, with the mouse button 0 pressed, any digit can be drawn by the user. As the mouse moves, line segments are created. All points of line segments created by this movement are stored. When the mouse button 0 is released, the set of points gathered are changed into set of radial unit vectors as shown in Sect. 3. This set is given as an input to a deep neural network. The network is then trained.

For the model, a dataset of digits 0–9 were used. Each digit was drawn 100 times in the program to get 100 different sets of vectors for each digit to train upon. Thus, a total of 1000 sets of vectors were created for training.

4.2 Neural Network

Feed forward neural network was used to train the set of vectors. The size of input was 50. Two hidden layers were used of size 30 and 20 respectively, followed by output layer of size 10. Learning rate was 0.05. The activation function used was sigmoid on all layers. Sigmoid function produces an output between 0 and 1 based on the input. It is represented by $\sigma(x)$, where x is the input. It is calculated using Eq. (4) as follows:

$$\sigma(x) = \frac{1}{1 + e^{-x}} \tag{4}$$

where e is Euler's number.

4.3 Forward Propagation

Each layer can be equated using Eq. (5)

$$L^n = \sigma\left(W * L^{n-1}\right) \tag{5}$$

where L^n is the vector corresponding to current layer, L^{n-1} is the vector corresponding to previous layer, and W is the weight matrix connecting them.

4.4 Backward Propagation

The error function used for the model is mean square error as shown in Eq. (6).

$$e = \sum_i (T_i - O_i)^2 \tag{6}$$

where T is target vector and O is output vector.

This error is then traversed back using backpropagation and weights are corrected for each layer as shown in Eqs. (7) and (8).

$$\frac{\delta e}{\delta W_{jk}^n} = \frac{\delta e}{\delta L_k^{n+1}} * \frac{\delta L_k^{n+1}}{\delta W_{jk}^n} \tag{7}$$

$$where, \frac{\delta L_k^{n+1}}{\delta W_{jk}^n} = \frac{\delta\left(\sum_i \sum_j W_{jk}^n * L_j^n\right)}{\delta W_{jk}^n} = L_j^n \tag{8}$$

W_{jk}^n is the weight in weight matrix of nth layer belonging to j^{th} row and k^{th} column.

$\frac{\delta e}{\delta W_{jk}^n}$ is error of W_{jk}^n.
$\frac{\delta e}{\delta L_k^{n+1}}$ is the backpropagated error of kth neuron of Layer n + 1.
L_j^n is jth element of nth layer.
Then the weights can then be corrected through the error using the Eq. (9)

$$W_{jk}^n = W_{jk}^n - \alpha * \frac{\delta e}{\delta W_{jk}^n} \tag{9}$$

where α is the learning rate.

Similarly, the error of the current nth layer can be calculated using the Eqs. (10) and (11) as follows:

$$\frac{\delta e}{\delta L_j^n} = \frac{\delta e}{\delta L_k^{n+1}} * \frac{\delta L_k^{n+1}}{\delta L_j^n} \tag{10}$$

$$where,\ \frac{\delta L_k^{n+1}}{\delta L_j^n} = \frac{\delta\left(\sum_i \sum_j W_{jk}^n * L_j^n\right)}{\delta L_j^n} = \sum_j W_{jk}^n \tag{11}$$

This error will not be used to correct the nth layer but will be used as the backpropagation error for (n−1)th layer's weight matrix.

5 Experimental Results

For the experiment, the model was compared with a deep artificial neural network. This deep neural network contains 1 input layer of size 28^2 (the image's dimensions), 2 hidden layers of sizes 50 and 20 respectively, and an output layer having a size of 10 (for 10 digits). It was trained with 60,000 images taken from the MNIST dataset and was trained for 100 epochs. The vector learning model was trained upon a dataset of 1000 vectorized digits and was trained for 1000 epochs. It contains 1 input layer of size 50, 2 hidden layers of sizes 50 and 20 respectively, and an output layer of size 10. For a fair comparison, no parallel processing was used.

It took 5 h for the deep neural network to train the model while vector learning model took only 15 min. This is because it requires less input.

5.1 Training Dataset

After training the deep neural network and vector learning model, these were first tested against the training dataset on which they were trained. The root mean square error for both were:

0.0154908 - Deep neural network
0.0059133 - Vector Learning Method

5.2 Test Dataset

Highly Deformed Digits

For both the models, two similar test datasets were made that contained some digits that were deliberately made to be highly deformed than their original shape. For example, a variation of digit '4' belonging to the test dataset contains a hugely elongated neck, while another has a very small neck but a very broad face. The test dataset for both models together with the result of each digit is shown in Fig. 2 and Fig. 3.

Vector learning model outperforms the traditional neural network method. The reason is that this model is trained from the abstract idea of curves and hence even though the dataset contains highly deformed digits, it recognizes them very accurately.

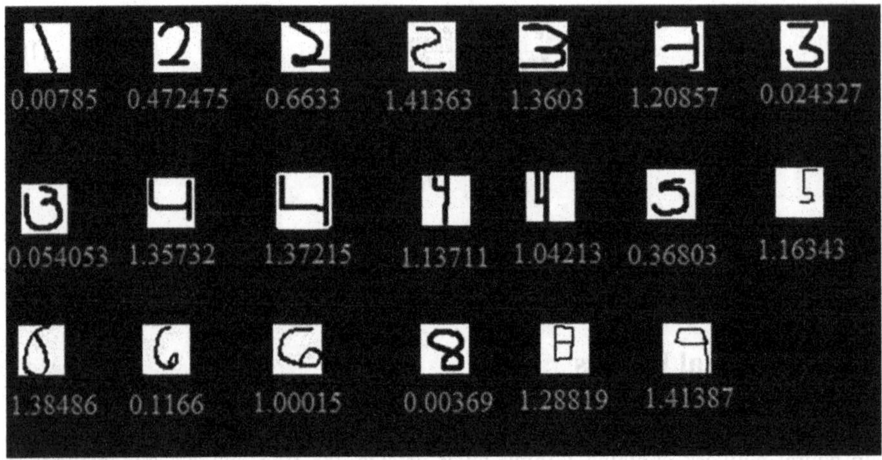

Fig. 2. Highly deformed dataset: root mean square errors given by deep neural network.

Variable Sized Digits

Vector learning model is independent the digit's size. This is because it requires the vectors as input or in other words, it requires the set of curves that forms the digits and not the pixel value of the image. The amount of curves required to make a digit remains the same no matter what the size of the digit is whereas the amount of pixels will change depending on the size of the image.

Position Independent Digits

Another perk of this method is that it is position independent too. The number of curves required to make a digit remains the same no matter where the digit is placed on the canvas, whereas the entire pixel color arrangement in the image changes with the change of position of the digit in traditional methods. Figure 4 and Fig. 5 shows some of the combined variable position and size test set along with the root mean square errors of deep neural network and vector learning method respectively.

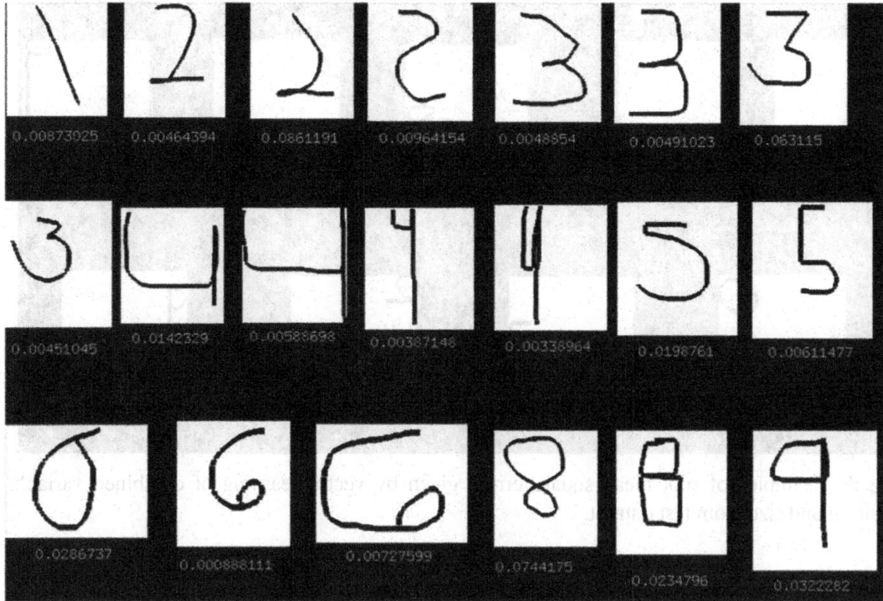

Fig. 3. Highly deformed dataset: root mean square errors given by vector learning method.

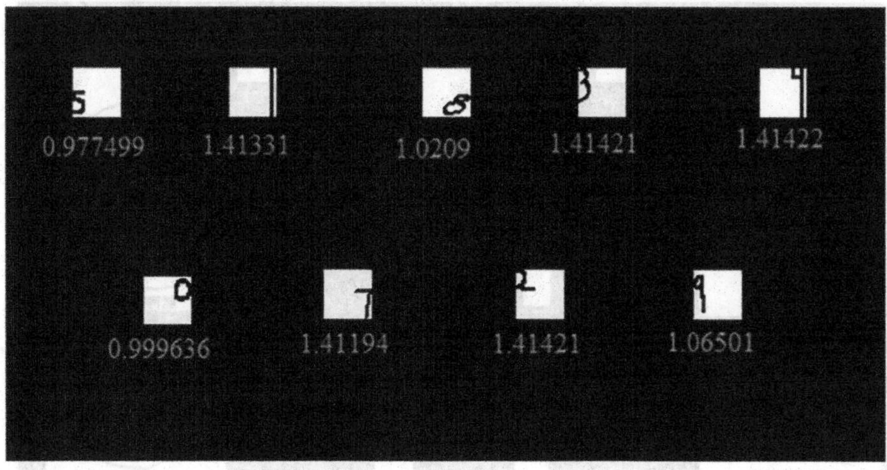

Fig. 4. Examples of root mean square errors given by deep neural network of combined variable position and size from test dataset.

Variable size of canvas

Vector learning model becomes independent of the size of the canvas. This test dataset is only made for vector learning model because variable canvas requires the variable size of weight matrices and hence the traditional method is incapable of solving this

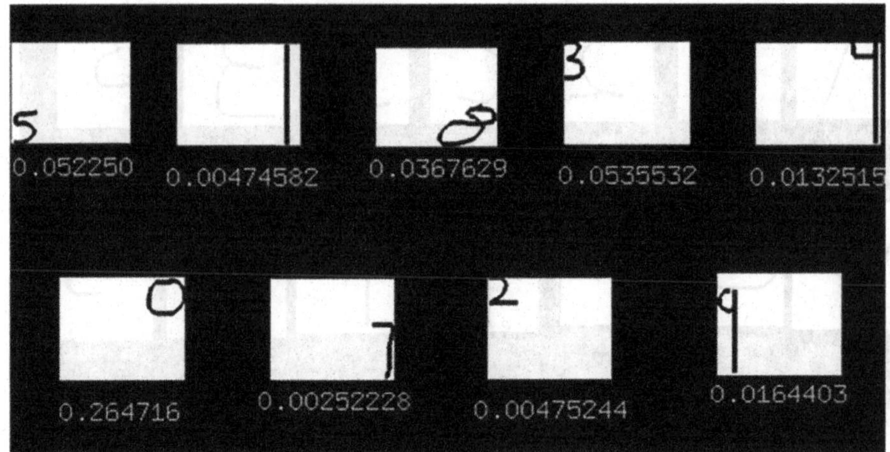

Fig. 5. Examples of root mean square errors given by vector learning of combined variable position and size from test dataset.

problem. Some of the examples from this dataset a with their root mean square error are displayed in Fig. 6.

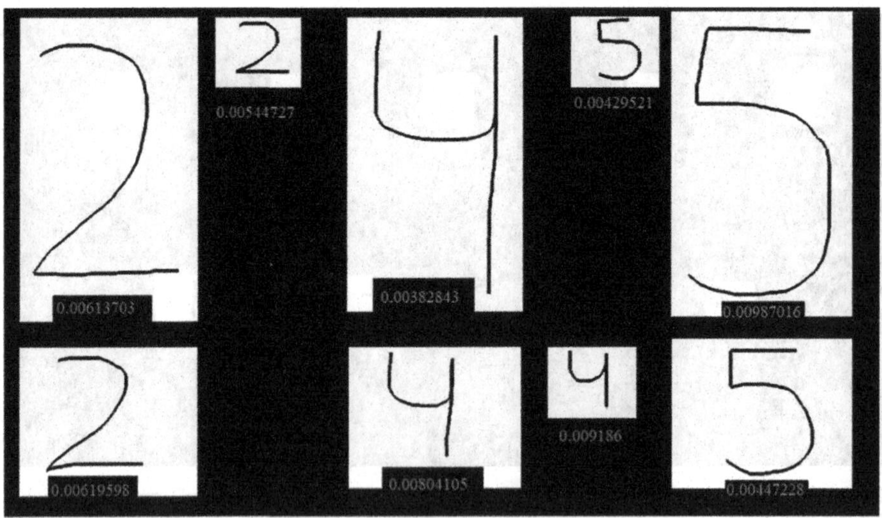

Fig. 6. Canvas independence dataset.

Figure 7 displays the root mean square error of the deep neural network and vector learning method in all the experiments.

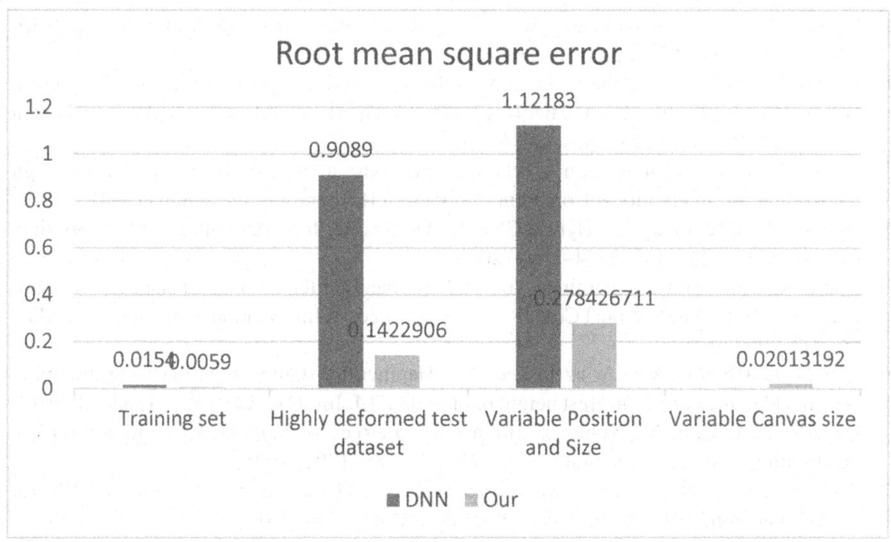

Fig. 7. Root mean square errors for deep neural network and vector learning method.

6 Conclusion

The vector learning model outperforms deep neural network. This demonstrates the power of learning from abstract ideas, which in this case was the idea of forming a digit from curves. This also makes the model capable of recognizing digits with variable sizes. Hence the neural networks should be fed with core ideas of how things work rather than providing mere pixel values of the image.

7 Future Work

Currently, the vectors are extracted from hand drawn images by the user with the help of mouse. This way vectors can be easily extracted but the whole process of drawing all the digits is very tedious and time consuming. For the future work, the set of vectors can be extracted from the image itself using path tracing algorithms.

References

1. McCulloch, W., Pitts, W.: A logical calculus of the ideas immanent in nervous activity. Bull. Math. Biophy. **5**, 115–133 (1943)
2. Rosenblatt, F.: The perceptron: a probabilistic model for information storage and organization in the brain. Psychological Review **65**(6), 386–408 (1958)
3. Ivakhnenko, A., Lapa, V.G.: Cybernetics and Forecasting Techniques. American Elsevier Publishing Company (1967).
4. Werbos, P.J.: Backpropagation through time: what it does and how to do it. In: Proceedings of The IEEE vol 78(10) (1990). October 1990.

5. Dreyfus, S.: The computational solution of optimal control problems with time lag. IEEE Trans. Auto. Cont. **18**(4), 383–385 (1973)
6. LeCun, Y., Bengio, Y.: Globally Trained handwitten word recognizer using spatial representation, space displacement neural network and hidden markov models. In: Advances in Neural Information Processing Systems, vol. **6** (1994)
7. Lecun, Y., et al.: Comparision of classifer methods: a case study in handwritten digit recognition. In: International Conference on Pattern Recognition. Jerusalem (1994)
8. Ahlawat, S., Choudhary, A.: Hybrid CNN-SVM classifier for handwritten digit recognition. Procedia Comp. Sci. **167**, 2554–2560 (2020)
9. Aneja, N.a.A.S.: Transfer learning using CNN for handwritten devanagari character recognition. In: 2019 1st International Conference on Advances in Information Technology (ICAIT), pp. 293–296 (2019)
10. Babu, U.R., Chintha, A.K., Venkateswarlu, Y.: Handwritten digit recognition using structural, statistical features and K-nearest neighbor classifier. I.J. Inf. Eng. Elec. Bus. **1**, 62–63 (2014)
11. Bengio, Y., LeCun, Y., Nohl, C., Burges, C.: LeRec: A NN/HMM hybrid for on-line handwriting recognition. Neural Computation **7**, 1289–1303 (1995)
12. Paul, S., Sarkar, R., Nasipuri, M., Chakraborty S.: Feature map reduction in CNN for handwritten digit recognition. Advan. Intel. Sys. Comp. **740** (2019)
13. Dutta, K., Krishnan, P., Mathew, M., Jawahar, C.V.: Improving CNN-RNN hybrid networks for handwriting recognition. In: 2018 16th International Conference on Frontiers in Handwriting Recognition (ICFHR), pp. 80–85 (2018)
14. Ebrahimzadeh, R., Jampour, M.: Efficient handwritten digit recognition based on histogram of oriented gradients and SVM. Int. J. Comp. Appl. **104**(9), 10–13 (2014)
15. Cakmakov, D., Gorgevik, D.: Handwritten digit recognition by combining SVM classifiers. In: EUROCON 2005 - The International Conference on "Computer as a Tool", pp. 1393–1396 (2005)
16. Turin, W., Hu, J., Brown, M.K.: HMM based online handwriting recognition. IEEE Trans. Pattern Anal. Mach. Intel. **18**, 1039–1045 (1996)
17. Lee, S.-W., Kim, Y.-J.: A new type of recurrent neural network for handwritten character recognition. In: Proceedings of 3rd International Conference on Document Analysis and Recognition, pp. 38–41 (1995)
18. Mahrishi, M., et al.: Video index point detection and extraction framework using custom YoloV4 darknet object detection model. In: IEEE Access, vol. 9, pp. 143378-143391 (2021). Print SSN: 2169-3536, https://doi.org/10.1109/ACCESS.2021.3118048
19. Niu, X.-X., Suen, C.Y.: A novel hybrid CNN–SVM classifier for recognizing handwritten digits. Pattern Recognition **45**, 1318–1325 (2012)
20. Wang, Y., Wang, R., Li, D., Adu-Gyamfi, D., Tian, K., Zhu, Y.: Improved handwritten digit recognition using quantum K-nearest neighbor algorithm. Int. J. Theor. Phys. **58**(7), 2331–2340 (2019)
21. Wigington, C., Stewart, S., Davis, B., Barrett, B., Price, B., Cohen, S.: Data augmentation for recognition of handwritten words and lines using a CNN-LSTM network. In: 2017 14th IAPR International Conference on Document Analysis and Recognition (ICDAR), pp. 639–645 (2017)
22. Zanchettin, C., Dantas Bezerra, B.L., Azevedo, W.W.: A KNN-SVM hybrid model for cursive handwriting recognition. In: The 2012 International Joint Conference on Neural Networks (IJCNN), pp. 1–8 (2012)
23. Su, J., Vargas, D.V., Sakurai, K.: One pixel attack for fooling deep neural networks. IEEE Transactions on Evolutionary Computation, 828–841 (2019)
24. Brown, T.B., Mané, D., Roy, A., Abadi, M., Gilmer, J.: Adversarial patch. In: 31st Conference on Neural Information Processing Systems. Long Beach, CA, USA (2018)

25. Cantareira, G.D., Mello, R.F., Paulovich, F.V.: Explainable adversarial attacks in deep neural networks using. Computer Graphics **xx**(200y), 1–13 (2021)
26. Chen, P.-Y., Sharma, Y., Zhang, H., Yi, J., Hsieh, C.-J.: EAD: elastic-net attacks to deep neural networks via adversarial examples. In: Thirty-Second AAAI Conference on Artificial Intelligence. New York (2018).
27. Clements, J., Lao, Y.: Backdoor attacks on neural network operations. In: 2018 IEEE Global Conference on Signal and Information Processing (GlobalSIP), pp. 1154–1158 (2018)
28. Goodfellow, I.J., Shlens, J., Szegedy, C.: Explaining and harnessing adversarial examples. In: ICLR (2015)
29. Kurakin, A., Goodfellow, I.J., Bengio, S.: Adversarial examples in the physical world. In: ICLR (2017)
30. Khronos: OpenGL - The Industry Standard for High Performance Graphics. 12 March 2021. [Online]. Available: https://www.opengl.org

An Efficient Classifier Model for Opinion Mining to Analyze Drugs Satisfaction Among Patients

Manish Suyal(✉) and Parul Goyal

Department of CA & IT, SGRR University, Dehradun, India
Suyal.manish922@gmail.com

Abstract. Now day's people opinion or sentiment matters a lot in the field of research. There are many patients who stay away from their family, friends and live in the place where hospital, facilities are not available. Then if they have any health related problems and they have to take a medicine in emergency then which medicine will be better for them. Therefore to overcome the problem developed the proposed predictive classifier model in which experience patients who used a medicine in the past, their reviews or feedback will be mined. In the developed model the worthless model has been removed by the stop word removing and stemming algorithms. The developed model will train the simplex decision tree classification algorithm rather than the complex classification algorithm employed in the model used in the previous years. The model used in the previous years were processed over the entire sentence of the reviews, which caused the model to take a long time to process the drug reviews because the worthless words exists in the sentence.

Keywords: Opinion · Sentiment · Drug reviews · Proposed predictive classifier model · Decision tree classification algorithm · Stop word algorithm · Stemming algorithm

1 Introduction

It is predictive classifier model that helps inexperienced patients find right medicine. The first step of the model is dataset collection and feature extraction, in the step dataset contains eight features in which reason, rating, comments, side effect, age, sex, date and duration are prominent but we will extract two features from the dataset in which side effect and comments are prominent. In classification, we will train the whole dataset created after preprocessing step with the simplex decision tree classification algorithm. The entire data of the dataset is divided in the form of binary tree on the basis of yes and no decision. The fourth step is testing in which we will pass the input query of the inexperienced patients to the train proposed model using decision tree classification algorithm. The final step is efficiency of the proposed model in which we will use the confusion matrix to calculate the efficiency of the proposed model by applying the precision, recall and f-measure mathematical formulas and will be compared with the efficiency of the model used in previous years.

V. E. Balas et al. (Eds.): ICETCE 2022, CCIS 1591, pp. 30–38, 2022.
https://doi.org/10.1007/978-3-031-07012-9_3

2 Related Work

We have study the review paper of the last 10 years to understand the work done in the previous year related to the proposed classifier model and after the study of the review papers of the last 10 years, we found the research gap.

Sinarbasarslan, Kayaalp [1] analyzed social media has become a major part of the human life; using the concept of sentiment analysis we can know about any social media post whether the post is positive or negative.

The study [2] proposes Manguri K.H, lot of people have posted their views on social media about covid-19 in which doctor, world health organization (WHO) and government are the head. Angkor, Andresen [3] used a python language to implement the concept of sentiment analysis.

The study [4] describes the supervised learning and unsupervised learning techniques. It is all machine learning techniques with to train a dataset, applying the concept of sentiment analysis to the dataset to make meaningful and then implementing these machine learning techniques on it.

Mohey [5] analyzed, sentiment analysis is a technique of natural language processing (NLP) with the help which, we can process the text and extract subjective information from it. The several machine learning algorithms applied to the dataset are described, among which support vector machine (SVM) and naive bias are prominent [6].

The purpose of this study is to explain the election of America which was held in 2016 in which Donald Trump and Hillary Clinton are prominent [7]. The study [8] proposes Kaur P, psychological disorders are identified in the human brain by supervised machine learning techniques. Mika, Graziotin [9] presented a concept of sentiment analysis is not only used for online product. Kyaing, Jin-Cheon Na [10] proposed the clause-level sentiment classification algorithm is applied on the drug-reviews.

The study [11], expected that extract the important features from the data of the dataset using the concept of sentiment analysis, which gives a complete meaningful dataset. Kechaou [12] analyzed the field of e-learning has carved a distinct identify in the world. Many students are studying online with the help of e-learning. Today's research focus for sentiment analysis is the amendment of level of detail at aspect level, representing two different goals: aspect extraction and sentiment classification of product reviews and sentiment classification of target-dependent tweets [13].

The study [14], expected the process is started from the root node of the decision tree and progresses by applying divided conditions at each non-leaf node resulting uniform subset. Liu, Zhang [15] have analyzed the aspect based sentiment classification mainly emphasis on recognizing the sentiment without thinking the relevance of such words with respect to the given aspects in the sentence. Opinion mining has resulted in the creation of a large variety of technologies for analyzing consumer opinion in the majority of key commercial contexts, including travel, housing, consumables, materials, goods, & services [16]. Shouman, Turner [17] analyzed, decision tree mainly depends on Gain Ratio and Binary Discretization. Information Gain and Gini Index are two other important types of decision trees that are less used in the diagnosis of heart disease.

3 Material, Methodologies, Tools

3.1 Data Collection and Feature Extraction

We plan to use two kinds of dataset gathered at www.askapatient to evaluate user satisfaction. Dataset I consist of customers reviews of Bactrim drug and dataset II consist of customer reviews of the Cymbalta drug. We will use the dataset in the proposed model in which an experience patients has review about a medicine.

3.2 Preprocessing

The second objective of the study is preprocessing face, so that we will make the dataset meaningful. The proposed classifier model will work on these 2 features of the dataset. We will extract the meaningful opinion from the sentence to make the dataset is meaningful for we use stop word removal and stemming algorithm.

- Stop Word Removal. This method laminates meaningful words. This method excludes all term that are not noun, verbs or adjective.
- Stemming Algorithm. The stemming algorithm enhances the system efficiency and reduces indexing size, example the words "workers", "working", or "works" can be minimized to stem.

3.3 Classification

The third objective of the study is classification in which we will train the whole dataset created after preprocessing stage with the decision tree classification algorithm. The decision tree classification algorithm keeps splitting the data of the dataset into binary tree form in two parts based on threshold value.

3.4 Testing

The fourth objective of the study is testing in which we will pass the input query of the inexperience patient to the train Proposed classifier model using decision tree classification algorithm. The proposed model very easily predict the result of input query on the leaf node of the tree, means the model predict the patients the medicine is better for him or not. We will use the 30% portion of the dataset for testing.

3.5 Model Efficiency and Comparison

And along with the, we will use the confusion matrix to get the efficiency of the proposed predictive classifier model. In the confusion matrix, we will compare the actual values with the predictive values of the reviews. The model efficiency is determined by applying the precision, recall, and f-measure. The proposed model derived from the precision, recall and f-measure will be compared with the efficiency of the model used in previous years.

4 Proposed Model

4.1 Proposed Approach

The proposed predictive classifier model responds to positive or negative about the medicine or drug and the patients satisfy his\her desire about the drug in the critical situation with the result given by the proposed predictive classifier model. The proposed model takes a very less time to predict the result of the input query of the inexperience patients (Fig. 1).

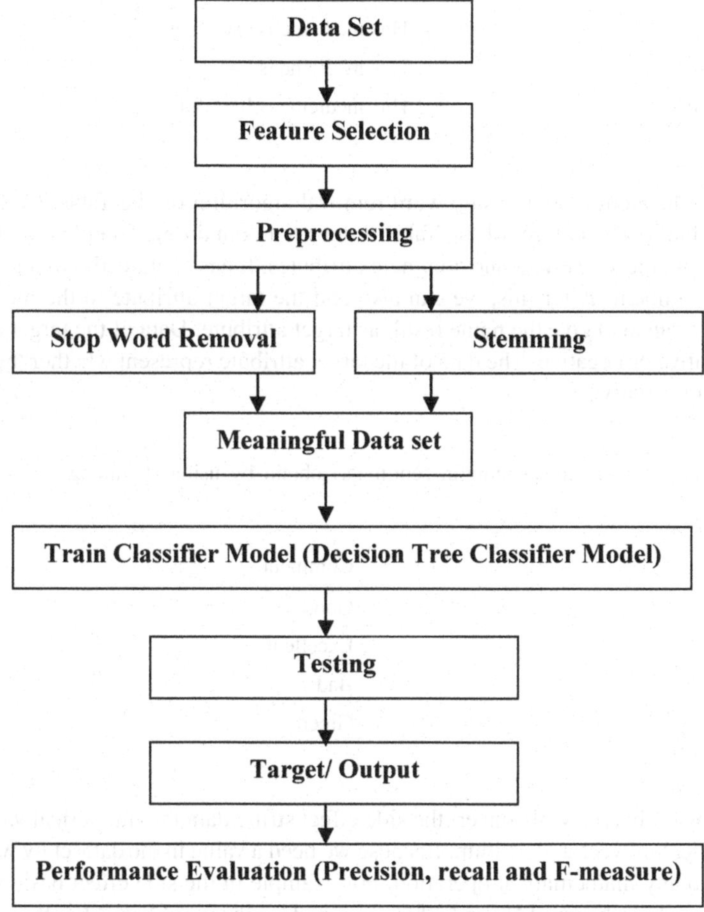

Fig. 1. Data flow diagram (DFD) of the proposed classifier model

5 Experimental Result

The Table 1 shows the dataset will contain two basic attributes, side effects and comments in which there will be complete sentence of reviews about the medicine, but we should not work on the complete sentence of reviews, because work has already been done on the model used in previous years. The complete sentence of reviews has meaningful words and worthless words, which takes a lot of time to process. So we will not work on the complete sentence of reviews in the proposed classifier model.

Table 1. Data set (complete sentence of reviews)

Side effect	Comment
No side effect	The medicine is good
No side effect	This medicine is excellent
Yes side effect	This medicine is bad
Yes side effect	This medicine is harmful

We will the stemming and stop word removal algorithm on the Table 2 which will give the Table 3. The dataset of the Table 3 does not contain the complete sentence of reviews within the side effect and comment attributes. It has contained only meaningful words or sentiment. After this, we can also add the target attribute to the meaningful data set. We can also give the name result as target attribute. Data of the target attribute will be positive or negative. The data of the target attribute represent whether the review is positive or negative.

Table 2. Data set (contain meaningful sentiment) obtain by using stemming and stop word removal

Side effect	Comment
No	Good
No	Excellent
Yes	Bad
Yes	Green

Along with the, we will convert the side effect string data into numerical value with the help of feature vector algorithm. Because we need a value in the dataset by which we can perform any mathematical operation. For example in the side effect basic attribute 200 is assigned for the word No and 300 is assigned for the word Yes. After that, with the help of decision tree classification algorithm, we have spitted the meaningful dataset at Table 4. Suppose we have taken the threshold value is 250. Decision Tree Classification algorithm split the meaningful data set at Table 4 according to the Fig. 2 based on the threshold value. Now we pass an input query to the train model (Fig. 2).

According to the query the proposed predictive classifier model predict the result is positive means review is positive because the side effect is 200 less than 250 taking yes decision and it will come on the comment: good Node, but the comment is not

Table 3. Data set (add target attribute to the meaningful data set, side effect, and comment are basic attribute, side effect data convert to the numerical values using feature vector technique)

Side effect	Comment	Result (Target)
200	Good	Positive
200	Excellent	Positive
300	Bad	Negative
300	Harmful	Negative

Table 4. Meaningful data set with result (target attribute)

Side effect	Comment	Result (Target)
200 (No)	Good	Positive
200 (No)	Excellent	Positive
300 (No)	Bad	Negative
300 (No)	Harmful	Negative

good, taking no decision and it will come on the comment: Excellent Node, because the comment is Excellent taking yes decision and it will come on the positive node. It is target attribute. Hence the proposed predictive classifier model predicts the given review is Positive.

We can find out the efficiency of the proposed model by placing the value of the confusion matrix in the precision, recall and f-measure formulas. We compare the actual values with the predictive values. We calculate the efficiency of the proposed model using the precision, recall and f-measure formulas of the confusion matrix. In the Table 5, we have total 165 reviews. If the review is negative in actual and its predictive value is also negative then it will be counted in TN. If the review is negative in actual and its predictive value is also positive then it will be counted in FP If the review is positive in actual and its predictive value is also negative then it will be counted in FN. Similarly, if the review is positive in actual and its predictive value is also positive then it will be counted in TP. In this way we create confusion matrix based on the actual values and the predictive values. In the example, we have the 50 TN and 10 FP, similarly 5 FN and 100 TP in the confusion matrix. The sum of all these, TN, FP, FN and TP should be equal to the 165 total reviews that means the confusion matrix is consisting the correct values. Now we can find out the efficiency of the proposed model by placing the value of the confusion matrix in the precision, recall and f-measure formulas.

Precision: - In the precision formula, the total values of TP is divided by the sum of total value of TP and total value

$$P = TP/TP + FP \tag{1}$$

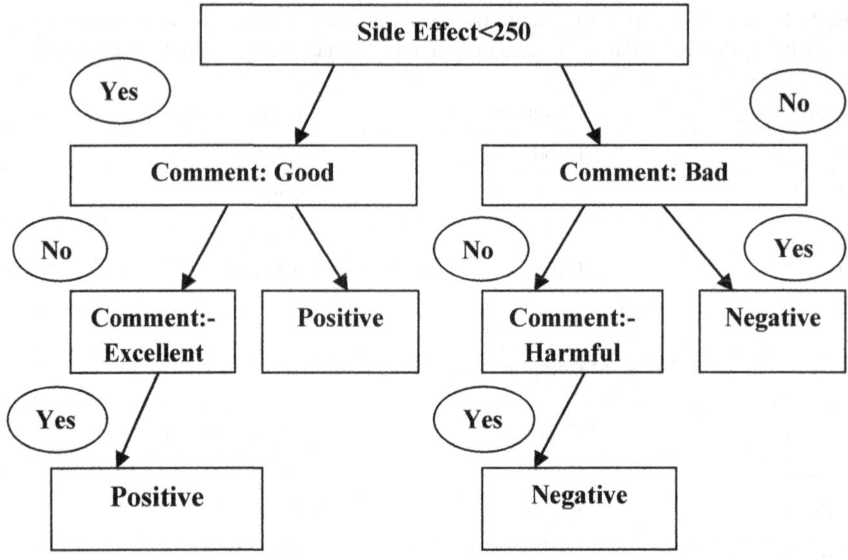

Input Query: Comment:-This medicine is Excellent, Side Effect: - No Side Effect

Meaningful Sentiment:-Comment: Excellent, Side Effect: No (200)

Decision Tree Classifier Model Predict:-Result: Positive (Review)

Fig. 2. Train data set of the proposed predictive classifier model using decision tree classification algorithm with the tested input query of the inexperience patients.

Table 5. Confusion matrix based on the actual values and predictive values of reviews for proposed model.

	Predictive values			
Actual values	165	Yes	No	
	No	TN 50	FP 10	60
	Yes	FN 5	TP 100	105
		55	110	

Note: - TP:-True Positive, FP:-False Positive, FN: - False Negative, TN:-True Negative

Recall: - In the recall formula the total value of TP, is divided by the sum of total value of TP and total value of TN.

$$R = TP/TP + FN \tag{2}$$

F-Measure: - from the f-measure formula, we find out the harmonic mean using precision (Fig. 3 and Table 6).

$$F = 2PR/P + R \tag{3}$$

Table 6. Comparison of proposed model efficiency with the previous model (support vector machine (SVM) and back propagation (BP) technique)

	Precision	*Recall*	*F-Measure*
Proposed model (decision tree)	0.64	0.95	0.76
Previous Model (support vector machine)	0.61	0.89	0.72
Previous model (back propagation)	0.63	0.85	0.71

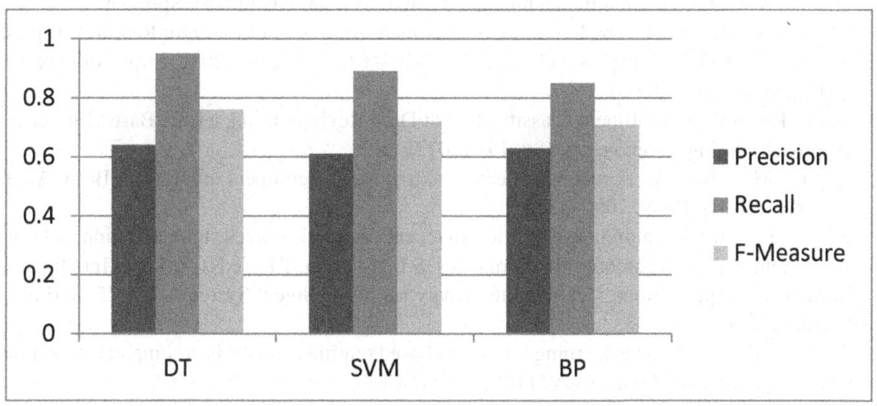

Fig. 3. Comparison chart of the proposed model efficiency (DT) with the previous model (support vector machine and back propagation technique)

6 Conclusion

From, conclude that we do not have to work on the noisy dataset of the model, used in previous years, means we do not have to process the entire sentence of the reviews we only have to work on the sentiment of the sentence in the proposed predictive classifier model. The second conclusion is that we do not have to train the proposed classifier model with the complex classification machine learning algorithm used in previous years for example support vector machine (SVM) and Back Propagation The model prediction growth rate is highly reduced with complex classification algorithm. So we will train the proposed model with the Decision Tree Classification algorithm as it splits the dataset into the form of a binary tree.

References

1. Sinarbasarslan, M., Kayaalp, F.: Sentiment analysis with machine learning methods on social media. J. Adva. Distrib. Artif. Intel. **9N**, 5–15 (2020)
2. Manguri, K.H., Amin, R.N.R.: Twitter sentiment analysis on word wide Covid-19 outbreaks. J. Kurdistan J. Appl. Res. (KJAR) **eISSN**, 2411–7706 (2020)
3. Angkor, M., Andresen, A.: Sentiment analysis using naive bays algorithm of the data crawler. Twitter (2021)
4. Madhoushi, Z., Hamdan, A.R.: Sentiment Analysis Technique in Recent Works. IEEE Xplore (2015)
5. Mohey, D.: A survey on sentiment analysis challenges. J. King Saud Univ.–Eng. Sci. 330–338 (2016)
6. Basra, H., Manglani, R.: Sentiment analysis of twitter data using python. Int. Res. J. Eng. Technol. (IRJET) **06**(eISSN), 2395–0056 (2019)
7. Nausheen, F., Begum, S.H.: Sentiment Analysis to Predict Election Result Using Python. IEEE Xplore (2021)
8. Kaur, P., Sharma, M.: Diagnosis of human psychological disorders using supervised learning and nature-inspired computing techniques. J. Med. Sys. **43**, 204 (2019). Springer
9. Mika, V., Graziotin, D.: The Evolution of Sentiment Analysis- a Review of Research Topics, Venues, and Top Cited Papers, vol. 27, ISSN 15704–0137. Elsevier (2018) https://doi.org/10.1016/j.cosrev.2017.10.002
10. Na, J., Kyaing, W.: Sentiment Classification of Drug Reviews Using a Rule Based Linguistic Approach. Springer-Verlay, pp. 189–198 (2012)
11. Asghar, M., Khan, A.: A review of feature extraction in sentiment analysis. J. Basic Appl. Sci. Res. (2014). ISSN: 2090–4304
12. Kumar, T., et al.: A comprehensive review of recent automatic speech summarization and keyword identification techniques. In: Fernandes, S.L., Sharma, T.K. (eds) Artificial Intelligence in Industrial Applications. Learning and Analytics in Intelligent Systems, vol. 25. Springer, Cham (2022)
13. Do, H.H., Prasad, P.: Deep learning for aspect- based sentiment analysis: a comparative review. Expert System with Applications **118**, 2 (2019). Elsevier
14. Tangirala, S.: Evaluating the impact of GINI index and information gain on classification using decision tree classifier algorithm*. Int. J. Adva. Comp. Sci. App. **11**, 2 (2020). University of Botswana
15. Liu, Q., Zhang, H.: Content attention model for aspect based sentiment analysis. In: Proceedings of the 2018 World Wide Web Conference, pp. 1023–1032. Switzerland (2018)
16. Sánchez-Rada, J.F., Iglesias, C.A.: Social context in sentiment analysis: formal definition, overview of current trends and framework for comparison. Inf. Fusion **52**, 344–356 (2019)
17. Shouman, M., Turner, T.: Using decision tree for diagnosing heart disease patients. In: Proceeding of the 9-th Australasian Data Mining Conference, vol. 12. Australia (2011)

Detection of Liver Disease Using Machine Learning Techniques: A Systematic Survey

Geetika Singh(✉), Charu Agarwal, and Sonam Gupta

Ajay Kumar Garg Engineering College, Dr. A.P.J. Abdul Kalam Technical University, Uttar Pradesh, Ghaziabad, India
singh.geetika93@gmail.com

Abstract. The rapid growth in count of patients suffering from liver disease is a major concern all over the globe. Identification of persons having liver disease is done through liver biopsy and by visual checking of MRI by trained experts which is a tedious and time-consuming process. Therefore, there is a need to develop automated diagnosis system which can provide results in less time and with high accuracy. Researchers worked on this domain and came up with various models for detection of liver disease and its severity using machine learning algorithms. This paper presents a systematic and comprehensive review of the work done in this domain focusing on various machine learning techniques developed by various authors for prediction of liver disease. The performance comparison of the various algorithms is also discussed. This study also explores the datasets used by the various authors for liver disease prediction. Finally, in the conclusion section the challenges involved in liver disease prediction and future scope is discussed.

Keywords: Liver disease · Grey level co-occurrence matrix (GLCM) · Decision tree · Naïve bayes · Support vector machine

1 Introduction

Liver originates from Greek word 'Hepar' and it weighs between 1.44 and 1.66 kg (kg). It is a reddish-brown organ which is vital part of digestive system that helps store sugar and vitamin, move toxin out, prepare bile for digestion to take place in intestine and protein for blood to clot. The texture for liver is rubbery and it is present on the right side of belly protected by rib cage, divided into two sections i.e., right and left lobe. Liver has the ability to regenerate or repair on its own when it is injured either by frequent consumption of alcohol or due to any disease which leads to formation of scar tissue. The food that we consume moves down from stomach to gut, from gut it easily flows with blood into liver through portal vein. The liver filters the blood before it flows out to other organs.

Due to change in lifestyle and eating habits which includes consumption of packaged food, alcohol, junk food which are low in nutrients and consist of high saturated fat, sugar/salt and this deteriorates the health conditions causing obesity, chronic liver disease, cancer, non-alcoholic fatty liver disease. Liver disease has been ranked as the

V. E. Balas et al. (Eds.): ICETCE 2022, CCIS 1591, pp. 39–51, 2022.
https://doi.org/10.1007/978-3-031-07012-9_4

fifth major cause of death in UK according to the National Statistics of UK [1] and second major cause of death in US out of all the digestive diseases [2].

Liver disease can be categorized into inflammation, fibrosis, cirrhosis, liver failure. Inflammation is the early sage in liver disease during which liver swells up and if this swelling of liver is not treated properly, it may cause scarring of liver (Fibrosis). The late stage scarring of liver is termed as "Cirrhosis" which cannot be reversed back.

Two stages of Cirrhosis are Compensated and Decompensated [3]. Compensated is an early-stage disease which is asymptomatic. Whereas, decompensated is advanced stage disease, symptomatic in nature and caused due to high pressure of blood in the portal veins as it becomes difficult for blood to seep into the liver because of hardening of liver called Portal Hypertension. The symptoms include fatigue, itchy skin, spider-like blood vessels on skin, Ascites. One must visit a hepatologist in case of any of the symptoms mentioned above arise for early detection of disease and to avoid further complications that are accumulation of fluid in the legs and abdomen called edema and ascites respectively, spleen gets enlarged, jaundice, risk of liver cancer, Portal Hypertension which may either make the blood flow to smaller veins, causing the veins to burst or cause the enlargement of veins either in the stomach (gastric varices) or esophagus (esophageal varices). To avoid the complications, early diagnosis of disease is important. Diagnosis can be done through various blood and liver tests. The blood tests that can help detect abnormalities in the functioning of liver are Liver function panel, Bilirubin, Albumin, Hepatitis (A, B, C), Alkaline phosphatase.

It is the need of an hour to work on providing computer aided automatic tools for prediction of liver disease which would help doctors diagnose disease in time and provide proper medication to reduce the chances of complications. In the last few years with the growth of machine learning techniques many researchers worked in this direction and come out with promising results. Researchers make use of different machine learning techniques such as Decision Tree [4, 6, 10, 19, 21], Naïve Bayesian [4, 6, 7, 17–19], logistic regression [20, 21], SVM [5, 11, 16, 18, 19, 21], K-NN [8, 11, 17, 18, 20] etc. for the detection of liver disease and its severity.

This paper presents a detail study of the research work conducted in the detection of liver disease and its severity using machine learning models. As per our knowledge, this is the first paper which presents a detail review in this domain. This paper explores the existing literature, technique and database used in them and their performance analysis. It also presents a comparative analysis of these techniques. At last, in the conclusion section the challenges involved in liver disease prediction and future scope is discussed.

2 Machine Learning Approaches

2.1 Decision Tree

It is a tree-structured algorithm in which internal nodes represent attributes of dataset, rules are represented by the branches and outcome of decision rule is represented by the leaves. Figure 1 depicts the general structure of the decision tree.

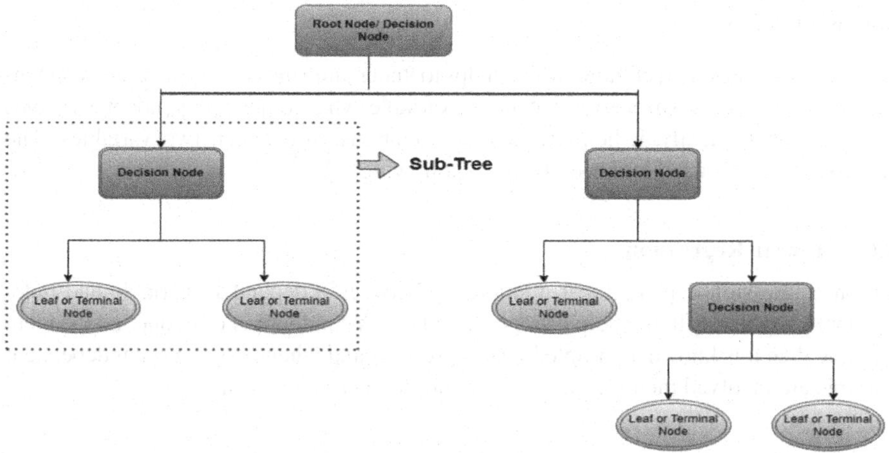

Fig. 1. Decision tree

2.2 Naïve Bayes (NB)

Naïve Bayes comprises of two words: Naïve and Bayes, where Naïve signifies that a feature's occurrence is not dependent on occurrence of other feature and the term Bayes signify the usage of Bayes Theorem. It is probability based supervised classification algorithm and helps to check if record belongs to a particular class or not.

2.3 Artificial Neural Network (ANN)

It is a statistical machine learning algorithm designed in a way that it replicates the functionality of neurons in human brain. As the brain reaches to a conclusion based on the past instance, similarly ANN provides output based on the dataset provided to the model.

2.4 Backpropagation Neural Network

Is also called "backward propagation of errors" and used for training artificial neural network. Backpropagation method calculates the gradient of loss function by chain rule for all the weights of a neural network and hence does fine tuning of the weights based on error.

2.5 Support Vector Machine (SVM)

SVM can classify features into appropriate classes using hyper-plane. Its main aim is to generate decision boundary so that it becomes easy to classify n-dimensional space into classes. Hence, new record generated can be easily put under right class.

2.6 Regression

A supervised learning technique which helps to understand the change in value of dependent variable with respect to independent variable when other independent variables are constant. Basically, it helps to generate relationship between two variables. The regression models used in the study are as following: SVM,

2.6.1 Linear Regression

Linear Regression depicts linear relationship between a dependent variable and independent variables with the help of regression line. When there is only one independent variable then it is known as simple linear regression and when two or more independent variable are involved then it is known as multiple linear regression.

2.6.2 Logistic Regression

Logistic Regression is used for speculating the categorical dependent variable with the help of set of independent variables already provided. The result is either categorical value or discrete value like 0 or 1, yes or no etc.

2.6.3 Random Forest

Random Forest relies on ensemble learning which means using combination of classifiers to solve the problem. Random Forest makes use of varied decision trees to reach to a conclusion instead of just using one decision tree algorithm. More will be the number of trees; more will be the accuracy.

2.7 K-Nearest Neighbor (K-NN)

KNN is a supervised machine learning algorithm used for classification and regression as well and works on the idea of proximity. KNN saves all the data and puts new data into category based on similarity. KNN is lazy algorithm as it doesn't learn from dataset provided for training.

3 Existing ML Based Schemes for Liver Disease Detection

This section presents a review of existing machine learning techniques-based liver disease detection algorithms and their features and characteristics. After exploring a lot of web space, we found that computer automated liver disease detection model generally makes use of machine learning algorithms for designing such models. Figure 2 depicts general block diagram of liver disease detection model.

Kumar et al. [4] proposed a model for analysis of liver disease using data mining algorithms. In this review, creators have used C4.5 Decision Tree (DT) and Naïve Bayesian (NB) classification algorithms to order the patient dependent on his/her information into any of the one class having liver infection or not having liver illness. They gathered 2453 records of patients with 15 attributes from Public Charitable Hospital,

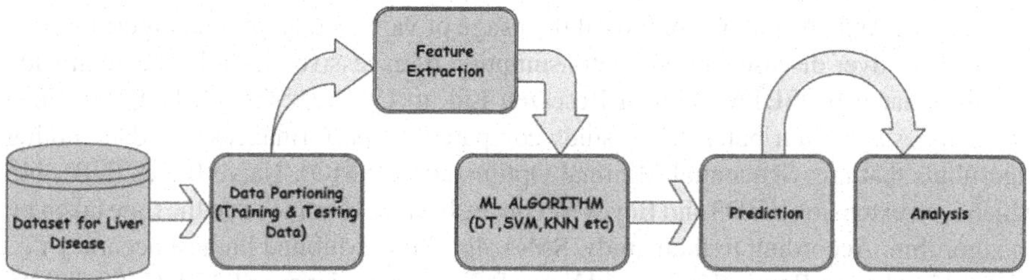

Fig. 2. General block diagram of liver detection model

Chennai out of which 2230 records were experiencing Liver illness and 223 individuals were not having liver sickness. The performance boundaries utilized by the creators to assess the exhibition of the supportive strategy are mean absolute error (MAE) and root mean square error (RMSE). guaranteed that they acquired 0.01 value for MAE and 0.08 for RMSE. They likewise guaranteed that the C4.5 Decision Tree gives most extreme exactness of 99.20% for 90:10 split (90% information is preparing information and 10 percent is Testing Data) of dataset and Naïve Bayesian classifier gives greatest precision of 89.60% for 90:10 split of dataset. They also analyzed their work in terms of computational time which shows that NB classifier takes less time to execute (0.03 s) in comparison to C4.5 DT which takes 0.16 s.

Prakash et al. [5] fostered a model "EP2CD (Efficient Prediction and Classification for Cirrhosis Disease)" for conclusion of phases of Cirrhosis. The dataset utilized for assessment of proposed model is taken from Kaggle Liver patient and MVISS 3.0 T which contains 101,950 records out of which 81,232 average liver picture tests, 9653 early cirrhosis tests and 11,065 progressed cirrhosis samples. The MRI of the liver goes through pre-handling for evacuation of commotion and the Region of interest is separated. The size of crude MRI is extremely enormous which is decreased utilizing Lloyd calculation. In the feature information discovery phase GLCM (Grey Level Co-occurrence Matrix) is used to perform LBP (Local Binary Pattern) filtering on the Region of Interest (ROI). During the final phase, SVM (Support Vector Machine) is used to classify the records of dataset into healthy, early stage and advanced stage cirrhosis. Authors compared the performance of proposed model with GLCM + SVM and GLCM + KNN and stated the accuracy of proposed model i.e., Efficient Prediction and Classification for Cirrhosis Disease to be 100% for healthy Liver,100% for early-stage liver cirrhosis, 99.5 for late-stage liver cirrhosis which is most elevated of the other two models.

Alfisahrin et al. [6] in this review compared the performances of C4.5 Decision Tree developed by J Ross Quinlan, Naïve Bayes (NB) and NB Tree algorithm developed by Kohavi in 1996 for building a productive and ideal liver illness detection model. Authors considered 10 attributes of the dataset to construct this model. The dataset utilized by creators for this review is taken from UCI vault which contains 583 records altogether. Out of which 416 patients were suffering from Liver Disease and 167 were solid. Authors concluded that NB Tree algorithm showcased highest accuracy which is 67.01% whereas accuracies for DT and NB are 66.14% and 56.14% respectively. But NB takes lowest computation time which is 0.04 s, whereas computation time of DT and NB Tree algorithms are 0.45 and 2.51 s.

Study of Arshad et al. [7] proposed the usage of various data mining algorithms for diagnosis of liver disease caused by consumption of excessive alcohol. They extracted dataset donated by BUPA Medical Research Ltd. to UCI repository which contained 345 samples and 8 attributes. Their study compared the performances of 4 data mining algorithms that are: Sequential Minimal Optimization (SMO), Naïve Bayes (NB), J48 which is an extension of ID3 and Bayes Net on the basis of accuracy and the time taken by an algorithm. According to their study, SMO algorithm exhibited highest accuracy i.e., 97.3913% whereas Bayes Net, J48 and Naïve Bayes accuracies are 81.1594%, 80.2899% and 70.7246% respectively.

Singh et al. [8] worked on building a model for identification of degree of liver damage i.e., A (Lowest Degree of Liver Damage), B, C and D (Highest Degree of Liver Damage). which could help doctors prescribe medication accordingly. They used liver function test dataset containing 606 instances and 4 attributes ALK, ALT, AST and TBL to find out the classifier that is best suited for their study. They analyzed LDA, DLDA, QDA, CART, DQDA and KNN on the basis on accuracy. From the study conducted, authors concluded K-NN to be the best classifier with accuracy of 92.53% whereas the accuracy for QDA, CART, DQDA, LDA, DLDA are 32.51%,32.34%,31.52%,28.38% and 27.06% respectively for test dataset. With that, author also focused on finding out the accuracies for predicting various degrees of liver disease using proposed KNN model based on Euclidean distance which are: Degree A – 93.23%, Degree B – 92.57%, Degree C – 92.57%, Degree D – 91.75%.

Ayeldeen et al. [9] designed a decision tree classifier based model for determination of level of liver fibrosis categorized into F_0 = no fibrosis, F_1 = portal fibrosis without septa, F_2 = fibrosis with few septa, F_3 = fibrosis with few septa and F_4 = cirrhosis. Record of 100 patients suffering from chronic Hepatitis C was taken from Kasr Alainy Hospital of Cairo University from which 67% were records of male patients with mean age = 38years and out of male records the percentage of patients with level F_0 = 6%, level F_1 = 29%, level F_2 = 16%, level F_3 = 7% and level F_4 = 9%. The study used 10 variables from which except 'age' other 9 variables were categorized into Routine Function Test and Serum Test. Authors used MySQL 5.2 for storage of data. The proposed decision tree model accurately classified 93.7% patients.

Nahar et al. [10] carried out this study to compare the performance of 7 decision tree algorithms i.e., J48, Logistic Model Tree (LMT), Random Forest (RF), Random Tree (RT), Decision Stump (DS), Hoeffding Tree (HT) and REP Tree for detection of liver disease on the basis of seven parameters (ACC%, MAE, PRE, REC, FME, Kappa Statistics and runtime). In this study, dataset was taken from UCI repository and it contained 538 records of patients where 167 records were negative and 416 records tested positive for liver disease. After the experiment, author came to the conclusion that Decision Stump was the best out of all with 70.67% accuracy while accuracies for J48, LMT, RF, RT, REP Tree and HT are 65.69%, 69.47%, 69.30%, 66.55%, 66.13% and 69.75% respectively. Also REP tree, Random Tree and J48 Decision Tree algorithms execute faster as compared to other algorithms and LMT takes longest execution time even when dataset is small.

Thirunavukkarasu et al. [11] compared three classification algorithms which are Linear Regression (LR), K-Nearest Neighbor (KNN) and Support Vector Machine (SVM) on the basis of accuracy and confusion matrix for prediction of Liver Disease. They used Indian Liver Patient Dataset which contained 567 instances and is taken from UCl repository. Authors used dataset in the ratio 70:30 for their study and came to the conclusion that accuracy for KNN and LR is 73.97 whereas accuracy for SVM is 71.97%. Also, the sensitivity of LR (Sensitivity = 0.952) is higher than KNN (Sensitivity = 0.904) which meant LR correctly diagnose the patients suffering from Liver Disease.

Abdar et al. [12] compared two decision tree-based classification algorithms i.e., Boosted C5.0 and CHAID for generation of rules that could be used to predict liver disease. For this study, authors extracted Indian Liver Patient Dataset (ILPD) from UCI repository which contained 583 records and 10 attributes out of which 416 suffered from liver disease and 167 did not. The dataset was divided into training, testing and validation set in the ratio 60:30:10. They compared the performance of Boosted C5.0 with CHAID for generation of rules that ease out detection of liver disease and to check if these rules can predict accurately. During the experiment, Boosted C5.0 generated 92 rules out of which 50 rules belong to class 1 (Liver Disease) and 42 rules belong to class 2 (Non Liver Disease).On the other hand, CHAID produced 9 rules out of which 6 belong to class 1 (Liver Disease) and 3 belong to class 2 (Non Liver Disease).The results obtained from experiment showcase that DB(Direct Bilirubin), ALB(Albumin), SGPT, TB(Total Bilirubin) and A/G attributes play important role in diagnosis of liver disease in both the algorithms. Three additional attributes Alkphos(Alkaline Phosphatase), Age and SGOT are important for diagnosis of liver disease in Boosted C5.0 algorithm. The authors compared the performance of algorithms and came to the conclusion that Boosted C5.0 has higher accuracy of 93.75% than CHAID algorithm which is 65%. Lastly, authors concluded that Boosted C5.0 was better in terms of accuracy and generated a greater number of rules than CHAID.

Hassoon et al. [13] used Genetic Algorithm (GA) to minimize the number of rules generated by Boosted C5.0 used in [14] for accurate prediction of liver disease in reduced amount of time. Genetic algorithm is based on Selection, Crossover, and Mutation. The Indian Liver Patient Dataset (ILPD) was taken from UCI repository for this study by the authors. The number of rules generated by Genetic Algorithm were 24, whereas Boosted C5.0 provided 92 rules. Accuracy achieved after conducting experiment was 92.93% for GA and 81.87% for Boosted C5.0 algorithm. Authors came to the conclusion based on various performance parameters, that GA correctly predicted existence of Liver Disease in less time.

P. Kumar et al. [15] proposed Fuzzy ANWKNN classification model for correctly predicting the liver infection using Liver Function Test (LFT). The model proposed by the authors is produced by mixture of fuzzy adaptive k-nearest neighbor (KNN) which is used for calculating membership of training data and Neighbor weighted KNN used for calculation of membership of test data. The authors used Indian Liver Patient Dataset (ILPD) with 10 attributes extracted from UCI repository and Madhya Pradesh Region Liver Patient Dataset (MPRLPD) with 12 attributes. The ILPD dataset contains 583 records out of which 416 are liver patients and 167 are not. The MPRLPD dataset contains 7865 records out of which 6282 are records of patients suffering from liver infection and 1583 are solid. The authors compared the performance of Fuzzy ANWKNN with Fuzzy ADPTKNN model proposed by Patel and G.S Thakur [23] using both the datasets and came to the conclusion that accuracies of Fuzzy ANWKNN are 84.29%, 91.12% for ILPD and MPRLPD respectively, which is better than that of Fuzzy ADPTKNN (Accuracies 76.34% - ILPD and 83.13% -MPRLPD).

Prajith et al. [16] proposed a model for classification of level of seriousness of liver fibrosis into normal, grade 1, grade 2, grade 3 and cirrhosis based on the texture properties extracted from imaging. In this experiment, First Order Statistic (FOS) and Gray Level Co-occurrence matrix (GLCM) used for extracting 8 texture features which are mean, skewness, kurtosis, standard deviation extracted using FOS and energy, entropy, homogeneity, correlation extracted using GLCM. The image database consists of 214 samples where 44 samples were normal, 42 for grade 1, 41 samples for grade 2, 42 – for grade 3, 45-cirrhosis samples. The classifiers used by then are ANN, SVM and GMM for classification of level of fibrosis. The dataset for classifiers contained 150 samples for training and 50 for testing. After conducting the experiment, authors came to the conclusion that accuracy of SVM is the highest which is 94% whereas accuracies for GMM and ANN are 86%, 88% respectively.

S. Kumar et al. [17] compared the performances of 5 data mining classification algorithms, namely Naïve Bayes (NB), Radom Forest (RF), K-Means, K Nearest Neighbor (KNN), C5.0 based on 3 performance parameters accuracy, precision, recall for correct prediction of liver disorder. The dataset used by authors have been collected from north eastern areas of India. It contains 416 records of patients suffering from liver disorder and 167 record of patients not suffering from liver disorder. After completing their study, authors concluded that out of 5 classification algorithms random forest predicts more correctly with accuracy of 72.18% while accuracies for C5.0, K-NN, K-Means and NB are 71.43%, 70.68%, 69.47%, 66.17%. Authors also stated that implementation of C5.0 with adaptive boosting gives the highest accuracy of 75.19%.

Ramana et al. [18] proposed a study for prediction of liver disease by comparing the performances of Naïve Bayes (NB), C4.5, Back Propagation (BP), K-NN and SVM classification algorithms for varying attribute combinations. For carrying out this study 2 datasets were considered, one was BUPA Liver Disorder dataset from UCI machine learning repository which has 345 records of patients with 5 attributes and the other dataset contains 751 records of liver patients with 12 attributes which is taken from Andhra Pradesh, India (AP dataset). Firstly, AP dataset was taken into consideration for analyzing the performances of classification algorithms with different combinations of attributes. Similarly, performance of classification algorithms was analyzed using dataset from UCI repository. Authors came to the conclusion that performance of AP dataset is higher than dataset extracted from UCI repository for NB, C4.5, BP, KNN and SVM algorithm as AP dataset has a greater number of attributes. Also, KNN, BP and SVM are providing better results for all combinations of attributes.

Auxilia [19] proposed a comparative study for predicting liver disorder in a person. Author compared the performances of various machine learning classification algorithms such as DT, NB, SVM, RF and ANN. To pursue her study, she used dataset from UCI repository and came to the conclusion that Decision Tree algorithm's performance is the best out of all which 81% accuracy while accuracies for NB, SVM, RF, ANN are 37%, 77%, 77%, 71% respectively.

Shaheamlung et al. [20] proposed a hybrid model based on machine learning classifiers for diagnosis of liver disease. Authors prepared a hybrid model which includes RF, K-NN and LR classifiers. They used Kaggle liver patient dataset containing 582 records and 10 attributes. They divided the dataset into 60% for training and remaining 40% for testing. In this paper authors used python along with Spyder tool for implementation and concluded that proposed model's performance was better than the existing model as proposed model's accuracy was 77.58%, execution time = 0.25 s and the accuracy for existing model = 73.27%, execution time = 0.4 s.

Gogi et al. [21] compared the performances of SVM, LR and DT classification algorithm for predicting liver disease. They used Liver Function Test records of 584 patients for carrying out the experiment. They came to the conclusion that performance of LR is better than that of SVM and DT with accuracy of LR = 95.8%, SVM = 82.7% and DT = 94.9%. Then authors tested 572 patients using linear regression classifier out of which 394 were predicted with 'YES' and 178 were predicted with 'NO'.

4 Dataset

As it is widely known that to implement any machine learning algorithms dataset is required to train and test the developed model. Very few public datasets related with liver disease detection are available till now for research purpose, so many researchers either worked on more than one dataset or worked on combined dataset from different sources. This section presents a detail of the various datasets used by the liver disease detection algorithms discussed in Sect. 3.

4.1 Public Charitable Hospital in Chennai

This database was designed by a charitable hospital located in Chennai [1]. It contains the records of 2,453 patients, of which 939 were female and 1,514 were male. This dataset gives the details regarding the various parameters such as age, sex, fever, DA, LD, LA etc.

4.2 Indian Liver Patient Records (ILPD)

It is most widely used dataset downloaded from UCI machine learning repository which contains records of 583 patients, of which 441 were male and 142 were female. The data is categorized into 'Liver Disease' and 'No Disease'. According to the dataset, the number of patients categorized under Liver Disease are 416 and 167 patients are put under 'No Disease' label. This dataset is available at [22].

4.3 Madhya Pradesh Region Liver Patient Dataset (MPRLPD)

MPRLPD is a dataset which records the patient's data from a hospital located in the city of Bhopal, Madhya Pradesh state in India [15]. It contains 7865 records in all with 12 features. Out of 7865 records, 6282 people were having liver disease and 1583 were healthy.

5 Performance Analysis

This section presents analysis of the various liver disease detection methods based on machine learning techniques. Table 1 tabulates the various parameters related with each scheme.

Table 1. Analysis of various liver disease detection schemes

Reference	Dataset	Size of dataset	ML technique	Accuracy
[4]	Public Charitable Hospital, Chennai	2453	a) NB b) C4.5 DT	C4.5 DT:99.20% NB:89.60%
[5]	Kaggle liver patient dataset, MVISS 3.0 T	101,950	SVM + GLCM + LBP	Healthy Liver: 100% Early: 100% Later:99.5%
[6]	UCI Repository	583	C4.5 DT, NB and NBTree	C4.5 DT: 66.14% NB:56.14% NBTree:67.01%
[7]	UCI Repository (BUPA Medical Research Ltd.)	345	SMO, NB, J48, Bayes Net	SMO: 97.3913% Bayes Net: 81.1594% J48:80.2899% NB:70.7246%
[8]	Liver Examination Data that signifies 4 drug doses (A,B,C,D)	606	LDA, DLDA, QDA, DQDA, CART, KNN	LDA:28.38%, DLDA:27.06%, QDA:32.51%, DQDA:31.52%, CART:32.34% and KNN: 92.53%
[9]	Kasr Alainy Hospital of Cairo University	100	Decision Tree	93.7%
[10]	UCI Machine learning repository	583	J48, LMT, RF, RT, REPTree, DS, Hoeffding Tree	J48:65.69%, LMT:69.47%, RF = 69.30%, RT = 66.55%, REPTree = 66.13%, DS = 70.67%, Hoeffding Tree = 69.75%
[11]	ILPD dataset	567	LR, KNN, SVM	KNN:73.97% LR: 73.97% SVM: 71.97%
[12]	ILPD dataset	583	a) Boosted C5.0 b) CHAID	Boosted C5.0: 92.33% CHAID: 76.14%

(*continued*)

Table 1. (*continued*)

[13]	UCI Database	583	Genetic Algorithm (compared with Boosted C5.0)	a) Genetic Algorithm: 92.93% b) Boosted C5.0: 81.87%
[15]	ILPD and MPRLPD	ILPD: 583 MPRLPD:7865	Fuzzy ANWKNN	a) Fuzzy-ANWKNN:91.12% b) Fuzzy-ADPTKNN: 83.13%
[16]	Self CollectedData from Private Hospital	214	ANN, GMM, SVM	ANN: 88% GMM: 86% SVM: 94%
[17]	UCI Repository	583	NB, RF, K-means, C5.0, KNN	C5.0 with Adaptive Boosting: 75.19%, RF: 72.18%, C5.0: 71.43%, KNN: 70.68% K-Means:69.47%, NB: 66.17%
[18]	AP And UCLA	345 + 751	NB, C4.5, BPN, SVM, KNN	NB:56.52%, C4.5:68.69%, BP:71.59%, KNN:62.89%, SVM:58.26%
[19]	ILPD from UCI	583	DT, NB, SVM, RF, ANN	DT:81%, NB:37%, SVM:77%, RF:77%, ANN:71%
[20]	Kaggle liver patient dataset	583	Proposed Model (KNN + RF + LR)	77.58%
[21]	LFT dataset	584	SVM, DT, LR	LR:95.8%, SVM:82.7%, ANN:94.9%

A study in which Gray Level Co-Occurrence Matrix and Local Binary Pattern is used for extracting features, then SVM is used as a classifier provides highest accuracy of 99.5% [2]. Decision Tree is most widely used data mining algorithm used by authors and out of all, C4.5 Decision Tree gives highest accuracy of 99.20%. Naïve Bayes (NB) algorithm has also been used by authors for conducting their study. The highest accuracy provided by NB is 89.60%.

6 Conclusion

As Liver is an integral part of digestive system, it become very important to detect abnormality at an early stage so that it does not turn into life threatening disease. Hence, to help doctors various machine models have been developed for detection of liver disease. This paper presents a comprehensive review of study already done by various authors for developing liver disease detection model using machine learning techniques. In the future, work can be done on enhancing the performance of classification algorithms by considering important attributes, combining two or more classification algorithms and huge dataset can be collected from health care sector for testing the performance.

References

1. http://www.statistics.gov.uk/
2. https://www.medicalnewstoday.com/articles/305075#structure
3. https://www.hepatitis.va.gov/cirrhosis/background/stages.asp
4. Kumar, A., Venkateswaran, J.: Estimating the surveillance of liver disorder using classification algorithms. Int. J. Comp. Appl. (57), 39–42 (2012)
5. Prakash, K., Saradha, S.: Efficient prediction and classification for cirrhosis disease using LBP. GLCM and SVM from MRI images (3), (2021)
6. Alfisahrin, S.N.N., Mantoro, T.: Data mining techniques for optimization of liver disease classification. In: International Conference on Advanced Computer Science Applications and Technologies, pp. 379–384 (2013)
7. Arshad, I., Dutta, C., Choudhury, T., Thakra, A.: Liver disease detection due to excessive alcoholism using data mining techniques. In: International Conference on Advances in Computing and Communication Engineering (ICACCE), pp. 163–168 (2018)
8. Singh, A., Pandey, B.: An euclidean distance based KNN computational method for assessing degree of liver damage. In: International Conference on Inventive Computation Technologies (ICICT), pp. 1–4 (2016)
9. Ayeldeen, H., Shaker, O., Ayeldeen, G., Anwar, K.M.: Prediction of liver fibrosis stages by machine learning model: A decision tree approach. In: Third World Conference on Complex Systems (WCCS), pp. 1–6 (2015)
10. Nahar, N., Ara, F.: Liver disease prediction by using different decision tree techniques. Int. J. Data Mining & Know. Manage. Proc. (8), 01–09 (2018)
11. Thirunavukkarasu, K., Singh, A.S., Irfan, M., Chowdhury, A.: Prediction of liver disease using classification algorithms. In: 4th International Conference on Computing Communication and Automation (ICCCA), pp. 1–3 (2018)
12. Abdar, M., Zomorodi, M., Das, R., Ting, I.-H.: Performance analysis of classification algorithms on early detection of Liver disease. Expert Systems with Applications (67), 239–251 (2017)

13. Hassoon, M., Kouhi, M.S., Zomorodi-Moghadam, M., Abdar, M.: Rule optimization of boosted C5.0 classification using genetic algorithm for liver disease prediction. International Conference on Computer and Applications (ICCA), pp. 299–305 (2017)
14. Patel, H., Thakur, G.: An improved fuzzy k-nearest neighbor algorithm for imbalanced data using adaptive approach. IETE J. Res. 1–10 (2018)
15. Kumar, P., Thakur, R.S.: Diagnosis of liver disorder using fuzzy adaptive and neighbor weighted K-NN method for LFT imbalanced data. International Conference on Smart Structures and Systems (ICSSS), pp. 1–5 (2019)
16. Mahrishi, M., Sharma, G., Morwal, S., Jain, V., Kalla, M. Chapter 7 data model recommendations for real-time machine learning applications: a suggestive approach. In: Kant Hiran, K., Khazanchi, D., Kumar Vyas, A., Padmanaban, S. (eds.) Machine Learning for Sustainable Development, pp. 115–128. De Gruyter, Berlin, Boston (2021). https://doi.org/10.1515/9783110702514-007
17. Kumar, S., Katyal, S.: Effective analysis and diagnosis of liver disorder by data mining. In: International Conference on Inventive Research in Computing Applications (ICIRCA), pp. 1047–1051 (2018)
18. Ramana, B., Surendra, P., Venkateswarlu, P.: A critical study of selected classification algorithms for liver disease diagnosis. Int. J. Datab. Manage. Sys. (IJDMS) (3), (2011)
19. Auxilia, L.A.: Accuracy prediction using machine learning techniques for indian patient liver disease. In: 2nd International Conference on Trends in Electronics and Informatics (ICOEI), pp. 45–50 (2018)
20. Shaheamlung, G., Kaur, H.: The diagnosis of chronic liver disease using machine learning techniques. Info. Technol. Indu. (9), (2021)
21. Gogi, V.J., Vijayalakshmi, M.N.: Prognosis of liver disease: using machine learning algorithms. In: International Conference on Recent Innovations in Electrical, Electronics & Communication Engineering (ICRIEECE), pp. 875–879 (2018)
22. https://www.kaggle.com/uciml/indian-liver-patient-records
23. Patel, H., Thakur, G.: A hybrid weighted nearest neighbor approach to mine imbalanced data. In: Proceedings of the International Conference on Data Mining (DMIN). The Steering Committee of The World Congress in Computer Science, Computer Engineering and Applied Computing (WorldComp), p. 106 (2016)

Identification of Dysgraphia: A Comparative Review

Dolly Mittal[1](✉), Veena Yadav[2], and Anjana Sangwan[3]

[1] Department of IT, SKIT, Jaipur, India
dolly.mittal@skit.ac.in
[2] Department of CSE, PCE, Jaipur, India
veena.yadav@poornima.org
[3] Department of CSE, SKIT, Jaipur, India
anjana@skit.ac.in

Abstract. Dysgraphia is a common learning disability in children worldwide. It is characterized as a disturbance or difficulty in the production of written language presented through visual graphics. Almost 10–20% of school going children do face this issue. The child's functional limitation in creating correct formation of letters or words, insufficient speed and legibility of written text is considered as Developmental Dysgrphia Disorder. Also the term developmental dysgraphia refers to the fact that a child is unable to get writing skills, in spite of the sufficient opportunities to learn in the absence of any neurological disorder. Because of dysgraphia children may have serious issues in their day to day life. It is proposed by various researchers that there might be serious consequences in a child's academic, social and emotional behaviour because of handwriting difficulties. It is therefore required to detect it in the earlier phase.

There are various scales which are developed to assess the handwriting quality. The Objective of this paper is to present various methods available for automatic detection of dysgraphia. We also presented a comparative study of existing research work for early detection of dysgraphia based on some already available measures.

Keywords: Dysgraphia · Handwriting quality · Developmental dysgraphia

1 Introduction

Handwriting is an important skill and in spite of the widespread use of digital tools, it is considered as an important communication medium. A child spent much of his time in various academic processes which involve handwriting activity, also handwriting is an important mean used by a child to convey his progress of learning to the teachers. Handwriting involves linguistics, perceptual, attention and fine motor skills [1,2]. It can be assessed either in qualitative (legibility) or in quantitative (speed) manner. Dysgraphia is a common learning disability

V. E. Balas et al. (Eds.): ICETCE 2022, CCIS 1591, pp. 52–62, 2022.
https://doi.org/10.1007/978-3-031-07012-9_5

in children worldwide. It is described by difficulty in writing words or letters presented through visual graphics. Hamstra-Bletz and Blote [3] defined dysgraphia or handwriting difficulty disorder as a disturbance or difficulty in the production of written language that is related to the mechanics of writing.

Children with dysgraphia do have problem in identifying words or letters, use inappropriate formations like mirror writing, letter reversal etc. Almost 10–20% of school going children do face this issue [4]. Dysgraphia is associated with dyslexia and dyspraxia, which are described as reading disorder and developmental coordination disorder respectively. All three of these are neuro-developmental disorders [5–7]. The child's functional limitation in creating correct formation of letters or words, insufficient speed and legibility of written text is considered as developmental dysgrphia disorder [8,9]. Children with neuro-developmental issues were found to have handwriting difficulties, in some cases dysgrphia was reported in children without any neurological disorder too [10].

In literature there are two meaning of dysgraphia are used, some researcher used it to refer to the problems in spelling, others used it to refer to deficits in the process of production of legible handwriting. Also the term developmental dysgraphia refers the fact that a child is unable to get writing skills, in spite of the sufficient opportunity to learn in the absence of any neurological disorder.

School is the best place, where teachers can help in early diagnosis of handwriting difficulties. In case, a child is not developing the required handwriting skills could be given some kind of handwriting practise sessions. While writing a paragraph, a child with this disorder will learn less about the paragraph composition or the topic as compared to the child with typical handwriting skills.

Because of dysgraphia children may have serious issues in their day to day life. It is therefore required to detect it in the earlier phase. It is proposed by various researchers that there might be serious consequences in a child's academic, social and emotional behaviour because of handwriting difficulties. These research therefore suggest to detect handwriting issues as early as possible [11–13].

The most common method to detect dysgraphia is by evaluating the letters or words written by a child on paper sheets [14–18]. Various test are used for assessing the quality of handwriting, some of them are based on evaluation of various features for the handwritten text like legibility, spacing between words, inclination of letters, handwriting speed etc. These tests cannot be used all over the world as there are differences in language alphabets. Another issue with these tests is that there is an expert of handwriting, who investigates the sentences written by a child and for this reason these tests are scaled poorly and are biased. Also, these tests do not take into account very important kinematic aspects like tilt and pressure of pen, writing dynamics, which can be used in the preliminary diagnosis of dysgraphia [19–22]. Nowadays with the easy availability of digital tablets, these ignored characteristics can be measured at an affordable cost and it can be scaled easily [23–25].

The Objective of this paper is to present various methods available for automatic detection of dysgraphia. Further this paper is organized as follows, Sect. 2 describes the motivation for this research, Sect. 3 and 4 discuss various handwriting analysis methods and the work already done for the detection of dysgraphia respectively, followed by the last section which concludes the paper.

2 Motivation

Very little attention has been paid relatively by researchers to the developmental deficits in the acquisition of writing skills. The motivation behind this research is to impart knowledge about these deficits to the people, also this will be a preventive and corrective measure towards these deficits. Research in this direction will enhance our understanding about the difference between lack of normal writing skills and developmental dysgraphic deficits.

3 Handwriting Analysis Methods

3.1 BHK Method

BHK, also known as Concise Evaluation Scale for Children's Handwriting was presented in the Netherlands in 1987 at the department of developmental psychology of the university of Leiden [3]. This method is basically used in French speaking countries to assess the quality of handwriting. This is the most common method for assessing the quality of writing in various clinical and academic activities.

In this method a standardized text, along with some tools is given to the child, who is the subject of this test. This text is copied by the subject using following tools: A4 size white paper without lines, a black pen, a timer and physical support. The standardized text for BHK method is dutch text, many researchers have re-adapted it to their own native languages.

The subject is instructed to copy this text on white paper in 5 min and first five lines at least. It consists of assigning scores to thirteen parameters to find out the features of a manuscript document, which helps in detection of dysgraphia. The most significant thirteen feature for this handwriting task are as follows:

1) Writing size, 2) non-aligned left margin, 3) skewed writing, 4) insufficient space between two words, 5) sharp angles, 6) broken links between two letters, 7) collision between two letters, 8) irregular size of the letters, 9)inconsistent height between letters with extension and without extension 10) atypical letters, 11) ambiguous letters, 12) traced letters, 13) unstable track.

All of these features except the 9th feature, are assigned a score between 0 and 5, feature 9 is assigned a value in between 0 and 4. If the final rating deviates below the average standard deviation, then the subject is considered as dysgraphic. BHK is not appropriate for screening purpose, as it needs a rigorous training and consumes approximately 15–20 minutes even if the evaluator is an experienced person.

3.2 Systematic Screening for Handwriting Difficulties (SOS)

This method is introduced by Smits-Engelsman, Stevens, Vrenken and van Hagen in 2005 [26]. It was developed as a short and efficient tool for handwriting screening. It is based on BHK, but it can evaluate the score in a shorter time. It was suggested that first SOS can be used for screening purpose and if detailed information is necessary, then complete BHK test can be performed. It was proved in a study that out of the thirteen features of BHK test, six features represents 65% of the variance. These six features are redesigned and simplified for the development of SOS test. The six features are as follows: letter formation fluency, letter connection fluency, height of letter, letter height regularity, word spacing and sentence regularity. First 5 lines of the written text are used to assign values to these 6 features. The total SOS assessment score can be in between 0 and 12, a high value of this indicates a poor handwriting.

3.3 Handwriting Proficiency Screening Questionnaire (HPSQ)

HPSQ is a method to identify handwriting difficulties in school going children using a questionnaire of 10 questions by their teachers. Some of the items in this questionnaire are legibility of handwriting, performance time i.e. time required to complete a writing task, physical and emotional state of the subject , inability to read his/her own handwriting, lack of interest in completing assigned homework etc. These items are scored on a scale of five point. A cut off score of 14 is used to designate a handwriting deficient children. The authors [27] well established the validity and reliability of this screening method, given good values for inter-rater, intra-rater and test-retest reliability and interclass coefficient. This is an appropriate screening method to identify childrens with handwriting difficulties for academic and clinical purpose. This method was developed for Israeli kids, now several countries are adapting this method for screening purpose.

3.4 Minnesota Handwriting Assessment Tool

It is a good handwriting assessment tool which requires very few minutes and assesses various good psychometric features. The words in the Minnesota Handwriting Assessment are written in print and as such. This test is not valid for children who are learning a cursive font.

3.5 Herbrew Handwriting Evaluation (HHE) Test

In this method the examiner observes the position of paper and pencil, posture of the child and fatigue etc. Inclusion of dynamic features of handwriting samples leads to improved accuracy in the diagnosis task of dysgraphic children, also these diagnoses are less biased and can be scaled.

3.6 Evaluation Tool of Children's Handwriting- Manuscript (ETCH-M)

This method is developed by Amundson in 1995. This is a standard tool, which is designed for the evaluation of legibility and speed test of handwriting in children of age group staring from 6 year and above. The pre requirement for this test is 10–12 weeks prior experience of manuscript writing. In this test 6 different tasks are assigned to the children of grade 1–2, these tasks are very much similar to the work they are assigned in the classroom. In first task children are required to write all alphabets in lowercase and uppercase using their own memory. In second task child is required to write numeral from 1 through 12 using memory. The third and fourth tasks are copying tasks named as near point copying and far point copying task respectively. In the third task child has to copy a short sentence from a pre printed sheet placed at a distance of 3 in. from the top of response sheet. In the fourth task child has to copy a short sentence from a pre printed sheet placed on a wall at a distance of 6 to 8 ft from the desk of the child. In fifth task 2 words of 5 letter each and 1 zip code of 5 numbers are recited by the examiner. In sixth task the child is expected to compose a sentence of at least 5 words.

3.7 Detailed Assessment of Speed of Handwriting (DASH)

This method is designed as per the United Kingdom norms for assessing the handwriting speed of children of age in the range 9–16 year [28]. This method includes various handwriting tasks that can be either copying a given text or writing some text using memory, it also includes a free handwriting task.

The 4 important tasks of DASH method are described as follows:

1. Copy best: In this task the child or subject is instructed to copy the following sentence in the best manner, he can for a duration of two minutes:
   "The quick brown fox jumps over the lazy dog"
   The observer records the number of word written in each minute, illegible word and punctuation marks are excluded from the total score.
2. Copy fast: This task is very much similar to the copy best task, but here the child is instructed to write the text as fast as possible. Also he has to make sure that the copied text must be legible.
3. Alphabet: In this task the child is instructed to write all alphabets repeatedly in the correct sequence from memory for duration of one minute. Also he has to write alphabets in lowercase and the alphabets must be in readable form. The alphabet letter is discarded if it is written in uppercase, illegible, out of order or reversed. The observer records the number of letters written per minute.
4. Free writing: In this task the child is instructed to write on a particular topic for a duration of ten minute. The child is given one minute to think of some ideas related to that topic prior to the task. The content of this task is not considered for evaluation, the observer has to record only the per minute words which is averaged for a 10 min period.

The observed scores for all of these tasks are then converted to the standardized value having a mean and standard deviation of 10 and 3 respectively. After this an overall score is also computed and converted to the standardized overall score with a mean and standard deviation of 100 and 15 respectively.

4 Comparative Analysis of Various Methods

It is proposed by various researchers that there might be serious consequences in a child's academic, social and emotional behaviour because of handwriting

Table 1. Comparative Analysis of work for Dysgraphia Detection

Year	Author	Data collection method	Data analysis approach	Participant	Features	Age or grade group	Performance analysis
2020	G. Dimauro et al.	BHK (manual as well as digital)	Document analysis algorithms are used	12 Dysgraphic and 92 Typically Developing (TD) children	13	2nd-5th grade, average age is 8.84 with std 0.94	96% Accuracy
2018	T. Asselborn et al.	BHK using a digital Tablet (wacom intuos)	K-fold cross validation on a random forest classifier	56 Dygraphic and 242 TD children	Total 53 out of them 8 are most discriminative features	1st–5th grade	F1-Score 97.98%, std of 2.68%
2012	V. Waelvelde et al.	SOS for screening purpose and BHK for detailed analysis	Interclass Correlation Coefficient (ICC) Model was used for the evaluation of inter-rater, intra-rater and test-retest reliability of the SOS Score	259 D and 603 TD Dutch children	6	7–12 age group	Inter and Intra-rater reliability is excellent, test-retest reliability is moderate, a correlation of 0.70 is found between SOS and BHK score
2019	K. Zolna et al.	Child has to write all the alphabets, all 10 digits (randomely dictated) on a tablet	Compared CNN vs RNN model	24 D, 971 TD Children	A probability score is assigned to each glyph, which is then averaged for all 36 glyphs, giving D statistic	Pre to 5th grade	More than 90% dysgraphic children were detected correctly. RNN model outperformed the CNN model
2020	T. Mavrea et al.	Data is collected from the archive of 2nd KEDDY of Thessaloniki	WISC-III (Wechsler Intelligence Scale third edition) Compared Intellectual profiles	104 children, distributed in 3 clinical (Dysgraphia, dyslexia, Attention Deficit Hyperactivity Disorder) and 26 TD	10 main scales out of 13 scales of WISC-III method, also verbal, performance and full scale intelligence quotients	8–12	ADHD group showed lower performance in comparison to other clinical and TD group
2020	P. Drotar et al.	Manuscript were created by writing various letters, words and sentences using a tablet	Adaptive boosting algorithm (AdaBoost) was found most effective for their experiments out of various machine learning algorithms	57 D, 63 TD	22 spatiotemporal, kinematic features	8–15	Accuracy 80%

(*continued*)

Table 1. (*continued*)

Year	Author	Data collection method	Data analysis approach	Participant	Features	Age or grade group	Performance analysis
2020	T. Asselborn et al.	Used digital tablet iPad for writing first 5 sentences of BHK Test	Multi-dimensional analysis of handwriting, used PCA to extract features, K-mean clustering is used	390 TD, 58 D	Total 63 features (static, kinematic, pressure and tilt aspects)	5–12	Sensitivity 91%, Specificity 90%
2019	V. Zvoncak et al.	A paragraph of 63 words, 371 characters is copied using a tablet	Tunable Q-factor wavelet transform (TQWT) vs HPSQ-C Sequential floating forward selection is used along with SVM/RF classifier	65 children divided in two groups on the basis of HPSQ-C score	Total 665 (kinematic, spatial, temporal, horizontal and vertical projection)	3rd to 5th grade	Accuracy 79.16% Sensitivity 86.22% Specificity 73.32%
2016	J. Mekyska et al.	HPSQ, subject is asked to write seven herbe letters in a sequence using a tablet	Used random forest classfier, intra writer normalization that increases estimation accuracies	27 D, 27 TD	Total 51 features (kinematic, non linear dynamic and others)	Age group 8–9 and 3rd grade	96% Sensitivity and Specificity
2020	P. Yogarajah et al.	14 hindi words and 3 conjoined consonant, 267 image samples (164 with D child and 103 TD) from regular classroom work	Used CNN, accuracy is estimated through 5-fold cross validation	54 children	Visual features of hindi handwriting	1st to 5th grade, mean age 8.39 year	Mean accuracy 86.14% std 1.02
2016	S. Rosenblum et al.	Copy two five word sentences and write two specific letters eight times using a tablet	Linear SVM was used for training and testing of machine learning model, performance is estimated using10-fold cross validation	Total 99, 50 Proficient and 49 no proficient	Total 20 from all 4 tasks	3rd grade and 8–9 year age group	Accuracy 89.9%

(*continued*)

Table 1. (*continued*)

Year	Author	Data collection method	Data analysis approach	Participant	Features	Age or grade group	Performance analysis
2020	T. Gargot et al.	BHK with a digitizing tablet	Linear model was constructed fromex-tracted features, K-mean Clustering is used to group children into different clusters according to the severity of dysgraphia	218 TD and 62 Dysgraphic (231 from different school and 49 from a learning impairment clinic)	12 features (kinematic, static, pressure, tilt etc.)	Average age of 9 year	A new classification method for Dysgraphia is established on the basis of features of handwriting
2018	P. D'Antrassi et al.	A sentence Containg all the Italian alphabet was given to copy (Manual and Digital)	Hand-motor performance analysis is used to evaluate writing elements stokes and component, PCA is used on kinematic and qualitative parameters	2 group of 20 students in each	16 qualitative and 12 kinematic parameters	10–11 Age group	70% to 100% PCA variance

difficulties. These research therefore suggest to detect handwriting issues in its early phase. A comparative analysis of the research work for identification of Dysgraphia is presented in Table 1.

Following are the major findings based on the analysis of existing work:

1. It has been found out that some methods llike BHK are specific to a particular language, they cannot be applied as it is in other parts of the world. For a nation like India where every region has its own native language, it would be unfair to use dysgraphia detection test specific to English language for a child who is more familiar to his own native language than english. Hence we need a dysgraphia detection test which is particular to child's language.
2. Consideration of fine motor skills like pen tilt, pressure and other aspects is important while identifying whether a child is dysgraphic or not.
3. Deep learning approach can also be incorporated along with machine learning for more accurate analysis of handwritten images.

5 Conclusion

A person's functional limitation in creating correct formation of letters or words, insufficient speed and legibility of written text is considered as Dysgrphia. The aim of this study is to impart knowledge about these deficits to the people, also awareness about these deficits will help to take preventive and corrective measure. The most common method to detect Dysgraphia is by evaluating the handwriting sample written by a child on paper sheets. Various test are used for assessing the quality of handwriting, some of them are based on evaluation of various features for the handwritten text like legibility, spacing between words, inclination of letters, handwriting speed etc. Dysgraphia diagnosis based on these features is biased and cannot be used uniformly all over the world as there are differences in language alphabets. Inclusion of dynamic features of handwriting samples leads to improved accuracy in the diagnosis task of dysgraphic children, also these diagnosis are less biased and can be scaled.

References

1. Feder, K.P., Majnemer, A.: Children's handwriting evaluation tools and their psychometric properties. Phys. Occup. Therapy Pediatrics **23**(3), 65–84 (2003)
2. Feder, K.P., Majnemer, A.: Handwriting development, competency, and intervention. Dev. Med. Child Neurol. **49**(4), 312–317 (2007)
3. Hamstra-Bletz, L., Blöte, A.W.: A longitudinal study on dysgraphic handwriting in primary school. J. Learn. Disabil. **26**(10), 689–699 (1993)
4. Mekyska, J., Faundez-Zanuy, M., Mzourek, Z., Galaz, Z., Smekal, Z., Rosenblum, S.: Identification and rating of developmental dysgraphia by handwriting analysis. IEEE Trans. Human-Mach. Syst. **47**(2), 235–248 (2016)
5. Asselborn, T., Gargot, T., Kidziński, Ł, Johal, W., Cohen, D., Jolly, C., Dillenbourg, P.: Automated human-level diagnosis of dysgraphia using a consumer tablet. NPJ Dig. Med. **1**(1), 1–9 (2018)
6. Yogarajah, P., Bhushan, B.: Deep learning approach to automated detection of dyslexia-dysgraphia. In: The 25th IEEE International Conference on Pattern Recognition (2020)

7. Mavrea, T., Malegiannaki, A.C., Apteslis, N., Kosmidi, M.H.: Comparison of intellectual profiles among children with different types of neurodevelopmental disorders and typically developing children. ENCEPHALOS **57**, 35–43 (2020)
8. Rosenblum, S., Dror, G.: Identifying developmental dysgraphia characteristics utilizing handwriting classification methods. IEEE Trans. Human-Mach. Syst. **47**(2), 293–298 (2016)
9. Döhla, D., Willmes, K., Heim, S.: Cognitive profiles of developmental dysgraphia. Front. Psychol. **9**, 2006 (2018)
10. Gargot, T., et al.: Acquisition of handwriting in children with and without dysgraphia: a computational approach. PLoS One **15**(9), e0237575 (2020)
11. D'Antrassi, P., Perrone, I., Cuzzocrea, A., Accardo, A.: A composite methodology for supporting early-detection of handwriting dysgraphia via big data analysis techniques. In: De Pietro, G., Gallo, L., Howlett, R.J., Jain, L.C. (eds.) KES-IIMSS 2017. SIST, vol. 76, pp. 241–253. Springer, Cham (2018). https://doi.org/10.1007/978-3-319-59480-4_25
12. Treatment of handwriting problems in beginning writers: Transfer from handwriting to composition. J. Educ. Psychol. **89**(4), 652–666 (1997)
13. Rosenblum, S., Weiss, P., Parush, S.: Product and process evaluation of handwriting difficulties. Educ. Psychol. Rev. **15**, 41–81 (2003)
14. Pagliarini, E.: Childrens first handwriting productions show a rhythmic structure. Sci. Rep. **7**(1), 1–10 (2017)
15. Schneck, C., Amundson, S.: Prewriting and handwriting skills. Occup. Therapy Child., 555–582 (2010)
16. Medwell, J., Wray, D.: Handwriting automaticity: the search for performance thresholds. Lang. Educ. **28**, 34–51 (2021). http://eprints.nottingham.ac.uk/id/eprint/45306
17. Rosenblum, S.: Relationships between handwriting features and executive control among children with developmental dysgraphia. Am. J. Occup. Therapy **70** (2015)
18. Asselborn, T., et al.: Bringing letters to life: handwriting with haptic-enabled tangible robots. In: Proceedings of the 17th ACM Conference on Interaction Design and Children, IDC 2018, p. 219230. Association for Computing Machinery, New York (2018). https://doi.org/10.1145/3202185.3202747
19. Zolna, K., Asselborn, T., Jolly, C., Casteran, L., Johal, W., Dillenbourg, P., et al.: The dynamics of handwriting improves the automated diagnosis of dysgraphia. arXiv preprint arXiv:1906.07576 (2019)
20. Dimauro, G., Bevilacqua, V., Colizzi, L., Di Pierro, D.: Testgraphia, a software system for the early diagnosis of dysgraphia. IEEE Access **8**, 19564–19575 (2020)
21. Mahrishi, M., Morwal, S., Muzaffar, A.W., Bhatia, S., Dadheech, P., Rahmani, M.K.I.: Video index point detection and extraction framework using custom yolov4 darknet object detection model. IEEE Access **9**, 143378–143391 (2021)
22. Asselborn, T., et al.: Reply: limitations in the creation of an automatic diagnosis tool for dysgraphia. npj Dig. Med. **2**, 1–2 (2019)
23. Asselborn, T., Chapatte, M., Dillenbourg, P.: Extending the spectrum of dysgraphia: a data driven strategy to estimate handwriting quality. Sci. Rep. **10**(1), 1–11 (2020)
24. Drotár, P., Dobeš, M.: Dysgraphia detection through machine learning. Sci. Rep. **10**(1), 1–11 (2020)
25. Zvoncak, V., Mekyska, J., Safarova, K., Smekal, Z., Brezany, P.: New approach of dysgraphic handwriting analysis based on the tunable q-factor wavelet transform. In: 2019 42nd International Convention on Information and Communication Technology, Electronics and Microelectronics (MIPRO), pp. 289–294. IEEE (2019)

26. Van Waelvelde, H., Hellinckx, T., Peersman, W., Smits-Engelsman, B.C.: Sos: a screening instrument to identify children with handwriting impairments. Phys. Occup. Therapy Pediat. **32**(3), 306–319 (2012)
27. Rosenblum, S.: Development, reliability, and validity of the handwriting proficiency screening questionnaire (HPSQ). Am. J. Occup. Therapy **62**(3), 298–307 (2008)
28. Barnett, A., Henderson, S., Scheib, B., Schulz, J.: Development and standardization of a new handwriting speed test: the detailed assessment of speed of handwriting. BJEP Monograph Ser. II, Number 6 - Teach. Learn. Writ. **1**, 137–157 (2009)

Using a Technique Based on Moment of Inertia About an Axis for the Recognition of Handwritten Digit in Devanagri Script

Diwakar Sharma(✉) and Manu Sood

Department of Computer Science, Himachal Pradesh University, Shimla, India
diwakarsharma1337@gmail.com

Abstract. A neural network for handwritten digit recognition is dependent on the dataset used to train it. Most of the techniques available in the literature recognize the handwritten characters of any script based upon the moment of inertia of the character uses a central point to it and suffers from certain drawbacks. In this paper, we develop methods that can be applied to any neural network, to make the dataset insensitive to brush size and insensitive to rotation applied to the digit based on the moment of inertia that the digit produces about an axis. All images in the dataset undergo pre-processing. First, the image is made independent of brush size. Then, the image is rotated to negate the rotation that is applied to the image according to the axis of the least moment of inertia. This pre-processing makes the dataset independent of brush size and rotation. The results show that when this preprocessing technique is applied to neural a network, it does makes it insensitive to rotation and brush size.

Keywords: Handwritten digit recognition · Image recognition · Machine learning · Moment of inertia · Rotation insensitivity · Brush size insensitivity

1 Introduction

Handwritten digit recognition is a basic task for a neural network. R.A. Fisher developed the first algorithm to do so [1]. He proposed a model of two normally distributed populations of n-dimensional vectors having mean vectors with co-variance matrices. He proved that the optimal Bayesian solution is a quadratic function. Then Rosenblatt developed the base of modern neural networks, the perceptrons [2]. The perceptron consisted of an array containing 400 photocells which were connected randomly to neurons. Weights were updated during learning using the electric motors and were encoded using potentiometers. Perceptron was not a program, it was a machine, specially constructed for recognizing images. Custom-built hardware was implemented for it and was named Mark 1 perceptron. Researchers started developing neural networks that were insensitive to change in translation or rotation. Reid et al. developed a system that was independent of translation, rotation, and scale [3]. It used a single-layer second-order neural network. The second-order neural network used invariance rules to make two or more

V. E. Balas et al. (Eds.): ICETCE 2022, CCIS 1591, pp. 63–78, 2022.
https://doi.org/10.1007/978-3-031-07012-9_6

weights to be constrained by equality. Hence, a fewer number of weights are required to train, reducing the risk of combinatorial explosion for the higher-order neural networks. Fukushima et al. developed neocognitron [4] in 1983. It was insensitive to translation and small deformation. Widrow et al. developed the MADALINE [5]. It was insensitive to translation and rotation by 90°. Lee et al. developed a method to make digits insensitive to translation, rotation, scale, and deformation [6], which used high order neural network as a preprocessor.

An axially symmetric architecture was proposed by H. Sawai [7]. It used weights arranged in a concentric circle for each layer. To recognize the rotation images, humans perform a mental rotation of images in their minds. Using this, Fukumi et al. created a rotational-invariant neural network that performs mental rotation [8]. Patil and Sontakke made Devanagri script digits insensitive to rotation, insensitive to scale, and insensitive to translations by using the general fuzzy networks [9]. Multiple instances of convolutional neural networks running together can also be used to achieve insensitiveness to rotation. This is shown by Ayushi Jain et al. in their neural network also known as RIMCNN [10]. These techniques either increases the input dataset by rotating the input images to various degrees or by increasing the intricacy and difficulty in training for the neural network.

1.1 Pattern Recognition Based on Moment Invariants

Ming-Kuei Hu [11] first presented the idea to use moments to extract features of a pattern. A total of seven-moment invariants were developed using geometric properties of the pattern. These patterns are known as geometric invariants Using a combination of these moments, a point can be generated in a 2D plane. The Euclidian distance of these points for two patterns can tell how similar these patterns are. Michael Read Teague [12] expanded upon the idea by using Zernike polynomials to generate Zernike moment invariants as well as using Legendre polynomials to generate Legendre moment invariants. Using moments generated by one of these techniques, any image can be reconstructed to a point in 41-dimensional space. The similarity of the two images is determined based on the values of the moments and how close they are.

Another use of Zernike moment invariants was found by Alireza Khotanzad and Jiin-Her Lu [13]. A multilayer perceptron or feed-forward neural network was used to classify patterns. The Zernike moment invariants were calculated for an image. These moments were then transformed into an input vector to be fed to the neural network. The results showed that the neural network performed better than other classifier techniques. Jan Flusser and Tomas Suk [14] used affine transformation to generate affine moment invariants. These transformations preserve lines and parallelism. For each pattern, three features are extracted based on affine moment invariants. Minimum distance on 3D Euclidean feature space is used to classify the patterns. Yet another polynomial-based technique was developed by R. Mukundan [15]. Discrete Tchebichef polynomials were used to generate Tchebichef moment invariants. It performed better than orthogonal moment invariants such as those based on Zernike polynomials or Legendre polynomials.

Yinan observed that the error from translation, rotation, and scaling invariants can further be reduced by using united moment invariants [16]. It imposed three conditions on geometric invariants. These conditions reduce the error in the invariants. Krawtchouk polynomials were used by Yap to generate Krawtchouk moment invariants [17]. These are used as with geometric invariants in a linear combination to classify patterns. The information redundancy in the moment-based pattern recognition was reduced by Zhou [18]. In this technique, discrete orthogonal Hahn polynomials are used to generate Hahn moment invariants. These are then used to classify patterns. The idea of Zhou was enhanced by Zhu by using Dual Hahn polynomials instead of Hahn polynomials [19]. These polynomials generated Dual Hahn moment invariants. The moments were generated using Dual Hahn polynomials on a non-uniform lattice. These moments perform better than Hahn polynomials for pattern recognition. Raj used Pollaczek polynomials to generate Pollaczek Moment invariants [20]. These polynomials ensured numerical stability.

To reduce error on rotation invariants, Xiao introduced Bessel Functions to generate Bessel-Fourier moment invariants [21]. Yang and Dai used weighted Hermite polynomials to generate Gaussian-Hermite moment invariant [22]. The Gaussian-Hermite moments proved to be more effective for translation and rotation invariant in a pattern. Sayyouri and Hmimid [23] introduced Meixner polynomials to generate Meixner moment invariants. The computational costs of these moments were less than that of orthogonal moment invariants. Hmimid yet again introduced Charlier polynomials to generate Charlier moment invariants [24]. The computational costs of these moments are low, at the same time they were invariant to translation, rotation, and scaling factors. Pandey used a type of orthogonal polynomials known as Bateman polynomials to generate Bateman moment invariants [25]. These polynomials performed better for shape analysis.

Karmouni [26] was able to develop a technique to use both Krawtchouk and Tchebichef polynomials together to generate Krawtchouk-Tchebichef moment invariants. Batioua on the other hand was able to use three polynomials in conjunction [27]. These polynomials were Krawtchouk polynomials, Tchebichef polynomials, and Dual Hahn polynomials. To further improve on translation and scale invariants, Zhi used Krawtchouk moments to generate a new technique [28]. It uses normalization for scale and translation invariance. Binod Prasad introduced two new techniques known as Pixel moment of Inertia and delta distance encoding to generate moments invariants [29]. These moments were fed to a hidden Markov model to classify digits in the MNIST database. All these techniques are summarized in Table 1.

All the methods for moment invariants discussed so far calculate moment from a central point, it is the first moment of distribution most of the time. Furthermore, these techniques feed the vector of these moments to the classifiers as the input and not the image of the pattern itself. The method proposed in this paper improves on the idea by taking the moment of inertia across the entire axis rather than a central point for the Devanagri script digits. The axis chosen is such that the moment of inertia around it is the lowest. The axis found lies in the plane of the image. The entire image is then rotated such that the axis of the least moment becomes the vertical axis. The rotated image is then fed to a feed-forward neural network, instead of the moment. The image is also made invariant to the brush size used for drawing the pattern.

Section 2 and 3 explain the task in detail. Section 2 describes how the image is made insensitive to brush size, while Sect. 3 describes how the image is made insensitive to

Table 1. Various moment invariant based techniques for pattern recognition.

Author	Types of moment	Work
Ming-Kuei Hu	Developed 7 moments (geometric invariants) for feature extraction based on the central moment, similitude moment and orthogonal moment invariants	A pair of moments invariants computed from the combination of central moments is used. The pair is a point in two-dimensional space. The Euclidian distances of these points for two patterns determine how similar they are
Michael Reed Teague	Uses Zernike & Legendre polynomials instead of geometric invariants to construct Zernike and Legendre moment invariants	Image is identified as a point in 41-dimensional space based on Zernike or Legendre moments invariant. The similarity of the two images is determined based on these invariants
Alireza Khotanzad	Uses Zernike moment invariants on MLP (multilayer perceptron)	Zernike moment invariants are extracted from the image. The set of moments so obtained are converted to a vector to be fed as input to a FFNN (MLP) and compared against other classifier techniques. NN performed better than other techniques even with fewer training samples
Jan Flusser and Tomas Suk	Uses affine transformation to generate affine moment invariants	Uses three features based on affine moment invariants and classification was done using minimum distance in 3D Euclidean feature space
R. Mukundan	Developed Tchebichef Moment invariants based on discrete Tchebichef polynomials	Proved to be better than orthogonal invariants such as Legendre or Zernike
Yinan	Developed United moment invariant to reduce the error from translation, rotation and scaling from geometric invariants	Uses three conditions based on geometric invariants to remove error generated by scaling factor in geometric invariants

(*continued*)

Table 1. (*continued*)

Author	Types of moment	Work
Yap	Uses Krawtchouk polynomials to generate Krawtchouk moment invariants	Uses weighted Krawtchouk polynomials with a linear combination of geometric invariants
Zhou	Uses discrete orthogonal Hahn polynomials to generate Hahn Moment Invariants	Was able to generate feature with minimum information redundancy
Zhu	Uses Hahn and Racah polynomials to generate Dual Hahn and Racah Polynomials respectively	Uses Dual Hahn polynomials on non-uniform lattice to define moments. It performs better than Hahn polynomials for pattern classification
Raj	Uses Pollaczek polynomials to generate Pollaczek Moment Invariants	Normalizes Pollaczek polynomials to ensure numerical stability
Xiao	Uses Bessel function to generate Bessel-Fourier moment	Performs good for rotation invariance in image analysis
Yang and Dai	Uses weighted Hermite polynomials to generate Gaussian-Hermite Moment invariant	Gaussian-Hermite moments are more invariant to translation and rotation in patterns
Sayyouri and Hmimid	Uses Meixner Polynomials to generate Meixner moment invariants	Moments reduced computational costs as compared to orthogonal moments
Hmimid	Uses Charlier Polynomials to generate Charlier moment invariant	Was invariant to rotation, translation and scale factors and improved computational cost
Pandey	Uses Bateman Polynomials to generate Bateman moment invariant	These polynomials perform better for shape analysis
Karmouni	Developed Krawtchouk-Tchebichef Moment invariant	Improved the method of image reconstruction
Batioua	Uses three moment invariant Tchebichef, Krawtchouk & Dual-Hahn moment to generate new set of moments	Improved performance of recognition and invariability
Zhi	Uses Krawtchouk moment to generate new technique to improve on translation and scale invariants	Not generated from polynomials and uses normalization for scale and translation invariance

(*continued*)

Table 1. (*continued*)

Author	Types of moment	Work
Binod Prasad Kumar	Uses Pixel moment of Inertia and delta distance encoding	Uses Multiple hidden Markov model to classify digits in the MNIST database by using Pixel moment of Inertia and Delta Distance Encoding

rotations. Section 4 tells us about the neural network and dataset used to train. Section 5 explains the results. Section 6 provides the conclusion.

2 Insensitiveness to Brush Size

Each image has a white background and the digit is drawn with a black brush. The first job of the preprocessor is to make the input image insensitive to the brush size used to draw the digit. For this, the input image is scanned three times.

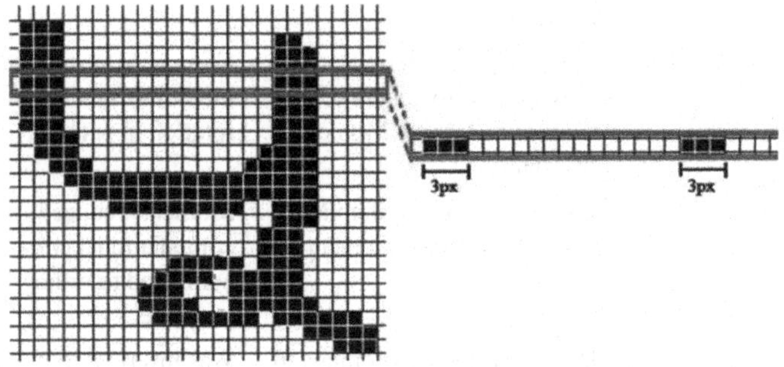

Fig. 1. Average brush size for a row for the digit '5' in Devanagri script.

2.1 The First Scan

The first scan is done row by row. This scan is used to obtain the brush size using which the digit was drawn. An array is used which counts the number of instances of possible brush size. The scan starts from the leftmost pixel of each row and traverses to the rightmost pixel. During the traversal, it checks for the stream of continuous black pixels. This is displayed in Fig. 1. The number of pixels in the continuous stream of black pixels is taken and then its value in the array is increased. When the scan finishes with the last row, the number in the array with maximum value corresponds to the brush size with maximum occurrences and is named as the average brush size.

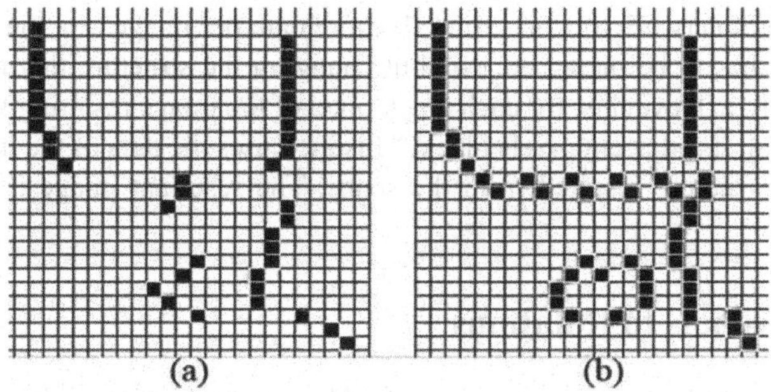

Fig. 2. The result of the second scan (a) without average brush size correction and (b) with average brush size correction for the digit '5' in Devanagri script.

2.2 The Second Scan

The second and the third scan makes the image insensitive to brush size. The second scan starts scanning row by row. Just like the first scan, the second scan checks for the continuous stream of black pixels. For each stream of black pixels, only the middle pixel in the stream is kept, all the other pixels are removed. This alone will work most of the time, but it will pick only the middle pixel when there is a long horizontal line, thus, it will remove too much information as displayed in Fig. 2(a). To overcome this problem average brush size is used. The continuous stream of black pixels is first divided into blocks of average brush size. Then the middle pixel is chosen from each block rather than the entire stream. This gives us a better result as displayed in Fig. 2(b).

2.3 The Third Scan

The third scan is done on the resultant image of the second scan. It is similar to the second scan. The difference being, that the scan is done column by column from top to bottom. The result after all the scans is given in Fig. 3. The image obtained after the third scan removes most of the unnecessary information and thus presents a more generalized form of the digit.

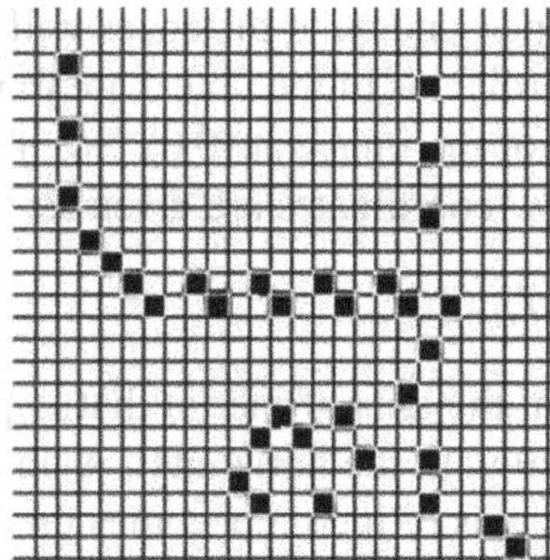

Fig. 3. The resultant image after all the scans.

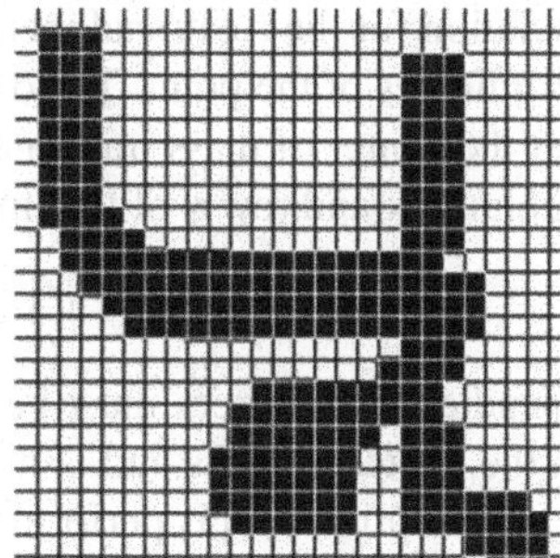

Fig. 4. Increasing the data in the image.

For the neural network to work, it needs more data per image. So, the total number of black pixels has to be increased, without changing the structure that resulted from the third scan. To achieve this, for each black pixel of the image resulting from the third scan, we color all its adjacent pixels black. This increases the data of the image without disturbing the structure resulted from the scans. The resultant image is displayed in Fig. 4.

3 Insensitiveness to Rotations

The image obtained in the previous step is used here to make it insensitive to rotations. To do this, first of all, the center of mass of the image is calculated.

3.1 Center of Mass

Only the black pixels contribute to the total mass of the image, while the white pixels are treated as empty space. Let the total number of black pixels in the image be n and let R_i and C_j represent a black pixel in the ith row and the jth column of the image. Then, coordinates of the center of mass are calculated as follows:

$$X_{com} = \frac{1}{n}\sum_{j=0}^{n} C_j$$

$$Y_{com} = \frac{1}{n}\sum_{i=0}^{n} R_i$$

3.2 Axis of Least Rotational Inertia

For any arbitrary axis which passes through the center of the mass, the rotational inertia is calculated around it by adding the square of perpendicular distances from the all the points to that axis. This is explained in Fig. 5.

To calculate the rotational inertia about any axis in the plane of the image, we transform the coordinate system such that the axis about which rotational inertia is to be calculated, becomes the y-axis in the transformed coordinate system. This will make the calculation for the rotational inertia easy, as the perpendicular distances will be just the x coordinates of the point. To do this, all the coordinates of the image are translated such that the center of mass translates to the center of the image at (0,0). So, if a pixel lies on column x and row y of the image, then its new coordinates are given as:

$$[x'y'] = [xy] - [X_{com}Y_{com}]$$

Perpendicular distances must be calculated from the axis of inertia that is passing through the center of mass and is making an angle θ with the horizontal. A change of basis transformation [30] is performed such that the axis of inertia becomes the Y-axis. This is done as:

$$[x^{\theta}y^{\theta}] = [cos\theta sin\theta - sin\theta cos\theta][x'y']$$

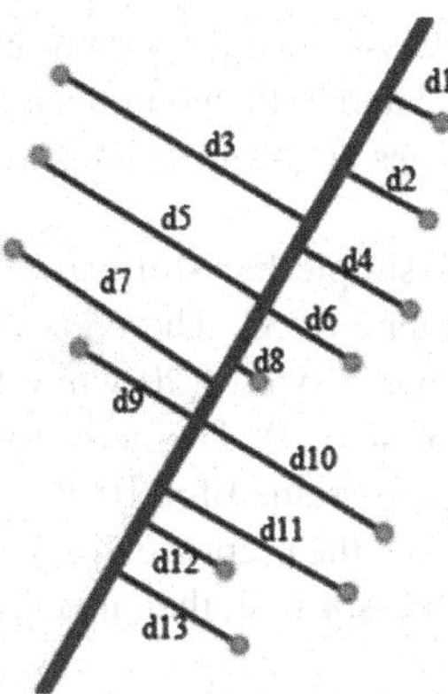

Fig. 5. Calculation of rotational inertia around an axis. The blue line is the axis of rotational inertia, the red dots represent points (black pixels in image) and the black lines are perpendicular distances from points to the axis.

Now the perpendicular distances to the axis are simply the x coordinates of the pixels in our new coordinate system. Rotational inertia about an axis at angle θ passing through the center of mass becomes:

$$RI^{\theta} = \sum_{i=0}^{n} \left(x_i^{\theta}\right)^2$$

where n is the number of black pixels in the image.

For the image, the rotational inertia is calculated at every π/8 rad starting from −π/2 radians and ending on π/2 radians. The rest of the angles yield symmetrical results. From all these angles, the angle which gives the least value for the rotational inertia is chosen. Let this angle be φ. All black pixels from the image are translated to the center of mass as done before, followed by a change of basis transformation at angle φ. After this, the coordinates of pixels are translated once again so that the top left corner becomes 0,0. This is done by adding coordinates of the center of mass to the coordinates of the pixel.

Some of the coordinates obtained might lie beyond the boundaries of the image. To overcome this, a new image is made. If the height of the original image is H and its width is W, the dimensions of the new image will be (L, L) where L is given by:

$$L = \sqrt{H^2 + W^2}$$

All the new coordinates fit in this new image size.

4 Dataset and the Neural Network

Devanagri digits were used to make the dataset. Devanagri has 10 digits from 0 to 9. For each digit, a set of hundred images were hand-drawn by the authors using Microsoft paint. Each image was 24 px by 24 px in size. This resulted in 1000 images for the dataset.

This dataset is fed to the preprocessor, where it randomizes the dataset, performs the calculations mentioned in Sects. 2 and 3, then outputs another dataset of brush size and rotation insensitive greyscale images of Devanagri digits, each having dimensions 33 px by 33 px.

The neural network used is a simple feed-forward neural network. The size of the input layer is 1089 which is a square of 33. There are two hidden layers in the neural network. The first hidden layer has a size of 20 while the second hidden layer has a size of 10. The output layer is of size 10. The activation function used in each layer is sigmoid. The neural network was trained for 1000 epochs. In each epoch, all 1000 images of the dataset generated from the preprocessor, were used. For testing, the image to be tested undergoes the preprocessor first, then it is fed to the trained neural network to obtain results.

5 Results

This section is divided into three sections. The first section tests the brush size insensitivity of the neural network with and without the proposed preprocessing. The second section tests the rotation insensitivity of the neural network with and without the proposed preprocessing. The third section tests both brush and rotation insensitivity combined. Other methods to compare the performance of the classifiers may be explored.

5.1 Brush Size Insensitivity

A new test set is created from the training set by selecting 5 images for each digit. Each of these images is further recreated two times, first by decreasing the brush size of the image and second, by increasing the brush size of the image. This results in a test set containing 150 images of different brush sizes.

Two feed-forward neural networks were created as described in Sect. 4. The first one uses the proposed preprocessing method and the second one does not do so. The test set is fed to both of them and the resulting average error for the networks is displayed in Fig. 6.

The results describe that our method is better than both feed-forward neural networks and convolutional neural networks. For a more visual evaluation, Table 2 displays some images from the test set and the resulting error of neural networks when the image was fed to them.

Table 2 shows two images per row. The first image is the train image and the second image is the modified version of the image, drawn with increased or decreased brush size. Rows 0 and 2 shows that for both feed-forward neural networks and convolutional neural networks, error increases when the brush size used to draw the image is reduced. While for our method, the error remains about the same. Rows 1, 3 and 4 shows that, for feed-forward neural network and convolutional neural network, the error is increased 10 to 100 times when the image was drawn with increased brush size. While with our method, the error did not change that much.

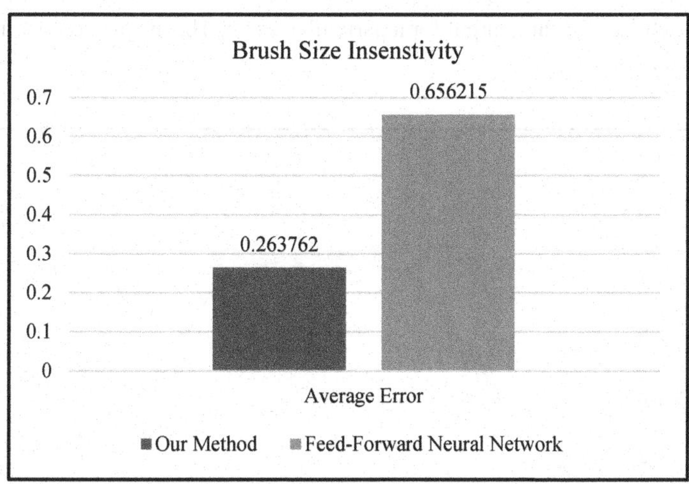

Fig. 6. Average error for brush size insensitivity for the digits in Devanagri script in the dataset.

Table 2. Comparing different brush sizes for same digit in Devanagri script in the dataset.

S. No.		Images	Error	
			Our Method	Feed-Forward Neural Network
1	Normal		0.0116026	0.00952600
	Modified		0.0108535	1.03056000
2	Normal		0.0149874	0.01928200
	Modified		0.0068879	0.63851200
3	Normal		0.00612053	0.02955100
	Modified		0.0112352	1.01531000
4	Normal		0.0126002	0.02515980
	Modified		0.00712356	0.22231900
5	Normal		0.0116026	0.00952609
	Modified		0.0108535	1.03056000

5.2 Rotation Insensitivity

The training set is rotated for every π/8 rad thus giving us 16 sets each corresponding to a particular angle. Each set is fed to neural networks to obtain output for a particular angle. The average error for each set is given in Fig. 7. The training set is rotated for every π/8 rad thus giving us 16 sets each belonging to a particular angle. Each set is fed to neural networks to obtain output for a particular angle. The average error for each set is given in Fig. 7.

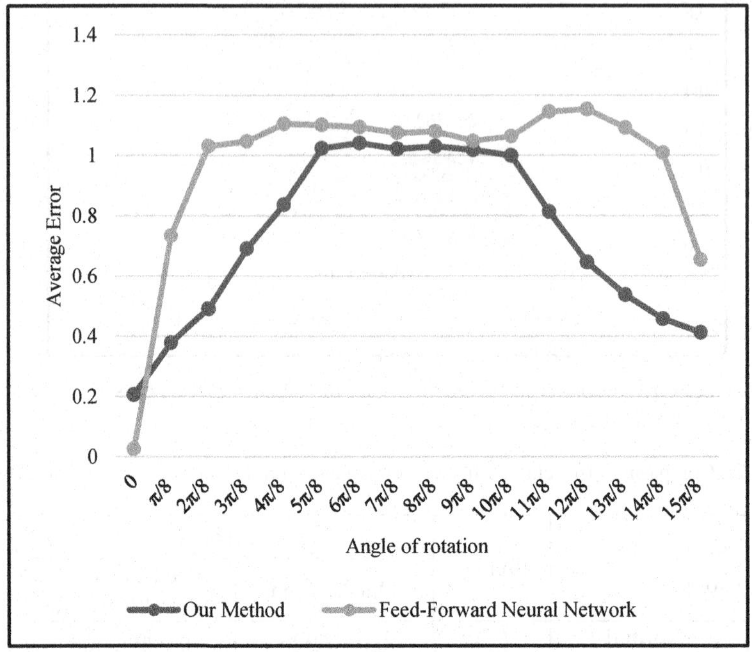

Fig. 7. Average error for each angle of rotation for the digits in Devanagri script in the dataset.

The results show that our method is insensitive to rotations from $-\pi/2$ radians to $\pi/2$ radians and on average, the error resulting from our method is less than that of the feed-forward neural network when input images are rotated. For visual comparisons, the best of the results is displayed in Table 3.

Table 3. Comparing rotation insensitivity of neural networks for the digits in Devanagri script in the dataset.

S. No.	Rotation angle	Modified Image	Error	
			Our Method	Feed-Forward Neural Network
1	π/2		0.00928173	0.999705
2	3π/2		0.0119288	0.999741
3	3π/8		0.013543	1.00002
4	15π/8		0.0135827	0.999831
5	15π/8		0.0152274	1.00005
6	5π/8		0.0242213	0.999712
7	3π/2		0.0272953	1.00028

5.3 Rotation Insensitivity with Brush Size Insensitivity

The test set described in Sect. 5.1 is used. This test set is rotated for every π/8 rad, thus giving us 16 sets. These sets are fed to the neural network to obtain the average error done in Sect. 5.2. The results are shown in Fig. 8.

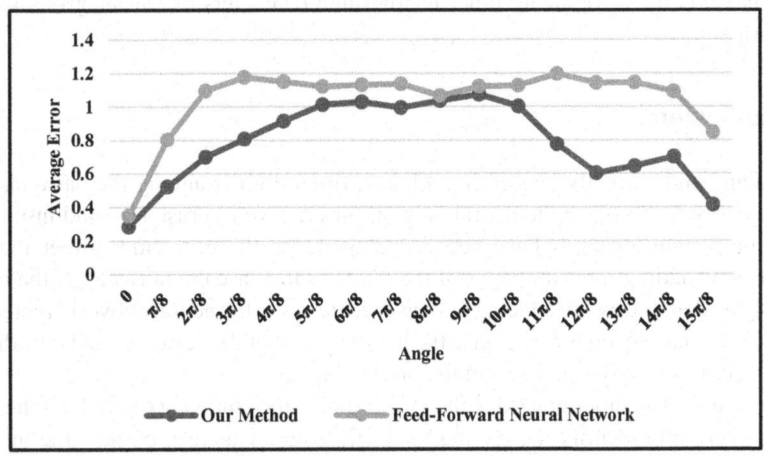

Fig. 8. Comparison of average error to test combined rotational and brush size insensitivity for the digits in Devanagri script in the dataset.

The results show that on average, the error resulting from our method is less than that of the feed-forward neural network when input images are rotated with a change in brush size. For more visual comparison, Table 4 shows some of the results obtained.

Table 4. Comparison of error when brush size and rotation insensitivity is combined for the digits in Devanagri script in the dataset.

S. No.	Rotation angle	Modified Image	Error	
			Our Method	Feed-Forward Neural Network
1	15π/8		0.0131309	0.995249
2	12π/8		0.0678723	0.999851
3	10π/8		0.0826384	0.995041
4	15π/8		0.0679514	0.970848
5	12π/8		0.122546	1.00611
6	12π/8		0.183466	1.01296
7	12π/8		0.182823	1.01195

6 Future Work

For future work, other techniques to generate moment of inertia around axis in the plane of the image can be developed. The works of previous authors for moments around a central point can be compared with the moments around the axis lying in plane of the image. Authors are at present in process of exploring concept of moment of inertia around the axis for recognizing other handwritten characters of Devanagri as well as other scripts.

7 Conclusion

Recognizing and correctly classifying a handwritten digit is among the basic tasks of a neural network. To make the neural network insensitive to brush size and insensitive to rotations, many methods have been developed, but these methods either increase the dataset by adding rotated images of the digits or increase the intricacy of the neural network, for example by introducing higher-order terms in the neural network. Increasing the size of the dataset tends to increase the learning time of the networks and introducing higher-order terms may lead to combinatorial explosion.

The proposed method shows that the axis around which moment of inertia is the least, is able to correctly identify the orientation of the digit. This method uses the intuitive concept of rotational inertia about an axis to make the image insensitive to rotations and uses the three-scan technique to get the basic shape of the digit in the image, thus making it insensitive to brush size. These methods modify the input image in such a way that the image becomes insensitive to brush size and rotations. Thus, there is no increase in the size of the dataset and no increase in the intricacy of the neural network.

References

1. Fisher, S.R.A.: 138: the use of multiple measurements in taxonomic problems. Ann. Eugen. **7**, 179–188 (1936)
2. Rosenblatt, F.: Principles of neurodynamics, perceptrons and the theory of brain mechanisms. Am. J. Psychol. **76**, 705 (1963)
3. Reid, M.B., Spirkovska, L., Ochoa, E.: Rapid training of higher-order neural networks for invariant pattern recognition. In: International Joint Conf. Neural Networks, 1989 (1982)
4. Fukushima, K., Miyake, S., Ito, T.: Neocognitron: a neural network model for a mechanism of visual pattern recognition. IEEE Trans. Syst. Man Cybernet. 826–834 (1983)
5. Widrow, B., Winter, R., Baxter, R.: Layered neural nets for pattern recognition. IEEE Trans. Acoust. Speech Signal Process. 9 (1988)
6. Lee, S., Choi, Y..: Translation, rotation, scaling and distortion invariant recognition of handwritten numerals. In: Proceedings of IJCNN'92, Beijing, Nov 1992
7. Sawai, H.: Axially symmetric neural network architecture for rotation-invariant pattern recognition. In: Proceedings of 1994 IEEE International Conference on Neural Networks (ICNN'94) (1994)
8. Fukumi, M., Omatu, S., Nishikawa, Y.: Rotation-invariant neural pattern recognition system estimating a rotation angle. IEEE Trans. Neural Netw. 568–581 (1997)
9. Patil, P.M., Sontakke, T.R.: Rotation, scale and translation invariant handwritten Devanagari numeral character recognition using general fuzzy neural network. Pattern Recog. 2110–2117 (2007)
10. Jain, A., Subrahmanyam, G.R.K.S., Mishra, D.: Rotation invariant digit recognition using convolutional neural network. In: Proceedings of 2nd International Conference on Computer Vision & Image Processing (2018)
11. M. K. Hu, "Visual pattern recognition by moment invariants," *IEEE Transactions on information Theory,* pp. 179–187, 1962
12. Teague, M.R..: Image analysis via the general theory of moments. J. Opt. 920–930 (1980)
13. Khotanzad, A., Lu, J.-H.: Classification of invariant image representations using a neural network. IEEE Trans. Acoust. Speech Signal Process. **38**, 1028–1038 (1990)
14. Flusser, J., Suk, T.: Pattern recognition by affine moment invariants. Pattern 167–174 (1993)
15. Mukundan, R., Ong, S., Lee, P.: Image analysis by Tchebichef moments. IEEE Trans. Image Process. 1357–1364 (2001)
16. Mahrishi, M., Hiran, K.K., Meena, G., Sharma, P. (eds.): Machine Learning and Deep Learning in Real-Time Applications. IGI Global (2020). https://doi.org/10.4018/978-1-7998-3095-5
17. Yap, P., Paramesran, R., Seng-Huat, O.: Image analysis by Krawtchouk moments. IEEE Trans. Image Process. 1367–1377 (2003)
18. Zhou, J., Shu, H., Zhu, H.: Image analysis by discrete orthogonal Hahn moments. In: International Conference Image Analysis and Recognition (2005)
19. Zhu, H., Shu, H., Zhou, J., Luo, L., Coatrieux, J.L.: Image analysis by discrete orthogonal dual Hahn moments. Pattern Recogn. Lett. 1688–1704 (2007)
20. Raj, P.A., Venkataramana, A.: Image reconstruction using Pollaczek moments. In: The First Biennial National Conference on Computer Vision, Pattern Recognition, Image Processing and Graphics (2008)
21. Xiao, B., Ma, J.F., Wang, X.: Image analysis by Bessel–Fourier moments. Pattern Recogn. 2620–2629 (2010)
22. Yang, B., Dai, M.: Image analysis by Gaussian–Hermite moments. Signal Process. 2290–2303 (2011)

23. Sayyouri, M., Hmimid, A.: A fast computation of novel set of Zeixner invariant. Circuits Syst. Signal Process. 875–900 (2014)
24. Hmimid, A., Sayyouri, M., Hassan.: Image classification using novel set of Charlier. In: International Conference on Image and Signal Processing, pp. 156–167 (2014)
25. Pandey, V.K., Bajaj, A., Singh, J., Parthasarathy, H.: Image analysis using Bateman. In: First India International Conference on Information Processing, pp. 1–4 (2016)
26. Karmouni, H., et al.: Image analysis using separable Krawtchouk-Tchebichef's moments. In: International Conference on Advanced Technologies for Signal and Image Processing (ATSIP), pp. 1–5 (2017)
27. Batioua, I., Benouini, R., Zenkouar, K., Zahi, A.: Image classification using separable invariants moments based on Racah polynomials. Procedia Comput. Sci. **127**, 320–327 (2018)
28. Zhi, R., Cao, L., Cao, G.: Translation and scale invariants of Krawtchouk moments. Inf. Process. Lett. **130**, 30–35 (2018)
29. Prasad, B.K.: Moment of inertia-based approach to recognize arabic handwritten numerals. In: Saini, H., Singh, R., Kumar, G., Rather, G., Santhi, K. (eds.) Innovations in Electronics and Communication Engineering. Lecture Notes in Networks and Systems, vol. 65. Springer, Singapore (2019)
30. Sanderson, G.: 3Blue1Brown, 15 Jul 2021 [Online]. Available: https://www.youtube.com/playlist?list=PLZHQObOWTQDPD3MizzM2xVFitgF8hE_ab

A Two-Phase Classifier Model for Predicting the Drug Satisfaction of the Patients Based on Their Sentiments

Manish Suyal(✉) and Parul Goyal

Department of CA and IT, SGRR University, Dehradun, India
suyal.manish923@gmail.com

Abstract. Since the corona virus has emerged, genuine clinical resources, such as a paucity of experts and healthcare workers, a lack of adequate equipment and medications, and so on, have reached their peak of inaccessibility. Several people have died as a consequence of the medical profession's concern. Individuals began self-medicating due to a lack of supply, which exacerbated an already precarious health situation. A rise in new ideas for automation is being spurred by machine learning's recent success in a varied variety of applications. In this paper, we have proposed a two-phase Decision Tree Classifier based on Artificial Neural networks (DTNN). The work is based on the satisfaction of the drugs among patients with the help of their comments as positive or negative polarity. The dataset of drugs used in this paper is Cymablta and Depopovera. The proposed results are compared with the existing methodology of Support Vector Machine Neural Network (SVMNN). The results are shown in graphical and tabular form which shows the efficiency of the proposed methodology.

Keywords: Sentiment · Opinion · Health · Drugs · Classification · Support vector machine · Decision tree classification · Preprocessing · Stop word removal algorithm · Stemming algorithm · Precision · Recall · F1-score

1 Introduction

Opinion mining has resulted in the creation of a large variety of technologies for analyzing consumer opinion in the majority of key commercial contexts, including travel, housing, consumables, materials, goods, & services [1–3].

Drug monitoring is critical for ensuring the safety of drugs after they are released for general use. Drug trials are often conducted on a small number of test individuals, with a low possibility of detecting unusual adverse effects. Additionally, volunteers or patients enrolled in drug studies vary from those getting commercially available pharmaceuticals in terms of age, co-morbidity, & polypharmacy. Thus, it is vital to conduct epidemiological studies on the safety of marketed drugs. Additionally, it is critical to learn how the general public utilizes a certain medicine, how they distinguish

V. E. Balas et al. (Eds.): ICETCE 2022, CCIS 1591, pp. 79–89, 2022.
https://doi.org/10.1007/978-3-031-07012-9_7

its security, effectiveness, & side effects. The purpose of this paper is to demonstrate how neural network approaches may be used to classify opinions in the health care domain [4, 5].

Previous work on medical sentiment analysis has focused on biomedical literature and clinical narratives (generally objective and technical, but includes information relevant to determining patient outcomes). Medical social media data presents particular difficulties for sentiment analysis due to the use of jargon, non-standard medical terminology, incorrect spelling, and incorrect grammar. In addition, medical texts commonly use sentiment words that may have the opposite meaning in the medical domain. For example, a lab test with a negative result may indicate a positive patient outcome. Thus, determining sentiment in medical social media presents challenges that stem from both the medical and general sentiment domains.

Advances in neural network methods, such as the quick training algorithm for deep multilayer neural networks, have been seen in recent years. Recent success with neural network approaches in numerous Natural Language Processing (NLP) applications, like Opinion Mining (OM), has resulted in the introduction of prediction architectures dependent on recursive NNs that may be utilized to offer a syntactic parser competition for sentences in natural language [6].

The literature, on the other hand, does not offer much to the field of opinion classification using neural networks and machine learning (ML) algorithms. For data classification in many other fields, however, it has been shown by several studies that Artificial Neural Network (ANN) and decision tree (DT) models are more successful than other methods.

2 Literature Review

Neural network approaches are being used in this work to mine social web opinions about health care. A total of two reviews have been retrieved from the Internet. The effectiveness of classification systems is examined experimentally by comparing the results of reviews of two different drugs. Using Neural Network (NN) for opinion mining outperforms support vector machine in standings of recall and f-score and precision, according to the results. A neural network with a radial basis function approach is proven to perform better than the probabilistic Neural Network (NN) technique in terms of the metrics used to evaluate performance [7].

The purpose of this paper is to offer a drug recommendation device that may significantly minimize the number of specialists. Using multiple factorization methods including Bow, Word2Vec, Term Frequency Inverse Document Frequency TF-IDF), as well as Manual Feature Analysis to forecast patient sentiment, we develop a pharmaceutical recommendation system that leverages patient reviews to recommend the best medicine for a certain illness employing various categorization algorithms. AUC (Area under the Curve), precision, recall, f1score, as well as accuracy all, played a role in determining how accurate our predictions were of actual feelings. There is a 93% difference between linear support vector classifier (SVC) and the other Term Frequency Inverse Document Frequency (TF-IDF) factorization classifier models in terms of accuracy [8].

The study [9] proposes Galen OWL, an online framework with semantic capability, to aid professionals in discovering information about drugs. According to this research, a patient's infection, sensitivity, and interactions with other drugs can be used to determine which drugs should be prescribed to the individual.

Leilei Sun [10] analyzed a vast number of treatment records to find the most appropriate treatment for each patient. The objective was to apply a semantic clustering method to estimate the similarities between treatment data.

In this study, [11] expected that Adverse Drug Reactions (ADR) mentions are connected with negative sentiment in online posts. They tested this theory by including sentiment-aware features into ADR Mine, a cutting-edge algorithm for extracting ADR mentions.

In this [12] paper, 2 Deep Fusion Models dependent on 3-way decision theory are suggested to assess the drug review data. The 1st 3-way fusion of one deep model with a traditional learning algorithm (3W1DT) used a deep learning approach as a primary classifier as well as a secondary classifier for classical learning whenever the deep method's confidence in test samples classification was low.

Manure [13] analyzed a lot of people have posted their views on social media about covid-19 in which doctor and world health organization (WHO) the head. The people got many benefits with the help of the post [13].

In this study, [14] describes how psychological disorders are identified in the human brain by the supervised machine learning techniques.

The study [15] proposes Hetu Bhavsar, that several machine learning algorithms applied to the dataset are described, among which SVM and Naive Bias are prominent.

In this [16] paper, the field of e-learning has carved a distinct identify in the world. Many students are studying online with the help of e-learning.

In this [17] paper, we collect customer reviews from healthcare field. After that, important features are removed from the review of each sentence and based on the extracted features with the help, we generate the feature vector. Then, we apply two-pass classifier, support vector machine and artificial neural network (SVMNN) in order to predict the given customer review is positive or negative.

3 Proposed Methodology

The two-phase classifier used in this study is a combination of the Decision Tree classifier as well as an artificial neural network (ANN). The system includes (1) feature selection (2) pre-processing (3) feature vector generation (4) classification modules.

3.1 Feature Selection

The features collection is used rating, rationale, side effects, remarks, date, sex, and age length. In this study, we use python to automatically extract the tree feature from each researcher, across with their rating score. Only comments, gender, age, side effects, and rating details are extracted from each review in the proposed approach. The rating ranges from 0 to 5. Furthermore, assigning a feature value of "0" to female reviewers and a feature value of "1" to male reviewers is our standard procedure. Non-numeric data is used to extract comments and side effects. There is no way to train the comments using string. Because of this, we must convert the string into a numerical value. These features are passed on to the preprocessing stage in this scenario. ANNs are supervised machine-learning approaches that use a known collection of observations (that is training points) as well as environmental variables to classify unknown and new data. In general, ANNs can handle enormous datasets, approximate non-linear connections, and generalize from somewhat imprecise input data.

3.2 Preprocessing

A key step in the development of opinion mining features is preprocessing. String features were transferred to the pre-processing step following feature selection. Dependent on the Senti WordNet dictionary, these sentences are divided into objective and subjective ones. The reviews in this work include data that is both numerical and non-numerical, as well as string and redundant data. We can't begin the prediction process correctly if the material is in numerous formats, and this hurts opinion mining.

The paper, therefore, includes several pre-processing procedures. In this preprocessing, we use stop word removal and stemming as two of the most important phases. Firstly, we extract rating, gender, as well as age features from the dataset and store them in a single text document, as shown in the figure. The next phase will use this text document as its input. Detailed descriptions of the preparation steps are provided below.

Stop Word Removal. Stop words are the words that are often filtered out when a natural language is processed. These are some of the most common terms in any language (including articles, pronouns, prepositions, conjunctions, etc.) and do not add much information to the text.

Reviewer comments, as well as side effects, are critical to determining whether a drug is safe. From one reviewer to the next, the side effects keep changing. Comments are made in light of any negative impacts. Many common words and some worthless ones can be found in the comments section. Remove these terms and you'll be able to focus more on the most crucial content in your paper. Such terms can be omitted from our training model and have no negative impact on the model's effectiveness.

Stemming. Following the removal of high-frequency terms, we put on the stemming method to remaining words. When a term is reduced to its root or stem, it is called "stemming." Send, sent, and sending are just a few examples of words that can be grouped. Sentence tenses are derived from one root word, which is sent. There is only one word left after we truncate the rest of the words: send.

3.3 Feature Vector Generation

We have only top words (technical words) after the stop word removal and stemming operation. There are a certain number of words associated with each statement and its effect on the body. We then assign a score value S_v to each word. As a result of this, we arrive at an average S_Av for each comment and side effect. To put it another way, S_Av stands for "satisfied attribute value." Five features are automatically selected for opinion mining in this study. Based on the feature's value, we create a feature vector. In the feature vector, the number of reviewers and the number of features are $n \times m$ respectively.

3.4 Artificial Neural Network

The training, as well as classification phases of ANN application, is intertwined. Three interconnected layers of architecture make up the examined neural networks. For starters, an input layer is made up of several nodes (that is. one for each variable in the environment that was tested). On another level, a hidden layer involves ANN that represents radial Basis function network (RBFs) (i.e., varied no of neurons were examined) (using a Gaussian activation function). Another layer has 5 artificial neurons (one for each expected soil drainage class), which transmits o/p values. A value between 0 and 1 (the lowest and greatest) is a probability value (Fig. 1).

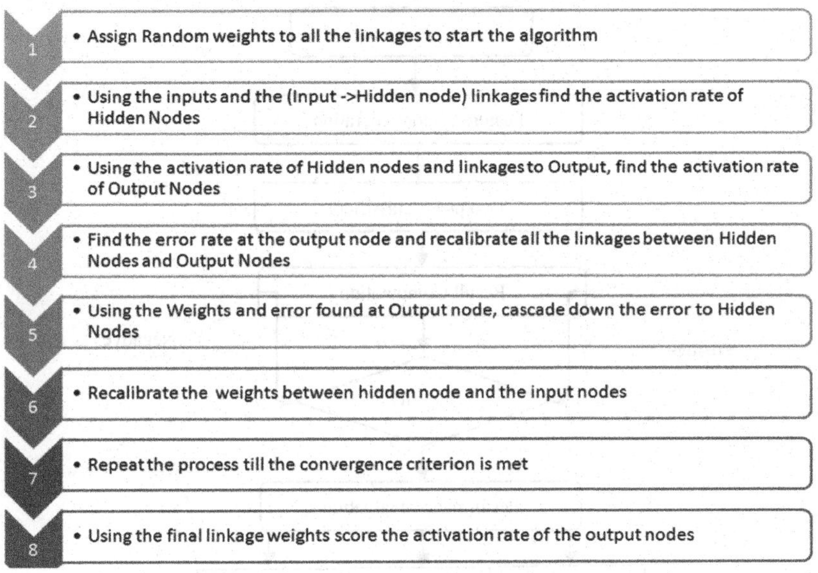

Fig. 1. ANN that represents radial basis function network (RBFs)

3.5 Decision Tree

A machine learning approach known as decision tree classification (DTC) makes use of the recursive division of a dataset to classify a target variable uniformly. As the data is divided up, the algorithm seeks to minimize the entropy of each variable by selecting the best split from a variety of possible splits.

- Step 1: Begin the tree by the root node, S, that involves the whole dataset.
- Step 2: Using the Attribute Selection Measure (ASM), locate the dataset's most valuable attribute.
- Step 3: Split S into sets that include the optimal possible values for attributes that you're interested in.
- Step 4: Create a decision tree node with the best attribute in it.

- Step 5: Recursively generate new DTs based on dataset subsets that were developed in step 3. A leaf node is the last node of a tree when you can no longer classify the nodes further.

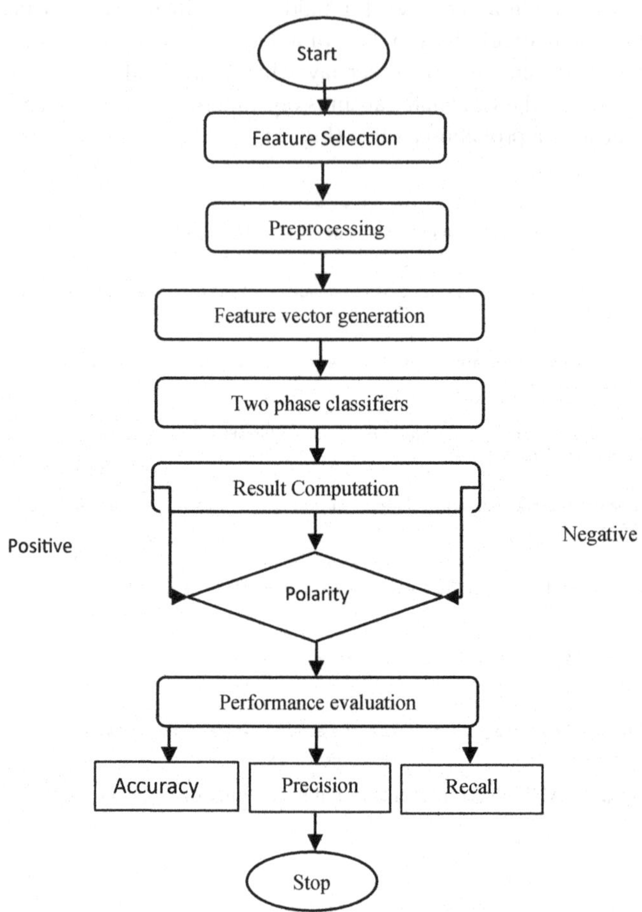

Fig. 2. Flowchart of proposed methodology

4 Findings and Discussions

The drug review dataset was derived from the well-known website www.askapatient.com. Figure 2 provides an example of a review format from the website asking a patient. Due to the nature of the language utilized, opinion mining is domain-specific. In contrast, this study focuses solely on opinion mining (OM) in the context of drug reviews. To create 2 datasets (Dataset I as well as Dataset II), we used two popular medications, Cymbalta and Depo-Provera. The antidepressant Cymbalta is prescribed to individuals with serious depressive disorder. Progesterone, a female hormone, is the active ingredient in Depo-Provera, a contraceptive pill prescribed to women who want to avoid getting pregnant. Both of these medicines are used in different ways. There are 2546 customer reviews in the dataset, which were collected between 2005 and 2018. The dataset includes eight features, including rating, rationale, side effects, comments, sex, age, length, & date added. There are also 1136 review comments in the second dataset, which was compiled from 2000 to 2018. The performance of the suggested feature-based opinion mining is evaluated using these two datasets.

The text reviews have been classified based on the feedback from the users, who have given their opinions. Ratings are explained in detail using words in the text review. Meaning of a 5-star review varies from person to person because reviews are supplied by a diverse group of people with varying backgrounds. Data format review scores are used to assign an opinion class to an extracted document. Class labeled positive if review score is greater than 3 & class labeled negative if review score is less than 3. The research work is implemented using a Python simulation tool using various libraries of Python (Figs. 3, 4, 5, 6, 7, 8 and 9).

Accuracy. The ratio of correctly predicted classes compared to the total number of samples assessed.

$$Accuracy = \frac{TP + TN}{TP + TN + FP + FN}$$

Recall. Used to calculate the percentage of positive patterns that are correctly classified.

$$Sensitivity = \frac{TP}{TP + FN}$$

Precision. To determine positive patterns that are accurately anticipated by all predicted patterns in a class of predicted patterns that are positive.

$$Precision = \frac{TP}{TP + FP}$$

The precision of positive reviews is described as a percentage of positive reviews that are correctly confidential as such. If precision is low, a large % of classes are being labeled as positive even when they are not. As a result, repairing them would be a waste of time and effort for developers. As a result, high precision is always demanded (Table 1).

F1-Score. Harmonic average of recall & precision rates is calculated.

$$F1 - score = 2 \times \frac{Precision \times Recall}{Precision + Recall}$$

Note: - TP:-True Positive, FP:-False Positive, FN:-False Negative, TN:-True Negative

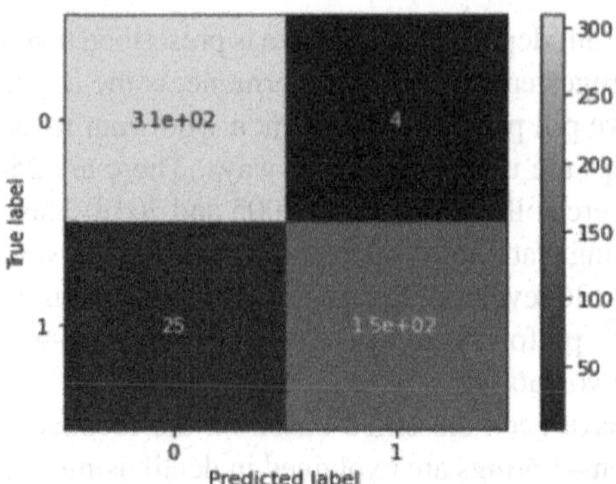

Fig. 3. Confusion matrix of the Cymbalta dataset

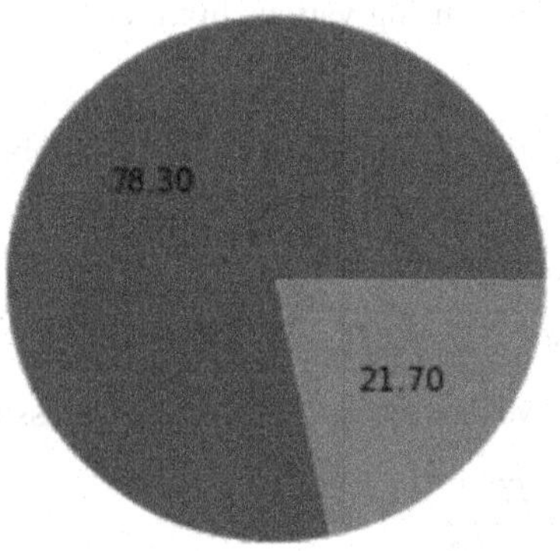

Fig. 4. Gender ratio (Female and Male) for Cymbalta dataset

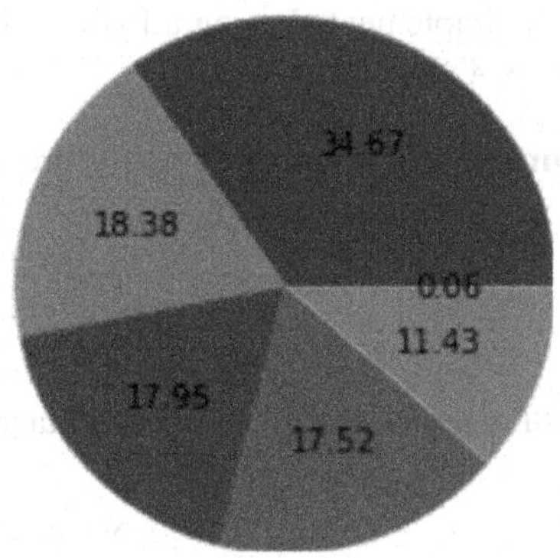

Fig. 5. Ratings (1–5) for Cymbalta dataset

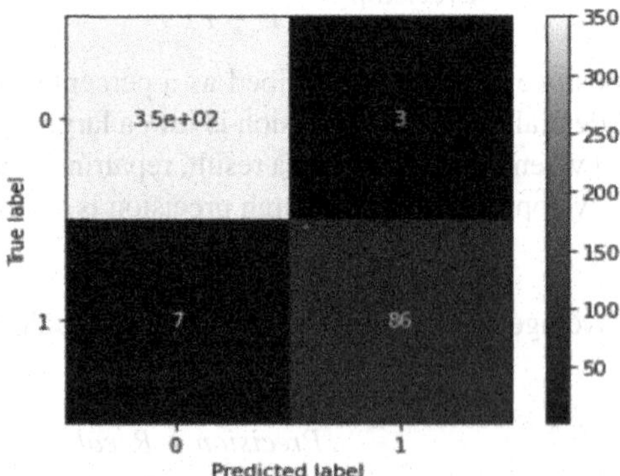

Fig. 6. Confusion matrix of the Depoprovera dataset

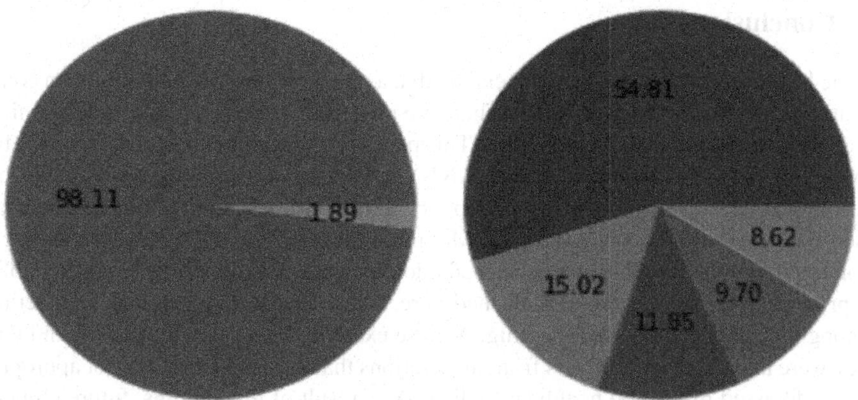

Fig. 7. Gender ratio (Female and Male) of the Depoprovera dataset

Fig. 8. Ratings (1–5) for Depoprovera dataset

Table 1. Comparison of various performance parameters

Parameters/Results	Accuracy	Precision	Recall	F1-score
Cymbalta SVMNN	0.812%	0.89	0.74	0.76
Depoprovera SVMNN	0.769%	0.88	0.53	0.49
Cymbalta DTNN	0.94%	0.95	0.92	0.93
Depoprovera DTNN	0.97%	0.97	0.96	0.97

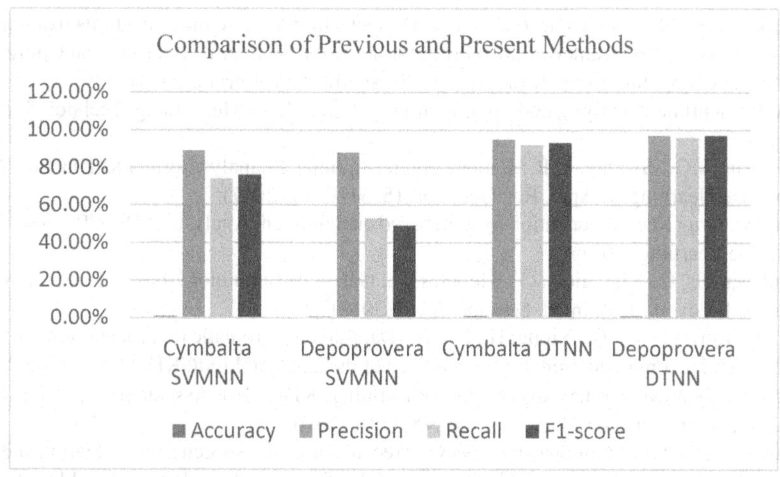

Fig. 9. Comparison graph of the previous and present methodology

5 Conclusion

In the last several years, the use of social media, a relatively new method of human communication, has grown significantly. There are numerous areas where machine learning is applied to analyze social media data. This paper proposes function approximation by employing Artificial Neural Network (ANN) based method combined with a decision tree. In the previous methodology, the support vector machine (SVMNN) technique was used but due to the noisier dataset SVM cannot outperform well. Therefore, a Decision tree has opted for this proposed methodology since it is more robust and reliable. A prior study's findings were confirmed here. There was a high level of satisfaction among the patients who used the drug. A close examination of their records found that they were having adverse effects from medications that they thought were not appropriately addressed by mental health specialists. As a result of the findings, future clinical practice research as well as development can use the information as a guide to what is currently considered optimal practise. From the outcomes, it is obvious that our proposed technique is greater to previous approach.

References

1. Sánchez-Rada, J.F., Iglesias, C.A.: Social context in sentiment analysis: Formal definition, overview of current trends and framework for comparison. Inf. Fusion **52**, 344–356 (2019)
2. Ramírez-Tinoco, F.J., Alor-Hernández, G., Sánchez-Cervantes, J.L., Olivares-Zepahua, B.A., Rodríguez-Mazahua, L.: A Brief Review on the Use of Sentiment Analysis Approaches in Social Networks, pp. 263–273 (2018)
3. Schuller, B., Mousa, A.E.D., Vryniotis, V.: Sentiment analysis and opinion mining: On optimal parameters and performances. Wiley Interdiscip. Rev. Data Min. Knowl. Discovery, vol. 5, pp. 255–263 (2015)
4. Chee, B.W., Berlin, R., Schatz, B.: Predicting adverse drug events from personal health messages. In: AMIA Annual Symposium proceedings/AMIA Symposium, pp. 217–226 (2011)
5. Sharif, H., Abbasi, A., Zafar, F., Zimbra, D.: Detecting adverse drug reactions using a sentiment classification framework. In: Proceedings of the Sixth ASE International Conference on Social Computing (SocialCom), pp. 1–10. Stanford, California (2014)
6. Liu, B.: Sentiment analysis and opinion mining. Synth. Lect. Hum. Lang. Technol. **5**, 1–167 (2012)
7. Krishnan, V.G., Swamy, C.R.: Patient opinion mining to analyze drugs satisfaction using supervised learning. J. Appl. Res. Technol. **15**, 311–319 (2017)
8. Garg, S.: Drug Recommendation System based on Sentiment Analysis of Drug Reviews using Machine Learning (2021)
9. Doulaverakis, C., Nikolaidis, G., Kleontas, A.: GalenOWL: ontology-based drug recommendations discovery. J. Biomed. Semant. **3**(14) (2012)
10. Sun, L., Liu, C., Guo, C., Xiong, H., Xie, Y.: Data-driven automatic treatment regimen development and recommendation. In: Proceedings of the 22nd ACM SIGKDD International Conference on Knowledge Discovery and Data Mining (KDD '16), Association for Computing Machinery, pp. 1865–1874. New York, NY, USA (2016)
11. Kumar T., et al.: A comprehensive review of recent automatic speech summarization and keyword ıdentification techniques. In: Fernandes S.L., Sharma T.K. (eds.) Artificial Intelligence in Industrial Applications. Learning and Analytics in Intelligent Systems, vol. 25. Springer, Cham (2022)

12. Basiri, M.E., Abdar, M., Cifci, M.A., Nemati, S., Acharya, U.R.: A novel method for sentiment classification of drug reviews using fusion of deep and machine learning techniques. Knowl. Based Syst. (2020)
13. Manure, K.H., Amin, R.N.R.: Twitter sentiment analysis on word wide Covid-19 outbreaks. J. Kurd. J. Appl. Res. (KJAR), eISSN. 2411–7706 (2020)
14. Kaur, P., Sharma, M.: Diagnosis of human psychological disorders using supervised learning and nature-ınspired computing techniques. J. Med. Syst. **43**, 204 (2019). Springer
15. Bhavsar, H., Manglani, R.: Sentiment analysis of twitter data using python, Int. Res. J. Eng. Technol. (IRJET) **06**, eISSN. 2395–0056 (2019)
16. Kechaou, Z., Ammar, M.B.: Improving e-learning with sentiment analysis of users opinions, IEEE Xplore 1032 (2011)
17. Padmavathy, P., Mohideen, S.P.: An efficient two-pass classifier system for patient opinion mining to analyze drugs satisfaction. Biomed. Signal Process. Control **57** (2019). Elsevier, Article. 101755

An Empirical Investigation in Analysing the Proactive Approach of Artificial Intelligence in Regulating the Financial Sector

Roopa Balavenu[1], Ahamd Khalid Khan[2], Syed Mohammad Faisal[2], K. Sriprasadh[3](✉), and Dharini Raje Sisodia[4]

[1] Department of MBA, K S School of Engineering and Management (Visvesvaraya Technological University), Bangalore, India

[2] Management, Jazan University, Jazan, Saudi Arabia
{akkhan,dfaisal}@jazanu.edu.sa

[3] Department of Computer Science and Engineering, Thirumalai Engineering College, Kilambi, Kanchipuram, India
srisaiprasadhhh@gmail.com

[4] Management, Army Institute of Management and Technology, Greater Noida, India

Abstract. The use of artificial intelligence (AI) within the finance industry can be considered as a transformative approach as it enables the financial institutions to enhance their performance capacity. The use of artificial intelligence within the finance sector helps the industries to streamline the processes and optimise their management efficiently for various types of operations pertaining to credit decisions-making, financial risk assessment and management and quantitative trading. The paper aims at analysing the proactive approach that can be taken with the use of AI in order to enhance effective management within the financial sector. The empirical study conducted in the paper utilizes various types of secondary materials with a qualitative approach. The findings of the study demonstrate the enhanced capacity of AI that can be used for a proactive approach, utilised for the assessment of risks or threats prior to any mismanagement incident. In this regard, fintech companies such as *Enova, Ocrolus, ZestFinance, and DataRobot* and so on have taken a predominant position in aiding the financial industries to use AI-based systems that aids the management process. However, the inclusion of AI within the financial sector is faced with certain challenges such as lack of knowledge regarding technological infrastructure, poor financial investment especially for government aided banks, unawareness of the employees and weak collaboration with the IT industry. Regardless, AI technologies in recent years have achieved great advancement, leading to the enhancement of its capacity to assist the effective management within the financial sector.

Keywords: Fintech · Artificial intelligence (AI) · Financial sector · Effective management · Proactive approach · Risk management · Organisational performance

V. E. Balas et al. (Eds.): ICETCE 2022, CCIS 1591, pp. 90–98, 2022.
https://doi.org/10.1007/978-3-031-07012-9_8

1 Introduction

In recent times, the use of artificial intelligence (AI) has been used widely within the financial sector that has positively transformed the services by enhancing the performance capacity of the organisations. Various financial technology (Fintech) companies have generated AI-based systems that can be used for the optimisation of management. Especially, with the constant growth of online banking services, the financial sectors face various challenges in management and regulating the excessive pressure of managing the user data, services management, risk assessment, management and troubleshooting and so on. The use of AI-based management systems or services for effective management has been recognised by the financial sectors on a global scale in order to gain a substantial competitive position as well. Hence, AI is being used to regulate the services in an automated manner that not only decreases service delivery timing but also helps in the recognition of risk factors and encounters with it.

The paper aims at conducting an empirical investigation of the proactive approach of AI based systems and services in order to regulate the operations and services for enhancing effective management capacity of the financial sectors. With the rise of customer-centric services approach on a global scale, the global financial sector has been inclined to adopt the advanced AI technology based systems for enhancing their performance and management capacity [1]. In recent years, the collaboration between the banking sector, retail and IT has led to the enhancement of cyber-attack rates [2]. The banking sector also faces difficulties in maintaining an advantageous competitive position due to lack of financial support and technological infrastructure. Merely 32% of the financial sector uses various AI based systems such as voice recognitions, proactive analytics and other [3]. Thus, the paper seeks to address the challenges faced in the implementation of proactive AI analytics in services along with the benefits attained by AI in regulating management operations in the financial sectors.

2 Materials and Methods

The study uses a qualitative approach for the development of comprehensive knowledge on the topic. Demonstration of knowledge in an in-depth manner aids in the process of discovering new ideas [4]. As the paper aims at an empirical investigation, it has benefited greatly from a deductive research approach in order to build new ideas based on existing information. The secondary data collection method has been conducted through a systematic and targeted search in electronic databases such as Google Scholar and ProQuest. Various industrial reports and information from official websites have also been collected. An inclusion-exclusion criteria had been set for the purpose of a systematic sampling process that ensures that all the resources used in the study holds authenticity, validity and relevance.

3 Results

The use of AI in the financial sector provides banks with a chance to utilise sophisticated technology for the management of various processes and operations. The use of AI based systems within the industry has been defined as beneficial due to multi-dimensional functionality. The use of AI has been able to make financial services cheap, automated, error-free, personalised and swift [5]. AI usage has found popularity among the developed countries more ardently as opposed to the developing countries such as India as financial support and infrastructural support is required for proper implementation. Proactive data analytics is an important feature of AI systems that analyses and monitors business data and prevents incidents from occurring or escalating [6]. In essence, the use of AI proactive analytics helps in the identification of potential issues and helps in generating accurate solutions.

AI-based analysis of data helps in regulating the potential causes of incidents that otherwise would be harmful to the banking sector. Within the financial sector, AI can be used for deep neural networks, clustering algorithms, price optimisation, customer sentiment analysis, fraud detection, loss forecasting and compliance monitoring [7]. Various AI tools such as EP, ML, NLP and topological data analysis are AI-based functions that help in quantitative analysis and process optimizations that are used in the widest capacity for businesses (28.7% segmentation for commercial and commercial banks) [7]. Among the fastest-growing fintech companies, *Enova, Ocrolus, Datarobot* and others provide AI services to the finance industry [8]. For example, Enova has developed an AI and machine learning platform named Colossus that provides advanced analytics technology used by various businesses and the banking sector [8]. The growth of fintech companies has played a critical role in the emerging necessity for using AI in the banking sector to attain and retain a beneficial competitive position [16] (Fig. 1).

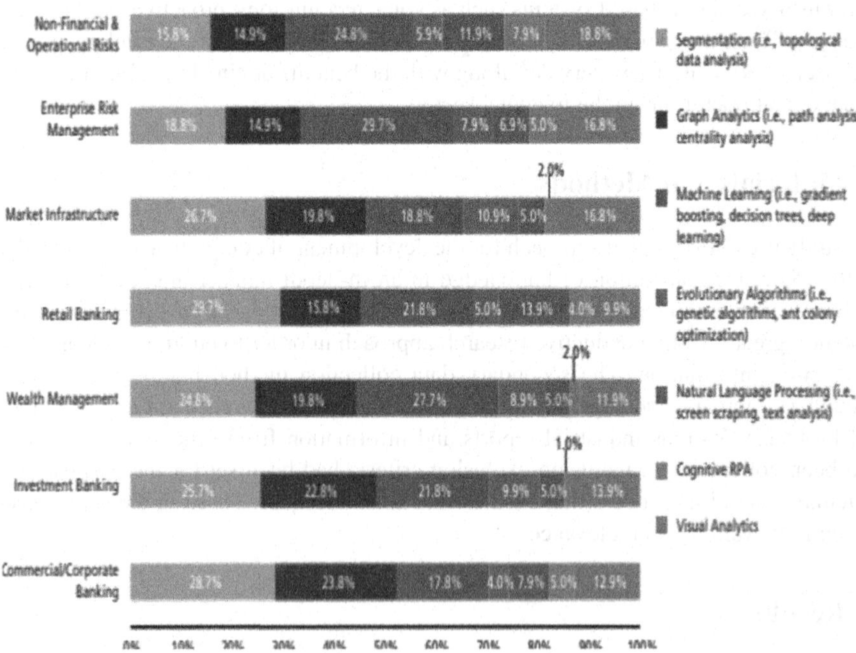

Fig. 1. Category wise analytics technology used in various business segments in India [7]

The above figure indicates the use of various types of analytics technology for banking, market infrastructure analysis, risk management, retail banking and so on. This provides a clear indication that various operations regarding data analysis and regulations are conducted through the help of big data analytics systems, machine learning and so on for the purpose of predictive analysis that reduces the threats of incidents in operations (Table 1 and Fig. 2).

Table 1. AI tool adoption rate within various industries in India [7]

AI tool for risk and compliance	Rate
Investment research	7.1%
WM	7.1%
Compliance	18.4%
ERM	22.4%
Retail analytics	25.5%
Corporate banking	2.0%
Others	17.3%

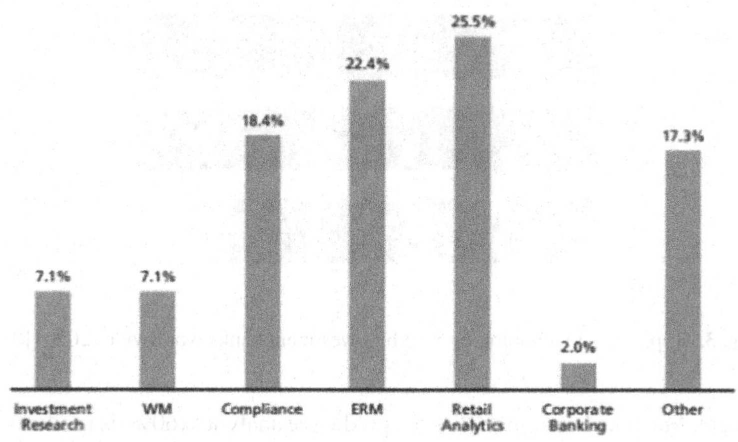

Fig. 2. AI tool adoption rate within various industries in India [7]

On the other hand, retail analytics (25.5%) and ERM (22.4%) are the specific areas in which AI technologies are used for risk analysis while corporate banking is seen to use the least AI tools (2.0%) [7]. Hence, the banking sector of India as a developing country exhibits the inability to implement proper AI tools for management regulations. On the other hand, within the global segment, investment banks have indicated a significant rise of AI use in 2020 (Table 2 and Fig. 3).

The use of data analytics systems in the banking sector has been the most common as the vast quantity of data processing, monitoring and management requires proper

Table 2. Types of AI technologies used by investment banks worldwide (2020) [9]

Types of AI tools used by investment banks globally in 2020	Rate
Machine learning	63%
Predictive analytics	60%
Virtual assistant	58%
Image analysis	52%
Natural language processing	52%
Robotic process automation	45%

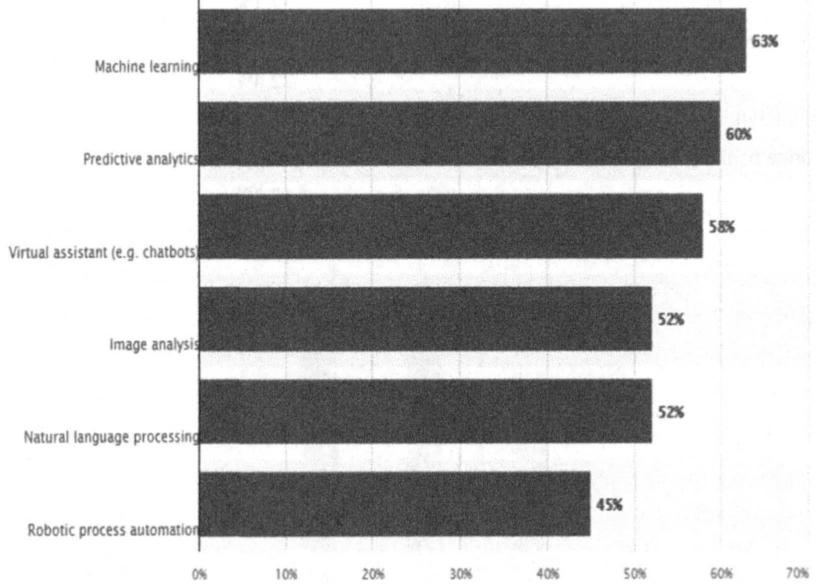

Fig. 3. Types of AI technologies used by investment banks worldwide (2020) [9]

management. Machine learning (63%) and predictive analytics (60%) is used most commonly by the investment banking sector [9]. It indicates the growing acknowledgement of the necessity to use AI in the management of data that may be prone to corruption and cyber-attacks in recent times [17]. The onset of the worldwide covid-19 pandemic has enhanced the number of customers using online banking services and has enhanced the necessity to ensure proper analysis and protection of data [10]. The proactive approach to analysis of data and prevention of incident may help in regulating the finance sectors to ensure that predictive analysis is conducted. There are various components involved in predictive analysis that is closely aligned with the proactive approach that is taken by an AI. Predictive analysis consists of machine learning and data exploration features that help in the analysis process through statistical model, consumer behaviour, machine learning, patterns, scoring, visualisation, predictive behaviour, decision-making and communication [11]. These are used further in the proactive approach to assess and monitor each financial service provided by the organisations and prevent mismanagement. Data theft, data loss and other incidents can be prevented beforehand with the use of AI in the finance sector (Fig. 4).

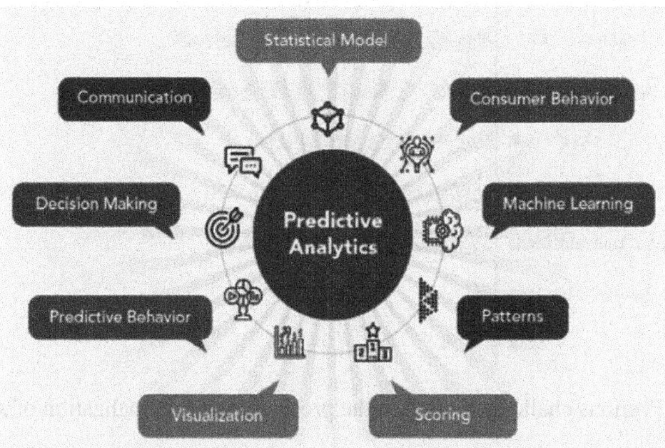

Fig. 4. Process of predictive analysis [11]

However, there are certain challenges that have been encountered in the application of AI for effective financial management. The implementation of AI based systems faces challenges in proper systematic and personalised modelling (36.6%), gathering relevant data for analysis and monitoring (60.4%), specific data strategies adopted by the organisations (44.6%), collaboration and linking with business (30.7%) and so on [7]. Additionally, the implementation is also faced with challenges due to the lack of proper knowledge regarding the use of AI. The qualms regarding algorithm biases and discriminatory data, uncertainty regarding the use of artificial codes, enhanced risks of cyber conflicts and threats have also impacted the positive reception of AI in the financial sector [12] (Fig. 5).

Regardless, the advancement of AI has been aided by the era of industrial digitalisation on a global scale and the banking sectors have been prompted to adopt AI systems for multidimensional management. In essence, the position of AI usage for the management of data with a proactive approach is being provided with importance in the financial sector that helps these finance institutes to conduct error-free and swift operations [18].

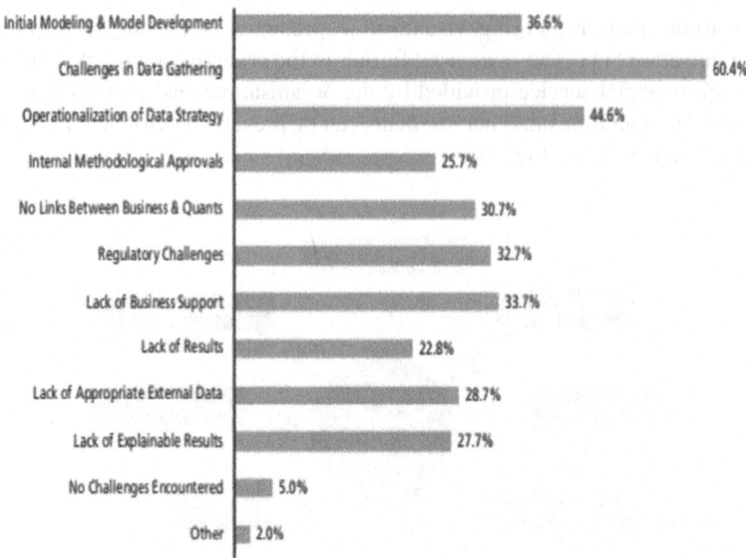

Fig. 5. Various challenges faced in the process of operationalization of AI [7]

4 Discussion

The growing necessity to include proactive AI approaches for the management of large-scale data in the finance industry can be indicated by the analysis of the information presented above. The critical implications of proper risk assessments and management have become one of the most common needs of AI implementation. The use of AI in various divisions in India is seen to be lower, especially in the financial sector, in comparison to the global data presented above. It can thus be stated that a leading position has been occupied by the economically developed countries in AI implementation. However, the growing threats of cyber-attacks, data theft, data displacement and so on have made the adoption of AI technologies in the finance sectors a rather necessary factor for the financial institutes in developing countries as well.

The proactive approach to data analysis and regulation also enables the organisational capacity to expand their online domain without causing higher risks of incidents. AI designs face operational issues as organisational demands change in accordance with the missions and visions of the other organizations [13]. Practical implementation of AI enhances the mode of control over the gathered data and aids the organisational decision-making process [14]. In essence, the mode of operations controlled in this manner lends the organisations with more scope for ensuring that sensitive data retrieved from the customers are stored and analysed for providing the organisations with accurate reports of segmentation, market changes and customer demands. This enables the banking organisations to manage their customers and reduce the number of complaints from customers. Decreased number of data-related mishaps is also attributed as one of the important advantages of AI based technology.

The lack of proper financial and technological infrastructure may be regarded with seriousness as it poses critical risks in the implementation process. The lack of adequate technological infrastructure in AI implementation exposes cyber conflicts and algorithm pattern mismanagement [12]. The AI, with proper analytics designs, helps in the creation of algorithm patterns that are used efficiently for the purpose of assessing risks and managing it beforehand. On the other hand, knowledge of any potential mishaps can be analysed automatically with the help of AI technology that lessens the burden of errors.

AI usage has the potential to transform the overall practices in the finance industry as it can be used in a multidimensional manner. The tasks conducted through sophisticated machine learning enable organisations to demonstrate the flexible management of operations [15]. This can be attributed to the capacity of the AI technology to complete multiple operations simultaneously. The system in itself is able to offer service efficiency that leads to a profitable future for these organisations. Additionally, with the implementation of sophisticated and proactive AI based systems, the organisations are able to collaborate with various third-party companies in a protective manner [19]. Thus, the overall analysis of the gathered information indicates the efficiency of a proactive AI analytical approach for data processing and management that leads to the enhancement of performance capacity of financial organisations. In addition, the multi-dimensional and wide-speared capacity of AI pertaining to data management can be used efficiently for regulatory-focused analytics and management.

5 Conclusion

The struggles of the financial sector, on a global scale, regarding the management of data is a potent driver behind the increasing demand for AI based systems. Various fintech companies such as Enova, Datarobots, and Ocrolus and so on have contributed to the development of AI based systems that help in optimizing the operational processes within the financial sector. The use of AI for proactive and predictive analysis in order to monitor data for the prevention of incidents has been identified as one of the most common demands fulfilled by these AI systems. Moreover, assistance in decision-making, customer segmentation, risk assessment and management are conducted efficiently and swiftly by AI technologies.

It helps the financial organisations to assume more control over the operations through a regulatory-focus analysis. However, lack of proper technological and financial infrastructure has been identified that hinders the process of implementation, especially among economically developing countries such as India. Additional support from the government of the developing countries can thus be recommended to aid AI implementation. Thus, it can be concluded that AI regulatory systems using proactive analytics with sophisticated technological design are able to enhance the effective management capacity of organisations of financial sectors.

References

1. Truby, J., Brown, R., Dahdal, A.: Banking on AI: mandating a proactive approach to AI regulation in the financial sector. Law Financ. Mark. Rev. **14**(2), 110–120 (2020)

2. Soni, V.D.: Role of artificial intelligence in combating cyber threats in banking. Int. Eng. J. Res. Dev. **4**(1), 7 (2019)
3. Wipro.com: Why banks need artificial intelligence. Available at: https://www.wipro.com/business-process/why-banks-need-artificial-intelligence/ (2021). Accessed 24 Oct 2021
4. Siponen, M., Klaavuniemi, T.: Why is the hypothetico-deductive (HD) method in information systems not an HD method? Inf. Organ. **30**(1), 100287 (2020)
5. Lui, A., Lamb, G.W.: Artificial intelligence and augmented intelligence collaboration: regaining trust and confidence in the financial sector. Inf. Commun. Technol. Law **27**(3), 267–283 (2018)
6. Kibria, M.G., Nguyen, K., Villardi, G.P., Zhao, O., Ishizu, K., Kojima, F.: Big data analytics, machine learning, and artificial intelligence in next-generation wireless networks. IEEE Access **6**, 32328–32338 (2018)
7. Tcs.com: The state of AI in risk management. Available at: https://www.tcs.com/content/dam/tcs/pdf/Industries/Banking%20and%20Financial%20Services/State-of-AI-in-Risk-Management.pdf (2019). Accessed 24 Oct 2021
8. Allayannis, G.Y., Becker, J.M.: A global fintech overview. Darden Case No. UVA-F-1860 (2019)
9. Norrestad, N.: Share of investment banks worldwide using artificial intelligence (AI) as of 2020, by types of application. Statista.com. Available at: https://www.statista.com/statistics/1246874/ai-used-in-investment-banking-worldwide-2020/ (2021). Accessed 24 Oct 2021
10. Kuhn, K., Bicakci, S., Shaikh, S.A.: COVID-19 digitization in maritime: understanding cyber risks. WMU J. Marit. Aff. **20**(2), 193–214 (2021). https://doi.org/10.1007/s13437-021-00235-1
11. Mahrishi, M., Hiran, K.K., Meena, G., Sharma, P. (eds.): Machine Learning and Deep Learning in Real-Time Applications. IGI Global (2020)
12. Kaur, D., Sahdev, S.L., Sharma, D., Siddiqui, L.: Banking 4.0: 'The influence of artificial intelligence on the banking industry & how AI is changing the face of modern day banks'. Int. J. Manag. **11**(6) (2020)
13. Golić, Z.: Finance and artificial intelligence: the fifth industrial revolution and its impact on the financial sector. Zb. Rad. Ekon. Fak. Istoč. Sarajevu **19**, 67–81 (2019)
14. Wójcik, D.: Financial geography II: the impacts of FinTech–Financial sector and centres, regulation and stability, inclusion and governance. Prog. Hum. Geogr. **45**(4), 878–889 (2021)
15. Mhlanga, D.: Industry 4.0 in finance: the impact of artificial intelligence (ai) on digital financial inclusion. Int. J. Finan. Stud. **8**(3), 45 (2020)
16. Jain, A., Yadav, A.K., Shrivastava, Y.: Modelling and optimization of different quality characteristics in electric discharge drilling of titanium alloy sheet. Mater. Today: Proc. **21**, 1680–1684 (2019)
17. Jain, A., Pandey, A.K.: Modeling and optimizing of different quality characteristics in electrical discharge drilling of titanium alloy (Grade-5) sheet. Mater. Today: Proc. **18**, 182–191 (2019)
18. Jain, A., Pandey, A.K.: Multiple quality optimizations in electrical discharge drilling of mild steel sheet. Mater. Today: Proc. **8**, 7252–7261 (2019)
19. Panwar, V., Sharma, D.K., Kumar, K.V.P., Jain, A., Thakar, C.: Experimental investigations and optimization of surface roughness in turning of EN 36 alloy steel using response surface methodology and genetic algorithm. Mater. Today: Proc. (2021). https://doi.org/10.1016/j.matpr.2021.03.642

Sentiment Analysis on Public Transportation Using Different Tools and Techniques: A Literature Review

Shilpa Singh(✉) and Astha Pareek

Department of CS & IT, The IIS University, Jaipur, India
shilpamca85@gmail.com, astha.pareek@iisuniv.ac.in

Abstract. With the rapid increase use of WWW people from all age group shares their views, thought, ideas, suggestions on the internet. This leads the internet to become a hub for user-generated data. If this extracted data from the web is analyzed properly, then this may become the largest platform for making decisions for new product reviews, political issues, public issues, etc. But manually extracting and analyzing this content becomes a drastic task.

So, for solving this problem a new area of research has been introduced that is Sentiment Analysis. It is the part of data mining where it extracts and analyzes the unstructured data. It is a domain where there is a huge scope of research. Various researchers already had researched different datasets with different Tools and Techniques on various domains like politics, sports, public issues and etc. This paper is focused on a literature survey for Sentiment Analysis on public transportation using different Tools and Techniques.

Keywords: Literature review · Polarity · Public transportation · Sentiment Analysis (SA) · Tools · Techniques

1 Introduction

With the huge popularity of social media, forums, microblogs, blogs, E-commerce sites, etc. Internet has become a bag full of crucial information like people's opinions, comments, and thoughts on almost everything. These resources are very helpful in understanding user choice for different products, events, and services. This opinion can be beneficial for both the provider and the customer because a customer can check other's experiences or thoughts before purchasing any product similarly producers may see the users or criticizer's thoughts about their goods or idea and may take further decisions. But, the critical task is to fetch and analyze data from the internet. So, this problem has been solved by using Sentiment Analysis. Sentiment Analysis is useful in reducing the complexity of unstructured data and Natural Language Processing.

It is a part of Natural Language Processing (NLP) and artificial intelligence that deals with the interaction between the human-computer language which automatically extracts the opinion or attitude of a text [14].

It is the process of finding the sentiments of the sentence. These sentiments can be classified into three different levels. The Fig. 1 shows levels of classification.

V. E. Balas et al. (Eds.): ICETCE 2022, CCIS 1591, pp. 99–110, 2022.
https://doi.org/10.1007/978-3-031-07012-9_9

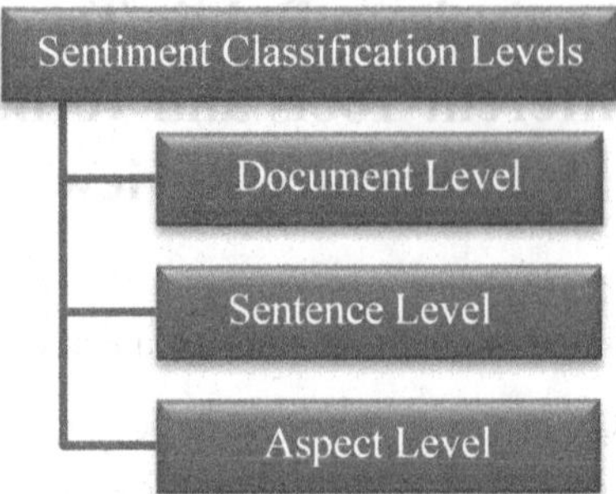

Fig. 1. Three levels of classification

Document Level: When only one product need to be revised as a single entity then document level sentiment analysis is used where whole document analyzed and classified as a single entity, which results in opinion of whole document expressed as positive, negative or neutral sentiment. It may not be applicable on single document that contains multiple product reviews.

Sentence Level: The main task of sentence level is to classify each sentence separately to find out whether the sentence are of positive, negative or neutral opinion. The level of analysis is nearly cognate to subjective classification and it discriminate subjective sentence from objective sentence because subjective sentence, manifest influenced information or expression about sentence whereas objective manifest factual information about sentence.

But both Document level and sentence level analysis do not discover what exactly people likes and dislikes.

Aspect Level: Earlier, it was known as Feature level sentiment analysis. It instates finer-grained analysis. Rather of looking different linguistic constraints like sentence, phrase, clause, document or paragraph it looks directly to the opinion itself as it can find out what people actually incline to say about any products. This level is more engrossing than the above two levels.

This polarity describes the nature of the sentence as positive, negative, or neutral sense. Sentiment Analysis consists of two tasks. First is the extraction of data and next involves analyzing data using various Sentiment Techniques. Data may be extracted from different sources like Twitter, Facebook, etc. using various tools available. And for analyzing the data there are various Techniques. These Techniques can be divided into two parts first is machine learning and the other is lexicon based. When both these types of techniques combine or work together then it becomes a Hybrid technique. Below Fig. 2 shows different techniques available for Sentiment Analysis.

Sentiment Analysis can be done for various datasets like politics, movies review, product reviews, trending topics, etc. [3] In this paper, Public Transportation has been taken for further investigation.

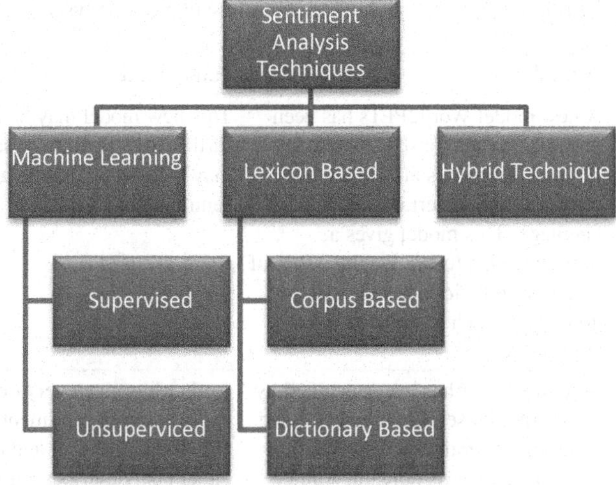

Fig. 2. Sentiment analysis techniques

This paper is well defined for work done by various researchers on Sentiment Analysis and various Tools and Techniques available. The rest of the paper is structured as follows. Section 2 gives an inspection on Sentiment Analysis. In Sect. 3 explanation and comparison of Tools and Techniques available for Sentiment Analysis on Public Transportation has been done. Lastly, a conclusion of the literature survey has been given in Sect. 4.

2 Sentiment Analysis

All material Innumerable research work has been bringing out for Sentiment Analysis. It is the study of attitudes, notions, emotions, and mindset for any topic or product. The process of analysis involves ascertaining, extracting, and an assortment of feelings. [15].

A lot of researches have been explored for Sentiment Analysis. Below Table 1, I show some research dimensions in Sentiment Analysis.

3 Tools and Techniques Used for Sentiment Analysis on Public Transportation

As Public Transportation has become a crucial topic for research, also an important part of human daily life as lots of people daily travel from Public Transportation. People prefer Public Transportation for safety, cost-saving, etc. Some authors like [7] done research on congestion caused by lots of vehicles on road. The author suggested that key solution of congestion and traffic jam is the increased use of public transportation but the reluctance of public vehicles are based on several factors like traveling time, cost, safety, and security as well as the pleasure and convenience of the users of private transport. Author [16] used an Odd-Even traffic scheme for analyzing the sentiments of people on

Table 1. Table research dimensions in sentiment analysis

S. No.	Author	Work done	Future facets
1	[22]	A new model Word2PLTs has been proposed by author that rooted on fuzzy mathematics also assess the fuzziness and uncertainty of language. This model gives a improved path to resolve the issues of variation of lexicon based. Also a good way for short text sentiment analysis	This new model may be used for further investigation or same model may be used for public transportation sentiment analysis
2	[23]	A survey paper has been specified by author that presents challenges faced while using sentiment classification in single and cross domain. Author also stated that contestation handling, classification of polarity and subjectivity analyzing are required to be conducted before doing sentiment analysis	Author has given very deep explanation of sentiment classifications level and techniques. Paper also describes various ways to collect data like on the bases of rating, likes, dislikes, emojis, and etc. All these may be used for further research
3	[24]	7000 tweets have been taken from twitter on McDonald's and KFC. Many machine learning approach have been used to find out accuracy, precision and recall	Work may be done in improvement in accuracy of these techniques by combine it with lexicon based techniques
4	[18]	A brief introduction has been given on sentiment analysis and their techniques. Their benefits and limitations have been described	Author gives good explanation for different approaches to find out good polarity result. This may be taken care while doing experiment
5	[9]	This research paper focused on a crucial analysis of literature from the year 2012 to 2017 on Sentiment analysis using SVM	Author provides good and comprehensive study on SVM that may be help in doing research work on the same
6	[13]	This paper gives the overview on sentiment analysis and their techniques	Different approaches has been described that may be used further
7	[4]	In this paper author proposed an algorithm for collecting a sample of expert tweets instead of following the conventional method of random sampling. These expert tweets are those tweets which are retweeted by large number of people	Only retweets has been considered. If all tweets has been taken may change the result

(*continued*)

Table 1. *(continued)*

S. No.	Author	Work done	Future facets
8	[2]	Author analyzed and tested an algorithm in which news headlines has been classified as positive or negative sentiments using NLTK tool. For each headline manually as well as experimental polarity using SentiWordNet has been calculated and compared	For a huge data, this manual classification task can be erroneous and a tedious job and can change the results. Work can be done using Vader

public transportation. NLTK toolkit has been used by the author for fetching data from Twitter. From all the tweets only those tweets have been taken for the further experiment which has retweets count of more than 100 and the experiment has been done on that tweets only.

Sentiment analysis has various methodologies and their advantages and disadvantages [4]. The author compared various Techniques used for Sentiment Analysis after analyzing various methodologies. She has given a comparative table of three approaches used in Sentiment Analysis that are Machine Learning, Lexicon Based and Hybrid approach with Rapid Miner tool based on classification, advantages, and disadvantages. These techniques can be classified as supervised and unsupervised methods [6, 20]. Supervised machine learning techniques like Naïve Bayes, Maximum Entropy (ME), and Support Vector Machines (SVM) have achieved great success in text categorization because in this approach each of the classifiers is trained on a collection of representative data. And then a test set is used to test the performance of the automatic classifier [14]. Figure 3 shows different Machine Learning techniques used by different authors.

Fig. 3. Various machine learning techniques

In contrast to supervised learning, unsupervised learning has no explicit target output associated with the input. The Lexicon Based Method is an Unsupervised Learning approach since it does not require prior training data sets. In this approach, a sentiment dictionary with sentiment words is used. The dictionary contains the polarity of each word as positive, negative, and neutral words. Polarity may be calculated for all words by matching them with dictionary words [1]. Some of the Lexicon Based Techniques shown in Fig. 4.

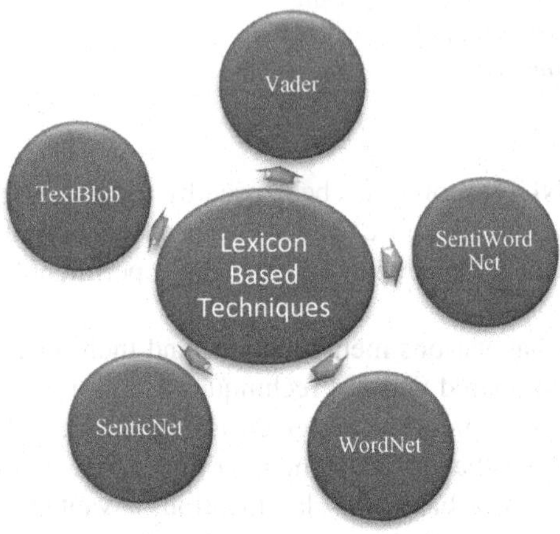

Fig. 4. Various lexicon based techniques

The development and application of Sentiment Analysis require the use of powerful software tools. As the quality of accessible tools keeps on developing the decision of the most suitable tool turns out to be progressively difficult [21] Some authors

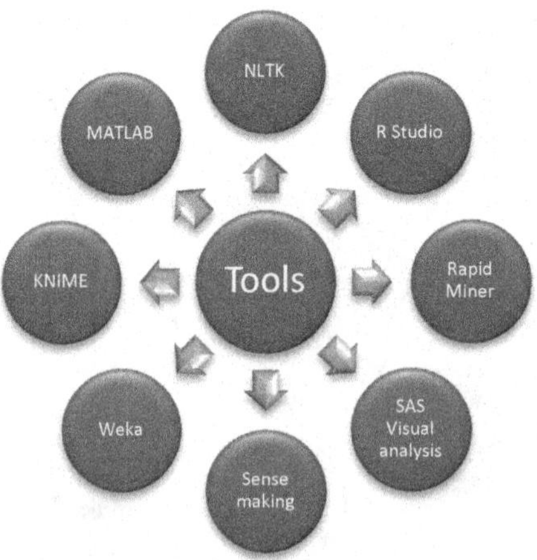

Fig. 5. Various tools for sentiment analysis

have researched the Rapid miner software tool [4] and some authors prefer the NLTK programming tool [10] Fig. 5 shows various tools used for Sentiment analysis.

So, below Table 2 shows some research work done on different tools and techniques for doing Sentiment Analysis on Public Transportation.

Table 2. Research dimensions on various tools and techniques

S. No.	Author	Method	Tool	Comment/Goal
1	[5]	SentiWordNet, SVM, Naïve Bayes and Random Forest	NLTK	**Observation:** Author used Telugu language tweets and used hybrid approach for finding sentiment analysis. This research paper shows that Naïve Bayes gives 85% accuracy while Random Forest gives 82% accuracy **Limitation:** work has been done only on Telugu language. Same method may be used for other datasets
2	[19]	SenticNet, SentiWordNet, Happiness Index, and SentiStrength, Decision trees, SVM, Naïve Bayes	iFeel utility software	**Observation:** Author used many techniques together. And a new tool that is iFeel utility software. Accuracy is in between 80–87% **Limitation:** Author used many methods together so, lack in clear picture of framework. The work done by author may be exploring more
3	[25]	SentiWordNet and SVM	CMU POS tagger Software	**Observation:** Hybrid technique has been used SVM gives 83% accuracy **Limitation:** Conjunctions and short form of words are not treated properly
4	[26]	Naïve Bayes and Vader	online YouTube scrapper, KNIME	**Observation:** Author used ruled based Vader approach with 85% accuracy **Limitation:** Rule based approach but the experiment may be done using NLTK toolkit for more accurate result

(*continued*)

Table 2. (*continued*)

S. No.	Author	Method	Tool	Comment/Goal
5	[30]	Machine Learning- C.4.5, Naïve Bayes, SVM, Random Forest	Rapid Miner	**Observation:** In this experiment SVM shows higher performance from the remaining three methods **Limitation:** In this paper how tool is used not clearly mentioned
6	[16]	Hybrid- Lexicon, SVM, Naïve Bayes, Logistic regression	NLTK	**Observation:** SVM gives highest accuracy result of 81.42% **Limitation:** This experiment lacks in removal of slag words
7	[27]	Machine Learning- Naïve Bayes	R Studio and Rapid Miner	**Observation:** Naïve Bayas gives good accuracy result of 83.98% with Rapid miner tool **Limitation:** Very few data has been taken. Only one technique has been used by author
8	[28]	Rules-Based approach	SAS Visual analytics	**Observation:** Static software approach has been used. This approach may be used for further research **Limitation:** One may not enhance the existing static model like may be with Lexicon methods
9	[12]	AVL systems, Rule Based approach	Wilcoxon Rank Sum test	**Observation:** A detailed comparison has been done by author may use for further research **Limitation:** Only a comparative study on taxi at airports. May improve the same using other Sentiment approaches
10	[29]	Naive Bayes, SVM and logistic regression	Sensemaking	**Observation:** An integrated approach has been used that gives good accuracy result of 95% **Limitation:** This approach can be further applied to other cities as well as to other forms of public transportation, e.g., subways and taxis

(*continued*)

Table 2. (*continued*)

S. No.	Author	Method	Tool	Comment/Goal
11	[8]	Hybrid and Rule base-Semantria	KNIME	**Observation:** Rule based technique is used. Microblogging site may use to collect data **Limitation:** Work done only on Airport services. This approach may be used for other transportation services like railways
12	[7]	SVM	Weka	**Observation:** SVM gives good accuracy result of 92.76% with Weka tool **Limitation:** Only one technique has been used so comparison not done with other techniques
13	[3]	Machine Learning approach- Support Vector Machine, Naïve Bayes, and Decision Tree	NLTK, Rapid Miner	**Observation:** Again SVM gives highest accuracy result of 72.97%. But experiment has been done on large and unbalanced data **Limitation:** Preprocessing and classification of data not done properly. This may be done using lexicon based to improve accuracy
14	[33]	Many machine learning and Lexicon based	NLTK	**Observation:** Author gives a deep study on various methods available for sentiment analysis in different languages
15	[31]	Naive Bayes (NB), Adaptive Boosting (AdaBoost), Support Vector Machines (SVM), Ridge Regression (RR), and Maximum Entropy (ME	NLTK and TF-IDF software	**Observation:** Author used NLTK and TF-IDF software in their experiment for finding sentiment analysis. Other approach may be used **Limitation:** Proper use of TF-IDF software has not been explained by author
16	[32]	SentiWordNet, TextBlob, W-WSD, Naïve Bayes and SVM	Weka	**Observation:** Weka has been used by author other tool may be used. TextBlob and WordNet has given highest accurate result **Limitation:** Author done experiment only for Urdu language. Author manually translate Urdu Language into English language

There are various datasets used by authors for doing sentiment analysis of public transportation. Data has been taken from twitter by many researches. So for doing further research, data may be fetched from twitter on public transportation. But the public transportation consists of different transportation like Railways, Bus, Cab and etc. Below Table 3 shows some of the datasets used by authors.

Table 3. Analysis of various datasets under public transportation

S. No.	Type of transportation/Corpus	Hashtag (#) for scraping raw data from twitter
1	Railways	railMinIndia
2	Metro	Metrotrains, metrorail, mumbaimetro, MetrorailF
3	Airlines	Aero plane, Spicejet, VitoAeroplane, airindiain, jetairways, Indigo6E
4	Bus	Redbus, diamondbuses, The_Green_Bus, redBus_in, VolvoBusIN, busrentalindia, MagicBusIndia, Roadwaysbuses, Rajasthanroadways, TransportDelhi, MORTHIndia
5	Taxi	carrentaldelhi, Taxiindia
6	Cab	Uberindia, olaindia,

From all the research done by us it has been observed that all authors used some model for doing sentiment analysis. Figure 6 shows basic model for sentiment analysis.

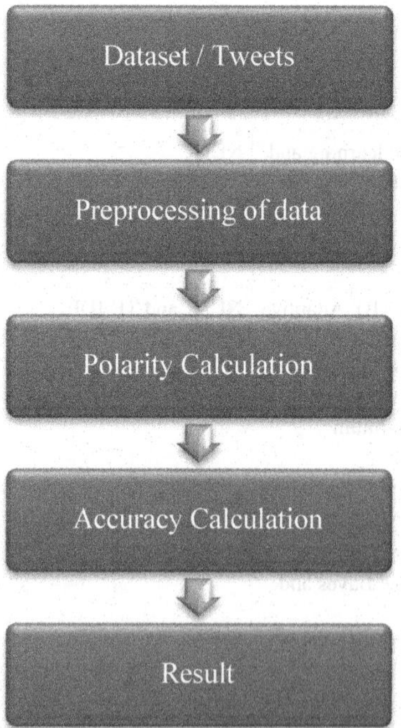

Fig. 6. Basic model for sentiment analysis

4 Conclusion

Sentiment Analysis on Public Transportation has a vast area to explore and research. This paper is fully focused on comparative study of various Sentiment Analysis Tools and Techniques. From the above survey, it has been observed that there are many tools available for Sentiment Analysis but NLTK is the best tool that may be used for further research. Also there are many Machine Learning and Lexicon Based Techniques available for Sentiment Analysis. The theoretical study shows that Machine learning techniques gives more accurate result than Unsupervised Lexicon based techniques. But performance of machine learning approach depends on domain set, quantity and quality of training data. Various authors researched Hybrid Technique which is the combination of Machine Learning and Lexicon based approaches. This Hybrid Technique combines the strength of both the techniques and avoids their weakness. So, the Hybrid Technique gives good accurate results and may be used for further research.

References

1. Adarsh, R., Patil, A., Rayar, S., Veena, K.M.: Comparison of VADER and LSTM for sentiment analysis. Int. J. Recent Technol. Eng. **7**, 540–543 (2019)
2. Agarwal A., Sharma, V., Sikka, G., Dhir, R.: Opinion mining of news headlines using SentiWordNet. IEEE **16**, 978-1-5090-0669-4 (2016)
3. Anastasia, S., Indra, B.: Twitter sentiment analysis of online transportation service providers. IEEE **16**, 978-1-5090-4629-4 (2016)
4. Apoorva, G., Chowdary, D.E., Vaishnav, N.R., Uddagiri, C.: An approach to sentiment analysis in twitter using expert tweets and retweeting hierarchy. IEEE, 971-1-4673-6621-2 (2016)
5. Tammina, S.: A hybrid learning approach for sentiment classification in telugu language. IEEE (2020). https://doi.org/10.1109/AISP48273.2020.9073109
6. Samal, B.R., Behera, A.K., Panda, M.: Performance analysis of supervised machine learning techniques for sentiment analysis. IEEE 978-1-5090-4929-5 (2017)
7. Effendy, V., Novantirani, A., Sabariah, M.K.: Sentiment analysis on twitter about the use of city public transportation using support vector machine method. IJOICT **2**, 57–66 (2016)
8. Gitto, S., Mancuso, P.: Improving Airport Services Using Sentiment Analysis of the Websites, vol. 22, pp. 132-136. Elsevier
9. Ahmad, M., Aftab, S., Bashir, M.S., Hameed, N.: Sentiment analysis using SVM: a systematic literature review. Int. J. Adv. Comput. Sci. Appl. **9**, 182–188 (2018)
10. Hanswal, S.S., Pareek, A., Sharma, A.: Twitter sentiment analysis using rapid miner tool. Int. J. Comput. Appl. **177**, 44–50 (2019)
11. Hanswal S.S., Pareek, A., Vyas, G., Sharma, A.: Sentiment Analysis on E-Learning Using Machine Learning Classifiers in Python. Springer. 1187, 978-981-15-6014-9_1 (2021)
12. Ji, Y., Cao, Y., Du, Y., Zhang, H.M.: Comparative Analyses of Taxi Operations at the Airport, vol. 25, pp. 2222–2232. Elsevier (2017)
13. Kaur, A., Kaur, J.: Sentiment analysis: approaches and methods. Int. J. Comput. Sci. Eng. **6**, 2347–2693 (2018)
14. Madhuri, K.: A machine learning based framework for sentiment classification: indian railways case study. IJITEE **8**, 441–445 (2019)
15. Ramírez F.J., Alor, G., Sánchez, J. L., Olivares, B. A., Rodríguez, L.: A Brief Review on the Use of Sentiment Analysis Approaches in Social Networks. Springer International Publishing. 1007, 978-3-319-69341-5_24 (2018)

16. Sharma, S.K., Hoque, X.: sentiment analysis for odd-even scheme in Delhi. Indian J. Sci. Technol. **24**, 0974–6846 (2018)
17. Singh, S., Pareek, A.: Improving public transport services using sentiment analysis of twitter data. J. Inf. Comput. Sci. **10**, 234–250 (2020)
18. Tyagi, P., Tripathi, R.: A review towards the sentiment analysis techniques for the analysis of twitter data. In: International Conference on Advanced Computing and Software Engineering (2019)
19. Khan, M.Y., Jinnah, M.A., Karachi, U., Khurum, P., Junejo, N.: Exerting 2D-space of sentiment lexicons with machine learning techniques: a hybrid approach for sentiment analysis. Int. J. Adv. Comput. Sci. Appl. **11**, 599–608 (2020)
20. Gupta, B., Negi, M., Vishwakarma, K., Rawat, G., Badhani, P.: Study of twitter sentiment analysis using machine learning algorithms on python. Int. J. Comput. Appl. **165**, 0975–8887 (2017)
21. Rahmath, P.H., Ahmad, T.: Sentiment analysis techniques – a comparative study. Int. J. Comput. Eng. Manag. **17**, 2230–7893 (2014)
22. Song, C., Wang, X.-K., Cheng, P.-F., Wang, J.-Q., Li, L.: SACPC: a framework based on probabilistic linguistic terms for short text sentiment analysis. **194**, 105572 (2020)
23. Kumar, R., Sarddar, D., Sarkar, I., Bose, R., Roy, S.: A literature survey on sentiment analysis techniques involving social media and online platforms. Int. J. Sci. Technol. Res. **9**, 2277–8616 (2020)
24. Kumar, T., Mahrishi, M., Meena, G.: A Comprehensive review of recent automatic speech summarization and keyword identification techniques. In: Fernandes, S.L., Sharma, T.K. (eds.) Artificial Intelligence in Industrial Applications. LAIS, vol. 25, pp. 111–126. Springer, Cham (2022). https://doi.org/10.1007/978-3-030-85383-9_8
25. Gupta, I., Joshi, N.: Enhanced twitter sentiment analysis using hybrid approach and by accounting local contextual semantic. J. Intell. Syst. **29**(1), 1611–1625 (2020). https://doi.org/10.1515/jisys-2019-0106
26. Chaithra, V.D.: Hybrid approach: naive bayes and sentiment VADER for analyzing sentiment of mobile unboxing video comments. In. J. Electr. Comput. Eng. **9**(5), 4452–4459 (2019)
27. Dutta Das, D., Sharma, S., Natani, S., Khare, N., Singh, B.: Sentimental analysis for airline twitter data. In: IOP Conference Series: Materials Science and Engineering, vol. 263, No. (4) (2017)
28. Bosznay, T.: Mind-map the Gap - Sentiment Analysis of Public Transport, vol. 1264. Amadeus Software Ltd. (2017)
29. Hoang, T., Cher, Hua (XU Peihua), P., Lim, E.-p., Lim, E.-P.: Crowdsensing and analyzing micro-event tweets for public transportation insights. In: 2016 IEEE International Conference on Big Data, pp. 2157–2166 (2017)
30. Madhuri, D.K.: A machine learning based framework for sentiment classification: Indian railways case study. Int. J. Innov. Technol. Exploring Eng. **8**, 2278–3075 (2019)
31. Gamal, D., Alfonse, M., El-Horbaty, E.-S.M., Salem, A.-B.M.: Twitter benchmark dataset for Arabic sentiment analysis. Int. J. Mod. Educ. Comput. Sci. **11**(1), 33–38 (2019). https://doi.org/10.5815/ijmecs.2019.01.04
32. Hasan, A., Moin, S., Karim, A., Shamshirband, S.: Machine learning-based sentiment analysis for twitter accounts. Math. Comput. Appl. **23**(1), 11 (2018). https://doi.org/10.3390/mca23010011
33. Kumar, T.S.S., Devi, N.T.D., Krishnendhu, T.K., Neethu, K.E, Radhakrishnan, S.C., Professor, A.: Review of sentiment analysis: a multilingual approach. Int. J. Adv. Res. Comput. Commun. Eng. **9**(1) (2020). https://doi.org/10.17148/IJARCCE.2020.9110

BigTech Befriending Circular Economy

Ruban Whenish and Seeram Ramakrishna(✉)

Department of Mechanical Engineering, National University of Singapore, 9 Engineering Drive 1, Singapore 119260, Singapore
mpesr@nus.edu.sg

Abstract. Nature has a great capability of utilizing waste in its ecosystem through recycling. Earth is a significant example of a circular economy. At the same time generation of waste is inevitable as well as it has association with various disposal principles. These days, by mimicking the various natural sources, the waste is treated as a possible resource, and the conversion of waste materials into a useful product, gaining a lot of focus. The growing awareness of usage of wastages, the depletion of limited natural sources, the consciousness of the environment of human health are key players by extending the life of waste materials and reusing them again after quality enhancements. The environmental, social, and governance (ESG) are the pillars of sustainability that encourages industries as well as nations to adopt circular economy concepts to achieve a zero-carbon economy in the coming years. Resources extraction and processing, and subsequent waste management are the major causes of carbon emissions. In this perspective, the adoption of circular economy principles by big technology players and their conscious shift towards zero carbon emissions pledge through 3R (reduce, reuse and recycling) principles were discussed.

Keywords: Circular economy · Zero-carbon strategies · Sustainability · UN SGDs

1 Introduction

Carbon dioxide (CO_2), methane (CH_4), nitrous oxide (N_2O), hydrofluorocarbons (HFCs), sulfur hexafluoride (SF_6) are the responsible GHGs for global warming and climate change. Among these, CO_2, CH_4 and N_2O are the predominant GHGs which accounts approximately 64%, 17% and 6% of human made global warming [1, 2]. CO_2 is more harmful and long-lived GHG gas which absorbs the heat and gradually releases over a period. The atmospheric CO_2 concentration leads to climate change and brings disaster to the globe. According to Mauna Loa Observatory in Hawaii, CO_2 concentration in atmosphere was calculated less than 320 ppm in 1960s and in 2020, it has reached to 412.5 ppm, a peak which was not witnessed in several past years. The CO_2 concentration has risen 2–3 ppm per annum may lead to more danger to earth's climate which would affect the millions of earth's habitants. The efforts and measures to be taken to return the CO_2 concentration level to 350 ppm [3]. The man-made damage to the nature leads to severe climate change that is rattling weather patterns, directing

V. E. Balas et al. (Eds.): ICETCE 2022, CCIS 1591, pp. 111–126, 2022.
https://doi.org/10.1007/978-3-031-07012-9_10

towards acute weather events and uncertain water availability. More than 50% of CO_2 emissions observed from resource extraction, processing, and its waste management. The constant efforts are needed at this hour to tackle the global climate change [4]. Production and process of materials account for 23% of global emissions. Processing and conversion of material from its natural form to products contribute a huge amount of global emissions [5]. 3.2% of global GHG emissions occurred due to waste generation and waste management [6]. Material efficiency strategies (MES) prevent a reasonable percentage of carbon emissions. A comparative study has been made to find out the life-cycle emissions of greenhouse gases made with and without MES in G7 countries, India, and China. MES reduce 60% life cycle emissions in India and China, 35% in G7 countries [7, 8].

An immediate shift towards zero carbon emission policies can make significant impacts on waste reduction, resource extraction and processing and lower the toxic relationship with natural resources [9]. Circular Economy (CE) is a pivotal tool to protect the environment by reducing waste formation, lowering the GHG emissions and avoiding depletion of natural resources by decreasing the dependence of natural resources such as water, material, and energy. CE concepts are building greener future through various frameworks and policies [10]. Zero-carbon/low-carbon development strategies, green and clean energy and sustainable living atmosphere for present and future generations are key constituents of CE principles [11]. Low carbon materials can produce low carbon products that can be applied for various applications which creates a circular economy framework with controlled CHG emissions without compromising their performance for creating a sustainable future [12].

Sustainability aims to reduce carbon emissions and circular solid waste management to avoid pollution and reduce waste generation [13, 14]. The best sustainable methods for reducing the negative environmental consequences of manufacturing processes must be found. About Cleaner Manufacturing and Production, SDG-12 focusses on the application of proactive environmental measures to processes (saving energy and raw materials, lowering the usage of harmful chemicals, lowering the intensity of emissions of waste generation and carbon dioxide emissions), products (reducing impact over the entire life cycle), and services to improve. By considering sustainable and cleaner manufacturing practices based on 17 UN SDGs, the review intends to present Sustainability manufacturing strategies towards net-zero Carbon-footprint, adopting Industry 4.0 principles for smart production and waste reduction [15]. CE concepts are recommended by UN SDG12 for a sustainable consumption and production through elimination of waste and pollution, recycle and reuse the materials and products as much as possible without compromising the quality and regenerate natural systems [16–18]. In addition, waste materials and pollutants also can be inputs to various recycling programs with an environmental, social, and economic effect that must be assessed [19].

Industries are going through various transitions throughout the centuries. The first industrial revolution introduced water and steam power to mechanize the production with zero date complexities in 1784. The operator-physical system was further advanced into following industrial revolutions 2.0 and 3.0 which adopted electric power and transformed into automated manufacturing systems. Industry 4.0 is revolutionizing the present manufacturing systems by transforming it into smart manufacturing systems [20, 21].

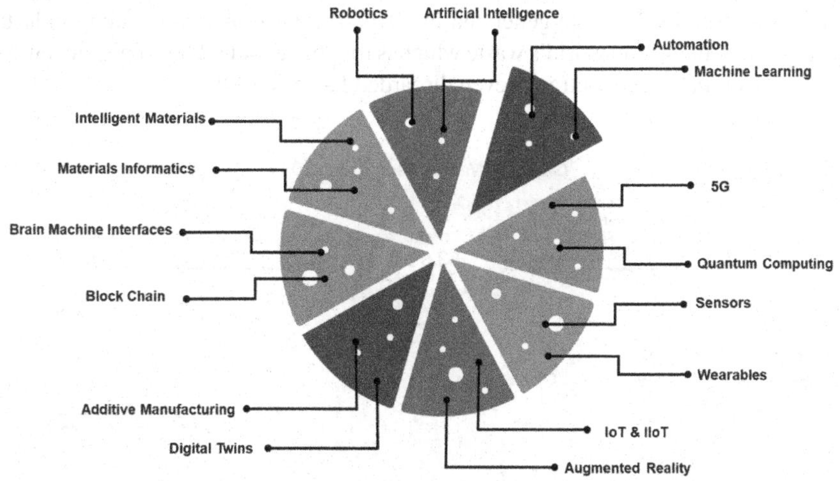

Fig. 1. Emerging landscape of Industry 4.0 technologies

There are various emerging technologies adopted by Industry 4.0 as depicted in Fig. 1 manufacturing environment with effective handling of resources and quality product outcomes. Industry 4.0 is known for amalgamating physical machine interface with information and communication technology as IP- enabled industrial cyber-physical systems [22]. Industry 4.0 will be considered as a forerunner to Industry 5.0 human-machine symbiosis which will be expected to arrive in 2035. The smart manufacturing market size expected to grow from $215 billion to $385 billion from 2020 to 2025 after the inclusion of Industry 4.0 [23]. Manufacturing operations equipped with technology driven ideas such as Big data-driven quality control, Robot-assisted production, predictive maintenance, optimized logistics and supply chain, etc. which enhanced the performance of smart factory [24, 25]. Industry 4.0 implementations in smart factory can generate a value worth efficiency improvement of 15% to 20% [26]. For instance, predictive maintenance offers early fault detection, improves reliability with reduced downtime, cost reduction, and better performance over traditional maintenance strategies [27]. The distinctive core principle of a smart factory is to meet the demands of various products in the same production line by interacting and reconfiguring the dynamic nature of the production [28]. Based on the literature study, by acquiring various circular economy principles in Manufacturing environment by major BigTech pave the way towards zero waste practices with enhanced sustainability.

2 The Need for Circular Economy Focusses in BigTech

The global E-waste studies outlined that, 53.6 Mt of E-waste generated in 2019, that is 21% hike in last 5 years. Out of 53.6 Mt, 9.3 Mt (less than 20%) were recycled, and remaining E-waste was discarded to landfill. The report further mentioned the prediction of global E-waste which will be expected to touch 74 Mt in 2030 which was double the amount compared with global E-waste collected in 2014 [29]. Figure 2 depicts the statistics of global E-waste generated in 2019. The fashion and modeling industry generates 4% of the whole world's waste which is roughly about 92 Mt of physical waste per year. This much waste outranks even discarded E-waste [30].

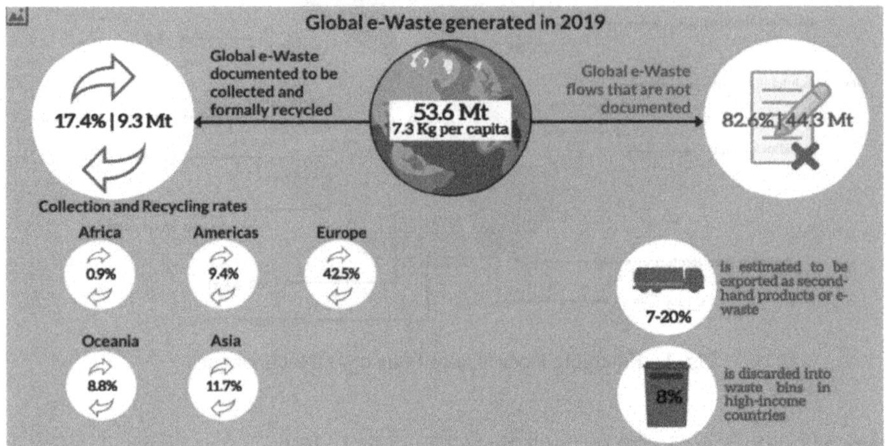

Fig. 2. Statistics on the amount of E-waste processed formally and informally across the globe [31] (Reprinted with permission from Science Advances from [29] under the Creative Commons Attribution 4.0 International License) (http://creativecommons.org/licenses/by/4.0/).

The 6R's are Reduce, Reuse, Recycle, Recover, Redesign, and Remanufacture. Reduce emphasis largely on the first three phases of product life cycles and relates to decreased resource consumption in pre-manufacture, reduced energy use and materials in production, and waste reduction during the usage phase. For reuse after its initial life cycle, a product or its components may be reused for successive life cycles to decrease the consumption of fresh raw materials in the manufacturing of such goods and components. Recovering involves gathering items at the end of their useful lives, dismantling, classifying, and cleaning them for reuse in later product life cycles. The redesign is changing goods to make future post-use procedures easier to make the product more sustainable. Remanufacture is restoring previously used items to their original form by reusing as many parts as feasible while maintaining functioning. This Idea was crucial for the development of the later CE concepts that focus on achieving sustainability dimensions [32–34]. Lean manufacturing or Lean production is a well-admired term among manufacturing industries by eliminating any kind of waste and making production/process more efficient. It is high time to transform the lean concepts as "Green lean" principles by addressing the reduction of GHG emissions and energy usage [35–38].

The growing importance of Industry 4.0 technologies, offers production cost reduction, boosting manufacturing agility and flexibility, efficient energy and resource consumption, reduction of harmful gas emission reduction, environmental responsibility development in the BigTech [39]. Though smart manufacturing is well developed through industry 4.0 technologies, sustainability and CE is compelling necessity now for BigTech [40]. Though, most of the studies described on smart factory characteristics, limited studies were done on the sustainability and circular economy attributes of BigTech [41]. Table 1 has shown mapping of Industry 4.0 onto UN SDGs and Economic Sectors.

Various countries pledged to achieve net-zero CO_2 equivalent emissions (all greenhouse gases) to the Earth's atmosphere by 2050. Motivation towards usage of renewable energy, Extended Producer Responsibility (EPR) for e-waste, plastics, packaging wastes, Green Mark Certification for Buildings, % Renewable energy targets, % Emissions reduction targets, % Recycled materials targets, % Renewable materials targets, Eco-labelling, Circularity Performance, Carbon Tax are few sustainability policies and regulations initiated by governments to implement carbon-neutral economy, low-carbon economy, and circular economy [42–47].

3 The Circular Economy of 3D Printing Industries

The adoption of the 3R principle in 3D printing can help in effective material utilization and avoid huge wastage. Sustainable dimensions are embraced well by 3D printing operations for efficient energy consumption, improved operational performance, no or fewer inventories, elimination of tooling cost, reduced carbon emission, economic balance and de-globalization, customization. 3D printing has the capability of acquiring circular materials using design for circularity principle with low waste production [48–50]. Approximately, 3D printing techniques generates 1/3rd of the waste compared than traditional building techniques. The approximate cost spent for 3D printed house roughly 20% of the cost spent for standard 2-bedroom apartment [51, 52]. The waste polymer powders converted into useful products through recycling [53]. Ocean-source plastic waste commercialization and conversion to useful 3D printed objects were done favorably by Plasticbank [54]. 'Waste for Life', a project in Argentina encouraged people to convert plastic into 3D printed parts and offered self-sustaining income to people who were involved with this initiative [55]. A waste feedstock material in one 3D printing used as recycled feedstock in another 3D printing technique provides a feasible way to support CE and sustainability in future 3D printing techniques [56]. The recycled waste polymer can be converted into useful AM feedstock such as polymer filaments for the FDM process and subsequently create 3D printed functional parts to build the circular economy [57–61] A low-cost, open-source, self-replicating 3D printer (Extrusion-based Additive Manufacturing (AM) technique) used the waste polymers as feedstock to promote the CE. The operating cost is found to be about 40 times lesser and is more economical than purchasing commercial filaments. The recycled plastic waste also contributed to valuable energy savings [62, 63]. An energy storage device comprising 40% recycled thermoplastics exhibits voltage potential that is par with commercial dry cells. In addition, the device showed better rheological and mechanical properties which ensures the capability to undergo severe loading conditions [64]. Used nylon powders in SLS L/PBF process are recycled and converted as a polymer filament reinforced with waste Mg powders in 2%, 4%, and 8% ratios. First and foremost, the waste nylon and Mg powders were identified as extrudable and printable as FFF filament. The thermal and mechanical performance of nylon-Mg polymer composites were also evaluated. The findings depicted that the waste nylon can be recycled and used as a viable filament material in the FFF technique. The printed parts depict desirable geometric and surface properties after slight postprocessing. Lignin, a biomass material from industrial

Table 1. Mapping of 4IRs onto UN SDGs and Economic Sectors

UN SDG	Sectors	Robotics & Automation	IoT, IIoT	Digital Twins	AI, Machine Learning	Block chain	Data Analytics	Sensors & 5G	Nanotechnology	3D Printing
1	Family income	☺☺	☺☺	☺☺	☺☺	☺☺	☺☺	☺☺	☺☺	☺☺
2	Food	☺☺	☺☺☺	☺☺	☺☺☺	☺☺☺	☺☺☺	☺☺☺	☺☺☺☺☺	☺
3	Health	☺☺☺	☺☺☺	☺☺☺	☺☺☺	☺☺☺	☺☺☺	☺☺☺	☺☺☺	☺☺☺
4	Education	☺☺	☺☺☺☺	☺☺☺	☺☺☺☺☺	☺☺	☺☺☺☺☺	☺☺☺☺	☺☺☺	☺☺☺
5	Gender equality	☺☺	☺☺	☺☺	☺☺	☺☺	☺☺	☺☺	☺☺	☺☺
6	Water and Sanitation	☺☺	☺☺☺☺	☺☺	☺☺	☺	☺☺☺☺	☺☺☺	☺☺☺☺☺	☺
7	Energy	☺☺	☺☺☺	☺☺☺☺☺	☺☺☺	☺	☺☺☺☺☺	☺☺	☺☺☺	☺
8	Economic Growth	☺☺☺	☺☺	☺☺	☺☺	☺☺☺	☺☺☺☺	☺☺	☺☺	☺
9	Infrastructure	☺☺☺	☺☺☺☺	☺☺☺☺☺	☺☺☺☺	☺☺	☺☺☺☺☺	☺☺☺☺	☺☺☺	☺☺
10	Social mobility	☺☺	☺☺	☺☺	☺☺	☺☺	☺☺	☺☺	☺☺	☺☺

(continued)

Table 1. *(continued)*

11	Urbanization	☺	☺ ☺ ☺ ☺	☺ ☺ ☺ ☺ ☺	☺ ☺	☺ ☺	☺ ☺ ☺ ☺ ☺	☺ ☺ ☺	☺ ☺ ☺ ☺ ☺	☺
12	Consumption & Production	☺ ☺ ☺	☺ ☺ ☺	☺ ☺	☺ ☺ ☺	☺ ☺ ☺	☺ ☺ ☺	☺ ☺ ☺	☺ ☺ ☺	☺
13	Combating Climate Impact	☺	☺ ☺	☺ ☺	☺ ☺ ☺ ☺	☺ ☺ ☺ ☺ ☺ ☺	☺ ☺ ☺ ☺ ☺	☺ ☺ ☺	☺ ☺ ☺	☺ ☺
14	Oceans & Water Resources	☺	☺	☺	☺ ☺	☺ ☺	☺ ☺	☺ ☺	☺ ☺ ☺ ☺	☺
15	Terrestrial Ecosystems	☺	☺ ☺ ☺	☺ ☺ ☺ ☺ ☺	☺ ☺ ☺	☺ ☺ ☺	☺ ☺ ☺ ☺ ☺	☺ ☺ ☺ ☺	☺	☺
16	Peace & Security	☺ ☺	☺ ☺	☺	☺ ☺ ☺	☺ ☺ ☺ ☺ ☺	☺ ☺ ☺ ☺ ☺	☺ ☺ ☺	☺ ☺	☺
17	Trade & Commerce	☺ ☺	☺ ☺	☺	☺ ☺ ☺	☺ ☺ ☺ ☺ ☺	☺ ☺ ☺ ☺ ☺	☺	☺	☺

feedstock waste (food, pulp & paper industries) was incorporated with synthetic polymers like acrylonitrile-butadiene-styrene (ABS) and nylon on a 40–60% weight basis to promote sustainable manufacturing. Titanium alloy waste powder (almost 100% waste) was recycled and converted into affordable wire for AM process. CP-Ti Gr 2 HDH and Ti-6Al-4V GA waste alloy powders continuously extruded through ConformTM and then were cold drawn. The mechanical performance demonstrated a strong potential for use as a feedstock in the wire AM process and microstructural analysis depicted a fine-grained structure. The effective usage of waste powders into a wire enables sustainable supply chains as an attractive circular economy practice [65].

In addition, Additive Manufacturing can provide a new paradigm in sustainable context on recycling of its feedstock and utilizing low carbon materials by reducing the ecological footprint [66–71]. Recycling of metal powders in AM depends on powder recovery, recycling optimization, recycled powder quality and lifecycle. In addition to recycling strategies, improved support structure and less waste generation should be mapped. Virgin AM feedstock is quite expensive and leads to deterioration of the resources. At the same time, the development high strength recycled materials without sacrificing their quality can reduce carbon footprint [72, 73] and reduce waste disposal and generate efficient material management [72, 73]. There are limitations such as the recycled polymer parts printed through AM techniques may not meet the application requirements. There is a need for more investigation and research in this direction to justify the effectiveness of recycled feedstock by enhancing circular economy through the improvement of the LCA of feedstock in AM [74].

4 The Adoption of Circular Economy by BigTech

World Economic Forum (WEF) appreciates the efforts of HP by embracing the fourth industrial revolution (4IR) manufacturing concepts to achieve profitable growth through a sustainable route [75]. Google officially launched Sycamore quantum processor in 2019 for quantum computing purpose. The quantum computing promises effective management of fast computational tasks with the support of high-fidelity processor that could be able to process quantum algorithms in a vastly large computational space. Sycamore quantum processor requires 200 s to perform calculation which cannot be complete in 10,000 years by conventional computers [76].

Power usage effectiveness (PUE) is defined as the ratio of total facility power to ICT equipment power. Ideal PUE is 1, and in practice it ranges from 1.2 (very efficient) to 3 (very inefficient). Other performance indicators are carbon usage effectiveness (CUE); on-site energy fraction, OEF, and on-site energy matching, OEM), energy reuse (energy reuse factor, ERF), and water usage (water usage effectiveness, WUE). Average internet user spends 1.5 GB per day whereas smart hospital requires 3000 GB/day, autonomous driving requires 4000 GB/day, airplane requires 40,000 GB/day and smart factory requires 1 million GB/day [77]. Data centers are large energy hogs which comprises servers, storage equipment and cooling infrastructures which demands huge appetite for electricity. The estimation of electricity consumption by data centers calculated by International Energy Agency was around 1% of the world's electricity. The electricity consumption could strike double-digits by 2030, making related emissions a problem [78, 79].

Singapore's rapid digitalization growth demands an appealing space to build centers and reviewed the balancing between green and business needs. The ideas behind the planting data centers in Singapore is for enhanced resource efficiency which could fulfill the objective of Singapore's economic and strategic developments. Globally Singapore has obtained the 5th place for top data center markets in 2021 and 3rd place in 2022, ties with Silicon Valley. In Asia-pacific region Singapore has achieved the feet, i.e., first position. A $23 million center at NUS to develop innovative and sustainable cooling solutions for data centers located in tropical locations. Singapore government prioritizes green economy-based skills in the job market to achieve their green economy targets. According to the Ministry of sustainability and environment estimation, in the next decade, Singapore requires 55,000 new and upgraded vacancies in the sustainability sector [80]. The National University of Singapore made Sustainability strategies and a sustainable resource management team to study and implement the action roadmap 2030 in renewable energy, waste management, water management, greening the campus, and integrated mobility [81]. The Smart Industry Readiness Index (SIRI) was developed by the Singapore Economic Development Board (EDB) in collaboration with a cluster of leading technology firms, industries, academic experts, and consultancies. SIRI has prepared various frameworks and tools which help manufacturers from small to large scale industries to transform their manufacturing transformation journeys in attaining sustainability. SIRI comprises the three major elements such as Process, Technology, and Organization of Industry 4.0. SIRI framework involves 3 layers, 3 building blocks, 8 pillars and 16 dimensions as shown in Fig. 3 [82].

DELPHI, an automotive company has come up with an optimized design from their prior design of a heat exchanger by reducing its mass. The optimized lightweight design increased the energy efficiency of the heat exchanger [83]. Famous shoemaker, ADIDAS unveils FUTURECRAFT.FOOTPRINT which is made of natural and recycled materials. FUTURECRAFT.FOOTPRINT shoe with a carbon footprint of ~ 3 kg of CO_2 per pair compared to ~ 14 kg carbon emissions of conventional shoes [84]. Epson's heat-free inkjet printer to minister for sustainability as their key focus and to make good environmental implications [85]. Recycling household appliances like refrigerators and washing machines from start to finish can do their functions efficiently [86]. The Submersible eLAB-ID concept watch made up of recycled steel, titanium, silicon, plastics by Panerai reached highest possible sustainability of 98.6% [87].

A portable healthcare system could be achieved through wearable health monitoring devices which also foster decentralized healthcare delivery which is contrary to traditional centralized healthcare practice. The Covid pandemic boosts the wearable gadgets market by encouraging many people to keep their interest in health[88]. Tech giant Google acquired the leading smartwatch firm Fitbit at $2.1 billion [89]. A wireless and fully implantable high-channel count brain-machine interface (BMI) was developed by Neuralink for the people affected with paralysis. This interface enables such people to use their neural activity directly as well as enable them to operate devices such as mobile phones and computers with speed and ease. The goal of Neuralink is to build a wireless and implantable clinical BMI system with safety and effective. [90]. A computer program developed by NTU researchers which can identify individuals who are at a higher risk of developing depression [91].

The tech sectors are regarded as the industry for future. Greenpeace, an independent global campaigning network for protecting the environment made a survey on climate commitments of Tech industries. 30 companies were selected, each 10 from China, Japan and South Korea which was listed in Forbes Top 100 Digital Companies. The highest score C+ was given to Japanese firm, Sony near the top in climate action. At the same time, Korean internet firm Kakao, failed by Greenpeace ratings for having low climate commitment and poor advocacy [92, 93].

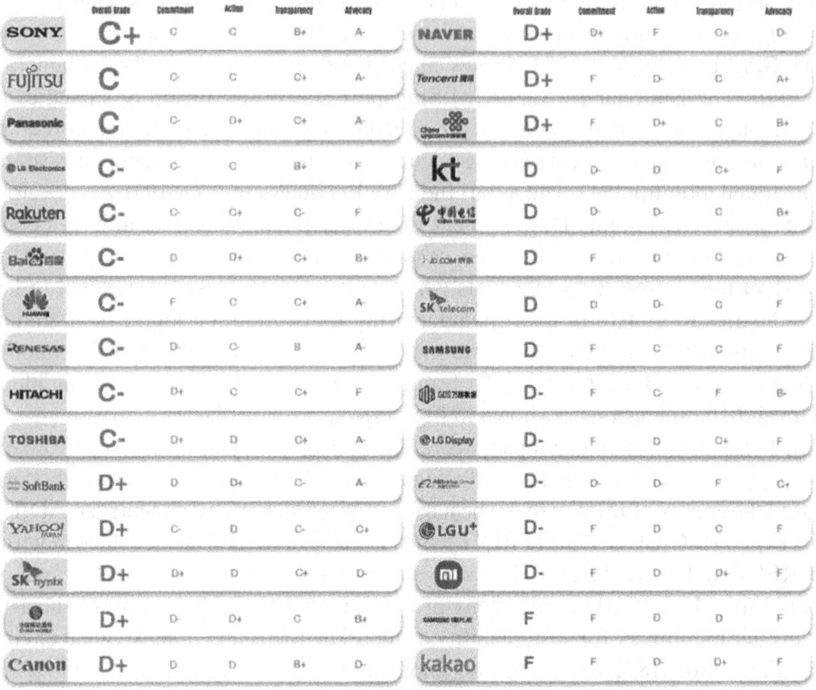

	Overall Grade	Commitment	Action	Transparency	Advocacy
SONY	C+	C	C	B+	A-
FUJITSU	C	C-	C	C+	A-
Panasonic	C	C-	D+	C+	A-
LG Electronics	C-	C-	C	B+	F
Rakuten	C-	C-	C+	C-	F
Baidu百度	C-	D	D+	C+	B+
HUAWEI	C-	F	C	C+	A-
RENESAS	C-	D-	C-	B	A-
HITACHI	C-	D+	C	C+	F
TOSHIBA	C-	D+	D	C+	A-
SoftBank	D+	D	D+	C-	A-
YAHOO! JAPAN	D+	C-	D	C-	C+
SK hynix	D+	D+	D	C+	D-
[illegible]	D+	D-	D+	C	B+
Canon	D+	D	D	B+	D-
NAVER	D+	D+	F	C+	D-
Tencent 腾讯	D+	F	D-	C	A+
China Unicom	D+	F	D+	C	B+
kt	D	D-	D	C+	F
中国电信	D	D-	D-	C	B+
JD.COM 京东	D	F	D	C	D-
SK telecom	D	D	D-	C	F
SAMSUNG	D	F	C	C	F
GDS万国数据	D-	F	C-	F	B-
LG Display	D-	F	D	C+	F
[illegible]	D-	D-	D-	F	C+
LG U+	D-	F	D	C	F
mi	D-	F	D	D+	F
SAMSUNG DISPLAY	F	F	D	D	F
kakao	F	F	D-	D+	F

Fig. 3. Greenpeace ranking of Asian technology firms [92]

4.1 Warnings

Greenwashing is the deceptive process of generating a false impression on the products about how environmentally they are sound than actual information advertised by companies [94]. Danish Environmental Protection Agency made a life cycle assessment and concluded that a bag produced by polypropylene, paper and cotton shall be utilized 37, 43, and 7100 times, respectively for the same environmental benefits. Greenwashing can be avoided through deepening of the knowledge of sustainability, ESG, and net-zero carbon [95]. Policies, regulations, standards, accreditation & compliance, and education are necessary to avoid sinkholes of greenwashing and to address the conundrum of climate actions and economic growth [96].

5 Conclusion

An economic development and investment focus made a subtle shift by fueling its demands on tech-driven based workforce. A tech-driven global economy activities involving technologies like AI, big data, block chain, FinTech, GreenTech cyber security, 5G, etc. For instance, an auditor needs to acquire skills to do 'Green audits and a lawyer need to shape his legal mind towards ESG – driven activities [97]. The International Labor Organization estimates there will be 24 million green jobs worldwide by 2030. Green economy and zero-carbon strategies should be backed up with relevant skills, workforce, and financial support for green technologies. Social Laboratories in the UK and Shikshantar from India designed the Gigaton Challenge course and delivered through online platform 'complexity university'. The faculty team encouraged the course participants through practicing prototypes and tests and adjusts proposed interventions which demonstrate in 2 weeks' time that each participant should lessen CO_2 emissions by one ton [98]. ISO TC 323 circular economy standards facilitate implementation, dialogue, and communication between stakeholders. These standards offer opportunities such as integrating eco-design tools in the process, integrating long-term vision, considering waste as a resource, cooperating with all value chain stakeholders, promote value networks [99, 100]. Low-carbon materials are purposely designed and/or selected with lower environmental footprint and social costs and higher circularity potential while satisfying the cost as well as functional requirements [101]. The strategies for low-carbon materials consist of materials, technology-intensive processing/ manufacturing technologies, optimized supply chain, Usage and End-of-Life Management, and Design Out Waste and Pollution [102]. The adoption of circular economy principles by BigTech and their conscious shift towards zero carbon emissions could enable the future industry towards sustainable focus and shall yield fruitful outcomes on balancing the ecological footprint.

Acknowledgment. The authors would like to acknowledge the support given by National Research Foundation (Sustainable Tropical Data Centre Test bed: A-0009465–01-00) for this research work.

References

1. European Commission: Causes of climate change (2014)
2. United States Environmental Protection Agency: Global Greenhouse Gas Emissions Data
3. Carbon dioxide peaks near 420 parts per million at Mauna Loa observatory
4. IPCC: Climate Change 2021 Working Group I contribution to the Sixth Assessment Report of the Intergovernmental Panel on Climate Change Summary for Policymakers (2020)
5. Krausmann, F., et al.: Global socioeconomic material stocks rise 23-fold over the 20th century and require half of annual resource use. Proc. Natl. Acad. Sci. **114**, 1880–1885 (2017)
6. Ritchie, H., Roser, M.: CO_2 and greenhouse gas emissions. Our world in data (2020)
7. Pao, H.-T., Chen, C.-C.: Decoupling strategies: CO2 emissions, energy resources, and economic growth in the group of twenty. J. Clean. Prod. **206**, 907–919 (2019)
8. Ekins, P., et al.: Resource efficiency: Potential and economic implications (2016)
9. Zaman, A.U.: A comprehensive review of the development of zero waste management: lessons learned and guidelines. J. Clean. Prod. **91**, 12–25 (2015)
10. Whenish, R., Ramakrishna, S., Jaiswal, A.K., Manivasagam, G.: A framework for the sustainability implications of 3D bioprinting through nature-inspired materials and structures. Bio-Design and Manuf. (2022). https://doi.org/10.1007/s42242-021-00168-x

11. Sen, A., Meini, L., Napoli, E., Napoli, C.: Beyond energy: incentivizing decarbonization through the circular economy
12. Das, O., et al.: Demystifying Low-Carbon Materials. Materials Circular Economy **3**, 1–11 (2021)
13. Singapore's path towards a zero waste nation
14. Kucukvar, M., Egilmez, G., Onat, N.C., Samadi, H.: A global, scope-based carbon footprint modeling for effective carbon reduction policies: lessons from the Turkish manufacturing. Sustain. Prod. Cons. **1**, 47–66 (2015). https://doi.org/10.1016/j.spc.2015.05.005
15. Giannetti, B.F., Agostinho, F., Eras, J.J.C., Yang, Z., Almeida, C.M.V.B.: Cleaner production for achieving the sustainable development goals (2020). https://doi.org/10.1016/j.jclepro.2020.122127
16. Delivering the Sustainable Development Goals
17. Ensure sustainable consumption and production patterns
18. Meena, G., et. al.: Traffic Prediction for Intelligent Transportation System using Machine Learning. In: 2020 3rd International Conference on Emerging Technologies in Computer Engineering: Machine Learning and Internet of Things (ICETCE), pp. 145–148 (2020). https://doi.org/10.1109/ICETCE48199.2020.9091758
19. Haapala, K.R., et al.: A review of engineering research in sustainable manufacturing. J. Manuf. Sci. Eng. 135 (2013)
20. Betti, F., de Boer, E., Giraud, Y.: Industry's fast-mover advantage: enterprise value from digital factories. In: World Economic Forum and McKinsey & Company (2020)
21. Zhong, R.Y., Xu, X., Klotz, E., Newman, S.T.: Intelligent manufacturing in the context of industry 4.0: a review. Engineering **3**, 616–630 (2017)
22. Hoosain, M.S., Paul, B.S., Ramakrishna, S.: The impact of 4ir digital technologies and circular thinking on the united nations sustainable development goals. Sustainability (Switzerland). **12**, 1–16 (2020). https://doi.org/10.3390/su122310143
23. Ramakrishna, S.: Fourth Industrial Revolution Technologies for Progress in UN SDGS (2021)
24. Tiwari, K., Khan, M.S.: Sustainability accounting and reporting in the industry 4.0. J. Clean. Prod. 258 (2020). https://doi.org/10.1016/j.jclepro.2020.120783
25. Warke, V., Kumar, S., Bongale, A., Kotecha, K.: Sustainable development of smart manufacturing driven by the digital twin framework: a statistical analysis. Sustainability. **13**, 10139 (2021)
26. Arnold, C., Kiel, D., Voigt, K.-I.: How the industrial internet of things changes business models in different manufacturing industries. Int. J. Innov. Manag. **20**, 1640015 (2016)
27. Daily, J., Peterson, J.: Predictive maintenance: how big data analysis can improve maintenance. In: Supply chain integration challenges in commercial aerospace, pp. 267–278. Springer (2017)
28. Cimini, C., Pezzotta, G., Pinto, R., Cavalieri, S.: Industry 4.0 Technologies Impacts in the Manufacturing and Supply Chain Landscape: An Overview. In: Borangiu, T., Trentesaux, D., Thomas, A., Cavalieri, S. (eds.) SOHOMA 2018. SCI, vol. 803, pp. 109–120. Springer, Cham (2019). https://doi.org/10.1007/978-3-030-03003-2_8
29. Murthy, V., Ramakrishna, S.: A review on global E-waste management: urban mining towards a sustainable future and circular economy. Sustainability (Switzerland). 14 (2022). https://doi.org/10.3390/su14020647
30. MINIWIZ: 13 fashion brands who show that upcycling can be fabulous. https://miniwiz.medium.com/13-fashion-brands-who-show-that-upcycling-can-be-fabulous-6cbed4c09d58. Accessed 26 Feb 2022
31. Baldé, C.P., van den Brink, S., Forti, V., van der Schalk, A., Hopstaken, F.: The dutch WEEE flows 2020. What happened between (2010)

32. Jawahir, I.S., Bradley, R.: Technological elements of circular economy and the principles of 6R-based closed-loop material flow in sustainable manufacturing. Procedia Cirp. **40**, 103–108 (2016)
33. Aljuneidi, T., Bulgak, A.A.: Carbon footprint for designing reverse logistics network with hybrid manufacturing-remanufacturing systems. J. Remanuf. **10**(2), 107–126 (2019). https://doi.org/10.1007/s13243-019-00076-5
34. Leino, M., Pekkarinen, J., Soukka, R.: The role of laser additive manufacturing methods of metals in repair, refurbishment and remanufacturing - Enabling circular economy. In: Physics Procedia, pp. 752–760. Elsevier B.V. (2016). https://doi.org/10.1016/j.phpro.2016.08.077
35. Dües, C.M., Tan, K.H., Lim, M.: Green as the new Lean: how to use Lean practices as a catalyst to greening your supply chain. J. Clean. Prod. **40**, 93–100 (2013)
36. Garza-Reyes, J.A.: Lean and green – a systematic review of the state of the art literature. J. Clean. Prod. **102**, 18–29 (2015). https://doi.org/10.1016/j.jclepro.2015.04.064
37. Fercoq, A., Lamouri, S., Carbone, V.: Lean/green integration focused on waste reduction techniques. J. Clean. Prod. **137**, 567–578 (2016)
38. Gholami, H., Jamil, N., Mat Saman, M.Z., Streimikiene, D., Sharif, S., Zakuan, N.: The application of green lean six sigma. Bus. Strateg. Environ. **30**, 1913–1931 (2021)
39. Viswanathan, R., Telukdarie, A.: The role of 4IR technologies in waste management practices-a bibliographic analysis. Proce. Comp. Sci. **200**, 247–256 (2022). https://doi.org/10.1016/j.procs.2022.01.223
40. Vrchota, J., Pech, M., Rolínek, L., Bednář, J.: Sustainability outcomes of green processes in relation to industry 4.0 in manufacturing: Systematic review (2020). https://doi.org/10.3390/su12155968
41. Kamble, S.S., Gunasekaran, A., Gawankar, S.A.: Sustainable industry 4.0 framework: A systematic literature review identifying the current trends and future perspectives. Proc. Saf. Environ. Prot. **117**, 408–425 (2018)
42. Latest climate pledges could limit global temperature rise, a new report: The COP26 SUMMIT
43. Our Waste, Our Resources: A Strategy for England 1 Our Waste, Our Resources: A Strategy for England Our Waste, Our Resources: A Strategy for England 2 (2018)
44. Waste: Materials Management and Circular Economy
45. Silva, A., Rosano, M., Stocker, L., Gorissen, L.: From waste to sustainable materials management: Three case studies of the transition journey. Waste Manage **61**, 547–557 (2017)
46. Tomaszewska, J.: Polish transition towards circular economy: Materials management and implications for the construction sector. Materials. **13**, 5228 (2020)
47. Recycling: Extended Producer Responsibility
48. Giurco, D., Littleboy, A., Boyle, T., Fyfe, J., White, S.: Circular economy: questions for responsible minerals, additive manufacturing and recycling of metals. Resources **3**, 432–453 (2014). https://doi.org/10.3390/resources3020432
49. Ponis, S., Aretoulaki, E., Maroutas, T.N., Plakas, G., Dimogiorgi, K.: A systematic literature review on additive manufacturing in the context of circular economy (2021). https://doi.org/10.3390/su13116007
50. Sauerwein, M., Balkenende, R., Doubrovski, Z., Bakker, C.: Exploring the potential of additive manufacturing for product design in a circular economy. J. Clean. Prod. (2019). https://doi.org/10.1016/j.jclepro.2019.04.108
51. All you need to know about India's first 3D-printed home
52. Park, B.K., Kim, C.J., Kwon, D.E., Lee, Y.W.: Design and fabrication of partially foamed grid structure using additive manufacturing and solid state foaming. Processes. **8**, 1 (2020). https://doi.org/10.3390/pr8121594

53. Gu, F., Guo, J., Zhang, W., Summers, P.A., Hall, P.: From waste plastics to industrial raw materials: A life cycle assessment of mechanical plastic recycling practice based on a real-world case study. Sci. Total Environ. **601–602**, 1192–1207 (2017). https://doi.org/10.1016/j.scitotenv.2017.05.278
54. Fox Business: One Man's Plastic Waste is Another's Business Mission (2016)
55. Baillie, C.: Waste for life. Mater. Today **11**, 6 (2008). https://doi.org/10.1016/S1369-7021(08)70187-1
56. Uddin, M., Williams, D., Blencowe, A.: Recycling of selective laser sintering waste Nylon powders into fused filament fabrication parts reinforced with Mg particles. Polymers **13** (2021). https://doi.org/10.3390/polym13132046
57. Atsani, S.I., Mastrisiswadi, H.: Recycled polypropylene filament for 3D printer: extrusion process parameter optimization. In: IOP Conference Series: Materials Science and Engineering, p. 012022. IOP Publishing (2020)
58. Mahrishi, M., Morwal, S., Muzaffar, A.W., Bhatia, S., Dadheech, P., Rahmani, M.K.I.: Video Index Point Detection and Extraction Framework Using Custom YoloV4 Darknet Object Detection Model. IEEE Access **9**, 143378–143391 (2021). https://doi.org/10.1109/ACCESS.2021.3118048
59. Zhao, P., Rao, C., Gu, F., Sharmin, N., Fu, J.: Close-looped recycling of polylactic acid used in 3D printing: An experimental investigation and life cycle assessment. J. Clean. Prod. **197**, 1046–1055 (2018)
60. Mikula, K., et al.: 3D printing filament as a second life of waste plastics-a review. https://doi.org/10.1007/s11356-020-10657-8/Published
61. Algarni, M., Ghazali, S.: Comparative study of the sensitivity of pla, abs, peek, and petg's mechanical properties to fdm printing process parameters. Crystals. **11** (2021). https://doi.org/10.3390/cryst11080995
62. Baechler, C., Devuono, M., Pearce, J.M.: Distributed recycling of waste polymer into RepRap feedstock. Rapid Prototyping Journal. **19**, 118–125 (2013). https://doi.org/10.1108/13552541311302978
63. Romani, A., Rognoli, V., Levi, M.: Design, materials, and extrusion-based additive manufacturing in circular economy contexts: From waste to new products (2021). https://doi.org/10.3390/su13137269
64. Singh, R., Singh, H., Farina, I., Colangelo, F., Fraternali, F.: On the additive manufacturing of an energy storage device from recycled material. Compos. B Eng. **156**, 259–265 (2019). https://doi.org/10.1016/j.compositesb.2018.08.080
65. Smythe, S.A., Thomas, B.M., Jackson, M.: Recycling of titanium alloy powders and swarf through continuous extrusion (Conformtm) into affordable wire for additive manufacturing. Metals **10**, 1–18 (2020). https://doi.org/10.3390/met10060843
66. Frăğilă, D., Rotaru, H.: Additive manufacturing – a sustainable manufacturing route. 03004 (2017)
67. Agrawal, R., Vinodh, S.: State of art review on sustainable additive manufacturing **6**, 1045–1060 (2019). https://doi.org/10.1108/RPJ-04-2018-0085
68. Diegel, O., Kristav, P., Motte, D., Kianian, B.: Additive Manufacturing and its Effect on Sustainable Design. 73–99 (2016). https://doi.org/10.1007/978-981-10-0549-7
69. Whenish, R., Antony, M., Balaji, T., Selvam, A.: Design and performance of additively manufactured lightweight bionic hand. 020028 (2021)
70. Whenish, R., Velu, R., Anand Kumar, S., Ramprasath, L.S.: Additive manufacturing technologies for biomedical implants using functional biocomposites. In: Praveen Kumar, A., Sadasivuni, K.K., AlMangour, B., Abdul bin Majid, M.S. (eds.) High-Performance Composite Structures: Additive Manufacturing and Processing, pp. 25–44. Springer Singapore, Singapore (2022). https://doi.org/10.1007/978-981-16-7377-1_2

71. Raheem, A.A., et al.: A review on development of Bio-inspired implants using 3D printing. Biomimetics 6 (2021). https://doi.org/10.3390/biomimetics6040065
72. Shanmugam, V., et al.: Polymer recycling in additive manufacturing: an opportunity for the circular economy. Mat. Circ. Eco. **2**(1), 1–11 (2020). https://doi.org/10.1007/s42824-020-00012-0
73. Patsavellas, J., Salonitis, K.: The carbon footprint of manufacturing digitalization: critical literature review and future research agenda. In: Procedia CIRP, pp. 1354–1359. Elsevier B.V. (2019). https://doi.org/10.1016/j.procir.2019.04.026
74. Colorado, H.A., Velásquez, E.I.G., Monteiro, S.N.: Sustainability of additive manufacturing: the circular economy of materials and environmental perspectives. J. Market. Res. **9**, 8221–8234 (2020). https://doi.org/10.1016/j.jmrt.2020.04.062
75. Factories of the Future Find Growth and Sustainability Through Digitalization
76. Arute, F., et al.: Quantum supremacy using a programmable superconducting processor. Nature **574**, 505–510 (2019)
77. Shah, V., Director, R.: IT transformation - an imperative for business outcomes (2017)
78. Manganelli, M., Soldati, A., Martirano, L., Ramakrishna, S.: Strategies for improving the sustainability of data centers via energy mix, energy conservation, and circular energy. Sustainability **13**, 6114 (2021)
79. Manganelli, M., Soldati, A., Martirano, L., Ramakrishna, S.: Strategies for improving the sustainability of data centers via energy mix, energy conservation, and circular energy, (2021). https://doi.org/10.3390/su13116114
80. SkillsFuture Singapore: Skills demand for the future economy (2021)
81. National University of Singapore: Nurturing passion & Expertise in sustainability, https://sustainability.nus.edu.sg/campus/campus-infrastructure/. last accessed 26 Feb 2022
82. Singapore Economic Development Board (EDB): The Smart Industry Readiness Index. https://siri.incit.org/about. last accessed 26 Feb 2022
83. le Bourhis, F., Kerbrat, O., Dembinski, L., Hascoet, J.Y., Mognol, P.: Predictive model for environmental assessment in additive manufacturing process. Procedia CIRP. **15**, 26–31 (2014). https://doi.org/10.1016/j.procir.2014.06.031
84. Adidas and Allbirds unveil low-carbon-footprint sneaker from their collaboration
85. Sustainability a key focus at Epson's new South-east Asia headquarters
86. More efficient recycling of large appliances like fridges at automated plant in Tuas
87. Panerai Explores New Reaches of Sustainability with 98.6%-Recycled Submersible eLAB-ID Concept Watch
88. Pandemic lifts sales of wearable gadgets Read more at: https://cio.economictimes.indiatimes.com/news/consumer-tech/pandemic-lifts-sales-of-wearable-gadgets/83984942
89. Google closes $2.1B acquisition of Fitbit as Justice Department probe continues
90. Elon Musk's Neuralink "shows monkey playing Pong with mind."
91. NTU scientists develop computer program that can detect those at higher risk of depression
92. How green is Asian tech? C+ at best: Greenpeace | News | Eco-Business
93. Race to Green Scoring Tech Companies from China, Japan and South Korea on their Climate Action and Renewable Energy Use. (2021)
94. de Freitas Netto, S.V., Sobral, M.F.F., Ribeiro, A.R.B., da Luz Soares, G.R.: Concepts and forms of greenwashing: A systematic review. Environ. Sci. Euro **32**, 1–12 (2020)
95. Sakthivel, T.S., Seal, S., Ramakrishna, S.: Sustainable materials and the circular economy: the circular economy and sustainability are multifaceted concepts and thus should be pursued through collective advancements across various scientific fields. Adv. Mater. Processes **179**, 28–32 (2021)
96. Gomez-Echeverri, L.: Climate and development: enhancing impact through stronger linkages in the implementation of the paris agreement and the sustainable development goals

(SDGs). Philosophical Transactions of the Royal Society A: Mathematical. Phys. Eng. Sci. **379**, 20160444 (2018)

97. Ven Sreenivasan: Role of EDB in ensuring Singapore's success in a new tech-driven world (2022)
98. Unlocking talent for the global green skills revolution
99. Singhal, R.: Recourse to the Circular Economy: The Path Ahead. In: Dasgupta, P., Saha, A.R., Singhal, R. (eds.) Sustainable Development Insights from India. ISBE, pp. 261–280. Springer, Singapore (2021). https://doi.org/10.1007/978-981-33-4830-1_13
100. Subramoniam, R., Sundin, E., Subramoniam, S., Huisingh, D.: Riding the digital product life cycle waves towards a circular economy. Sustainability **13**, 8960 (2021)
101. Ramakrishna, S., Pervaiz, M., Tjong, J., Ghisellini, P., Sain, M.M.: Low-carbon materials: genesis, thoughts, case study, and perspectives. Circ. Econ. Sustain. 1–16 (2021)
102. Das, O., et al.: Demystifying low-carbon materials. Mat. Circ. Econ. **3**(1), 1–11 (2021). https://doi.org/10.1007/s42824-021-00044-0

Internet of Things (IoT)

Emerging Role of Artificial Intelligence and Internet of Things on Healthcare Management in COVID-19 Pandemic Situation

G. S. Raghavendra[1](✉), Shanthi Mahesh[2], and M. V. P. Chandra Sekhara Rao[3]

[1] CSE, RVR & JC College of Engineering, Chowdavaram, India
raghavendragunturi@gmail.com

[2] Information Science and Engineering, Atria Institute of Technology, Bengaluru, India
shanthi.mahesh@atria.edu

[3] Computer Science and Business Systems, RVR & JC College of Engineering, Chowdavaram, India

Abstract. The tremors caused by the COVID-19 epidemic have ushered in a new era of healthcare problems for people all across the globe. Significant problems and challenges have been observed in areas such as victim assistance, remote monitoring, health resources, and healthcare staff, among others. The goal of this study is to provide a comprehensive perspective of digital healthcare during the COVID-19 pandemic, which is now underway. Various efforts such as mobile applications, digital sites, and sophisticated analytics are being used in order to improve early diagnosis and overall healthcare management. While briefly describing the major aspects leading to widespread implementation of e - healthcare concepts, This research also sheds light on some key aspects of artificial intelligence and the Internet of Things, as well as their practical applications such as clinical decision support systems and predictive risk modelling, which are particularly relevant in the context of combating the COVID-19 pandemic's practical difficulties. The potential uses of artificial intelligence and Internet of Things technology in the battle against the COVID-19 pandemic are thoroughly examined in this paper. The present and future applications of artificial intelligence and the Internet of Things are covered in depth, as well as a thorough assessment of the supporting tools and methodologies. An in-depth analysis.

Keywords: Covid-19 pandemic · Artificial Intelligence (AI) · Internet of Things (IoT) · Healthcare · Deep learning

1 Introduction

In the past decade it is noted that there is a exponential increase on the usage of highly sophisticated devices like Internet of Things (IoT). It has been regarded that the healthcare industry has been stated as the most significant industry which has adopted the technology in a more effective manner.. The combination of new technological breakthroughs such as AI and deep learning has propelled the field of electronic health records

V. E. Balas et al. (Eds.): ICETCE 2022, CCIS 1591, pp. 129–137, 2022.
https://doi.org/10.1007/978-3-031-07012-9_11

(eHR) into the stratosphere in recent years. Because of the impending pandemic situation like COVID-19 pandemic, the whole industry around the globe has been shaken in a variety of ways, such as its capacity and deliverability, promptness in reaction, linked information, and assessment of the situation. The implementation of these technologies in the healthcare industry has made the area of eHealth completely trans disciplinary, and the COVID-19 pandemic, in particular, has created new horizons of problems. Aspects of eHealth, such as virtual healthcare solutions, have been strengthened in the face of this pandemic, on the one hand, and this has been made possible by the rapid evolution of virtual healthcare solutions. For their part, technological areas related to the AI and other technologies have received a significant amount of attention in order supply quick and enhance the services in a efficient manner, particularly from the perspective of COVID-19 [1]. These areas are frequently used to automate and simplify some many tasks performed by healthcare professionals.

This study is aimed on emphasizing some of the most important elements of these technologies in healthcare, which are particularly pertinent during the current COVID-19 epidemic. It has been shown what the main elements of health care will be during COVID-19 [2]. Following that, it has been done a country evaluation of the latest technologies and solutions, such as mobile applications and virtual consultations. Throughout the COVID-19 epidemic, there has been a significant increase in the use of technology in the healthcare field.

2 Critical Issues in Healthcare Industry During Pandemic

The current pandemic situation due to COVID-19, has impacted the global healthcare industry adversely, many nations has closed down their borders, imposed lockdowns and restricted the movement of men and materials so as to reduce the impact of the diseases. An additional burden was placed on hospitals and healthcare facilities by the very stringent standards related to overall sanitation, enable in proper disposal of waster and other aspects were focused on. Implementing these procedures included new difficulties in terms of patient care in hospitals and healthcare facilities. Furthermore, these difficulties during times of pandemic opened the door to new opportunities for IoT and artificial intelligence in the healthcare sector. The application of IoT is considered as the key element as it enables in enhancing the connectivity of various systems. Due to the outbreak of the deadly novel coronavirus, the healthcare industry is facing various challenges in treating the patients, provide better awareness to the individuals so that they can protect their life and their loved ones. There has been scarcity of individuals in supporting the health care industry, hence there needs better technologies in coordinating and providing the treatments for the patients, this will enable an increase in the demand for various medical services and supplies [3] (Table 1).

Table 1. The critical element in the healthcare industry

Element	Description
Patient Follow-up	Because to the increased danger of infection and the reduction of collections in medical centres, the issue of various remote services which was related to non-critical patients with various illnesses were recommended of avoiding hospital visits to free up health related aspects due to Covid 19. Clearly, this entailed disrupting current non-COVID patients' normal treatment and monitoring, leading to additional difficulties in preserving the essential degree of therapy and follow-up, on the one hand, while on the other, keeping provisions for COVID-19 individuals
Healthcare Personnel	Due to the impact of the pandemic, the situation has resulted in the lack of medical personnel available to treat the normal patients. Due to the scarcity, many nations have made particular efforts to meet the unexpected increase in the demand of healthcare professionals. Several nations have cooperated with the faculty members of medicine and governmental health agencies, inviting and involving medical students in the fight against COVID 19 and summoning retired doctors in the epidemic to join the current health staff
Medical Resource Shortage	As a result of the uneven distribution of the corona cases around the world, there is a overall sharing and cross-provincial transport of medical supplies between various healthcare facilities was often required in dynamic terms, in order to avoid shortages at a specific facility that had high demands, based on the progression of COVID-19 cases. A number of partnerships were formed between the university and the industry in order to avoid a shortage of vital medical equipment such as ventilators and quick test kits for devices such as artificial respirators or UV sterilizers

3 IoT Applications During COVID-19

The development of IoT-based methods has created a rising interest amongst developers of healthcare apps. There is no question that IoT technologies are a very feasible alternative for pandemic management in COVID-19. The IoT technologies may be used to trace patterns of infection spreading via the COVID-19 system, to diagnose and integrate COVID-19 patients into wearables and to provide telemedicine services. Various IoT applications for COVID-19 illness management are described in the following paragraphs [4] (Fig. 1).

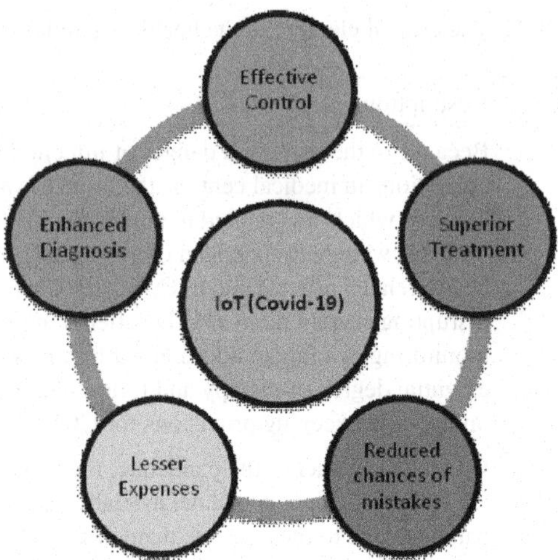

Fig. 1. IoT application during Covid-19

3.1 COVID-19 Diagnosis Though the Application of Internet of Things (IoT)

The IoT posses more potential in accelerating, increase more profitability and achieve better accuracy of the COVID-19 diagnostic and treatment process. The current pandemic situation has resulted in implementing the different program, developed by the standards specified, has been incorporated into a cloud-based software. Automated diagnosis is achieved via the use of electronic medical records and machine learning methods in the application. It may help the department of health by collecting data from patients, managing coordination, and allowing patients to self-diagnose themselves, among other things. The implementation of the 5G technologies, which take use of the greater bandwidth and better communication efficiency that such networks provide, among other things. By using WeChat, it is possible to improve coordination between assessment and treatment, as well as provide a single platform for physicians, medical scientists, and other health professionals [5].

3.2 Internet of Things-Enabled Wearable Technologies for COVID-19 Prediction

Several smartphone apps for analysing the physiological indicators, such as measuring the blood glucose levels, measure the temperature on a real time, measure the sleep-wake cycle has developed in response to the development of Internet of Things-enabled wearables (IoT). In conjunction with machine learning technology, these gadgets or sensors may be utilized to anticipate the various stages of COVID-19 infection. Alarms may be generated by automated analysis and overall monitoring, enable in measuring the oxygen level etc. [6].

3.3 Internet of Things (IoT) for Telemedicine Services at COVID-19

In order to prevent the reduce the overall pandemic related illness, individuals afflicted with COVID-19 must be able to get appropriate healthcare services from a distance. Many medical organizations are concentrating their efforts on creating mobile applications that will simplify the screening of the illness and the scheduling of follow-up appointments. Robotic telemedicine carts may be utilized for remote patient monitoring under quarantined without the need for appearance of health care professionals on the premises. A handful of hospitals utilize bidirectional audio and video communications to monitor patients in critical care units, and this technology is becoming more popular [7].

4 Application of Artificial Intelligence (AI) in Medical Imaging

The critical application and breathing issues caused by COVID-19 is characterized by alveolar oedema with major inflammatory aspect and interstitial thickness may also develop as a result of the process of healing in the most extreme instances various phases of condensation are characterized, although the most severe cases involve alveolar congestion, bleeding, necrosis, and ultimately lung fibrosis, among other symptoms and complications [8] (Fig. 2).

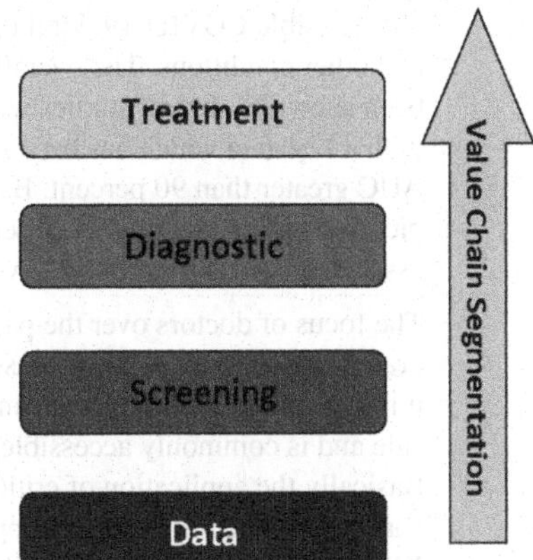

Fig. 2. Applications of AI in medical imaging

Because of the characteristic diagnostic patterns of COVID-19, lung imaging (LI) has become a critical tool in the diagnosis of the disease (Table 2).

Table 2. Three types of imaging methods are regularly employed.

Imaging Method	Description
Chest Computed Tomography (CT)	It can be regarded as the CT scans can generate more specific images, they can identify GGO and pulmonary participation previously in COVID-19 than other imaging techniques. CT, on the other hand, requires substantial radiation doses and requires patient handling that can be difficult. Similar to XR, CT automatic analysis makes use of transmission learning and the pictures are divided into three categories based on the similarities they contain. Furthermore, using cross sectional lung analysis, it is possible to distinguish between impacted regions, allowing for the measurement of pulmonary compromise. The overall accuracy related to COVID-19 disease surveillance were higher with AI assisted CT when compared to conventional X-ray imaging
Chest X-Rays (XR)	Ground-glass opacities (GGO) and pleural consolidations are often identified by X-ray techniques in COVID-19 patients who have lung impairment, according to the literature (9). Traditional artificial intelligence (AI) methods use pre-trained deep neural networks to categories clinical pictures into different patterns: normal XR, possible COVID-19, viral or bacterial pneumonia, and other conditions. The overall accuracy aspects as been shown in several studies was more than 86 percent, with a key area which has been focused in addressing the AUC greater than 90 percent. Based on the clarity of the pictures and the tagging, each sequence may be associated with a greater AUC value
Lung Ultrasound (LUS)	The focus of doctors over the past 15 years has been drawn to Point - Of - care Ultrasound, primarily because it is a portable technique that can be carried out at the bed side and is commonly accessible in medical institutions. Basically the application of critical stages is used to analyse the pneumothorax as opposed to conventional XR, thus acquiring the field of diagnosis, treatment and monitoring of lung diseases in other medical areas (pulmonary and internal medicine)

5 The Use of Intelligent Techniques May Pose a Risk

Although the implementation of the stated technologies like the Ai and IoT tend to offer tremendous promise to enhance and automate COVID tracking, management, and related healthcare applications, the dangers connected with its deployment must not be ignored [9]. This section addresses the possible vulnerabilities and consequences of erroneous choices made by artificial intelligence and Internet of Things technology.

5.1 The Risks of Artificial Intelligence

Artificial intelligence (AI) methods are data-driven strategies in which judgments are made depending on a model that has been built entirely from data and information. There is a phrase that goes "trash in, rubbish out." Now, every judgement made based on artificial intelligence will be skewed if the data used to make the conclusion is incorrect or false. During COVID-19, knowledge dissemination was identified as a key issue due to the large amount of incorrect and misleading information that is being propagated in the media at the moment. Consequently, policy choices, patient monitoring, and diagnosis may be skewed or incorrectly performed as a result of insufficient or incorrect information [10]. As a result, there is a considerable degree of risk involved with the automating of artificial intelligence-based systems. It is highly important to analyse the different dataset which will enable in analysing the overall characteristics through the application of deep learning and other AI related methods. The learning model may thus be used by an adversary to create fictitious computed tomography (CT) scan or other medical imaging data and pictures [11].

6 Internet of Things (IoT) Risks

In terms of security and privacy, present IoT applications are unquestionably lacking. In terms of security issues, the contact tracking mechanisms now used by many countries pose significant challenges. For example, the approach used in trace together necessitates a high level of trust in government agencies, as they may be able to identify sensitive information about COVID-19-infected persons. as a result, while the trace together programme cannot provide privacy-preserving data solutions for the affected group, it can keep that information safe from third parties in general. DP3decentralized T's method may bring additional types of privacy.

6.1 Improving AI and Iot Technology Privacy and Security

Given the nature of information disclosed and related consequences, privacy and security are the major difficulties of AI and IoT methods and it is extremely essential to provide the proper safeguards for data protection and security in AI and IoT applications [14]. These safeguards should generally ensure the privacy of the trusted authority, attackers and other users of the program me. In the case of COVID-19 apps for contact tracking, polling may be a privacy measure for COVID-19 infected users from non-Infected users where non-Infected users may periodically query the health authority to check whether they are in close touch with a contaminated user [15]. Privacy may be provided via private messages systems and private intersection procedures from the health authority. These methods may nevertheless be computationally costly, thus a trade-off between security and computational efficiency has to be devised. Adding authentication credentials that integrate a report memo with a cryptography may improve the integrity of the COVID positive reports. The note may include an overview of user symptoms and health services certifications [16].

It may be very beneficial to integrate new technology to protect information exchange during a pandemic like COVID-19. For its smooth integration in healthcare systems, suitable standardized techniques for new communication and network protocols are needed in this respect [17]. As IoT applications allow broad access to system surveillance data, new safety methods powered by data may be created to construct predictive ML models to identify IoT threats. For safe information exchange across many healthcare providers, distributed ledger technologies like block chain may protect information integrity without engaging a third party, allowing for intelligent healthcare systems to maintain their privacy [18].

7 Conclusion

The healthcare industry has been one of the most rapidly adopted for IoT and AI. The whole health care industry faced significant difficulties, in particular during the COVID-19 epidemic. In this regard, many possibilities have been suggested for the use of IoT and AI. It comprises efficient virtual consultations and remote patient monitoring, intelligent diagnosis, the exchange of EHRs, and prioritized patient planning. Many nations confront difficulties in areas such as digital division and demographic inequality with access to digital technology in healthcare. In addition, despite other challenges, such as the limited power with which big data are handled by health data interoperability among heterogeneous interested parties, as well as a lack of a unified eHealth, AI and IoT implementation structure, the health care sector poses huge potential, particularly during or after the COVID-19 pandemic. Furthermore, current developments in IoT have been examined for tracking and reporting COVID-19 infected patients and the prospective uses for automated wearable technology monitoring. During COVID-19 scenarios, uses of block chain technology have been presented for decentralised decision-making in health systems. Security and privacy elements of the AI and IoT technologies and their effect on crucial policy-making were also developed.

References

1. Rahman, M., Peeri, N., Shrestha, N., Zaki, R., Haque, U., Hamid, S.: Defending against the novel coronavirus (COVID-19) outbreak: how can the internet of things (IoT) help to save the world?, Health Policy and Technol. **9**(2), 136–138 (2020). https://doi.org/10.1016/j.hlpt.2020.04.005
2. Bai, L., Yang, D., Wang, X., Tong, L., Zhu, X.: Chinese experts' consensus on the internet of things-aided diagnosis and treatment of coronavirus disease 2019 (COVID-19). Clinical eHealth **3**, 7–15 (2020). https://doi.org/10.1016/j.ceh.2020.03.001
3. Yang, T., Gentile, M., Shen, C., Cheng, C.: Combining point-of-care diagnostics and internet of medical things (IoMT) to combat the COVID-19 pandemic. Diagnostics **10**(4), 224 (2020). https://doi.org/10.3390/diagnostics10040224
4. Torous, J., Keshavan, M.: COVID-19, mobile health and serious mental illness. Schizophrenia Research **218**, 36–37 (2020). https://doi.org/10.1016/j.schres.2020.04.013
5. Kapoor, A., Guha, S., Kanti Das, M., Goswami, K., Yadav, R.: Digital healthcare: the only solution for better healthcare during COVID-19 pandemic?. Indian Heart Journal **72**(2), 61–64 (2020). https://doi.org/10.1016/j.ihj.2020.04.001

6. Zhang, M., Chow, A., Smith, H.: COVID-19 contact-tracing apps: analysis of the readability of privacy policies (preprint). J. Med. Inter. Res. (2020). https://doi.org/10.2196/21572
7. Chu, L., Anandkumar, A., Shin, H., Fishman, E.: The potential dangers of artificial intelligence for radiology and radiologists. J. Amer. Coll. Radiol. **17**(10), 1309–1311 (2020). https://doi.org/10.1016/j.jacr.2020.04.010
8. She, A., et al.: Healthcare data breaches: insights and implications. Healthcare **8**(2), 133 (2020). https://doi.org/10.3390/healthcare8020133
9. Allam, Z., Jones, D.: On the coronavirus (COVID-19) outbreak and the smart city network: universal data sharing standards coupled with artificial intelligence (AI) to benefit urban health monitoring and management. Healthcare **8**(1), 46 (2020). https://doi.org/10.3390/healthcare8010046
10. Wheelus, C., Zhu, X.: IoT network security: threats, risks, and a data-driven defense framework. IoT **1**(2), 259–285 (2020). https://doi.org/10.3390/iot1020016
11. Liu, H., Crespo, R., Martínez, O.: Enhancing privacy and data security across healthcare applications using blockchain and distributed ledger concepts. Healthcare **8**(3), 243 (2020). https://doi.org/10.3390/healthcare8030243
12. Jain, A.Y., Shrivastava, Y.: Modelling and optimization of different quality characteristics in electric discharge drilling of titanium alloy sheet. Mat. Today: Proc. **21**, 1680–1684 (2020). https://doi.org/10.1016/j.matpr.2019.12.010
13. Meena, G., Sharma, D., Mahrishi, M.: Traffic prediction for intelligent transportation system using machine learning. In: Proceedings of 3rd International Conference on Emerging Technologies in Computer Engineering: Machine Learning and Internet of Things, ICETCE 2020, pp. 145–148 (2020)
14. Jain, A., Pandey, A.: Multiple quality optimizations in electrical discharge drilling of mild steel sheet. Mat. Today: Proc. **4**(8), 7252–7261 (2017). https://doi.org/10.1016/j.matpr.2017.07.054
15. Panwar, V., Kumar Sharma, D., Pradeep Kumar, K., Jain, A., Thakar, C.: Experimental investigations and optimization of surface roughness in turning of en 36 alloy steel using response surface methodology and genetic algorithm. Mat. Today: Proc. (2021). https://doi.org/10.1016/j.matpr.2021.03.642 Accessed 27 July 2021
16. Fang, Y., et al.: Sensitivity of chest CT for COVID-19: comparison to RT-PCR. Radiology **296**(2), E115-E117 (2020). https://doi.org/10.1148/radiol.202020043
17. Huang, L., et al.: Serial quantitative chest CT assessment of COVID-19: a deep learning approach. Radiology: Cardiothoracic Imaging **2**(2), e200075 (2020). https://doi.org/10.1148/ryct.2020200075
18. Ying, S., Zheng, S., Li, L., Zhang, X., Zhang, X.: Deep learning enables accurate diagnosis of novel coronavirus (COVID-19) with CT images (2020). https://doi.org/10.1101/2020.02.23.20026930

Intelligent Smart Waste Management Using Regression Analysis: An Empirical Study

Abinash Rath[1(✉)], Ayan Das Gupta[2], Vinita Rohilla[3], Archana Balyan[4], and Suman Mann[5]

[1] School of Commerce, Finance & Accountancy, CHRIST (Deemed to be University), Lavasa, Pune, India
abi.atlantis@gmail.com

[2] Department of Geography, WBES, Chandernagore Government College affiliated to the University of Burdwan, Hooghly, West Bengal, India

[3] Department of Computer Science, Maharaja Surajmal Institute of Technology, New Delhi, India

[4] ECE Department, Maharaja Surajmal Institute of Technology, New Delhi, India
Archanabalyan@msit.in

[5] Department of IT, Maharaja Surajmal Institute of Technology, New Delhi, India
sumanmann@msit.in

Abstract. The term deep learning is seen as an important part of artificial intelligence that allows the system to understand and make decisions without special human intervention. In-depth learning uses a variety of statistical models and programs that allow different computational properties to reach the highest point. It is estimated that the market development of artificial intelligence and technology for deep learning will amount to USD 500 billion by 2026. The use of advanced technology, such as neural networks, enables better image recognition and the use of automated processes for deep operations. The main purpose of the study is to understand the critical determinants of Deep Learning in Creating a better City through Intelligent Smart Waste Management, the major determinants cover: System usability scale, Implementation of RFID sensors and Optimizing route selection. The proposed work is that implementation of advanced tools like deep learning methodologies and machine learning tools can support in managing the waste in a smart way, this will enable in creating better cities, enhance the environment and support sustainable living. Smart cities today need to use tools like deep learning and other artificial intelligence to effectively manage waste. Smart vessels are mainly controlled and implemented, which makes it easier for users to open vessels, it is also suitable for storing solid and dry waste, but provides information on the total degree of filling, can share data and information with central waste management service, you can collect waste quickly and avoid flooding. To achieve this, governments, administrators and communities are introducing sensors that transmit data and information to the waste management company in real-time and take appropriate action.

Keywords: Deep learning · Smart waste management · Regression analysis · Analysis of variance

V. E. Balas et al. (Eds.): ICETCE 2022, CCIS 1591, pp. 138–148, 2022.
https://doi.org/10.1007/978-3-031-07012-9_12

1 Introduction

The increase in population and relocation of citizens from rural and semi-urban to smart cities has posted an increase in a greater number of individuals being accumulated in a specified location which creates an increase in tons of waste. The term waste management intends to focus on end-to-end activities which is involve in monitoring of waste from the source point when the companies, individuals and other produce waste. There exist numerous ways through which the waste can be controlled and disposed, recycled in an effective manner [1]. In the present day, waste management is considered to be a challenge as the waste is increasing due to usage of products and services. The waste management tend to refer to those action which need to dominate from the inception to the demolish of an activity. The waste can be categorized into solid aspect, liquid and gas, there are numerous processes tend to deal with different type of waste which includes biological, household, industrial aspects. The household waste tends to include the plastics, papers, glasses etc.

The term deep learning is considered as part of the significant part of Artificial Intelligence (AI) which will allow the system to ability to understand and make decision without much human intervention. The deep learning tends to apply different statistical models and program which will allow different features in computing that has reached the highest aspect [2]. It is estimated that the market growth of AI and deep learning technologies tend to grow to nearly $500 billion by 2026 [3]. The implementation of advanced technologies like neural network supports in enhanced image recognition, support in using automated procedures for making profound actions.

The traditional waste management system involves in the collection of waste through manual effort directly from the public waste bins. The manual disposal is mainly made through defined process, through which it involves in recycling through defined process. Hence, it involves more human efforts and there may be significant errors as many processes are handled manually. However, it also creates various other issues which includes application of waste bins, the overfilled waste bins can enable in creating toxic hazes and increase foul smell which causes allergic and ecological imbalances [4]. The waste collection from the underfilled waste bins tend to increase inefficiency in the process. The waste collection vehicles tend to collect the waste in sequence so that it takes more time to collect which will increase excess fuel consumption.

In order to avoid the traditional waste management system, smart cities are now using Deep learning tools and other artificial intelligence (AI) technologies in order to manage the waste in an efficient manner. The AI technologies are Smart trash bins are mainly tested and applied which will enable the users to open the bin more conveniently, this trash can also store the dry and solid waste, but offer the overall fill level information, this supports in sharing the data and information to the central waste management agency which can collect the waste quickly and can avoid spillovers. In order to accomplish this the governments, administrators and communities implement sensors which will transmit real time data and information to the waste management company and appropriate actions are taken [5].

Ultrasonic Sensors are mainly used to measure the fill levels of the trash bins through different technologies, these sensors can also estimate the exact location of the bins so that the refill can be made quickly. The information from these sensors is monitored and communicated trough deep learning tools, Internet of Things (IoT) techniques etc. The data are then stored so as to take informed action and decision making. The next aspect considered in the implementation of smart waste management is through Radio frequency identification (RFID) sensors which are highly compact and can be fixed in any place [6]. These tags act as transponders which can be installed in monitor the trash bins, also provide critical information about the usage of these bins and number of other bins located in the area.

Smart waste management involves in using image processing and neural network which supports in classifying the waste which are available in the bins. The research has stated that the neural network and related tools is used for managing the solid waste, this tool support in detecting the plastic and non-plastic materials, support in organizing the waste so that they can be recycled and managed effectively [7]. Furthermore, smart waste management server is considered as the main component in the entire process, it is mainly work on the cloud server and be easily accessed by the waste management company, government administrators and others through remote access [8]. This application is mainly responsible in communicating with the smart bins for implementing appropriate control, track the number of bins reaching its maximum and send the delivery truck to collect the garbage. Furthermore, it analyses the information from the truck control system, recycling plant and other aspect for forecasting the data and information.

The communication status with the stated controllers and report the communication failure for the suitable action. These systems also run the waste management application and carry out the required data analytics based on the waste management. These systems also store the data related to waste management update information from trash bins and controllers for analyzing it. Support in creating alerts and notifications during emergency situations related to waste disposal. Provide information on the type of waste disposal in the locality and provide necessary information to the individuals and citizens to act responsibly [9]. Also analyses the fill level stats data for each bin and instruct to make profound actions. Provide the optimized route information for garbage collection so that the fuel and other expenses can be curtailed effectively.

Hence, this study is focused in analyzing the key factors of deep learning tools in Creating better City through Intelligent Smart Waste Management. The major factors are system usability scale, implementation of Radio frequency identification (RFID) sensors for effective tracking and optimizing the route selection for effective waste management.

2 Review of Literature

The author has proposed a unique allocation model which will enable in identifying different type of waste through deep learning mechanisms. It has been stated that these tools can also be enabled in enhancing the recycle of different type of waste. The research has stated a scheme through which the deep learning tools can automatically identify the type of garbage and propose immediate actions [10].

The researcher has noted that the deep learning model was applied in the overall classification of the waste, it also supports in measuring the trash bins in each location and analyses the total capacity. The Convolutional neural network (CNN) model is considered as one of the effective mode which supports in extracting higher image content also enable in effective management by identifying different type of waste. Furthermore, the authors have stated that the optimal planning algo can support in solving sustainability issues, enable in identifying optimal transportation so that the waste can be collected at lower cost.

Due to population growth, urbanization and industrial treatment, a lot of waste is generated daily. If this waste is not treated and handled properly, there is a great risk of health risks, pollution and contamination [11]. In order to avoid such problems in advance, waste management is considered a priority that needs to be addressed for several years. Researchers, researchers and engineers around the world are working to improve existing intelligent waste management systems and identify new features and functions that will be integrated into the new intelligent waste management system that is under development [12]. Many research articles and important reading are available on the Internet on the Intelligent Waste Management System. Let's take a look at some of the existing research articles to get enough information on this research topic.

The authors claim that this ideal design algorithm can calculate the best garbage collection routes for smart and sustainable cities with the right software platform. An ideal programming algorithm that runs on an open source programming tool called Net2Plan. Net2Plan is often used to model and design communication networks. Net2Plan as well as the Net2Plan-GIS library make it easy to enter city layout information into the algorithm, including smart storage locations. The calculated optimal waste collection routes minimize the number of trucks used for waste collection and thus fuel consumption, which in turn reduces the impact of CO_2 emissions and noise. The authors also discussed a practical case from the city of Cartago (Spain), where the ideal way to collect plastic waste is discussed [13].

An intelligent network based on smart sensors, which can communicate with any original server located in the communication area, is also becoming a practical idea and requires broad coverage in waste collection and management [14]. The authors discussed a solution based on an RFID tag that provides information about garbage. This information was then used to improve waste management, enabling automatic and rapid identification of waste at the waste level. The authors confirm that no other external information system is required to track waste with RFID-tagged Smart Bins, and this solution enables better waste sorting [15]. In addition, this RFID tag allows the residual contents of the Smart Bin to be returned to the recycling system.

Today, based on the coverage areas, the municipality and the smart city administration organize garbage collection vehicles to collect garbage from different parts of the city and geographical areas. This manual waste management is a long and tedious job., hence automating them tend to save more time and enable in automating the whole process in a better manner.

3 Methodology

The researcher has used descriptive research design as this supports the researchers in analyzing the key determinants in an effective manner, this also enable the researchers for measuring the main aspects of the deep learning aspects in enhancing the waste management in city. The study involves in sourcing the information from primary and secondary sources, which involves in using questionnaire to collect the information from the respondents, nearly 156 respondents were chosen based on purposive sampling method. The secondary data is collected through online data base like EBSCO, Google scholar and published research articles from Scopus indexed journals.

The research is mainly intended to understand the key components of deep learning in enhancing the smart waste management in cities. The government, policy makers and others are focusing in using the digital technologies to carefully plan and manage the waste in an efficient manner [16]. The application of deep learning-based garbage system support in efficient tracking of waste, forecast the bins required for each area, track the data on a real time basis and provide necessary information to the law makers for efficient decision making. The implementation of different router and sensors enable in analyzing the key information, support in enhancing the smart garbage system efficiently[17].

Critical Determinants of the Study

There exist no major differences among system usability scale and creating better city with smart waste management.

There exist no major differences among implementation of RFID sensors and creating better city with smart waste management.

There exist no major differences among optimizing the route selection and creating better city with smart waste management.

4 Analysis and Discussion

The data analysis covers frequency table analysis of critical variables, regression analysis and Analysis of variance for testing the hypothesis.

The authors intend to understand the importance of Deep learning supporting in smart waste management, as the government and others are increasingly using these tools in order to enhance the implementation process and manage the waste better (Table 1).

Table 1. Deep learning supporting in smart waste management

Particulars	Frequency	Percent
Strongly disagree	9	5.8
Disagree	15	9.6
Neutral	26	16.7
Agree	57	36.5
Strongly agree	49	31.4
Total	156	100

The above table states that the deep learning is highly supportive in managing the waste in making smart city more sustainable, it is noted from the literature that waste disposal and management is one of the critical issues faced by government to transform into smart city, from the table it is noted that 31.4% have strongly agreed to the statement and 36.5% have agreed to the statement that deep learning is highly supportive in smart waste management whereas 16.7% of the respondents mentioned as neutral, 9.% were disagreeing to the statement.

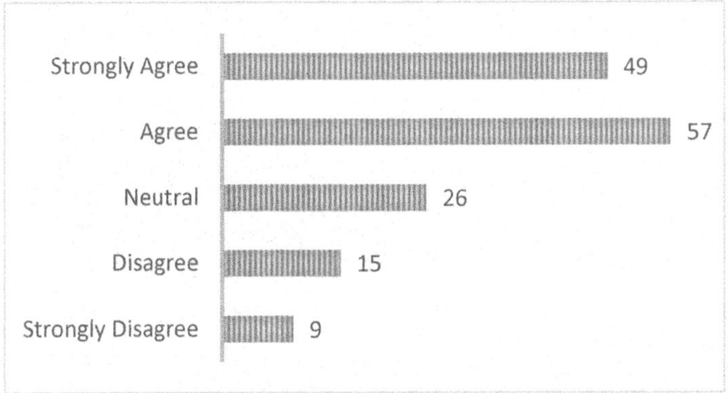

Fig. 1. Deep learning supporting in smart waste management

The Fig. 1 confirms that most of the respondents agree that the deep learning support in smart waste management, hence this confirms with the literature review. Hence, the respondents were stating the deep learning tools Neural designer, Microsoft cognitive toolkit, Pytorch etc. support in monitoring the fullness of torch, this allows the municipalities to optimize the waste collection routes, times etc.

The second aspect to be analyzed is the identification of waste management as critical aspect in making smart city, with the increase in population climate changes has become huge concern for government and Nongovernmental Organisation hence creating smart city can enable in using the resources efficiently and support in reducing the increase in temperature (Table 2).

Table 2. Waste management is key concern for developing smart city

Particulars	Frequency	Percent
Strongly disagree	11	7.1
Disagree	14	9
Neutral	25	16
Agree	53	34
Strongly agree	53	34
Total	156	100

The table shows that the waste management is considered as the critical component in creating and sustaining smart city by the authorities. With the use of digital tools and intelligent systems the policy makers can enable in creating better waste management and thereby enable in creating sustainable smart city. From the above table nearly 68% of the respondents have agreed to the statement that waste management is the key concern in developing smart city, whereas 16% are neutral and remaining are disagreeing with the statement.

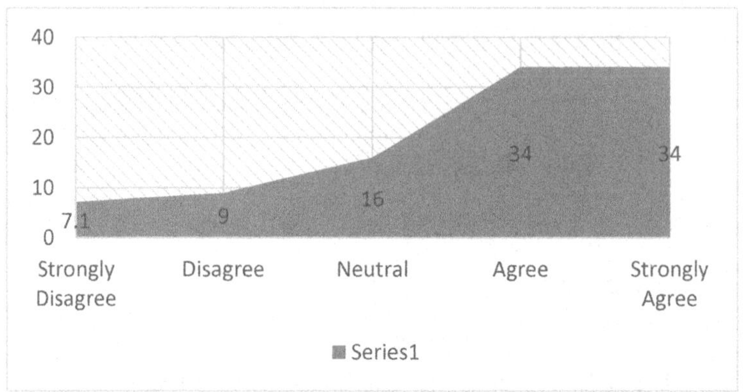

Fig. 2. Waste management is key concern for developing smart city

Furthermore, Fig. 2 states that the 68% of the agreed to the statement the use of digital tools and intelligent systems the policy makers can enable in creating better waste management and thereby enable in creating sustainable smart city. Hence, this is confirming that government and others are focusing in stating that the waste management is more concern in creating a smart city.

Regression Analysis

The critical part in the data analysis is to understand the overall relationship between the independent variables and dependent variable (Table 3).

Table 3. Regression analysis

Regression analysis	B	SE	P val.
(Constant)	0.2	0.18	0.269
System usability scale	0.431	0.098	**0.00**
Implementation of RFID sensors	0.384	0.095	**0.00**
Optimising route selection	0.073	0.083	0.381
ANOVA	SS	F	P
Regression	**163.88**	153.66	.000b
Residual	**54.04**		
Coefficient of determination	0.752	75.20%	

Based on the above table the determination coefficient is 0.752 or 75.2% which shows that the model is regarded as good fit. The sum of squares of the ANOVA table states that the regression coefficient is 163.38, F value is at highest level of 153.66 and the significance value is 0.00 which is less than 5% hence the variables are statistically significant.

Furthermore, the regression expression is stated as

Smart waste management = 0.20 + 0.431 × System usability scale + 0.384 × implementation of RFID sensors + 0.073 × Optimizing route selection.

Hence stated that all the variables possess positive association towards creating smart waste management.

Testing of Hypothesis Using Analysis of Variance

The last step is to test the hypothesis using ANOVA.

Hypothesis 1 – Null: There exist no major differences among system usability scale and creating better city with smart waste management (Table 4).

Table 4. ANOVA between system usability scale and creating better city with smart waste management

ANOVA	SS	F	P
Between groups	165.8	143.29	0
Within groups	43.67		
Total	209.4		
Levene's	Value	df1	P
Levene's test	6.647	4.00	0

The Analysis of variance (ANOVA) analysis states that the Sum of squares (SS) between groups is 165.8 and within groups is 43.67, whereas the F value is 143.29 and significance value is 0.00, also the Levene test shows a value of 6.64 and significance value of 0.00, which is less than 0.05. Hence concluded that there exist major differences among system usability scale and creating better city with smart waste management.

Hypothesis 2 – Null: There exist no major differences among implementation of RFID sensors and creating better city with smart waste management (Table 5).

Table 5. ANOVA between RFID sensors and creating better city with smart waste management

ANOVA	SS	F	P
Between groups	195.5	165.98	0
Within groups	44.47		

(continued)

Table 5. (*continued*)

ANOVA	SS	F	P
Total	240		
Levene's	Value	df1	P
Levene's test	42.89	4.00	0

The ANOVA analysis states that the SS between groups is 195.5 and within groups is 44.47, whereas the F value is 165.98 and significance value is 0.00, also the Levene test shows a value of 6.64 and significance value of 0.00, which is less than 0.05. Hence concluded that there exist major differences among implementation of RFID sensors and creating better city with smart waste management.

Hypothesis 3 – Null: There exist no major differences among optimizing the route selection and creating better city with smart waste management (Table 6).

Table 6. ANOVA between optimizing the route selection and creating better city with smart waste management

ANOVA	SS	F	P
Between groups	145.7	82.36	0
Within groups	66.78		
Total	212.5		
Levene's	Value	df1	P
Levene's test	6.247	4.00	0

The ANOVA analysis states that the SS between groups is 145.7 and within groups is 66. 78, whereas the F value is 82.36 and significance value is 0.00, also the Levene test shows a value of 6.64 and significance value of 0.00, which is less than 0.05. Hence concluded that there exist major differences among optimizing the route selection and creating better city with smart waste management.

5 Limitations

The major limitations of the study are focused in analyzing the key determinants like System usability scale; implementation of RFID sensors and Optimizing route selection in measuring the Intelligent Smart Waste Management to create better city. The researcher has not applied other factors in managing the waste management are considered. The time available to perform the study is short and hence only 156 respondents were chosen. The responses received from the sample population may have biased information.

6 Conclusion

The Intelligent waste management consists of image processing and a neural network that classifies waste that is thrown in the trash. Research has shown that neural networks and related devices are used to treat solid waste, this tool helps to detect plastic and non-plastic materials, helps to organize waste so that it can be recycled and managed efficiently. The implementation of a coherent neural network achieved 96% accuracy in the detection of solid waste and provided important data and information. In addition, the intelligent waste management server is a key component of the entire process, runs primarily on a cloud server and is easily accessible to the waste management company, government administrators and others with remote access. This application is primarily responsible for communicating with the smart landfills to implement the correct control, monitor the amount of landfills that reach the maximum and send the truck for garbage collection. It also analyzes the truck's control system, recycling facility information and other aspects to predict data and information. The researcher noted that the deep learning model has been used to classify waste in general, supports the measurement of trash at each site and analyzes total capacity.

References

1. Dipak, S.G., Aithal, P.S.: Smart city waste management through ICT and IoT driven solution. Int. J. Appl. Eng. Manag. Lett. **5**(1), 51–65 (2021)
2. Pardini, K., Rodrigues, J.J.P.C., Diallo, O., Das, A.K., de Albuquerque, V.H.C., Kozlov, S.A.: A smart waste management solution geared towards citizens. Sensors **20**(1), 1–15 (2020)
3. Meena, G., Sharma, D., Mahrishi, M.: Traffic prediction for intelligent transportation system using machine learning. In: 2020 3rd International Conference on Emerging Technologies in Computer Engineering: Machine Learning and Internet of Things (ICETCE), pp. 145–148 (2020).https://doi.org/10.1109/ICETCE48199.2020.9091758
4. Branco, P., Bolton, K., Eriksso, M.: Environmental impacts of waste management and valorisation pathways for surplus bread in Sweden. Waste Manag. **117**(1), 136–145 (2020)
5. Muncipal Solid Waste Management Manual Part II: Central Public Health and Environmental Engineering Organisation (CPHEEO). Ministry of Urban Development, Government of India, pp. 3–41 (2020)
6. Gade, D.: Introduction to smart cities and selected literature review. Int. J. Adv. Innov. Res. **6**(2), 7–19 (2019). Part 4
7. Bueno-Delgado, M.-V., Romero-Gázquez, J.L., Jiménez, P., Pavón-Mariño, P.: Optimal path planning for selective waste collection in smart cities. Sensors **19**(1), 1–14 (2019)
8. Esmaeilian, B., Wang, B., Lewis, K., Duarte, F., Ratti, C., Behdad, S.: The future of waste management in smart and sustainable cities: a review and concept paper. Waste Manag. **81**, 177–195 (2018)
9. Pardini, K., Rodrigues, J.J.P.C., Kozlov, S.A., Kumar, N., Furtado, V.: IoT-based solid waste management solutions: a survey. J. Sens. Actuator Netw. **8**(1), 1–25 (2019)
10. Khoa, T.A., et al.: Waste management system using IoT-based machine learning in university. Wirel. Commun. Mob. Comput. **2020**(1), 1–13 (2019)
11. Jain, A., Yadav, A.K., Shrivastava, Y.: Modelling and optimization of different quality characteristics in electric discharge drilling of titanium alloy sheet. Mater. Today Proc. **21**, 1680–1684 (2019)

12. Jain, A., Pandey, A.K.: Modeling and optimizing of different quality characteristics in electrical discharge drilling of titanium alloy (Grade-5) sheet. Mater. Today Proc. **18**, 182–191 (2019)
13. Abba, S., Light, C.I.: IoT-based framework for smart waste monitoring and control system: a case study for smart cities. Eng. Proc. **2**(1), 1–10 (2020)
14. Jain, A., Pandey, A.K.: Multiple quality optimizations in electrical discharge drilling of mild steel sheet. Mater. Today Proc. **8**, 7252–7261 (2019)
15. Panwar, V., Sharma, D.K., Pradeep Kumar, K.V., Jain, A., Thakar, C.: Experimental investigations and optimization of surface roughness in turning of EN 36 alloy steel using response surface methodology and genetic algorithm. Mater. Today: Proc. **46**, 6474–6481 (2021)
16. Pathak, N., Deb, P.K., Mukherjee, A., Misra, S.: IoT-to-the-rescue: a survey of IoT solutions for COVID-19-like pandemics. IEEE Internet Things J. **8**(17), 13145–13164 (2021)
17. Rathod, P., Usoro, A.: Benchmarking machine learning approaches to predict the uncertainties of pandemic disease: an explorative study. In: Proceedings of the 7th Annual International Conference on Information Technology and Economic Development, Gregory University, Uturu, Nigeria, 16–18 Nov 2020

An Investigative Analysis for IoT Based Supply Chain Coordination and Control Through Machine Learning

K. Veerasamy[1(✉)], Shouvik Sanyal[2], Mohammad Salameh Almahirah[3], Monika Saxena[4], and Mahesh Manohar Bhanushali[5]

[1] Department of Computer Science, Karpagam Academy of Higher Education, Coimbatore, India
veerasamyca@kahedu.edu.in

[2] Department of Marketing and Entrepreneurship, Dhofar University, Salalah, Sultanate of Oman

[3] Business Administration Department, Isra University, Amman, Jordan
mohammad.almahirah@iu.edu.jo

[4] Computer Science, Banasthali Vidyapith, Aliyabad, Rajasthan, India

[5] Management Studies-Operations, VPM's Dr. V.N. Bedekar Institute of Management Studies, Thane, Maharashtra, India

Abstract. The use of the Internet of Things (IoT) has brought about radical changes in the construction and business sectors, and companies are now using technology to remain competitive, support the exploitation of competitive advantages and increase growth and profitability. The use of the next generation of computers facilitated industrial change in all areas, and IoT helped shape the production structure, build an efficient value chain and achieve economic growth points. It can be argued that the introduction of IoT has changed the way we create value in the supply chain, which creates better opportunities for companies to improve scalability, make faster decisions and achieve better profits and growth. Although there are few challenges for the company, such as optimizing resources, investing in IoT and related digital technology, changing the production process and supply chain, etc., these new problems tend to change the organization's bases and change the traditional way of doing things. Business. digital environment for effective customer engagement.

Keywords: Internet of Things · Machine learning · Supply chain management · Regression analysis · Mediation analysis

1 Introduction

The enhanced application of the Internet of Things (IoT) has resulted in making profound changes in the manufacturing and business sectors, companies are now using the technologies in order to stay competitive, support in realizing competitive advantage and enhance growth and development sustainably. The application of next generation

V. E. Balas et al. (Eds.): ICETCE 2022, CCIS 1591, pp. 149–159, 2022.
https://doi.org/10.1007/978-3-031-07012-9_13

IT has supported in reaching industrial changes across sectors, the IoT has supported in enabling the formation of production modulation, support in creating efficient value chain process and achieving economic growth points [1]. The critical aspect of the IoT is the overall ability to connect things to things, which is considered as the main element in the supply chain process (SCPs). The Organisation intend to get the real time information related to the availability of the resources in the production facilities, estimate the capacities and tack the movement of goods from one location to another, hence big data has become an integral part of the supply chain process [2]. It is also necessary to understand that implementation of IoT not only provides the data but also analyze the data effectively so as to support the organizational decision making. Hence, IoT is not becoming critical in making decisions, devise strategies and support in economic growth of the company.

The SCP is stated to be different from the normal production chain as the are more focused on physical components, however IoT enable in connecting with different systems, understand the components and provide real time information. The intelligent quotient of IoT supports in strengthening the value chain aspects of the supply chain, enhance the interconnectivity among systems and people located in different geographical locations and support in creating dashboard reports for quick decision making which again acts as source of information for the future, hence it forms the virtuous cycle in creating value promotion [3]. The application of IoT based things to things (T2T) supports in creating better and efficient network of different terminals covering products and services which forms part of the supply chain process, this assists the managers in possessing effective and mutual communication, data collection and transmission, analysis of the data for making decisions and control the overall operations. This nature of information support in enhanced data interaction which make the overall diverse functions and values which are integrated, support in adding value chain for all the stakeholders. Hence IoT expands from the individual dimension value of conventional product system to multi-dimensional value-added system in the supply chain process.

Furthermore, it is considered that in the IoT environment, the application of technologies, tools and algorithms has become integral part of the process. In SCP, the application of sensors, processors, collaborating and connecting the systems, real time data tracking and synchronizing various storage options has enhanced the production and supply chain process, support in enhancing the overall performance, reduce cost and attain competitive edge over the competitors. Hence, compared with the traditional model of operating the supply chain, the application of IoT offer new function and facilities to the management through higher reliability of data access, utilization of the resources, deploy the assets effectively and support cross border tracking and movement in an efficient manner. The SCP can replace the physical accessories and manual tracking into real time IoT based automated tracking of product movement and also sustain the business process. The major association of applying IoT in the manufacturing process enables in exchange of critical data and information among the products, support in creating new digital functions so as to enhance the value chain in the supply chain process [4].

It has been further stated that the overall composition and characteristics of the SCP has transformed the overall nature of organizational structure, enhance competition between the companies located in both domestic and international markets, support in identifying new opportunities, eliminate risk and leverage the existing resources for enhancing the product and service offerings. The IoT architecture has supported in improving the economic efficiencies of new product and services. The manufacturing of such products support in making new design, identifying creative tools in manufacturing, logistics and after sales services through the implementation of digital technologies like IoT, Big data and Robotics [5]. This triggers new value chain-based improvement in different domains covering manufacturing, supply chain, marketing, finance and operations. The scope and deployment of resources also changes substantially through the implementation of digital technologies, which will promote large degree of change in the production process.

Hence, in summary it can be stated that the IoT has radically transformed the business environment where the supply chain system operates. The identification and implementation of real time data collection has supported the process in identification of efficiency, optimize the usage of existing resources, collarbone with other systems for efficient decision making. In the dynamic business environment, the real time massive accumulation and transfer of data is being made faster in the SCP, hence efficiency and optimization need to be improved for better allocation of resources. However, the overall visualization of the supply chain systema and usage of big data tend to depend on the company's own decision and interactions with the vendors and suppliers. Hence, IoT offers better environment through laying technical foundation for the creation of better economic system, which supports in creating essential changes in market demand, production function and supply chain process [6]. Therefore, the IoT enables ecosystem brings in emerging management protocols and issues like addressing the inefficiencies, reduction of mundane task and optimize cash flows.

It can be stated that implementation of IoT has changes the manner of value creation in the supply chain operations, create better opportunities for the companies to enhance scalability, support in quick decision making and achieve better profit and growth. Though there are few challenges like the ability of the company to optimize the resources, invest in IoT and related digital technologies, alteration of the production and supply chain process etc. these emerging questions tend to transform the organizational foundation and move from the traditional mode of managing the business to digital environment so as to connect with customers in an efficient manner.

The main purpose of the study is to analyze the extent of implementing IoT in the supply chain process for efficient coordination and control through machine learning approaches. It is understood that the IoT enables machine learning approaches involve in data collection, analysis and embed them into the production process so as to trigger new opportunities, deploy the resources effectively, support in value chain, attain economic efficiency and lead to superior cost management. This study is more focused in understanding the critical determinants of using IoT approaches for attaining Supply Chain Coordination and Control through Machine Learning.

2 Review of Literature

The authors have stated that when compared with the non-IoT SCP, the implementation of IoT based machine learning has enabled in restructure the value chain of the SCP in many organizations. This resulted in competing and proposition of new product development, create profound effect on the manufacturing and delivery of goods and services. The IoT enables environment has resulted in creating different product lines, deploy the resources effectively and support in organizational efficiencies [7].

It is stated that IoT enables environment has placed variety of interconnection among the systems for better interaction and taking intelligent decision making. [8]. The overall function and value of these aspect are mainly forming the customer value chain process. IoT support in better interaction among the heterogenous and complimentary services, hence the customers get better value for the price that paid, enable in reducing the cost of the company and thereby support sustainable growth and development.

Through the IoT enables environment, the companies focus in moving from a single product offering to multiproduct offerings, use the existing system to cross sell the products, forecast the demand for the products and plan the supply chain functions effectively. The IoT enables service-based product portfolio critically understand the system collaborating [9]. Machine learning approaches assort different components of the SCP which will help in finding the value in the existing business process. This supports the supply chain companies to understand the process, use the digital technologies for better growth and development.

There is extensive literature on IoT's impact on supply chain management that is closely related to this study. The authors said that the hybrid manufacturing and recycling system is the ideal optimization of production in the closed supply chain in the IoT environment. In addition, an IoT-compliant information architecture is being developed to meet the challenge of providing effective emergency information for seniors. A conceptual model for a vulnerable food chain with IoT capacity and validation of the proposed model to improve the supply chain's performance through a case study [10]. New, robust information structure and control system with cost-effective IoT solutions to meet the challenge of managing a dynamic IoT-compliant logistics production system. Similar work is being done to design an IoT system that automates data collection and processing to optimize port operations and multimodal transhipment, logistics and supply chain performance.

Some researchers have mentioned the concept of IoT-based resource allocation for cloud computing. IoT development perspective and its impact on resource allocation. In addition, the ideal strategy for transferring construction resources between companies on the IoT cloud platform. However, research in this area is still in its infancy and requires a lot of research. Develop more realistic approaches to supply chain measurement, supply chain cyclical structures and quality assessment made possible by the introduction of IoT technology [11]. The IoT's impact on supply chain management, through literature review and bibliometrics, classifies existing research and proposes future research directions to facilitate the implementation of IoT.

IoT is driving a change in the hierarchy of consumer behavior and value propositions, forcing companies in the supply chain to change their product design models. With its powerful and ubiquitous sensors and controls, IoT has become part of consumers 'daily lives and has a direct impact on consumers' lifestyles and habits. Although the means of connection itself form part of the service provider's network, consumer behavior, the concept of consumption and values can change drastically as a result.

The IoT big data service is changing the value chain and the structure of the supply chain, leading to new problems in coordinating the operation of the supply chain and the control mechanism for sharing data between these centers. In an IoT environment, there are basically two types of big data companies: big data owners and big data providers. These latter companies have professional expertise in the processing and analysis of big data. Big data providers create value by processing data, utilizing knowledge and possibly providing effective decision support. The presence of large data providers fundamentally changes existing value chains and structures for supply chains, which raises new questions in the choice of structure for supply chains and the design of incentive contracts. In addition, even if real-time data transfer over IoT increases the display level of production resources and the capacity of nodes in the supply chain.

3 Methodology

The main purpose of the study is to analyses in Applying Internet of Things (IoT) based Supply Chain Coordination and Control through Machine Learning, hence the researches are focused in using the descriptive design for this purpose. This design supports in understanding the critical aspects which has been carried in the area and enable in critically analyze the factors which contribute in implementing IoT towards supply chain process. The descriptive design is the most common applied tool for performing the study involving social research [12]. Furthermore, the researcher intends to collect the data from secondary sources and primary sources, the secondary source of information is collated through the digital libraries like ProQuest, Springer published journal, Google scholar etc. The tool to collect the primary data is through the closed ended questionnaire, the researcher has focused in preparing critical constructs for the questionnaire and choses the sample respondents from the available population [13]. Convenience sampling method is used by the researcher to choose the respondents as it provides the researcher in choosing the respondents easily. The data is sources from 152 respondents, the data are then organized through Likert scale (1 – Strongly Disagree to 5 – Strongly Agree).

The IoT enables things to things interconnection of the SCP lead to linking of different nodes from the foundational level to the resources end. This aspect allows the business enterprise to optimize the allocation of the resources at different levels. The accurate and personalized data process increases the customer experiences in the dynamic market environment. Hence, the study analyses the determinates of IoT enabled machine learning in enhancing the coordination and control of supply chain management process [14]. The researchers use SPSS package in performing the data analysis which covers correlation analysis, regression analysis and mediation analysis. The mediation analysis enables in quantifying the overall extent to which the stated variable in the transmittance of change from the cause to the effect, the mediation analysis is manly comprised of three different set of analysis covering (Fig. 1).

Hypothesis Testing

There is no statistical association between IoT enables value chain efficiencies and supply chain coordination and control through machine learning approaches.

There is no statistical association between achieving economic efficiencies and supply chain coordination and control through machine learning approaches.

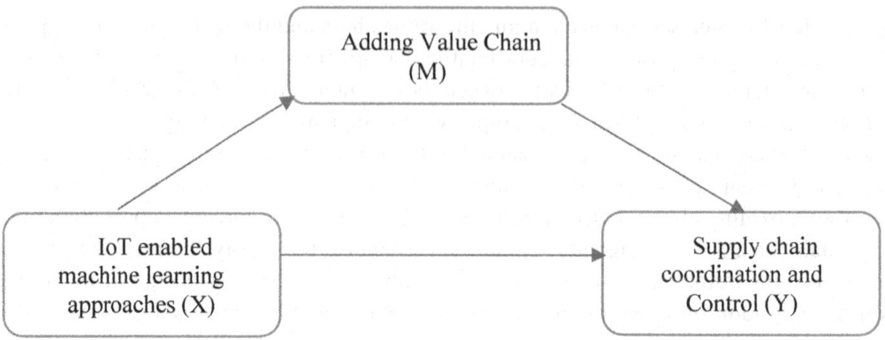

Fig. 1. Mediation analysis framework

There is no statistical association between overall cost control and supply chain coordination and control through machine learning approaches.

4 Analysis and Discussion

This part provides the critical aspect in analyzing the data collected from the sample respondents, the major analysis covers correlation analysis, regression analysis and mediation analysis.

The first step is to measure the frequency analysis of two critical construct which is focused in understanding the importance of IoT in the supply chain management process (Table 1).

Table 1. IoT supports in supply chain management process

IoT supports in supply chain management process	Frequency	Percent
Strongly disagree	11	7.2
Disagree	9	5.9
Neutral	20	13.2
Agree	41	27
Strongly agree	71	46.7
Total	152	100

From the analysis it is noted that nearly 46.7% of the residents have strongly agreed to the statement that IoT enables in supporting the supply chain management process effectively, this is in line with the discussion made in literature review, also 27% of the respondents have agreed to the statement. On the other hand, 13.2% were neutral to the mentioned aspect and remaining have disagreed to it (Fig. 2).

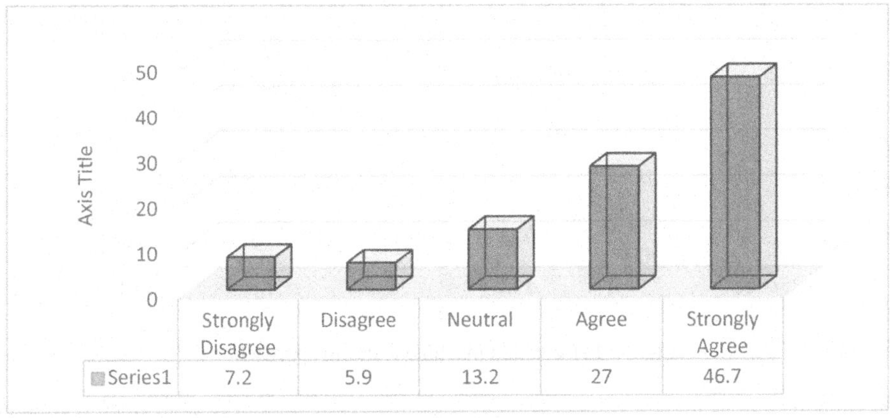

Fig. 2. IoT and supply chain management

The next factor is focused in analyzing the importance of IoT in enabling machine learning approaches for better control and coordination of supply chain process (Table 2).

Table 2. IoT enabled machine learning process

IoT enabled ML offers better control and coordination	Frequency	Percent
Strongly Disagree	11	7.2
Disagree	17	11.2
Neutral	18	11.8
Agree	63	41.4
Strongly Agree	43	28.3
Total	152	100

The analysis states that nearly 28.3% of the respondents have strongly agreed to the statement that the IoT enabled machine learning approaches tend to provide better control and coordination to the supply chain process effectively. [5]. This is also stated by the researchers as mentioned in the literature review, also 41.4% of the respondents have agreed to the statement. Whereas, 11.8% were neutral to the statement, 11.2% were disagreeing to it and remaining were strongly disagreeing to the statement (Fig. 3).

4.1 Correlation Analysis

The next step in the analysis is the estimation of degree of association between the variables, the researchers has considered the key independent variables as: Enhancing value

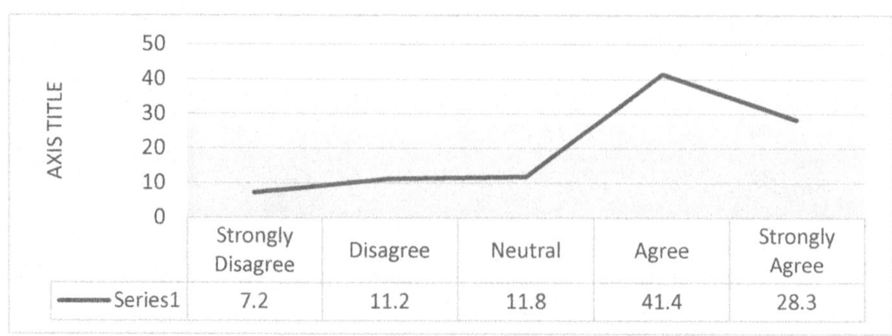

Fig. 3. IoT enabled machine learning approach

chain; Achieving economic efficiencies and Overall cost control whereas the dependent variable is stated to be Supply chain coordination and control. The coefficient of correlation lies between −1 to + 1 (Table 3).

Table 3. Correlation analysis

Correlation analysis	Enhancing value chain	Achieving economic efficiencies	Overall cost control	Supply chain coordination and control
Enhancing value chain	1	.887**	.820**	.864**
Achieving economic efficiencies	.887**	1	.863**	.871**
Overall cost control	.820**	.863**	1	.831**
Supply chain coordination and control	.864**	.871**	.831**	1

Based on the analysis it is noted that the correlation analysis between all the independent variables and dependent variable is more than +0.80 hence there is a positive association between the variables. The highest association is noted between Achieving economic efficiencies and Supply chain coordination and control with coefficient of + 0.871, Enhancing value chain and Supply chain coordination and control with coefficient +.864 and lastly Overall cost control and Supply chain coordination and control is .831.

Hence, it is concluded that all the variables possess higher association and is considered to possess significant impact on the supply chain process in the Organisation.

4.2 Regression Analysis

The next analysis is focused in understanding the extent of relationship between the variables, the regression equation can support in understanding the major drivers influencing the dependent variable chosen in the study (Table 4).

Table 4. Regression analysis

Correlation analysis	B	t Value	P Coeff.
(Constant)	0.274	1.746	0.083
Enhancing value chain	0.371	4.663	0.00
Achieving economic efficiencies	0.313	3.743	0.00
Overall cost control	0.226	3.215	0.002
F value	211.4		
P sig	0.00		
R squared	0.811		

From the regression analysis it is noted that the R squared is 0.811 hence the model is considered to be a good fit and further analysis is being made. The regression coefficient can bee formed to make the equation as follows.

Supply chain coordination and control = 0.274 + 0.371 × Enhancing value chain + 0.313 × Achieving economic efficiencies + 0.226 × Overall cost control.

4.3 Mediation Analysis

The mediation analysis is one of the effective tools in understanding the impact of mediator variable in influencing the dependent variable (Table 5).

Table 5. Analysis of variables

Analysis	B	t Value	P Coeff.
(Constant)	0.4676	2.6979	0.07
Value chain	0.6477	7.325	0.00
ioT enabled machine learning	0.2235	2.7365	0.00
R squared	0.759		

The analysis shows that the R squared is at 0.759 which is more than 0.70, hence specified that the data is considered to be the best fit. Also, the p value is 0.00 which is less than 0.00, hence is statistically significant (Table 6).

Table 6. Mediation analysis

Mediation effect	B	t Value	P Coeff.
ioT enabled machine learning	0.8633	21.039	0.07
Value chain	0.6477	7.325	0.00
R squared	0.7469		

From the analysis it is noted that the independent variable and mediator variables possess positive effect on the deponent variable. The independent variable has highest factor of nearly 0.8633 when compared to the mediator variable of 0.6477.

Therefore, the analysis shows that the null hypothesis is rejected and alternate hypothesis is accepted [15]. Hence, the variables possess higher and critical significance towards building supply chain coordination and control in the Organisation and the management need to focus in implementing the necessary tools so as to harness the potential of IoT in the supply chain management domain [16].

5 Conclusion

The IoT helps to improve aspects of the supply chain, improves the connection between systems and people in different geographical areas and produces dashboard reports for quick decision-making, which again serves as a source of information for the future. the good cycle of value creation. The use of objects based on IoT (T2T) helps to create a better and more efficient network of different terminals, covering products and services that are part of the supply chain, which helps managers to communicate, collect and transfer data in an efficient and reciprocal way.. data analysis for decision making and control of all functions. In addition, the application of technologies, tools and algorithms in an IoT environment is considered an integral part of the process. Within SCP, the use of sensors and processors, interoperability and interconnection of systems, real-time data monitoring and synchronization of several storage options have improved the production and supply chain, and committed to improving overall efficiency while reducing costs and a competitive advantage over competitors… It was also claimed that the general composition and characteristics of SCP changed the holistic nature of the organizational structure, strengthened competition between companies in the domestic and international market, helped to identify new opportunities, eliminate risks and utilize existing resources.

In short, the IoT has radically changed the business environment in which the supply chain operates. The discovery and implementation of real-time data collection has supported the exploration of efficiency, optimization of the use of existing resources, which is fundamental to effective decision-making with other systems. In a dynamic business environment, bulk data collection and real-time data transfer are accelerated in SCP. Therefore, efficiency and optimization need to be improved to better allocate resources.

References

1. Longfei, H., Mei, X., Bin, G.: Internet-of-things enabled supply chain planning and coordination with big data services: certain theoretic implications. J. Manag. Sci. Eng. **5**, 1–22 (2020)
2. Zhang, Y., Wang, Z.: Optimal RFID deployment in a multiple-stage production system under inventory inaccuracy and robust control policy. IEEE Trans. Ind. Inf. **15**(6), 3230–3242 (2019)
3. Li, M., Wang, Z.: An integrated robust replenishment/production/distribution policy under inventory inaccuracy. Int. J. Prod. Res. **56**(12), 4115–4131 (2018)
4. Govindan, K., Cheng, T.C.E., Mishra, N., Shukla, N.: Big data analytics and application for logistics and supply chain management. Transp. Res. Part E: Logist. Transp. Rev. **114**, 343–349 (2018)
5. Celia Garrido-Hidalgo, F., Ramirez, J., Olivares, T., Roda-Sanchez, L.: The adoption of internet of things in a circular supply chain framework for the recovery of WEEE: the case of Lithium-ion electric vehicle battery packs. Waste Manag. **103**, 32–44 (2020)
6. Cui, L., Deng, J., Liu, R., Dongyang, X., Zhang, Y., Maozeng, X.: A stochastic multi-item replenishment and delivery problem with lead-time reduction initiatives and the solving methodologies. Appl. Math. Comput. **374**, 125055 (2020). https://doi.org/10.1016/j.amc.2020.125055
7. Wang, T., Hu, B., Chang, S., Ding, L.: Inventory inaccuracies and radio frequency identification technology: risk analysis and coordination. Comput. Ind. Eng. **125**, 9–22 (2018)
8. Tu, M., Lim, M.K., Yang, M.-F.: IoT-based production logistics and supply chain system - Part 1 Modeling IoT-based manufacturing IoT supply chain. Ind. Manag. Data Syst. **118**, 65–95 (2018)
9. Jain, A., Yadav, A.K., Shrivastava, Y.: Modelling and optimization of different quality characteristics in electric discharge drilling of titanium alloy sheet. Mater. Today Proc. **21**, 1680–1684 (2019)
10. Jain, A., Pandey, A.K.: Modeling and optimizing of different quality characteristics in electrical discharge drilling of titanium alloy (Grade-5) sheet. Mater. Today Proc. **18**, 182–191 (2019)
11. Munuzuri, J., Onieva, L., Cortes, P., Guadix, J.: Using IoT data and applications to improve port-based intermodal supply chains. Comput. Indus. Eng. **139**, 105668 (2020)
12. Kamble, S.S., Gunasekaran, A., Parekh, H., Joshi, S.: Modeling the internet of things adoption barriers in food retail supply chains. J. Retail. Consum. Serv. **48**, 154–168 (2019)
13. Mahrishi, M., Morwal, S., Muzaffar, A.W., Bhatia, S., Dadheech, P., Rahmani, M.K.I.: Video index point detection and extraction framework using custom YoloV4 darknet object detection model. IEEE Access **9**, 143378–143391 (2021)
14. Choi, T.-M., Wallace, S.W., Wang, Y.: Big data analytics in operations management. Prod. Oper. Manag. **27**(10), 1868–1883 (2018)
15. Jain, A., Pandey, K.: Multiple quality optimizations in electrical discharge drilling of mild steel sheet. Mater. Today Proc. **8**, 7252–7261 (2019)
16. Panwar, V., Sharma, D.K., Pradeep Kumar, K.V., Jain, A., Thakar, C.: Experimental investigations and optimization of surface roughness in turning of EN 36 alloy steel using response surface methodology and genetic algorithm. Mater. Today: Proc. **46**, 6474–6481 (2021). https://doi.org/10.1016/j.matpr.2021.03.642

Comparative Analysis of Environmental Internet of Things (IoT) and Its Techniques to Improve Profit Margin in a Small Business

Khongdet Phasinam[1], Mohammed Usman[2], Sumona Bhattacharya[3](✉), Thanwamas Kassanuk[1], and Korakod Tongkachok[4]

[1] Faculty of Food and Agricultural Technology, Pibulsongkram Rajabhat University, Phitsanulok, Thailand
{phasinam,thanwamas.k}@psru.ac.th

[2] Department of Management, College of Economics, Management and Information Systems, Thaksin, Thailand
usman@unizwa.edu.om

[3] Department of Commerce and Financial Studies, Atal Bihari Vajpayee University, Bilaspur, C.G., India
drsumona.abvv@gmail.com

[4] Faculty of Law, Thaksin University, Thaksin, Thailand
tokorakod@tsu.ac.th

Abstract. The Internet of Things (IoT) is a computer concept in which common items are enhanced with computational and wireless communication capabilities, generally through the inclusion of resource-constrained components such as sensors and actuators that allow them to access the internet. The Internet of Things (IoT) is regarded as a critical component in the implementation of intelligent environments. Nonetheless, the present IoT ecosystem provides a plethora of different connectivity options with varying performance characteristics. This circumstance makes determining the best IoT connectivity solution for a certain intelligent environment extremely difficult. In this article, we look at the specific requirements of major smart settings, such as the home automation, smart healthcare, smart urbanization, and advanced manufacturing, and compare them to modern Iot communication solutions. We define the key features of these smart settings before providing a detailed assessment of applicable IoT communication technologies and systems. The Internet of Things (IoT) has the potential to change organizations by automating operations ranging from inventory management to robotics and automation, therefore resulting in cost savings. Can, meanwhile, tiny businesses profit from IoT? This study investigated the increasing importance of the Internet of Things (IoT) in small companies, its influence on their capacity to engage in a fast-changing digital world, and their awareness, attitudes, opinions, and desire to embrace it. An initial exploratory method is used in the research, which is predicated on a study of instance studies in the literature, interviews with many economic development employees, and several small and medium-sized business executives. The adoption of IoT can lead to greater operational efficiency and cost savings in enterprises. These advantages have been confirmed by the medium-sized firms surveyed.

V. E. Balas et al. (Eds.): ICETCE 2022, CCIS 1591, pp. 160–168, 2022.
https://doi.org/10.1007/978-3-031-07012-9_14

Keywords: Internet of Things (IoT) · Business · Environment · Development · Management · Benefits · Communications

1 Introduction

The emergence of IoT has been regarded as one of the critical elements in the business environment, IoT is often referred as network of interlinked systems that often use key attributes and systems to gather and distribute data. It is mainly stated as an outgrowth of the interconnection to various sensors and devices or things, providing sometimes simplified objects, such as light bulbs, locks, and vents, with a greater degree of computing and analytical functionality. The Internet of Things (IoT) is made up of a network of material things with the integrated capability to converse, communicate, and exchange information. The Internet of Things (IoT) may bring substantial assistance to businesses companies by automating fundamental operations such as inventories management and supply chain management. This gives merchants real-time knowledge on where their inventory is at any given time, avoiding product loss, and automating accounting and replenishing. Intelligent monitors for temperature and lighting management may be used in any office to save money on energy expenditures. Wearable technology is being used by medical practitioners to monitor various patient health, with automatic monitoring and notifications if patients display warning signals such as cardiac issues or hypotension. To increase food quality and safety, small eateries can employ sensors for monitoring food temperature and expiration dates. This tool is considered as the key system which supports the management in connecting with different things to the internet which will enable in collecting the data from the different means, analyzing them, and acting on it to achieve competitive advantage.

Furthermore, dealing with a wide range domain (for example, smart home/health/city/factory) have varying needs and hence technological options, which impacts the strategies of how and where information is transmitted and how to rely on the data within a given environment. Moreover, the evolution of different types of smart settings occurs at varying rates: Some vertically domains (e.g., smart homes and smart health) can assess and implement new technologies much more quickly, but others (e.g., smart factories and smart cities) cannot because they must cope with old systems. This need challenges the selection of communications technology and the accessibility of solutions for certain intelligent devices. Moreover, few studies have been conducted to determine if small companies comprehend what these systems tend to offer the related benefits, the critical impact which they create on the business and how the managers can focus in embracing them. This paper investigates these variables and provides a review of the status of small company perspectives, IoT technology adoption and deployment, and the ability to affect performance or competitive advantages.

2 What Does IoT Mean?

The Internet of Things (IoT) refers to how gadgets communicate with one another in a network setting. These devices may perform a wide range of tasks, from sensors in thermometers and manufacturing machinery to printers, televisions, and even refrigerators. It can be stated that the overall applicability of IoT in business context are highly diverse in nature, hence the implementation of IoT has seen a great increase among the entrepreneurs, business management and leaders for enhancing their competitive edge. [1].

3 What are the Advantages of the Internet of Things (IoT) for End Users?

The major objective of this previous article is to investigate the advantages of IoT technologies for businesses, especially in terms of business operations and efficiency. As we go any further, it is important to note that Internet of Things technology provides substantial benefits for end users that should not be overlooked [2]. Considering the flexibility that the Internet of Things provides to businesses, particularly for employees who must spend time away from the office, such as salesmen or engineers. IoT is a windfall for firms with employees in these sectors, and employees are the first one to benefit [3]. Based to one study, large-scale IoT deployment contributed to higher levels of staff engagement, resulting in:

- Improved productivity.
- Improved job satisfaction.
- Improved performance and customer service.

4 The Nature and Advantages of Implementing IoT in Business

Many enterprises are now focusing in leveraging the Internet of Things to gather, handle, analyses, and apply enormous volume of critical information to automate the overall processes, enable in analyzing the changing behaviors of the consumers, and track goods and services, devices, and focus in deploying the resources efficiently and cost cutbacks, gives the company a competitive benefit [4]. The challenge is the ability of SME sector in implementing the technologies so that they can compete effectively against the MNE (Multinational Enterprises).

5 The Critical Application of Implementing IoT in Small Businesses

In the recent business environment, small business enterprises are focusing in implementing various technologies to compete effectively against the peers and major companies. It would allow them to optimize time-consuming, manual procedures and engage with consumers more efficiently through tailored interactions [5]. He suggests many strategies for small enterprises to adopt IoT technology at a minimal price. Using a digital assistant like Google Assistant is a simple way to get started.

By notifying a small company about equipment issues before they occur and reducing production downtime, adding sensors to industrial equipment may save time. Incorporating IoT devices like Nest to monitoring temperature, lighting, or when equipment is turned on has the potential to save a lot of money on energy. A couple more ideas on how smaller companies may efficiently employ IoT devices. Small merchants, restaurants, and wholesalers, for example, can substantially enhance efficiency by putting RFID tags on products for tracking inventory. This might significantly cut the time required to detect incoming and outgoing items and improve inventory accuracy [6]. Each item must be scanned separately using standard bar codes. RFID tags, on the other hand, enable firms to monitor all items along the supply chain, from original purchase through processing and distribution to intermediaries.

6 Using the Internet of Things, How are big Businesses Using it?

He implies several significant aspects that these applications of these technologies tend to impact the overall process, such as using automatons to complete labor shortages, enabling sensors which will be connected to different systems through cloud to monitor and measure temperature, analyses the pressure, and other aspects; and instantly notifying technicians of difficulties or maintenance. These systems tend to support the management in tracking the movement of goods, deploy the resources effectively and support in reducing the cost in an effective manner [7]. RFID tags are currently used by most significant corporations. These are tiny electrical gadgets that can store and transfer data and may be incorporated in goods. Walmart, for instance, has been applying these tools in realizing the efficiency in the logistics and supply chain process, while other large retailers have embraced it to enhance the supply chain activities and many managers have used it to automate operations and increase inventories accuracy.

Worn sensors that measure heart rhythm, environmental variables, air quality in hazardous situations, and other aspects are mainly focused in enhancing the working environment, support for collaboration among the employees etc. Likewise, wearable sensors are used efficiently in telemedicine to monitor patients and can also interact with Artificial Intelligence technology to understand patient patterns, measure the current health conditions on real time basis, enable in predicting the future complications based on lifestyle, food habits etc. to offer health care to the individuals [8]. Other advantages include the potential to schedule the appointments in real time basis, understand the vitals, and even manage office equipment using machine learning, which gathers information, learns trends, and produces models to enhance efficiency and save costs.

7 Responses of Small and Medium Sized Enterprises

Most answers from the top leaders who manage the SME sectors were brief and sent by e-mail, in response to inquiries from the overall development personnel described in the preceding section [9]. In Table 1 we were, moreover, welcomed in offering an overview of the medium-sized manufacturing plant, which gave us with a great deal of data on the application of different systems related to Internet of Things which is focused in enhancing the overall productivity, enable in achieving better efficiency, improve quality, and cost-cutting. We'll keep that till the end of this part.

Table 1. Responses of small and medium-sized enterprises

Type	Description
Small nursery	We don't utilize sensors very much when expanding. Temperature sensors are commonly utilized to aid with airflow
Small environmental consulting company	This little company, which employs around 32 people, conducts environmental services. Most of the staff have PhDs or master's qualifications in difficult fields such as statistics or biology. They do complex study on agriculture, water, and other topics and are required by the EPA to gather certain information in specified methods
Pet Clinic	It has been noted that the employees in the procurement department have stated that extensive study related to RFID however they were not mainly adopted for the clinic. The manager states that these technologies might not provide the necessary inventory information. Hence, it can be stated that may of the owners and employees in the industry are not having comprehensive understanding on the application of technologies to manage the inventory better
Infrastructure business	It can be stated that the company is substantial in the New England and employs more than 1200 individuals. It features cutting-edge technology and IT skills. As per the CIO, the construction sector is making strides in a variety of technological sectors. Sensors have been around for a while and are becoming more capable and useful. The technologies are mainly applied to collate and forecast data to take proactive actions and achieve competitive edge

8 IoT Adoption Barriers for SMEs

It's not a good idea to approach IoT adoption with the mindset of "build it and they will come" [10]. Small businesses must be aware of the technology's complexities and challenges. According to a survey of small business perspectives on IoT adoption, most respondents (69%) believe that implementing IP-based automation systems within the next three years is crucial for their businesses. However, network administration and monitoring limits are seen by half of those questioned as a key hindrance to expansion in that sector [11]. Smaller firms, more than any other type of organization, have the most difficulty understanding and implementing IoT. For example, in retail, they must develop a computerized system (Fig. 1).

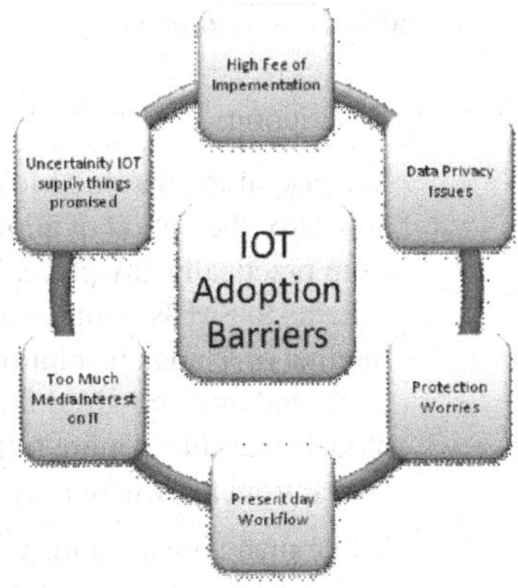

Fig. 1. IOT adoption barriers for SMEs

9 Business Advantages of the Internet of Things (IoT)

In Table 2 Let's look at some of the ways SME sectors may use the Internet of Things and its cutting-edge technology to enhance crucial parts of their operations [14] (Fig. 2).

Table 2. Business advantages of IoT [12, 13]

Type	Description
Efficiency & productivity	Efficiency is the foundation for the type of increased productivity that increases income for SMBs. The management can focus in unleashing the potential of the technologies to enhance productivity through automating the repetitive and non-core activities. An example IoT function for this technique would be an autonomous PDF translation and generation tool that reduces barriers to PDF editing and archiving, speeding up communications and documenting
It's all about the customer experience	While IoT technology has primarily affected back-end operations due to its newness, today's IoT has resulted in universal strategies that have fundamentally changed the way customers perceive their relationships with enterprises. Front-end consumer interaction has become a focus for businesses as they begin to interact their customers more effectively, typically through IoT gadgets

(*continued*)

Table 2. (*continued*)

Type	Description
Mobility & agility	Because of the nature of IoT technology, businesses now have the option of allowing their staff to work from practically any place, which may bring significant benefits to SMBs. Office rents aren't cheap, and the Internet of Things revolution has resulted in a rush of small and medium-sized firms altering their business models, recruiting more full-time remote employees in "work from anywhere jobs
Opportunities for business	While many organizations try to tap into the revenue-generating capabilities of online services, the majority without a clear plan for doing so. In this regard, the technology can be regarded as the key attribute in collecting the large volume of information and use them to take informed decisions. These analytics give fresh data that are being used to not only build new business models but are also fundamentally altering existing sectors
Cost-cutting	One of the critical aspects of implementing the sophisticated technologies is to enhance productivity and efficiency which will lead to reducing the overall operational cost effectively. This boost profitability of the business effectively. Furthermore, the IoT can support in protecting the company data and information from cyber-attacks, reducing the spending on IT infrastructure as they can be stored in cloud computing which also benefits the business and managers largely

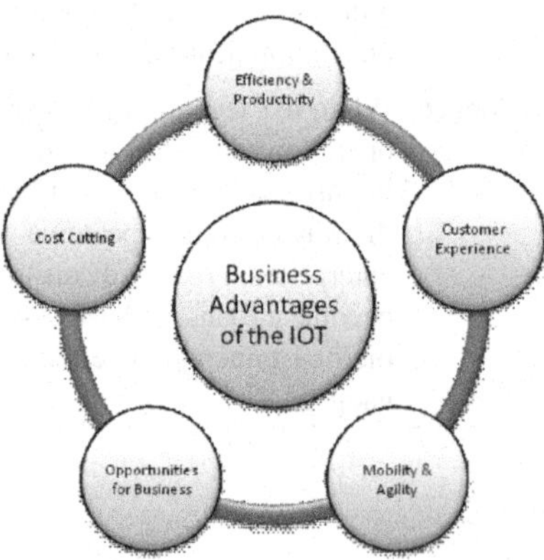

Fig. 2. Business advantages of IoT

10 IoT perceptions of Small Businesses

SME's awareness and grasp of IoT technology will be assessed. Moreover, 78% of respondents employed for small businesses, 41% were proprietors, and 39% of the were related to production. It has been identified that the managers and owners a comprehensive understanding on the technologies and were stated to utilize them, if they were interested in them, and if they were important to their business [15]. The majority of those who did not use RFID were interested in doing so. The capacity to gather and analyses information for enhanced decision making was recognized as the most significant benefit of adopting IoT. In terms of views, 34% of these participants said that the IoT will unquestionably add value to the company. Surprisingly, owners seemed to be more optimistic about IoT technologies than managers. In perspective of acceptance difficulties. Return on investment (46.3%), a lack of internal IT experience (37%), integrating IoT solutions with existing systems (29.6%), and security threats were the top hurdles identified (25.9%). She did, meanwhile, indicate that unfavorable impressions may be linked to a lack of knowledge about the technology and their advantages. The traditional technology acceptance paradigm in relation to SMEs' adoption of IoT technologies.

11 Conclusion

IoT technology may dramatically increase quality and efficiency while lowering costs. While most big and many medium-sized firms have embraced them and experienced considerable benefits, small businesses tend to be unaware of them or the benefits they may get. The case studies back up the results that construction and manufacturing businesses appreciate their employees more than other industries. We realize, though, that education level and experience also play a significant role in awareness and acceptance. It should also be mentioned that when asked why SMEs are not embracing IoT technology, the site manager of the medium-size components firm for the automotive sector gave a succinct and to-the-point response: It can be a big capital outlay. As a result, we reach the conclusion that there is potential for small businesses to enhance their competitiveness in most industries by trying to adopt these IoT sensor technologies, and we propose that economic growth organizations or similar organizations in most towns could assist with awareness, education, acceptance, and development.

References

1. Gazis, A.: What is IoT? The Internet of Things explained. Academia Let. **1003**, 1–8 (2021). https://doi.org/10.20935/al1003
2. Alaslani, M., Nawab, F., Shihada, B.: Blockchain in IoT systems: end-to-end delay evaluation. IEEE Internet Things J. **6**(5), 8332–8344 (2019). https://doi.org/10.1109/jiot.2019.2917226
3. Panwar, V., Sharma, D.K., Pradeep Kumar, K.V., Jain, A., Thakar, C.: Experimental investigations and optimization of surface roughness in turning of en 36 alloy steel using response surface methodology and genetic algorithm. Mater. Today: Proc. **46**, 6474–6481 (2021). https://doi.org/10.1016/j.matpr.2021.03.642
4. Kilari, N.: Internet of things for business. Int. J. Eng. Comput. Sci. **6**(3), 20738–20741 (2017). https://doi.org/10.18535/ijecs/v6i3.60

5. Meena, G., Sharma, D., Mahrishi, M.: Traffic prediction for intelligent transportation system using machine learning. In: Proceedings of 3rd International Conference on Emerging Technologies in Computer Engineering: Machine Learning and Internet of Things, ICETCE 2020, pp. 145–148 (2020)
6. Jeong, Y.S.: A study on improving manufacturing environment using IoT technology in small business environment. Convergence Soc. SMB **7**(2), 83–90 (2017). https://doi.org/10.22156/CS4SMB.2017.7.2.083
7. Lee, I.: The internet of things for enterprises: an ecosystem, architecture, and IoT service business model. Internet of Things **7**, 100078 (2019). https://doi.org/10.1016/j.iot.2019.100078
8. Jain, A., Pandey, A.: Multiple quality optimizations in electrical discharge drilling of mild steel sheet. Mater. Today: Proc. **4**(8), 7252–7261 (2017). https://doi.org/10.1016/j.matpr.2017.07.054
9. Khuja, M.S.A.A.L., Mohamed, Z.A.B.: Investigating the adoption of E-business technology by small and medium enterprises. J. Adm. Bus. Stud. **2**(2), 71-84 (2016) https://doi.org/10.20474/jabs-2.2.3
10. Gordon, H.W.: The Meaning of dreams: don't be a freud. Contemp. Psychol.: A J. Rev. **34**(8), 790–791 (1989). https://doi.org/10.1037/031025
11. Jain, A., Pandey, A.K.: Modeling and optimizing of different quality characteristics in electrical discharge drilling of titanium alloy (Grade-5) sheet. Mater. Today: Proc. **18**, 182–191 (2019). https://doi.org/10.1016/j.matpr.2019.06.292
12. Dinka, S.T.: Assessment of critical challenges facing small business enterprises: a survey of small businesses in Woliso Twon, Ethiopia. Res. J. Finance Accounting **10**(3), 22–31 (2019). https://doi.org/10.7176/rjfa/10-3-04
13. Šmeringaiová, A., Pavlenko, S., Wittner, M.: Industrial internet of things - strategic way to optimize the production environment in small enterprises. EAI Endorsed Trans. Internet Things **4**(15), 156718 (2018). https://doi.org/10.4108/eai.5-3-2019.156718
14. Alam, M., Khan, I.: Internet of things (IOT) as key enabler for efficient business processes. SSRN Electron. J. (2021) https://doi.org/10.2139/ssrn.3806408
15. Jain, A.Y., Shrivastava, Y.: Modelling and optimization of different quality characteristics in electric discharge drilling of titanium alloy sheet. Mater. Today: Proc. **21**, 1680–1684 (2020). https://doi.org/10.1016/j.matpr.2019.12.010

A Theoretical Aspect on Fault-Tolerant Data Dissemination in IoT Enabled Systems

Vishnu Kumar Prajapati(✉), T. P. Sharma, and Lalit Awasthi

National Institute of Technology, Hamirpur, H.P., India
vishnu08jec@gmail.com, {teek,lalit}@nith.ac.in

Abstract. The Internet of Things (IoT) field has grown exponentially and, it is a significant emerging research field in this era. IoT consists of sensors, actuators, radio frequency identifier (RFID), and machine-to-machine communication. A failure may occur in IoT at layering architecture such as applications: sensor and actuator nodes can be lost, missed, and failed, network links can be down processing, and storage components of IoT can fail to perform appropriately. That is the reason for which data dissemination fault-tolerant (F.T.) has become a significant role for IoT-enabled systems to provide Quality of service (QoS). It can reduce the end-to-end delay, low energy consumption, and maximize throughput. The authors aim to identify and systematically review data aggregation, data dissemination, and F.T. mechanisms. Data dissemination has become effective when data aggregation, fault detection, and fault-tolerant has done at the same level of communication. The authors also discuss the theoretical aspect of efficient and fault-tolerant data dissemination using a group formation framework for IoT-enabled systems and different fault-tolerant techniques with a comparative survey.

Keywords: Data aggregation · Data dissemination · Fault-tolerant · Internet of Things · Quality of Service (QoS)

1 Introduction

The term "Internet of Things" was coined by Kevin Ashton in 1999. AT that time, Kevin Ashton worked as assistant director of the Auto-ID Center at the Massachusetts Institute of Technology (MIT) [24, 25]. The IoT has slightly changed, and the organization International and Telecommunication Union (ITU) is defined as "Internet of Things "are the global Infrastructure for the information society, enabling advanced services by interconnecting things based on existing and evolving interoperable information and communication technology [25].

The Internet of things is becoming the most popular field in 2020 for researchers and real-world implementation [23]. IoT has four pillars which are M2M, SCADA, RFID, and WSN. Machine to machine (M2M) is direct communication between two or more machines with wired or wireless communication channels [24]. Supervisory control and data acquisitions are hardware elements and system software to control the

V. E. Balas et al. (Eds.): ICETCE 2022, CCIS 1591, pp. 169–181, 2022.
https://doi.org/10.1007/978-3-031-07012-9_15

industrial process locally and in remote locations [23, 24]. Radio Frequency identifier is a technology that allows the identification of an object by using radio waves. A wireless sensor network (WSN) is a self-configured and infrastructure-less wireless network that monitors physical or environmental conditions, i.e. temperature, sound, pressure, motion, vibration, etc. [23–25].

Internet of Things (IoT) enabled systems are very complex and massive heterogeneity which consists of sensor and actuator enable devices with deferent geographical locations. IoT devices have connected with wire/warless and fetching information from environmental or required information from different places. Using aggregation techniques, they are aggregating data from ends devices [21] and data that can be disseminated over the networks so that design of IoT architecture is less faulty. Fault-tolerant provides the ability of a system to continue operating without interruption when one or more (hardware/ software) components fail [14]. IoT technology is used in billions of devices, and each device collects data which is sensors or actuators. Sensors and actuators have limited power. These sensors and actuators may collect duplicate or redundant data. Data aggregation is one of the most influential techniques to eliminate data redundancy and improve energy efficiency, a lifetime of the IoT system [5]. The data dissemination is distributing or transmitting statistical or other data to the end-users. The routing process of data and queries in a wireless sensor network is called data dissemination. It is a routing process of transferring desired data from active sensor nodes to sink nodes that collect data in the sensor network [2]. In an IoT environment, data dissemination is first addressed by a framework, which reduces the consumption of device energy and the amount of data that has to be disseminated over the network [7]. IoT device distributions data based on run-time knowledge and interest node enhances the efficiency of data dissemination [4]. So overall the data dissemination is a massive field for IoT-enabled system and need to disseminate data over the network with minimum data overhead.

The architecture of data dissemination has shown in Fig. 1, IoT applications, IoT gateways, and IoT devices. IoT devices use the hierarchical network architecture data transmission at cluster head level and further decrease the data overhead and optimize the routing so that IoT applications take fruitful data dissemination in IoT enabled systems.

Fault detection is a process for industry monitoring IoT, but it is a challenging and complicated task because many factors influence data and could cause faults; the fault is application and sensor type [22]. Fault-tolerant provides the ability of a system such as networks, cloud, Cluster, and computer to continue operating without interruption when one or more IoT system components fail [17]. Fault detection is a part of communication over the network. It provides to take care of faults. A fault may be tolerable or fixed or take appropriate action so that data dissemination becomes more effective.

Data aggregation, data dissemination, fault detection, and fault tolerance techniques have significant applications to maximize lifetime, minimize overhead, availability, and reliability of IoT-enabled systems [15]. The healthcare system, agriculture monitoring system, home appliance, transportation system, etc. that is IoT enabled systems have a great chance to address these techniques in over architecture.

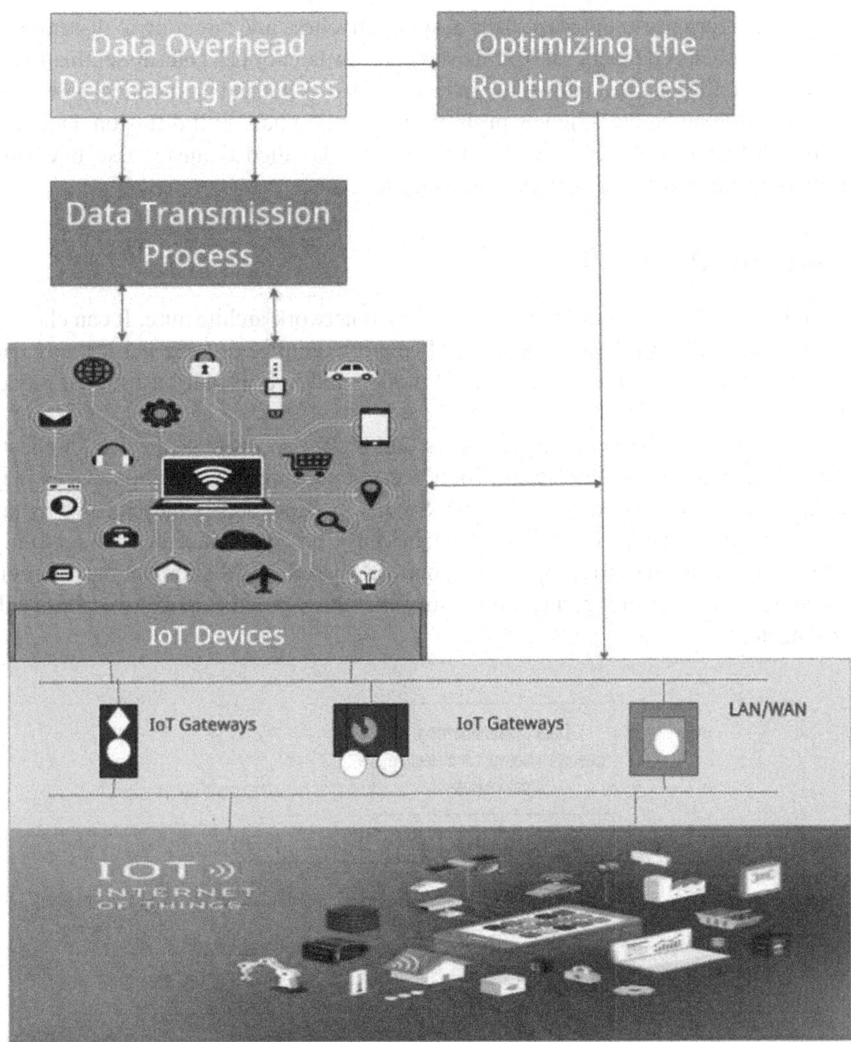

Fig. 1. Data dissemination architecture for IoT

2 Data Dissemination

The IoT-enabled systems are needed to handle highly heterogeneous devices and, during transferring data, do not achieve energy efficiency, reliability, and higher delivery ratio. Hence, appropriate data dissemination to IoT devices is critical to achieving their goals. It is the process of disseminating and monitoring information, tasks that perform and reported to a goal such as a sink. So, designing appropriate data dissemination issues is essential for different-different application scenarios [2]. Because most of the data dissemination techniques/frameworks in IoT are use monitoring base systems, at run time environment. These monitoring frameworks are used to structure-based approaches in various data formats to optimize their cost-effectiveness and targeting to disseminate information with max delivery ratio, min overhead, low latency [1]. Data dissemination in structure-based IoT-enabled networks can be classified as flat and hierarchical networks. A flat network may be classified as push diffusion, two-phase pull diffusion, and one-phase pull diffusion. The hierarchical network can be classified as cluster base, tree base, chain-based, and grid-based networks, shown Fig. 2.

2.1 Flat Network

IoT-enabled systems generally use structure-based network architecture. It can classify as flat network architecture and hierarchical network architecture. the flat network can classify as a pull-based model, push-based model, and hybrid pull-push-based model [33].

IoT systems have become very complex and highly heterogeneous. The duplicate message is standard for data dissemination because sensing or actuating devices may send the copy to its neighbor or sink node. So that multiple copies may receive by the receiver. Only the first copy is acknowledged; in a push-based scheme, they can send only the last copies acknowledged [33, 34]. Flat network-based data dissemination protocols are flooding, SPIN, gossiping, direct diffusion, one-phase pull diffusion, two-phase pull diffusion, etc.

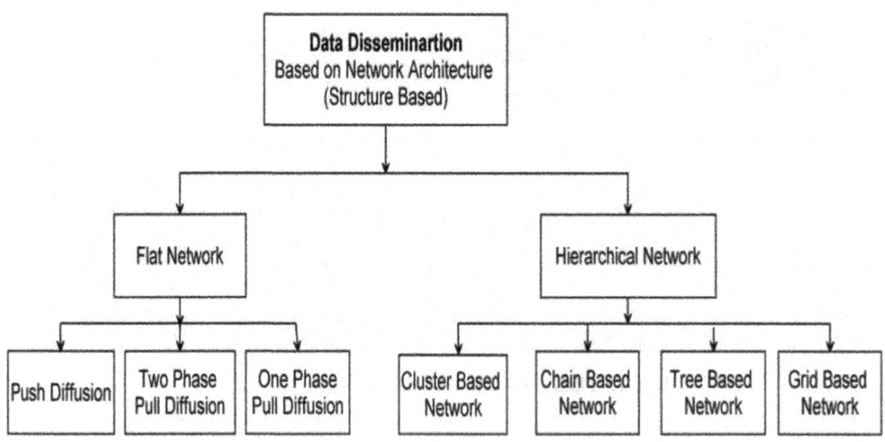

Fig. 2. Classification of data dissemination based on structured IoT system

2.2 Hierarchical Network

IoT industry 4.0 has based on a hierarchical network-based and data can be disseminated over the network. It can classify as grid base, cluster base, tree base, and chain base data dissemination. In grid-based data, the dissemination-based protocol is designed using a grid of square size and crossing point using greedy geographical forwarding phenomenon. If there is no IoT device at the crossing point, select the nearest IoT device to the crossing point as a dissemination node [30]. The Crosspoint approach is used layer base clustering and does not support network mobility to enable anytime and anywhere data accessibility at IoT -based networks [31]. Now clustering approach is used in a heterogeneous environment. It is based on layered IoT frameworks: LEACH, two-level-LEACH, power-efficient and adaptive clustering hierarchy, and pragmatic two-layer IoT architecture [34]. Tree-based data dissemination approach formatted a tree network which is a root node, enter median nodes and leaf nodes. Data dissemination takes place from a root node to a leaf node or leaf node to the root node in such a way so that energy consumption is minimum [33]. Many tree-based routing protocols are used to minimize energy consumption that is low power and lossy networks. The neighbor-based probabilistic broadcast protocol is an example of a tree base scheme. Chain-based data dissemination and aggregation apply. If the sink node is the farthest distance from the source node, it increases the energy consumption for further communication between them (Table 1).

Table 1. Review on data dissemination

S. No.	Name of protocols	Finding/contribution	Research gaps
1	Neighbor Based Probabilistic Broadcast Protocol [3]	1. Reduces End-to-end delay, 2. adaptive connectivity for packet delivery ratio and network overhead	Energy consumption
2	Energy Optimal Data Aggregation and Dissemination Protocol [1]	1. Problem of energy optimal routing and multiple sink aggregation by wireless mesh network using mixed-integer programming formulations and algorithm 2. Network lifetime optimization by min-max per node energy optimization for both 1K, nK models 3. Box-plot for 1K and nK model suing approximation algorithm	1. IoT reliability 2. redundant paths 3. using only uni-cast aggregation 4. min-max energy optimization method is less efficient

(*continued*)

Table 1. (*continued*)

S. No.	Name of protocols	Finding/contribution	Research gaps
3	Mobile Edge Aided Data Dissemination Protocol [6]	1. Data dissemination consists of - installation process for loading data in IoT nodes; data propagation is used for efficiently and reliably forwarding the whole data block in all IoT nodes 2. Channel resolution scheme reduces channel collisions 3. It reduces the overall delay 4. Adaptive protocol reduced total transmission count	1. Mobile server and IoT node distance calculated by Euclidean distance 2. slow starting because of mobile edge server travel
4	Direct diffusion	1. No implosion 2. Low energy consumption as compared to simple flooding protocol	It is applicable if the number of sink nodes is less
5	Secure Data Dissemination Techniques (SDDT) for IoT applications: Research challenges and opportunities [7]	1. focused on SDDT in IoT environment 2. Discussed efficiency, security, reliability of data aggregation, and data dissemination-based techniques 3. Data dissemination schemes for the Internet of drones, vehicles, and battlefield things	Do not cover secure execution of delay-sensitive applications, Cluster-based data dissemination techniques
6	Edge Mesh a New Paradigm to enable distributed intelligence in IoT [2]	1. Edge Mesh focused distributed processing, fault-tolerant, low latency, energy efficiency, and better scalability 2. Task allocation, reduced redundancy of data from end devices 3. Load balancing 4. It has focused on data shared between end devices that includes converting the data to the appropriate format 5. Adaptive protocol reduced total transmission count	1. Speed of data dissemination may affect due to intra-channel 2. do not cover the Data Delivery ratio

Based on the above data dissemination techniques and framework, most authors focus on end-to-end delay, packet delivery ratio, network overhead, low energy consumption, task allocation, and reduced data redundancy.

3 Fault-Tolerant

A fault is defined as the defect of the System [23], which is an error of the System it may lead to failure, and it may affect the system functionality. The fault may occur in different components of the System. Fault-tolerant is the process of the ability of the System to continue operating without interruption when one or more components fail. Fault tolerance techniques provide reliability and availability, maximize lifetime and energy efficiency, increase the delivery ratio, and reduce the System's overhead []. Fault tolerance routing approach is essential for IoT applications since actuator or sensors devices monitor and track important events. Fault tolerance tasks are fault prediction, fault detection, and fault recovery (Table 2).

Table 2. Type of fault detected in iot enabled system

S. No.	Type of fault detected	Fault detection techniques	Category
1	Application failure	Task resubmission, Check pointing, S-Guard	Reactive
2	Node failure	Replication, Task resubmission, Job migration, S-guard, Preemptive migration	Reactive, Proactive
3	Process failure	Replication, Job migration	Reactive
4	Host failure	Job migration, Retry, self-healing	Reactive, Proactive
5	Network failure	Retry, Self-healing	Reactive, Proactive

Effective and fault-tolerant data dissemination has a significant challenge in the Internet of Things. When data is disseminated over the networks, they should detect a fault, which is done using reactive and proactive approaches. There are application failure, nodes failure, process failure, host failure, and network failure kinds of fault. These faults should be detected using fault detection techniques such as to ask Resubmission, Check Pointing, S-Guard, Replication, Task resubmission, Job migration, S-guard, Preemptive migration, Replication, Job migration, Job migration, Retry, self-healing, Retry, Self-healing.

There are several approaches for achieving fault tolerance in IoT-enabled systems. These are.

Table 3. Review on Fault-Tolerant in IoT

S. No	Name of protocol	Finding/contribution	Research Gaps
1	Self-learning sensor fault detection Framework [8]	The authors apply group-based fault-detection. It provides an online sensor faul- detection framework. Statistics Sliding Windows contains the recent sensor data and regresses each window by Gaussian distribution	1. They do not focus on Statistics Sliding Windows (SSW) size 2. Not Covered status transform windows size
2	Layered fault management scheme for end-to-end transmission in IoT [9]	1. They provide uniform fault managing procedure control 2. Layered fault management structure provides the unified fault countermeasure and maximizes existing facilities utilities 3. Advanced fuzzy cognitive maps (FCM) provide efficient fault detection in end-to-end	Do not cover 1. Energy efficiency 2. Delivery ratio 3. Duplicate and redundancy 4. Throughput
3	Optimizing Multipath Routing With Guaranteed Fault Tolerance in Internet of Things [10]	1. PMSO routing algorithm to construct, recover, and select k-disjoint paths, which tolerate the failure while satisfying the QoS parameters 2. It provides fast recovery from path failure 3. It uses both reactive and proactive routing mechanisms 4. It uses personal-best position and global position for velocity updating to that enhance the performance of the routing algorithm	Do not focus 1. Delivery ratio 2. Redundancy 3. Other than ring and mesh network topology 4. routing done by hop (sensor) count strategy

(*continued*)

Table 3. *(continued)*

S. No	Name of protocol	Finding/contribution	Research Gaps
4	A decentralized latency aware task allocation and Group Formation [11]	1. They provide task reliability in the distributed edge computing environment 2. Increase the overall application availability by providing fault tolerance 3. Fair resource allocation in this environment 4. Select a backup node based on a reliability model to provide redundancy	Focused on 1. Delivery ratio 2. Decreasing overall network traffic 3. Lower latency
5	A Microservices architecture for reactive and proactive F.T. [12]	They used fault-tolerance with two micro-services, 1. Real-time fault-tolerant are used for complex event processing to analyze stream data for rapid error recovery. 2. Predictive fault-tolerant is used as a learning mechanism to learn fault patterns and mitigate future faults	Do not cover 1. Delay 2. Delivery ratio 3. Aggregation 4. Task allocation
6	Device clustering for fault monitoring [13]	1. Constricting a monitoring system are done by using Heuristic algorithms for both clusters monitoring and route monitoring within each Cluster are done 2. Fault route monitoring is done by using m-TSP and compared to N.N. and used CLNN 3. Use the greedy solution for Identifying cluster heads	1. m-TSP are complex 2. We have limited power, and computation, we need a less complex algorithm 3. NP Hard Problem

(continued)

Table 3. *(continued)*

S. No	Name of protocol	Finding/contribution	Research Gaps
7	Fault Tolerance Techniques and Architectures in Cloud computing: a comparative analysis [14]	1. Hardware fault-tolerance (structure system can work with other faulty Systems) and software fault-tolerance may be static or dynamic, 2. It manages scalability, throughput, reliability, response time, performance, overhead, usability, and availability 3. F.T. techniques are checking point, replication, job migration, task resubmission, rescue workflow, and preemptive migration	1. Not covered comparison between these fault-tolerance techniques 2. do not focusing framework based suitability fault-tolerance techniques
8	Reliable and Fault-Tolerant IoT- dge Architecture [15]	1. They proposed a four-level fault-tolerance architecture cloud-fog-mist-dew 2. Reliability is achieved by replication, redundancy, and a backup system 3. Mobile agent works as a resource and network monitoring agent 4. This algorithm work with reactive and proactive fault-tolerance process	Do not cover 1) Delay 2) Delivery ratio 3) Aggregation 4) Task allocation

The above Table 3 shows that the fault-tolerant literature review provides fruitful information regarding IoT-enabled systems. The authors focused on finding or contributions and research gaps so that the fault-tolerant techniques and implementation become more effective and routing performs minimum delay, maximum delivery ratio, low energy consumption, and System become less faulty. There are several fault-tolerant techniques to provide fast recovery from path failure like Particle Swarm Optimization (PSO) based routing scheme to find out the optimal loop-free path for solving the multipath problem [27], it also chooses the appropriate path for load scheduling to distribute routing load for resolving multipath load balancing [28]. PSO-based schemes use to minimize the power consumption, like the EPSO-CEO scheme to minimize the power consumption for each node and optimize the cluster head selection [30, 32].

4 Efficient and Fault-Tolerant Data Dissemination

In this paper, the authors propose effective and fault-tolerant data dissemination techniques based on the literature review. Effective data dissemination refers to optimizing delay, reducing data overhead, and optimizing energy consumption. Fault-tolerant data dissemination refers to choosing the best routing path in real-time applications, real-time routing load distribution and load balancing at cluster head, path recovery at the multipath scheme, and replication are done at IoT layer architecture. The proposed data dissemination framework is based on a group formation approach, and routing is done based on the PSO scheme. In group formation framework refers to the task management process, group formation at forming step, algorithms design at norming and storming and simulation are done at performing steps.

Group formation framework consists of four major septs task management process, group formation, algorithm, and simulation.

4.1 Task Management Process

Independent sub-task management is performed at this stage. Data collection by IoT-enabled end devices and distribution processes are done at the cluster head. This stage is focused on managing a task that is supported to optimize delay, minimize power consumption and reduce data overhead by applying effective data aggregation techniques (Fig. 3).

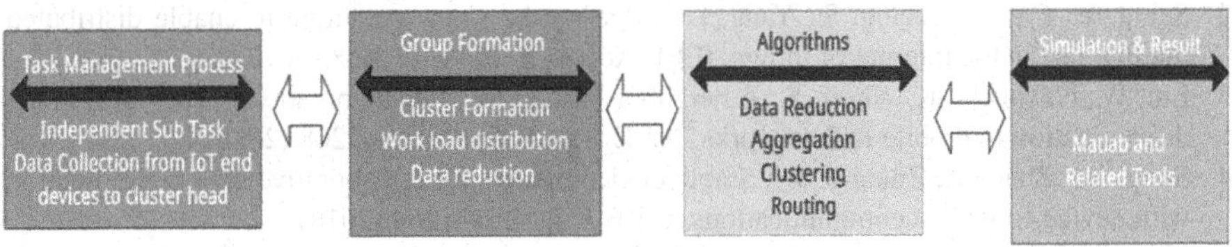

Fig. 3. Group formation framework for IoT enabled system

4.2 Group Formation

Forming and storing are done at this stage. Cluster formation, load balancing, load distribution in real-time, multipath routing, data transmission, and data reduction are made at this step. It is an essential step for fault-tolerant because load balancing, replication, and fast recovery from multipath failure are made. Node failure, host failure, process failure, and network failure will manage at this step.

4.3 Algorithms

The proposed group formation framework chooses the best and appropriate algorithm to perform the real-time application. Home automation, smart home, smart cities, smart health care, smart traffic control system, etc., are real-time applications. The appropriate algorithms are applied as needs of applications.

4.4 Simulation

This is the final step of the proposed group formation framework. IoT enabled systems to support what kind of simulators are used as application-dependent. MATLAB with Simulink simulator may choose for implementation.

5 Conclusion and Future work

Data dissemination over the network is a massive field for IoT-enabled systems. When data dissemination is done without fault or less faulty in the network with less energy consumption, the IoT enables systems to become more reliable than normal data dissemination over the networks. They should focus the review on data aggregation, fault detection, fault-tolerant in data dissemination over the networks. Researchers can model the problem and propose the objectives to achieve their goals. The authors proposed an efficient and fault-tolerant data dissemination using a group formation framework for IoT-enabled systems. Group formation frameworks are applied to implement efficient and fault-tolerant data dissemination techniques to achieve their goal.

References

1. Emma, F., Micha, L.P., Artur, T.: Energy optimal data aggregation and dissemination for the internet of things. IEEE J. **5**, 955–969 (2018)
2. Sahni, Y., Cao, J., Zhang, S., Yang, L.: Edge mesh: a new paradigm to enable distributed intelligence in the internet of things. IEEE Access **5**, 16441–16458 (2017)
3. Liu, W., Nakauchi, K., Shoji, Y.: A neighbor-based probabilistic broadcast protocol for data dissemination in mobile IoT networks. IEEE Access **6**, 12260–12268 (2018)
4. Song, W., Zhao, Y., Zhuang, W.: Stable device pairing for collaborative data dissemination with device-to-device communications. IEEE J. **5**, 1251–1264 (2018)
5. Fitzgerald, E., Pi'oro, M., Tomaszewski, A.: Energy-optimal data aggregation and dissemination for the internet of things. IEEE J. **5**, 955–969 (2018)
6. Shu, C., Zhao, Z., Min, G., Chen, S.: Mobile edge aided data dissemination for wireless healthcare systems. IEEE Trans. **6**, 898–906 (2019)
7. Umesh, B., Sudeep, T.: Secure Data Dissemination Techniques for IoT Applications: Research Challenges and Opportunities. Wiley publication, pp. 1–23 (2020)
8. Liu, Y., Yang, Y., Lv, X., Wang, L.: A self-learning sensor fault detection framework for industry monitoring IoT. Math. Probl. Eng. **2013**, 1–8 (2013)
9. Li, X., Ji, H., Li, Y.: Layered fault management scheme for end -to-end transmission in IoT. IEEE 1021–1025 (2011)
10. Hasan, M.Z., Al-Turjman, F.: Optimizing multipath routing with guaranteed fault tolerance in internet of things. IEEE J. **17**, 6463–6473 (2017)
11. Mudassar, M., Zhai, Y., Liao, L., Shen, J.: A decentralized latency-aware task allocation and group formation approach with fault tolerance for IoT applications. IEEE **8**, 49212–49223 (2020)
12. Power, A., Kotonya, G.: A microservices architecture for reactive and proactive fault tolerance in IoT systems. In: IEEE 19th International conference, pp. 588–599 (2018)
13. Zhou, S., Lin, K., Shih, C.: Device clustering for fault monitoring in Internet of Things systems. In: IEEE 2nd WF-IoT, pp. 228–233 (2015)

14. Kaur, P.D., Priya, K.: Fault tolerance techniques and architectures in cloud computing - a comparative analysis. In: IEEE International conference, pp. 1090–1095 (2015)
15. Grover, J., Garimella, R.M.: Reliable and fault-tolerant IoT-Edge architecture. In: 2018 IEEE SENSORS, pp. 1–4 (2018)
16. Sanket, T., Ashwini, P., Ashraf, S.: The Internet of Things – Emerging Technologies, Challenges and Applications. Int. J. Comput. Appl. **149**, 21–24 (2016)
17. Zhou, H.: The internet of things in the cloud. In: A Middleware Perspective. CRC Press, Taylor and Francis Group (2013) ISBN 978-1-4398-9302-9
18. Pan, Y., et al.: CDD: coordinating data dissemination in heterogeneous IoT networks. IEEE Commun. Mag. **58**, 84–89 (2020). https://doi.org/10.1109/MCOM.001.1900473
19. Cheraghlou, M.N., Khadem-Zadeh, A., Haghparast, M.: A survey of fault tolerance architecture in cloud computing. J. Netw. Comput. Appl. **61**, 81–92 (2016)
20. Ramakrishnan, J., Shabbir, M.S., Kassim, N.M., Nguyen, P.T., Mavaluru, D.: A comprehensive and systematic review of the network virtualization techniques in the IoT. Int. J. Commun. Syst. **33**(7), e4331 (2020). https://doi.org/10.1002/dac.4331
21. Ray, P.P.: A survey on Internet of Things architectures. J. King Saud Univ. – Comput. Inform. Sci. **30**(3), 291–319 (2018)
22. Liu, Y., Yang, Y., Lv, X.P., Wang, L.: A self-learning sensor fault detection framework for industry monitoring IoT. Math. Probl. Eng. **2013**, 1–8 (2013)
23. Madakam, S., Ramaswamy, R., Tripathi, S.: Internet of Things (IoT): a literature review. J. Comput. Commun. **3**, 164–173 (2015)
24. Shafique, K., Khawaja, B.A., Sabir, F., Qazi, S., Mustaqim, M.: Internet of Things (IoT) for next-generation smart systems: a review of current challenges, future trends and prospects for emerging 5G-IoT scenarios. IEEE Access **8**, 23022–23040 (2020)
25. Somayya Madakam, R., Ramaswamy, S.T.: Internet of Things (IoT): a literature review. J. Comput. Commun. **03**(05), 164–173 (2015)
26. Mahajan, A., Ahmed, S., Gupta, S., Suri, A.: Comparative analysis of different data dissemination strategies in wireless sensor networks. Int. J. Adv. Res. Comput. Commun. Eng. **3**(4), 6350–6354 (2014)
27. Hasan, M.Z., Al-Turjman, F.: Optimizing multipath routing with guaranteed fault tolerance in internet of things. IEEE Sensors J. **17**(19), 6463–6473 (2017)
28. Shu, C., Zhao, Z., Min, G., Chen, S.: Mobile edge aided data dissemination for wireless healthcare systems. IEEE Trans. Comput. Soc. Syst. **6**(5), 898–906 (2019)
29. Pal, Y., Awasthi, L.K., Singh, A.J.: Maximize the lifetime of object tracking sensor network with node-to-node activation scheme. In: 2009 IEEE International Advance Computing Conference (2009)
30. Hasan, M.Z., Al-Turjman, F.: SWARM-based data delivery framework in the Ad Hoc internet of things. In: GLOBECOM 2017 - 2017 IEEE Global Communications Conference (2017)
31. Bodkhe, U., Tanwar, S.: Secure data dissemination techniques for IoT applications: research challenges and opportunities. Software: Pract. Exper. **51**, 2469–2491 (2020)
32. Maiti, P., Sahoo, B., Turuk, A.K.: Service oriented fault monitoring in internet of things device management. In: 2016 International Conference on Signal Processing, Communication, Power and Embedded System (SCOPES), pp. 243-248. (2016)
33. Kaur, P.D., Priya, K.: Fault tolerance techniques and architectures in cloud computing - a comparative analysis. In: 2015 International Conference on Green Computing and Internet of Things (ICGCIoT) (2015)
34. Ghafoor, H., Koo, I.: CR-SDVN: a cognitive routing protocol for software-defined vehicular networks. IEEE Sensors J. **18**(4), 1761–1772 (2018)
35. Rajagopalan, R., Varshney, P.K.: Data-aggregation techniques in sensor networks: a survey. IEEE Commun. Surv. Tutorials **8**(4), 48–63 (2006)

Security and Privacy in Internet of Things

Md. Alimul Haque[1(✉)], Shameemul Haque[2], Kailash Kumar[3], Moidur Rahman[4], Deepa Sonal[5], and Nourah Almrezeq[6]

[1] Department of Computer Science, Veer Kunwar Singh University, Arrah 802301, India
shadvksu@gmail.com

[2] Al-Hafeez College, Arrah 802301, India

[3] College of Computing and Informatics, Saudi Electronic University, Riyadh, Kingdom of Saudi Arabia

[4] Department of Computer Science and Information Technology, Jazan University, Jizan, Kingdom of Saudi Arabia

[5] Department of Computer Science, Patna Women's College, Patna, India

[6] Deanship of the Common First Year, Jouf University, Sakakah, Kingdom of Saudi Arabia

Abstract. The Internet of Things (IoT) is a worldwide network of physically interconnected devices that communicate with one another over an online platform. The Internet of Things envisions the connectivity of a few billion and billions of smart things around each other, each completely unique and accessible on a regular basis. Those very same objects will be able to collect, process, and share information with themselves and their surroundings. Medical services, developing smart city projects with the latest Industry 5.0 systems, civil and military monitoring, and collection of data are some of the biggest examples of IoT technologies and their applications. Intelligent sensors and actuators, as well as RFIDs, have been developed recently, resulting in a huge number of wireless networks with smart and intelligent equipment (objects, or things) interconnected to the Internet continually transmitting data. As a result, providing privacy and security for all this data in the IoT is a highly difficult challenge that must be prioritised for numerous present and future applications. Poor equipment updates, a shortage of effective security measures, consumer low awareness, and well-known active device surveillance are just a few of the difficulties that IoT is struggling with. IoT security solutions prevent unauthorised access by securing unwanted modifications or damage. Privacy methods retain the right to restrict the acquired information that is available for utilisation and objective. This article presents a comprehensive survey of IoT systems' security and privacy with countermeasures.

Keywords: Security · Privacy · Internet of Things · Computing · Unauthorized access

1 Introduction

The Internet of Things (IoT) is a network of interconnected products and devices that may be connected to the Internet via a wired or wireless connection. The Internet of Things is gaining increasing prominence as these technologies are utilised for various

V. E. Balas et al. (Eds.): ICETCE 2022, CCIS 1591, pp. 182–196, 2022.
https://doi.org/10.1007/978-3-031-07012-9_16

purposes such as data communication, education, transmission, and economic expansion. There are currently around 26.66 billion Internet of Things devices in the world [1]. The rapid expansion of the Internet of Things has benefitted businesses in a variety of ways, including improved market research and commercial strategy. People's lives have also been enhanced by the introduction of automated services through the Internet of Things. However, such unchecked expansion has raised worries about privacy and security.

By providing hostile apps access to sensitive data in IoT systems, unintentional password usage, password failure, and device failure have all exacerbated cybersecurity problems. Security criteria are not properly stated, although numerous security solutions have been created to safeguard IoT devices from cyber assaults [2, 3]. Devices enabled with IoT have been used in many business contexts as well as in industrial settings [4]. Figure 1 presents the growth rate of Internet-connected equipment worldwide, and Fig. 2 predicts that by 2025, IoT is expected to be the leading application. These applications help these businesses obtain a competitive advantage. Due to the rapid deployment of numerous smart devices with data interchange and integration, companies are concerned about their privacy and data breaches, which interrupt the entire network activation. These threats should be addressed, and policies to secure the network must be kept in mind at the time of designing a framework to secure the company assets.

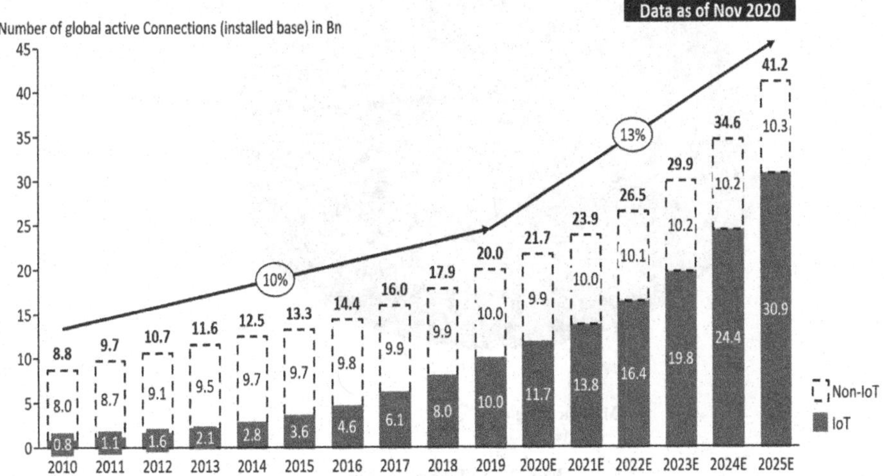

Fig. 1. Number of internet connected devices by 2025 [5].

Every day, new technologies are developed and existing ones are improved. Take a look at the most recent 5G network developments. IoT technology will undoubtedly play an important role in the 5G network. The short wavelength, on the other hand, necessitates infrastructure changes since it needs more base stations to cover the same area as previous wireless technologies. Additional risks, such as fake base stations, are introduced by the new structure. It is important to comprehend security dangers and potential remedies [6, 7].

We hope to offer an overview of IoT applications, advantages, and dangers in this paper. Furthermore, to provide a framework for studying and further improving optimal security practices, either by applying and evaluating present schemes or by inventing new ones. We offer suggestions for avoiding such risks and fixing security vulnerabilities. This study will assist regulatory authorities in continuing to enforce regulations, educate users and organizations, and help in developing and implementing more suitable security and privacy safeguards.

The paper is arranged accordingly: Sect. 2 provides the literature review with deep analysis. Section 3 explains security and privacy in the IoT. Section 4 addresses the classification of security attacks in IoT systems. Section 5 explores the vulnerabilities and countermeasures of these threats. Finally, the paper is concluded with future research challenges.

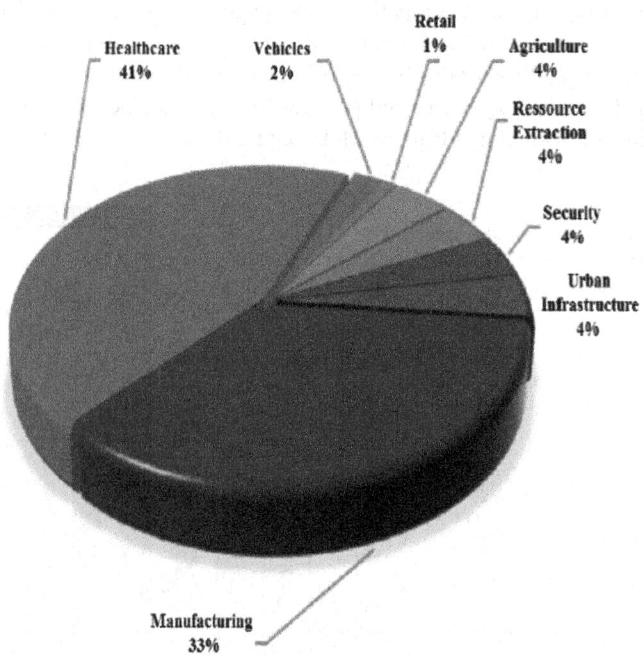

Fig. 2. IoT applications.

2 Literature Review

According to the authors of [8], a number of challenges, such as blocking and spoofing cyberattacks, as well as other unwanted access, have damaged the integrity of the user's data. A user may use a variety of solutions to help them implement various steps to safeguard their IoT devices. Many privacy risks have arisen in recent years, according to [9], and they have the potential to undermine IoT technologies and their related networks.

Interceptors and analyzers aid in the detection and investigation of a wide range of cyber threats [10]. Multiple platforms are available to test this security protocol, as well as numerous simulation tools, modelers, and other tools that can aid in the development of a unique IoT security protocol can also help. It's fair to say that IoT security research has advanced fast, and our work has benefited from a variety of simulation tools and modelers. The implications will be severe if the IoT devices fail. Despite the numerous advantages that the Internet of Things offers consumers, the authors of chapter [11] believe that there are challenges that need to be addressed. Two of the most critical themes mentioned are cybersecurity and privacy concerns. Businesses and government organizations are having a lot of trouble with these two. The vulnerabilities in IoT technology have been exposed by recent high-profile cybersecurity attacks. Novel security measures are necessary since the interconnectedness of networks in the Internet of Things allows access to the Internet from an anonymous and untrusted source. According to [12], ending a contract that involves many devices with various communication protocols is one of the most significant things to consider. Different protocols make it difficult to create separate service contracts, which are critical components of any Internet of Things cybersecurity architecture. Furthermore, the authors of [13, 14] have shown that scalability is a key criterion for the cybersecurity Internet of Things architecture's success. Experts believe that the Internet-of-Things ecosystem must be scalable enough to manage a billion Internet-related and cybersecurity issues. The authors of [15] highlighted some of the existing IoT cybersecurity solutions in a similar spirit. The company takes some basic security precautions and claims that providing high-quality solutions is not viable for them. Companies are unlikely to give the greatest Internet of Things cybersecurity solution. Furthermore, [16] describes today's smart phones and cyber-physical systems as pervasive, ranging from industrial control systems to contemporary vehicles to vital infrastructure. Through strong connectivity and effective utilization of new generations of embedded devices, current advancements and initiatives such as Industry 4.0 and the Internet of Things (IoT) enable new business models and user experiences. These systems generate, process, and distribute a lot of information. Attacks against the Internet of Things system, which are appealing targets because of security and confidentiality issues, cause bodily injury and disrupt people's lives. Because they can be a threat, cybersecurity and privacy are essential. Due to the complexity of these systems and the possible effect of cyber-attacks, linked industrial IoT systems are also facing additional threats. General security frameworks can be used to solve security and privacy issues in industrial IoT systems. The current state of IoT systems makes it impossible to offer the needed capabilities.

As a result, it's more important than ever to explore and examine numerous IoT security issues. One of the major goals of IoT security is to provide all users with privacy and confidentiality, as well as enhanced protection, infrastructure, and the assurance that various IoT ecosystem services will be available. As a result, with the aid of various modeling tools and a variety of computer platforms, research in varied IoT security is getting the necessary traction [17].

3 Security and Privacy in IoT

The Internet of Things has provided significant benefits to consumers, but it has also some technical issues and challenges. The major concerns of academics and security experts are cybersecurity and privacy issues. The vulnerabilities in IoT technology have been exposed by recent high-profile cybersecurity attacks. The interconnectedness of networks in the Internet of Things provides a vulnerability that allows access from the untrusted and anonymous Internet, necessitating innovative security measures [18, 19].

Security and privacy are the most significant known barriers to IoT adoption. There were several issues with today's automobile systems.

3.1 Security

Due to the huge amount of connectivity, the Internet of Things is more vulnerable to security threats [20]:

- IoT deployments often consist of a set of equipment that is similar or nearly equivalent in terms of characteristics. Because of this similarity, any security issue that affects a significant number of them has a greater impact.
- Several institutions have established risk assessment standards in a similar spirit. It goes without saying that many of these gadgets will form haphazard links and interact with one another. This demands a review of the existing IoT security technologies, methods, and tactics. Because IoT devices are so interconnected, one that isn't properly protected or linked has the potential to threaten global Internet security and resilience. This is only one sign of the problem that has arisen as a result of the widespread deployment of similar IoT devices. The Internet of Things [21] emphasises the necessity of collaborating to find a solution that is both effective and suitable. Figure 3 presents the matrix of security issues in the IoT.

3.2 Privacy

The Internet of Things' efficacy is influenced by how well it respects people's privacy concerns. Issues related to privacy and other risks linked to the IoT may play a significant role in delaying the IoT's successful adoption. It is vital to recognize that customer trust and faith in the Internet of Things, smart devices, and associated services are founded on respect for user privacy and privacy rights. Substantial work is being done to ensure that the Internet of Things (IoT) redefines privacy issues such as eavesdropping and spying [22].

3.3 Interpretability

A fragmented ecosystem of proprietary IoT technology implementations has been identified as a hindrance to users' value. Poorly constructed IoT devices might harm the networking resources to which they connect [23].

Security characteristic	Device/Hardware	Network	Cloud/Server-side
.1. **Confidentiality**	Hardware attacks	Encryption with low capability devices	Privacy data leaks fingerprinting
2. **Integrity**	Spoofing; Lack of attestation	Signatures with low capability devices Sybil attacks	No common device identity
3. **Availability**	Physical attacks	Unreliable networks, DDoS, radio jamming	DDoS (as usual)
4. **Authentication**	Lack of UI, default passwords, hardware secret retrieval	Default passwords, lack of secure identities	No common device identity, insecure flows
5. **Access Control**	Physical access; Lack of local authentication	Lightweight distributed protocols for access control	Inappropriate use of traditional ACLs, device shadow
6. **Non-Repudiation**	No secure local storage; No attestation, forgery	Lack of signatures with low capability devices	Lack of secure identity and signatures

Fig. 3. Matrix of security issues in the IoT

Another important component is cryptography, which has been used in numerous applications for many years to protect against security flaws. It is difficult to construct an effective defensive system against attacks with only one security application. As a result, combating threats to IoT authentication necessitates many levels of protection.

Hacking might be avoided if more sophisticated security measures were developed and integrated into products. Customers would buy items that already have enough security features in place to protect against flaws. So this evasion happens. Cybersecurity frameworks are one of the methods offered to assure IoT security [24].

Furthermore, several elements and concerns may influence attempts to protect Internet of Things devices, such as:

- Periodic updates: IoT makers generally release security fixes every three months. Similar upgrades are made to OS versions and security patches [25]. As a consequence, cybercriminals have plenty of time to exploit security weaknesses and steal sensitive information.
- Support personnel can use embedded passwords to debug OS problems or remotely install needed upgrades on IoT devices. Hackers, on the other hand, might exploit the functionality to undermine device security.
- Remote access: Wi-Fi, Z-Wave, and ZigBee are some of the network protocols used by IoT devices for remote access. Typically, precise limitations on what can be utilised to deter hackers are not stated. As a result, using these remote access protocols, hackers may easily establish a hostile connection [26, 27].
- A wide range of third-party apps: There are several software programs accessible on the Internet that companies may employ to execute certain tasks. However, it was difficult to determine the legitimacy of these applications.
- Inadequate Device Monitoring: All IoT suppliers generally create unique device IDs to monitor and trace devices. On the other hand, some manufacturers do not follow

security requirements. As a result, it may be difficult to keep track of problematic Internet behavior.

4 Classification of Security Attacks in IoT System

This section attempts to gather some information concerning attacks and vulnerabilities on IoT systems. Assaults on the Internet of Things system might come from within the system's network, via the physical layer, from the system's applications, and, finally, from attacks on the IoT network's encryption system methods. As we all know, IoT is implemented through the use of current technologies such as WSN, RFIDs, sensors, actuators, the internet, and so on. As a result, it is necessary to classify threats/attacks on IoT systems in a way that encompasses all the threats/attacks so that countermeasures may be created and deployed to mitigate these attacks/threats on these IoT systems..

4.1 Physical Attacks

Physical assaults target IoT systems' physical or perceptual layers. For these assaults to succeed, the attacker must be isolated from the IoT system. This category includes assaults that endanger the hardware of IoT systems by reducing their lifespan and usefulness [11]. Figure 4 presents attacks and security approach in Physical layers.

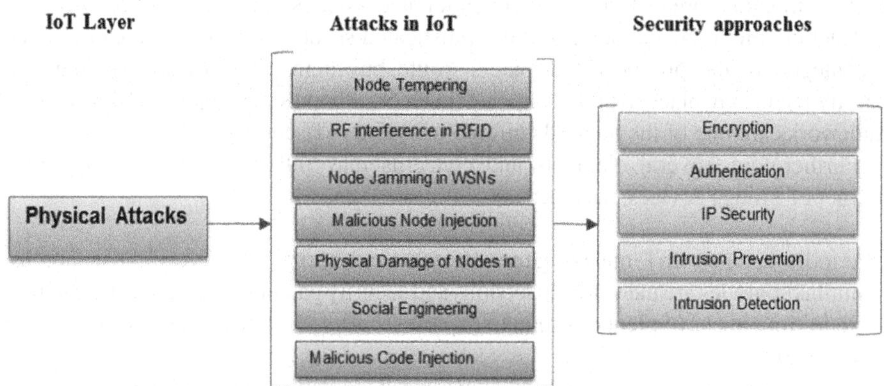

Fig. 4. Physical layers Attacks and Security approaches

Node Tempering
"An attacker can damage a sensor node by physically altering it or a piece of it, to obtain access to and modify sensitive information like shared cryptographic keys (if any) or routing tables [11]."

RF Interference in RFID
By generating and transmitting noise signals across the Radio Frequency signals used by RFIDs for communication, a "Denial of Service" attack against any RFID tag may be carried out. Noise will interrupt RFID signals, making communication impossible [11].

Node Jamming in WSNs
"This attack is similar to the physical Radio Frequency Interference attack for RFIDs mentioned earlier, except it uses WSNs instead. The attacker can jam the signals and impede connection by interfering with the wireless sensor nodes' radiofrequency. If an attacker succeeds in jamming important sensor nodes, he will be able to quickly disable IoT services [11]."

Malicious Node Injection
Between the other IoT nodes in the IoT system, an attacker can insert a malicious node. As a result, the attacker can use this rogue node to take control of the other nodes, as well as the information flow between them [11].

Physical Damage of Nodes in IoT System
For personal benefit, the attacker can physically destroy the nodes/sensors in the IoT systems. The security of the building and the region where the physical sensor nodes are installed are addressed in this area of IoT System security.

Social Engineering
"The attacker uses an IoT device to trick people into giving up personal information or doing actions that will help him achieve his objectives. Because the attacker must establish physical contact with IoT network users to fulfill his objectives, this type of attack is classified as a physical attack".

Malicious Code Injection
"Consider an attacker physically introducing malicious code into a node to get access to the IoT system. It means an attacker can take total control of the terminal or may be complete network".

4.2 Network Attacks

The focus of assaults is on IoT network systems. In these types of assaults, the attacker utilizes the internet and a radio frequency network to target IoT systems. Figure 5 presents attacks and security approach in Network layers.

Traffic Analysis Attacks
Because RFID technology is wireless, an attacker may take out critical information or any other data flowing through it.

RFID Spoofing

"To access and collect data broadcast by an RFID tag, an attacker spoofs RFID signals. It can then transmit his data with the original tag ID, mimicking the original source and gaining total access to the system".

RFID Cloning

An attacker copies data from the victim's RFID tag and pastes it into different RFID tags to create a clone. Even though the two RFID tags contain similar data, unlike the RFID spoofing attack, this technique does not duplicate the RFID's original ID, allowing users to identify between the original and the compromised.

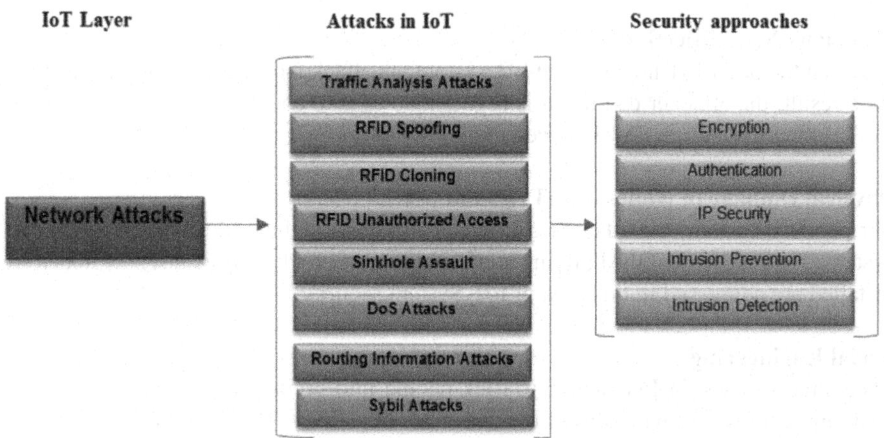

Fig. 5. Network layers Attacks and Security approaches

RFID Unauthorized Access

RFID Unauthorized Access: "Due to a lack of sufficient authentication mechanisms in the majority of RFID systems, tags can be accessed by anybody. As a result, the attacker gets access to the RFID nodes' data and may read, alter, and delete it".

Sinkhole Assault

In this sort of attack, the attacker builds a sinkhole in the network, drawing all IoT and WSN node traffic to a single location. In the IoT network, this sort of attack causes data forwarding secrecy issues as well as denial-of-service issues.

DoS Attacks

A Denial of Service (DoS) attack occurs when an attacker floods the IoT network with junk data or information. As a result, the IoT system is jammed, resulting in a DoS situation in the network [11].

Routing Information Attacks
"Hello Attack and Black hole attack are examples of direct attacks in which the adversary spoofs, modifies, or replays routing information to confuse the network and create routing loops, allow or drop traffic, send false error messages, shorten or extend source routes, or even partition the network".

Sybil Attacks
"A malicious node is a single node that takes on the identities of several other nodes. Nearby WSN nodes accept fake data as a result of this type of assault".

4.3 Software Attacks

IoT systems, like any other computerized system, are subject to software assaults. The attackers use the application layer of IoT systems to develop malicious software applications that cause mayhem. Figure 6 presents attacks and security approach in Software layers

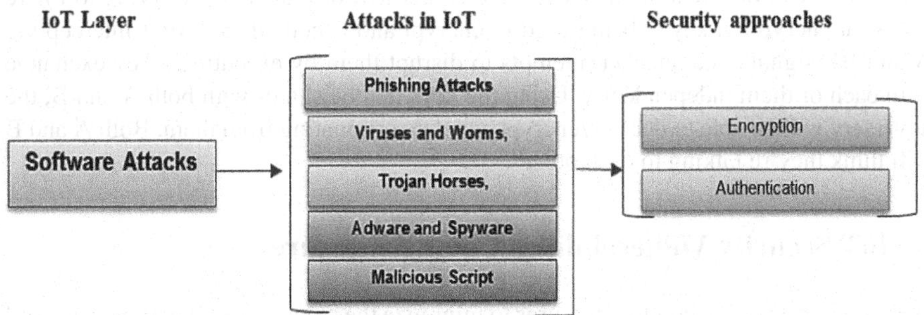

Fig. 6. Software layers Attacks and Security approaches

Phishing Attacks
In this attack, the attacker spoofs the credentials of the machines or users in the IoT network, generally using phishing messages, to obtain access to confidential data in the IoT network.

Viruses, Worms, Trojan Horses, Adware, and Spyware
These malicious software programs may be installed in the IoT system via the application layer, and then the attacker can use them to do various tasks such as DoS, DDoS, data tampering, data integrity and security, and so on.

Malicious Script
Once the IoT system is linked to the network, a malicious script may be executed. A malicious script, such as Active-X, can be fooled into executing on the computer managing the gateway. This might lead to data theft, manipulation, and other problems.

4.4 Encryption Attacks

It is an attack against IoT devices that is primarily centered on cracking the encryption mechanism in place. Figure 7 presents attacks and security approach in Encryption layers.

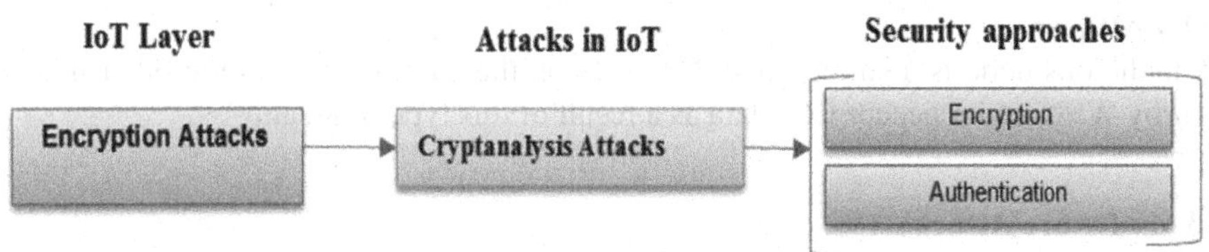

Fig. 7. Encryption Attacks and Security approaches

Cryptanalysis Attacks
These attacks work by assuming you have cipher text or plaintext and trying to figure out what encryption key is being used to encrypt and decode data. "After intercepting A and B's signals, the attacker attempts to disrupt them by executing a key exchange with each of them independently. Using the keys that he shares with both A and B, the adversary will be able to decrypt/encrypt any data originating from them. Both A and B will think they're talking to each other".

5 IoT Security Vulnerabilities Countermeasures

We'll demonstrate the countermeasures to minimize the risks we just mentioned, as well as the security needs for any IoT system.

5.1 Hashed Based Encryption

The facility of encryption provided by hash-based encryption security is cipher text, which converts a message into an anonymous form. When a message is transmitted from a sender, it is transformed into a different format using a key that only legitimate users can decipher. The length of the message is used to produce a key. It has a double-length key from the message at all times. As a result, breaking a key is a difficult process. The recipient receives the key as well. The recipient to transform the encrypted text into an original message can use the key.

5.2 PKI Protocol

A Public Key Infrastructure (PKI)-like protocol mechanism combines all of the processes outlined above, including authorization, authentication, and intrusion detection, and is implemented in the IoT architecture's cognitive layer. It is preferable to employ several methods separately. Many nodes are linked together to form a network. It is responsible for ensuring security. As a result, it has no faith in anybody sending a message. The public key and private key are both encrypted using the RSA method. The public key is kept at the base station, whereas the base station to each node provides the private key [25].

5.3 Security Mechanisms for NL

To defend the NL from a hacker assault, the following preventive steps should be considered:

Identity Management Framework
For each device to communicate with each other, authentication is required. It allows users to verify the validity of devices before sharing data. The identity management framework is a solution developed to meet this need. The environment, sensors, and receiver, as well as the network, make up the system. The environment refers to the area where sensors and things are linked. The sensors are in charge of gathering data from things. The network, which is utilized to transport information, is the last component. Sensors send information to computers, while computers send information to receivers [17].

SDN with IoT
In many application domains such as business, smart homes, and e-health systems, software-defined networking is the current network security management trend. The data plane runs on hardware, while the control plane runs on software, and both are logically centralized. SDN can keep track of network traffic and identify malicious activity. It locates the infected nodes and separates them from the rest of the network.

Cooperation of Node Communication Protocols
Security-conscious ad-hoc can protect against attacks from within the network; the goal of the proposed protocol is to identify nodes that can disrupt the entire network due to their misbehavior. The monitor, the reputation system, the path manager, and the trust manager are the four components. If any node in the network exhibits questionable activity, the node's information and suspicious behavior are transmitted to a trust manager. A trust manager sends out an ALERT message to notify all other nodes in its range. The ALERT message contains information on the address of a reporting node, the attacker node's address, and packet loss. Through a reputation system, the node that gets an ALERT message analysis whether the reporting node is authorized or not, as well as packet loss. If the information given is valid, the route manager creates a new path from the source node to the destination node to ensure consistency.

Security Mechanisms for AL
To defend the application layer from malicious attacks, use the following security measures.

Special Policies and Permissions
For accessing and controlling the IoT framework, specific regulations and permissions must be followed. Policies, like XACML terms, are made up of a collection of rules. They're made up of an effect (for example, allow/deny) and a condition. Access control lists can allow or disallow outgoing/incoming traffic as well as system access requests.

Anti-virus, Anti-adware and Anti-spyware
All of these tools can ensure the IoT territory's security, consistency, secrecy, and dependability.

Risk Assessment Techniques
These techniques detect vulnerabilities to the IoT network; therefore risk assessment should be used to safeguard the application layer. Update the firmware of the system devices to strengthen security measures in this case.

6 Research Challenges

This study aims to provide a comprehensive overview of the various security issues that arise on the Internet of Things. Although many studies have been published, their findings are still not complete. Some critical issues regarding security and privacy in IoT will be investigated and solved in future research, such as:

- Some of the factors that need to be considered when designing an IDS-capable network are the complexity of the factors that need to be considered when designing an IDS-capable network.
- Due to the nature of the connectivity and processing requirements of Internet of Things devices, they require low power to operate efficiently. The use of light weight encryption algorithms is necessary to secure these devices.

7 Conclusion

The internet has transformed our lifestyles, electronically changing people's relationships in a variety of situations ranging from intelligent life to social bonds. With the development of IoT devices in various corporate domains and in people's lives, security and privacy concerns are rising. As a result of resource constraints, a wide range of vulnerabilities have emerged. The majority of such issues may result in framework dissatisfaction in the IoT industry. Additionally, this study critically examines contemporary research on IoT security and privacy problems. The proposed approach examined in this article gives a cutting-edge overview of existing IoT cybersecurity threats. This article attempts to provide a thorough overview of IoT system security concerns, vulnerabilities, and solutions. It is hoped that this article will serve as a wake-up call to future IoT security scholars and researchers to perform appropriate research in order to minimise the necessary IoT security and privacy concerns.

Funding. This research received no external funding.

Conflict of Interest. The authors declare that they have no conflict of interest.

References

1. Tawalbeh, L., Muheidat, F., Tawalbeh, M., Quwaider, M.: IoT privacy and security: challenges and solutions. Appl. Sci. **10**(12), 4102 (2020)
2. Conti, M., Dehghantanha, A., Franke, K., Watson, S.: Internet of Things security and forensics: challenges and opportunities (2018)
3. Aldwairi, M., Tawalbeh, L.: Security techniques for intelligent spam sensing and anomaly detection in online social platforms. Int. J. Electr. Comput. Eng. **10**(1), 275 (2020)
4. Makhdoom, I., Abolhasan, M., Lipman, J., Liu, R.P., Ni, W.: Anatomy of threats to the Internet of Things. IEEE Commun. Surv. Tutor. **21**(2), 1636–1675 (2018)
5. Lueth, K.L.: State of the IoT 2020: 12 billion IoT connections, surpassing non-IoT for the first time, November 2020
6. Haque, M.A., Faizanuddin, M., Singh, N.K.: A study of cognitive wireless sensor networks: taxonomy of attacks and countermeasures (2012)
7. Haque, M.A., Amola, Y., Singh, N.K.: Threat analysis and guidelines for secure WiFi and WiMAX network (2011)
8. Meng, Y., Zhang, W., Zhu, H., Shen, X.S.: Securing consumer IoT in the smart home: architecture, challenges, and countermeasures. IEEE Wirel. Commun. **25**(6), 53–59 (2018)
9. Siby, S., Maiti, R.R., Tippenhauer, N.O.: IoTScanner: detecting privacy threats in IoT neighborhoods. In: Proceedings of the 3rd ACM International Workshop on IoT Privacy, Trust, and Security, pp. 23–30 (2017)
10. Hassan, W.H.: Current research on Internet of Things (IoT) security: a survey. Comput. Netw. **148**, 283–294 (2019)
11. Leloglu, E.: A review of security concerns in Internet of Things. J. Comput. Commun. **5**(1), 121–136 (2016)
12. Haque, A., Sinha, A.K., Singh, K.M., Sing, N.K.: Security issues of wireless communication networks. IJECCE **5**(5), 1191–1196 (2014)
13. Haque, M.A., Bokhari, M.U., Sinha, A.K., Singh, N.K.: Comparative study on wireless threats and their classification. In: INDIACom-2017; IEEE Conference ID: 40353 2017 4th International Conference on "Computing for Sustainable Global Development", 01st–03rd March 2017, pp. 5057–5059. BVICAM (2017)
14. Ali, S., Bosche, A., Ford, F.: Cybersecurity is the Key to Unlocking Demand in the Internet of Things. Bain Co., Boston (2018)
15. Sadeghi, A.-R., Wachsmann, C., Waidner, M.: Security and privacy challenges in industrial Internet of Things. In: 2015 52nd ACM/EDAC/IEEE Design Automation Conference (DAC), pp. 1–6 (2015)
16. Alsmadi, I., Easttom, C.: The NICE Cyber Security Framework. Springer, Heidelberg (2020)
17. Haque, M.A., Sonal, D., Haque, S., Nezami, M.M., Kumar, K.: An IoT-based model for defending against the novel coronavirus (COVID-19) outbreak. Solid State Technol. 592–600 (2020)
18. Haque, M.A., Haque, S., Sonal, D., Kumar, K., Shakeb, E.: Security enhancement for IoT enabled agriculture. Mater. Today Proc. (2021). https://doi.org/10.1016/j.matpr.2020.12.452
19. Alaba, F.A., Othman, M., Hashem, I.A.T., Alotaibi, F.: Internet of Things security: a survey. J. Netw. Comput. Appl. **88**, 10–28 (2017)
20. Haque, S., Zeba, S., Alimul Haque, M., Kumar, K., Ali Basha, M.P.: An IoT model for securing examinations from malpractices. Mater. Today Proc. (2021). https://doi.org/10.1016/j.matpr.2021.03.413
21. Sharma, G., et al.: Reverse engineering for potential malware detection: Android APK Smali to Java. J. Inf. Assur. Secur. **15**(1), 26–34 (2020)

22. Zaldivar, D., Lo'ai, A.T., Muheidat, F.: Investigating the security threats on networked medical devices. In: 2020 10th Annual Computing and Communication Workshop and Conference (CCWC), pp. 488–493 (2020)
23. Lo'ai, A.T., Somani, T.F.: More secure Internet of Things using robust encryption algorithms against side channel attacks. In: 2016 IEEE/ACS 13th International Conference of Computer Systems and Applications (AICCSA), pp. 1–6 (2016)
24. Dalipi, F., Yayilgan, S.Y.: Security and privacy considerations for IoT application on smart grids: survey and research challenges. In: 2016 IEEE 4th International Conference on Future Internet of Things and Cloud Workshops (FiCloudW), pp. 63–68 (2016)
25. Sharma, S., Mahrishi, M.: Implementation of trust model on CloudSim based on service parametric model. In: 2015 IEEE International Conference on Research in Computational Intelligence and Communication Networks (ICRCICN), pp. 351–356 (2015). https://doi.org/10.1109/ICRCICN.2015.7434263
26. Haque, M.A., Haque, S., Kumar, K., Singh, N.K.: A comprehensive study of cyber security attacks, classification, and countermeasures in the Internet of Things. In: Digital Transformation and Challenges to Data Security and Privacy, pp. 63–90. IGI Global (2021)

Implementation of Energy Efficient Artificial Intelligence-Based Health Monitoring and Emergency Prediction System Using IoT: Mediating Effect of Entrepreneurial Orientation

Mintu Debnath[1], Joel Alanya-Beltran[2], Sudakshina Chakrabarti[3](✉), Vinay Kumar Yadav[4], Shanjida Chowdhury[5], and Sushma Jaiswal[6]

[1] Department of Physics, Chakdaha College, Chakdaha, West Bengal, India
[2] Electronic Department, Universidad Tecnológica del Perú, Lima, Peru
C18121@utp.edu.pe
[3] Chettinad Hospital and Research Institute, Chettinad Academy of Research and Education, Chennai, Tamil Nadu, India
sudhianat@gmail.com
[4] Department of Business Management, Indira Gandhi National Tribal University, Amarkantak, Madhya Pradesh, India
[5] Department of General Educational Development, Daffodil International University, Dhaka, Bangladesh
shan_chydiu.ged@daffodilvarsity.edu.bd
[6] Department of Computer Science and Information Technology (CSIT), Guru Ghasidas Vishwavidyalaya, (A Central University), Koni, Bilaspur, CG, India

Abstract. The healthcare industry is developing rapidly, and innovations are now considered as the significant game-changer. IoT (Internet of Things) is shaping the healthcare industry in a new form with promising advances in testing, monitoring processes. Monitoring the health issues of the patients, organizing the treatment initiatives, and empowering the physicians it is providing superlative measures. The invention of the IoT through internet based artificial intelligence is determining the bright future of the medical field. Whether IoT is diagnosing the disease, or analyzing the past history of a certain disease the implementation of artificial intelligence is great. Here in this study the roles of internet based artificial intelligence are illustrated. Furthermore, it has described the current working features in health monitoring. Key aim of this study is to analyze this new innovative implementation in health monitoring. The article is developed including *secondary qualitative analysis.* Data collection, diagnosing health issues, and in monitoring the preventive care of IoT is compared with the traditional way of heath monitoring. Many experts see that artificial intelligence is more able than the conventional method to work in a more organized way. This study targets to analyze both the advantages, and disadvantages of implementation of artificial intelligence. Various components are addressed along with the gap to predict the increasing use of it in the near future. Comparing with the traditional; ways in giving better service experience is discussed. Including both the gaps, and benefits this study would be beneficial to give a better and effective understanding about the chosen topic.

V. E. Balas et al. (Eds.): ICETCE 2022, CCIS 1591, pp. 197–204, 2022.
https://doi.org/10.1007/978-3-031-07012-9_17

Keywords: Internet of Things (IoT) · Health monitoring · Artificial intelligence · Entrepreneurial orientation · Internet of Medical Things (IoMT)

1 Introduction

The Internet of things (IoT) is the developing technology that improves every individual in using the systems of wireless communication. Several health care systems are based on Artificial intelligence and these systems are monitored by the Internet of Medical Things (IoMT). It is used for analyzing and collecting the data of the health sectors with the least operation cost. The most significant challenge in this application is storing the data of the health care on the secured cloud for and it makes the system of disease diagnosis more energy efficient. Moreover, there is a rapid discharge of the IoMT technology that has involved several cyber-criminals and the continuous endeavor to compromise the medical devices by generating fake certificates and information loss. Therefore there is an increase in the modern and latest technologies for the health applications that are based on IoMT. These technologies offer trusted communication and secure health data against trespassers is gaining more attention.

The IoT plays a significant role in gathering information with the help of sensors and transmits the data of the patients to the end-users [1]. The aim of this research is to analyze the AI based health monitoring and the emergency prediction system with the help of IoT. Entrepreneurial Orientation (EO) also includes innovation, proactiveness and risk taking approach of a system that helps in making strategic decisions effectively. In the present context EO is prevalent in the implication of IoT in health monitoring practices.

1.1 Benefits of Energy Efficient AI Model for Health Monitoring

AI based Internet of Things (IoT) plays an important role in health monitoring and helps in gathering information with the help of sensors that are battery-powered. It, therefore, transmits the data of the patients to the end-users [2]. IoT has participated with the computing devices for the diagnosis of several infectious diseases and public health issues like malaria, typhoid, blood pressure, and many others. The biosensor collects the data related to health from the surgical devices that are implanted, mobile devices, wearable bands, and many other devices. Moreover, these biosensors are limited in terms of power transmission, processing units, storage, and battery powers and these biosensors check the patient parameters such as heartbeat, pulse rate and blood pressure. Monitoring the health issues of the patients, organizing the treatment initiatives, and empowering the physicians it is providing superlative measures. The invention of the IoT through internet based artificial intelligence is determining the bright future of the medical field. Therefore, the efficient utilization of these devices is one of the most common factors for the improvement of the medical system [3]. Many solutions are available regarding the problems of the health system that are presented for analyzing the medical symptoms by using the constraint-oriented biosensors of the Internet of Medical Things (IoMT). These sensors support several services in emergency cases and the medical biosensors are communicated directly to the local coordinator and collaborate with the sink nodes for sending the health data to the public cloud [4]. In corresponding to EO, innovation in medical practices makes the entire process more fast and accurate (Fig. 1).

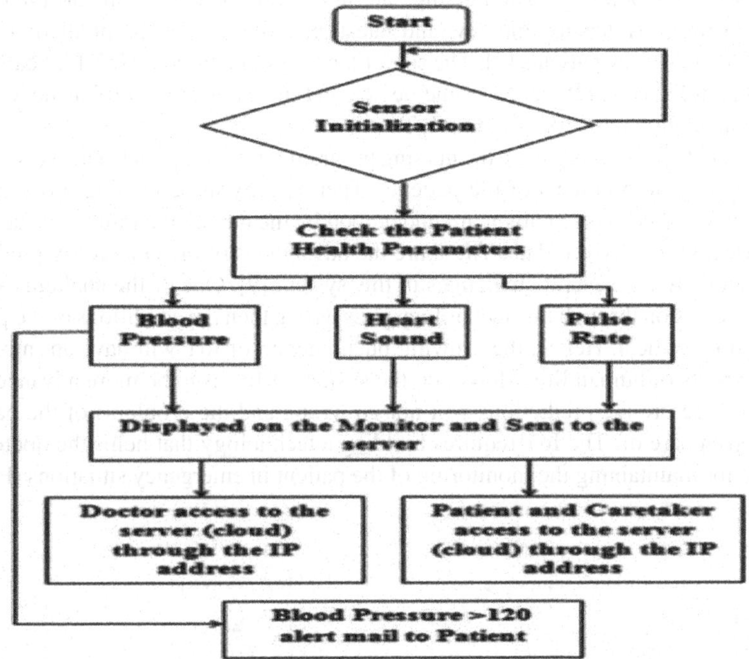

Fig. 1. Flow chart of Health Monitoring System of a patient (Source: [4])

Nowadays, the health monitoring of the costly sensors of IoT, vehicles, and the data analytic is used that requires special training of a technician for elaborating on the problem of the machine. Moreover in the era of IoT technology innovations, through sensors, the data is collected and then transmitted to the main server. It is done for reducing the effect of the requirement for the expert by developing AI diagnostics [5]. The technology of IoT and the requirement for storage, transmission, and processing bid data are the subjects of the research efforts for increasing improvement, further optimization, and high efficiency in cloud computing. The development of the efficient system of Health Monitoring and Diagnosis (HM&D) is done in the current scenario. Big data and IoT are the two sides of the system that decides the efficiency and smooth flow of the data.

2 Impact of IoT on Emergency Prediction System

IoT is rapidly increasing the interest nowadays in the health care system, particularly in personalized healthcare. For example, the body area sensor network (BASN) is broadly applied under the system of IoT for monitoring universal health. The system of ECG testing is broadly accepted as the critical method for diagnosing cardiovascular diseases [6]. The IoT has a novel system named WISE that is wearable to cloud and AI based health monitoring systems for the real time monitoring of personal healthcare. For improving the quality of the patient and care safety, the monitoring system of health care is used for promoting basic nursing care in the environment of the hospitals. The medical institutions develop the new and latest technologies for the prediction of the disease for emergency patients [7]. The surveillance system based on IoT for health care achieves simultaneous efficiency of the device and the transmission robustness within the communication network of public IoT.

Medical institutions organize the nursing program for tracking each and every patient and predicting the problems of the patients. Therefore by these nursing programs the critical conditions of every patient are understood by the nurses and immediate action is taken based on the derived data. These are the advantages of the emergency prediction system but there are several challenges in this system [9]. One of the challenges is, in case the prediction through this technology goes wrong then the conditions of the patient may be more critical. Hence, the growing phenomenon of IoT will have an impact on several aspects of human life. Moreover, the system of IoT is to be more advanced and new, and therefore the predictions will not go wrong and the problems of the patients will be taken care of. The IoT requires building a technology that helps the doctor and the nurse for maintaining the monitoring of the patient in emergency situations (Fig. 2).

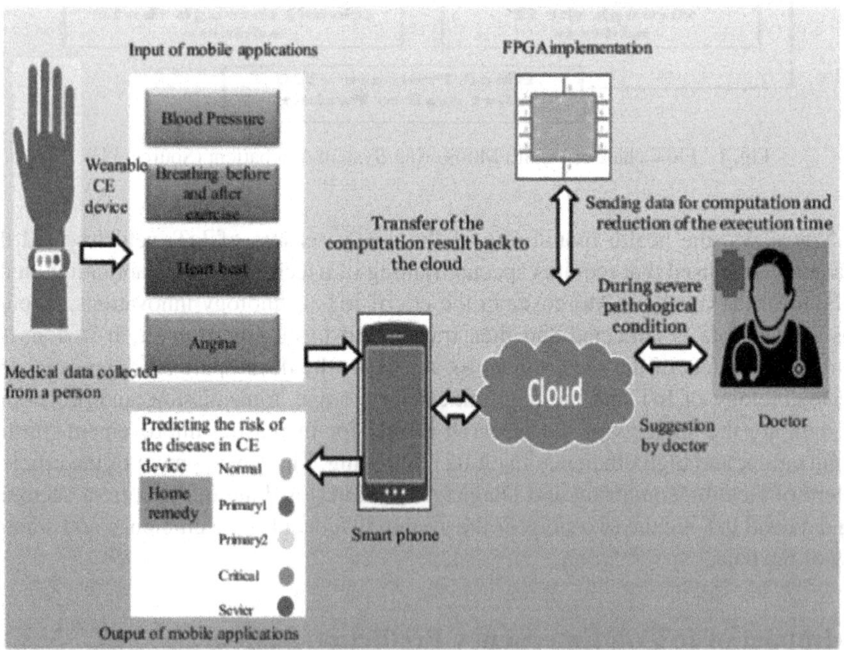

Fig. 2. Flow process of IoT based system of smart health care (Source: [10])

The analysis of the IoT based system of health care is proposed to be used for designing consumer electronic devices. The process of this system is to collect the medical data from the person through the sensors and then the risk of that person is predicted and the data is sent through a smartphone to the cloud [10]. Moreover, the transfer of the data of computation result back to the cloud again through FPGA implementation for reducing the time of implementation and the doctor predicts the disease.

3 Methods and Materials

The reliable and effective data analysis and data collection methods play an important role in this research. This research conducts the ***Secondary Qualitative data*** collection method for collecting information about the AI based Health monitoring and emergency prediction system by using the technology of IoT. The authentic data is taken from the secondary resources like articles, newspaper journals, different websites, and graphs. Moreover, the method of secondary data collection helps the researcher to collect reliable data regarding the Entrepreneurial Orientation of IoT technology in health monitoring and the system of emergency prediction.

4 Findings and Discussions

The AI based health care resources are having slower medical systems and these resources are having lower response time from the medical experts and it is one of the challenges in this IoT technology. The medical systems are most sensitive for compromising the data and the legal attempts on the medical data and it gains adequate interest from the advanced technology. The proposed AI balances the heuristic decisions of the energy loads and decreases the communication overheads by communicating the smart technologies of healthcare. It offers cognitive performance over the network of the public and it transmits the reputation of the systems of medical detection [11]. Therefore, it establishes trust among the medical sensors and avoids unauthorized access to the medical data (Table 1).

Table 1. Revenue of AI systems in health sectors (Source: [11])

Revenue in millions	Years
1438.4	2017
2002.7	2018
2882	2019
4298.2	2020
6662.2	2021

The system of healthcare monitoring in the hospitals and several health centres has an experience on the significant growth and systems of portable healthcare monitoring

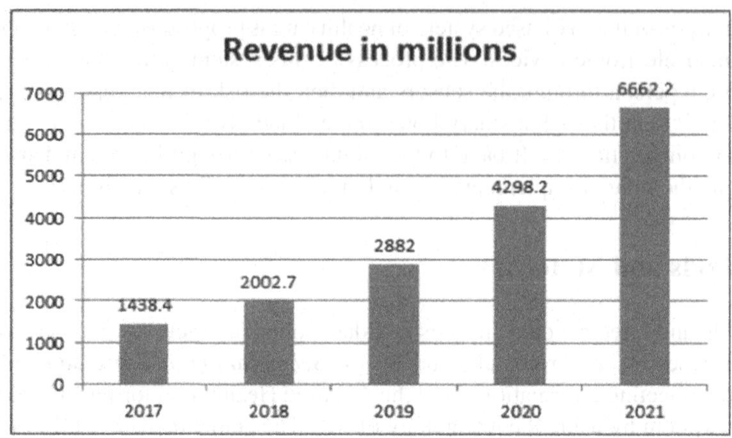

Fig. 3. Revenue of AI systems in health sectors (Source: [11])

with all the emerging technologies. These technologies are becoming a great concern to several countries nowadays. IoT technologies facilitate the actual progress of healthcare from consulting face to face to telemedicine [12]. The smart system of healthcare in the environment of IoT monitors the basic health signs of the patient and also the conditions of the rooms where the patients will stay. For this, the healthcare system has its own expenditure worldwide. Given in the above graph in Fig. 3 is the revenues of the AI system in health care globally, it is seen that every year the revenue is increasing day by day as the advanced technology is growing. The healthcare system is coming up with new and latest AI based technology for monitoring the patients. This system consists of five different types of sensors that are capturing the data from the environment of the hospital [13]. The sensors are capturing the body temperature, room temperature, heartbeat, Co level, carbon dioxide level. The conditions of all the patients are conveyed via a portal to the medical staff and the staffs then analyze and process the current situation and condition of the patient (Table 2).

Table 2. Predictions on the impact of AI in healthcare systems (Source: [13])

Predictions of AI in Healthcare	Better	Worse	No impact
AI programs will be capable of providing the real healthcare services	67%	17%	16%
AI programs will be capable of an accurate diagnosis for several diseases	69%	14%	16%
Medical appointments are booked through the AI	54%	16%	29%

IoT is making the objects that are internally connected in the current decade and it is considered as the upcoming technological revolution. The monitoring of the health care systems is one of the applications of IoT and this application has a tremendous use

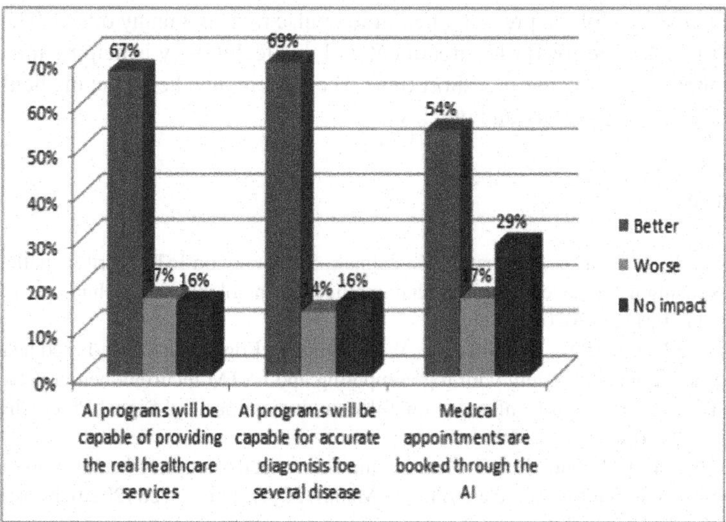

Fig. 4. Predictions on the impact of AI in healthcare systems (Source: Developed based on [13])

of IoT in the healthcare management that provides the tracking facilities to environment and health conditions. The above Fig. 4 is given the predictions on the impact of AI in the healthcare systems and this figure analyses the better technology predictions of AI in the healthcare sectors. The graph also states that the lids can be improved by the AI programs for providing the healthcare services like the robot assisted surgeries, early diagnosis of the patients, and many others [14]. The IoT is linking the computers to the internet utilizing the networks and the sensors. These components that are connected are used on the devices for monitoring the heath. The sensors are then used for forwarding the information to distant locations. It is an energy efficient, simple, scalable, interoperable, and smarter way of optimizing and tracking the care of any health problems. Nowadays, modern systems are providing a flexible interface, management of mental health, and assistant devices for leading the smart life of human beings.

5 Conclusion

The system of IoT in the healthcare systems is to monitor the important signs of the patients like the blood pressure, body temperatures, heart rate, and many other measures. The condition of the room like level of carbon dioxide, CO, room humidity is also taken care of by this technology. This system is beneficial for the doctors and nurses in the situations of crises or the epidemics for analyzing the medical data easily in a short time. The system of IoT is too useful in case of various critical conditions and the developed system improves the system of current health care that protects several lives from death. The system looks bulky but it is a tiny process for manufacturing and upgrading the technology for the future. The video feature might be added for consulting face to face between the patients and the doctors. Many features that can be significant for determining the condition and monitoring the diseases of the patients like the level of respiration, diabetes, blood pressure, heart rate, pulse rate, and many others. The data of the patient is collected for further treatment and this technology is helping the medical staff to complete the process in a short time. The system also helps the medical staff in taking necessary steps easily and quickly.

References

1. Mohanta, B.K., Jena, D., Satapathy, U., Patnaik, S.: Survey on IoT security: challenges and solution using machine learning, artificial intelligence and blockchain technology. Internet of Things **11**, 100227 (2020)
2. Queralta, J.P., Gia, T.N., Tenhunen, H., Westerlund, T.: Edge-AI in LoRa-based health monitoring: fall detection system with fog computing and LSTM recurrent neural networks. In: 2019 42nd International Conference on Telecommunications and Signal Processing (TSP), pp. 601–604. IEEE, July 2019
3. Gupta, N., et al.: Economic IoT strategy: the future technology for health monitoring and diagnostic of agriculture vehicles. J. Intell. Manuf. **32**(4), 1117–1128 (2020). https://doi.org/10.1007/s10845-020-01610-0
4. Ghazal, T.M., et al.: IoT for smart cities: machine learning approaches in smart healthcare—a review. Future Internet **13**(8), 218 (2021)
5. Popli, S., Jha, R.K., Jain, S.: A survey on energy efficient narrowband internet of things (NBIoT): architecture, application and challenges. IEEE Access **7**, 16739–16776 (2018)
6. Huang, D., Wang, S., Liu, Z.: A systematic review of prediction methods for emergency management. Int. J. Disaster Risk Reduction 102412 (2021)
7. Sahoo, S., Borthakur, P., Baruah, N., Chutia, B.P.: IoT and machine learning based health monitoring and heart attack prediction system. In: Journal of Physics: Conference Series, vol. 1950, no. 1, p. 012056. IOP Publishing, August 2021
8. Yacchirema, D.C., Sarabia-Jácome, D., Palau, C.E., Esteve, M.: A smart system for sleep monitoring by integrating IoT with big data analytics. IEEE Access **6**, 35988–36001 (2018)
9. Bodur, G., Gumus, S., Gursoy, N.G.: Perceptions of Turkish health professional students toward the effects of the internet of things (IOT) technology in the future. Nurse Educ. Today **79**, 98–104 (2019)
10. Kadhim, K.T., Alsahlany, A.M., Wadi, S.M., Kadhum, H.T.: An overview of patient's health status monitoring system based on Internet of Things (IoT). Wirel. Pers. Commun. **114**(3), 2235–2262 (2020). https://doi.org/10.1007/s11277-020-07474-0
11. Meena, G., Sharma, D., Mahrishi, M.: Traffic prediction for intelligent transportation system using machine learning. In: Proceedings of 3rd International Conference on Emerging Technologies in Computer Engineering: Machine Learning and Internet of Things, ICETCE 2020, pp. 145–148 (2020)
12. Jain, A., Pandey, A.K.: Modelingand optimizing of different quality characteristics in electrical discharge drilling of titanium alloy (Grade-5) sheet. Mater. Today Proc. **18**, 182–191 (2019)
13. Jain, A., Pandey, A.K.: MultipleQuality optimizations in electrical discharge drilling of mild steel sheet. Mater. Today Proc. **8**, 7252–7261 (2019)
14. Panwar, V., Sharma, D.K., Kumar, K.V.P., Jain, A., Thakar, C.: Experimental investigations and optimization of surface roughness in turning of EN 36 alloy steel using response surface methodology and genetic algorithm. Mater. Today: Proc. (2021). https://doi.org/10.1016/j.matpr.2021.03.642

Artificial Intelligence Empowered Internet of Things for Smart City Management

Abinash Rath[1], E. Kannapiran[2(✉)], Mohammad Salameh Almahirah[3], Ashim Bora[4], and Shanjida Chowdhury[5]

[1] School of Business, The Assam Kaziranga University, Jorhat, India
[2] Department of Electrical and Electronics Engineering, Dr. N.G.P. Institute of Technology, Coimbatore, Tamilnadu, India
kannapiran@drngpit.ac.in
[3] Business Administration Department, Isra University, Amman, Jordan
mohammad.almahirah@iu.edu.jo
[4] Department of Mathematics, Diphu Government College, Diphu, Assam, India
ashim.bora@rgu.ac.in
[5] Department of General Educational Development, Daffodil International University, Dhaka, Bangladesh
shan_chydiu.ged@daffodilvarsity.edu.bd

Abstract. The Research on the Internet of Things (IoT) has paved the way for a revolution in community services. It was found that the application of IoT in a smart city is mainly carried out without major human intervention. The different uses of IoT devices allow interfaces to available systems, improve communication and perform a variety of tasks. In addition, the availability of multiple IoT devices has been shown to be compatible to protect integrity, collect and analyze information, and improve functionality. The study is focused in understanding the critical role Artificial Intelligence empowered Internet of Things (IoT) for creating Smart city. The researchers have used extensive data analysis by collecting the data from the respondents to understand the importance of AI empowered IoT in creating better smart city. The analysis is focused in using regression tools and other key data analysis to test the hypothesis. Based on the overall analysis it is concluded that enhanced security and privacy; implementing smart sensors; implementation of Intelligent analytics and better collaboration and Networking has supported the organization and other stakeholders in creating better smart city.

Keywords: Internet of Things · Artificial Intelligence · Smart city management · Regression analysis · Chi square test analysis

1 Introduction

The population of the world is increasing with the passage of time, based on the report released by World Health Organisation (WHO), it is noted that the total population of the world will increase by 2.3 billion by 2050, moreover it has also been stated that nearly 54% of the population are currently living in cities and this will increase to 66% in the next

V. E. Balas et al. (Eds.): ICETCE 2022, CCIS 1591, pp. 205–216, 2022.
https://doi.org/10.1007/978-3-031-07012-9_18

30 years [1]. Hence, an incredible increase in the population focuses in giving ways to new challenges, issues and opportunities for managing the city effectively. These aspects tend to include the overall sustainability of the cities, support in greater development. The research in the area of Internet of Things (IoT) has paved the way to revolutionize the services across communities. It has been noted that in smart city, the implementation of IoT is mainly applied without major human intervention. The diverse uses of IoT devices support in interconnecting with the available systems, enhance communication and perform various tasks [2]. Moreover, it has been regarded that the availability of higher number of IoT devices as supported in protecting the privacy, collection and analysis of information and support in enhancing the functions. Billions of systems are being connected through the usage of IoT resulting in large volume of information, process them for better decision making.

On a general parlance, the city which focuses in becoming smarter and more sustainable entitles to be highly attractive and sustainable. Furthermore, it offers better living atmosphere by having better air quality, drinking water quality, sewage system, reduction in traffic congestion and much more [3]. This will enable in making the city a better place to live for the current and future generations. The smart city enables the individuals to have a better life style, support in attracting more citizens, tourist and business enterprises, focus on value aspects like ethical aspects, overall social responsibility, diversity and inclusion etc. Hence, it is regarded that the application of IoT can support the government and communities in creating better and effective city which supports great place to live.

The different aspect which needs to be concentrated in creating IoT enables smart city covers four key aspects: Enhanced security and privacy; Implementing smart sensors; Intelligent analytics and better collineation and Networking [7].

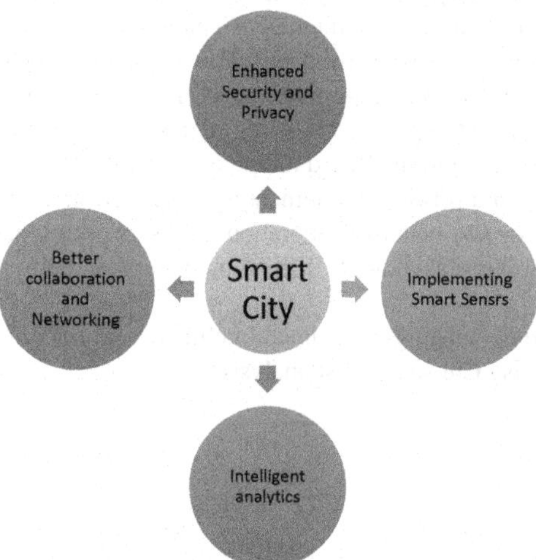

Fig. 1. IoT in Smart city implementation (Source: Jin et al. [8])

The Fig. 1 presents in detail the critical factors influencing the smart city through implementation of IoT, the major factors are Enhanced security and privacy; Implementing smart sensors; Implementation of Intelligent analytics and better collaboration and Networking. These are the major drivers of creating smart city through IoT as they support in enhancing traffic management, address better connectivity; enable in implementing smart waste management etc. [9].

2 Review of Literature

The researchers have mentioned that the building of smart cities involve enhanced security and privacy for the individuals. The smart cities need to possess essential infrastructures, enable in gathering data and information on the process of creating better aspects so as to attain sustainable development.

It is further stated by that [10], the smart cities need to support in acting against cybercrimes, support in delivering government policies to the citizens, enable in safer living and attract more tourist and business for better enhancement. The implementation of tools like AI, IoT and other digital technologies support in collecting more data, hence these needs to be stored in safer location so that it can be retrieved for better planning and management of smart cities. The implementation of cloud computing and blockchain support the government and administrators in safeguarding the data effectively.

Furthermore, various business enterprises and nongovernmental organizational may use these IoT enabled technologies in collecting information about the customers with consent approval for complying with regulators. Banks and financial institutions need to gather the know your customer (KYC) and anti-money laundering (AML) for complying with the regulatory requirements. Hence, these will enable the government to collate the necessary information in tracking the progress, tax liabilities and compliance of the individuals effectively [11]. Moreover, these data need to be stored as they are highly confidential and can be used only the government and authorities for better decision making.

Security and integrity are key factors for smart cities. Smart cities are the existence of basic urban infrastructure on the Internet, and any differences in the operation of urban services will cause inconvenience to citizens and put people's lives and property at risk. Therefore, safety is an important factor in smart cities. At a time when cybercrime and warfare have become widespread in world politics, smart cities risk becoming more and more targets for these malicious attacks. This scenario requires encryption of data transmitted over the network. To be successful, smart city projects need the trust and participation of citizens [12]. The proliferation of sensors in smart cities, which continuously collect data on human activity, can expose citizens to the everyday activities of citizens. Furthermore, companies and companies within the IoT network can use citizen data for purposes such as targeted advertising and engage in acts such as espionage, etc. without the citizen's consent [13].

Table 1. Comparison of previous work

Authors	Source	Criticality of implementation
Jin et al. [8]	Overall city administration	IoT based digital model is suggested for better implementation
Hussain et al. [11]	Environmental protection	Critical indicators like quality, affordability, riskiness and cost structure are major drivers
Liu et al. [12]	Transport management	Socio – economic activities are mainly focused
Santos et al. [13]	Waste management	Significant strategies are implemented for better management

3 Methodology

The objective of the study is to understand the critical determinants of Artificial Intelligence Empowered Internet of Things (IoT) Security for better Smart City Management. The researchers focus in apprehending the usage of explanatory research design as it supports the study in effectively analyzing the various aspects related to the study. The study involves in using different key measures in the study, the study is intended to collect information from the respondents who are part of the government, nongovernment enterprises where they are more involved in supporting and collaborating for implementation of smart cities [14]. The concept of smart cities involves in addressing various government and societal goals of creating better place to live in, adhere to environmental goals and protect the safety and security of the individuals.

The researcher enables in collecting the data from both primary and secondary data sources, the primary source of information is collected through questionnaire which will be issued to the respondents who are chosen through purpose sampling method. The data will be sourced from nearly 148 respondents and these will be used for analyses using SPSS package. The researchers have used online library like EBSCO, ProQuest and other leading Scopus indexed journals in order to source the information for preparing the past literature and to understand the critical aspects of AI in creating smart city. The researchers have identified that four main variables viz., Enhanced security and privacy; Implementing smart sensors; Implementation of Intelligent analytics and better collaboration and Networking tend to play critical role in the implementation of smart cities.

It is regarded that the application of AI based tools are playing a crucial role in the development and enhancement of smart city management. These tools are supporting the communities, administrators and others in forecasting and estimating the cluster of tasks from the available data [15]. The machine learning approaches are considered as the overall set of application which the system can be used to process from the available empirical data. Moreover, AI enabled IoT can support in enhancing the smart agriculture, support in creating smart services and harness the effective use of renewable energy thereby reducing the carbon footprint.

3.1 Mathematical Model

Smart City Management = Constant + a1 × Enhanced security and privacy + a2 × Implementing smart sensors + a3 × Implementation of Intelligent analytics + a4 × better collaboration and Networking.

3.2 Hypothesis Testing

There are no major differences among the IoT enables Enhanced security and privacy and creation of better smart city.

There are no major differences among the IoT enabled Implementing smart sensors in creation of better smart city.

There are no major differences among Intelligent analytics and smart city creation.

There are no major differences among IoT influences in better collaboration and Networking and smart city creation.

4 Analysis and Discussion

This section enables in presenting the critical data analysis based on the information collected from the respondents, the analysis covers correlation analysis, regression analysis and chi square test analysis.

The part of the analysis involves in providing descriptive analysis of the critical components which states role of AI in enhancing the security system for smart city (Table 2).

Table 2. Role of AI in enhancing security

Role of AI in enhancing security	Frequency	Percent
Strongly disagree	11	7.4
Disagree	14	9.5
Neutral	16	10.8
Agree	31	20.9
Strongly agree	76	51.4
Total	148	100

Based on the Table 1 it is identified that nearly 51.4% of the respondents have stated that the AI supports in enhancing the security and led to better smart city management, as stated in literature this is confirmed by the respondents, moreover 20.9% have agreed to the statement, 10.8% of the respondents have stated that they are neutral to the statement, 9.5% have disagreed and remaining have strongly disagreed.

The Fig. 2 mentions the overall role of AI in enhancing the security and majority of the respondents are agreeing to the statement that AI supports in enhancing security and hence it is highly essential in creating smart city.

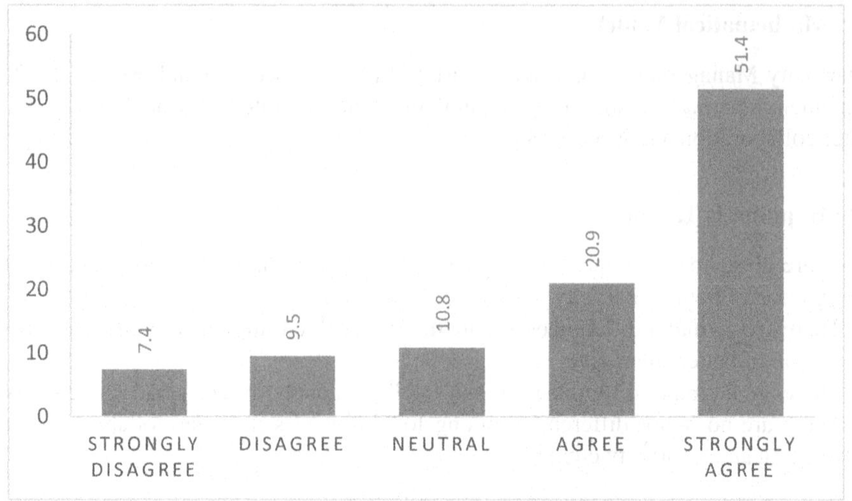

Fig. 2. Role of AI in enhancing security

The second aspect analyzed is the application of IoT in creating smart city management so that better control mechanism can be ascertained (Table 3).

Table 3. IoT support in smart city management

IoT support in smart city management	Frequency	Percent
Strongly disagree	14	9.5
Disagree	14	9.5
Neutral	25	16.9
Agree	65	43.9
Strongly agree	30	20.3
Total	148	100

Though the above table it is noted that 20.3% have strongly agreed to the statement that IoT support in smart city management which confirms with the literature review stated in previous section. [16]. Furthermore, it is noted that 43.9% of the respondents have agreed to the statement, whereas 16.9% have been neutral and 19% of them have disagreed and strongly disagreed.

The Fig. 3 has mentioned that the respondents opinion in implementing IoT for smart city management and it has been regarded that most of the respondents has mentioned that the IoT supports in enhancing the implementation of smart city management.

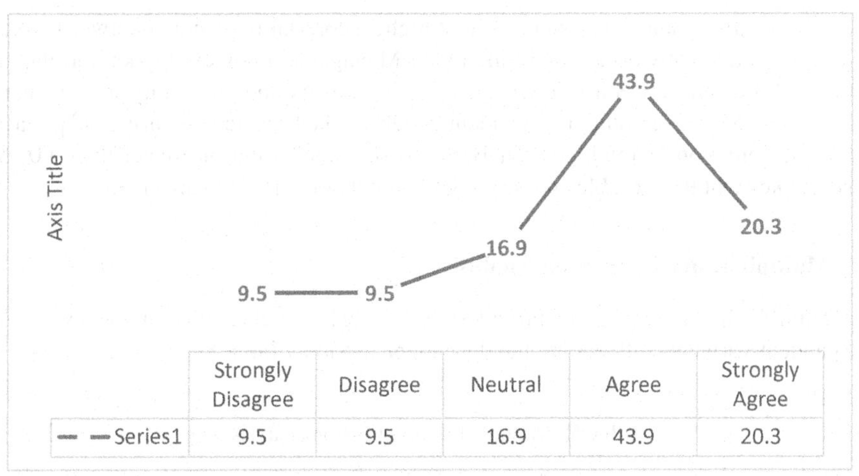

Fig. 3. IoT support in smart city management

4.1 Correlation Analysis

The second step on the analysis part is to perform correlation analysis between the independent variables and dependent variable: Enhanced security and privacy; Implementing smart sensors; Intelligent analytics and better collineation and Networking with the creation of better smart city (Table 4).

Table 4. Coefficient of correlation

Correlation analysis	Enhanced security and privacy	Implementing smart sensors	Implementation of Intelligent analytics	Better collaboration and networking	Smart City management
Enhanced security and privacy	1	.893**	.833**	.881**	.837**
Implementing smart sensors	.893**	1	.855**	.906**	.843**
Implementation of Intelligent analytics	.833**	.855**	1	.870**	.757**
Better collaboration and Networking	.881**	.906**	.870**	1	.878**
Smart City Management	.837**	.843**	.757**	.878**	1

From the above analysis it is noted that highest correlation is noted between better collaboration and Networking and Smart City Management possess higher correlation of 0.878, followed between Implementing smart sensors and smart city management with +0.843. Moreover, other independent variables like Enhanced security and privacy and Implementation of Intelligent analytics possess coefficient of greater than +0.75. Hence, states that the variables possess significant association between them.

4.2 Multiple Liner Regression Analysis

The multiple linear regression provides the critical apprehension on the extent of independent variables influencing the dependent variable (Table 5).

Table 5. Multiple linear regression analysis

Regression analysis	B	t	P Coeff.
(Constant)	0.005	0	1.00
Enhanced security and privacy	0.255	2.698	0.008
Implementing smart sensors	0.179	1.807	0.07
Implementation of Intelligent analytics	−0.152	−1.833	0.07
Better collaboration and Networking	0.647	6.031	0
F value	140.28		
P Coeff	0.00		
R sqd.	0.797		

Based on the above table it is identified that the coefficient of determination is 0.797 which is more than 0.60, hence the model is stated as the best fit and used for regression analysis. The F value 140.28 and p coefficient is 0.00, which is less than 0.05 hence it is statistically significant. The regression equation is

Smart City Management = 0.005 + 0.255 × Enhanced security and privacy + 0.179 × Implementing smart sensors –0.152 × Implementation of Intelligent analytics + 0.647 × better collaboration and Networking.

4.3 Testing of Hypothesis Analysis

The last section involves in testing the hypothesis which has been mentioned by the researcher, for this purpose chi square test is used.

Hypothesis 1
Null: There is no major differences among the IoT enables Enhanced security and privacy and creation of better smart city (Table 6).

The above analysis reveals that the p coefficient between Enhanced Security & Privacy and Smart city management is 0.00 hence it is concluded that major differences

Table 6. Cross tab between Enhanced Security & Privacy and Smart city management

Hypothesis 1	Value	df	P Coeff
Pearson Chi-Square	217.058a	16	0.00
Likelihood Ratio	172.145	16	**0.00**
Linear-by-Linear Association	102.952	**1**	0

exist among the IoT enables Enhanced security and privacy and creation of better smart city.

Hypothesis 2
Null: There is no major differences among the IoT enabled Implementing smart sensors in creation of better smart city (Table 7).

Table 7. Cross tab between Implementing smart sensors and Smart city management

Hypothesis 2	Value	df	P Coeff
Pearson Chi-Square	224.950a	16	0.00
Likelihood Ratio	170.72	16	**0.00**
Linear-by-Linear Association	104.343	**1**	0

The above analysis reveals that the p coefficient between IoT enabled Implementing smart sensors in creation of better smart city is 0.00 hence it is concluded that major differences exist among the IoT enabled Implementing smart sensors in creation of better smart city.

Hypothesis 3
Null: There is no major differences among Intelligent analytics and smart city creation (Table 8).

Table 8. Cross tab between Implementing Intelligent and Smart city management

Hypothesis 3	Value	df	P Coeff
Pearson Chi-Square	188.320a	16	0.00
Likelihood Ratio	149.777	16	**0.00**
Linear-by-Linear Association	84.258	**1**	0

The above analysis reveals that the p coefficient between Intelligent analytics and smart city creation is 0.00 hence it is concluded that major differences exist among Intelligent analytics and smart city creation.

Hypothesis 4

Null: There is no major differences among IoT influences in better collaboration and Networking and smart city creation (Table 9).

Table 9. Cross tab between better collaboration & Networking and Smart city management

Hypothesis 4	Value	df	P Coeff
Pearson Chi-Square	246.195a	16	0.00
Likelihood Ratio	192.047	16	**0.00**
Linear-by-Linear Association	113.389	**1**	0

The above analysis reveals that the p coefficient between better collaboration & Networking and Smart city management is 0.00 hence it is concluded that major differences exist among IoT influences in better collaboration and Networking and smart city creation.

4.4 Implementation Approach

It has been identified from the analysis that there exist better association between the independent variables like Enhanced security and privacy; Implementing smart sensors; Implementation of Intelligent analytics and better collaboration and Networking tends to support the organization and other stakeholders in implementing smart city. The transformation in the digital environment has resulted in enhancing the usage of critical tools like AI, IoT and Robotics in creating better smart city. Hence, it is critical for the companies to focus in unleashing the overall potential of the AI and IoT tools for creating better city and thereby support the community in achieving sustainable growth and development.

5 Conclusion

A smart city must improve the overall quality of life for all citizens, support the use of IoT-based technology to improve the quality of services and support the efforts of all individuals, systems and processes to achieve sustainable growth and development. Key aspects of digital transformation are the collection of information from various sources, real-time analysis for efficient urban planning and the use of more predictive models for better planning and management. To create smart cities, the digital network needs to be developed with the introduction of tracking centers in all parts of the city. The purpose of this study is to focus more on the technical aspects that urban planners, authorities, administrators and communities need to focus on using IoT to build and maintain the smart cities of the future. The various aspects of creating an IoT to enable a smart city include four main aspects: improved security and data protection. Smart sensor application. Smart resolution, better negotiation and networking. Implementing smart city management through improved artificial intelligence IoT often also covers security and integrity as it is very important to make the city safer and better as the population grows by supporting the use of smart sensors to handle traffic efficiently, which helps to reap rainwater and snow removal and other considerations. In addition, the implementation of IoT enables more information to be collected on a daily basis, so it is important that we use this information to anticipate future needs and plan effective action. Therefore, it includes the application of technical tools in various areas such as governance, housing design and development, traffic management, sustainable housing, attracting more tourists, supporting entrepreneurship and better housing. Researchers say that building smart cities means more security and integrity for individuals. Smart cities need to have a basic infrastructure that allows them to gather data and information about the development process of the best aspects to achieve sustainable development.

References

1. Syed, A.S., Sierra-Sosa, D., Kumar, A., Elmaghraby, A.: IoT in smart cities: a survey of technologies, practices and challenges. Smart Cities **4**, 429–475 (2021)
2. Majumdar, S., Subhani, M.M., Roullier, B., Anjum, A., Zhu, R.: Congestion prediction for smart sustainable cities using IoT and machine learning approaches. Sustain. Cities Soc. **64**, 102500 (2021)
3. Sánchez-Corcuera, R., et al.: Smart cities survey: technologies, application domains and challenges for the cities of the future. Int. J. Distrib. Sens. Netw. **15** (2019)
4. Trakadas, P., et al.: An artificial intelligence-based collaboration approach in industrial iot manufacturing: key concepts, architectural extensions and potential applications. Sensors **20**, 5480 (2020)
5. Thakker, D., Mishra, B.K., Abdullatif, A., Mazumdar, S.: Explainable artificial intelligence for developing smart cities solutions. Smart Cities **3**, 1353–1382 (2020)
6. Jan, B., Farman, H., Khan, M., Talha, M., Din, I.U.: Designing a smart transportation system: an internet of things and big data approach. IEEE Wirel. Commun. **26**, 73–79 (2019)
7. Meena, G., et al.: Traffic prediction for intelligent transportation system using machine learning. In: 2020 3rd International Conference on Emerging Technologies in Computer Engineering: Machine Learning and Internet of Things (ICETCE), pp. 145–148 (2020)
8. Jin, X.B., Yang, N.X., Wang, X.Y., Bai, Y.T., Su, T.L., Kong, J.L.: Hybrid deep learning predictor for smart agriculture sensing based on empirical mode decomposition and gated recurrent unit group model. Sensors **20**, 1334 (2020)
9. Jain, A., Yadav, A.K., Shrivastava, Y.: Modelling and optimization of different quality characteristics in electric discharge drilling of titanium alloy sheet. Mater. Today Proc. **21**, 1680–1684 (2019)
10. Jain, A., Pandey, A.K.: Modeling and optimizing of different quality characteristics in electrical discharge drilling of titanium alloy (grade-5) sheet. Mater. Today Proc. **18**, 182–191 (2019)
11. Hussain, A., et al.: Waste management and prediction of air pollutants using IoT and machine learning approach. Energies **13**, 3930 (2020)
12. Liu, X., Xiao, Z., Zhu, R., Wang, J., Liu, L., Ma, M.: Edge sensing data-imaging conversion scheme of load forecasting in smart grid. Sustain. Cities Soc. **62**, 102363
13. Santos, G.L., et al.: Accelerometer-based human fall detection using convolutional neural networks. Sensors **19**, 1644 (2019)
14. Jain, A., Pandey, A.K.: Multiple quality optimizations in electrical discharge drilling of mild steel sheet. Mater. Today Proc. **8**, 7252–7261 (2019)

15. Panwar, V., Sharma, D.K., Kumar, K.V.P., Jain, A., Thakar, C.: Experimental investigations and optimization of surface roughness in turning of EN 36 alloy steel using response surface methodology and genetic algorithm. Mater. Today: Proc. (2021)
16. Qu, Y., Ming, X., Qiu, S., Zheng, M., Hou, Z.: An integrative framework for online prognostic and health management using internet of things and convolutional neural network. Sensors **19**, 2338 (2019)

Critical Analysis of Intelligent IoT in Creating Better Smart Waste Management and Recycling for Sustainable Development

Joel Alanya-Beltran[1], Abu Md. Mehdi Hassan[2], Akash Bag[3](✉), Mintu Debnath[4], and Ashim Bora[5]

[1] Electronic Department, Universidad Tecnológica del Perú, Lima, Peru
C18121@utp.edu.pe
[2] Department of Physics, Government City College, Chittagong, Bangladesh
[3] Amity Law School, Amity University, Raipur, Chhattisgarh, India
aakashbagh06@gmail.com
[4] Department of Physics, Chakdaha College, Chakdaha, West Bengal, India
[5] Department of Mathematics, Diphu Government College, Diphu, Assam, India
ashim.bora@rgu.ac.in

Abstract. The Internet of Things (IoT) has found extensive use in areas such as water management, waste management, and sustainable development. With its broad connectivity, the Internet of Things is a new and promising technology that has the potential to positively transform human existence globally. IoT allows low-energy devices to exchange information and interact with one another. Waste management is a daily chore that necessitates a huge number of labour resources and has an impact on natural, fiscal, efficient, and social elements. The rate of garbage generation has been magnified at an alarming rate as a result of fast urbanisation and expanding population. As a result, as the world faces global environmental problems, it is necessary to develop changes in waste management systems and technology to address issues that have never been addressed in such innovative ways. Many applications throughout the world have been using IoT to conduct various activities to provide unique services for wastage handling and maximise energy efficiency. These improvements allow IoT technologies to serve as a bridge between basic network-based systems and technologies that scan and gather data from the real environment, as well as deliver new services and applications that help people in a variety of ways. As a result, this research conducts a study of existing IoT-enabled waste management solutions and sustainable development. The goal is to get an understanding of the strengths and weaknesses so that changes and innovations may be made to effectively and efficiently manage waste while also maintaining a healthy environment in the communities. The impact of waste management on long-term economic, social, and environmental sustainability is also depicted in the literature review. The secondary data has been used in the research to gain a reliable conclusion of the study.

Keywords: Internet of Things · Waste management · Sustainable development · Circular economy

V. E. Balas et al. (Eds.): ICETCE 2022, CCIS 1591, pp. 217–225, 2022.
https://doi.org/10.1007/978-3-031-07012-9_19

1 Introduction

The Internet of Things (IoT) is a cutting-edge new communication system that envisions a global number of physical devices and gadgets that can communicate with one another [1]. It is a network of physical items, devices, machines, homes, and eventually, all of which are connected with electronics, sensors, software, and network access, as well as appropriate protocol stacks, allowing them to gather and exchange data. In the area of waste management, IoT plays a major part in improving the quality of life and achieving sustainability. If waste is correctly treated with innovative technologies, it may be renewed into a valued resource.

The vision of attaining sustainable development that improves people's quality of life while also allowing for more efficient resource management necessitates radical adjustments. The growing population of countries has resulted in a growth in waste volume, which is a major concern. Waste management systems in cities are inefficient and ineffective for a multitude of reasons, including collection, disposal, vehicle routing, and pollution, all of which must be controlled [2]. The technologies such as IoT are planned to change the current state of economic renewal and give light to a more well-organized circular economy. The term "circular economy" refers to the efficient employment of natural resources. The circular economy aims to reduce wastage and develop a confined ecosystem for optimal resource utilisation and utilization.

IoT technologies enable both design and consumption perspective shifts, resulting in objects that may be reused, refurbished, and recycled. Smart technology can aid in the realisation of the vision by assisting with infrastructure and management decision-making [3]. A clear framework with immediate optimum trash collection can be realised using IoT embedded technology and a rising number of internet users in urban and rural locations. The Internet of Things enables the garbage collection system to become interconnected and efficient, lowering prices and decreasing waste disposal problems. With the capacity to communicate data over the Internet, an IoT-based model embedded sensors to continuously monitor the volume of waste in a container or trash cans. The data is acquired and accessed via the internet, where innovative and streamlined calculations are used to optimise the waste is collected in real time.

Therefore, solving environmental concerns in overall, and waste management in specifically, is a crucial issue that must be investigated [4]. Based on a comprehensive evaluation of the literature, the study highlights the likely of green infrastructure and associated groups to support waste management initiatives. Researchers present a unique strategy for waste management that is both vigorous and efficient by predicting the likely level of waste in the garbage can.

2 Literature Review

A. Jain et al. [5] states that Waste management is an essential topic that is usually addressed when it comes to the implementation of IoT in smart cities. As the world's population nurtures, so does the amount of garbage produce daily. However, in cities, one of the most significant issues is the failure to efficiently manage this waste. The methods for dealing with garbage are extremely limited and necessitate a great deal of human work or assistance. As a result, some governments or organisations lack the capacity to track and measure the daily wastage in cities, forcing people to live in harmful conditions.

Table 1. Barriers to adoption of waste management

S/no.	Barriers	Description
1.	Lack of awareness among community	One of the primary constraints in the implementation of a smart waste management platform is lack of information, i.e. lack of awareness in the subject and experience in various waste management technologies. Therefore, in this era of new technology, citizens' participation is crucial to achieving an efficient solid waste management system
2.	Implementing innovative technology is challenging	Municipalities with ever-increasing garbage are faced with the task of improving their current technology. Every area of information and communication technology (ICT) that can be used to create an effective waste management system should be prioritised. As a result, the task of determining the best combination of ICTs while taking into account robustness, cost, complexity, and environmental implications remains unsolved
3.	People's resistance to change	Waste and garbage in public places is a big problem, with serious consequences for the environment and public health. There is no systematic and scientifically designed waste separation at either the household or societal level. People's attitudes on trash segregation, as well as their attitudes toward labour, must alter
4.	Dynamic scheduling and routing are not available	Improved garbage collection and transport routing can lessen the environmental impact of open dumps while also increasing the efficiency of the waste collection system, lowering costs and time
5.	Lack of proper financial planning	One of the main hindrances to improving existing systems is the absence of financial analysis and resource management, as well as the establishment of cost-effective waste management systems [7]

Barnabas SG et al. [6] argues that individual decisions have an impact on the waste generation rate. People's decisions are chiefly focused on mechanical variables such as price rather than social or environmental objectives. As a result, waste management has evolved into a complex metropolitan issue. The citizens' relationship with technology is a two-way road. While people decisions have an impact on the waste management strategy, its infrastructure that surrounds residents has an impact as well. Cities need creative, cross-industry keys to make trash collection and removal smoother. These solutions should be reproducible, scalable, and flexible.

It has thoroughly examined specific limits from a variety of perspectives to identify the major roadblocks to the circular economy's goal of smart solutions. Table 1 lists the barriers discovered after exhaustive literature research and expert validation, as well as their brief descriptions.

3 Methodology

In the context of methodology, this research employs a case study approach based on a real-world execution of the changing technology. The generated resolution (a real sample of a smart bin and waste management software coupled with the IoT) is exhibited, tested, and confirmed by real-world testing. A detailed literature review was conducted by reviewing scholarly journals as part of the Secondary data collection process.

The role of technology in attaining sustainable development goals has been the subject of heated debate in recent years. IoT is an unique generation of technology that provides coordination across cyber and physical infrastructures and serves as the industry's backbone [8]. Due to the high expense of waste collection and the environmental risks involved with landfills, modern civilizations are concerned about garbage output. Waste engendered by municipal actions, both industrial and household, can pose health problems and damage the atmosphere if there isn't a solid waste management plan in place that is effective and efficient [9]. If waste is not adequately handled can become a major cause of pathogens, contribute to global climate change by emitting greenhouse gases, and even promote urban violence by degrading urban surroundings.

The key to an effective waste management model necessitates an awareness of the waste generated, resource availability, and societal environmental circumstances. In this context, the Internet of Things appears to be a potential alternative for waste collection and recovery operations [10]. A specific form of garbage collection and treatment is required to solve the waste management challenge. Solid waste management includes a wide range of activities such as waste generation, storage, collection, transportation, processing and disposal [11]. The circular economy involves much more than just physical recovery. Hence, in the circular economy, composting has gotten more attention.

It has been noticed that several IoT-enabled waste management solutions have been proposed, each with its own set of strengths and weaknesses. For example, Low-power wide-area networks (LPWANs) leveraging long-range (LoRa) technology have emerged as feasible IoT solutions. LoRa is an IoT technology that is wireless, long-range, low-power, and uses unlicensed radio airwaves in the industrial sector [12]. Several IoT-enabled solutions for waste management in cities have been created using LoRa. Further, a cloud-based system has also been found to be effective for waste management. In this system, waste bins are fitted with sensors that detect and alert of waste level status, which is then sent and stored in the cloud. Furthermore, route optimization and path selection for waste collection can be done depending on the received statuses of trash bins for effective and efficient town administration and waste management. The drawback is that it cannot be used for door-to-door rubbish collecting because it needs each sort of a waste to have its own recycling bin [13]. In this situation, Big Data analysis can be used to analyse data from various locations.

In addition, IoT Smart and wireless waste management offered a smart garbage bin with sensors [14]. When the garbage bin is about to be full, the system can notify the necessary authorities. Informing them of the impending collection of waste in bins to keep the environment clean. In this system, the bin can be emptied only when a user reports the status of the bin to the server through the mobile application. As a result, the downside is that the authorities concerned cannot monitor waste levels in real-time and must instead rely on the information. Furthermore, if a user is unable to transmit the message, it implies that when the bin is full, the environment will be polluted with waste (Fig. 1).

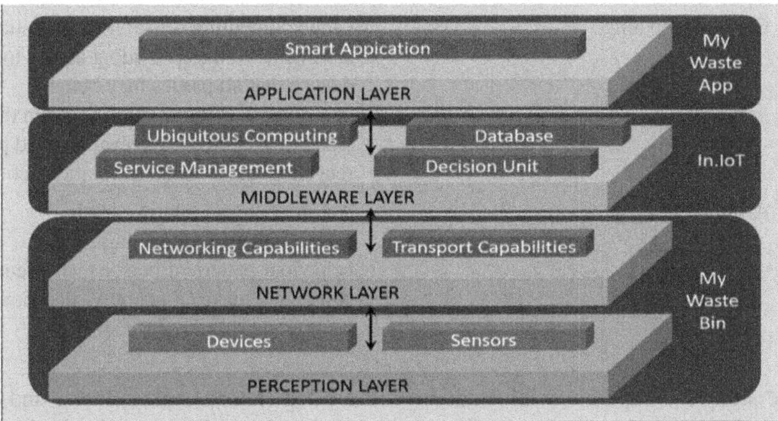

Fig. 1. Layered construction of the waste management system (Source: Anusree PS, Balasubramanian P)

Moreover, the work presented in [15] is likewise concerned with garbage management via smart bins. This smart garbage disposal system is based on a strategy that usages solar energy to operate the device and exist sensors to keep track of how much garbage has accrued inside the insertion. If essential, the container can condense the trash and reduce its volume by up to ten times before it is collected Wi-fi connection is used to send data about the fill level to a cloud server, where it is deposited. The automated bin can be used as a Wi-Fi hotspot and can be simply converted to any form of container, from little trash cans to giant garbage cans, even underground containers. In addition, this smart technology aids utilities in reducing truck fleets, lowering fuel usage and increasing pick-up time, all while lowering operating expenses. Table 2: Depict the table that lists the smart waste management technologies.

Table 2. Smart waste management with smart technologies

S/no.	Smart technologies	Application in smart waste management
1.	NFC sensors, RFID tags, and GPS sensors are forms of IoT sensors	By involving a physical or virtual device to the cyberspace, data can be collected, detected, stored, and distributed. IoT sensors can aid in the immediate monitoring of automated waste bins, that can deliver data on the progress of each bin as well as its location
2.	Ultrasonic sensors are used in solar-powered trash gatherers	Solar-powered bins compressed garbage mechanically into little packets, while sonar sensors aid in bin maintenance and monitoring
3.	Big Data Analytics	Big Data Analytics can help administration in refining its competence by merging data from socio-economic issues and their physical aspects
4.	Cloud Computing	Cloud computing is useful for statistics storing, as well as providing access to this data so that all parties may better decisions. The information saved in the database can be used to evaluate and plan recycling and revenue restoration actions
5.	Artificial Intelligence	Automation can improve trash categorization and recognise the behaviour pattern of persons who generate garbage to make an operative decision

The study's comparative information will aid legislators, stakeholders, and government specialists in framing policies and initiatives to eliminate the highlighted impediments to a more efficient and innovative waste management system.

4 Findings/Result

It has been evaluated that Waste management is a critical environmental concern that IoT is attempting to address. All of the research included in this paper established a waste collection system based on IoT devices that works by sensing the waste bin's cutoff point and sending notifications or warnings to the proper authorities when the container is full. This paper presents a waste management strategy for cities that is both efficient and real-time, using a citizen-centred approach. In the study, various models for managing waste have been explained. The proposed system combines sensor technologies that collect trash data in real time from smart bins (things) and broadcast it to an online podium where anyone can browse and monitor the status of chambers located throughout the city. Data is collected and transmitted via a wireless mesh network in the suggested system.

The verdicts assist in the collection of a more effectual garbage collection path, along with ensure that trash will be composed without destructive the atmosphere and distributed to appropriate discrimination plants for discarding and reprocessing. This true smart container prototype is built, tested, verified and authenticated and ready-mapped into a real solution.

Figure 2 depicts the three basic components of a waste management system. The first part discusses the automated bin, the second the IoT middleware integration, and the third and final block is the handler's application.

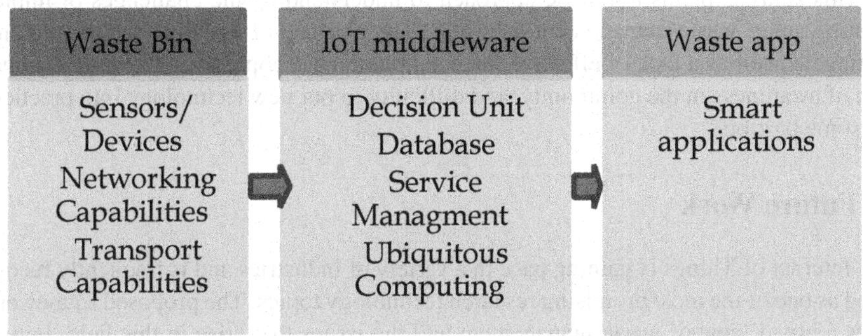

Fig. 2. Illustration of the system architecture

The waste bin is made up of a container with a lid and a sensor-equipped inclusion. An ultrasonic sensor that measures the amount of garbage filling in the chamber. This is crucial inside the resolution as it allows for the avoidance of waste overflow or excess waste deposit through its operation. A load cell module (load sensor) is also included in the solution, which measures the mass of the leftovers in the bin. In addition, this system contains an implemented solution in which the containers are monitored by sensors that provide real info on each one's filling level.

5 Conclusions

The current need in society is for the progress of a sustainable environment and economy. Garbage in various forms such as solid junk, gaseous surplus, and liquid waste amasses as an outcome of population growth, urbanisation, and industrialisation, impacting the whole world. Emissions, air smog, soil pollution, and water trash are all challenges caused by improper trash disposal. As a result, for proper waste management, smart technology, information and communication technology (ICT), and smart waste management approaches are necessary.

The integration of technology to waste management procedures that involve sensing results in a smart solution to waste problems. The report examines existing research on IoT-enabled waste management approaches and presents a conceptual framework for overcoming the waste recovery gap. The Internet of Things communication paradigm has enabled objects to communicate and share information over large distances while using less electricity. Researchers conducted and presented a study of smart waste collection systems employing IoT technology in this research. The system employs an IoT technology in which sensors continuously monitor the dumped rubbish from the smart bin, providing real-time information on the full level of each chamber. These data are saved and analysed in an IoT. Building a real smart container prototype, developing a waste management application, can effectively change people's waste disposal habits while maximizing economic and material resources. It was determined that the projected model may effectively improve people's rubbish disposal while also maximising economic and material resources.

This study explains a holistic approach to understanding the challenges of implementing smart waste management solutions. The absence of dynamic scheduling and routing, as well as a lack of effective financial planning, People's resistance to change, lack of awareness in the community, and difficulty to put new technology into practice, are some barriers.

6 Future Work

The Internet of Things is gaining pace in a variety of industries and is frequently recognised as one of the most promising research technology topics. The proposed framework takes a broad view of waste management and the issues that arise in this field. Future work could be done to examine the effects of other aspects, such as legislation, law, product development strategy, and technology, which could have an impact on waste management. In addition, a study of the solution's investment and operating expenses will be quite interesting and can be done as future study.

References

1. Alqahtani, F., Al-Makhadmeh, Z., Tolba, A., Said, W.: Internet of things-based urban waste management system for smart cities using a Cuckoo Search Algorithm. Clust. Comput. **23**(3), 1769–1780 (2020). https://doi.org/10.1007/s10586-020-03126-x
2. Anusree, P.S., Balasubramanian, P.: IoT-enabled logistics for E-waste management and sustainability. In: Gunjan, V.K., Suganthan, P.N., Haase, J., Kumar, A. (eds.) Cybernetics, Cognition and Machine Learning Applications. AIS, pp. 71–77. Springer, Singapore (2021). https://doi.org/10.1007/978-981-33-6691-6_9
3. Jain, A., Pandey, A.K.: Modeling and optimizing of different quality characteristics in electrical discharge drilling of titanium alloy (grade-5) sheet. Mater. Today Proc. **18**, 182–191 (2019)
4. Jain, A., Pandey, A.K.: Multiple quality optimizations in electrical discharge drilling of mild steel sheet. Mater. Today Proc. **8**, 7252–7261 (2019)
5. Jain, A., Yadav, A.K., Shrivastava, Y.: Modelling and optimization of different quality characteristics in electric discharge drilling of titanium alloy sheet. Mater. Today Proc. **21**, 1680–1684 (2019)
6. Barnabas, S.G., Vasantha Geethan, K.A., Ganesh, S.V., Rajakarunakaran, S., Kumar, P.S.: Role of modern technologies and internet of things in the field of solid waste management. Int. J. Comput. Commun. Control **16**(5), 1–12 (2021)

7. Dogo, E.M., Salami, A.F., Nwulu, N.I., Aigbavboa, C.O.: Blockchain and internet of things-based technologies for intelligent water management system. In: Al-Turjman, F. (ed.) Artificial Intelligence in IoT, pp. 129–150. Springer, Cham (2019). https://doi.org/10.1007/978-3-030-04110-6_7
8. Khan, S., Venkat, S., Jayavel, K.: Autonomous smart waste collection system using internet of things. In: 2017 IEEE International Conference on Power, Control, Signals and Instrumentation Engineering (ICPCSI), 21 September 2017, pp. 1181–1186. IEEE (2017)
9. Meena, et al.: Traffic prediction for intelligent transportation system using machine learning. In: Proceedings of 3rd International Conference on Emerging Technologies in Computer Engineering: Machine Learning and Internet of Things, ICETCE 2020, pp. 145–148 (2020)
10. Maksimovic, M.: Leveraging the internet of things to revolutionize waste management. Int. J. Agric. Environ. Inf. Syst. (IJAEIS) **9**(4), 1–3 (2018)
11. Nahr, J.G., Nozari, H., Sadeghi, M.E.: Green supply chain based on artificial intelligence of things (AIoT). Int. J. Innov. Manag. Econ. Soc. Sci. **1**(2), 56–63 (2021)
12. Nasiri, M., Tura, N., Ojanen, V.: Developing disruptive innovations for sustainability: a review on Impact of Internet of Things (IOT). In: 2017 Portland International Conference on Management of Engineering and Technology (PICMET), 9 July 2017, pp. 1–10. IEEE (2017)
13. Ramanujam, V., Napoleon, D.: IoT based green computing: an attempt to delineate e-waste management (2020)
14. Saha, H.N., et al.: Waste management using internet of things (IoT). In: 2017 8th Annual Industrial Automation and Electromechanical Engineering Conference (IEMECON) 16 August 2017, pp. 359–363. IEEE (2017)
15. Panwar, V., Sharma, D.K., Kumar, K.V.P., Jain, A., Thakar, C.: Experimental investigations and optimization of surface roughness in turning of EN 36 alloy steel using response surface methodology and genetic algorithm. Mater. Today Proc. (2021). https://doi.org/10.1016/j.matpr.2021.03.642

A Compressive Review on Internet of Things in Healthcare: Opportunity, Application and Challenges

Pankaj Dadheech[1], Sanwta Ram Dogiwal[1], Ankit Kumar[2](✉), Linesh Raja[3], Neeraj Varshney[2], and Neha Janu[1]

[1] Swami Keshvanand Institute of Technology, Management and Gramothan, Jaipur, Rajasthan, India

[2] GLA University, Mathura, UP, India
kumar.ankit@gla.ac.in

[3] Manipal University Jaipur, Jaipur, Rajasthan, India

Abstract. This article presents a review of the Internet of Things in the health area, focusing on the solutions that currently exist in home-oriented health. A very promising future is predicted with the appearance of portable smart devices, using protocols such as 6LoWPAN, which will allow the development of many applications for solving every-day problems in the health sector and the rapid implementation of the home-centered health model. In the various works and solutions consulted, the use of wireless technologies pre-dominates, such as: WPAN, WBAN, MBAN, Wi-Fi, WiMAX, ZigBee, Bluetooth, ANT, ultrawideband, ingestible sensors, and epidermal electronics, smart bandages, smartphone applications, RFID, RTLS and IPS. These technologies are widely used for biomedical census systems. This reflects a broad advance at the level of technologies and network architecture based on IoT, which provide specific solutions to problems in the health sector, especially in scenarios focused on home-centered health, which allows the maximum use of technology IoT in this area, very commonly called IoT Health. The above sounds promising for the health sector and ICT industry in general because it allows personalizing the health service, and accelerating its evolution.

Keywords: e-Health · Internet of Things · Telemedicine · Healthcare

1 Introduction

Internet of Things (IoT) is about connecting devices and objects of everyday life over the Internet [1]. IoT allows sensors and devices to be integrated with objects that are connected to the Internet through fixed and wireless networks [2–4]. It should be noted that the objects that will make up the network are varied: household appliances, industrial machinery, vehicles, among others. In IoT each object would have its own address of its Internet provider. This indicates that if we have objects such as car parts and cameras connected to the Internet, it will be easier to locate and identify them, it would be easier to carry out inventories, to know if the object is on or off at any moment. The Internet

V. E. Balas et al. (Eds.): ICETCE 2022, CCIS 1591, pp. 226–235, 2022.
https://doi.org/10.1007/978-3-031-07012-9_20

of things will allow a better quality of life, since it has the ability to collect and analyze such data that together can be converted into important information and knowledge.

1.1 The Internet in Health

The Internet of things applied in the health area will allow many people, regardless of their social class, to use the services that could be offered through the IoT and that are already being offered in many countries. It will serve to constantly monitor our health, taking into account that there are many diseases in which symptoms are silent and that an early diagnosis would allow prevention and possible solutions to diseases that can be fatal. All of the above sounds very convenient, considering that the global health system tends to change, focusing on home care. Figure 2 presents different applications of IoT in health.

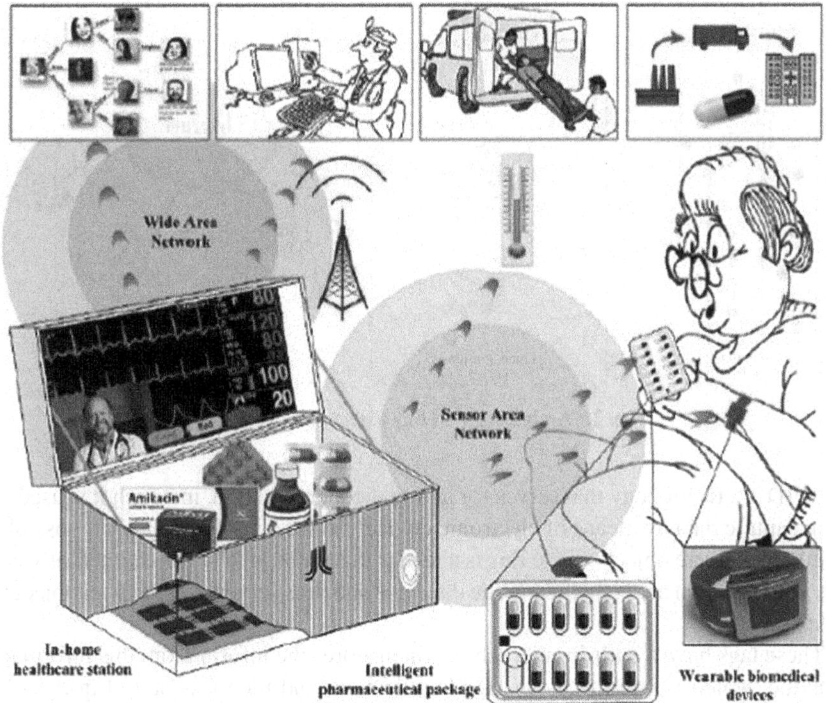

Fig. 1. Application of IoT in health

The different solutions/applications of the IoT in health can be sectorized in the services of Telemedicine, Emergency, Medication, Social networks for health, home health, smart pharmaceutical packages, biomedical devices, as shown in Fig. 1 [6].

1.2 IoT Technologies in Health

Among the fundamental characteristics offered by the Internet of Things is the increase in nodes connected to each other, which at the same time becomes a great challenge for the development of new communication protocols and updating of topologies. Sensor networks (WSN, Wireless Sensor Network) [7, 8] meet many of the requirements proposed by the technological solution of the Internet of Things. WSNs are made up of a set of nodes scattered in a given area, communicated in an ad-hoc manner [9] and can work cooperatively, this allows their applicability to be increased in almost all fields of the industry.

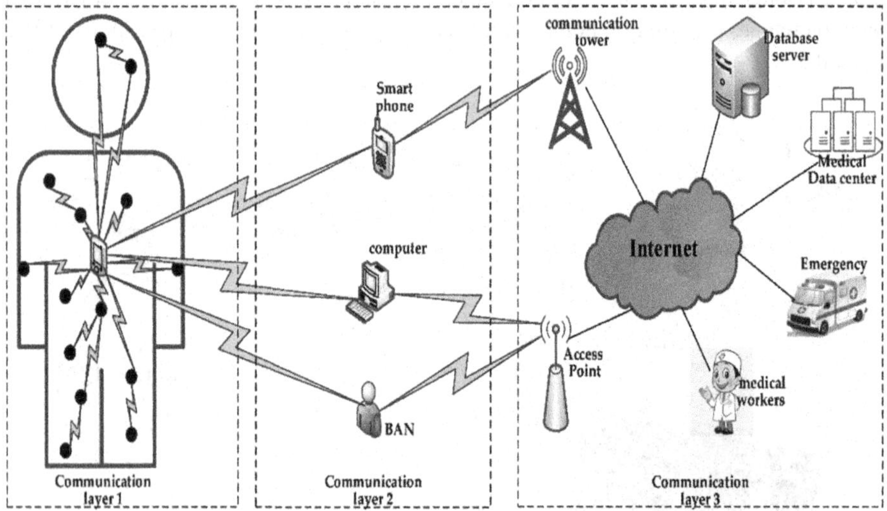

Fig. 2. Architecture of bio wireless network [14]

RFID is a technology that serves for automatic identification, in which it is used [10, 11] to capture data by means of electronic devices. This technology makes it possible to identify and locate objects, since this is a reader that reads labels, which internally carry data, which in turn are consulted in the database and thus serve to identify an object that contains it.

These tags have a built-in microchip, which stores the information, this information is unique, which is why it has an ID identification, and then it is stored in the RFID memory; these labels are of different size according to the system. Then the RFID is incorporated into the object to be identified and located. This is why in the health area it would be of great interest, because what is wanted is that a patient or worker be monitored and controlled to have better care.

2 Common IoT Application Architectures in Healthcare

The IoT makes good use of wireless technologies. In [12] an architecture is presented that serves as a test bed for applications on body area networks (BAN) (see Fig. 3) or body area sensor networks (WSAN), it is a network designed with devices (sensors) of low power incorporated in the body, through which its movements and its vital parameters are controlled, and in turn these will be connected to a wireless network to be able to transmit the data to a base station, and that data is sent on time real to hospital and clinics. As said, this architecture is used to implement test applications; many solutions worldwide that are based on BAN use architectures similar to this.

After all the above, it is not difficult to realize that the ubiquity of the IoT almost requires us that most of the everyday objects can be accessible through the famous IP (Internet Protocol) addresses. IPV6 (IP version 6) over 6LoWPAN [13] personal area sensor networks (IPv6 over Low power Wireless Personal Area Networks) is a protocol that provides IPv6 compatibility with sensor networks. In [14] an architecture that integrates hardware and software is presented, based on heterogeneous 6LoWPAN networks. It is motivated by the need for the fusion of a body area network and an environment control network, for health monitoring applications. Internet connectivity was added through the use of an edge router.

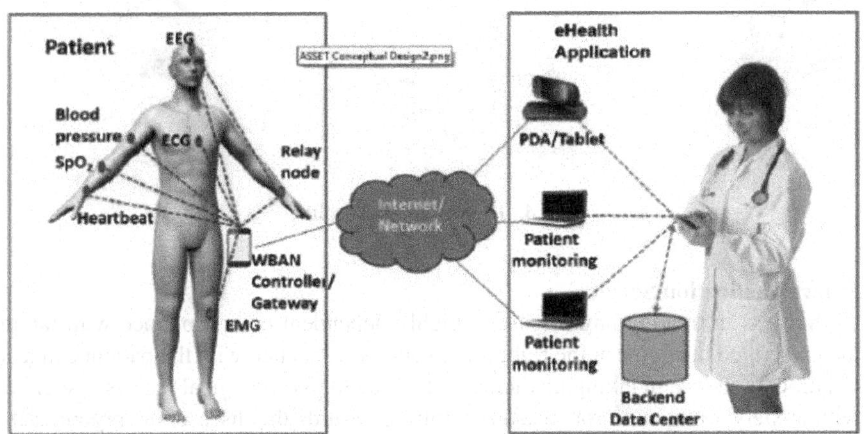

Fig. 3. Human network in eHealth [32]

3 Typical Scenarios of Application of the Internet of Things in Health Smart Emergency Service

Emergency services (ES) are also joining the evolution and new paradigm of patient-centered medical service from home. The SE is extremely important, since many lives depend on its good and timely management [15–17]. Many times, the responsibility falls on the prehospital service and the quality of the emergency medical service. So far, there are several solutions that help in the timely management of the pre-hospital service, through wireless technologies and the use of IoT as a basis. In [18, 19], time is described as a purpose to improve communication between the ambulance, the dispatch control center and the hospital emergency service (Fig. 4).

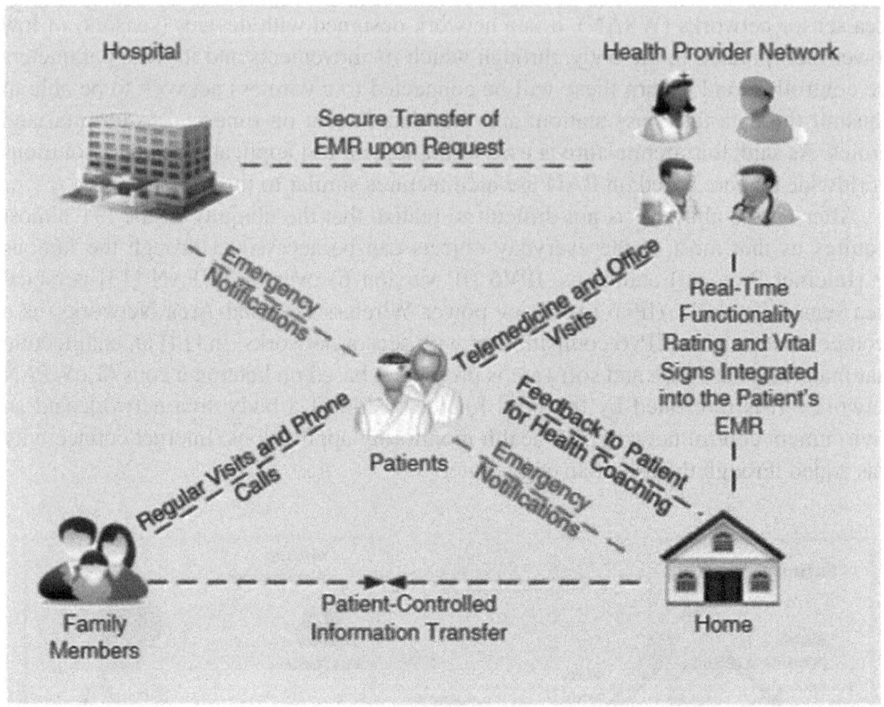

Fig. 4. Home-centered health

- **Smart Medication Service**
  Success in treating any disease is highly dependent on compliance with taking the prescribed medicine at the right time. Factors associated with the patient can lead to non-compliance in taking medicine, either due to psychological factors associated with anxiety, motivation for recovery, attitude towards the disease, or forgetting the schedule in the elderly [20–23].

Biomedical Devices

The research and development of wearable smart devices (SWDs) for personalized services (health being one such example) has motivated governments and research agencies around the world in recent years. Europe, the United States and Asia have put a notable interest in funding research projects in which topics such as E-Health and ICT are the main ones. An example of the numerous projects developed in the United States is the "Oregon Center for Aging and Technology", a project focused on the development and deployment of sensor technology to monitor older adults at home.

A network architecture based on 3G, GSM, Internet infrastructure and wireless body area sensor networks (WBASN) is proposed, called Community Physicians Networks, CMN for its acronym in English. (Community Medical Network). The architecture proposed by is divided into 5 main parts: WBASN, mobile devices, mobile communication networks, backbone and servers in local hospitals. Regarding the implementation of the proposed system, WBASN is responsible for both monitoring the health status of patients and the acquisition of medical information; this management plays an important role in the CMN. According to its authors, CMN does not need large costs for its deployment, it is not altered by the mobility of patients, a device called AMI (Active Monitoring Insole) is proposed for monitoring lower limb loads. According to its authors, the AMI formulated system is trained to be used by a doctor or a therapist in order to analyze the progress of patients and take the necessary corrective measures for their recovery.

On the other hand, MuSa (MUlti Sensor Assistant) is proposed by Bianchi et al. MuSa is a portable multisensory device, trained in detecting drops in heart rate and respiration rates.

- **Social Media Service focused on Health**

  With the appearance of the World Wide Web, the Internet has been the protagonist of numerous advances in society, even with the explosion of the so-called Web 2.0, the phenomenon of social networks began. Social media sites (social networks), such as Facebook, Twitter, YouTube, among others, have become powerful tools through the Internet, content generation applications, blogs, wikis, have largely strengthened the way of do business on the Internet, being the source of income for many companies and families worldwide.

Currently with the IoT, communities on the Internet are growing even more, as for example, in a community of social support for weight loss is described, which is shared through an Internet social network. These people can share their motivations and experiences. Support for these people is as personalized as if they were face-to-face with the doctor, but offers unique aspects of comfort, anonymity, and non-judgmental interaction. Within the methodology used, they conduct surveys to analyze information.

It is proposed that social support within Internet communities for weight loss deserves more attention and better evaluations; these communities can improve, a study is also carried out in young adults between 18 and 29 years old with the purpose of controlling their weight loss. The purpose was to offer guidance regarding their eating habits and analyze their weight management behavior. The work yielded good results and contributed to the understanding of the strengths and limitations of the use of Facebook in research and health promotion in general presents a not very common statistical study, in which social networks are used for medical use and research related to control, trust and intimacy. This study uses data from Europe, which contains information on behavior and perceptions on social issues such as the media, privacy and social networks. The results show that there is some evidence of a digital divide due to age, while income does not significantly affect access to social networks in the health area. Perceptions of privacy appear to be a deterrent, while trust in health care providers is not an important factor for social development and network use.

The area of radiology has also benefited claim that social networking services on the Internet have changed the way we communicate as a society and offer opportunities to improve the way radiology is practiced today. This study proposes some areas where radiologists can benefit from the new tools provided by IoT and concludes with several steps they can take to become more involved with social networks through the Internet.

According to, the accelerated change in health technology is rooted in trends. In their work they review the activities in the training of consumers and technology, and found that social trends are visible in the integration of information and communication technologies in the health area, both in search and exchange of information on the Internet, in the use of social media to create new types of interactions with family, providers and colleagues, and in the e-patient, which integrates these new functions and new technologies.

Another important example is found, where social media and file-sharing websites have brought together MS patients from all over the world and facilitated the sharing of personal experiences and information derived from medical research and its relationship with CCSVI (chronic cerebrospinal venous insufficiency). This paper carefully reviews the resources available on the Internet for MS patients seeking information on CCSVI, and the responsibilities of clinicians as they participate in these online discussions. As a result, there has been an accelerated growth in the number of patients seeking treatment for this syndrome, contemplating the possibility of improving their current condition.

The IoT through social media has also had an impact on mental health. In a study with this purpose is presented, in which the development and current uses of social networks on mental health are reviewed, ethical and legal dilemmas, as well as practical guidelines for mental health practitioners interested in developing a social networking site to support their professional practice. In conclusion, the ethical implications that clinicians should consider when using personal social networking sites are discussed.

- **Telemedicine Service**

  There is more than a one analyst who thinks that the IoT has the potential to revolutionize civilization and also increase the standard of living and increase industry performance of the economy on a lot more than the internet and IT can. Expanding the advantages of IoT [24–26] system-level qualities including scalability, dependability, inclusiveness, and general protection combine to generate a real opportunity for designers to develop efficient, highly successful and resilient IoT infrastructures. The paper shows the current state of the field of functionality in relation to academic institutions and researchers as well as its implementations which make things like the principles of cloud-based, operational, and safe all at the same time-integrated pillars in the coming. This paper gives an overview of the design of IoT, discussing the current state of the architecture of IoT architecture. Furthermore, highlighted to be ready for future IoT implementation are the IoT system-related problems that will allow for more real-time applications to be developed. There are millions of devices connected to the network and various protocols are required for the interconnection to be deemed relevant. This paper reveals the existing state of the communication requirements and implementation protocols for IoT as well as well as a systematic review of their actual and possible future use. Various service capabilities including data offloading, resource and system management are present in the computing paradigms including Cloud, Fog, Cloudlet, and Edge computing that help IoT move forward [27–29]. The main theme of this report is the search for better ways to make use IT to link devices and systems and operate things at the edge of an organization. As more and more people begin to use IoT, there may be problems of privacy and protection. This paper stresses how the protection concerns, risks, security strategies, and traditional mitigating tactics are on top of this subject [30, 31]. The additional scope that this address is presenting is concerning privacy. Scheduling is enabled by Real-Time Operating Systems (RTOS) [32, 33] simplifies the use of time-critical IoT applications because of their reduced memory footprints, and in-time execution, which is why it is an advantage for those working on them. To see the latest features and networking infrastructure specifications of the review also gives an overview of the RTOS's abilities with respect to IoT and networking. In the final section, it addresses open research problems in IoT device production.

4 Conclusion and Future Work

One of the widely used applications of telemedicine is the electrocardiogram (ECG); there are several published works. For example, in (60) the design of a portable cardiology system is presented that allows the electrocardiographic signal to be transmitted in real time via mobile telephony. A similar but more applied solution is presented in a study carried out in Bangladesh (South Asia), where the majority live in rural areas that lack specialized care, the need for Internet-based telemedicine systems is much greater, which would allow a large number of doctors and hospitals to collectively provide health services to entire populations. A scalable architecture is proposed, based on the Internet for telemedicine (ECG), which integrates multiple hospitals; is a distributed information system based on mobile telemedicine that allows connecting patients in rural locations with specialist doctors. Regarding connectivity, the solution indicates that the portable ECG equipment communicates through the mobile device to a server (robust computer equipped with sophisticated software and conditioned for the solution) located in the rural medical center. This, in turn, communicates through the network with a specialist who can verify and process the information to diagnose through their mobile device or a compute. The solution indicates that the portable ECG equipment communicates through the mobile device to a server (robust computer equipped with sophisticated software and conditioned for the solution) located in the rural medical center. This, in turn, communicates through the network with a specialist who can verify and process the information to diagnose through their mobile device or a computer. The solution indicates that the portable ECG equipment communicates through the mobile device to a server (robust computer equipped with sophisticated software and conditioned for the solution) located in the rural medical center. This, in turn, communicates through the network with a specialist who can verify and process the information to diagnose through their mobile device or a computer.

References

1. Swamy, S.N., Kota, S.R.: An empirical study on system level aspects of Internet of Things (IoT). IEEE Access **8**, 188082–188134 (2020). https://doi.org/10.1109/ACCESS.2020.3029847
2. Neshenko, N., Bou-Harb, E., Crichigno, J., Kaddoum, G., Ghani, N.: Demystifying IoT security: an exhaustive survey on IoT vulnerabilities and a first empirical look on internet-scale IoT exploitations. IEEE Commun. Surv. Tutor. **21**(3), 2702–2733 (2019). https://doi.org/10.1109/COMST.2019.2910750

3. Meneghello, F., Calore, M., Zucchetto, D., Polese, M., Zanella, A.: IoT: Internet of Threats? A survey of practical security vulnerabilities in real IoT devices. IEEE Internet Things J. **6**(5), 8182–8201 (2019). https://doi.org/10.1109/JIOT.2019.2935189
4. Said, O., Al-Makhadmeh, Z., Tolba, A.: EMS: an energy management scheme for green IoT environments. IEEE Access **8**, 44983–44998 (2020). https://doi.org/10.1109/ACCESS.2020.2976641
5. Frustaci, M., Pace, P., Aloi, G., Fortino, G.: Evaluating critical security issues of the IoT world: present and future challenges. IEEE Internet Things J. **5**(4), 2483–2495 (2018). https://doi.org/10.1109/JIOT.2017.2767291
6. Hwang, J., Aziz, A., Sung, N., Ahmad, A., Le Gall, F., Song, J.: AUTOCON-IoT: automated and scalable online conformance testing for IoT applications. IEEE Access **8**, 43111–43121 (2020). https://doi.org/10.1109/ACCESS.2020.2976718
7. Gutiérrez-Madroñal, L., La Blunda, L., Wagner, M.F., Medina-Bulo, I.: Test event generation for a fall-detection IoT system. IEEE Internet Things J. **6**(4), 6642–6651 (2019). https://doi.org/10.1109/JIOT.2019.2909434
8. Said, O., Albagory, Y., Nofal, M., Al Raddady, F.: IoT-RTP and IoT-RTCP: adaptive protocols for multimedia transmission over internet of things environments. IEEE Access **5**, 16757–16773 (2017). https://doi.org/10.1109/ACCESS.2017.2726902
9. Samaila, M.G., Sequeiros, J.B.F., Simões, T., Freire, M.M., Inácio, P.R.M.: IoT-HarPSecA: a framework and roadmap for secure design and development of devices and applications in the IoT space. IEEE Access **8**, 16462–16494 (2020). https://doi.org/10.1109/ACCESS.2020.2965925
10. Zarca, A.M., Bernabe, J.B., Skarmeta, A., Alcaraz Calero, J.M.: Virtual IoT HoneyNets to mitigate cyberattacks in SDN/NFV-enabled IoT networks. IEEE J. Sel. Areas Commun. **38**(6), 1262–1277 (2020). https://doi.org/10.1109/JSAC.2020.2986621
11. Bhayo, J., Hameed, S., Shah, S.A.: An efficient counter-based DDoS attack detection framework leveraging software defined IoT (SD-IoT). IEEE Access **8**, 221612–221631 (2020). https://doi.org/10.1109/ACCESS.2020.3043082
12. Muñoz, R., et al.: Integration of IoT, transport SDN, and edge/cloud computing for dynamic distribution of IoT analytics and efficient use of network resources. J. Lightwave Technol. **36**(7), 1420–1428 (2018). https://doi.org/10.1109/JLT.2018.2800660
13. Alsaedi, A., Moustafa, N., Tari, Z., Mahmood, A., Anwar, A.: TON_IoT telemetry dataset: a new generation dataset of IoT and IIoT for data-driven intrusion detection systems. IEEE Access **8**, 165130–165150 (2020). https://doi.org/10.1109/ACCESS.2020.3022862
14. Jain, P., et al.: Impact analysis and detection method of malicious node misbehavior over mobile ad hoc networks. Int. J. Comput. Sci. Inf. Technol. (IJCSIT) **5**(6), 7467–7470 (2014)
15. Condry, M.W., Nelson, C.B.: Using smart edge IoT devices for safer, rapid response with industry IoT control operations. Proc. IEEE **104**(5), 938–946 (2016). https://doi.org/10.1109/JPROC.2015.2513672
16. Sathyadevan, S., Achuthan, K., Doss, R., Pan, L.: Protean authentication scheme – a time-bound dynamic KeyGen authentication technique for IoT edge nodes in outdoor deployments. IEEE Access **7**, 92419–92435 (2019). https://doi.org/10.1109/ACCESS.2019.2927818
17. Shafiq, M., Tian, Z., Bashir, A.K., Du, X., Guizani, M.: CorrAUC: a malicious bot-IoT traffic detection method in IoT network using machine-learning techniques. IEEE Internet of Things J. **8**(5), 3242–3254 (2021). https://doi.org/10.1109/JIOT.2020.3002255
18. Simiscuka, A.A., Markande, T.M., Muntean, G.-M.: Real-virtual world device synchronization in a cloud-enabled social virtual reality IoT network. IEEE Access **7**, 106588–106599 (2019). https://doi.org/10.1109/ACCESS.2019.2933014
19. An, J., et al.: Toward global IoT-enabled smart cities interworking using adaptive semantic adapter. IEEE Internet Things J. **6**(3), 5753–5765 (2019). https://doi.org/10.1109/JIOT.2019.2905275

20. Cheng, Y., Xu, Y., Zhong, H., Liu, Y.: Leveraging semisupervised hierarchical stacking temporal convolutional network for anomaly detection in IoT communication. IEEE Internet of Things J. **8**(1), 144–155 (2021). https://doi.org/10.1109/JIOT.2020.3000771
21. Ngu, A.H., Gutierrez, M., Metsis, V., Nepal, S., Sheng, Q.Z.: IoT middleware: a survey on issues and enabling technologies. IEEE Internet Things J. **4**(1), 1–20 (2017). https://doi.org/10.1109/JIOT.2016.2615180
22. Chen, S., Xu, H., Liu, D., Hu, B., Wang, H.: A vision of IoT: applications, challenges, and opportunities with China perspective. IEEE Internet Things J. **1**(4), 349–359 (2014). https://doi.org/10.1109/JIOT.2014.2337336
23. Xu, D., Zhu, H.: Secure transmission for SWIPT IoT systems with full-duplex IoT devices. IEEE Internet Things J. **6**(6), 10915–10933 (2019). https://doi.org/10.1109/JIOT.2019.2943377
24. Azari, A., Stefanović, Č., Popovski, P., Cavdar, C.: On the latency-energy performance of NB-IoT systems in providing wide-area IoT connectivity. IEEE Trans. Green Commun. Netw. **4**(1), 57–68 (2020). https://doi.org/10.1109/TGCN.2019.2948591
25. Petrov, V., et al.: Vehicle-based relay assistance for opportunistic crowdsensing over narrowband IoT (NB-IoT). IEEE Internet Things J. **5**(5), 3710–3723 (2018). https://doi.org/10.1109/JIOT.2017.2670363
26. Hafeez, I., Antikainen, M., Ding, A.Y., Tarkoma, S.: IoT-KEEPER: detecting malicious IoT network activity using online traffic analysis at the edge. IEEE Trans. Netw. Serv. Manag. **17**(1), 45–59 (2020). https://doi.org/10.1109/TNSM.2020.2966951
27. Metzger, F., Hoßfeld, T., Bauer, A., Kounev, S., Heegaard, P.E.: Modeling of aggregated IoT traffic and its application to an IoT cloud. Proc. IEEE **107**(4), 679–694 (2019). https://doi.org/10.1109/JPROC.2019.2901578
28. Wang, J., et al.: IoT-praetor: undesired behaviors detection for IoT devices. IEEE Internet of Things J. **8**(2), 927–940 (2021). https://doi.org/10.1109/JIOT.2020.3010023
29. Meena, G., Sharma, D., Mahrishi, M.: Traffic prediction for intelligent transportation system using machine learning. In: 2020 3rd International Conference on Emerging Technologies in Computer Engineering: Machine Learning and Internet of Things (ICETCE), Jaipur, India, pp. 145–148 (2020). https://doi.org/10.1109/ICETCE48199.2020.9091758
30. Sultania, A.K., Mahfoudhi, F., Famaey, J.: Real-time demand response using NB-IoT. IEEE Internet Things J. **7**(12), 11863–11872 (2020). https://doi.org/10.1109/JIOT.2020.3004390
31. Fernández-Caramés, T.M.: From pre-quantum to post-quantum IoT security: a survey on quantum-resistant cryptosystems for the Internet of Things. IEEE Internet Things J. **7**(7), 6457–6480 (2020). https://doi.org/10.1109/JIOT.2019.2958788
32. Xu, L., Yin, W., Zhang, X., Yang, Y.: Fairness-aware throughput maximization over cognitive heterogeneous NOMA networks for industrial cognitive IoT. IEEE Trans. Commun. **68**(8), 4723–4733 (2020). https://doi.org/10.1109/TCOMM.2020.2992720
33. Chettri, L., Bera, R.: A comprehensive survey on Internet of Things (IoT) toward 5G wireless systems. IEEE Internet Things J. **7**(1), 16–32 (2020). https://doi.org/10.1109/JIOT.2019.2948888

Internet of Things Based Real-Time Monitoring System for Grid Data

Sanwta Ram Dogiwal[1], Pankaj Dadheech[1], Ankit Kumar[2](✉), Linesh Raja[3], Mukesh Kumar Singh[2], and Neha Janu[1]

[1] Swami Keshvanand Institute of Technology, Management and Gramothan, Jaipur, Rajasthan, India

[2] GLA University, Mathura, UP, India

kumar.ankit@gla.ac.in

[3] Manipal University Jaipur, Jaipur, Rajasthan, India

Abstract. Climate change, as well as the continuous and increasing exploitation of the planet's natural resources, have forced future generations to reconsider their behavior. Because the availability of fossil fuels is finite, and their usage as an energy source causes pollution and harms the ozone layer, it is critical to take steps to mitigate or eliminate the environmental effect. Improving the network's operation, lowering electrical losses, and incorporating a substantial share of renewable energy sources are just a few of the options available. These are the key reasons for the introduction of smart grids as a solution to the old power grid's architecture, which have sparked international interest and support. Information was kept in a SQLite database for this project. The measuring nodes' communications network was built utilizing the low-power, fault-tolerant ZigBee protocol. The developed supervision system allows for the reduction or deletion of third-party-managed communication nodes in some circumstances, resulting in increased independence and dependability. Through the growth of the National Software Industry and the usage of open-source platforms, it also provides for technical sovereignty.

Keywords: Smart grid · Smart energy meters · Supervision system · Grid data

1 Introduction

Using the infrastructure of communications, computing, electronics, and mathematical algorithms, the new intelligent network has diagnostic capabilities, decision-making, and efficiency that has never been seen in the history of the human race. However, there are additional challenges to contend with in the new millennium. Consumers are becoming increasingly concerned about network security, and this concern now extends to all of the entities involved. Risks of breaching client privacy must also be considered. It's impossible to ignore the rise in popularity of electric vehicles and the benefits they provide when used as a source or store of electricity during critical points in a grid's functioning. There is a need to rethink system stability, frequency management, and

V. E. Balas et al. (Eds.): ICETCE 2022, CCIS 1591, pp. 236–247, 2022.
https://doi.org/10.1007/978-3-031-07012-9_21

voltage control at nodes in light of electric automobiles and the Nano generation. It has been implemented at the substation and distribution level, with sophisticated methods for the identification and removal of defects. One final option is to rearrange the electrical supply circuit to ensure that as many customers as possible are supplied with electricity in the case of a permanent failure. These are just a few examples of how the new "smart" trend is reshaping society. It's the real-time monitoring capacity that allows the network to transition to a new design and accommodate all these new developments. Using free software, a monitoring system for smart electrical grids has been built that is tailored to the unique needs of Cuban grids. Flexibility, minimal system costs, and the establishment of a fundamental infrastructure are the foundations upon which its primary attributes are built. It's important to keep in mind that importing a system with such advantages comes at a great price. This massive transfer of technology may not be compatible with the country's technological, social, and political conditions, which must also be understood. Flexibility, minimal system costs, and the establishment of a fundamental infrastructure are the foundations upon which its primary attributes are built. It's important to keep in mind that importing a system that provides so many advantages come with a significant investment and upkeep expense. A technology transfer of this size may not be compatible with the country's technological, social, and political conditions. Flexibility, minimal system costs, and the establishment of a fundamental infrastructure are the foundations upon which its primary attributes are built. Consider that bringing in a system with this many advantage would demand a significant outlay of capital as well as ongoing maintenance expenses. A technology transfer of this size may not be adaptable to the country's technological, social, and political contexts.

2 Literature Review

There is a lot of focus on security, latency, fault tolerance, and communication protocols when it comes to the monitoring and management of smart grids. As more and more devices become interconnected, the amount of data that must be monitored and controlled grows exponentially. Flexible management, virtually limitless growth, and data flow monitoring are all requirements for these systems. Developing countries that are at the forefront of smart grids have a robust communications infrastructure that is constantly being improved. We have a unique challenge in the United States, where we lack a fully-deployed network capable of handling the additional data volumes necessary. It's common for the monitoring system to be linked to the internet, which necessitates extra security precautions and takes advantage of a third-party communications network. It is necessary to have a communication network with high fault tolerance for this system due to its importance. This necessitates that the communications network be guaranteed to function. It is suggested to use open-source software technologies to create a national-level monitoring system. Python was chosen because of its versatility across platforms and paradigms. This programming language has grown in popularity in recent years and may now be found in nearly every aspect of modern computing. The ZigBee protocol is utilized by connecting devices to link the terminals and their deployment [1]. Low power consumption, reasonable data transmission speed and the ability to connect a large number of devices are all advantages of this protocol.

2.1 Predecessor Architecture of Smart Grids

One of the most distinguishing features of the older, more conventional electrical networks is the one-way flow of electricity from the generators or big centres to the consumers. Communication and computation among the components can be extensive, although isolation between them predominates [2]. A comprehensive system of management and decision-making is not available since there is no centralised system or "intelligence." For a long time, the architecture of electrical power systems [3] may be shown in Fig. 1.

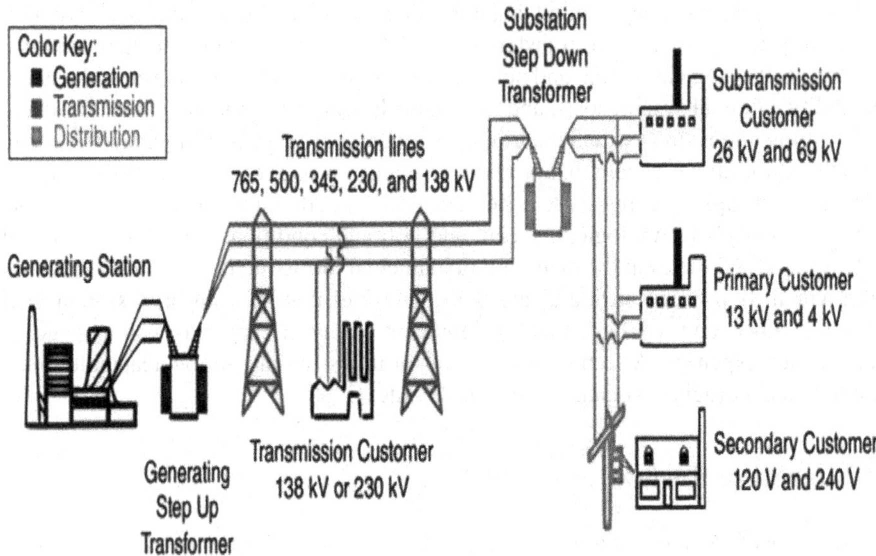

Fig. 1. Electrical power systems predecessors of smart grids

There are several drawbacks [4] to this architecture, including the need to transmit energy across large distances and via the whole electrical system, which results in significant losses. Fossil fuels, which contribute to climate change, are the primary source of energy in the United States. Due to their size and complexity, large power plants are only meant to operate at full capacity. There's also the lost revenue from not using it to its full potential. Because energy storage capacity is rare or nonexistent, it is required to continually rebalance supply and demand.

The capacity to remotely monitor all or the majority of energy users in a reliable and structured manner is probably what encouraged the new structure of renewable energy producing centers, where there may even be a high degree of communication between its parts. This enables the generation [5] to be as near to the loads as feasible, on a big scale and with a high level of penetration into the electrical system. Loads should be able to generate their own energy, which gives birth to the term "prosumer" (producer and consumer of energy at once).

2.2 Need for Change in the Electricity Grid

The power grid may now go into a new phase of growth thanks to additional technical support infrastructures [6, 7]. Electronics, communications, and large-scale computing in everyday life supplied this assistance [8, 9]. The electrical network might change as a result of the following things in this context of technological advancement:

1. There is a large portion of the energy consumed on the creation and delivery of electricity. The network's conventional architecture does not allow for bringing the generation to the loads directly as a solution.
2. Second, the traditional network has severe issues with the integration of all forms of generation and storage, including wind, solar, hydro, and nuclear.
3. There is a lack of ability to manage and dispose of resources effectively.
4. In order to protect the conventional network, its primary anticipatory response is to respond. In order to prevent and minimise the negative effects on customers, it is vital to have the ability to automatically recognise and respond to issues of concern.
5. A sluggish and susceptible reaction to terrorist attacks and natural calamities.
6. The integration of energy sources at the nano-generation scale is constrained since consumers are passive and unaware of the network's condition.
7. It's difficult to participate in energy markets since the services provided are restricted to energy delivery.
8. Climate change, fossil fuel depletion, a growth in energy use (and the number of people using it, which increases the network's complexity), new energy sources, and more are all connected in some way or another. renewable energy sources and the cycle of technological innovation.

2.3 Architecture of Smart Grids

The electrical grid specialist's defining criterion for the smart grid, or at least the weight of the relevance it gives, will be established in this way, depending on his or her vision [10]. Incomplete definitions or areas of higher importance include: distributed generation, substation automation and sophisticated metering infrastructure with energy metres. Every single one of the above-mentioned criteria is somewhat correct [11]. Intelligence-capable networks are those capable of incorporating a wide range of modern measuring technologies as well as communication methods into the electrical network.

2.4 Consumer Participation

1. There are options to handle a variety of power production sources. Set up a new power market that includes new services and goods [12].
2. Excellent quality energy that can support an economy built on electronic communication, computation, and network functioning in the digital era.
3. Efficient use of existing and newly acquired resources.
4. Network self-healing by predicting and responding to network disruptions.
5. Self-defence against both natural and man-made threats

Figure 2 depicts the further development of the brand-new architecture. The smart grid is becoming more complicated as new applications and technology are added. In addition to the internet of things existing in the demotics of buildings and smart homes, there is a rise in internet support for the smart grid [13–16].

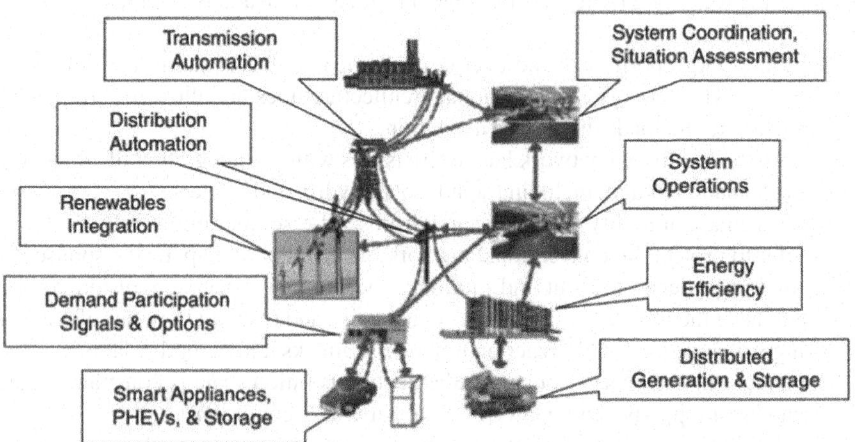

Fig. 2. Representative architecture of smart grids

3 Proposed Work

3.1 Tailored Supervision System Developed for Cuban Conditions

The Advanced Metering Infrastructure (AMI) is at the heart of all smart grid projects [17]. The smart energy meter is the centerpiece of this system. In 2007, Italy became the first country to use this particular device. It is a digital electronic device that can process and store data digitally, as well as display it visually and communicate it. For this reason, the intelligent network's bandwidth and dispersed processing capacity must also be upgraded to accommodate new methods of transmitting encrypted data. Unlike the more developed countries, not all developing countries have a communications infrastructure or Internet connection that meets the standards of speed, operational assurances, expansion capacity and technologies, appropriate or current. In the past, mobile networks have been used to link smart network terminals to each other, as well as to the Internet, and this method has been widely accepted.

The existence of a legal framework must be taken into consideration because, if all or most of the remote measuring equipment is connected to the communication networks of another specialized firm, the legal framework must be taken into consideration. When there is a problem with the data network, the electrical provider takes the lead. There must thus be regulations and legislation in place to penalize and encourage the corporations in charge of maintaining the energy grid's stability and security. Because they can better control their own infrastructure and services, electricity firms often choose not to rely on a third party for their critical infrastructure.

3.2 Communication System Proposal Must Take into Account the Following Aspects

- The highest number of nodes and connection complexity is in the last mile.
- The physical layout of the terminal measurement equipment and the architectural barriers for their deployment and operation.
- Technical flexibility for the wide variety of existing equipment and conditions.
- Support for the volume of data and number of terminals.
- Low operating consumption.
- Transmission and reception distance.

The Internet's communications infrastructure connects the nodes in the last mile of a path. The system's running expenses might be exorbitant in less developed countries. Reducing the number of nodes in this infrastructure and utilizing network hubs to provide a two-way communication channel is one solution to this problem.

Licenses or patents are another technical and financial issue. There is a wide range of equipment vendors for the key infrastructure of smart grids. It's a problem because the vast majority of them were created under non-public permission. In a technologically backward country, reliance on this sector might paralyses and badly hinder your progress, not to mention the licensing charge. The United States of America's economic policies have recently had an impact on the Chinese corporation Huawei. The temporary ban on the use of certain technology and electronic gadgets makes it impossible to quantify the company's economic and public image losses.

According to Fig. 3, an improved measuring infrastructure that is essential for an intelligent network may be managed, decreasing technology reliance and foreign policy. With regard to its development, open platforms have been utilized as much as possible. The use of free software in the development process is essential for reducing costs and preserving autonomy.

Clusters or blocks can be used to deploy the system, which is built in layers. It is necessary to divide the network into three distinct layers: the top layer is in charge of overseeing everything, the middle layer is in charge of setting up the primary communication network and is equipped with the necessary routers to do so, and the bottom layer is in charge of connecting all of the final terminal units (intelligent energy meters but not exclusively). The server-side application used to keep track of the infrastructure is an intriguing aspect of this project. The Python programming language is used to create the software. Code readability is prioritized in the syntax of this language. Due of its support for object-oriented, imperative, and a lesser extent functional programming, it is multi-paradigm. The GNU General Public License [18] is compatible with version 2.1.1 of this open source project. A database management that adhered to the principles of free software was chosen.

3.3 Supervision System Operation Method

Unlike other computer programs, a monitoring system must have scheduled tasks that operate at well-defined time intervals. Professional SCADA systems have the ability to handle a complex structure of timed tasks. The main difference with a general-purpose

SCADA system is that the developed system has a specific application, which means that a less complex task call time control system can be created [19, 20].

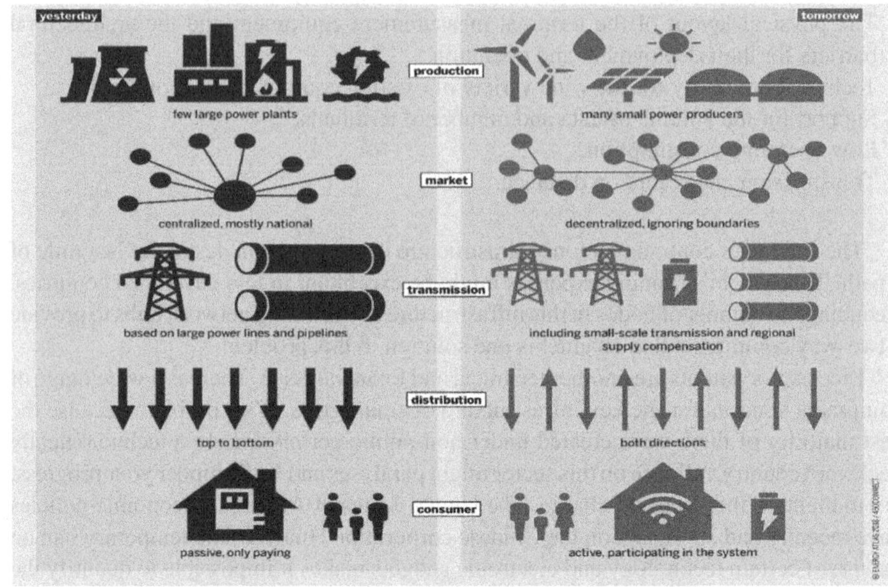

Fig. 3. Advanced communications infrastructure of an intelligent network

The Python implementation for a multitasking system with execution in well- defined times can be done by two methods: using threads (Threading) or by multiple processes (Multiprocessing). Both options have their advantages and disadvantages. The last option being the one selected to implement in this work, the timed tasks.

There are several high priority timed tasks:

Request and storage of the data presented by the terminal devices. Monitoring and updating of the system database [21, 22]. Updating of all the information shown in the program. Status of the network nodes. There are other tasks in the system that are not mentioned so as not to make it difficult to explain the basic operation of the system. System alarms is an example of the tasks in the previous case. There are tasks that can be seen as a service to a more complex and higher order task.

For all of the above, we are not in the presence of a sequential program in its operation, but it is based on events with and without timing. In Fig. 4, there is a block diagram representing the simplified algorithm of the high-level functions of the super- vision system.

When executing the program code, the necessary libraries are initially loaded, followed by some predefined variables (parameters). The variables allow setting the basic configuration of not only the graphical environment, but also the needs of the database.

The rest of the algorithm is event-based. There are three processes of utmost importance: the timing of the data capture of the measurement devices, the operations in the database and the operation of the supervisory system in charge of managing all the other

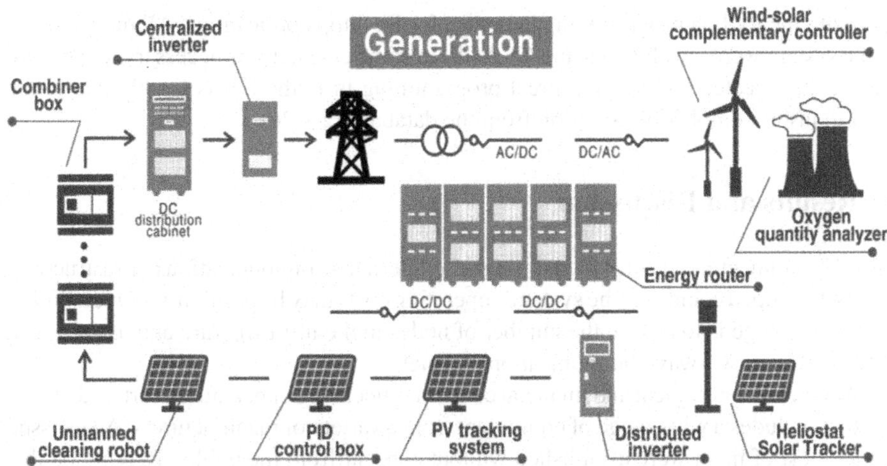

Fig. 4. Proposed model developed using artificial supervision system

processes [23, 24]. Although it is not represented in the diagram, the processes have an information exchange mechanism between them.

3.4 Design of the Graphical User Environment

The application's graphical windows were implemented through a set of multiplateform libraries developed for the creation of graphical user interfaces (GUI). The GTK+ set has excellent features and facilitates design using the GLADE tool, which allows the interfaces to be designed graphically and independently of the programming language. See Fig. 5. The tool is multiplatform and open software, contributing to the philosophy of keeping the code free of proprietary patents. It integrates in a simple and very comfortable

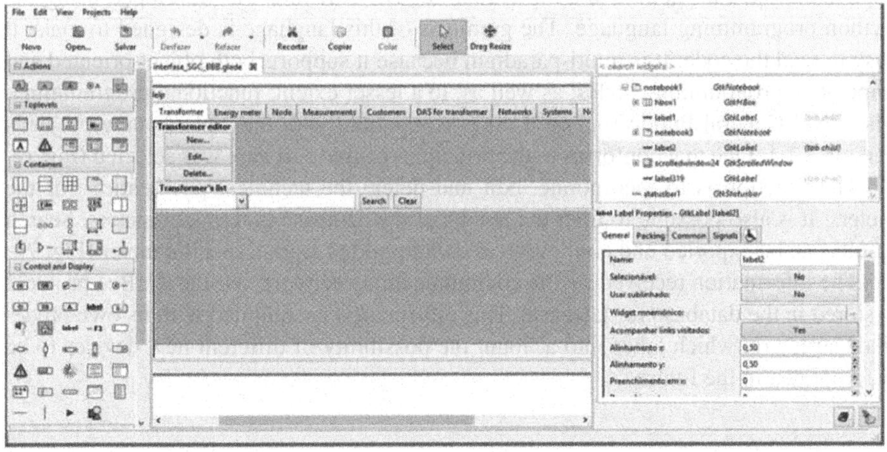

Fig. 5. Design environment of the graphical user interface

way with the Python programming language for Desktop applications. All the windows of the system were developed using GLADE. In some cases, it was necessary to generate part of the selected window by direct programming in Python. It is largely due to the need to insert complex information from the database [25, 26].

4 Results and Discussion

Last mile connectivity is provided through the Internet's communications infrastructure. In less developed countries, the system's operating costs may be prohibitive. One solution to this challenge is to reduce the number of nodes in the infrastructure and use network hubs to offer a two-way communication channel.

Another technological and financial concern is licensing or patents. Smart grid infrastructure includes a wide range of equipment from a variety of manufacturers. A key issue is that most of them were established without consent from the public. Relying on this sector in a technologically underdeveloped nation might paralyze and severely restrict your growth, not to mention the licensing fee. Economic policies in the United States have recently had an influence on Huawei, the Chinese multinational. The company's economic and public image losses cannot be quantified because of a temporary prohibition on the usage of certain technologies and electronic gadgets.

To lessen dependency on foreign policy and foreign technology, it is possible to strengthen the measurement infrastructure necessary for an intelligent network, as shown in Fig. 3. To the greatest extent feasible, open platforms have been utilized in the creation of this software. In order to keep expenses down and maintain control, free software must be used in the development process.

It is possible to install the system in stages using clusters or blocks. the top layer is in charge of monitoring everything and is equipped with routers to establish up a primary communication network whereas layer two connects every terminal unit, layer three is in charge of setting up a secondary communication network (intelligent energy metres but not exclusively). An noteworthy component of this project is the server-side application used to monitor the infrastructure. The software was developed using the Python programming language. The grammar of this language is designed to make it easy to read the code. It is multi-paradigm because it supports both object-oriented and imperative programming styles, as well as, to a lesser extent, functional programming. The GNU General Public License [18] is compatible with this open source project's version 2.1.1. One that conforms to the principles of free software was selected (Fig. 6).

The function is to incorporate, edit and delete the technical pacifications of the meters. It is also possible to filter the results of the database in terms of meters. Search results can be exported and configuration data imported from a specific meter (Fig. 7).

The information received by the communications network and the ZigBee protocol is stored in the database for later use. This information is compiled with its own predefined structure, which takes into account the possibility of different new devices to be incorporated in the future.

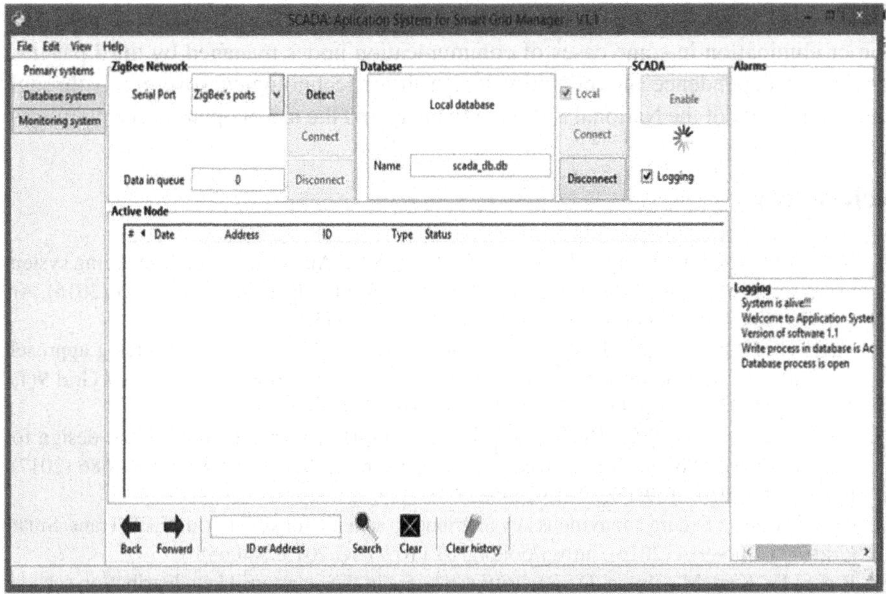

Fig. 6. Monitoring system of the metering infrastructure of a smart grid using open software

Energy meter editor
New...
Edit...
Delete...
Tools
Import...
Export...
Energy meter's list
All
Search
Clear

#	Identification code	Manufacturer	Fases	Power	Voltage	Current	Version	Date of build	Xbee address
0	LAT_metro_5555	Latino	3	15.0 kVA	240.0 V	100.0 A	V1.01	15/August/2013	0B32A45C456
1	LAT_metro_4587559	Latino	1	10.0 kVA	120.0 V	100.0 A	V1.84	15/April/2014	AA45C34D560
2	LAT_metro_5887145	Latino	3	10.0 kVA	240.0 V	100.0 A	V1.84	27/February/2016	0B32A45C125
3	LAT_metro_5887155	Latino	1	25.0 kVA	440.0 V	100.0 A	V1.5	27/February/2015	0B32A45D857
4	LAT_metro_258147	Latino	3	15.0 kVA	440.0 V	100.0 A	V2.10 - M1	20/August/2014	0A12B23D323

Fig. 7. Capture of the administration of smart meters belonging to the supervision system.

5 Conclusion

A system capable of performing the supervision and monitoring tasks of an intelligent network of less developed structure was obtained, using open programming, through free Python software. The information obtained was stored in a SQLite database. The communications network for the measurement nodes was deployed using the low-power, fault-tolerant ZigBee protocol. The implemented supervision system allows the reduction or elimination in some cases of communication nodes managed by third par-ties, providing independence and reliability. It also allows technological sovereignty through the development of the National Software Industry and the use of open source platforms.

References s

1. Xing, Y.Q., Jin, J.X., Wang, Y.L., Du, B.X., Wang, S.C.: An electric vehicle charging system using an SMES implanted smart grid. IEEE Trans. Appl. Supercond. **26**(7), 1–4 (2016). Art no. 5701504. https://doi.org/10.1109/TASC.2016.2602245
2. Yang, H., Zhang, J., Qiu, J., Zhang, S., Lai, M., Dong, Z.Y.: A practical pricing approach to smart grid demand response based on load classification. IEEE Trans. Smart Grid **9**(1), 179–190 (2018). https://doi.org/10.1109/TSG.2016.2547883
3. He, X., Ai, Q., Qiu, R.C., Huang, W., Piao, L., Liu, H.: A big data architecture design for smart grids based on random matrix theory. IEEE Trans. Smart Grid **8**(2), 674–686 (2017). https://doi.org/10.1109/TSG.2015.2445828
4. Tsai, J., Lo, N.: Secure anonymous key distribution scheme for smart grid. IEEE Trans. Smart Grid **7**(2), 906–914 (2016). https://doi.org/10.1109/TSG.2015.2440658
5. Park, J.H., Kim, M., Kwon, D.: Security weakness in the smart grid key distribution scheme proposed by Xia and Wang. IEEE Trans. Smart Grid **4**(3), 1613–1614 (2013). https://doi.org/10.1109/TSG.2013.2258823
6. Sha, K., Alatrash, N., Wang, Z.: A secure and efficient framework to read isolated smart grid devices. IEEE Trans. Smart Grid **8**(6), 2519–2531 (2017). https://doi.org/10.1109/TSG.2016.2526045
7. Spanò, E., Niccolini, L., Pascoli, S.D., Iannacconeluca, G.: Last-meter smart grid embedded in an internet-of-things platform. IEEE Trans. Smart Grid **6**(1), 468–476 (2015). https://doi.org/10.1109/TSG.2014.2342796
8. Kim, J., Cho, S., Shin, H.: Advanced power distribution system configuration for smart grid. IEEE Trans. Smart Grid **4**(1), 353–358 (2013). https://doi.org/10.1109/TSG.2012.2233771
9. Moghbel, M., Masoum, M.A.S., Fereidouni, A., Deilami, S.: Optimal sizing, siting and operation of custom power devices with STATCOM and APLC functions for real-time reactive power and network voltage quality control of smart grid. IEEE Trans. Smart Grid **9**(6), 5564–5575 (2018). https://doi.org/10.1109/TSG.2017.2690681
10. Tiwari, A., Sharma, V., Mahrishi, M.: Service adaptive broking mechanism using MROSP algorithm. In: Kumar Kundu, M., Mohapatra, D.P., Konar, A., Chakraborty, A. (eds.) Advanced Computing, Networking and Informatics- Volume 2. SIST, vol. 28, pp. 383–391. Springer, Cham (2014). https://doi.org/10.1007/978-3-319-07350-7_43
11. Ahn, M.C., Ko, T.K.: Proof-of-concept of a smart fault current controller with a superconducting coil for the smart grid. IEEE Trans. Appl. Supercond. **21**(3), 2201–2204 (2011). https://doi.org/10.1109/TASC.2010.2091386
12. Malpani, P., et al.: A novel framework for extracting geospatial information using sparql query and multiple header extraction sources. In: Afzalpulkar, N., Srivastava, V., Singh, G., Bhatnagar, D. (eds.) Proceedings of the International Conference on Recent Cognizance in Wireless Communication & Image Processing. Springer, New Delhi (2016). https://doi.org/10.1007/978-81-322-2638-3_56
13. Uludag, S., Lui, K., Ren, W., Nahrstedt, K.: Secure and scalable data collection with time minimization in the smart grid. IEEE Trans. Smart Grid **7**(1), 43–54 (2016). https://doi.org/10.1109/TSG.2015.2404534

14. Morello, R., De Capua, C., Fulco, G., Mukhopadhyay, S.C.: A smart power meter to monitor energy flow in smart grids: the role of advanced sensing and IoT in the electric grid of the future. IEEE Sens. J. **17**(23), 7828–7837 (2017). https://doi.org/10.1109/JSEN.2017.2760014
15. Wan, Y., et al.: An integrated cyber-physical simulation environment for smart grid applications. Tsinghua Sci. Technol. **19**(2), 133–143 (2014). https://doi.org/10.1109/TST.2014.6787366
16. Monteiro, V., Pinto, J.G., Afonso, J.L.: Operation modes for the electric vehicle in smart grids and smart homes: present and proposed modes. IEEE Trans. Veh. Technol. **65**(3), 1007–1020 (2016). https://doi.org/10.1109/TVT.2015.2481005
17. Ye, F., Qian, Y., Hu, R.Q., Das, S.K.: Reliable energy-efficient uplink transmission for neighbourhood area networks in smart grid. IEEE Trans. Smart Grid **6**(5), 2179–2188 (2015). https://doi.org/10.1109/TSG.2015.2392130
18. Gope, P., Sikdar, B.: Privacy-aware authenticated key agreement scheme for secure smart grid communication. IEEE Trans. Smart Grid **10**(4), 3953–3962 (2019). https://doi.org/10.1109/TSG.2018.2844403
19. Ullah, R., Faheem, Y., Kim, B.: Energy and congestion-aware routing metric for smart grid ami networks in smart city. IEEE Access **5**, 13799–13810 (2017). https://doi.org/10.1109/ACCESS.2017.2728623
20. Koutitas, G.: Control of flexible smart devices in the smart grid. IEEE Trans. Smart Grid **3**(3), 1333–1343 (2012). https://doi.org/10.1109/TSG.2012.2204410
21. Cassidy, A., Strube, M., Nehorai, A.: A framework for exploring social network and personality-based predictors of smart grid diffusion. IEEE Trans. Smart Grid **6**(3), 1314–1322 (2015). https://doi.org/10.1109/TSG.2014.2366729
22. Gong, Y., Cai, Y., Guo, Y., Fang, Y.: A privacy-preserving scheme for incentive-based demand response in the smart grid. IEEE Trans. Smart Grid **7**(3), 1304–1313 (2016). https://doi.org/10.1109/TSG.2015.2412091
23. Akaber, P., et al.: CASeS: concurrent contingency analysis-based security metric deployment for the smart grid. IEEE Trans. Smart Grid **11**(3), 2676–2687 (2020). https://doi.org/10.1109/TSG.2019.2959937
24. Zhang, N., Hu, Z., Dai, D., Dang, S., Yao, M., Zhou, Y.: Unit commitment model in smart grid environment considering carbon emissions trading. IEEE Trans. Smart Grid **7**(1), 420–427 (2016). https://doi.org/10.1109/TSG.2015.2401337
25. Tushar, W., et al.: Energy storage sharing in smart grid: a modified auction-based approach. IEEE Trans. Smart Grid **7**(3), 1462–1475 (2016). https://doi.org/10.1109/TSG.2015.2512267
26. Yassine, A., Shirehjini, A.A.N., Shirmohammadi, S.: Smart meters big data: game theoretic model for fair data sharing in deregulated smart grids. IEEE Access **3**, 2743–2754 (2015). https://doi.org/10.1109/ACCESS.2015.2504503

Machine Learning and Applications

Identification and Classification of Brain Tumor Using Convolutional Neural Network with Autoencoder Feature Selection

M. S. Hema(✉), Sowjanya, Niteesha Sharma, G. Abhishek, G. Shivani, and P. Pavan Kumar

Department of Information Technology, Anurag University, Hyderabad, India
hemait@anurag.edu.in, {sowjanyait,nitishasharma}@cvsr.ac.in

Abstract. The brain tumor is deadly disease. The correct and early identification is required to cure and improve the brain tumor patients' quality of life. The manual identification of brain tumor is very difficult. Sometimes, it leads to wrong prediction also. Computer aided identification and classification helps the doctors to identify and classify tumors accurately. To mitigate the above issues, Convolutional Neural Network (CNN) along with autoencoder based feature selection was proposed. The auto encoder was used to extract the relevant feature and eliminate the noise. The CNN was correct choice for prediction of brain tumor. The MRI brain images were used for classification and prediction of brain tumors. The CNN has divided into e layers namely pooling layer, convolutional layer and dense layer. The Convolutional layer divides the images in to segments and extract the relevant features. The pooling layer downsize the image to reduce the computational complexity. The tumors are identified and classified in the dense layer. The BRATS 2013 and BRATS 2015 data set is taken for experimentation. The accuracy, sensitivity, specificity, precision and Dice Similarity Coefficient matrices is considered for performance assessment. The proposed methodology performed better when compared to traditional CNN, Decision tree and Bayesian classification.

Keywords: Brain tumor · Auto encoders · Convolutional neural network · MRI brain images · Image segmentation

1 Introduction

The brain tumor is one among the deadliest tumor among all kinds of tumors. Fourteen thousand deaths are happening every year due to brain tumor. The brain tumor is divided into three categories. The category one patients have prolonged survival. The tumor will form slowly. The category two patients' tumor growth is slow but it will spread in the neighboring tissues. The category three patients' tumor tissues will duplicate quickly and it causes the death quickly [1]. So, early and accurate prediction of brain tumor is essential. The correct prediction of brain tumor is used not only for treatment of the patients and also for further investigations and follow ups. It will improve the patients' quality of life. There are so many methods such as MRI, X-ray and ultrasound which are the inputs

V. E. Balas et al. (Eds.): ICETCE 2022, CCIS 1591, pp. 251–258, 2022.
https://doi.org/10.1007/978-3-031-07012-9_22

used for predicting tumor in brain. Among these methods MRI brain image is widely used to identify and classify the brain tumor [2]. The dissection of MRI images is an important step in identification and classification of tumor. The image segmentation is done either manually or automatically. The manual segmentation of image will take more time consuming and also typing error may occur. Automatic segmentation is very effective and less prone. The computer aided automatic segmentation will improve the accuracy. There are two categories of learning algorithms namely supervised and unsupervised the main usage of them is for segmentation and classification [3]. The supervised learning methods are Support Vector Machines (SVM), Artificial Neural Network (ANN), Decision tree and Bayesian classification. The unsupervised algorithms are clustering and self-organizing maps [9]. Now days CNN supervised learning method is gaining popularity because of its automatic feature extraction and efficient classification. In CNN, the important features from the input image are extracted by convolutional layer using hyper parameters. The pooling layer is downsizing the image to reduce computational complexity. The dense layer classifies the image.

The contributions of the proposed methodology are as follows

- A methodology is proposed to extract relevant features, identify and classify brain tumor automatically.
- The autoencoder eliminates noise and extract relevant features from the image
- CNN segments MRI image, identifies and classifies the tumor in brain automatically.

The organization of the article is as follows. In Sect. 2, the previous works related to proposed methodology is summarized. In Sect. 3, the proposed methodology includes extraction of the feature, image segmentation, classification of brain tumor is presented. In section, the results are discussed. The conclusion is presented in Sect. 5.

2 State of Art

An automatic segmentation method based on convolutional Neural Networks is proposed. Usage of 3 × 3 kernels for image segmentation was done. Thus, usage of small kernels gave an effective overfitting and also the less no. of weights in the network. Also, the authors have used the intensity normalization as one of the preprocessing steps in their study which was proved to be effective for MRI images in brain tumor segmentation. The Computation time was reduced by tenfold [1]. A methodology was proposed to identify tumor objects in MRI. A segmentation method based on color which uses K-means clustering was developed. The main use of introducing K-means clustering in this was to convert a MR image from gray level into a RGB Color and then into a CIElab color model. To prove the above as a better segmentation, model the analysis was done on two datasets. 1) MRI Gray features and 2) RGB from color image [2]. The study has proposed a fully automatic segmentation model for brain tumor. U- Net based Networks is used for developing the model. BRATS2015 dataset is used which had both HGG and LGG Patients data [3]. A methodology used for image segmentation to identify brain tumor was proposed. For this DNN architecture was used. The proposed model used three convolutional layers with ReLU activation function. In dense layer the activation function- softmax was used. BRATS 2012, 2013, 2014, 2015, and 2017 data set was taken for experimentation [4]. A comparison of automatic detection methods in MRI images of tumor in brain was discussed. The authors based on LVQ have implemented modified probabilistic neural network model on image and data analysis. The experiment performed showed that modified PNN had an accurate classification [5]. The authors proposed an automated method on MRI of the brain tumor which can easily differentiate cancerous and non-cancerous disease. SVM methodology was proposed to categorize the brain tumor. Different preprocess steps were used to segment the images. Linear, Gaussian and cubic kernel sum variations were used. The local, Harvard and rider data sets were taken for experimentation. Area Under Curve (AUC) and accuracy performance metrics were taken for performance assessment [6]. A multi model for image segmentation and classifying the tumor was proposed. Fractural feature extraction and wavelet analysis methods were combined. The various types of MRI brain images were taken for experimentation. The self-organized map was used for feature fusion. The tumor regions from non-tumor regions were classified with the help of feedforward neural network [7]. The authors have developed a segmentation method for tumor in brain using two-D and three-D MRI data. The authors to assess a tumor have dome manual segmentation, 2D & 3D Visualizations. The study was first restricted to area of the tumor but was later extended to find the volume of the tumor [8]. A method for brain tumor detection based on Machine learning was proposed. The categorization of work is divided into the following phases: MRI images preprocessing, Texture features classification is done through muti layer perception, Naïve Bayes algorithm [9]. An approach was proposed to identify the brain tumor from MRI brain images. The input image quality was enhanced and the noise was suppressed. Potential field clustering was used for finding the subsets of tumor pixels [10]. A deep learning architecture [DLA] was developed which automatic detects the brain tumor. The DLA's are implemented using deep features, classification is done using KNN, SVM and a customized VGG19 network is developed to improve accuracy [11]. Various methods like Median Filtering for Noise Removal, De-noising Filters, Image Enhancement, Edge Detection, Threshold, Morphological Operation, segmentation and many more were discussed [12]. An effective method was developed for detecting tumor tissues. The segmentation was accomplished using clustering algorithm - k-means. The method proposed concentrated more on reducing the intensive distance computation between data point and previous nearest cluster. Also, it provides an efficient method to extract the Region of Interest [ROI] [13]. The authors proposed two approaches called identification and classification for brain tumor identification. The approaches were based on image processing algorithms and probabilistic ANN. The implementation is done using MATLAB and Canny Edge detection algorithm is used [14]. Proposed a methodology in which similarity of PD patients was identified. The above method is implemented using following steps: collection of data, Patient's similarity analysis, identification of subgroup and prediction of outcome [15]. A machine learning technique to identify the portion of metamorphic changes affected by Parkinson's disease was proposed [16]. A CNN based methodology to identify brain tumor was discussed. Method was used for accurate separation of MRI brain images. The features are optimized using entropy. The optimized features ensure the accurate and fast classification [17]. Delivered a detailed comprehensive study on related work on brain tumor detection. Brain MRNET: A new CNN model was proposed. The attention model was used for selecting important area in the image. Then convolutional layer is used to extract the features [18]. Auto encoder model was used to identify slice as tumor. The segmented slices were given to stacked sparse auto encoder model. Two hidden layer was used. 200 neurons and 400 neurons were used respectively in two layers [19]. Big Data technologies as well as various ML algorithms to predict the PD disease were used. The various features based on movement disorder are used for the prediction model. The work aimed at identifying the biomarkers used for cause of disease using big data., analyze data needs, visualization techniques [20].

3 Proposed Methodology

The aim of the methodology proposed is to classify the tumor in brain. The CNN is proposed for classification. The auto encoder is used to remove noise from the input and also used to select relevant features from the image. The architecture is illustrated in Fig. 1.

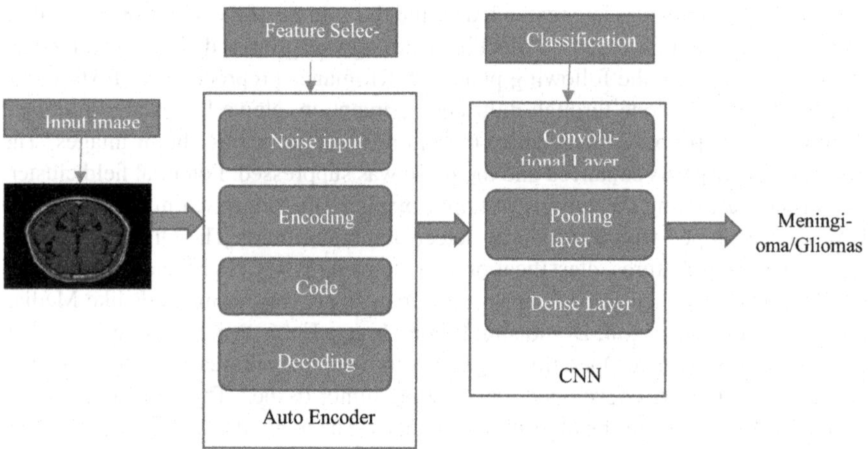

Fig. 1. Brain tumor classification approach

The input image is given to auto encoder to extract required features. The feature selection is essential step to improve the classification performance. The single hidden layer autoencoders with hyperbolic tangent activation function is used for feature selection. Each feature is considered as one neuron. The weight is associated with each feature. The number features are represented as n. The output (O_{ij}) of the input layer is calculate using formula shown Eq. 1.

$$O_{ij} = \sum_{n=1}^{N} h_{in} \times o_{nj} \tag{1}$$

h_{in} represents the contribution of ith input feature to the nth neuron in the hidden layer. It is calculated using formula shown in Eq. 2. The total number of input is represented as n

$$h_{in} = \frac{|W_{in}|}{\sum_{i=1}^{n} |W_{i*n}|} \tag{2}$$

where W_{in} represents the weight of the corresponding input feature.

o_{nj} is contribution of nth neuron in the hidden layer to jth output layer. It is calculated similar to h_{in}. The final output (O_i) is calculated. The formula is shown in Eq. 3

$$O_i = \sum_{j=1}^{M} \frac{O_{ij}}{\sum_{i=1}^{M} O_{i*J}} \tag{3}$$

Ranked the input features according to O_i values. Then top 50% of the features taken for classification.

The CNN model consists of different number of layers. The first layer of CNN is convolutional layer, which divides the input images into number of sub images and then extracts the features from the image. The 3×3 or 5×5 kernels are applied against the input image and features are extracted. It will extract all features from the images even in the edge of the image. The output of this layer is given as input for the activation function to obtain nonlinear output. In CNN, Rectified Linear Unit (ReLU) activation function is used in this layer. The second layer of CNN is pooling layer, which is used to downsize the image to reduce computational complexity of the model. Pooling layer is categorized into two: Max and Average pooling layer. The prior is used to downsize the image in proposed approach. The final layer of CNN is dense layer, which is used to classify the image as tumor image or health image. The softmax activation function is used in dense layer.

3.1 CNN Steps

1. Start with encoder output as input to the convolutional layer.
2. Apply filter in convolutional layer to extract the feature map.
3. The features are given to ReLU function to get non linearity.
4. The output activation function is given to pooling layer to down size the image.
5. The steps 2 to 4 is repeated until specified number of convolutional and pooling layer.
6. Flatten the output of the pooling layer into single vector.
7. The single vector is given as input to dense layer.
8. The dense layer performs classification.
9. The network is trained through forward propagation and backward propagation for specified number of epochs.

4 Results and Discussion

BRATS 2013 and BRATS 2015 datasets are taken for experimentation. The noise removal and feature extraction is done using autoencoders. The CNN model is used for classification. It is a multi-class classification. The metrics used to assess the proposed work and comparison are shown in Eqs. 4–8

$$DSC = 2True\,Positive/False\,Positive + 2True\,Positive + Falese\,Negative \quad (4)$$

$$SE = True\,Positive/True\,Positive + False\,Negative \quad (5)$$

$$Accuracy = True\,Positive + True\,Negative/True\,Positive + True\,Negative + False\,Positive + False\,Negative \quad (6)$$

$$SP = True\,Negative/True\,Negative + False\,Positive \quad (7)$$

$$Precison = True\,Positive/True\,Positive + False\,Positive \quad (8)$$

where DSC is Dice Similarity Coefficient. SE is sensitivity and SP is specificity.

The proposed approach performance is shown in Table 1. The performance of the proposed approach and existing methodology is shown Table 2. The methodology performs better when compared to existing methodology.

Table 1. Proposed approach-performance

Dataset	DSC	SE	SP	Precision	Accuracy
BRATS 2013	98.8	97.3	99.8	98.2	99.28
BRATS 2015	97.34	96.03	98.3	97.8	98.5

Table 2. Proposed approach and existing methodologies - comparison

Methodology	DSC	SE	SP	Precision	Accuracy
Proposed	98.53	97.59	99.00	97.50	98.99
CNN	96.5	94.39	97.25	95.20	97.34
Naïve bayes	75.00	7134	73.56	75.23	80.34
Decision tree	78.56	75.12	77.23	70.23	79.12

The proposed methodology improved 3%, 23%, and 20% when compared to traditional CNN, Naïve Bayes and decision tree classifier respectively.

5 Conclusion

The convolutional neural network with autoencoder feature selection was proposed and implemented. The autoencoder is used to select the appropriate features from the image and it removes the noise from input mage. The CNN is used for image segmentation and classification. The convolutional layer automatically segments the MRI brain images and extract the relevant features. The pooling layer downsized the image to reduce the complexity. The dense layer is used for automatic classification. BRATS 2013, BRATS 2015 dataset is taken for experimentation. The results of the proposed approach is compared with traditional CNN, decision tree and Bayesian classification. The proposed methodology performed better when compared to existing methodology.

References

1. Pereira, S., Pinto, A., Alves, V., Silva, C.A.: Brain tumor segmentation using convolutional neural networks in MRI images. IEEE Trans. Med. Imaging **35**(5), 1240–1251 (2016)
2. Wu, M.N., Lin, C.C., Chang, C.C.: Brain tumor detection using color-based k-means clustering segmentation. In: Proceedings of the Third International Conference on Intelligent Information Hiding and Multimedia Signal Processing (IIH-MSP 2007), vol. 2, pp. 245–250, November 2007. IEEE
3. Dong, H., Yang, G., Liu, F., Mo, Y., Guo, Y.: Automatic brain tumor detection and segmentation Using U-Net based fully convolutional networks. In: Valdés Hernández, M., González-Castro, V. (eds.) Medical Image Understanding and Analysis. MIUA 2017. CCIS, vol. 723, pp. 506–517. Springer, Cham (2017). https://doi.org/10.1007/978-3-319-60964-5_44
4. Amin, J., Sharif, M., Yasmin, M., Fernandes, S.L.: Big data analysis for brain tumor detection: deep convolutional neural networks. Futur. Gener. Comput. Syst. **87**, 290–297 (2018)
5. Sapra, P., Singh, R., Khurana, S.: Brain tumor detection using neural network. Int. J. Sci. Mod. Eng. (IJISME) ISSN 2319–6386 (2013)
6. Amin, J., Sharif, M., Yasmin, M., Fernandes, S.L.: A distinctive approach in brain tumor detection and classification using MRI. Pattern Recogn. Lett. **139**, 118–127 (2020)
7. Iftekharuddin, K.M., Zheng, J., Islam, M.A., Ogg, R.J.: Fractal-based brain tumor detection in multimodal MRI. Appl. Math. Comput. **207**(1), 23–41 (2009)
8. Ratan, R., Sharma, S., Sharma, S.K.: Brain tumor detection based on multi-parameter MRI image analysis. ICGST-GVIP J. **9**(3), 9–17 (2009)
9. Sharma, K., Kaur, A., Gujral, S.: Brain tumor detection based on machine learning algorithms. Int. J. Comput. Appl. **103**(1), 7–11 (2014)
10. Amin, J., Sharif, M., Raza, M., Saba, T., Anjum, M.A.: Brain tumor detection using statistical and machine learning method. Comput. Methods Programs Biomed. **177**, 69–79 (2019)
11. Rajinikanth, V., Joseph Raj, A.N., Thanaraj, K.P., Naik, G.R.: A customized VGG19 network with concatenation of deep and handcrafted features for brain tumor detection. Appl. Sci. **10**(10), 3429 (2020)
12. Borole, V.Y., Nimbhore, S.S., Kawthekar, D.S.S.: Image processing techniques for brain tumor detection: a review. Int. J. Emerg. Trends Technol. Comput. Sci. (IJETTCS) **4**(5), 2 (2015)
13. Vijay, J., Subhashini, J.: An efficient brain tumor detection methodology using K-means clustering algorithm. In: Proceedings of the 2013 International Conference on Communication and Signal Processing, pp. 653–657, April 2013. IEEE

14. Dahab, D.A., Ghoniemy, S.S., Selim, G.M.: Automated brain tumor detection and identification using image processing and probabilistic neural network techniques. Int. J. Image process. Vis. Commun. **1**(2), 1–8 (2012)
15. Hema, M.S., Maheshprabhu, R., Guptha, M.N., Sampathkumar, A., Amudhavel, J.: Similarity learning of Parkinsonism elicited from genetics and syndrome for pharmacotherapy decision making. Materials Today: Proceedings (2021)
16. Mary, G., et al.: Predicting metamorphic changes in parkinson's disease patients using machine learning algorithms. J. Biosci. Biotechnol. Res. Commun. Spec. Issue **13**(11), 147–152 (2020)
17. Mahrishi, M., Hiran, K.K., Meena, G., Sharma, P. (Eds.): Machine Learning and Deep Learning in Real-Time Applications. IGI Global, Harrisburg (2020). https://doi.org/10.4018/978-1-7998-3095-5
18. Amin, J., et al.: Brain tumor detection by using stacked autoencoders in deep learning. J. Med. Syst. **44**(2), 1–12 (2020)
19. Logeswari, T., Karnan, M.: An improved implementation of brain tumor detection using segmentation based on hierarchical self organizing map. Int. J. Comput. Theory Eng. **2**(4), 591 (2010)
20. Kanagaraj, S., Hema, M.S., Gupta, M.N.: Machine learning techniques for prediction of Parkinson's Disease using big data. Int. J. Innov. Technol. Explor. Eng. **8**(10), 3788–3791 (2019)

Machine Learning Based Rumor Detection on Twitter Data

Manita Maan[1(✉)], Mayank Kumar Jain[2], Sainyali Trivedi[1], and Rekha Sharma[3]

[1] Department Computer Science and Engineering, Swami Keshvanad Institute of Technology, Jaipur 302017, Rajasthan, India
manitamaan@gmail.com

[2] Department Computer Science and Engineering, Malaviya National Institute of Technology, Jaipur 302017, Rajasthan, India

[3] Computer Science Department, Kanoria PG Mahila Mahavidyalaya, Jaipur 302004, Rajasthan, India

Abstract. Rumors are misleading information that are not sustained at the time of circulation and are not true at the time of verification. In other words, Rumors are set of linguistic, symbolic or tactile propositions whose veracity is not quickly or ever confirmed. As the use of social media platform has grown in recent years, incorrect information and rumors have circulated widely causing a significant influence on people's lives. Rumors spreads faster than righteous news and spreads through social media. Because of the expansion of Internet and web technologies, it is now possible for anybody to post anything on online platforms such as blogs, comments on articles, post on social media, and so on, where false news, rumors, and true news are swiftly conveyed. This rapid and expansive spread of rumors has encouraged researchers to differentiate between rumors and non-rumors data. In this work, we have used stylometric and word vector features and put them into machine learning models. These features are extracted from the twitter-16 dataset and by applying SVM, we have attain the highest accuracy in compare to existing state-of-the-art studies.

Keywords: Rumors · Twitter · Machine learning

1 Introduction

The notion of rumor existed long before the Internet and other computer technologies. "Rumors can disseminate misinformation or disinformation, both of which are types of misleading information". There are several outlets for accessing rumors, such as social media, blogs, comments, and so on. Anyone who uses these sites has the ability to generate and distribute false news or rumors. Microblogging websites like Twitter, Instagram, Facebook, Telegram, etc., may quickly spread news, rumor, and authentic information. Moreover, rumors may

V. E. Balas et al. (Eds.): ICETCE 2022, CCIS 1591, pp. 259–273, 2022.
https://doi.org/10.1007/978-3-031-07012-9_23

not be safe for anyone [1–4]. False or unconfirmed information travels on the internet the same way that truthful information does, potentially going viral and affecting public opinion and choices.

Researching on rumours Sites such as snopes.com and factcheck.org which report on rumours, are examples of collaborative endeavours. Nevertheless, because such initiatives entail manual verification phases, these websites do not cover all topics and might take a long time to debunk [5,6]. Fake news and rumours are the most common types of misleading and unverified information, and they should be caught as quickly as possible to avoid drastic consequences. Nowadays, quickly Anyone may post anything on the Internet with the help of microblogs or social media platforms. Reach of social media platforms for Anyone may cause some problems for society sometimes due to spreading rumors on the microblogs. To identify rumors on a ny platform, we need to classify actual news and verify news authentication. To identify actual news, we need to check the authentic website or authentic news. Social media channels have been used to acquire information and news, and they are valuable in a variety of applications. They do, however, aid in the dissemination of rumours and fake news [7,8]. Many efforts have been made to detect and refute rumours on social media using machine learning algorithms to analyse their content and social environment. The majority of users on social media do not check material before spreading it. Manually spotting bogus news is a huge challenge for everyone. As a result, there is a need for an automated system that can recognise phoney news or rumours quickly. After changing the text, image, audio, and video, publish misleading material on social media platforms to deceive people.

Rumors study on multimedia content is an important area due to the following reasons:

- People spread wrong information, create controversies of that wrong information.
- Some users submit material on a blog, but the information does not verify the title or facts, degrading user experiences.
- False information emotionally stymies the public [9] emotions.
- Amplification in network interconnection leads to high-risk [10–12] of danger, such as rumors, various and fake news, which have a negative influence on society. The earthquake in Chile in 2010 is one example, where the spread of falsehoods on social media caused widespread disorder and confusion.
- Social networking platforms enable users to share their emotions, ideas, and sentiments about any event, opinions and occasions directly without any trusted external control. To encourage rumors or misinformation it is one of the stepping stones.

Identification of rumors, we apply machine learning approaches to datasets and identify how many rumors are in a dataset [13,14]. Rumor is defined as an unconfirmed remark spread over multiple networks that originates from a single or multiple sources. Because of the omnipresence of online platforms, a massive amount of multimedia data is created in social networks. The openness and simplicity with which information on social media platforms may be disseminated stimulates the spread of information across the network, regardless of its veracity. This sort of disinformation is typically disseminated in the context of breaking news. As a result of unconfirmed information, such information, sometimes referred to as rumors, has the potential to do catastrophic damage. Despite their popularity, the chaotic nature of social media platforms usually leads to the propagation of rumours [2,3]. Using supervised and unsupervised machine learning algorithms, we may identify rumors. Some machine learning methods are defined for rumor detection, such as classification, regression, clustering, and decision making. Machine Learning models may apply to the datasets, and these data sets are available on the internet.

In this work, we have used stylometric and word vector features and put them into machine learning models. These features are extracted from the twitter-16 dataset and by applying SVM, we have attain the highest accuracy in compare to existing state-of-the-art studies.

This paper includes previous related works in Sect. 2, rumor detection process, data collection, feature extraction, ML models, and metrics are discussed in Sect. 3. Moreover, in Sect. 4 the outcomes of experimental results and in Sect. 5 the summary of the paper is discussed.

2 Related Work

In recent years most of the work have been done on rumors detection [6,15]. Microblogging systems are perfect for propagating rumors, and automatically debunking myths is a critical issue. Existing methods for detecting rumors have relied on handmade features rather than machine learning algorithms. As shown in the Fig. 1, access data from social media approaches, like: APIs, scrapping the web and Selenium web driver.

In [16] authors recognize the rumor on Twitter data using propagation structure with kernel learning, where they achieve the accuracy of 75%. They also find the results on false rumor, true rumor, non-rumor, and verified rumor. In [17] authors recognize the fake news on Twitter data using across several platforms and languages, where they achieve the accuracy of 79%. They also detect fake news in multiple languages(Latin, Germanic, and Slavic). In [18] authors recognize the rumor on Twitter data using dual convolutional neural networks, where they achieve approximately 80% accuracy. They also detect features like behavioral patterns and propagation tree patterns. In [19] authors recognize the rumor

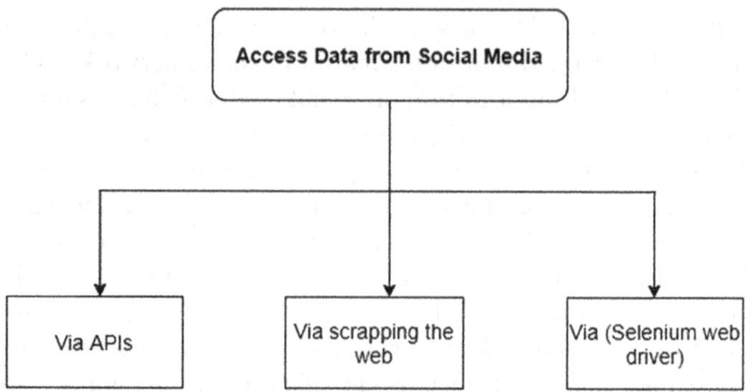

Fig. 1. Access data from social media

on Twitter data using Bidirectional Long Short-Term Memory (BiLSTM) with Convolutional Neural Network, where they achieved 86.12% accuracy. They also find the results on rumor and non-rumor. In [20] authors recognize the rumor on Twitter (PHEME dataset) data using deep learning, where they achieve 64.3% accuracy and they also find the results on rumor and non-rumor. In [14] authors recognize the rumor on Arabic Tweets using semi-supervised and unsupervised expectation-maximization (E-M), where they achieved 78.6% accuracy using Gaussian Naive Bayes (NB). In [13] authors recognize the rumor on Twitter data using one-class classification (OCC) approach, where they achieve 74% using Zubiagaset approach and 93% using Kwonset accuracy without using Non-Rumor.

As seen in the Fig. 2, certain critical criteria were taken into account during the rumour data gathering procedure.

3 Methodology

3.1 Rumor Detection Process

It's a procedure that starts with data collection from social networking sites through API or an actual (Open Source) data set. To extract significant characteristics, the acquired data must be in a standard organised manner. Cleaning, transformation, consolidation, and reduction are all part of the preprocessing process. The dataset is then categorised as rumor or non-rumor using various machine learning approaches once relevant characteristics are extracted. The rumor detection process is shown in Fig. 3 as:

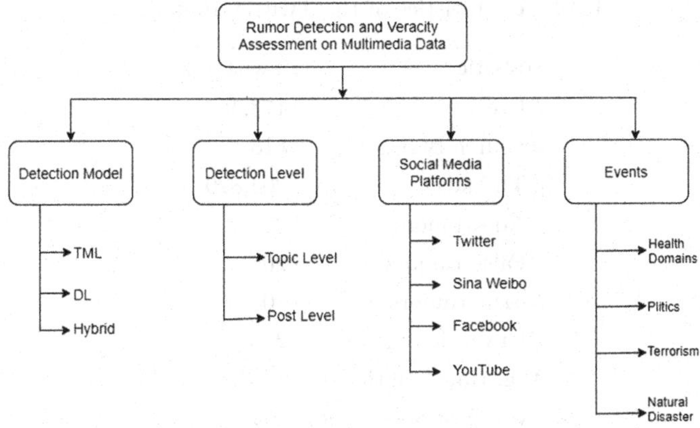

Fig. 2. The factors about the rumor data collection

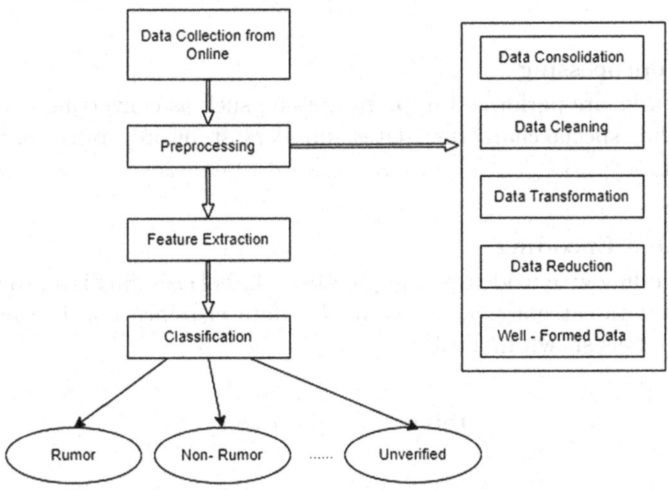

Fig. 3. Rumor detection process

3.1.1 Data Collection

In this work, data used is Twitter-16, and it is an open-source data set [16]. Twitter-16 data set have a total of 818 twists. Statistics of the Twitter-16 dataset as shown in Table 1:

Table 1. Statistics of the Twitter-16 dataset

Statistic	Twitter-16
#Users	173,487
#Source tweets	818
#Threads	204,820
#Non-rumors	205
#False rumors	205
#True rumors	205
#Unverified rumors	203
Avg. time length/tree	848 h
Avg. # of posts/tree	251
Max # of posts/tree	2,765
Min # of posts/tree	81

3.1.2 Preprocessing

Following tasks are performed in preprocessing such as converting text in lower case, removing special characters, dates, numbers, stopwords, punctuations, emojis, and urls.

3.1.3 Label Encoding

In label encoding we encode text into numbers. Label encoding is a process where we identify data categories and forward for data classification. In this data set we encode labels shown in Table 2.

Table 2. Label encoding

0	False
1	Non-Rumor
2	True
3	Unverified

3.1.4 Feature Extraction

Stylometry examines the semantic highlights of a content fragment, and is commonly used to assess the commencement of material or authenticity based on the linguistic style of writing. Linguistic characteristics assist distinguish the deceptive material used to conceal the writer's unique writing style. The suggested technique was used to extract stylometric highlights from the dataset's text field [21]. The stylometric highlights are shown in Table 3:

Word Vector Features [22–24]**:** To use data in classifiers, raw data must first be converted into a fixed-size vector. The phrase vector model refers to documents that are represented in a vector format. Remove all stop words and punctuation marks from the lexicon before changing the data in vector. There were two different types of vectors used:

Table 3. Content feature set

Feature no	Feature category	Feature name
F1	Message - based	Word count
F2	Message - based	Character count
F3	Message - based	Stop words
F4	Message - based	Average words in a sentence/average sentence length
F5	Content - based	Hash tag (#)
F6	Content - based	Count of '@'
F7	Content - based	Count of '?'
F8	Content - based	Upper case word
F9	Readability	Flesch reading ease
F10	Readability	Coleman liau index
F11	Readability	Flesch Kincaid score
F12	Readability	Lexicon count
F13	Readability	Lexical diversity
F14	Vocabulary	Syllables count
F15	Vocabulary	Unique word
F16	Vocabulary	Big word (≥6 characters)
F17	Word	Short word (<4 characters)
F18	Word	Long word (>15 words)

Bag of Words Term Frequency (BOW TF): Bag of words (count vector) has known as Bag of words term frequency (BOW TF), which has the formula 1. The document term matrix (DTM) is represented here, with rows representing documents and columns representing words. Each document versus word cell in the Table 4 represents the frequency of the phrase in the document. Term frequency

$$Termfrequency(TF) = \frac{freq\_T}{T\_o} \tag{1}$$

where, freq_T = No. of time token present in the document
T_o = Total tokens in documents

Table 4. BOW TF

	Token1	Token2	....	Token M
Rumor 1	count (token 1 in rumor 1)	count (token 2 in rumor 1)		count (token M in rumor 1)
Rumor 2	....	....	....	.....
Rumor N	(token 1 in rumor N)	count (token 2 in rumor N)		count (token M in rumor N)

Bag of Words Term Frequency - IDF: Each column of the document text matrix (DTM) in the BOW TF-IDF technique shown in Table 5. The Bag of words term frequency - Inverse document frequency formula shown in 2.

$$Inverse\_document\_frequency(IDF) = \log \frac{Doc}{T\_Doc} \tag{2}$$

where, Doc = number of total Documents and
T-Doc = in number of documents token present
TF-IDF = Term Frequency (TF)× Inverse document frequency (IDF).

Table 5. BOW TF-IDF

	Token1	Token2	.....	Token M
Rumor 1	tf-idf (token 1 in rumor 1)	tf-idf (token 2 in rumor 1)		tf-idf (token M in rumor 1)
Rumor 2	....	....	....	.....
Rumor N	tf-idf (token 1 in rumor N)	tf-idf (token 2 in rumor N)		tf-idf (token M in rumor N)

3.1.5 Metrics

Highlights use F1-Score, accuracy, precision, recall [21,25] to evaluate the classifier's exposition in the event of stylometric and word vector features: The classification is performed on the train set (80% of the dataset) and then put to the test on the test set (20% percent of the dataset). The model's performance is measured using a test set and is based on recall, precision, F1-score, and accuracy. To analyze the performance of various machine learning approaches, a variety of evaluation metrics are utilized. The following are the most frequent metrics:

True Positive (TP) [21]**:** The number of examples that were positive (+) and correctly classified as positive (+v).

False Negative (FN) [21]**:** The number of examples that were positive (+) and incorrectly classified as negative (−v).

True Negative (TN) [21]**:** The number of examples that were negative (−) and correctly classified as (−v).

False Positive (FP) [21]**:** The number of examples that were negative (−) and incorrectly classified as (+v). TP, TN, FP, and FN detect fake, genuine, misclassified, and unverified rumors.

Formulas for calculating these metrics are:

$$Precision = \frac{TP}{TP + FP} \tag{3}$$

$$Recall = \frac{TP}{TP + FN} \tag{4}$$

$$F1 - score = \frac{2 * Pre * Re}{Pre + Re} \tag{5}$$

$$Accuracy = \frac{TP + TN}{TP + FP + TN + FP} \tag{6}$$

3.2 Feature Selection

To reduce the dataset's dimension and improve the model's performance, this method is applied to stylometric and word vector features, with word vector features adjusted to reject a few words from the lexicon using wordnet lemmatizer and porter stemmer [21]. Apply a classifier after executing the feature reductions step on BOW TF (Beg of Words word frequency) and BOW TF-IDF vectors.

3.3 Parameters and Methods

To identify rumor differently using machine learning models based on stylometric and word vector parameters for classification work as shown in the Fig. 4.

These tests utilized simple classifiers like: Gaussian Naive Bayes (GNB), Support Vector Machine (SVM), Random Forest (RF), K-Nearest Neighbors (KNN), and Decision Tree (DT).

Gaussian Naive Bayes (GNB): GNB simple classifier to classify the rumor. Within the experiment python library, GNB applies and obtains good results. GNB apply to stylometric features and word vector feature.

Random Forest (RF): The text is classified by RF, and because of the randomization, it reduces the risk of overfitting. In this experiment, the RandomForestClassifier was utilised with n_estimators = 100 and the python library was employed, applied on both feature sets and get good result.

Support Vector Machine (SVM): This is another classification method. SVM stands for supervised machine learning classifier, and it is used to classify data. To improve the result, a leaner Kernel classification function was used.

K - Nearest Neighbors: The KNN classifier is used in this method to apply the python library function KNeighborsClassifier with n_neighbors = 5 and Euclidean distance using stylometric features. According to the elbow method, the optimal value is n neighbors = 5, which gives a satisfactory result.

Decision Tree (DT): This is another classifier method to perform python library function DecisionTreeClassifier. It provides good result.

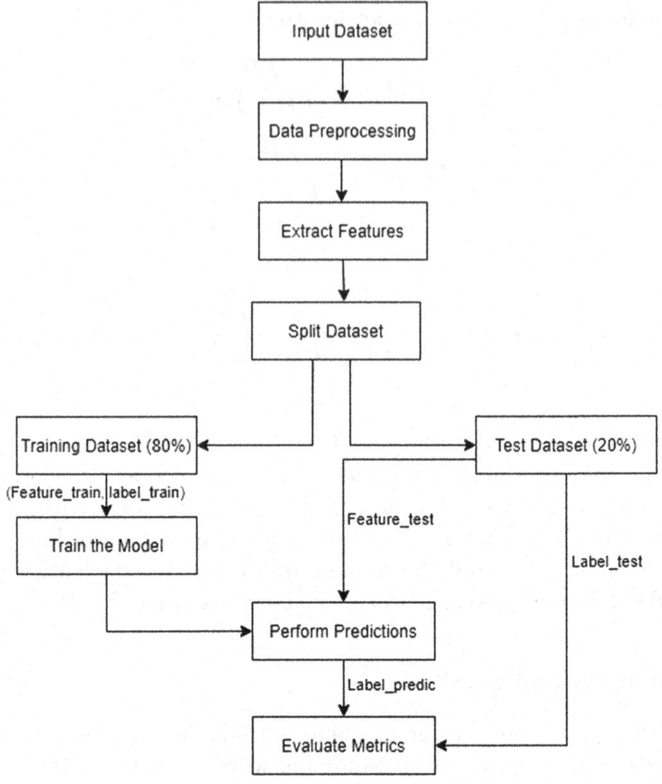

Fig. 4. Machine learning classification process flow chart [21]

4 Results

In this experiment, we take the help of stylometric and word vector features to detect rumors. In Table 6 and 7, we compare some machine learning classifiers like RF, DT, KNN, GBC, and SVM models on TF and TF-IDF features with the accuracy of 85.36% and 86.58% respectively. A similar comparison is made with Features obtained through Table 8 of Content features + TF and Table 9 of Content features + TF-IDF for all machine learning models. Among all comparisons, the SVM model gives the best accuracy.

Graphical representation of all machine learning classifier on Term Frequency graph shown in Fig. 5, TF-IDF graph shown in Fig. 6, Content Feature + Term Frequency graph shown in Fig. 7 and Content Feature + TF-IDF graph shown in Fig. 8.

Table 6. Results of all machine learning classifier on Term Frequency

CL	ACC	Pr	R	F-1
RF	85.36	.95	.95	.95
GNB	81.09	.80	.95	.87
SVM	84.75	.95	.90	.92
KNN	65.24	.87	.80	.84
DT	77.43	.87	.95	.91

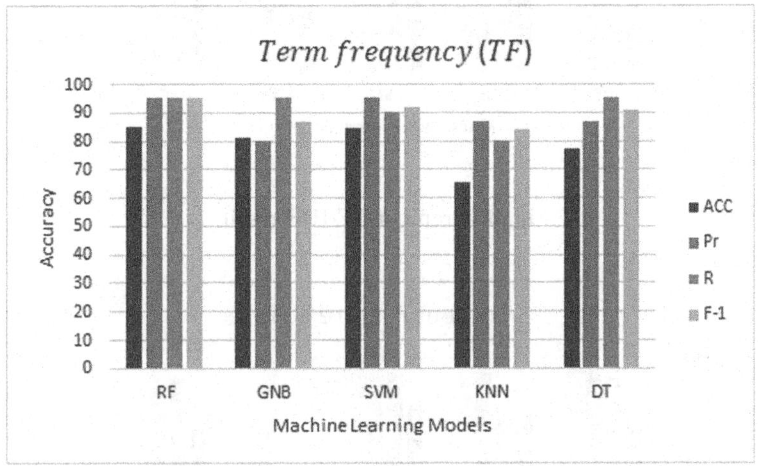

Fig. 5. Graph of term frequency result

Table 7. Results of all machine learning classifier on TF-IDF

CL	ACC	Pr	R	F-1
RF	83.53	.95	.95	.95
GNB	79.87	.86	.90	.88
SVM	86.58	1.0	.95	.97
KNN	73.78	.79	1.0	.88
DT	73.78	.95	.90	.92

Table 8. Results of all machine learning classifier on Content Feature + Term Frequency

CL	ACC	Pr	R	F-1
RF	82.92	.93	.95	.99
GNB	81.09	.80	.95	.87
SVM	83.53	.95	.88	.99
KNN	58.53	.62	.76	.68
DT	77.43	.93	.90	.91

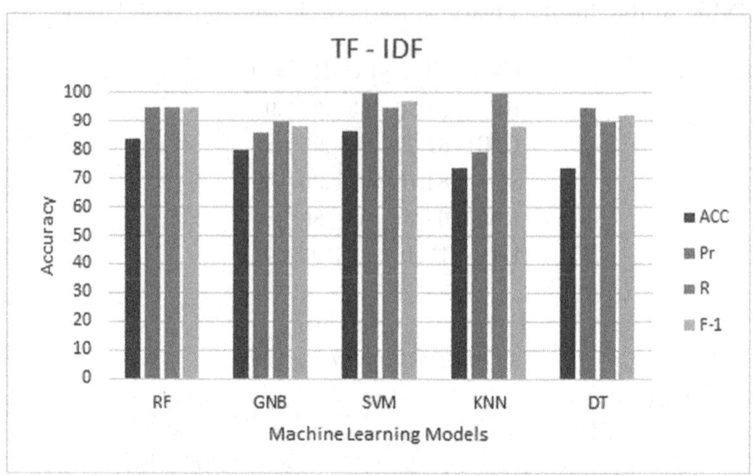

Fig. 6. Graph of TF-IDF result

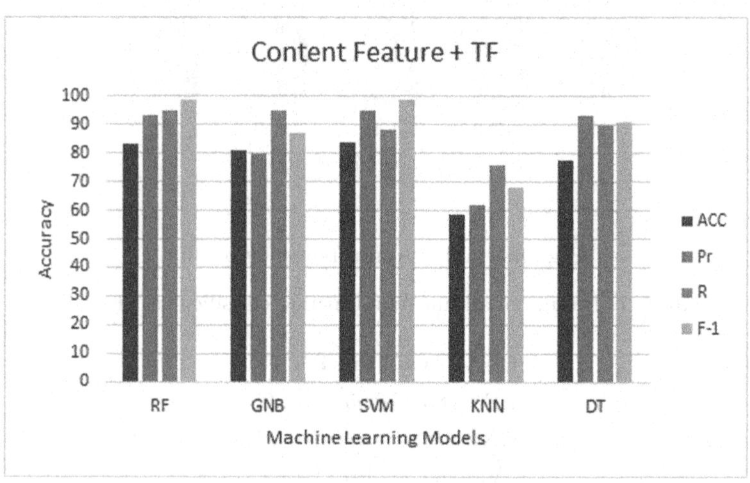

Fig. 7. Graph of Content Feature + Term Frequency Result

Table 9. Results of all machine learning classifier on Content Feature + TF-IDF

CL	ACC	Pr	R	F-1
RF	82.31	.93	.95	.94
GNB	80.48	.86	.93	.99
SVM	87.80	1.0	.98	.99
KNN	64.61	.62	.54	.58
DT	76.21	.86	.93	.89

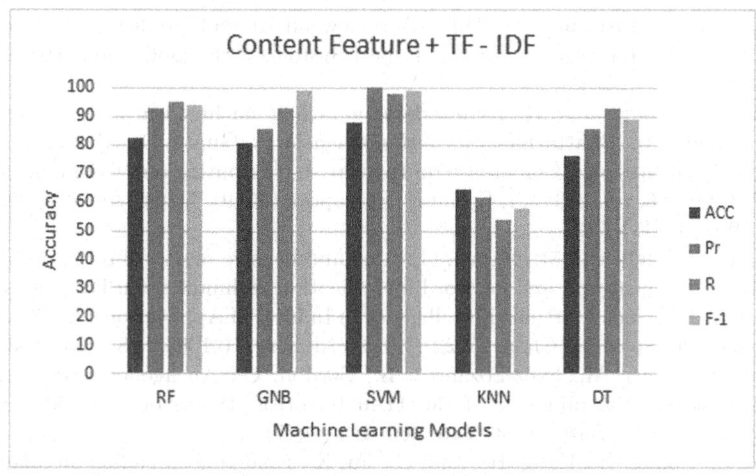

Fig. 8. Graph of Content Feature + TF-IDF Result

5 Conclusion

In this work, we have use only text field of the tweet messages and draw out the stylometric and word vector features to classify rumor. The result of these experiments are acquired in terms of accuracy, precision, recall, and F-1 score. In this experiment, features set contains 2909 features of content feature and word vector feature and after that these features are fed into various machine learning models like RF, GNB, SVM, KNN, and DT. We have attained the maximum accuracy of 87.80% through SVM on content feature and TF. In future, we will integrate extended deep learning models for our proposed work and also include large datasets that contain rumors and non-rumors.

References

1. Jabir, H.M., Naser, M.A., Al-mamory, S.O.: Rumor detection on Twitter using features extraction method. In: 2020 1st Information Technology To Enhance E-learning and Other Application (IT-ELA), pp. 115–120. IEEE (2020)
2. Pathak, A.R., Mahajan, A., Singh, K., Patil, A., Nair, A.: Analysis of techniques for rumor detection in social media. Proc. Comput. Sci. **167**, 2286–2296 (2020)
3. Bharti, M., Jindal, H.: Automatic rumour detection model on social media. In: 2020 Sixth International Conference on Parallel, Distributed and Grid Computing (PDGC), pp. 367–371. IEEE (2020)
4. Floos, A.Y.M.: Arabic rumours identification by measuring the credibility of Arabic tweet content. Int. J. Knowl. Soc. Res. **7**(2), 72–83 (2016)
5. Ma, J., et al.: Detecting rumors from microblogs with recurrent neural networks (2016)
6. Bondielli, A., Marcelloni, F.: A survey on fake news and rumour detection techniques. Inf. Sci. **497**, 38–55 (2019)

7. Varshney, D., Vishwakarma, D.K.: A review on rumour prediction and veracity assessment in online social network. Expert Systems with Applications **168**, 114208 (2020)
8. Kumar, A., Sangwan, S.R.: Rumor detection using machine learning techniques on social media. In: Bhattacharyya, S., Hassanien, A.E., Gupta, D., Khanna, A., Pan, I. (eds.) International Conference on Innovative Computing and Communications. LNNS, vol. 56, pp. 213–221. Springer, Singapore (2019). https://doi.org/10.1007/978-981-13-2354-6_23
9. Kumar, T., Mahrishi, M., Meena, G.: A comprehensive review of recent automatic speech summarization and keyword identification techniques. In: Fernandes, S.L., Sharma, T.K. (eds.) Artificial Intelligence in Industrial Applications. LAIS, vol. 25, pp. 111–126. Springer, Cham (2022). https://doi.org/10.1007/978-3-030-85383-9_8
10. Moya, I., Chica, M., Saez-Lozano, J.L., Cordon, O.: An agent-based model for understanding the influence of the 11-m terrorist attacks on the 2004 Spanish elections. Knowl.-Based Syst. **123**, 200–216 (2017)
11. Sun, M., Zhang, H., Kang, H., Zhu, G., Fu, X.: Epidemic spreading on adaptively weighted scale-free networks. J. Math. Biol. **74**(5), 1263–1298 (2016). https://doi.org/10.1007/s00285-016-1057-6
12. Fu, F., Christakis, N.A., Fowler, J.H.: Dueling biological and social contagions. Sci. Rep. **7**(1), 1–9 (2017)
13. Ebrahimi Fard, A., Mohammadi, M., Chen, Y., Van de Walle, B.: Computational rumor detection without non-rumor: a one-class classification approach. IEEE Trans. Comput. Soc. Syst. **6**(5), 830–846 (2019). https://doi.org/10.1109/TCSS.2019.2931186
14. Alzanin, S.M., Azmi, A.M.: Rumor detection in Arabic tweets using semi-supervised and unsupervised expectation-maximization. Knowl.-Based Syst. **185**, 104945 (2019). https://doi.org/10.1016/j.knosys.2019.104945
15. Cao, J., Guo, J., Li, X., Jin, Z., Guo, H., Li, J.: Automatic rumor detection on microblogs: a survey. arXiv preprint arXiv:1807.03505 (2018)
16. Ma, J., Gao, W., Wong, K.-F.: Detect rumors in microblog posts using propagation structure via kernel learning. Association for Computational Linguistics (2017)
17. Faustini, P.H.A., Covoes, T.F.: Fake news detection in multiple platforms and languages. Expert Syst. Appl. **158**, 113503 (2020)
18. Santhoshkumar, S., Dhinesh Babu, L.D.: Earlier detection of rumors in online social networks using certainty-factor-based convolutional neural networks. Soc. Netw. Anal. Min. **10**(1), 1–17 (2020). https://doi.org/10.1007/s13278-020-00634-x
19. Asghar, M.Z., Habib, A., Habib, A., Khan, A., Ali, R., Khattak, A.: Exploring deep neural networks for rumor detection. J. Ambient. Intell. Humaniz. Comput. **12**(4), 4315–4333 (2019). https://doi.org/10.1007/s12652-019-01527-4
20. Kotteti, C.M.M., Dong, X., Qian, L.: Ensemble deep learning on time-series representation of tweets for rumor detection in social media. Appl. Sci. **10**(21), 7541 (2020)
21. Jain, M.K., Gopalani, D., Meena, Y.K., Kumar, R.: Machine learning based fake news detection using linguistic features and word vector features. In: 2020 IEEE 7th Uttar Pradesh Section International Conference on Electrical, Electronics and Computer Engineering (UPCON), pp. 1–6. IEEE (2020)
22. Reddy, H., Raj, N., Gala, M., Basava, A.: Text-mining-based fake news detection using ensemble methods. Int. J. Autom. Comput. **17**(2), 210–221 (2020)

23. Bali, A.P.S., Fernandes, M., Choubey, S., Goel, M.: Comparative performance of machine learning algorithms for fake news detection. In: Singh, M., Gupta, P.K., Tyagi, V., Flusser, J., Ören, T., Kashyap, R. (eds.) ICACDS 2019. CCIS, vol. 1046, pp. 420–430. Springer, Singapore (2019). https://doi.org/10.1007/978-981-13-9942-8_40
24. Khan, J.Y., Khondaker, M.T.I., Afroz, S., Uddin, G., Iqbal, A.: A benchmark study of machine learning models for online fake news detection. Mach. Learn. Appl. **4**, 100032 (2021)
25. Li, Q., Zhang, Q., Si, L., Liu, Y.: Rumor detection on social media: datasets, methods and opportunities. arXiv preprint arXiv:1911.07199 (2019)

Personally Identifiable Information (PII) Detection and Obfuscation Using YOLOv3 Object Detector

Saurabh Soni(✉) and Kamal Kant Hiran(✉)

Sir Padampat Singhania University, Udaipur, India
saurabh.soni@jkcement.com, kamal.hiran@spsu.ac.in

Abstract. We live in an era of smart phones, and the number of smart phone users is growing by the day, resulting in a rapid increase in the number of people with access to the internet and social media. According to reports, the average person spends about five hours per day on his or her mobile phone. This extensive use of phone and internet had a significant impact on the amount of data exchanged on social media platforms such as Instagram, Facebook, Whatsapp, Snapchat, and others. Images are one type of data. Every day, millions and billions of images are shared, and many of them may contain information that compromises an individual's privacy. Hackers can use such information for malicious purposes and personal gain. They can even use this person's personal information to blackmail or threaten them. Cybercriminals can also use that information to open a bank account in the victim's name, create a forged driving licence, and other forms of identification. Although many researchers have addressed this issue by using tools such as Tensorflow and OpenCV to detect and obfuscate sensitive information in images. In this paper, we will use the YOLOv3 object detector to solve this problem. First, we'll label the data that needs to be blurred, and then we'll train our object.

Keywords: You Only Look Once (YOLO) · Convolutional Neural Network (CNN) · Face-detection

1 Introduction

We all know that new technology related to smart phones follow a J-curve or exponential advancement. Earlier generations utilised reel cameras, but with the introduction of camera phones, the first thing people look for when purchasing a new phone is the camera and its quality. The picture quality on new phones has greatly increased, and even the tiniest elements in the environment are clearly evident in the images we take. These days, features like twin cameras are fairly prevalent. Front cameras with 32 Mega Pixels are also rather prevalent. New smart phones with such outstanding picture quality are also quite affordable, resulting in a tremendous increase in the number of individuals using social media platforms. Even if we only consider Instagram, the statistics shows that 95 million photos are submitted every day. These data demonstrate how information

V. E. Balas et al. (Eds.): ICETCE 2022, CCIS 1591, pp. 274–282, 2022.
https://doi.org/10.1007/978-3-031-07012-9_24

sharing has become ingrained in people's daily lives. Whether they are going to the bank or getting their driver's licence, new generations provide status updates on everything they do in their everyday lives. On social media, information such as a person's full name, age, and occasionally their current location or the location of their residence puts them exposed to crimes such as burglary. The Data Privacy Law is used to protect sensitive information from falling into the hands of criminals by censoring or removing it from the internet (if necessary).

Information such as a person's name, date of birth, and Social Security Number (SSN) (sometimes known as the "Holy Trinty") is sufficient for a hacker or cybercriminal to gain access to his or her personnel account. And by answering simple security questions, fraudsters can gain access to that person's financial information.

Personally Identifiable Information (PII) is information about a person that allows us to determine his or her identity. Examples of PII include name, date of birth, location, phone number, driving licence number, account number, pan id, aadhar card number, and so on. If they have a photo of the victim's pan card, aadhar card, driving licence, credit card, or debit card, they can access this information. And the amount of damage that can be done with it is limitless. There was one such case, it was March 2021, and it was discovered using a third party that there was unauthorised access to data from payment cards at one of the stores of the company Forever 21. This could be because credit card information of customers was leaked from one of their stores.

Object detection is used to address the problem of spreading personnel information about any individual via photos or videos. Object detectors are used to detect objects in images [6, 11]. When we look at the big picture, we can see that today's object detectors detect objects not only in images, but also in videos.

2 Problem and Objectives

To detect and obfuscate personally identifiable or other sensitive information in image.

- This Find and collect images that contain sensitive information, then use tools like labelImg to label that information (this step is known as Data Annotation).
- Secondly, that use YOLOv3 (You Only Look Once) object detector to get coordinates of bounding box around that information.
- Third finally censor it using OpenCV (Open Source Computer Vision Library) and Python.

3 Theoretical Background

3.1 Object Detection

Object detection is a type of computer technology that detects objects in images or video [2, 3].

3.2 Convolutional Neural Network (CNN)

CNN is well-known for challenges requiring image processing or classification problems [12]. CNN is a more advanced version of the Artificial Neural Network (ANN), which is used to address image-related problems by finding patterns in images that make sense. It's quite different because if you try to tackle an image-related problem with an ANN, you'll end up with a lot of computation that's wasteful in terms of time and memory usage. CNN, on the other hand, employs convolutional layers, or hidden layers. When we give CNN an image, it constructs filters to recognise features (like eyes, nose, loopy pattern, etc.). Then, based on their location in the image, those discovered features will be stacked into a feature map.

3.3 YOLOv3

You Only Look Once, or YOLO, is a state-of-the-art object recognition algorithm that is so quick that it has practically become a standard method of detecting things in the field of computer vision [9, 10]. Previously, people relied on sliding window object detection, although faster versions such as RCNN (Region Based Convolutional Neural Network), Fast-RCNN, and Faster-RCNN were developed. However, in 2015, YOLO emerged, outperforming all prior object identification methods. We're utilising YOLOv3, which is the most recent and fastest version available as per the mentioned Fig. 1.

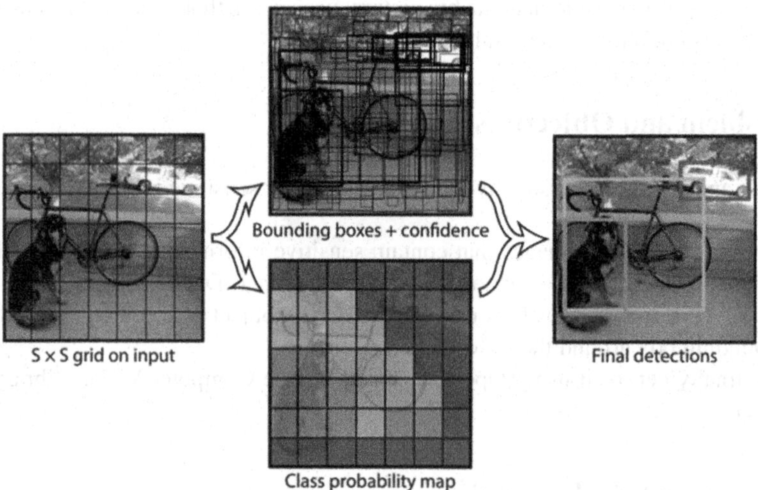

Fig. 1. YOLO (You Only Look Once)

The reasons why YOLOv3 is best suited for this project are:-

- As seen in Fig. 2, it is the quickest of all the options.

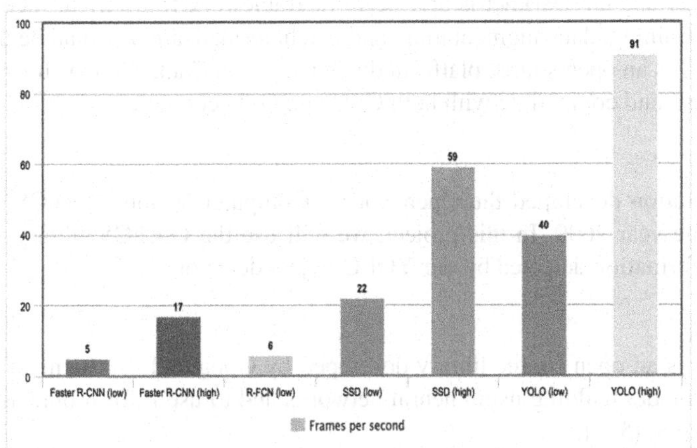

Fig. 2. Speed comparison (YOLO vs others)

- It has accuracy is good enough accuracy for our project. As shown Fig. 3.

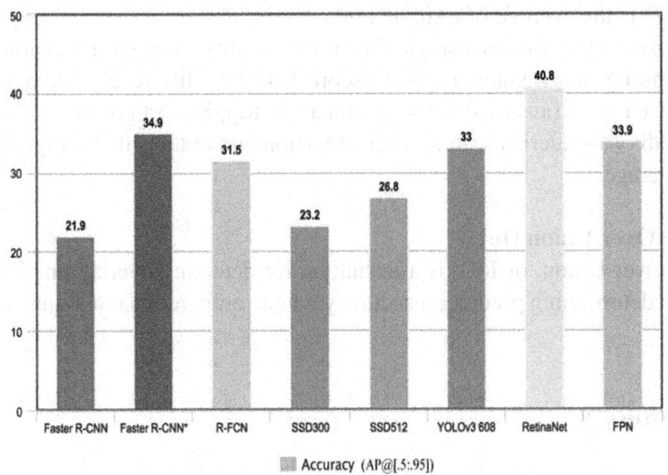

Fig. 3. Comparison of our model

- It recognizes small objects in images using the Feature Pyramid Network (FPN).

Darknet-53 is used by YOLOv3 to extract features from images. Darknet-53 is the moniker given to it since it uses 53 hidden layers or convolutional layers. It has 3X3 and 1X1 convolutional layers (hidden layers) in succession, as well as some short connections.

We'll also use the Darknet neural network framework to train our YOLO model. Multi scale training, data augmentation, and batch normalisation will all be used in this framework. It's an open source platform that's written in C and CUDA. It's also simple to set up, fast, and compatible with both CPU and GPU computing.

OpenCV

Intel Corporation developed the Open Source Computer Vision (OpenCV) library in C/C++ in the year 2000. In this project, we will use the OpenCV library to blur the sensitive information detected by our YOLO object detector [4].

TensorFlow

TensorFlow is an open source library developed by Google [1, 8]. Its main purpose is to simplify model building using neural networks and to use it for problems involving object detection [5, 7].

Mean Average Precision (mAP)

Graphs are extremely useful and convenient when analysing the scope of improvement for a specific thing it was used for. However, in the case of computers, we only need a single numerical value to determine whether one value is greater than another. We calculate mean precision by calculating the average of different queries, and average precision (AP) is the average of a single rank.

The precision measured is a single threshold quantity. Average Precision examines the entire ranking and assigns a quality score based on the results. And this step is crucial because rank 1 (awarded to the person at the top) is worth twice as much as rank 2. It obtains all of its queries with average precision and obtains its average value using arithmetic average.

Intersection Over Union (IoU)

Intersection Over Union, or IoU, is a technique for detecting overlapping in a dataset, which aids in determining prediction accuracy when compared to real object boundaries [13, 14].

4 Methodology

Data Gathering

Data is critical in machine learning problems. Images serve as our data in this project. We were able to collect over 50 images containing sensitive information. In future research, we will also use the Data Augmentation technique to improve the accuracy and reliability of our model, as well as to increase the volume of our dataset.

Data Annotation

Computer systems primarily use Data Annotation technology to assign keywords to image or video files. The term Annotating Image refers to the process of locating an object in an image by box-bounding and labelling it. This is our second step, and it is required before we can train and test the object detection model. In this step, we will use an open-source tool called labeling.

Training

We will use **pre-trained darknet-53 weights** to train our model because they will help our model converge quickly.

Censoring Information with OpenCV

The most important step follows training and testing. Obfuscate sensitive information or personally identifiable information (PII) detected in an image. Our object detection model (YOLOv3) will generate a bounding-box containing the coordinates (top, left, right, and bottom) of our region of interest. Following the extraction of our region of

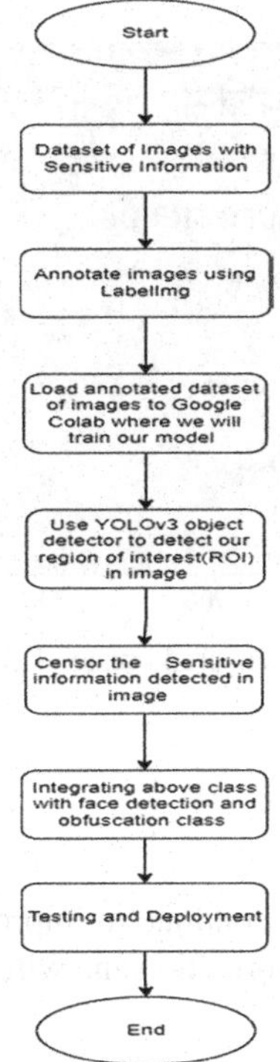

Fig. 4. Model workflow

interest, we would crop that portion and pass it as an argument to our blur function, which would replace the sensitive information portion of the original image with the blurred region, thereby censoring the information.

Integrating with Face-Detection Function
In this project, we also use a Python face detection library called "face-recognition", which was created with dlib and deep learning. It is quite accurate and can detect small faces in images. First, we used this library to determine whether or not there is a face in an input image. If a face is detected, we use its location (top, right, bottom, left) as an argument to our blur function, and after it is censored, the image is returned to our YOLO-based class where other sensitive information will be obfuscated.

Work Flow
The work flow of our model is depicted in Fig. 4.

5 Result

As shown in Fig. 5, after the preceding steps, we run our class, and the output of our code is shown below in the figure, where in a PAN card image, the Date of Birth, Permanent Account Number, and face of that individual are obfuscated.

Fig. 5. Censored image of pan card

6 Conclusions

We've taken our first step and will continue to improve its performance by collecting more data that contains sensitive information and will help our model be more accurate.

Using this research, we can censor any individual's sensitive information. Furthermore, this research can serve as a foundation for moving forward with the concept of privacy in images or videos available on the internet. Many platforms, such as YouTube, where creators blur their car registration number and hide their bank details because cyber criminals can use these details for blackmailing, earning money, or even political gain. In this day and age of bio-warfare and cyber-warfare, data privacy is a major concern.

References

1. Morales, J., Huliganga, V., Pasaoa, J., Melad, N.: Detecting and blurring potentially sensitive personal information containers in images using faster R-CNN object detection model with TensorFlow and OpenCV (2019)
2. Bochkovskiy, A., Wang, C.-Y., Liao, H.: YOLOv4: optimal speed and accuracy of object detection (2020)
3. Daoud, E., Vu Nguyen Hai, D., Nguyen, H., Gaedke, M.: Enhancing fake product detection using deep learning object detection models. **15**, 13–24 (2020). https://doi.org/10.33965/ijcsis_2020150102
4. Chen, P.-Y., Hsieh, J.-W., Gochoo, M., Wang, C.-Y., Liao, H.: Smaller object detection for real-time embedded traffic flow estimation using fish-eye cameras, pp. 2956–2960 (2019). https://doi.org/10.1109/ICIP.2019.8803719
5. Junayed, M.S., Jeny, A.A., Neehal, N., Ahmed, E., Hossain, S.A.: Incept-N: a convolutional neural network based classification approach for predicting nationality from facial features. In: Santosh, K.C., Hegadi, R.S. (eds.) RTIP2R 2018. CCIS, vol. 1036, pp. 466–475. Springer, Singapore (2019). https://doi.org/10.1007/978-981-13-9184-2_41
6. Kallam, S., Basha, M., Rajput, D., Patan, R., Balamurugan, B., Basha, S.; Evaluating the performance of deep learning techniques on classification using tensor flow application, pp. 331–335 (2018). https://doi.org/10.1109/ICACCE.2018.8441674
7. Shafiq, F., Yamada, T., Vilchez, A., Dasgupta, S.: Automated flow for compressing convolution neural networks for efficient edge-computation with FPGA (2017)
8. Janahiraman, T.V., Shahrul, M.S.M.: Traffic light detection using tensorflow object detection framework, pp. 108–113 (2019). https://doi.org/10.1109/ICSEngT.2019.8906486
9. Galvez, R.L., Bandala, A.A., Dadios, E.P., Vicerra, R.R.P., Maningo, J.M.Z.: Object detection using convolutional neural networks. In: TENCON 2018 - 2018 IEEE Region 10 Conference (2018)
10. Yanagisawa, H., Yamashita, T., Watanabe, H.: A study on object detection method from manga images using CNN. In: 2018 International Workshop on Advanced Image Technology (IWAIT), pp. 1–4 (2018). https://doi.org/10.1109/IWAIT.2018.8369633
11. Zhang, N., Luo, J., Gao, W.: Research on face detection technology based on MTCNN. In: 2020 International Conference on Computer Network, Electronic and Automation (ICCNEA), pp. 154–158 (2020). https://doi.org/10.1109/ICCNEA50255.2020.00040
12. Sharma, S., Shanmugasundaram, K., Ramasamy, S.K.: FAREC—CNN based efficient face recognition technique using Dlib. In: 2016 International Conference on Advanced Communication Control and Computing Technologies (ICACCCT), pp. 192–195 (2016). https://doi.org/10.1109/ICACCCT.2016.7831628

13. Hiran, K.K., Doshi, R., Fagbola, T., Mahrishi, M.: Cloud Computing: Master the Concepts, Architecture and Applications with Real-World Examples and Case Studies. Bpb Publications (2019)
14. Mahrishi, M., Hiran, K.K., Doshi, R.: Selection of cloud service provider based on sampled non-functional attribute set. In: Abraham, A., Siarry, P., Ma, K., Kaklauskas, A. (eds.) ISDA 2019. AISC, vol. 1181, pp. 641–648. Springer, Cham (2021). https://doi.org/10.1007/978-3-030-49342-4_62

Performance Analysis of Machine Learning Algorithms in Intrusion Detection and Classification

R. Dilip[1(✉)], N. Samanvita[2], R. Pramodhini[2], S. G. Vidhya[3], and Bhagirathi S. Telkar[4]

[1] Global Academy of Technology, Bengaluru, India
dilipr@gat.ac.in
[2] Nitte Meenakshi Institute of Technology, Bengaluru, India
[3] BGS Institute of Technology, Mandya, India
[4] Acharya Institute of Technology, Bengaluru, India

Abstract. Computer security is defined as the defense of computing systems against external threats in order to protect resource confidentiality, integrity, and availability. When an intrusion occurs, both network resources and the victim server are put at risk. When an intrusion occurs in a computer system or network, the Intrusion Detection System (IDS) tracks it and notifies the system administrator, allowing the appropriate action to be taken. People's faith in the Internet has declined as the frequency of cyber-attacks has increased. A security attack known as Denial of Service (DoS) is quite successful (DoS). When an intrusion detection system (IDS) detects external attacks as well as system abuse or internal attacks, it sends a signal to a central monitoring station. In terms of functioning, an intrusion detection system is similar to a burglar alarm. This article provides a machine learning based intrusion detection system. NSL KDD data set is used as input for experimental work. ANN, SVM and ID3 algorithms are used in analytical investigation.

Keywords: Security · Intrusion detection · Classification · Machine learning · Accuracy · SVM

1 Introduction

Defending computing systems against outside attacks to maintain resource confidentiality, integrity, and availability is the definition of computer security [1]. Network resources and the victim server are both at risk when an intrusion occurs [2]. When an intrusion occurs in a computer system or network, the Intrusion Detection System (IDS) tracks it and alerts the system administrator so that necessary action may be done. As the frequency of cyberattacks has risen, so has people's trust in the Internet. A security attack known as Denial of Service (DoS) is quite successful (DoS).

When an intrusion detection system (IDS) detects assaults from outside the organization, as well as system misuse or attacks from within, it sends a signal to a central

V. E. Balas et al. (Eds.): ICETCE 2022, CCIS 1591, pp. 283–289, 2022.
https://doi.org/10.1007/978-3-031-07012-9_25

monitoring station. An intrusion detection system resembles a burglar alarm in terms of functionality. For example, the lock system on an automobile helps keep it safe from being burglarized.

But if someone manages to break the lock mechanism and steal the car, the burglar detector will raise an alarm and alert the owner. An IDS will act as an alarm in a network or system to identify occurrences and warn users if any malicious conduct is detected.

Cybercriminals are always coming up with new techniques to get into a host or network and use the information for their own purposes. In addition to the operating systems on end hosts, the Internet's size and sophistication increase its vulnerability to flaws. As a result of these issues, current Internet best practices rely on evidence of recognizing attack trends, monitoring security vulnerabilities, and quickly resolving them. False alarms are being raised by existing intrusion detection systems. Components of IDS's Computational Intelligence (CI) can be simplified to reduce their impact. The researchers used a wide range of CI techniques and assessed their accuracy by comparing the results to industry standards.

As a multi-colored method, the Intrusion Detection System (IDS) looks for anomalous patterns in both inbound and outgoing network data before eliminating them. Data bases, analysis engines, and reaction managers are the three main components of IDS [4].

2 Literature Survey

Computational intelligence is a term used to describe a new type of information system built with the help of soft computing. Another goal is to construct intelligent machines that can give answers for complicated real-world issues that cannot be analytically modeled due to their complexity. In order to attain a comparable similarity to human-like decision making [5], it utilizes tolerance for approximation, uncertainty, imprecision, and partial reality.

If you're looking for anything to mirror the wisdom of nature, you've discovered an intelligent system (IS). Examples of these include awareness, flexibility, reasoning, and the capacity to deal with confusing information [6]. An overview of soft computing approaches used in intrusion detection is provided in this section. An artificial immune system is an artificial system that can detect and respond to foreign invaders using soft computing methods such as fuzzy logic and neural networks, rather than hard computational methods such as genetic algorithms. Detection of intrusions using Coupled Map Lattices is still a work in progress.

With his genetic algorithm (GA), Holland created an adaptable search method. It's a computer simulation designed to mimic the natural evolution process. The GA is a stochastic global search method that uses the survival of the fittest principle to find better and better approximations to a solution in a population of alternative solutions.

Each generation generates a new set of estimated solutions by picking individuals based on their level of fitness in the issue area and producing children from those individuals. Through this technique, people that are more problem-domain suited than the ones they evolved from are generated. An indicator of how individuals acted in the issue area is provided by the fitness function here [7].

It was inspired by the social behavior of flocking birds and schooling fish and was created in 1995 [8]. To understand PSO, you should know about its roots in fields such as artificial life, psychology, physics, and computer science. a "population" of particles travels through the problem hyperspace at predefined velocities in this approach. Stochastic calibration is used for the particle velocities, which are determined by the historical best location of a given particle and the best position in its immediate vicinity.

A user-defined fitness function is used to determine the particle's best and the neighborhood's best [9]. The movement of each atom leads to an ideal or nearly optimal result. "Swarm" describes the random movement of particles in the issue room, which resembles a swarm of insects more than a flock of birds or a school of fish these days.

Fuzzy logic (FL) is a logic of approximation that is used in problem solving and reasoning. There are several operations on fuzzy sets that are similar to classical (two-valued or "crisp") logic that are also included in fuzzy logic. Biological genetics and natural selection are employed in evolutionary computation as machine learning optimization and classification paradigms. evolutionary computation encompasses a wide range of concepts including genetic algorithms and evolutionary programming. The most often utilized algorithms in applications are genetic algorithms [10].

An improved variant of the naive Bayesian classifier, the secret naive Bayes (HNB) classifier reduces the restrictions. According to the HNB paradigm, a new layer is added to define the secret parent of each attribute. The structural characteristics of HNB are derived using naive Bayes. For each attribute, a hidden parent is generated, allowing the forces of all other characteristics to be integrated into one cohesive whole. It is possible to characterize hidden parents by averaging weighted one-dependence estimators [11].

This approach uses support vector machines (SVMs) and statistical learning theory to classify data (SLT). It is built on the idea of a linear separability hyperplane classifier. An optimal hyperplane to locate with SVM is one that maximizes the difference between two groups.

A deep belief network (DBN) is a sort of deep neural network in machine learning that is made up of multiple layers of latent variables ("hidden units") and has relationships between the layers but not within each layer [12]; it is a generative graphical model.

3 Methodology

Methodology consists of three machine learning algorithm. These algorithms are ID3, support vector machine and artificial neural network.

The ID-3 algorithm (Iterative Dichotomiser-3) was developed by J. Ross Quinlan and is the first evolving decision tree-based system. The entropy and information gain measurements are used in this algorithm's computations. For each iteration of the dataset, the entropy measure for the functional characteristics is computed. Datasets are divided into subsets based on which characteristics have lower error rates (entropy) and larger information gains (information gain). Until the data is accurately categorized to its target classes, the procedure is repeated on every subset. When building a decision tree, start with a non-terminal node and work your way down to the subset of terminal nodes at the end of the branch. The nonterminal node is defined by the split property, whereas the class labels are represented by the terminal node. Authors [13] developed an ID-3-based decision tree model to better categorize and forecast cardiac disease in the early stages. Using well-known decision tree techniques like CART and ID-3, a prediction model with a large health dataset is built. Validation is carried out using a 10-fold cross validation approach. Results show that decision tree classification approaches may be used to create an accurate and efficient model for predicting outcomes. Using the ID-3 approach, you may get better results using less datasets and a reduced computation measure. It becomes computationally costly, however, and the performance metric is reduced while dealing with continuous and huge data.

In their study, researcher [14] explain that the brain's information processing is carried out by densely linked neurons. Pattern recognition and data categorization are two common uses for this approach. Neurons are nodes or linked units that make up the network. Connection nodes are used to change signals as they go from one neuron to another. Artificial neural networks are fed real numbers as input and nonlinear inputs as output. The intensity of the signals at the linked edges steadily grows or decreases as the weighted edge size changes. The nodes set a threshold value, and the neurons can only send signals if they are higher than or equal to that number. Artificial neurons are often portrayed as a layered structure. Each layer transforms the inputs it receives in some way. It is common for signals to pass through multiple intermediary layers before reaching the final destination. Artificial neural networks are based on a notion that may be used to address issues by simulating the human brain. Using artificial neural networks for medical diagnosis, video and image identification, among other things, is becoming increasingly popular.

A non-probabilistic binary linear classification method is what it is: It creates a training model that categorizes the samples into one or more target classes based on their characteristics. Points in space represent the data items. There appears to be a chasm between the items of various categories, and this causes the gap's breadth to grow. The new instances are mapped to the appropriate side of the gap based on their target classes. When the input datasets are unlabeled, the support vector machine may do non-linear classification as well. The support vector machine uses an unsupervised learning strategy to cluster the data because there are no target classes to assign the instances to. New instances are added to existing clusters based on the functionality. Using a non-linear support vector machine, an efficient recommendation system is developed [9]. When dealing with unlabeled data, non-linear support vector machine techniques are the most commonly employed methodology. They are also used in several industrial applications. It generates an ideal hyperplane for each given collection of data and labeled training samples. In this way, the new instances of the input data model are further classified. One portion of the hyperplane is separated from the other by a line called the hyperplane. The classes are divided into two groups and are located on each side of the dividers. SVM separates classes, to put it simply. An suitable hyperplane is found to categorize the target classes. The hyperplane should be chosen so that it classifies the target class effectively. The choice of hyperplane changes depending on the situation. Three hyperplanes are required to accurately distinguish the target classes if there are any. The term "distance margin" is used in this situation. Better outcomes are obtained by varying the margin distance measurements according to different conditions. Using margin distance measurements, SVM classification algorithms such as the maximal margin classifier and the soft margin classifier may be used. It uses this to efficiently classify things. In a variety of fields, support vector machines (SVMs) are the most efficient classifiers. It also has a number of uses in the prognosis and treatment of cardiac disease. In high-dimensional environments, this approach is quite successful and works well with a little margin of separation between the two objects. In addition, it works effectively with more dimensions than just samples. In addition, this method is computationally and memory-intensive. Additionally, while dealing with noisy and ambiguous data, the performance metrics might worsen. The shortcomings aside, SVM has many useful applications because of its suitable and economical categorization features.

4 Result Analysis

NSL KDD [16] was the data set utilized in the experiment. The NSL-KDD dataset contains 24 different forms of assaults, and the data is either normal or one of them. Probe, DoS, R2L, and U2R are all attacks that fit into one of these four categories. The result comparison of these three classifiers is shown below in Fig. 1.

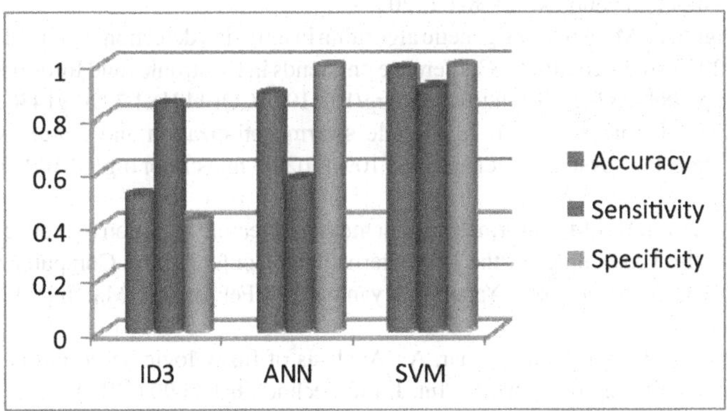

Fig. 1. Result comparison of classifiers

5 Conclusion

The protection of computing systems against external attacks in order to maintain resource confidentiality, integrity, and availability is characterized as computer security. When an intrusion occurs, it jeopardizes both network resources and the victim server. When an intrusion occurs in a computer system or network, the Intrusion Detection System (IDS) detects it and alerts the system administrator, allowing necessary action to be done. People's trust in the Internet has dwindled as the number of cyberattacks has grown. A security attack known as Denial of Service (DoS) is quite successful (DoS). When an intrusion detection system (IDS) detects external or internal threats, it transmits a signal to a central monitoring station. An intrusion detection system functions similarly to a burglar alarm. This article describes an intrusion detection system based on machine learning. The NSL KDD data set is utilized as the experimental input. In analytical research, the ANN, SVM, and ID3 algorithms are employed. SVM outperforms ANN and ID3 algorithms as far as accuracy is concerned.

References

1. Othman, S.M., Alsohybe, N.T., Ba-Alwi, F.M., Zahary, A.T.: Survey on intrusion detection system types. Int. J. Cyber Secur. Digit. Forensics **7**, 444–463 (2018)
2. Zarpelão, B.B., Miani, R.S., Kawakani, C.T., de Alvarenga, S.C.: A survey of intrusion detection in internet of things. J. Netw. Comput. Appl. **84**, 25–37 (2017)
3. Fragkiadakis, A.G., Siris, V.A., Petroulakis, N.E., Traganitis, A.P.: Anomaly-based intrusion detection of jamming attacks, local versus collaborative detection. Wiley Online Library, February 2013. wileyonlinelibrary.com, https://doi.org/10.1002/wcm.2341
4. Yan, S., Chung, Y.: Improved ad hoc on-demand distance vector routing (AODV) protocol based on blockchain node detection in ad hoc networks. Int. J. Internet Broadcast. Commun. **12**(3), 46–55 (2020)
5. Patel, A., Jain, A.: A study of various Black Hole Attack techniques and IDS in MANET. Int. J. Adv. Comput. Technol. **4**(3), 58–62 (2015)
6. Jain, V., Agrawal, M.: Applying genetic algorithm in intrusion detection system of IoT applications. In: 2020 4th International Conference on Trends in Electronics and Informatics (ICOEI) (48184), pp. 284–287 (2020). https://doi.org/10.1109/ICOEI48184.2020.9143019
7. Kunhare, N., Tiwari, R., Dhar, J.: Particle swarm optimization and feature selection for intrusion detection system. Sādhanā **45**, 109 (2020). https://doi.org/10.1007/s12046-020-1308-5
8. Win, T.Z., Kham, N.S.M.: Information gain measured feature selection to reduce high dimensional data. In: Proceedings of the 17th International Conference on Computer Applications (ICCA 2019), Novotel hotel, Yangon, Myanmar, 27 February–1 March 2019, pp. 68–73 (2019)
9. Chaudhary, A., Tiwari, V.N., Kumar, A.: Analysis of fuzzy logic based intrusion detection systems in mobile ad hoc networks. Int. J. Inf. Technol. **6**(1), 690 (2014)
10. Gaurav, M., Babita, D., Mehul, M., Hiran, K.K.: Performance comparison of network intrusion detection system based on different pre-processing methods and deep neural network. In: Proceedings of 3rd International Conference on Emerging Technologies in Computer Engineering: Machine Learning and Internet of Things, ICETCE 2020 pp. 145–148 (2020)
11. Nagar, P., Menaria, H.K., Tiwari, M.: Novel approach of intrusion detection classification deep learning using SVM. In: Luhach, A., Kosa, J., Poonia, R., Gao, X.Z., Singh, D. (eds.) First International Conference on Sustainable Technologies for Computational Intelligence. Advances in Intelligent Systems and Computing, vol. 1045, pp. 365–381. Springer, Singapore (2020). https://doi.org/10.1007/978-981-15-0029-9_29
12. Rajeshkanna, A., Arunesh, K.: ID3 decision tree classification: an algorithmic perspective based on error rate. In: 2020 International Conference on Electronics and Sustainable Communication Systems (ICESC), pp. 787–790 (2020). https://doi.org/10.1109/ICESC48915.2020.9155578

13. Harumy, T.H.F., Zarlis, M., Effendi, S., Lidya, M.S.: Prediction using a neural network algorithm approach (a review). In: 2021 International Conference on Software Engineering & Computer Systems and 4th International Conference on Computational Science and Information Management (ICSECS-ICOCSIM), pp. 325–330 (2021). https://doi.org/10.1109/ICSECS52883.2021.00066
14. Huang, J., Zhou, J., Zheng, L.: Support vector machine classification algorithm based on relief-F feature weighting. In: 2020 International Conference on Computer Engineering and Application (ICCEA), pp. 547–553 (2020). https://doi.org/10.1109/ICCEA50009.2020.00121

Fruit Classification Using Deep Convolutional Neural Network and Transfer Learning

Rachna Verma[1(✉)] and Arvind Kumar Verma[2]

[1] Department of CSE, Faculty of Engineering, J.N.V. University, Jodhpur, Rajasthan, India
rachnaverma@jnvu.edu.in

[2] Department of PI, Faculty of Engineering, J.N.V. University, Jodhpur, Rajasthan, India
akverma.pi@jnvu.edu.in

Abstract. Identification and classification of fruits of different qualities are vital for fruit industries. Traditional techniques, such as visual inspection of fruits by handpicking, are time-consuming, tiresome, and error-prone. To automate the fruit inspection process and segregating them into different classes, computer vision and machine learning approaches has been applied and researched. For segregating fruits into different classes or qualities, Transfer Learning is one of the popular technique to build a fruit classifier. This paper evaluates the performance of VGG16, InceptionV3, Xception, ResNet152V2, and DenseNet by training and testing models using transfer learning. The experiment is conducted on two fruits datasets. Further, to improve the accuracy of fruit classification, the pre-trained model DenseNet is partially unfreezed and retrained. The results show that this model archives accuracy of 99.61% for fruit classification.

Keywords: CNN · Deep learning · Transfer learning · Fruits classification

1 Introduction

Indian agriculture industries play a significant role in the economy of India. India is primarily an agriculture-based economy and has tremendous opportunities in exporting fruits and vegetables. As per APEDA (Agricultural and Processed Food Products Export Development Authority), Ministry of Commerce and Industry, India [1], India stands second in producing fruits and vegetables globally, after China. During 2020–2021, India exported 956961.00 metric tons of fruits, which were worth Rs 5647crores. Fruits such as pomegranates, bananas, oranges, and mangoes are exported in large amounts from India to different countries.

Farmers produce fruits and vegetables and sell them either in local markets or to fruit industries. The fruit industries process them to segregate into different grades (quality of fruits) before packaging for export. The high quality of fruit is the primary requirement for exporting. Hence, accurately and efficiently selecting high-quality fruits is vital for the fruit industry.

Fruits sorting in major fruits industries in India is mainly done by handpicking and inspecting fruits manually, which is time-consuming, tiresome, and error-prone. Further,

V. E. Balas et al. (Eds.): ICETCE 2022, CCIS 1591, pp. 290–301, 2022.
https://doi.org/10.1007/978-3-031-07012-9_26

rotten fruits need to be sorted and removed immediately to avoid spoiling other fruits. The quality of fruit is decided by many parameters, such as ripe, unripe, and rotten. Thus, there is a need to automate the fruit sorting process to reduce labor costs and accurately sort fruits into different qualities.

In literature, some works are reported to automate the fruit classification problem using various machine learning and computer vision techniques. The major steps used in fruits and vegetables classification are data acquisition, pre-processing, feature extraction, and classification [2]. Images are the primary input data in the fruit classification system, acquired using the camera, ultrasound, MRI, infrared, Lidar [2, 3]. The digital images of various fruits and vegetables of different qualities and grades are collected to create a training dataset.

The images acquired through sensors contain noises and distortions. In the pre-processing step, the noises and distortions are removed or minimized before the feature extraction process. Further, pre-processing enhances the image data and various features of the images, which is essential for obtaining discriminative features for classification. Sometimes segmentation is required as a pre-processing step that separates the foreground object of interest from the background. Segmentation techniques [3], such as thresholding, clustering are primarily used for segmentation. An improper segmentation degrades the classifier's performance.

After pre-processing, features such as texture and color, are extracted for further processing as these features play a vital role in discriminating one object from the other. These extracted features are used to train a fruit classifier. Various machine learning algorithms, such as KNN, SVM, ANN, deep learning networks, have been used in the literature [3–5] to design fruit quality classification algorithms.

Development in deep Convolutional Neural Networks (CNN) has proven to be very effective in image classification tasks [6, 7]. CNN is trained to identify and classify objects. Building and training a CNN from scratch requires a huge amount of data and time. Hence, transfer learning is used. Transfer learning is a way to transfer knowledge from a similar domain to a specific domain. It is used when the training data is limited [8]. In literature, transfer learning is also used fruit classification [9, 10]. Therefore, this work proposes a fruit classifier to automate the visual inspection of some Indian fruits using transfer learning. The major contributions are:

i) Performance evaluation of various popular pre-trained networks for fruit image identification and classification.
ii) Fine-tuning of Desenet121 by partially unfreezing a few layers of the convolutional network.

2 Related Work

Many research works on fruit and vegetable classification and their quality grading, based on color, texture and shape, have been reported using machine learning-based computer vision [2–5]. Four different machine learning classifiers, KNN, SVM, SRC, and ANN, have been used in [4] to classify fruits. A classification technique based on KNN, SVM, linear discriminates, and regression classifier is used by Singh and Singh

[11] to classify healthy and rotten apples. SVM classifier is used by Moallem et al. [12] to detect defects in apples. They used statistical, texture, and geometrical features and achieved 92.50% and 89.20% accuracy for healthy and defective apples. Long and Thinh [13] used external features, such as length, width, weight, and defect, to classify mango into three different grades: good, medium, and bad. The classifiers used by [13] are Random Forest, Linear discriminant analysis, KNN, and SVM. The random forest has outperformed with a precision of 98.1% compared with the other three models.

Developments in the deep convolutional neural network is found effective in image identification and classification tasks [6, 7]. In [14], authors used two models - a six layered CNN model and a customized VGG16 model for automatic fruit classification. A nine-layer CNN is proposed by [15] for fruit classification in uncertainty conditions. They used YOLO V3 to generate a bounding box around the apple in the original images. They reported an average accuracy level of 99.73%.

Full training of a deep CNN is very expensive in terms of time and resource. Hence, many researchers used the transfer learning approach, which uses weights of pre-trained generic models to build new specific models. Transfer learning is widely used in various computer vision tasks and helps efficiently fine-tune a generic pre-trained network for a specific purpose. Further, it has been found that transfer learning effectively transfers learned knowledge from the general domain to a specific domain even when data is limited. It not only saves training time but also reduces generalization error. Hence, many researchers in fruit and vegetable classification and grading have used transfer learning.

Zilong et al. [16] have investigated the effectiveness of various configurations of CNN to detect damaged apples. They also explored different fusion strategies and achieved an accuracy of 97.67%. They used VGG-19 and Inception-v3 CNN architecture for feature extraction. Vishal et al. [9] used DenseNet161, InceptionV3, and MobilenetV2 architectures to study the misclassification problem. Further, they proposed MNet architecture based on Inception-v3 to classify fruits into different classes and achieved an accuracy of 99.92%.

A real-time automatic visual inspection system for grading of apple and banana using pre-trained CNN architectures - ResNet, DenseNet, MobileNet, NASNet, and EfficientNet is studied by Nazrul and Malik [10]. They found EfficientNet to be the best model for fruit grading and achieved an accuracy of 99.2% and 98.6% on apple and banana test sets, respectively. Shih-Lun et al. [17] also used pre-trained networks AlexNet, VGG, and ResNet to classify Mangoes of different grades. A CNN architecture based on MobileNetV2 for the classification of fruits inside a plastic bag is proposed by Rojas Aranda et al. [18]. Besides fruit images, they also input additional features, such as color and centroid, to improve accuracy.

From the above, it can be concluded that fruit and vegetable classification and grading is an emerging area of research. There are many fruits and vegetables for which classifiers have to be developed. Further, there are many generic networks whose efficacy for fruit and vegetable classification has to the evaluated. In this paper, we have evaluated the performances of various networks.

3 Methodology

This work evaluates the performance of generic VGG16 [22], InceptionV3 [23], Xception [24], DenseNet [25] and ResNet152V2 [26] networks for fruit identification and classification. Further, we have attempted fine-tuning of Desenet121 by partially unfreezing a few layers of the convolutional network to enhance the accuracy of the fruit classifier, using two fruit image datasets [19, 20]. This section briefly describes the datasets, CNN networks, and transfer learning concepts used in the paper.

3.1 Datasets

In this paper, we have used two different fruit datasets for training fruit classifiers. The first dataset (dataset-1) is fruits fresh and rotten for classification [19]. It consists of three types of fruits - Apple, Banana, and Oranges, having 13599 healthy and rotten images; Of the total images, 7308, 2698 and 3593 images are used for training, testing and validation. The dataset consists of six classes: Fresh Apples, Bananas, and Oranges; and Rotten Apples, Bananas, and Oranges. The image sizes in dataset-1 are variable. Hence, they need to be properly resized before using them in a network. We have used resized images to 150 $\times$ 150 $\times$ 3 in this paper.

The second dataset (dataset-2) consists of images of six types of Indian Fruits: Apples, Bananas, Guavas, Limes, Oranges, and Pomegranates. The dataset consists of 12000 healthy and rotten images. The dataset is available at [20]. The dataset is divided into twelve classes, with each class having 1000 images. The twelve fruit classes in the dataset are Fresh Apples, Fresh Bananas, Fresh Guavas, Fresh Limes, Fresh Oranges, Fresh Pomegranates, Rotten Apples, Rotten Bananas, Rotten Guavas, Rotten Limes, Rotten Oranges and Rotten Pomegranates. The images are colored RGB images of sizes 256 $\times$ 256 $\times$ 3. To build the classifier, dataset-2 is partitioned into training, testing and validation in 70%,15%, and 15%, respectively.

Pre-trained networks used in this study require all input images to be of some specific sizes during their training. For example, the Xception network requires the input image dimensions of size 299 $\times$ 299 $\times$ 3 for its training. However, these networks accept input image dimensions to fall within a range. For example, the Xception network can take images from 299 $\times$ 299 $\times$ 3 to 71 $\times$ 71 $\times$ 3. The number of channels is fixed at 3. Similarly, DenseNet accepts the input image from 224 $\times$ 224 $\times$ 3 to 32 $\times$ 32 $\times$ 3. Therefore, we have used input images of size 224 $\times$ 224 $\times$ 3 for consistent comparisons of results for all networks for dataset-2. Dataset-1 contains images of variable sizes. Therefore, we have used input images of size 150 $\times$ 150 $\times$ 3 for training, testing, and validation of the classifier.

3.2 Convolution Neural Network

CNN has outperformed traditional technics for various computer vision tasks, including the identification and classification of images. A CNN is an artificial neural network specifically designed to process rectangular image pixel data and has produced superior results than other neural networks for image processing tasks. A typical CNN consists of two parts (see Fig. 1): A convolutional base and a classifier. A Convolutional base consists of a stack of layers comprised of convolutional and pooling layers. The convolutional base is used to learn and extract various features. The classifier consists of a fully connected layers to classify images. The initial layers, i.e., the head layers of the convolutional base, learn the general features, whereas the features extracted by the tail layers are specific to the chosen dataset and task.

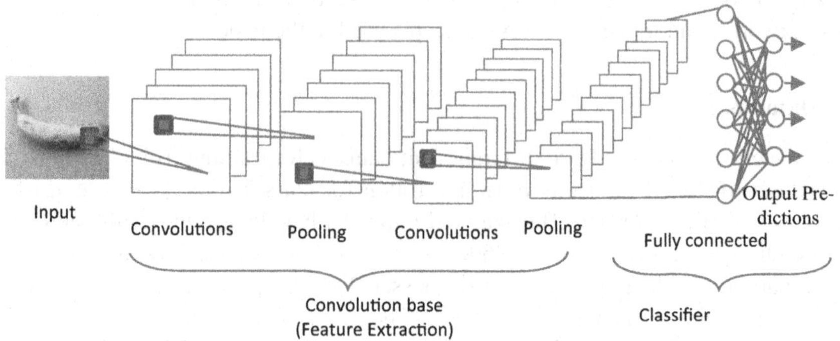

Fig. 1. Typical convolutional neural network architecture

3.3 Transfer Learning

Deep CNN networks consist of many layers in their convolutional base, sometimes more than 100 layers. Transfer learning is a way to build accurate deep CNN models efficiently [7]. Interested readers can refer to [8] for more details. In transfer learning, we use weights and parameters of pre-trained networks that have been trained earlier using generic datasets of similar types. In this way, the network does not have to learn from scratch and require a huge amount of training data. The advantage of transfer learning is that it is more efficient and requires less data than training the same network from scratch.

Various pre-trained networks are explored and compared to develop automatic fruit identification and classification systems [5, 9, 10]. The list of pre-trained models is large, and we have used a limited number of pre-trained networks ranging from classical to modern architectures. All the pre-trained networks used in this paper are trained on the ImageNet dataset of the ImageNet Large Scale Recognition Challenge (ILSVRC) [21] to classify 1000 generic classes, and their pre-trained weights are available. The pre-trained networks used in this paper for fruit classification are VGG16 [22], InceptionV3 [23], Xception [24], DenseNet [25] and ResNet152V2 [26].

3.4 Training of Networks

This paper adopts two strategies to train the above pre-trained networks using transfer learning to identify and classify fruits from dataset-1 and dataset-2. In strategy 1, we have used the frozen CNN (the base network) as a feature extractor, and the network is fine-tuned by adding a new classifier to the base network. This strategy is implemented by removing the pre-trained network's original classifier, and a new block of fully-connected layers are added on the top of the existing base network. The new block of fully connected layers consists of a dense layer of 1024 neurons (ReLU activation function), and a dropout layer to avoid overfitting. Finally, a dense softmax output layer of 6 or 12 neurons (depending on the dataset used) is added, (see Fig. 2). The probability of dropout is set to 0.2. We call this model as Model-I. Model-I is fine-tuned by freezing all the layers of the convolutional base network, using dataset-1 and dataset-2. Model-I is trained with both fruits datasets. In this strategy, during transfer learning, the original weights of the convolutional network are preserved.

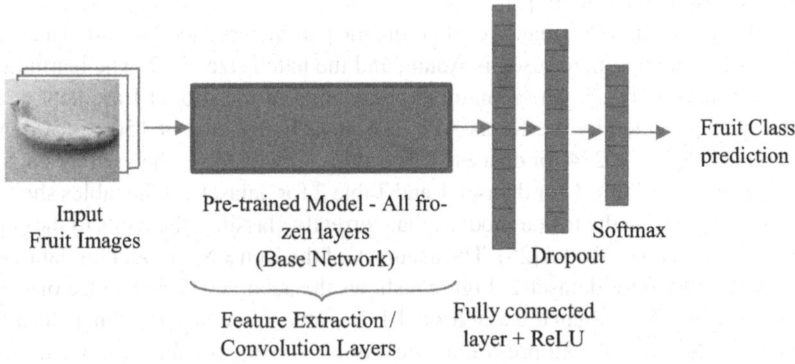

Fig. 2. Block diagram of Model-I

In the second strategy, the classifier remains the same as used in the first strategy. Here, a few layers of the convolutional base network are unfreezed for training (see Fig. 3). The unfreezed layers and a new block of fully-connected layers (shown in blue in Fig. 3) are retrained with dataset-2. We call it Model-II. During retraining, we have to be very careful with the learning rate that controls the weights of the pre-trained network. A small learning rate is preferred for pre-trained networks because a high learning rate may distort the CNN weights too early.

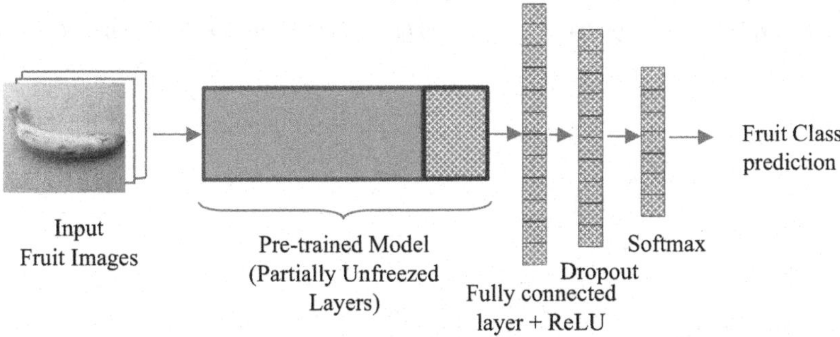

Fig. 3. Block diagram of Model-II

4 Results and Discussion

Our results for VGG16 [22], InceptionV3 [23], Xception [24], DenseNet [25] and ResNet152V2 [26] pre-trained networks (model-I and model-II), which are trained and evaluated with dataset1 and dataset-2, are reported in this section. These networks have been selected because these networks have achieved excellent performance on the Imagenet challenge [21]. All these pre-trained networks are freely available in the Keras API for use.

The experiments are performed using Google Colab with its GPU. We used Keras APIs in Python for building the fruit classifiers. Further, to reduce the training time, we used transfer learning for both datasets. Standard formulas are used to calculate accuracy, precision, recall, and F1-score [6].

For consistency in performance evaluation, the parameters used for both datasets are kept the same. The optimizer used is Adam, and the batch size = 32. The learning rate used for Model-I is 1e−3. The remaining parameters of the Model-I are kept at their default values provided by the Keras API. The input image size of 150 × 150 is used for dataset-1 and 224 × 224 for dataset-2. The values obtained for the various statistical metrics are given in Table 1 for dataset-1 and Table 2 for dataset-2. The tables show that the best model, among the tested models, that correctly classifies the fruits to the correct classes is the DenseNet Model [25]. The accuracy of the DenseNet model for dataset-1 is 99.03% and 99.11% for dataset-2. Figure 4 shows the accuracy performance of various pre-trained networks evaluated using Model-I. In Table 2, we observed that Recall and Accuracy are the same for all pre-trained networks. The equal values for accuracy and recall are because dataset-2 is highly balanced and contains an equal number of images for each class, i.e., 1000 images per fruit class.

Table 1. Performance metrics of pre-trained networks using Model-I for dataset-1

Model	Accuracy	Precision	Recall	F1-score
VGG16 [22]	96.62	96.75	96.50	96.59
InceptionV3 [23]	96.81	96.74	96.69	96.61
Xception [24]	96.62	96.94	96.52	96.69
ResNet152V2 [26]	97.14	97.13	97.08	97.10
DenseNet [25]	99.03	99.00	98.99	98.98

Table 2. Performance metrics of pre-trained networks using Model-I for dataset-2

Model	Accuracy	Precision	Recall	F1-score
VGG16 [22]	95.22	95.53	95.22	95.23
InceptionV3 [23]	97.05	97.14	97.05	97.02
Xception [24]	97.00	97.12	97.00	96.99
ResNet152V2 [26]	97.61	97.73	97.61	97.62
DenseNet [25]	99.11	99.13	99.11	99.11
Partial unfreezed DenseNet [25]	99.6111	99.6158	99.6111	99.6098

From Table 1 and Table 2, it can be observed that DenseNet is the best. The confusion matrices obtained for the DenseNet architecture for Model-I are shown in Fig. 5 (for dataset-1) and Fig. 6 (for dataset-2). The test set of dataset-1 contains 2672 images, and dataset 2 contains 1800 images (150 images for each class).

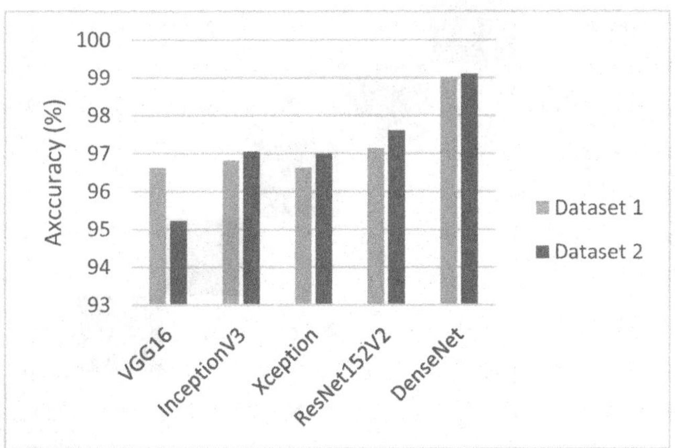

Fig. 4. Accuracy performance of various pre-trained networks using Model-I for dataset-1 and dataset-2.

To further increase the accuracy of the DenseNet model on fruit datasets, it is further evaluated using Model-II. As stated earlier, in Moel-II, we partially unfreezed the convolutional network of DenseNet and retrained the network (Fig. 3). The input image size for training DenseNet is 224 × 224; hence we trained and tested it only on dataset 2.

The convolutional base of DenseNet is a huge network and consists of five convolutional blocks, and each convolutional block consists of many sub-blocks of convolutional and pooling layers. We experimented with unfreezing a few sub-blocks of convolutional block-5. Convolutional block-5 consists of 16 sub-blocks. We experimented and evaluated the network's performance by unfreezing sub-blocks 16 to 12 of convolutional block-5. We found that the performance of the network improved till unfreezing sub-block 13 but degraded when further unfreezed to sub-block-12. When we unfreeze a sub-block, all the downline sub-blocks from that block are unfreezed until the network's end. We started by unfreezing one subblock of conv5_block from the bottom at a time for partial retraining and evaluated the network.

We have used a learning rate of 1e−04 with the optimizer Adam. Further, the early stopping technique is used to avoid overfitting the network. The evaluation metrics are reported in Table 2 (the last row). The accuracy of the DenseNet model evaluated using Model-II for dataset-2 is 99.61%, and its confusion matrix is shown in Fig. 7. From the confusion matrix (Fig. 7), we can see that the number of misclassified fruits is 7 out of 1800 fruits, and the only misclassified fruits are Fresh Apples and Rotten Oranges. All other fruits (fresh and rotten) are 100% correctly classified.

Since the proposed model, Model-II using DenseNet architecture, achieves an accuracy of 99.61%, therefore, DenseNet architecture when compared with other architectures considered in this paper can be applied in fruits industries for classification of fresh and rotten fruits.

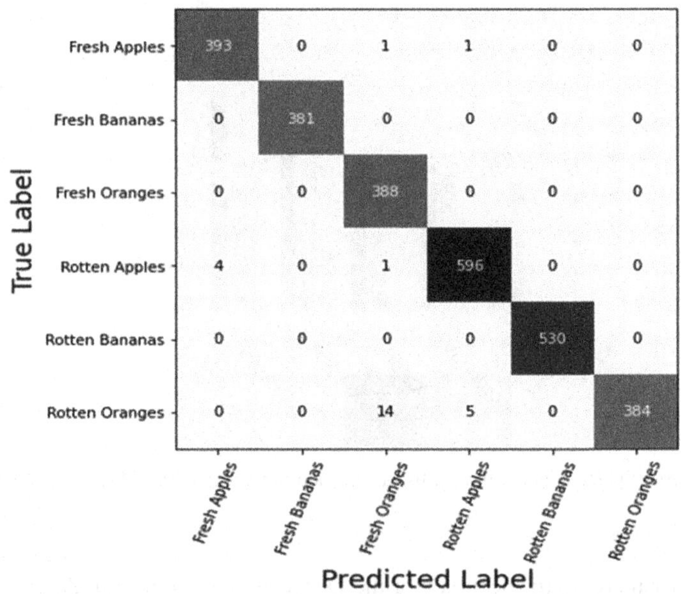

Fig. 5. Confusion matrix obtained for the DenseNet pre-trained network using Model-I for dataset-1

5 Conclusions

This paper proposes fruit classification systems based on generic image classifiers: VGG16, InceptionV3, Xception, DenseNet, and ResNet152V2. We have partially retrained these networks using transfer learning for two fruits datasets. We experimented with changing the classifiers and unfreezing a few sub-blocks of base convolutional networks of DenseNet network. We adopted two training strategies for transfer learning. In the first strategy, the base network is frozen, the old classifier is removed, and a new classifier is added and fine-tuned. Results show that the DenseNet model is superior than other considered networks on both datasets. The accuracy achieved is 99.03% (for dataset-1) and 99.11% (for dataset-2). We applied the second strategy to DenseNet to further improve its accuracy. In the second strategy, the base network of the DenseNet model is partially unfreezed and fine-tuned, which results in an accuracy of 99.61%. It can be concluded that an accurate fruit classification can be developed using these generic image classifiers. In a future study, these classifiers will be tested on other fruits and fruit datasets.

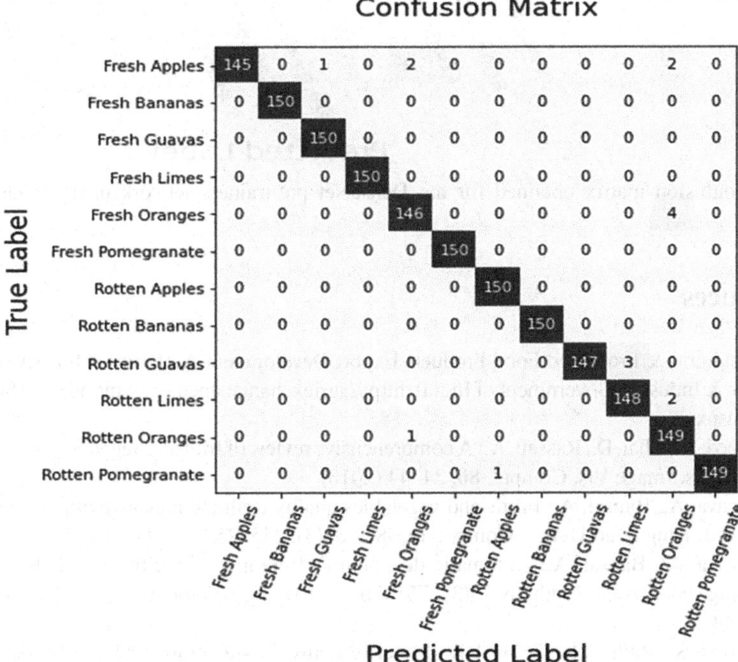

Fig. 6. Confusion matrix obtained for the DenseNet pre-trained network using Model-I for dataset-2

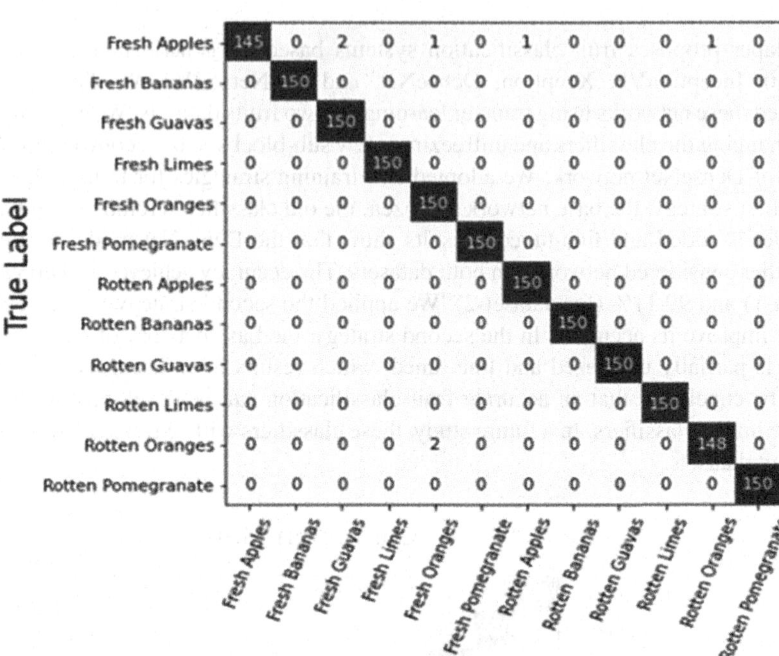

Fig. 7. Confusion matrix obtained for the DenseNet pre-trained network using Model-II for dataset-2

References

1. Agricultural & Processed Food Products Export Development Authority (Ministry of Commerce & Industry, Government of India). http://agriexchange.apeda.gov.in/indexp/18headproduct.aspx
2. Hameed, K., Chai, D., Rassau, A.: A comprehensive review of fruit and vegetable classification techniques. Image Vis. Comput. **80**, 24–44 (2018)
3. Bhargava, A., Bansal, A.: Fruits and vegetables quality evaluation using computer vision: A review. J. King Saud Univ. – Comput. Inf. Sci. **33**(3), 243–257 (2021)
4. Bhargava, A., Bansal, A.: Automatic detection and grading of multiple fruits by machine learning. Food Anal. Methods **13**(3), 751–761 (2019). https://doi.org/10.1007/s12161-019-01690-6
5. Behera, S.K., Rath, A.K., Sethy, P.K.: Maturity status classification of papaya fruits based on machine learning and transfer learning approach. Inf. Process. Agric. **8**(2), 244–250 (2021)
6. Geron, A.: Hands-On Machine Learning with Scikit-Learn, Keras, and TensorFlow, 2nd edn. O'Reilly Media, Inc. (2019). ISBN 9781492032649
7. Rawat, W., Wang, Z.: Deep convolutional neural networks for image classification: a comprehensive review. Neural Comput. **29**(9), 2352–2449 (2017)
8. Weiss, K., Khoshgoftaar, T.M., Wang, D.: A survey of transfer learning. J. Big Data **3**, 9 (2016). https://doi.org/10.1186/s40537-016-0043-6

9. Meshram, V.A., Patil, K., Ramteke, S.D.: MNet: a framework to reduce fruit image misclassification. Ingén. Syst. d'Inf. **26**(2), 159–170 (2021)
10. Ismail, N., Malik, O.A.: Real-time visual inspection system for grading fruits using computer vision and deep learning techniques. Inf. Process. Agric. **9**, 24–37 (2021)
11. Singh, S., Singh, N.P.: Machine learning based classification of good and rotten apple. In: Khare, A., Tiwary, U., Sethi, I., Singh, N. (eds.) Recent Trends in Communication, Computing, and Electronics. Lecture Notes in Electrical Engineering, vol. 524, pp. 377–386. Springer, Singapore (2019). https://doi.org/10.1007/978-981-13-2685-1_36
12. Mahrishi, M., Morwal, S., Muzaffar, A.W., Bhatia, S., Dadheech, P., Rahmani, M.K.I.: Video index point detection and extraction framework using custom YoloV4 darknet object detection model. IEEE Access **9**, 143378–143391 (2021). https://doi.org/10.1109/ACCESS.2021.3118048
13. Long, N.T.M., Thinh, N.T.: Using machine learning to grade the mango's quality based on external features captured by vision system. Appl. Sci. **10**, 5775 (2020)
14. Hossain, M.S., Al-Hammadi, M., Muhammad, G.: Automatic fruit classification using deep learning for industrial applications. IEEE Trans. Ind. Inf. **15**(2), 1027–1034 (2018)
15. Katarzyna, R., Paweł, M.: A vision-based method utilizing deep convolutional neural networks for fruit variety classification in uncertainty conditions of retail sales. Appl. Sci. **9**(19), 3971 (2019)
16. Hu, Z., Tang, J., Zhang, P., Jiang, J.: Deep learning for the identification of bruised apples by fusing 3D deep features for apple grading systems. Mech. Syst. Signal Process. **145**, 106922 (2020)
17. Wu, S.-L., Tung, H.-Y, Hsu, Y.-L.: Deep learning for automatic quality grading of mangoes: methods and insights. In: 19th IEEE International Conference on Machine Learning and Applications, pp. 446–453 (2020)
18. Rojas-Aranda, J.L., Nunez-Varela, J.I., Cuevas-Tello, J.C., Rangel-Ramirez, G.: Fruit classification for retail stores using deep learning. In: Figueroa Mora, K., Anzurez Marín, J., Cerda, J., Carrasco-Ochoa, J., Martínez-Trinidad, J., Olvera-López, J. (eds.) Pattern Recognition MCPR 2020. Lecture Notes in Computer Science, vol. 12088, pp. 3–13. Springer, Cham (2020). https://doi.org/10.1007/978-3-030-49076-8_1
19. https://www.kaggle.com/sriramr/fruits-fresh-and-rotten-for-classification
20. https://ieee-dataport.org/open-access/top-indian-fruits-quality
21. Russakovsky, O., et al.: Imagenet large scale visual recognition challenge. Int. J. Comput. Vis. **115**(3), 211–252 (2015)
22. Simonyan, K., Zisserman, A.: Very deep convolutional networks for large-scale image recognition. arXiv preprint arXiv:1409.1556 (2014)
23. Szegedy, C., Vanhoucke, V., Ioffe, S., Shlens, J., Wojna, Z.: Rethinking the inception architecture for computer vision. arXiv:1512.00567v3 [cs.CV] (2015)
24. Chollet, F.: Xception: deep learning with depthwise separable convolutions. In: Proceedings of the IEEE Conference on Computer Vision and Pattern Recognition, pp. 1251–1258 (2017)
25. Huang, G., Liu, Z., Maaten, L.V.D., Weinberger, K.Q.: Densely connected convolutional networks. In: Proceedings of IEEE Conference on Computer Vision and Pattern Recognition. Honolulu, USA, pp. 2261–2269 (2017)
26. Kaiming, H., Zhang, X., Ren, S., Sun, J.: Deep residual learning for image recognition. In: Proceedings of IEEE Conference on Computer Vision and Pattern Recognition. Nevada, USA, pp. 770–778 (2016)

Analysis of Crime Rate Prediction Using Machine Learning and Performance Analysis Using Accuracy Metrics

Meenakshi Nawal[1,2](✉), Kusumlata Jain[3], Smaranika Mohapatra[3], and Sunita Gupta[1,2]

[1] Swami Keshvanand Institute of Technology, Management and Gramothan, Jaipur, India
{meenakshi.nawal,sunita}@skit.ac.in
[2] RTU, Kota, Rajasthan, India
[3] Manipal University Jaipur, Jaipur, Rajasthan, India
{kusumlata.jain,smaranika.mohapatra}@jaipur.manipal.edu

Abstract. One of the most serious and overarching issues in our society is crime, which involves many crimes daily. Crime represents today's greatest threat to humanity. The crime rate is steadily increasing; it is impossible to predict because crimes are not systematic or random but grow and spread rapidly. The victims of crimes are all small towns and large cities. It is the police department's duty to control and reduce the crime that has increased the rate of crime. To do that, we need to deal with criminal matters much faster. This dataset includes the date and crime rate for each year. The crime rate of this project is based on murder, violence against women and auto theft. We use historical data to predict the percentage of crime in the future using a variety of regression algorithms using the accuracy scores.

Keywords: Crime rate · Machine learning · RMSE · R2-score · Linear regression · Polynomial regression

1 Introduction

Crime represents today's greatest threat to humanity. Criminology is a field which concentrates on science on crime and crime and enforcement, and a process aimed at identifying characteristics of crime [7]. The rate of crime is steadily increasing and cannot be predicted because of the lack of systemic or random nature of the crimes, but they are expanded and spread rapidly and widely. In an analysis that has varying degrees of success, criminologists and statistics use their expertise and knowledge [3]. Though promising, the social & web media context which surrounds events of interest does not take into account the rich and fast expanding [4]. The objective of this project is to predict crime using the data set. The data set is taken from the official websites. The aim is to develop a model for prediction. The training would be done via a validated dataset of training data. Multi-linear regression (MLR) crime prediction is used. Every year and based on population and number of crimes, the data set is visualized to analyze potential

V. E. Balas et al. (Eds.): ICETCE 2022, CCIS 1591, pp. 302–309, 2022.
https://doi.org/10.1007/978-3-031-07012-9_27

crimes. Data is displayed as Crime analyses; crime prevention and the analysis of crime patterns and trends is a systemic approach [2]. Sleepless nights and unforgettable days for common citizens were the result of repeated and frequent thieves, robberies, killings, rapes, shopping, pocket raids and drug-related abuse, illegal trafficking, smuggling, car-theft etc. NARX model, a nonlinear flexible approach to predict the next crime and suspect biography [5]. Crime is unpredictable in the classical sense. It is not necessarily random, nor does it persist in space or time [5]. A variety of documentation and cases are used to facilitate and speed up working on machine learning and data science. In particular, the paper is intended to predict the crime rate and analyze the factors leading to future crimes. The officials can take care of this information and attempt to lower the crime rate. To analyze crimes in the country, regardless of the crime, community and city pressures are being applied. Please note that the first paragraph of a section or subsection is not indented. The first paragraphs that follows a table, figure, equation etc. does not have an indent, either. The costs involved in police crime and prosecuting criminals are common monetary costs (Fig. 1).

Out[243]:

	P-CA	P-CB	P-CC	ACA	ACB	ACC
422	16616.589990	1365.031412	10712.241379	11812.0	1284.0	5267.0
312	1070.397596	214.125823	795.313599	177.0	145.0	687.0
120	5525.661338	1368.344984	6909.273216	5805.0	1280.0	5537.0
103	-3548.707618	-43.477570	-1275.090981	132.0	42.0	125.0
155	8054.006666	925.429361	10874.693722	9839.0	1057.0	15634.0
...	...	...	...	...	...	...
52	17170.611801	1247.731694	3504.730677	17449.0	1354.0	3729.0
395	13865.475591	1413.723079	3898.867847	17837.0	1328.0	3204.0
463	2539.838001	1674.828880	2366.933977	9332.0	1544.0	4005.0
286	17242.526835	3146.834966	16474.884186	14924.0	2793.0	13910.0
549	21976.435535	1887.301089	3452.955135	26125.0	2415.0	4305.0

168 rows × 6 columns

Fig. 1. Predicted and actual crime - prediction results

The research paper has the main requirement and findings to predict the exact crime rates using old data used to train and build the model. The forecast should be as precise as possible [10]. This article includes master learning, the methodology used in Fig. 2, to develop a model which can correctly predict predictive modeling. It includes machine-learning algorithms, which learn certain features from the training data to forecast the test data set [8]. It includes different regression models, the model to use the exact results.

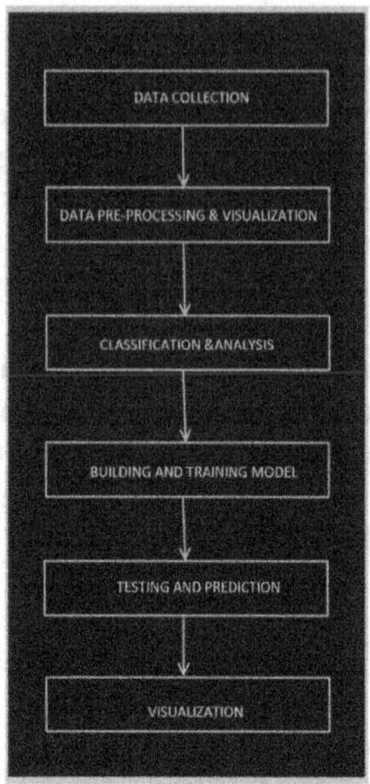

Fig. 2. Design flow of the process of prediction

2 Data Collection

The Internet bought all the available data for the ordinary man, it was difficult to find data for this project yearly, since things such as the population, gender ratio, literacy rate, etc. are calculated officially only once every 10 years. We have received annual data from many private and government survey companies and human rights organizations. This is the process by which the collected data can be made functional and useful. For the pre-processing of our dataset, we used python libraries like NumPy and Pandas. For 20 years from 2001 to 2020, we collected and cleaned all data from 28 countries. This included the management of null values (took the medium for arithmetic), outliers (removal) and values that affected system accuracy. In this phase, we identify and analyses and target the important variables or categories or classes. We used the selection of features to consider the correlation matrix and their coefficients to help us to combine the target variables for each crime and for each crime type. Few of the features were redundant and few of them affected our target variable very little or no.

3 Training and Testing a Model

As we predict the crime rate or the number of crimes that take place, our output is a numerical value predicted. This means that we use prediction regression models. Regression is a large topic, with countless algorithms for provision, we have chosen linear regression, multiple regression, lasso, and polynomial degree 2 & 3, as well as time spectrums, because it depends on time here year, after a detailed search of our target class and considered factors for building a model. We have all the known and suitable algorithms and have checked for consistent and accurate results, and then we have developed the model for training and testing the results and we have "r-2 scores" and "RMSE" which construct the results of all regression results. Tested and developed models using different algorithms to work together and separately on our three crimes. Using the comparison of the actual and forecast results and having a good r-2 score and a less RMSE (root mean squared value). Work again on this when the score is better, and less error is needed. This stage cleared our ready assumptions about few approaches, algorithms, and models when we put our data into various algorithms. The most accurate and closest results were predicted separately by lasso polynomial regression.

4 Visualization

Presenting the results through graphical and pictorial representation in a better and easier way to understand. We were able to show our results in much clearer ways with libraries such as Matplotlib, Plotly, Seaborn and Graphy. This made it easy for all to understand our results and predictions.The Fig. 3 shows the visualization of the collinearity between various parameters chosen like population, gender radiopurity rate,literacyrate, GDSP (at constant prices), Religious Bifurcation, crimes (for top three crimes in India). Crime A states for Crime against Women, Crime B for Murder and in Crime C for automobile theft.

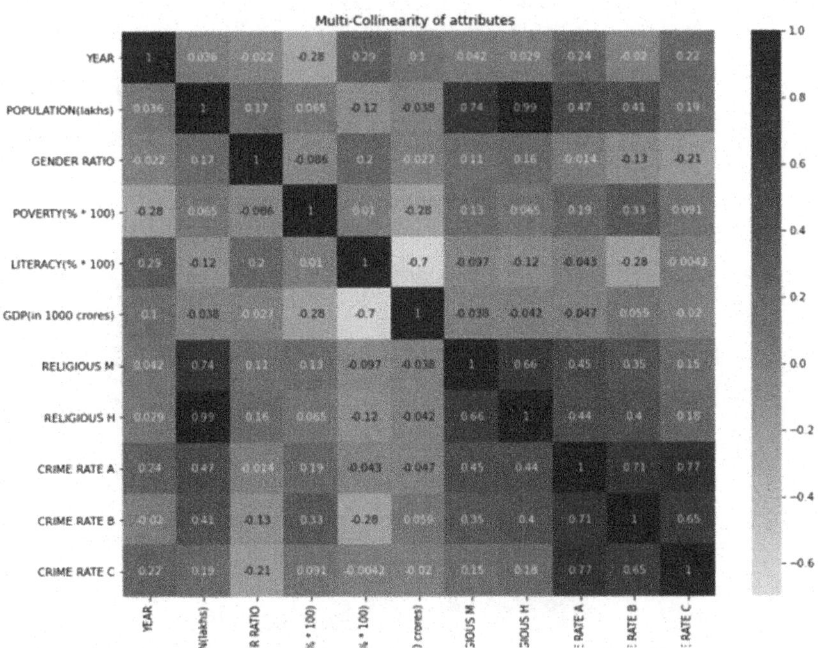

Fig. 3. Multi collinearity of attributes

5 Results and Discussion

Machine learning is a data analysis method which automates the development of an analytic model. The branch is based on the idea that systems can learn from data and identify patterns and make decisions with minimal human action. The way to create a model able to make predictions is predictive modeling. The process includes an algorithm to learn certain features from a training dataset to make these forecasts. Linear Regression is a supervised learning machine algorithm. It carries out a reversal task. Regression models a value based on independent variables for the target prediction. The

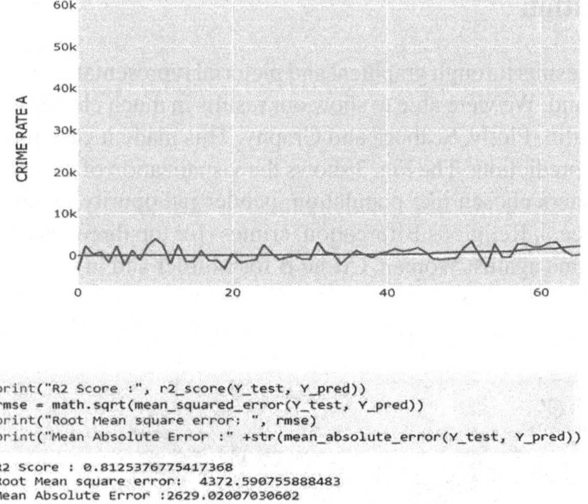

Fig. 4. Error and accuracy scores

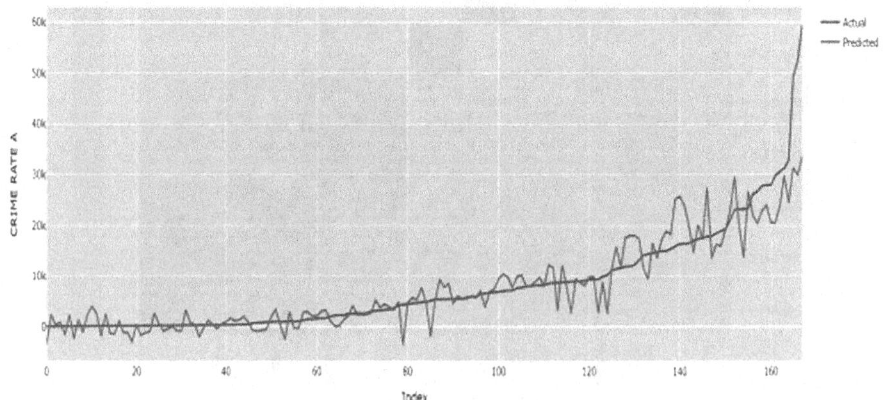

Fig. 5. Crime rate A-crime against women using linear regression

relationship between variables and forecasting is mainly determined. The type of relationship between dependent and independent variables is different, and the number of independent variables is considered.

The Fig. 4 describes the accuracy scores for analyzing he polynomial and multiple linear regression for the crime rates prediction. The r2 score is 0.81 which suggests that the accuracy of the system is 81%. Figure 5 we can analyze the rate of crime against women which has helped to give insights that there can be suggestive measures which should be taken for declining the rate of crime against women and Fig. 6 and 7 depicts the crime rate prediction for murder and automobile theft which has a rise in cases in crime rate.

The Figs. 5, 6 and 7 analyses the crime rate prediction using linear regression analysis. The Fig. 8, 9 and 10 uses polynomial regression analysis to predict the crime rate for women, murder and automobile theft.

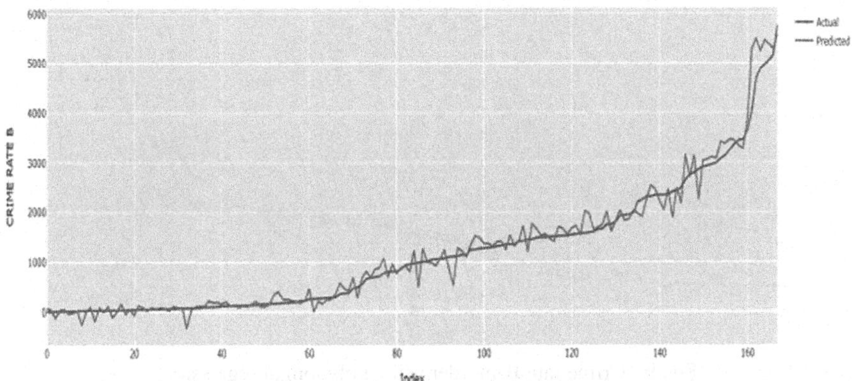

Fig. 6. Crime rate B-murderusing linear regression

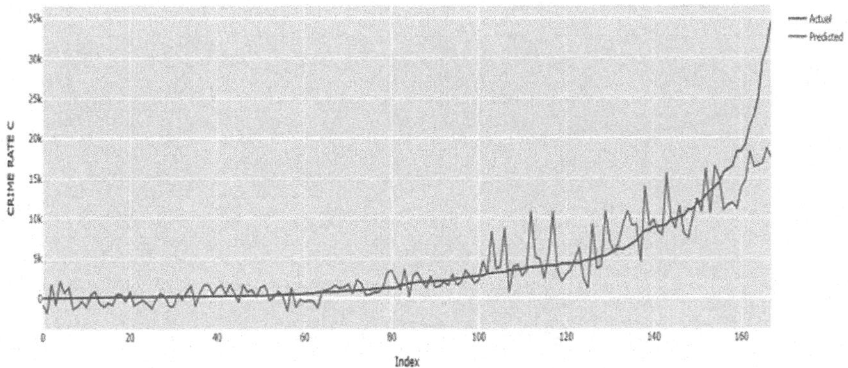

Fig. 7. Crime rate C-automobile theft using linear regression

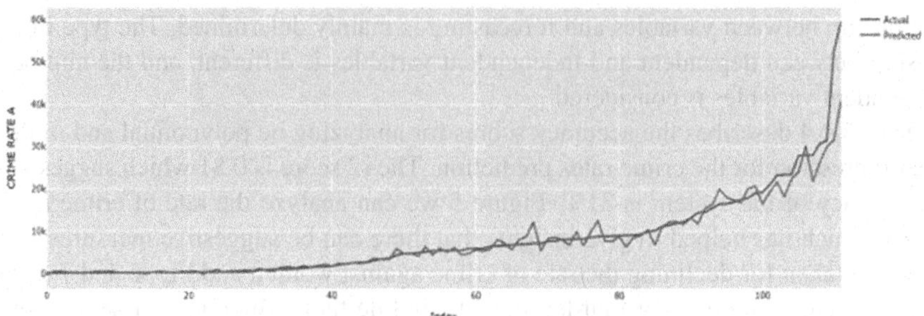

Fig. 8. Crime rate A-crime against women using polynomial regression

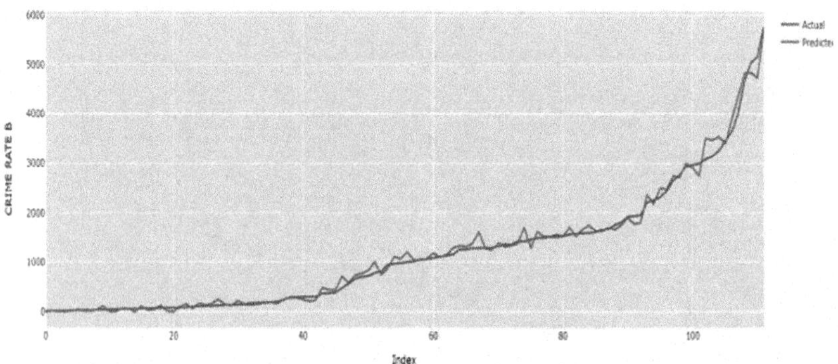

Fig. 9. Crime rate B-murder using polynomial regression

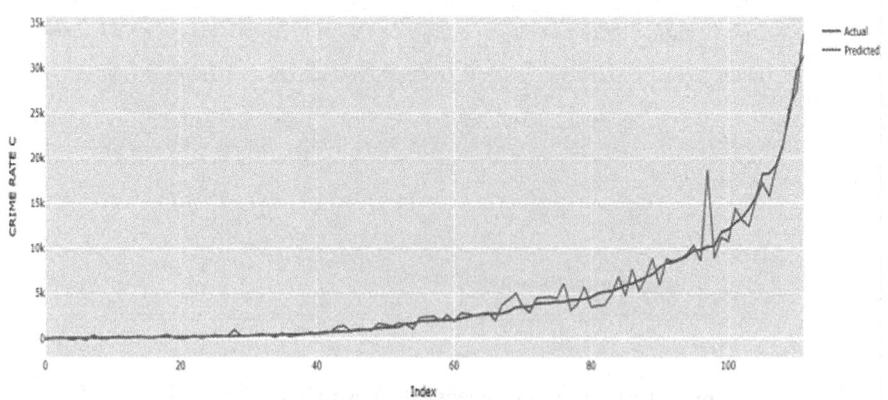

Fig. 10. Crime rate A-crime against women using polynomial regression

6 Conclusion and Future Scope

Every society's aim should not only be to catch criminals, but also to avoid crime. Future crime predictions: By using history and monitoring where recent crimes have occurred, we can predict where future crimes are likely to occur. For example, a burglary rush could correlate soon with more burglary in the vicinity. The system points to potential hotspots in a map that the police should consider more patrolling. The above results which were derived using the linear and polynomial regression for three categories of crime like murder, crime against women and automobile theft gives an insight that the government needs to make more strict policies for curbing the crime rate that's increasing day by day.

References

1. Sathyadevan, S., Devan, M., Gangadharan, S.: Crime analysis and prediction using data mining. In: First International Conference on Networks & Soft Computing (2014)
2. Babakura, A., Sulaiman, M.N., Yusuf, M.A.: Improved method of classification algorithms for crime prediction. In: International Symposium on Biometrics and Security Technologies (ISBAST) (2014)
3. Azeez, J., Aravindhar, D.J.: Hybrid approach to crime prediction using deep learning. In: International Conference on Advances in Computing, Communications and Informatics (ICACCI) (2015)
4. Dubey, N., et al.: A survey paper on crime prediction technique using data mining. Int. J. Eng. Res. Appl. (2014)
5. Chandra, B., Gupta, M., Gupta, M.P.: A multivariate time series clustering approach for crime trends prediction, pp. 892–896. IEEE (2008)
6. Malathi, A., Baboo, S.S.: An enhanced algorithm to predict a future crime using data mining. Int. J. Comput. Appl. **21**(1), 1–6 (2011)
7. Wang, B., et al.: Deep learning for real time crime forecasting. arXiv preprint arXiv:1707.03340 (2017)
8. Chun, S.A., et al.: Crime prediction model using deep neural networks. In: Proceedings of the 20th Annual International Conference on digital government research (2019)
9. Stalidis, P., Semertzidis, T., Daras, P.: Examining deep learning architectures for crime classification and prediction. Forecasting **3**(4), 741–762 (2021)
10. Kumar, K.K., Venkateswara, R.H.: Crime activities prediction system in video surveillance by an optimized deep learning framework. Concurr. Comput.: Pract. Exp. **1**, e6852 (2022)

Assessment of Network Intrusion Detection System Based on Shallow and Deep Learning Approaches

Gaurav Meena(✉), Babita, and Krishna Kumar Mohbey

Central University of Rajasthan, Ajmer, India
Gaurav.meena@curaj.ac.in

Abstract. This extensive review aims to classify the Intrusion Detection System (IDS) and various machine learning and deep learning (ML/DL) approaches used for IDS. The survey also addresses security, which is a concern with the Internet of Things. Several types of intrusion detection systems (IDSs), including shallow and deep learning methods and various learning algorithms to aid intrusion detection, are also categorized. This research expands on Network Intrusion Detection Systems and investigates techniques for improving their performance. It provides a more comprehensive understanding of deep and shallow learning methodologies with their benefits and drawbacks. The study component examines IDS classification, feature extraction techniques, machine learning, deep learning, and examples of how these may be applied. The essence of this review will establish a viable approach to assist professionals in modeling trustworthy and powerful IDS based on real-time requirements. Because the methods of intrusions and cyberattacks in networks are constantly evolving, it attracted the interest of many scholars and industrial professionals. However, cyber specialists struggle to develop an accurate and effective Intrusion Detection System (IDS). In addition, an increasing number of devices has resulted in more complicated network topology, raising security risks. As a result, a lengthy and exhaustive review is indispensable while developing a secure communication system.

Keywords: Artificial neural networks · Deep learning · Information security · Internet of Things (IoT) · Intrusion detection system · Knowledge discovery in databases · Machine learning

1 Introduction

We have seen extensive networking growth over a couple of years, contributing majorly to economic and social development. Many sectors increasingly depend on networks such as trade, defense, healthcare, finance, and banking. Hence, these sectors become more vulnerable to cyber-attacks, which have attracted many security experts to build an intrusion detection system (IDS). It is a crucial component of computer networks, which are more prone to internal and external intrusions [1, 2].

V. E. Balas et al. (Eds.): ICETCE 2022, CCIS 1591, pp. 310–335, 2022.
https://doi.org/10.1007/978-3-031-07012-9_28

Intrusion detection system (IDS) [3] that monitors the network's suspicious activities, keeps a check on network traffic, and alerts the network admin. In addition, it can block malicious traffic at its discretion by blocking the source IP address. These have been in use since 1980. However, cyber threats can be disastrous beyond imagination, such as DoS, which ceases genuine users' access to network resources and creates redundant traffic in the network [4]. Trojans, worms, and malware are more examples of attacks where attackers compromise the users' system using malicious executable codes [6].

The subsequent significant technological development on the Internet is the Internet of Things (IoT). Everything from your kitchen to your wardrobe will be linked to the Internet, providing users access to everything from anywhere, at any time, in any fashion and with little effort ("5 A's"). More and more hazards and threats are emerging in the operating environment because of IoT. While many mitigating techniques are implemented on IoT components, these include things like authentication, encryption, and traffic analysis. In the context of authentication, there are millions of different methods. This might consist of using a password of any type, a visual password, or applying a biometric system [2].

Intrusion detection systems emerged as a combating solution to cyber threats. In recent years, many scholars have focused on building IDS, which has a higher accuracy rate and low false performance parameters [5]. However, most IDS's are suffering from a high rate of false positives and negatives, which only incorrect recognition of intrusion, which makes network admin's job worse while handling intrusions. As the attacks gradually become advanced and less detectable, existing detection systems cannot recognize the latest attacks because they are all based upon signature-based detection methods. They require manual feeding of signatures of known attacks.

To overcome these issues, IDS which use machine learning, are seen as an alternative to existing IDS and do not require human intervention. First, we need to train ML IDS about genuine and malicious traffic using datasets like KDD CUP'99. After training, it can predict attacks by using machine learning algorithms like classification. Unfortunately, although ML algorithms are proved to be good, they also suffer from a high false positives and negatives rate, and efficiency is significantly less [5].

The research and study presented in this paper can be considered most pertinent in the current context because of the gradually evolving nature of the attack, systems, networks, and traffic data [8, 9]. Therefore, this comprehensive study has two primary aims; first, to have a clear picture of the existing situation of IDS and where we should shift our focus to build more reliant IDS [10, 11].

This survey paper includes an elaborative discussion upon IDS and comparative performance analysis based upon many experimental studies. This paper will give a more comprehensive picture of deep and shallow learning IDS and address each's merits and demerits.

This study aims to correctly classify IDS and various shallow and deep learning algorithms useful in intrusion detection.

The rest of the paper is organized as Sect. 2 discusses the IDS classification and some methods of detection, Sect. 3 discusses various machine learning techniques and extracted features, Sect. 4 discusses deep and artificial neural networks as well as deep learning algorithms, and Sect. 5 we will conclude our discussion with some future scopes.

2 What is Intrusion Detection System?

A computer network prone to attacks, IDS play is the role of this kind of place. It acts as a surveillance and alarming device. It monitors the traffic, and whenever it detects suspicious activity, it alerts the network admin. The basic functionality of IDS includes reporting threats, taking preventive steps whenever it identifies some intrusion, and recording all significant incidents happening in the network [6]. The following Fig. 1 shows a modular diagram of IDS.

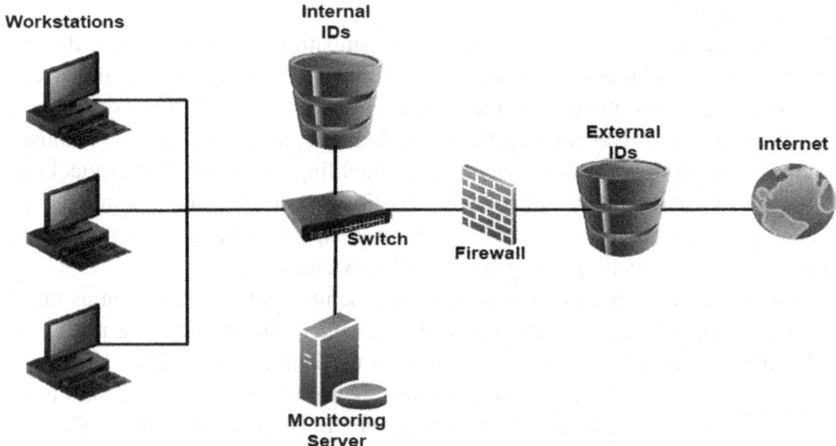

Fig. 1. Model of intrusion detection system [7]

2.1 IDS Classification

Following Fig. 2 is an illustrative categorization of IDS. A lot of cyber experts have developed the categorical representation of IDS. Researchers like H. Debar [15], S. Axelsson [8], S. Amer [9], C. Xenakis [10], Hung-Ten Lia [11], and more others had carried out a comprehensive as well as comparative study and survey of an intrusion detection system.

The survey paper is based on the works mentioned earlier and brings deep learning methods that did not discuss. As the volume of data increases with the increased use of the Internet in every sector, Deep learning becomes an essential player in intrusion detection. The survey presented in this paper provides every little detail about machine learning and discusses each term mentioned in Fig. 2, i.e., IDS classification.

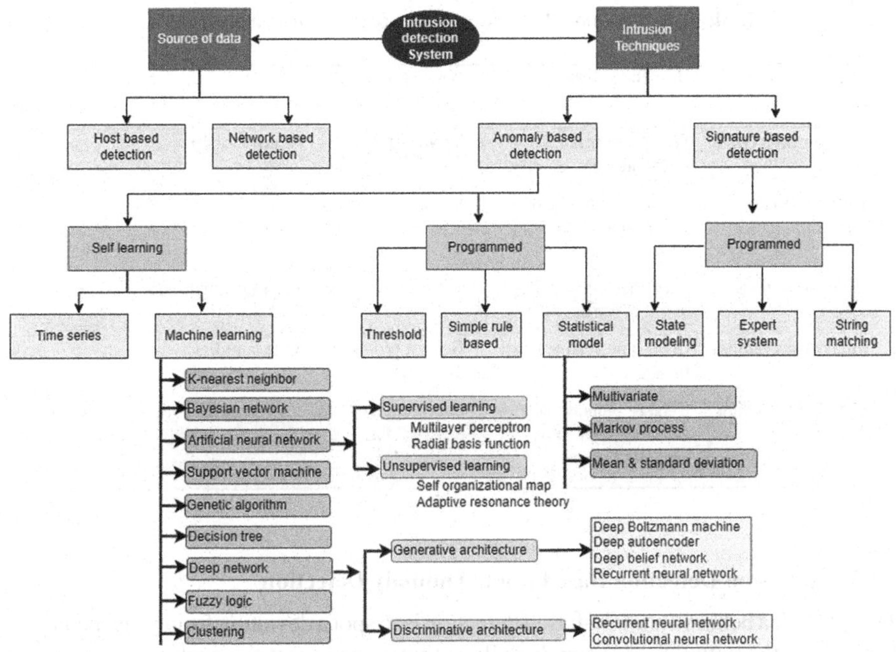

Fig. 2. IDS classification [13]

The intrusion detection system requires a dataset and some intrusion detection algorithm. The data used in detection can be some standard dataset or data collected from some network node. The source of data can be network or host, or both. We will further talk about this in the incoming Sect. 2.1.1. From Fig. 2, the intrusion detection technique into the anomaly and signature-based techniques, which are further divided into self-learning and programmed. These will be discussed in another section separately. Nevertheless, for some overview, self-learning is what a system learns while in programmed class; an external actor coaches the system.

2.1.1 Comparative Study of Network-Based IDS and Host-Based IDS

Host-based IDS (HIDS) was the first type of IDS to be employed [12]. An application-layer IDS, such as Host Intrusion Detection (HIDS), is installed on the host's computer, monitoring network traffic on system files and the operating system [21, 22]. Operating system audit trails contain data related to all the different types of traffic generated by the application and strategy [13]. HIDS is significantly disadvantaged because it only works with one system, which must be installed in every system [26].

On the other hand, Network-based IDS are present in computer networks to gather and analyze the traffic data passing through links [27, 28]. Therefore, it benefits from HIDS because a single NIDS is sufficient for monitoring the whole network, thus saving the installation cost and time in every host [14]. On the other hand, one of the demerits of NIDS is that it cannot identify internal intrusions arising in the network. Following Table 1 will make an exact crystal comparison between NIDS and HIDS based upon some essential criteria.

Table 1. HIDS and NIDS comparative performance analysis [15]

Performance in terms of:	Host-Based IDS	Network-Based IDS
Intruder deterrence	Strong deterrence for inside intruders	Strong deterrence for outside intruders
Threat response time	Weak real time response but performs better for a long term attack	Strong response time against outside intruders
Assessing damage	Excellent in determining extent of damage	Very weak in determining extent of damage
Intruder prevention	Good at preventing inside intruders	Good at preventing outside intruders
Threat anticipation	Good at trending and detecting suspicious behavior patterns	Good at trending and detecting suspicious behavior patterns

2.1.2 Intrusion Detection Based upon Anomaly Detection

The main idea behind anomaly-based detection is to spot a deviation from usual behavior. It monitors the network or system's daily activities and creates a profile for the site or system being watched [31, 32]. One of its benefits is that it can identify new system assaults [16]. Anomaly-based detection is further subdivided into two forms based on the course's normal behavior.

i) **Self-learning:** These systems learn independently, and a baseline for standard activities must be established. Time series models and machine learning are two types of self-learning systems. In Sect. 3, we'll go into machine learning in-depth.

   a) *a) Time series model*: It considers the sequence of observations that occur in uniform intervals. When the likelihood of a new observation is small, a shift in usual behavior is considered. This model has the advantage of observing changes in behavior and sounding an alarm if it detects any deviation from normal behavior. It is a resilient model against sequential attacks over time [17], but it is also computationally expensive [18]. Auto-Regressive Moving Average is an example of a time series model IDS (ARMA). [19] et al. used generalized ARMA and ARMA time series models. Four types of assaults are detected using predictive modeling: DoS, U2R, L2R, and probe. Although the models were able to predict the attack, generalized ARMA was more effective in detecting it.

ii) **Programmed:** These are the model where the system needs external help to detect behavioral changes. Here the external user sets the limit for abnormal behavior, and an alarm should be raised if a threat is detected [17]. Programmed models are categorized into threshold, simple rule-based, and statistical models.

a) ***Threshold models*** are the simplest programmed statistical detector [7]. A user can program the system at a known bottleneck of alarm on a statistical variable by analyzing and correlating the statistical data. To minimize the false alarm rate, one should carefully set the threshold. Very high threshold limits lead to missing an intrusion alarm [20]. For example, raise the alarm after three unsuccessful login attempts [38]. However, it cannot detect a threat not programmed as a rule in the system [21].
b) ***Simple rule-based models*** monitor events in a system that can raise/trigger suspicion other than the expected behavior. One such simple rule-based model is Repeated Incremental Pruning to Produce Error Reduction, i.e., RIPPER. R. Naidu et al. [40] made a performance comparison of the decision tree(C5), RIPPER rule, and SVM using the KDDCup '99 dataset. The recorded detection rate was 98.69% of the RIPPER rule, 98.75% of the decision tree, and 98.63% of the SVM.
c) ***Statistics models*** collect and analyze the data by storing it in a profile. The statistical analysis is compared to typical behavior and helps determine whether an action is abnormal or normal [13, 14]. In addition, this model creates a distance-vector for the profile and traffic observed, and if this distance is large enough, then an alarm is generated by the system [35, 41]. These models are further categorized in multivariate, operational, Markov, and mean/standard deviation models.

2.1.3 Intrusion Detection Based upon Signature

When a set of rules are used in network traffic to match patterns, and if a mismatch occurs, then an alarm is raised; this is known as Signature-based detection [42]. It detects attacks while giving a low false-positive detection rate [22], but it can only see those attacks in the database [12].

Signature-based models are further programmed in state modeling, string matching, and expert systems. In-state modeling, attacks are encoded as no. of various states of a finite automaton. As a result, every assault detected in the traffic profile will be classified as an intrusion. On the other hand, expert systems use rules to describe attack situations known to the system. They are, however, more vulnerable to assaults that are unrelated to the restrictions stated [2]. Finally, string matching learns the same information as an expert system but approaches that knowledge differently [7]. It does not participate in the decision-making process, but it deals with pattern matching generated by attack [23].

Nonetheless, this method is insufficient for IoT devices. Another option is to employ Two-Factor Authentication. This allows for two-tier password security solutions. They are, however, problematic, unmanageable, time-consuming, and easily counterfeited. As a result, that is not a practical method of security in IoT design. As a result, the third approach may be to employ biometric authentication because biometric is what we are and what we have. It cannot be stolen, forged, or misplaced. Alternatively, the most crucial aspect is that we do not need to remember or carry it out.

3 Machine Learning Techniques

We have discussed that anomaly detection systems detect intrusions by exposing abnormal behavior in a system, and these systems are independent of humans. However, their effectiveness depends upon one condition, i.e., how effectively they can differentiate abnormal behavior from ordinary.

A better alternative to detect existing and new attacks without human intervention is machine learning-based IDS. It is based upon a group of methods that can detect intrusion on its own and predict future trends [45, 46]. Approximately all the machine algorithms rely upon optimized feature selection. These features play a critical role in finding out trends and patterns. We will discuss the importance of feature selection and categorization (refer to Fig. 3) of various machine learning algorithms in the coming sections [7].

The hybrid scheme combines the benefits of deep learning and machine learning techniques. For latent feature extraction, a stacked autoencoder network is used first, followed by several classification-based intrusion detection methods such as support vector machine, random forest, decision trees, and naive Bayes, which are used to detect intrusion in massive network traffic data quickly and efficiently.

Rapid advancements in the internet and communication areas have resulted in a massive expansion of network size and data. As a result, many new attacks are being developed, making it difficult for network security to detect breaches effectively. Furthermore, intruders with the intent of launching various assaults within the network cannot be overlooked. An intrusion detection system (IDS) is a tool that inspects network traffic to assure its confidentiality, integrity, and availability and thereby protects the network from possible invasions. Despite the researchers' best efforts, IDS continues to encounter difficulties in boosting detection accuracy while lowering false alarm rates and detecting fresh intrusions. Machine learning (ML) and deep learning (DL)-based IDS systems have recently been implemented as viable solutions for quickly detecting intrusions across the network [70].

3.1 Feature Selection in Intrusion Detection System

Machine learning is a 2-step procedure as follows: first, classification, and second, training phase. To begin, the training phase feature distribution is to be taught, and when it comes time to identify abnormalities, the trained features are utilized as predefined attributes [47, 48]. Next, the testing and training phases clean the data to remove any noise. Training a classifier finally extracts features that were extracted during the training phase. After classifying the data, the trained classifier assigns it to the training set's specified features [25].

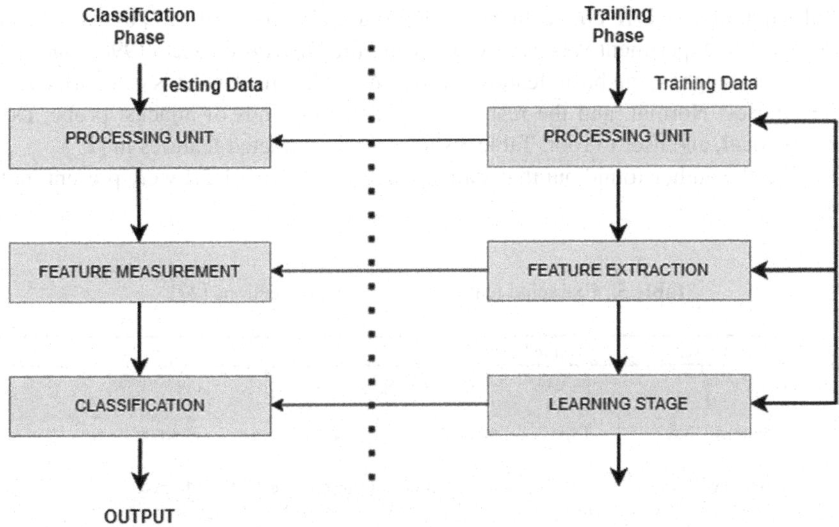

Fig. 3. Classification process of machine learning [7]

To achieve good results in machine learning IDS, we need a helpful quality training dataset. Both standard and malicious patterns are present in the training dataset [24]. First, we need to talk about features. Important information extracted from the raw dataset is known as the features, and they play a crucial role in classification and intrusion detection. They also affect the quality of efficiency of machine learning-based IDS. V. Mahoney et al. Analysed all network packets and flows' header-field and extracted features [26]. The proposed methodology detects intrusion based on UDP, ICMP, TCP, IP, and Ethernet's 33 fields. Table 2 shows some features extracted in [26].

Table 2. Header field analysis providing extracted features [26]

No.	Feature Description	No.	Feature Description
1	IP Source	10	IP protocol
2	IP Destination	11	IP Length
3	TCP Source Port	12	IP Type of Service
4	TCP Destination Port	13	IP Header Length
5	UDP Source Port	14	IP Time to Live
6	UDP Destination Port	15	Ether Source
7	UDP Length	16	Ethernet Size

Subsequently, TCP/IP connections were also used by Lee et al. [27] to extract characteristics. The experiment was carried out using the DARPA dataset (1998), and from domain knowledge, some basic features were removed. This dataset is categorized into five categories: Normal, and the rest are four different kinds of attacks: probe, DoS, remote to local, and user to root. Table 3 shows some extracted features in [27].

In [27], the author found out that some attacks like R2L & U2R were present in the packets'.

Table 3. Extracted features from TCP connections [27]

No.	Feature Name	Feature Description	Type
1	Duration	Length of connection in seconds	continuous
2	protocol_type	Type of the protocol used, e.g.. up,tcp, etc.	discrete
3	Service	Network service on the destination, e.g.http, telnet, etc.	discrete
4	scr_bytes	Number of data bytes from source to destination	continuous
5	dst_bytes	Number of data bytes form destination to source	continuous
6	Flag	Normal or error status of network	discrete
7	Land	1 if connection is from/to same host/port; 0 otherwise	discrete
8	wrong_fragment	Number of wrong fragments	continuous
9	Urgent	Number of wrong fragments	continuous

Payload. This information led to the formation of a new methodology where domain knowledge can be combined with features that are extracted from the payload of packets.

3.2 Categorization of Machine Learning Algorithms

In this section, we will discuss the various machine learning algorithms in detail. The following Fig. 4 will provide a proper classification of the different machine learning algorithms.

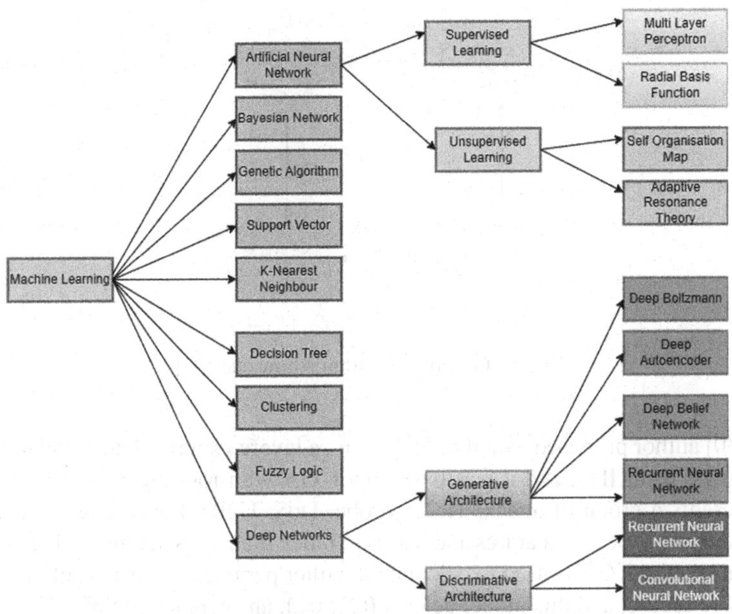

Fig. 4. Various techniques of machine learning [7]

a) ***Bayesian Networks:*** These are used to model the probability of variables using graphical modeling. These networks are directed graphs where each node represents a distinct random variable and its state in table form, i.e., conditional probability table (CPT). The direction of causality in the Bayesian network results from the relation between connected and domain variables. Hence, causal dependence exists between connected and domain variables.
An experiment was done in [28] on the NSL-KDD dataset using Bayesian networks. A filter approach was used to reduce the number of features, and features decreased from 41 to 16. Here, the Bayesian model performed well with an accuracy rate of 97.27% with a low rate of false-positive, i.e., 0.008. Other models were inferior to the Bayesian model, like Naive Bayes with 84.86% accuracy, decision stamp with 83.31% accuracy, and K-means clustering with an accuracy of 80.75%.
Another experiment was done in [29] with another dataset, i.e., the KDD Cup'99 dataset, using a Bayesian classifier to analyze network traffic by adopting a risk matrix approach to detect risk zones of attacks. Here Bayesian network performed better than the Genetic algorithm in the probe and U2R attack detection, but it acted the same as association rule mining in DoS detection.

b) ***Genetic Algorithm:*** This algorithm is based upon the principle of stochastic global search method [26], which initializes with random chromosome generation known as population. As shown in Fig. 5, they go through selection, crossover, and mutation, where each chromosome is an unsolved problem encoded in the form of a string. Their position is generally represented as an integral list or binary. During each generation, a solution is created based on the evaluation function. So, their (Chromo) selection is based on their fitness level [53, 54].

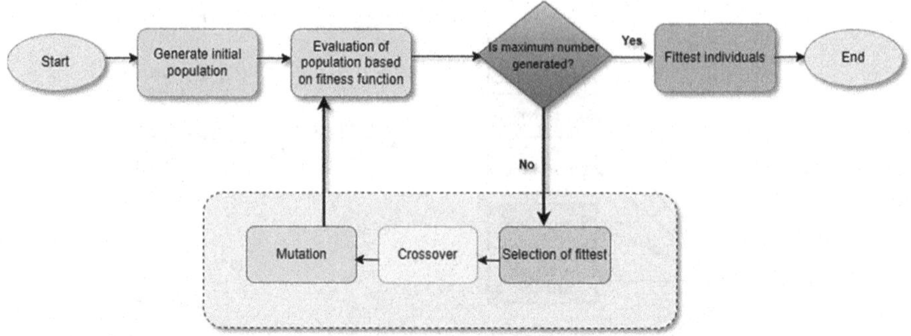

Fig. 5. Genetic algorithm's flow chart [7]

In [30] author proposed a methodology, i.e., a layered approach to combat the flaws of single layer IDS, and it was based upon GA with four layers where each layer represents a group of attacks (R2L, probe, DoS, U2R). Every layer gets separate training with specific features and learns to block malicious actions and act as filters. In this method, GA is used to calculate the filter parameters. As a result, the system is more secure, and this model detects R2L with an accuracy of 90%.
Following Fig. 6 will clearly show how SVM segregates data distinctly in N-dimensional space by finding a hyperplane, where N is the no. of features.

c) ***Support Vector Machines:*** This algorithm learns to classify the data using labeled training data into one or two classes. This algorithm builds a model that can identify the new data in which category [31, 32].

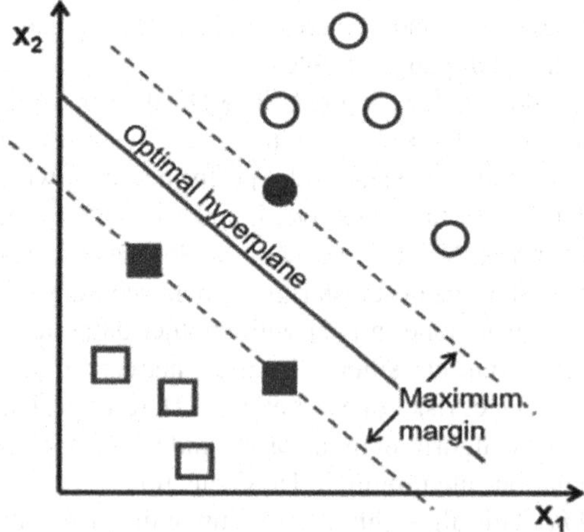

Fig. 6. SVM classifying data in two classes with a maximum margin

In [33], the author performed a performance analysis compared with four other classifiers using all features (41) and 10 features from the KDD Cup'99 dataset. GA was used to select ten of the 41 features. As a result, 90% of the dataset was utilized for training, while the remaining 10% was used for testing. Following Table 4 will present the performance analysis.

Table 4. SVM, GMM, Naive, MLP, and linear algorithms performance analysis [33]

Classified Algorithm	Accuracy% (41 feature)	Accuracy% (10 features)
Gmm	30.5	45.07
Naive	45.17	55.75
MLP	55.24	67.87
Linear	65.47	76.75
SVM	82.45	95.26

d) ***K-Nearest Neighbour (K-NN):*** It is the most basic algorithm for sample classification, which is also known for its highly efficient classification, and it is non-parametric [34]. Most test sample neighbors evaluate the class labels for test samples [35]. Here, the user will determine the parameter the 'k's. Furthermore, test sample data determines the 'k's no. of training points by considering the nearest distance to the test data. Thus, K nearest neighbor is the prediction of the test sample [36].
In [35] author proposed a hybrid method using K-NN + GA to find the intrusion in the KDD Cup'99 dataset, where GA is used to pick k nearest neighbor for the K-NN classifier. The proposed methodology outperformed the old K-NN in all values of k that were used in the experiment.

e) ***Decision Tree:*** It uses what it has learned from past data to understand and classify a new dataset [37]. It employs well-defined criteria in the selection of the finest feature. This model generates the path for each node using a root node linked to distinct nodes in attribute data [38]. The decision tree algorithm C4.5 is the most extensively utilized.
S. Sahu et al. [39] used a decision tree (J48) and a labeled dataset, Kyoto 2006 + , to develop an intrusion detection system with 24 features, 14 derived from the KDD Cup'99 dataset ten additional essential features. This experiment was carried out using the WEKA tool. It accurately classified 97.23% of the time and wrongly classified 2.67% of the time.

f) ***Fuzzy Logic (FL):*** The concept [40] of Fuzzy logic comes from the theory of fuzzy sets, which deal with imprecision and uncertainty [41]. For example, IDS works on the proper classification between normal and abnormal, and fuzzy logic helps build this classification.
Work done in [42] analyzed neuro-fuzzy networks, fuzzy inference, and GA using a 10% KDD Cup'99 dataset where classification was done with neuro-fuzzy classifiers. Its output became the input for the fuzzy inference system for finding the abnormal actions out of the dataset. It performed its best when a fuzzy decision tree was optimized with GA.

g) ***Clustering:*** It is the grouping of data based upon two or more parameters, and each group is known as a cluster having no similarity with others, and their labels are not known always. The most common example of clustering is the K-means algorithm [43, 44], which calculates the distance between data and cluster center using Euclidean distance. The focus here is to maximize the distance between clusters and minimize within the cluster [45].

3.3 Merits and Demerits of Machine Learning Algorithm

Table 5 will represent the advantages and disadvantages of ML algorithms.

Table 5. Pros and Cons of ML methods [45]

ML technique	Advantages	Disadvantages
Bayesian network	The graphical representation gives it an advantage of breaking complex problem into different smaller models	Slow in classifying data sets with many features
Genetic algorithm	• Uses a technique which is inspired by convolutional biological process • It has the capability of solving optimization problems during classification	Gets stack in local optima (overfitting)
Support vector machine	• The algorithm is simple to analyse mathematically • All computations are performed in space using kernels giving it an edge to be used practically	Slow in training and requires more memory space
K-nearest neighbour	East to implement and can solve multi-class problems	• Slow in training and requires large memory space • It is computationally complex because to classify a test sample involve the consideration of all training samples
Decision tree	• It has a unique structure therefore easy to interpret • It has no limitation in handling high dimensional data sets	• If trees are not pruned back, it causes overfitting • Types of data must be considered when constructing tree (i.e., Categorical, or numerical)
Fuzzy logic	• It is based on human reasoning concepts which are not precise • It gives a representation of uncertainty	Its construction has a high level of generality there by high consumption of resource
K-means algorithm	Simple to implement and effective	• The outcome of clustering depends on how cluster centres are initialised to specify K value • The algorithm works for only numerical data

3.4 Performance Metrics and Confusion Matrix

Any ML algorithm's performance is measured using True Negative (TN), True Positive (TP), False Negative (FN), and False Positive (FP) performance measures, which are all arranged in a matrix called the Confusion matrix [70]. A confusion matrix is depicted in Fig. 7.

		Predicted Class	
		Negative Class(Normal)	Positive Class(Attack)
Actual Class	Negative Class(Normal)	True Negative(TN)	False Positive(FP)
	Positive Class(Attack)	False Negative(FN)	True Positive(TP)

Fig. 7. Confusion matrix [47]

Now we will discuss the performance measures mentioned above.

1. True Positive (TP): When an assault is successfully identified as an attack, it is called a True Positive (TP).
2. True Negative (TN): When normal is correctly classified as normal, it is called True Negative (TN).
3. False Positive (FP): False Positive (FP) occurs when anything that appears to be normal is mistakenly classified as an assault.
4. False Negative (FN): when attack incorrectly classified as normal.

By using above mentioned measures we calculate the performance of machine learning IDS in the following manner:

1. Accuracy = (TP + TN)/(TP + TN + FP + FN)
2. Precision = TP/(TP + FP)
3. Recall = TP/(TP + FN)
4. False alarm = FP/(FP + TN)
5. F-score = 2 * (Precision * Recall/(Precision + Recall))

4 Deep and Artificial Neural Network-Based Intrusion Detection System

This section will have an elaborative discussion on IDS, which uses deep learning and ANN for attack detection.

4.1 Artificial Neural Network

It works on the same concept of how the human brain works. Its processing elements resemble the neurons which are present in the human brain. Figure 8 shows the further categorization of ANN.

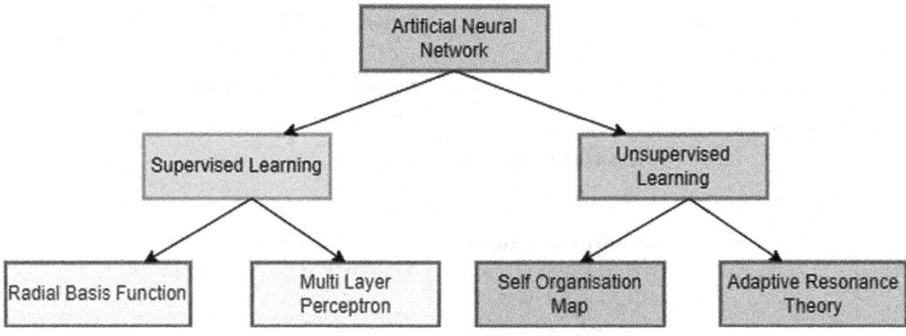

Fig. 8. ANN classification [7]

4.1.1 Supervised Learning

When the learning is supervised, we train the machine using a labeled training dataset where the dataset is already labeled with the correct answer.

There are two types of supervised learning methods used to train the neural network to detect intrusion.

i) **Radial basis function (RBF):** First, we will see how it works. The central concept is a classification based on the distance between the center of hidden neurons and input, and it is a kind of feed-forward neural network [46]. Figure 9 shows the architecture of RBF, which comprises input nodes, output nodes, hidden nodes, input, and output. The network output would be the linear combination of radial basis function because, with information, RBF has different parameters. Weights of hidden and input nodes are always one because the network's transfer function is RBF, and it helps adjust the weights between output and hidden nodes [48].

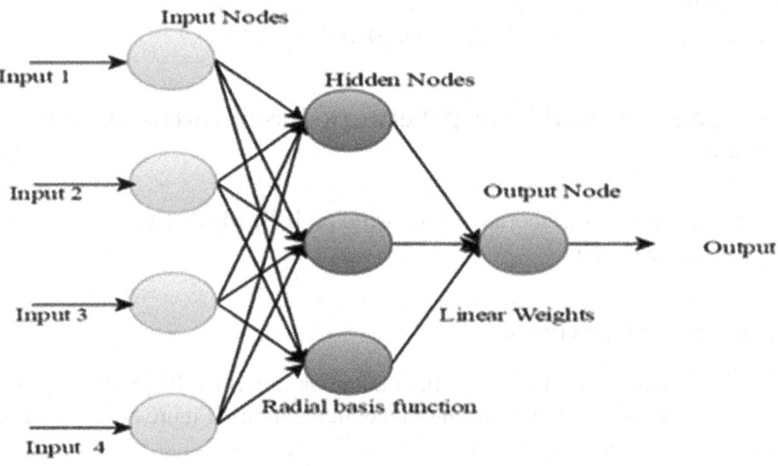

Fig. 9. Architecture of RBF [7]

In [49], the author tried to compare RBF and Multilayer perceptron with back-propagation (MLP-BP) with the KDD Cup'99 dataset to determine four different kinds of attacks. The output proved that RBF outperformed MLP-BP with an accuracy of 99.2% and a false positive rate of 1.2%, whereas MLP-BP achieved about 93.7% with a 7.2% false-positive rate. In [48], the author also tried to perform the same experiment mentioned above, but with the processed data that had less no. of features, and this time, RBF outperformed MLP-BP.

ii) **Multilayer perceptron (MLP):** These are the type of feed-forward neural networks, but if they are trained with the Backpropagation method, they are known as MLP-BP. The architecture of MLP consists of input, output nodes, and one or more nodes in the hidden layer, as shown in Fig. 10. MLP-BP architecture involves layers of non-linear elements to make an advanced hypothesis, and if we want to make hypotheses more advanced, we need to add more such layers.
Each node of MLP-BP takes an element of the feature vector, and output is represented as two classes, i.e., attack and regular.
We can evaluate hypotheses by setting the feedback process on input nodes, and the values keep on propagating to the output through the network. Gradient descent is used to push back the network's output error via backpropagation and determine the error of hidden nodes, which is calculated as given in [39].

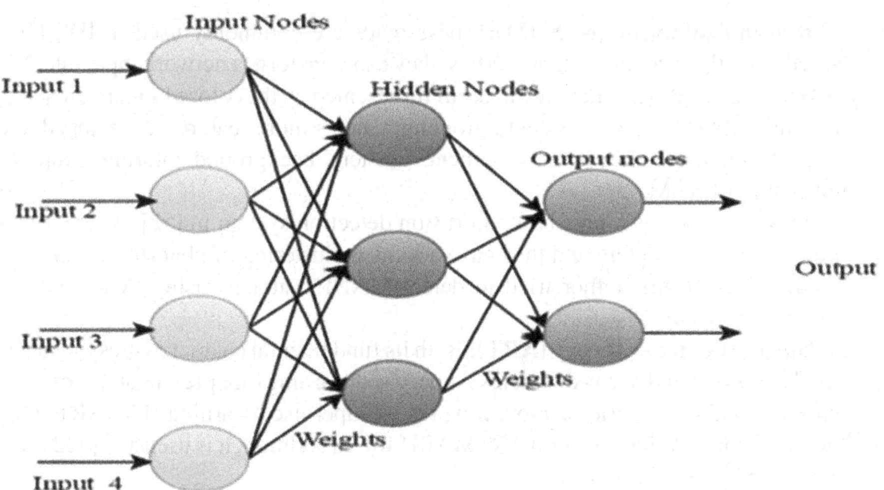

Fig. 10. 3-layer MLP [7]

In [50] author performed an intrusion detection experiment using a neural network that consisted of one input, a hidden, and an output layer. In 1999, the experiment was performed on the KDD Cup dataset, which had 41 features. The result will contain two classes: "attack" and "normal". This label is applied in the above scenario if the output was 0, and when the output is 1, it is branded as an attack. Following Table 6 will show the outcome of the experiment.

Table 6. Performance of MLP for different attacks [50]

Attack type	Detection rate %	False positive rate %	False negative rate %
DoS	99.75	4.78	0.24
Probe	98.16	1.33	1.83
U2R	87.09	31.03	12.9
DoS-Probe	99.33	18.5	0.66
R2L	98.99	9	1.01
Overall	81.96	8.51	18.03

4.1.2 Unsupervised Learning

This neural learning network is provided with only unlabelled data [51], and it must find the patterns on its own without having prior knowledge of output. There are two types of unsupervised learning algorithms. Kohonen's unsupervised learning is used for the computation of these feature maps.

i) **Self-organization maps (SOM):** These maps are commonly used in IDS [70]. Based upon their topological properties, they can transform a network input into 2-D feature maps, and these maps are neurons represented by the colored square. In SOM, the similarity is the basis for input grouping, and as more features are mapped, the colored square becomes more prominent. Clusters' background coloring computes the quality of SOM.
   SOM was applied as a host-based intrusion detection system in [52], where it was trained with data coding and thus gave a refined clustering of abnormal behaviors. Finally, in [53], the author tried to detect MANET attacks using SOM and got satisfactory results.
ii) **Adaptive resonance theory (ART):** Both its fundamental characteristics (unsupervised learning), and derived features (supervised learning) are present at its core, but when applied in a hybrid fashion, it works as supervised learning (like ARTMAP, Fuzzy ARTMAP, and Gaussian ARTMAP) [46]. In addition, it is used as a predictive and pattern recognition tool.

4.2 Deep Networks

Deep networks may look like MLP, but their architecture is different and training procedure. Here classification is done with unsupervised learning with multiple layers in the hierarchy by training data. These networks resemble the human brain. In [54], proposed the concept of Deep Belief Network (DBN). Deep networks IDS can be categorized based on the architecture and techniques used, as shown in Fig. 11 as follows:

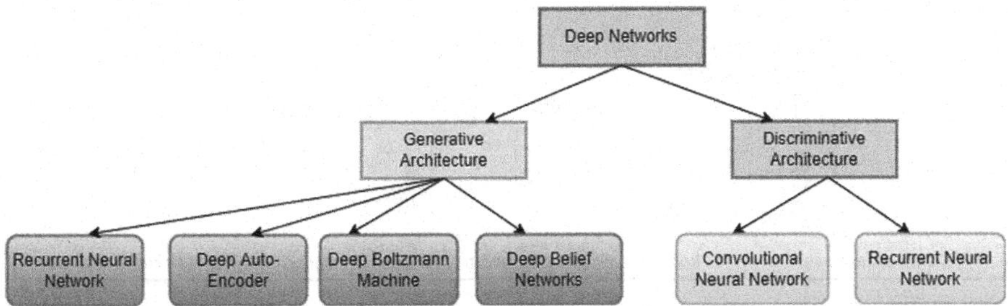

Fig. 11. Deep learning IDS categorization [7]

4.2.1 Discriminative Architecture

These architectures use a discriminative approach for classification, and convolutional and recurrent neural networks are 2 examples of this architecture.

a) **Recurrent neural networks:** These are designed to accept a series of inputs with no prior limit set on the input size. RNN can produce information and output vectors, respectively, and input weights and hidden state vector influence outputs. Different results are obtained from the same input. Figure 12 will show a simple RNN. It memorizes the past, and its decision is influenced by what it has learned in the past.

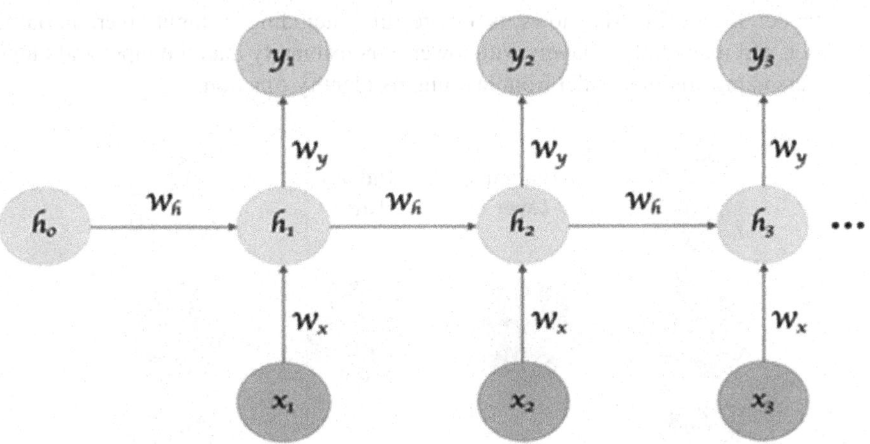

Fig. 12. A simple RNN

b) **Convolutional neural networks:** These were initially designed for the analysis of the image. They can take an image as input and assign weights to various objects and differentiate one from another. In CNN, as compared to others, less pre-processing is required [55]. CNN architecture resembles the connectivity pattern of neurons in the human brain, as shown in Fig. 13.

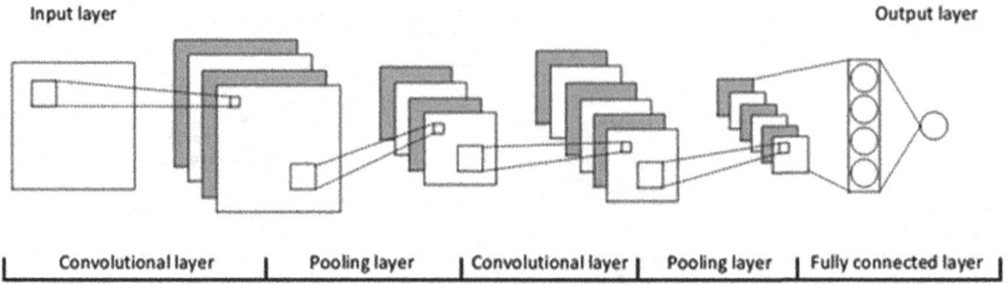

Fig. 13. CNN structure [60]

4.2.2 Generative Architecture

Due to the display of dependence or independence for distribution, these models are often known as graphical models. The nodes in the graph represent random variables, while the edges reflect their relationships. Supervised learning is linked to these models. RNN, which we've already examined, Deep autoencoder, Deep belief network, and Deep Boltzmann machine are all subclasses of generative models.

a) **Deep auto-encoder:** They are classed as generative models in their original form and as stacked auto-encoder and denoising auto-encoder in various forms [57]. An auto-principal encoder's task is to reconstruct output as closely as possible to input. When an auto-encoder has more than one buried layer, it is classified as a deep auto-encoder. Figure 14 [51] shows its architecture, including an input layer, an output layer, and many hidden layers with lower dimensionality than the input and output layers to prevent the model from learning its identity function.

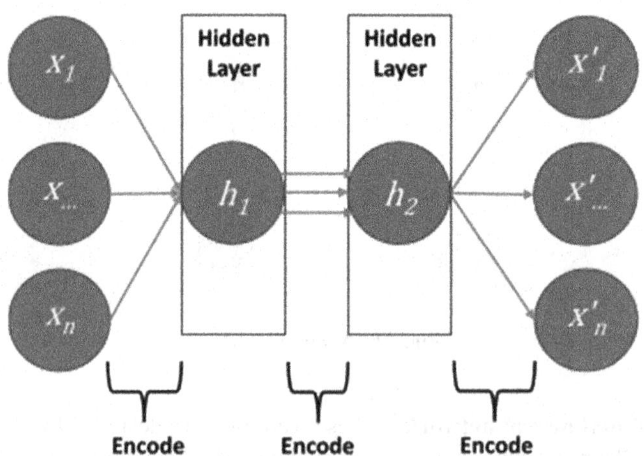

Fig. 14. A deep auto-encoder [58]

One issue with deep auto-encoder, i.e., when it gets trained with backpropagation, it gets stuck in local minima and does not give good results. In [57], the author suggested a new approach to detect intrusion using a deep auto-encoder. This experiment was done on the NSL-KDD dataset and compared the result with other PCA approaches, factor analysis, and Kernel/PCA. Results showed that deep auto-encoder beat all three in terms of accuracy.

b) ***Deep Boltzmann machine (DBM):*** The model is unidirectional and is suited for data collection that includes many unlabelled examples, as well as training and fine-tuning using labeled examples [56]. Formation of DBM is done by training a stack of RBM with multiple hidden layers where one RBM's output is input to another, as shown in Fig. 15.

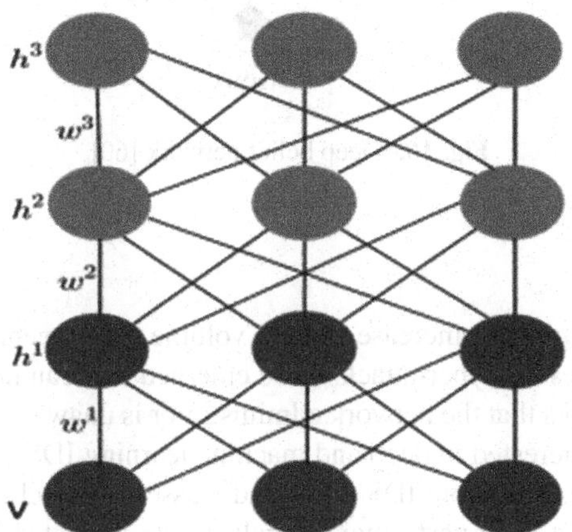

Fig. 15. A deep Boltzmann machines [7]

c) **Deep belief network (DBN):** These networks use supervised fine-tuning and unsupervised pre-training to build the models. Figure 16 will present how a deep belief network looks like RBM's top layer, and lower layers form a sigmoid belief network. It is made up of an RBM stack, and it then follows a greedy approach of layer-by-layer training in which each layer of the stack of RBM learns at a time.
N. Gao et al. [59] trained DBN as a classifier to detect intrusion and compare performance with SVM and ANN. The results showed that DBN outperformed SVM and ANN with the best accuracy, i.e., 93.49% and 0.76% false positive.

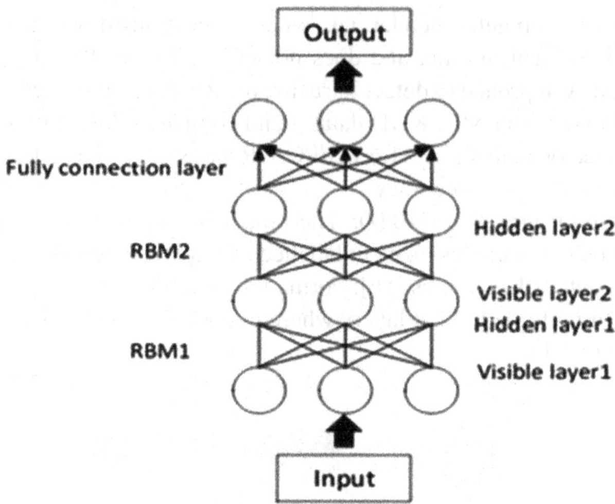

Fig. 16. Deep belief network [60]

5 Conclusion

We've observed a significant increase in data volume, heterogeneity, and noise in the last few years. As a result, cyber-attacks have emerged that can take advantage of new weaknesses in networks that the network administrator is unaware of. That is why cyber specialists are very interested in deep and machine learning IDS.

This study explored multiple IDS types and classed all machine learning and deep learning methodologies with performance analysis. As indicated in Table 7, the major goal of this paper is to compare the performance of several Shallow and Deep learning algorithms in the field of intrusion detection. Deep learning methods surpass machine learning techniques in intrusion detection, according to the results of the experiments. However, some deep networks, such as CNN, have yet to be properly investigated and used in intrusion detection.

Table 7. Performance table of ML and deep learning IDS

Algorithm	Authors	Dataset used	IDS performance	
			Accuracy %	False positive rate
Bayesian network	M.A. Jabbar et al. [69]	Kyoto dataset	99.9	0.015
Genetic algorithm	Md. Abdul Mukit et al. [68]	KDD Cup'99	95	0.3046
Support vector machine	B. Senthilnayaki et al. [33]	KDD Cup'99	95.26	–
K-nearest neighbor	Q.Zeng et al. [61]	KDD Cup'99	97.26	0.0603
Decision tree	S.Sahu et al. [39]	Kyoto 2006+	97.23	0.0267
Clustering	Li Jun Tao et al. [62]	KDD Cup'99	90	0.15
Fuzzy logic	Gomez and Dasgupta et al. [63]	KDD Cup'99	98.95	0.07
Multi-layer perceptron	P. Barapatre et al. [50]	KDD Cup'99	81.96	0.0851
Radial basis function	C.Zhang et al. [49]	KDD Cup'99	99.2	0.012
Self-organization map	H. Gunes Kayacik et al. [64]	KDD Cup'99	90.4	0.0138
Adaptive resonance theory	P. Somwang et al. [65]	KDD Cup'99	96.13	0.0386
Deep auto-encoder	Shone et al. [58]	KDD Cup'99	97.85	0.0215
Deep belief network	N. Gao et al. [59]	KDD Cup'99	93.49	0.0076
Recurrent neural network	Yin et al. [66]	NSL-KDD	83.28	–
Convolutional neural net	Yang and Wang [67]	KDD Cup'99	95.36	0.0076

References

1. Chand, N., Mishra, P., Krishna, C.R., Pilli, E.S., Govil, M.C.: A comparative analysis of SVM and its stacking with other classification algorithm for intrusion detection. In: 2016 International Conference on Advances in Computing, Communication, & Automation (ICACCA) (Spring), pp. 1–6. IEEE, April 2016
2. Mukherjee, B., Heberlein, L.T., Levitt, K.N.: Network intrusion detection. IEEE Netw. **8**(3), 26–41 (1994)
3. Jonnalagadda, S.K., Reddy, R.P.: A literature survey and comprehensive study of intrusion detection. Int. J. Comput. Appl. **81**(16), 40–47 (2013)
4. Mitchell, R., Chen, I.R.: A survey of intrusion detection techniques for cyber-physical systems. ACM Comput. Surv. (CSUR) **46**(4), 1–29 (2014)
5. Masduki, B.W., Ramli, K., Saputra, F.A., Sugiarto, D.: Study on implementation of machine learning methods combination for improving attacks detection accuracy on Intrusion Detection System (IDS). In: 2015 International Conference on Quality in Research (QiR), pp. 56–64. IEEE, August 2015
6. Hodo, E., Bellekens, X., Iorkyase, E., Hamilton, A., Tachtatzis, C., Atkinson, R.: Machine learning approach for detection of nontor traffic. In: Proceedings of the 12th International Conference on Availability, Reliability and Security, pp. 1–6, August 2017
7. Debar, H., Dacier, M., Wespi, A.: A revised taxonomy for intrusion-detection systems. Ann. Télécommun. **55**(7), 361–378 (2000)
8. Axelsson, S.: Intrusion detection systems: a survey and taxonomy. 2000. Chalmers University of Technology, Goteborg, Sweden (2005)
9. hafez Amer, S., Hamilton Jr, J.A.: Intrusion detection systems (IDS) taxonomy-a short review. This is a paid advertisement. STN 13-2 June 2010: Defensive Cyber Secur.: Policies Procedures **2**, 23 (2010)

10. Xenakis, C., Panos, C., Stavrakakis, I.: A comparative evaluation of intrusion detection architectures for mobile ad hoc networks. Comput. Secur. **30**(1), 63–80 (2011)
11. Liao, H.J., Lin, C.H.R., Lin, Y.C., Tung, K.Y.: Intrusion detection system: a comprehensive review. J. Netw. Comput. Appl. **36**(1), 16–24 (2013)
12. Debar, H., Dacier, M., Wespi, A.: Towards a taxonomy of intrusion-detection systems. Comput. Netw. **31**(8), 805–822 (1999)
13. Mounji, A., Le Charlier, B.: Continuous assessment of a unix configuration: Integrating intrusion detection and configuration analysis. In: Proceedings of SNDSS 1997: Internet Society 1997 Symposium on Network and Distributed System Security, pp. 27–35. IEEE, February 1997
14. Liu, S., et al.: A flow-based method to detect penetration. In: The 7th IEEE/International Conference on Advanced Infocomm Technology, pp. 184–191. IEEE, November 2014
15. Kozushko, H.: Intrusion detection: host-based and network-based intrusion detection systems. Independent Study **11**, 1–23 (2003)
16. Modi, C., Patel, D., Borisaniya, B., Patel, H., Patel, A., Rajarajan, M.: A survey of intrusion detection techniques in cloud. J. Netw. Comput. Appl. **36**(1), 42–57 (2013)
17. Qayyum, A., Islam, M.H., Jamil, M.: Taxonomy of statistical based anomaly detection techniques for intrusion detection. In: Proceedings of the IEEE Symposium on Emerging Technologies, pp. 270–276. IEEE, September 2005
18. Garcia-Teodoro, P., Diaz-Verdejo, J., Maciá-Fernández, G., Vázquez, E.: Anomaly-based network intrusion detection: techniques, systems and challenges. Comput. Secur. **28**(1–2), 18–28 (2009)
19. Pillai, T.R., Palaniappan, S., Abdullah, A., Imran, H.M.: Predictive modeling for intrusions in communication systems using GARMA and ARMA models. In: 2015 5th National Symposium on Information Technology: Towards New Smart World (NSITNSW), pp. 1–6. IEEE, February 2015
20. Debar, H., Becker, M., Siboni, D.: A neural network component for an intrusion detection system. In: Proceedings 1992 IEEE Computer Society Symposium on Research in Security and Privacy, p. 240. IEEE Computer Society, May 1992
21. Lunt, T.F.: A survey of intrusion detection techniques. Comput. Secur. **12**(4), 405–418 (1993)
22. Poston, H.E.: A brief taxonomy of intrusion detection strategies. In: 2012 IEEE National Aerospace and Electronics Conference (NAECON), pp. 255–263. IEEE, July 2012
23. Kuperman, B.A.: A categorization of computer security monitoring systems and the impact on the design of audit sources. Doctoral dissertation, Purdue University (2004)
24. Ghorbani, A.A., Lu, W., Tavallaee, M.: Network attacks. In: Ghorbani, A.A., Lu, W., Tavallaee, M. (eds.) Network Intrusion Detection and Prevention. Advances in Information Security, vol. 47, pp. 1–25. Springer, Boston (2010). https://doi.org/10.1007/978-0-387-88771-5_1
25. Nguyen, H.T., Franke, K., Petrovic, S.: Feature extraction methods for intrusion detection systems. In: Threats, Countermeasures, and Advances in Applied Information Security, pp. 23–52. IGI Global (2012)
26. Mahoney, M.V., Chan, P.K.: An analysis of the 1999 DARPA/Lincoln laboratory evaluation data for network anomaly detection. In: Vigna, G., Kruegel, C., Jonsson, E. (eds.) RAID 2003. LNCS, vol. 2820, pp. 220–237. Springer, Heidelberg (2003). https://doi.org/10.1007/978-3-540-45248-5_13
27. Lee, W., Stolfo, S.J.: A framework for constructing features and models for intrusion detection systems. ACM Trans. Inf. Syst. Secur. (TiSSEC) **3**(4), 227–261 (2000)
28. Onik, A.R., Haq, N.F., Mustahin, W.: Cross-breed type Bayesian network based intrusion detection system (CBNIDS). In: 2015 18th International Conference on Computer and Information Technology (ICCIT), pp. 407–412. IEEE, December 2015

29. Bode, M.A., Oluwadare, S.A., Alese, B.K., Thompson, A.F.B.: Risk analysis in cyber situation awareness using Bayesian approach. In: 2015 International Conference on Cyber Situational Awareness, Data Analytics and Assessment (CyberSA), pp. 1–12. IEEE, June 2015
30. Padmadas, M., Krishnan, N., Kanchana, J., Karthikeyan, M.: Layered approach for intrusion detection systems based genetic algorithm. In: 2013 IEEE International Conference on Computational Intelligence and Computing Research, pp. 1–4. IEEE, December 2013
31. Wang, G., Yeung, D.Y., Lochovsky, F.H.: A kernel path algorithm for support vector machines. In: Proceedings of the 24th International Conference on Machine Learning, pp. 951–958, June 2007
32. Burges, C.J.: A tutorial on support vector machines for pattern recognition. Data Min. Knowl. Disc. **2**(2), 121–167 (1998)
33. Senthilnayaki, B., Venkatalakshmi, K., Kannan, A.: Intrusion detection using optimal genetic feature selection and SVM based classifier. In: 2015 3rd International Conference on Signal Processing, Communication and Networking (ICSCN), pp. 1–4. IEEE, March 2015
34. Shi, K., Li, L., Liu, H., He, J., Zhang, N., Song, W.: An improved KNN text classification algorithm based on density. In: 2011 IEEE International Conference on Cloud Computing and Intelligence Systems, pp. 113–117. IEEE, September 2011
35. Canbay, Y., Sagiroglu, S.: A hybrid method for intrusion detection. In: 2015 IEEE 14th International Conference on Machine Learning and Applications (ICMLA), pp. 156–161. IEEE, December 2015
36. Zhang, H., Chen, G.: The research of face recognition based on PCA and K-nearest neighbor. In: 2012 Symposium on Photonics and Optoelectronics, pp. 1–4. IEEE, May 2012
37. Mahrishi, M., Hiran, K.K., Meena, G., Sharma, P. (eds.): Machine Learning and Deep Learning in Real-Time Applications. IGI Global (2020)
38. Sebastiani, F.: Machine learning in automated text categorization. ACM Comput. Surv. (CSUR) **34**(1), 1–47 (2002)
39. Sahu, S., Mehtre, B.M.: Network intrusion detection system using J48 decision tree. In: 2015 International Conference on Advances in Computing, Communications and Informatics (ICACCI), pp. 2023–2026. IEEE, August 2015
40. Rajasekaran, S., Pai, G.V.: Neural Networks, Fuzzy Logic and Genetic Algorithm: Synthesis and Applications (with CD). PHI Learning Pvt. Ltd. (2003)
41. Wahengbam, M., Marchang, N.: Intrusion detection in manet using fuzzy logic. In: 2012 3rd National Conference on Emerging Trends and Applications in Computer Science, pp. 189–192. IEEE, March 2012
42. Toosi, A.N., Kahani, M.: A new approach to intrusion detection based on an evolutionary soft computing model using neuro-fuzzy classifiers. Comput. Commun. **30**(10), 2201–2212 (2007)
43. Om, H., Kundu, A.: A hybrid system for reducing the false alarm rate of anomaly intrusion detection system. In: 2012 1st International Conference on Recent Advances in Information Technology (RAIT), pp. 131–136. IEEE, March 2012
44. Gupta, S.: An effective model for anomaly IDS to improve the efficiency. In: 2015 International Conference on Green Computing and Internet of Things (ICGCIoT), pp. 190–194. IEEE, October 2015
45. Ayed, A.B., Halima, M.B., Alimi, A.M.: Survey on clustering methods: towards fuzzy clustering for big data. In: 2014 6th International conference of soft computing and pattern recognition (SoCPaR), pp. 331–336. IEEE, August 2014
46. Wu, S.X., Banzhaf, W.: The use of computational intelligence in intrusion detection systems: a review. Appl. Soft Comput. **10**(1), 1–35 (2010)
47. Dong, S., Zhou, D., Ding, W.: The study of network traffic identification based on machine learning algorithm. In: 2012 Fourth International Conference on Computational Intelligence and Communication Networks, pp. 205–208. IEEE, November 2012

48. Bi, J., Zhang, K., Cheng, X.: Intrusion detection based on RBF neural network. In: 2009 International Symposium on Information Engineering and Electronic Commerce, pp. 357–360. IEEE, May 2009
49. Zhang, C., Jiang, J., Kamel, M.: Comparison of BPL and RBF network in intrusion detection system. In: Wang, G., Liu, Q., Yao, Y., Skowron, A. (eds.) RSFDGrC 2003. LNCS, vol. 2639, pp. 466–470. Springer, Heidelberg (2003). https://doi.org/10.1007/3-540-39205-X_79
50. Barapatre, P., Tarapore, N. Z., Pukale, S.G., Dhore, M. L.: Training MLP neural network to reduce false alerts in IDS. In: 2008 International Conference on Computing, Communication and Networking, pp. 1–7. IEEE, December 2008
51. Murphy, K.P.: Machine Learning: A Probabilistic Perspective. MIT Press, Cambridge (2012)
52. Meena, G., Dhanwal, B., et al.: Performance comparison of network intrusion detection system based on different pre-processing methods and deep neural network. In: Proceedings of the International Conference on Data Science, Machine Learning and Artificial Intelligence, pp. 110–115. ACM (2021)
53. Kumar, V.D., Radhakrishnan, S.: Intrusion detection in MANET using self organizing map (SOM). In: 2014 International Conference on Recent Trends in Information Technology, pp. 1–8. IEEE, April 2014
54. Hinton, G.E., Osindero, S., Teh, Y.W.: A fast learning algorithm for deep belief nets. Neural Comput. **18**(7), 1527–1554 (2006)
55. Ng, A., et al.: Unsupervised feature learning and deep learning tutorial. CS294A Lecture (2013)
56. Bengio, Y.: Learning Deep Architectures for AI. Now Publishers Inc. (2009)
57. Deng, L.: A tutorial survey of architectures, algorithms, and applications for deep learning. APSIPA Trans. Signal Inf. Process. **3**, 1–29 (2014)
58. Shone, N., Ngoc, T.N., Phai, V.D., Shi, Q.: A deep learning approach to network intrusion detection. IEEE Trans. Emerg. Top. Comput. Intell. **2**(1), 41–50 (2018)
59. Gao, N., Gao, L., Gao, Q., Wang, H.: An intrusion detection model based on deep belief networks. In: 2014 Second International Conference on Advanced Cloud and Big Data, pp. 247–252. IEEE, November 2014
60. Liu, H., Lang, B.: Machine learning and deep learning methods for intrusion detection systems: a survey. Appl. Sci. **9**(20), 4396 (2019)
61. Zeng, Q., Wu, S.: Anomaly detection based on multi-attribute decision. In: 2009 WRI Global Congress on Intelligent Systems, vol. 2, pp. 394–398. IEEE, May 2009
62. Tao, L.J., Hong, L.Y., Yan, H.: The improvement and application of a K-means clustering algorithm. In: 2016 IEEE International Conference on Cloud Computing and Big Data Analysis (ICCCBDA), pp. 93–96. IEEE, July 2016
63. Gomez, J., Dasgupta, D.: Evolving fuzzy classifiers for intrusion detection. In: Proceedings of the 2002 IEEE Workshop on Information Assurance, vol. 6, no. 3, pp. 321–323, June 2002
64. Kayacik, H.G., Zincir-Heywood, A.N., Heywood, M.I.: A hierarchical SOM-based intrusion detection system. Eng. Appl. Artif. Intell. **20**(4), 439–451 (2007)
65. Somwang, P., Lilakiatsakun, W.: Intrusion detection technique by using fuzzy ART on computer network security. In: 2012 7th IEEE Conference on Industrial Electronics and Applications (ICIEA), pp. 697–702. IEEE, July 2012
66. Yin, C., Zhu, Y., Fei, J., He, X.: A deep learning approach for intrusion detection using recurrent neural networks. IEEE Access **5**, 21954–21961 (2017)
67. Yang, H., Wang, F.: Wireless network intrusion detection based on improved convolutional neural network. IEEE Access **7**, 64366–64374 (2019)
68. Hoque, M.S., Mukit, M., Bikas, M., Naser, A.: An implementation of intrusion detection system using genetic algorithm. arXiv preprint arXiv:1204.1336 (2012)

69. Jabbar, M.A., Aluvalu, R., Reddy, S.S.S.: Intrusion detection system using Bayesian network and feature subset selection. In: 2017 IEEE International Conference on Computational Intelligence and Computing Research (ICCIC), pp. 1–5. IEEE, December 2017
70. Ahmad, Z., et al.: Network intrusion detection system: a systematic study of machine learning and deep learning approaches. Trans. Emerg. Telecommun. Technol. **32**(1), e4150 (2021)

Deep Learning Application of Image Recognition Based on Self-driving Vehicle

Stuti Bhujade[1(✉)], T. Kamaleshwar[2], Sushma Jaiswal[3], and D. Vijendra Babu[4]

[1] Analyst in Artificial Intelligence, UT Administration of DNH and Daman & Diu, Daman 396210, India
stutiasthana@gmail.com

[2] Department of Computer Science and Engineering, Vel Tech Rangarajan Dr.Sagunthala R&D Institute of Science and Technology, Chennai, Tamil Nadu, India

[3] Department of Computer Science and Information Technology (CSIT), Guru Ghasidas Vishwavidyalaya, (A Central University), Koni, Bilaspur 495009, Chhattisgarh, India

[4] Department of Electronics and Communication Engineering, Aarupadai Veedu Institute of Technology, Vinayaka Mission's Research Foundation, Paiyanoor 603104, Tamil Nadu, India

Abstract. A CNN (Convolutional Neural Network) is an artificial neural network used to evaluate visual pictures. It is used for visual image processing and is categorised as a deep neural network in deep learning. So, using real-time image processing, an AI autonomous driving model was built using a road crossing picture as an impediment. Based on the CNN model, we created a low-cost approach that can realistically perform autonomous driving. An end-to-end model is applied to the most widely used deep neural network technology for autonomous driving. It was shown that viable lane identification and maintaining techniques may be used to train and self-drive on a virtual road.

Keywords: Deep NN · Less cost · Autonomous car · Lane tracking · Lane keeping

1 Introduction

Images have recognized, and natural language processing has been improved with its usage in recent times. Because CNN (Convolutional Neural Network) performs better than conventional techniques, there is a lot of work being done in image recognition. The development of self-driving automobiles is a prominent area of investigation. Several instruments and cameras are necessary to detect the nearby atmosphere and instantaneously identify acceleration, deceleration. They halt to acquiring and analyzing information both within and outside the autonomous vehicle. Another benefit of using continuous video or data feeds for autonomous driving is tackling this problem much more rapidly. An autonomous driving system for automobiles is being developed and tested in this study using a deep learning algorithm. This approach, which has a significant influence on picture recognition, can easily handle the development of autonomous driving systems.

S. Bhujade—Independent Researcher.

V. E. Balas et al. (Eds.): ICETCE 2022, CCIS 1591, pp. 336–344, 2022.
https://doi.org/10.1007/978-3-031-07012-9_29

2 Deep Learning Method for Self-driving Car

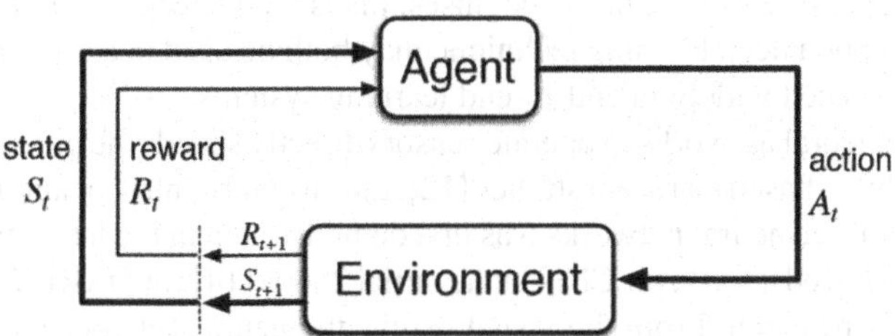

Fig. 1. Structure of reinforcement learning

As a result of this, convolutional artificial neural networks (CNNs) have been developed. Neural networks are the basic building blocks of CNNs. 1) Neurons between layers remain "local" and disconnected in each layer. Neuron input from a lower-layer neuron located in a rectangular space next to the neuron's upper layer. The neural receptive field may be found here. Neural cells in the same layer share the weights of "local" connections. The number of CNN model parameters is reduced because of the immutability of visual input. CNN is regarded as "implicit previous knowledge" in computer vision; CNN has proved to be a viable paradigm for resolving cv difficulties. AlexNet [3] surpassed ImageNet [4] in 2012, which prompted CNN and its newest algorithms to be adopted immediately. As a result, CNN has become a symbol of self-driving vehicles. An RNN is a deep learning method that works well with large datasets. Being able to deal with time. Natural language processing, as well as video, are two of its strong points. Streaming. An accurate depiction of self-driving vehicles is essential. Unpredictable and ever-changing environmental conditions Thus, driving data from the past may be used to create more accurate representations of the environment. Reinforcement learning may be modelled using the Markov Decision Process. In the same way as in a When the robot correctly determines the present state of the environment, it gets rewarded. Reinforcement, The optimum technique is learned via this process of trial and error. To maximise the cumulative Reward, the policy recommends an approach that maximises the sum of all rewards. Laws like this. Figure 1 depicts a conceptual framework for reinforcement learning. Learners should be rewarded for their efforts. In self-driving cars, technology is used in a variety of ways.

Tries to identify a continuous action strategy that maximizes the agent's advantage environment. Action, Reward, State, Agent, and Environment are the five components. a subject that learns and acts in the environment. When Agent Acts, Agent recognizes its own condition as Reward. The agent so interacts with reinforcement learning continues [6–9].

3 Deep Learning-Based Acknowledgement of Autonomous Driving

Figure 2 displays two self-driving car channels based on deep learning. However, (b) is an all-encompassing system that recognises, plans, and executes all at once. A non-learning method or a deep learning technique may both be used in a sequential pipeline. Deep learning is used widely in end-to-end learning systems.

Using deep neural networks to operate sensors directly speeds the process. Deep neural networks This is based on recent studies [12]. End-to-end control of autonomous vehicles using layer three neural networks was first demonstrated in the late 1980s [13]. This experiment employed a six-layer CNN in the initial 2000s [14] for DARPA's autonomous vehicle (DAVE) research. From the visual pixels, the neural network model generates steering control instructions. All intermediary steps are skipped in the sequential pipeline approach.

Cameras in automobiles take 30 frames per second of roadside images, according to NVIDIA research. This information was also obtained, which is how many degrees to the right or left. A cause and a result must be linked in supervised learning. Lane-maintaining AI uses the road image as a cause value (X) and the handle value. As a result, value (Y) (Y). The lane-maintenance function cannot be implemented using a modest linear equivalence. Because of this, an artificial neural network is needed. In additional arguments, as predictive models, artificial neural networks have taken the role of straight lines. The predicted handle value is returned as before when the highway copy is input into the artificial neural network. To minimise the overall number of mistakes, an artificial neural network is trained to provide the smallest possible difference between the predicted and actual values of the steering wheel.

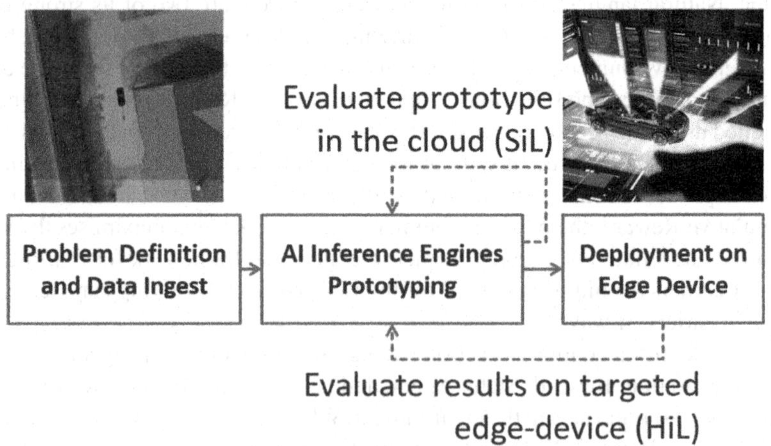

Fig. 2. A self-driving automobile powered by deep learning.

4 Low-Cost Autonomous Vehicle Architecture Design

4.1 Preparation of the Data for Analysis

Before gathering data for self-driving vehicles, the route must be prepared. Small or large, the road was created and used in an unsuitable manner. 1.5 to 2 times the width of an automobile is ideal. On a small path, it's problematic to ambition and get enough drill information. If the road is also varied, the camera may not be able to detention both left and right sides at the same time. Moderate curves and straight lines would be included in the roads. A discrepancy in rotational speed between the left and suitable motors in prototype studies makes it difficult to handle curving roads. Accomplished A incorrect turn might be difficult to navigate, even on a straight route. If an artificial intelligence system is operating in a safe mode, direct driving enables the driver to act like a human. The prototype road model is seen in Fig. 3.

Fig. 3. Prototypes of self-driving vehicles.

It is essential that the data proportions be equalised. As an example, for a left turn, the ratio is 14%; for a straight-ahead, it is 76%; and for a right turn, it is 20%. There is a 76% to 18% shift in the correct turn ratio after running decalcompy.py. Learning to turn left and right is improved by reducing the left/right balance. Scaling data may be done in two ways when it is not equal. Down Left/right turn counts are equalised by selecting the number of left/right angles from the total data of straight forward, which has an 18%, 76%, 18% ratio. Reduces the number of categories to a few from many with many classes, in other words. An artificial intelligence system may not function as well as it might if the missing knowledge is critical. When the ratio of left, straight, and right turns reaches 18%, it begins to replicate the data. It develops the data volume of a wide variety with a high data rate by continually copying and using the data of a category with a low data rate. Overfitting on repeated data might be a drawback of this approach, which benefits from increasing the number of data and altering the data rate. The data rate may be adjusted by using sampling in this study.

4.2 Prototype Training for the Self-driving Car

Figure 4 depicts the System hardware structure. Using Tensorflow to classify images is covered in more depth later. In this case, the picture and the arrow keys associated with it are preserved in the Orange Pie light internal folder. After gathering the training data, the system regulates the speed of the DC motor and runs using image classification.

This study employed a categorization model. Classification is the process of categorizing a result and predicting it. The system flow is as follows. First, using VNC Viewer to connect to Orange-pi remotely, the user collects learning data and saves it as an image. After gathering training data, a classification model is developed. Figure 5 depicts the model's classification process. The road picture is sent to the input layer of the artificial neural network, which predicts and expresses a value in the output layer of four neurons (stop, go straight, turn right, turn left). It also stops at a crosswalk on the road. Predicted values for four categories are shown as likelihood based on the current road picture. Among these, the most likely value is chosen as the AI's output value.

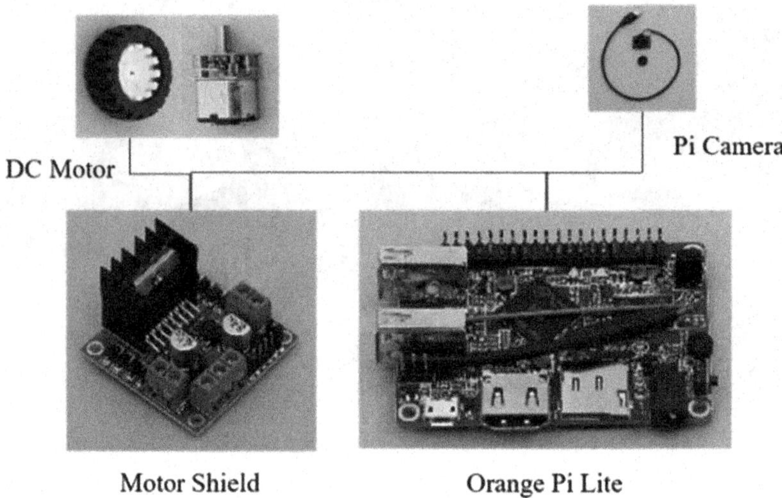

Fig. 4. Overall system structure

4.3 Network Architecture

As illustrated in Fig. 6, the learning technique employed in this study has four convolutional layers and three fully linked layers. Figure 7 also shows the internal structure of the convolutional and fully linked layers. The convolution layer uses a 3 $\times$ 3, and a 5 $\times$ 5 kernel in the sequence of convolution and activation functions indicated in Fig. 7. More steps in the convolutional layer improve feature detection but takes a long time to learn. However, if it gets above a particular point, the picture becomes distorted, and the performance suffers. Conversely, a low amount of convolutional layer limits feature extraction, resulting in poor accuracy.

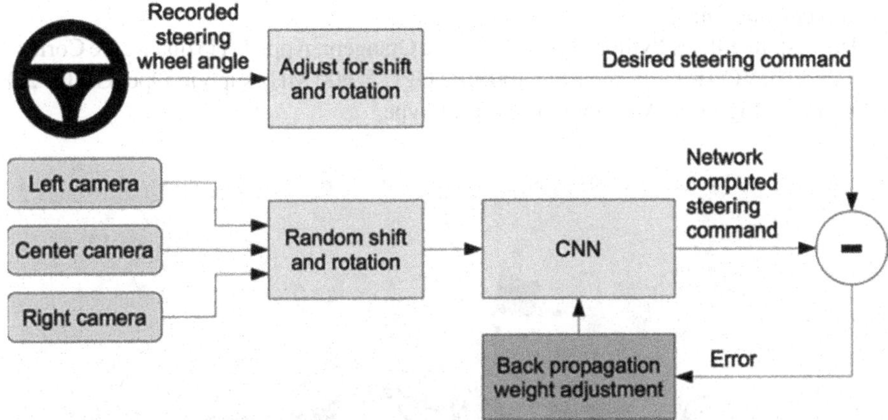

Fig. 5. Images of road signs are used to classify them.

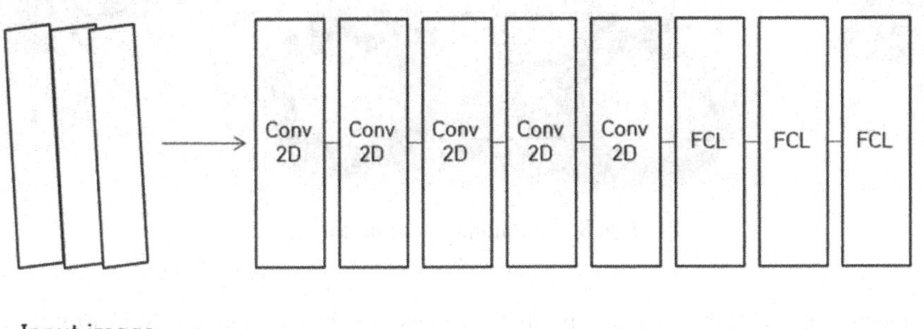

Fig. 6. CNN network structure

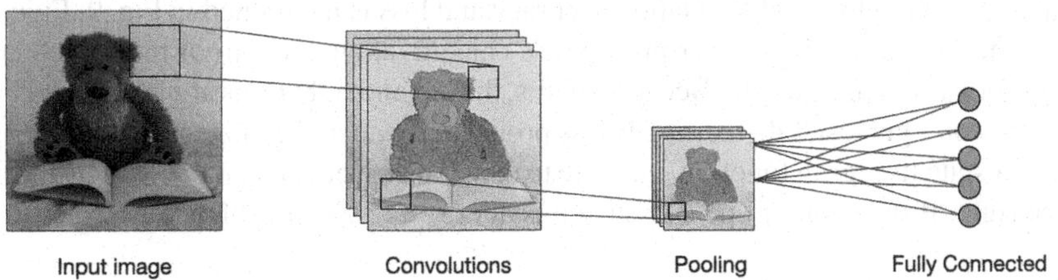

Fig. 7. Fully linked layer structure with a convolutional layer.

The kernel derives the image's features and checks if the detected feature exists in the area based on the learnt part. This work employed three layers for the ultimately linked layer. After flattening using 1D data, dropout is used to the first completely connected layer to avoid over-fitting. The second layer is dropped out again after employing the Relu activation algorithm. The final layer uses Relu and Dropout. The last layer connects the photographs to the same four layers as the experiment's image kinds.

Implementation:

The system utilises Python, Tensorflow, and Otangepi. Apple's H3 quad-core Cortex-A7 H.265/HEVC 4K CPU and Mali 400MP2 GPU @600 MHz support OpenGL ES 2.0 on Orangepi. Figure 8: Autonomous car prototype.

Fig. 8. Autonomous car prototype

During the learning process, prediction rates are shown graphically in Fig. 9. The final learning model has a prediction rate of 87%, which is the most significant. Tertiary learning models' loss values decrease as learning progresses in Fig. 9. Afterwards, the model is ready for self-driving operations. More iterations are mentioned training accuracy, training loss and Validation accuracy and loss is mentioned in Fig. 9. Figure 8 shows the camera on the toy car prototype's camera taking a photo of the vehicle as it moves forward, left, and right. Because of this, the left and right turns are 100% correct so that you won't get lost. Off-roading is less probable if you make more accurate left/right turns. Despite its zigzag function, it is safe to drive since it can only be steered by turning left or right. If the straight line's precision is at or over 50%, it is blended correctly.

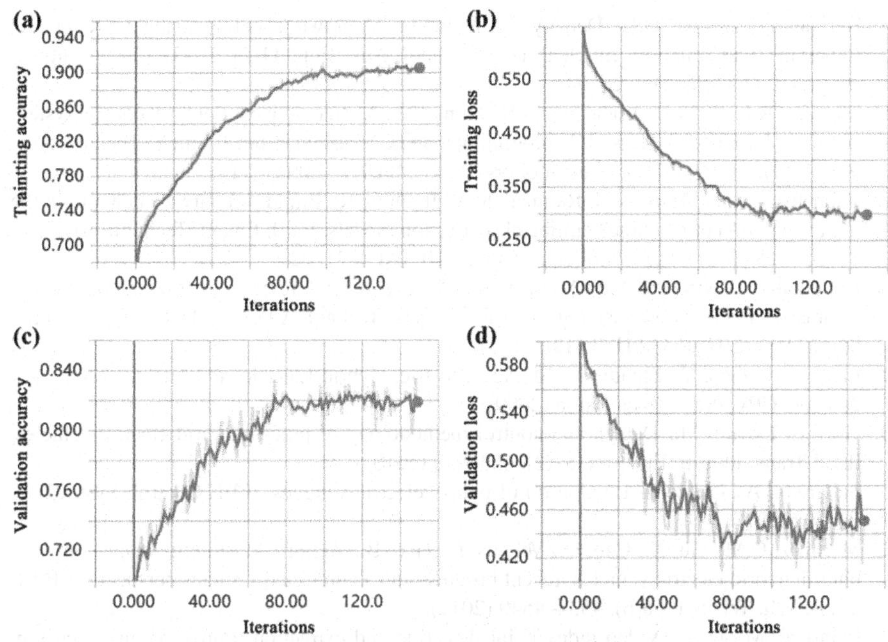

Fig. 9. Each experiment's self-driving prediction accuracy (loss value)

5 Conclusion

A CNN-based deep learning algorithm was used to construct an active vehicle autonomous driving system to examine low-cost autonomous driving based on picture categorisation. This article's discussion of self-driving cars relies on a technique known as convolutional neural networks (CNN). This might be done in the future with the creation of a low-cost testbed. By explaining the responsibilities of other subsystems in autonomous driving, it is feasible to investigate object identification and navigation.

References

1. Kurnia, R.I., Girsang, A.S.: Classification of user comment using word2vec and deep learning. Int. J. Emerg. Technol. Adv. Eng. **11**(5), 1–8 (2021). https://doi.org/10.46338/IJETAE0521_01
2. Adytia, N.R., Kusuma, G.P.: Indonesian license plate detection and identification using deep learning. Int. J. Emerg. Technol. Adv. Eng. **11**(7), 1–7 (2021). https://doi.org/10.46338/ijetae0721_01
3. Rahman, R.A., Masrom, S., Zakaria, N.B., Halid, S.: Auditor choice prediction model using corporate governance and ownership attributes: machine learning approach. Int. J. Emerg. Technol. Adv. Eng. **1**(7), 87–94 (2021). https://doi.org/10.46338/ijetae0721_11
4. Dela Cruz, L.A., Tolentino, L.K.S.: Telemedicine implementation challenges in underserved areas of the Philippines. Int. J. Emerg. Technol. Adv. Eng. **11**(7), 60–70 (2021). https://doi.org/10.46338/ijetae0721_08

5. Hermanto, Kusuma, G.P.: Density estimation-based crowd counting using CSRNet and Bayesian+ loss function. Int. J. Emerg. Technol. Adv. Eng. **11**(7), 19–27 (2021). https://doi.org/10.46338/ijetae0721_04
6. Mustapa, R.F., Rifin, R., Mahadan, M.E., Zainuddin, A.: Interactive water level control system simulator based on OMRON CX-programmer and CX-designer. Int. J. Emerg. Technol. Adv. Eng. **11**(9), 91–99 (2021). https://doi.org/10.46338/IJETAE0921_11
7. Rahman, A.S.A., Masrom, S., Rahman, R.A., Ibrahim, R.: Rapid software framework for the implementation of machine learning classification models. Int. J. Emerg. Technol. Adv. Eng. **11**(8), 8–18 (2021). https://doi.org/10.46338/IJETAE0821_02
8. Khotimah, N., Wibowo, A.P., Andreas, B., Girsang, A.S.: A review paper on automatic text summarization in Indonesia language. Int. J. Emerg. Technol. Adv. Eng. **11**(8), 89–96 (2021). https://doi.org/10.46338/IJETAE0821_11
9. Ehsani, M., Gao, Y., Emadi, A.: Modern Electric, Hybrid Electric, and Fuel Cell Vehicles, 2nd edn. CRC Press, Boca Raton (2009)
10. Gao, Y., Ehsani, M.: Design and control methodology of plug-in hybrid electric vehicles. IEEE Trans. Industr. Electron. **57**(2), 633–640 (2010)
11. Sciarretta, A., Guzzella, L.: Control of hybrid electric vehicles. IEEE Control. Syst. **27**(2), 60–70 (2007)
12. Martinez, C.M., Hu, X., Cao, D., Velenis, E., Gao, B., Wellers, M.: Energy management in plug-in hybrid electric vehicles: recent progress and a connected vehicles perspective. IEEE Trans. Veh. Technol. **66**(6), 4534–4549 (2017)
13. Mahrishi, M., et al.: Video index point detection and extraction framework using custom YoloV4 Darknet object detection model. IEEE Access **9**, 143378–143391 (2021)
14. Feng, T., Yang, L., Gu, Q., Hu, Y., Yan, T., Yan, B.: A supervisory control strategy for plug-in hybrid electric vehicles based on energy demand prediction and route preview. IEEE Trans. Veh. Technol. **64**(5), 1691–1700 (2013)
15. Li, L., Yang, C., Zhang, Y., Zhang, L., Song, J.: Correctional DP-based energy management strategy of plug-in hybrid electric bus for citybus-route. IEEE Trans. Veh. Technol. **64**(7), 2792–2803 (2015)
16. Manikyam, S., Kumar, S.S., Pavan, K.V.S., Kumar, T.R.: Laser heat treatment was performed to improve the bending property of laser welded joints o low-alloy ultrahigh-strength steel with minimized strength loss. **83**, 2659–2673 (2019)
17. Siddam, S., Somasekhar, T., Reddy, P., Kumar, T.: Duplex stainless steel welding microstructures have been engineered for thermal welding cycles & nitrogen (N) gas protection. Mater. Today Proc. (2021). https://doi.org/10.1016/j.matpr.2020.11.091
18. Kumar, T., Mahrishi, M., Meena, G.: A comprehensive review of recent automatic speech summarization and keyword identification techniques. In: Fernandes, S.L., Sharma, T.K. (eds.) Artificial Intelligence in Industrial Applications. LAIS, vol. 25, pp. 111–126. Springer, Cham (2022). https://doi.org/10.1007/978-3-030-85383-9_8
19. Dhotre, V.A., Mohammad, H., Pathak, P.K., Shrivastava, A., Kumar, T.R.: Big data analytics using MapReduce for education system. Linguist. Antverpiensia 3130–3138 (2021)
20. Gurugubelli, S., Chekuri, R.B.R.: The method combining laser welding and induction heating at high temperatures was performed. Design Eng. 592–602 (2021)
21. Pavan, K.V.S., Deepthi, K., Saravanan, G., Kumar, T.R., Vinay, A.V.: Improvement of delamination spread model to gauge a dynamic disappointment of interlaminar in covered composite materials & to forecast of material debasement. PalArch's J. Archaeol. Egypt/Egyptol. **17**(9), 6551–6562 (2020)

Using Artificial Intelligence and Deep Learning Methods to Analysis the Marketing Analytics and Its Impact on Human Resource Management Systems

Vinima Gambhir[1], Edwin Asnate-Salazar[2], M. Prithi[3](✉), Joseph Alvarado-Tolentino[4], and Korakod Tongkachok[5]

[1] SVIMS Business School, Wadala, Mumbai, Maharashtra, India
[2] Computer Science and Engineering, Universidad Nacional Santiago Antunez de Mayolo, Huaraz, Peru
edwin_johny@unasam.edu.pe
[3] Computer Science, Periyar University, Vellore, Tamil Nadu, India
prithimadhavan@gmail.com
[4] Science and Engineering with a Major in Audit and Computer Security, Universidad Nacional Santiago Antunez de Mayolo, Huaraz, Peru
jalvaradot@unasam.edu.pe
[5] Faculty of Law, Thaksin University, Songkhla, Thailand
tokorakod@tsu.ac.th

Abstract. In recent years, all firms have been concerned with marketing analytics. They're using numerous advanced technologies to analyse marketing analytics. Artificial intelligence (AI) and Deep Learning (DL) technology are highly capable of examining large databases for patterns and insights. These technologies enable the marketing function to encompass its reach and analytics empower a deeper understanding of how the market responds to actions. With the help of Deep learning, businesses are now able to connect a wide range of datasets to better understand what customers want with greater sophistication and analytic capacity, and then use that information to gain a competitive advantage. In addition, Deep learning uses numerous technical tools that excel at extracting perceptions and patterns from huge amounts of data and then predicting the future for marketing. Thus, Deep learning can potentially be used to create products that are tailored to what customers want. It has been found that marketing analytics play a significant role in HRM. It helps in examining the employees' skill sets and developing a training programme based on the market demands. AI assists firms in determining target audiences and devising a strategy to meet its objectives. Also, AI technology adapts and learns from data to make data-driven decisions. Many time-consuming and managerial chores will be automated by HR software that includes artificial intelligence. A lot of administrative activities are automated and speeded up using AI. Moreover, to gain in-depth knowledge of Artificial Intelligence and Deep learning this paper is conducted. Furthermore, the article examines how artificial intelligence and deep learning technologies are utilised to assess marketing statistics, as well as their impact on human resource management systems. For this paper,

V. E. Balas et al. (Eds.): ICETCE 2022, CCIS 1591, pp. 345–353, 2022.
https://doi.org/10.1007/978-3-031-07012-9_30

descriptive research methodology has been used for this study, and secondary data has been used to obtain reliable conclusions.

Keywords: Deep learning · Artificial intelligence · Marketing · HRM · Human Resource Management System (HRMS)

1 Introduction

In today's corporate world, many companies have increased their usage of advanced technologies. These technologies aid in the transformation of existing business processes to increase customer engagement, employee productivity, and company resiliency. The tempo in which management jargon shifted from big data to deep learning to artificial intelligence (AI) is staggering. The term "artificial intelligence" is used to cover a range of technologies that enable a system to execute tasks that would ordinarily need human intellect, such as decision-making [1]. In some data-rich environments, deep learning utilising neural networks has grown more widespread, bringing us closer to real AI, which refers to robots' ability to emulate dynamic human decision-making.

Nonetheless, few firms have even achieved the big data stage when it comes to staff management, where the promise of more informed judgments has been clearly proclaimed. Human resources are regarded as a significant factor impacting all of a company's activities at all stages of its growth and human resource policy, and all companies are interested in examining human resources-related data. Hence, Data analysis (HR analytics) increases the profitability of the firm dramatically. Further, with the help of AI, HRM is increasingly tailoring HR processes to account for the unique characteristics of each target group. The objective is to assist managers in improving intergenerational collaboration. Human Resources Analysis enables businesses to make the best use of their employee datasets to make better decisions and improve operational performance [2]. AI assists HR departments in making strategic decisions. The HR staff does not have to spoon-feed the information to employees. AI chatbots can manage all of an employee's questions and provide a suitable response. On the other side, companies employ chatbots to engage with customers directly over the internet. Chatbots helps to address the issues and complaint of the customers.

In addition to this, Artificial intelligence and deep learning are widely used developing technology that enables businesses to track real-time data to understand and respond quickly to client needs [3]. AI provides crucial consumer insight into consumer behaviour, which is critical for customer attraction and retention. The customer's next move is prompted by AI, which redefines the total experience. In artificial intelligence, data ingestion is a critical component. Artificially intelligent systems have to deal with a lot of data. The artificial intelligence system collects and analyses large amounts of data based on the requirements (Fig. 1).

2 Literature Review

A. Jain, A. K. Pandey, [4] states that Companies will profit from AI integration in their HRMS to train staff in their respective domains. Being at the top of the competition,

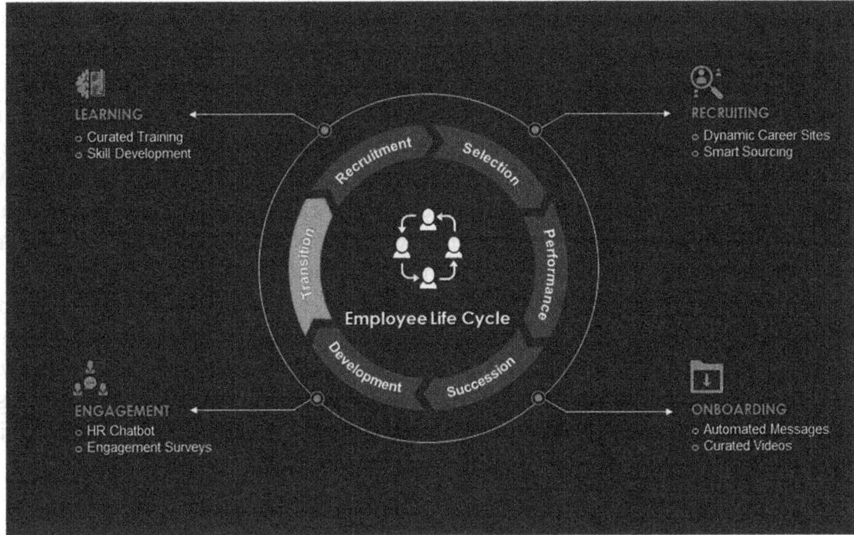

Fig. 1. AI integration in their HRMS (Source: Parry, Battista)

employees constantly improve their skill set. The HR department use AI to evaluate their employees' skill sets and offer a training module for employees' job. It compiles all of the information and assists the HR team in making informed decisions about what training to undertake in which field to increase abilities. HRMS benefit from AI-integrated systems when it comes to teaching and training personnel in their respective fields. It creates a learning route based on the employee's grasping power to help them acquire knowledge. Additionally, Artificial intelligence assists businesses in making data-driven conclusions. Data-driven decisions will aid in the firm's progress. Apart from their work inputs, the HR department is responsible for the overall well-being of personnel. A. Jain, et al. [5] argues that AI requires proper data storage for effective operation. It collects and analyses all relevant statistics, allowing HR to make data-driven choices. Thus, HR can make long-term and significant data-driven decisions.

Borges AF, et al. [6] reports that, in marketing, artificial intelligence recommends the proper information or items of customers based on their previous purchasing habits. Companies can monitor their clients' virtual footsteps using cookies and promote a certain commodity, or similar ones, on other online platforms, such as Facebook. Marketers have been able to personalise their client experiences thanks to artificial intelligence. People's internet search histories can be collected and utilised in marketing tailored to that customer. Artificial intelligence also helps organizations clearer grasp their target market and promote sales strategies that are focused on performance. Moreover, Advertisers may use AI to evaluate more data than a human psyche can grasp, better adapt to a customer's tastes, and understand the target market through natural language processing. The help of artificial intelligence enables a more in-depth analysis of data. A Deep-learning algorithm could analyse large data sets, connect them with other data, and provide more detailed insights. For a marketer, this means actually being able to integrate more data into campaign planning. Also, being able to act on those results much more quickly.

Further, marketing campaigns must be completely relevant to the target demographic. However, many marketers lack the information and analytics necessary to design a campaign that would engage their target demographic. They can get all the information they need and enhance the productivity of their endeavours with AI. However, De Bruyn A, et al. [7] claims that, despite numerous endeavours, AI cannot still be creative. Machine learning algorithms are unable to act on data in the same way that humans can. Yet, as a result, their potential to build something from those insights is severely limited.

Goralski MA, Tan TK [8] suggest that Deep learning has enormous marketing potential because of its capacity to crunch large amounts of data. This can benefit marketers and businesses in a variety of ways. Hyper-personalization is one of the most effective uses of Deep learning in marketing. In addition, DL can help marketers interpret vast amounts of data, which would take time for employees to do, and marketers can use this data to develop highly personalized marketing campaigns and content (Table 1).

Table 1. Areas where artificial intelligence put a substantial effect on marketing

S/no	Domain	Its effectiveness
1	Delivering the right message on the right channel at the right time	61%
2	Dynamic landing pages and websites	60%
3	Hyper-personalization of content	59%
4	Hyper-personalized product recommendations	60%
5	Predictive journey	58%
6	Programmatic advertising and media buying	57%
7	Business insights across data and systems	55%
8	Campaign analytics	58%
9	Digital asset management	54%
10	Productivity of marketers	57%
11	Customer segmentation	56%
12	Lead scoring	58%

Research Methodology

A descriptive research methodology was used in this study. It allows for a more thorough analysis of the research. In addition, secondary data sources were employed to get responses to queries and review research methodologies to help establish crucial information to answer the study's aims' criteria. Data was gathered from secondary sources such as books, research reports, journals, and the internet, and more. Intelligent technologies and systems such as Deep Learning and AI are increasingly making their way into the business world thanks to technological breakthroughs.

Artificial Intelligence is transforming the way things are done in every facet of the business [9]. The future of the HRM and marketing depends on the unique blend of digital AI and human work, with advancements in machine-learning technologies and Deep learning constantly being developed. Human resources experts believe AI can provide possibilities for learning new skills and freeing extra time, allowing Hr managers to expand their current roles and become more strategic within their organisations. Candidate rediscovery has been aided by artificial intelligence. AI technology may assess the existing pool of applicants and identify individuals who would be a good fit for new opportunities when they become available by maintaining a database of former applicants [10]. HR professionals may use this technology to identify qualified individuals more quickly and easily than ever before, rather than wasting time and resources seeking new talent. Candidate rediscovery has been aided by artificial intelligence. AI technology assesses the existing applicant pool and targets specific who would be a better match for new opportunities when they become available by maintaining a database of former applicants. Hr managers use these techniques to identify qualified employees more rapidly and effortlessly than ever before, rather than wasting time and money seeking new talent. Additionally, HR practitioners use artificial intelligence to enhance employee empowerment and employee satisfaction in addition to improving the hiring process. This is especially useful given how crucial it is to understand employees' overall needs, but there are also some major organisational advantages to having this knowledge.

Further, with the use of chatbots and remote assistance applications, AI technology enables new hires to get human resources assistance at any time and from any place [11]. This move not only allows workers to accomplish the onboarding at their own pace but also minimises operational effort and usually leads to faster integration. As a result, artificial intelligence will certainly continue to have a favourable impact on the area of human resources management in the future years (Table 2).

Table 2. How deep learning used in HR & recruitment

S. no	Areas	Benefits
1	Recruitment	Candidate sourcing, hiring process automation, Predicting new hire performance
2	Learning and development	Onboarding, reference information, personal training
3	Performance management	Objective setting, Capacity predictions, Forecasting performance, Team fit psychometrics
4	Engagement and recognition	Payroll and benefits automation, Engagement monitoring
5	Risk management	Retention and attribution analytics, Absence and accident prediction, Fraud detection
6	Compliance and reporting	Monitoring bias, Supervision of employee activities
7	Separation/Retirement	Outplacement, Pension administration, Alumni and boomerang hire sourcing

Apart from this, for marketers, AI opens up a world of possibilities. Marketers can run campaigns targeted at a specific demographic. As well as do so with a message that is both tailored to the needs of those clients and appropriately delivered. In-depth data analysis could also disclose the motives, desires, and pain points of each section. It might even assist forecast customers' future moves. For example, having a better knowledge of the data could also aid in the prediction of future campaign performance. It might even help forecast growth and create realistic, attainable goals.

In addition, by 2022, AI will handle 85% of all customer service queries. For example, Chatbots have risen to the forefront of user experience. According to the market, 39% of customers would prefer to utilise a chatbot to handle a problem or complaint. The key advantages of AI-powered customer service are the 24-h service and the ability to receive immediate assistance [12]. Amazon's use of AI to propose products to consumers is by far the most well-known example of artificial intelligence in marketing. Amazon's AI examines a user's previous purchases and viewing habits to determine which things they are most prone to shop next.

Furthermore, many e-commerce stores use AI to track competitive pricing as well as internal aspects (such as expenditures, etc.) AI is used by manufacturing firms, hotels, and a variety of other businesses to track purchasing trends and establish the most modest rates [13]. Hence, increasingly companies are turning to artificial intelligence to help them save money on customer support. AI in customer service can be used for augmented messaging, channelling support enquiries to the appropriate personnel, and improving phone assistance, to mention a few examples. Eventually, it is apparent that AI and deep learning will aid marketers in their profession. With the help of AI, marketers have a very clear vision for the future of content creation, from analytics to forecasting and unlimited refining opportunities [14] (Fig. 2).

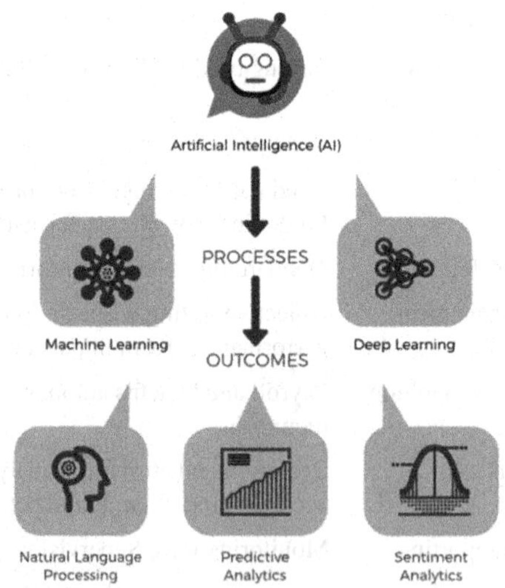

Fig. 2. Use of AI in marketing (Source: Vishnoi et al.)

3 Result/Findings

The fact that leads to this conclusion in marketing analytics is a crucial and necessary activity for every company. Marketing analytics isn't merely for developing strategies. It can track the effectiveness of marketing activities and make necessary adjustments to help the company expand faster. The purpose of marketing analytics is to improve and measure marketing campaigns [15]. It goes without saying that optimization is based on the power of analytics, with AI and Deep learning at the forefront. In this way, artificial intelligence can help at every stage in marketing. By automating insights, AI makes things simpler to link and bring diverse data together, automate datasets for scale, and shorten optimization cycles. Additionally, using the advantages of automation enables more rapid testing and optimization. AI technologies assist marketing analysts in making their jobs easier while also enhancing their efficacy. AI can learn about a client's prior behaviour and predict the likelihood of a customer churning with the use of past records.

In addition, the finding also shows that Although AI and DL are crucial for forecasting and predicting performance statistics over time, the HR industry is progressively using them. Deep learning adoption is benefiting a vast number of organisations. Across a spectrum of key HR practises, it has decreased manual participation, human errors, and improved predictive accuracy and decision-making [16]. Innovation, technology advancement, and their evolving impact on company vision and goals are the technology of the future.

4 Conclusion

It has been concluded from the study that from data integration to insight generation, artificial intelligence plays a key role. Marketers may scale up faster with AI's support while never losing out on an insight. AI has reduced optimization cycles to daily intervals, which were previously weeks or even months long. Marketing analytics is all about obtaining the most out of the data to make marketing campaigns more efficient and productive. This involves analysing and assessing marketing initiatives in order to avoid wasting marketing dollars and keep marketing programmes nimble and adaptive. In addition, marketers do this by focusing on consumer behaviour, customer insights, and marketing automation with advanced analytics. AI approaches such as deep learning assist businesses in filtering through enormous and diverse collections of data to come at exact constraint specifications. Companies can assess data analytics supported by deep learning to acquire a better knowledge of how to optimise marketing to intended results. Therefore, because of the efficiency and versatility they provide to marketing analytics and reporting systems, AI and deep learning are tremendously helpful to marketing. This is especially helpful when creating a basic layer on which to build a breath-taking view of client touchpoints.

Apart from this, Artificial intelligence is also used in several ways in the human resource management system to streamline procedures and increase productivity. Artificial intelligence is demonstrating its ability to help people manage and grow their skills. While there is always the potential for improvement in this area, systems that can provide calibrated assistance without the use of humans and allow more employees to advance in their jobs and stay engaged. Integration of artificial intelligence into HR operations will improve firms because these apps can evaluate, anticipate, and diagnose to assist HR professionals in making better decisions. Along with this, Deep learning has risen to prominence in the field of human resources. It can not only assist personnel in doing their jobs, but it can even replace them when necessary, freeing up HR workers to concentrate on more vital activities. Deep learning has the potential to improve the future of HR.

5 Future Work

The goal of this study was to explain the most important applications of Deep Learning and AI in marketing analytics. Regardless of the various benefits that deep learning and Artificial intelligence provide to marketers. As a result, there's no guarantee that the research won't need to be broadened in the future. However, this study fails to highlight the dangers of using Artificial intelligence in marketing. As a result, more research should be done in the future to detect the risks of using Artificial intelligence in advertising.

References

1. Abubakar, A.M., Behravesh, E., Rezapouraghdam, H., Yildiz, S.B.: Applying artificial intelligence technique to predict knowledge hiding behavior. Int. J. Inf. Manage. **49**, 45–57 (2019)
2. Agrawal, A., Gans, J.S., Goldfarb, A.: Artificial intelligence: the ambiguous labor market impact of automating prediction. J. Econ. Perspect. **33**(2), 31–50 (2019)
3. Jain, A., Pandey, A.K.: Modeling and optimizing of different quality characteristics in electrical discharge drilling of titanium alloy (Grade-5) sheet. Mater. Today Proc. **18**, 182–191 (2019)
4. Jain, A., Pandey, A.K.: Multiple quality optimizations in electrical discharge drilling of mild steel sheet. Mater. Today Proc. **8**, 7252–7261 (2019)
5. Jain, A., Yadav, A.K., Shrivastava, Y.: Modelling and optimization of different quality characteristics in electric discharge drilling of titanium alloy sheet. Mater. Today Proc. **21**, 1680–1684 (2019)
6. Borges, A.F., Laurindo, F.J., Spínola, M.M., Gonçalves, R.F., Mattos, C.A.: The strategic use of artificial intelligence in the digital era: systematic literature review and future research directions. Int. J. Inf. Manage. **57**, 102225 (2021)
7. De Bruyn, A., Viswanathan, V., Beh, Y.S., Brock, J.K., von Wangenheim, F.: Artificial intelligence and marketing: Pitfalls and opportunities. J. Interact. Mark. **51**, 91–105 (2020)
8. Mahrishi, M., et al.: Video index point detection and extraction framework using custom YoloV4 Darknet object detection model. IEEE Access **9**, 143378–143391 (2021)
9. Huang, M.-H., Rust, R.T.: A strategic framework for artificial intelligence in marketing. J. Acad. Mark. Sci. **49**(1), 30–50 (2020). https://doi.org/10.1007/s11747-020-00749-9
10. Lengnick-Hall, M.L., Neely, A.R., Stone, C.B.: Human resource management in the digital age: big data, HR analytics and artificial intelligence. In: Management and Technological Challenges in the Digital Age, 3 January 2018, pp. 1–30. CRC Press (2018)
11. Ongsulee, P.: Artificial intelligence, machine learning and deep learning. In: 2017 15th International Conference on ICT and Knowledge Engineering (ICT&KE), 22 November 2017, pp. 1–6. IEEE (2017)
12. Parry, E., Battista, V.: The impact of emerging technologies on work: a review of the evidence and implications for the human resource function. Emerald Open Res. **1**, 5 (2019)

13. Sterne, J.: Artificial Intelligence for Marketing: Practical Applications. Wiley, Hoboken (2017)
14. Tambe, P., Cappelli, P., Yakubovich, V.: Artificial intelligence in human resources management: challenges and a path forward. Calif. Manage. Rev. **61**(4), 15–42 (2019)
15. Panwar, V., Sharma, D.K., Kumar, K.V.P., Jain, A., Thakar, C.: Experimental investigations and optimization of surface roughness in turning of EN 36 alloy steel using response surface methodology and genetic algorithm. Mater. Today: Proc. (2021). https://doi.org/10.1016/j.matpr.2021.03.642
16. Vishnoi, S.K., Bagga, T.E., Sharma, A.A., Wani, S.N.: Artificial Intelligence enabled marketing solutions: a review. Indian J. Econ. Bus. **17**, 167–177 (2018)

A Case Study on Machine Learning Techniques for Plant Disease Identification

Palika Jajoo[1](✉), Mayank Kumar Jain[2], and Sarla Jangir[1]

[1] Swami Keshvanand Institute of Technology Management and Gramothan, Jaipur 302017, Rajasthan, India
palika@skit.ac.in

[2] Malaviya National Institute of Technology, Jaipur 302017, Rajasthan, India

Abstract. Plant diseases are the significant elements impacting food supply and minimizing production losses; hence crop illnesses must be detected and recognized quickly. Deep learning approaches have recently expanded their applicability in plant disease identification, providing a comprehensive instrument with accurate findings. In this paper, we present a thorough assessment of the literature to determine the art of present state in the application of deep learning techniques for noticing and classification of diseases of plants and identifying trends and gaps.

Keywords: Image processing · Leaf disease · Machine learning · Deep neural network

1 Introduction

All living beings rely heavily on plants as their primary source of nutrition. The spread of plant diseases is disastrous to society and jeopardizes global economic stability [1]. Pests and diseases wreak havoc on a third of all crops. Monitoring sanitary conditions and detecting weeds is a crucial task [2]. Plant diseases are affected mainly by fungi, bacteria, nematodes and cause drastic yield losses and quality limitations. It is possible to diagnose the plant leaf to monitor and detect symptoms early and predict disease at an early stage [3]. Detecting symptoms can be done mechanically or through direct observation with the naked eye. The most common method for identifying disease in plants is to observe it with the naked eye. However, this process necessitates people as well as ongoing monitoring.

It is possible to halt manual observation labour using an automatic detection system. Automatic illness identification has several advantages, including saving time, effort and providing accurate predictions [4]. The need is for techniques such as a system capable of accurately detecting diseases and classifying them accurately with the slightest error. Many types of research have been done. Crop protection has been accomplished using a variety of image processing approaches[5]. Plant diseases can also be classed depending on their chemical and morphological characteristics. Each plant's visible and infrared spectral reflectance is unique. Plant diseases are classified using a various methods such as fuzzy logic, artificial neural networks, and machine learning approaches [6].

V. E. Balas et al. (Eds.): ICETCE 2022, CCIS 1591, pp. 354–372, 2022.
https://doi.org/10.1007/978-3-031-07012-9_31

A decision tree and the nearest k-method can be used to identify diseases based on spectral reflectance. [17], PCA algorithms are utilized to identify weeds in the field of rice [8]. Based on the characteristics extracted during the leaf segmentation process, the picture segmentation technique primarily aids in the disease detection process. In some cases, for segmentation of color image, a genetic algorithm (GA) is used to make use of the SVM classifier, k-mean and radial basis function [9].

Performing manual feature extraction for disease prediction is a time-consuming task. The deep learning technique is a promising solution. Recently, deep learning and CNN have been employed to make predictions [10]. As demonstrated in Fig. 1, an extensive plant image library is employed for disease identification and analysis in deep learning. Numerous studies show the effectiveness of their constructed intelligent models, yet they all have flaws. Existing studies face issues such as inefficient performance, inability to adapt to a new context, and erroneous forecasts. This paper give insight into the state-of-the-art ML approaches used to detect leaf disease and present recommendations for future work.

The paper is further categorized as follows. Section 2 discusses the previous research works related concepts mentioned above in brief. Section 3 discusses various methodologies used in plant disease detection, and Sect. 4 gives some information about freely available datasets. Section 5 shows the limitations of the previous methodology. Section 6 concludes the research and presents the direction for upcoming research. Plant diseases are the significant elements impacting food supply and minimizing production losses; hence crop illnesses must be detected and recognized quickly. Deep learning approaches have recently expanded their applicability in plant disease identification, providing a comprehensive instrument with highly accurate findings. This paper presents a thorough assessment of the literature to determine the art of present state in the application of deep learning techniques for identify and classification of diseases of plants and identifying trends and gaps.

2 Literature Review

The rate of crop production is directly proportional to the health of the crops. The most crucial step of a crop production system is the timely diagnosis and treatment of crop diseases. Automated plant disease identification systems based on machine learning techniques and image processing can be developed for efficient results. The automated system can be trained to recognize the onset of disease by analyzing leaf images and observing the development of the disease. Diseases damage the parts of the plant. The most common symptoms include a change in color, texture, and size of leaves. As a result, image processing techniques can be used to identify plant disease diagnosis and treatment recommendations. Many researchers from around the world experimented with various algorithms of Machine Learning to automate the diagnosis of plant diseases. This section looks at several cutting-edge technologies for automatic spotting of disease in plants that are present in the literature.

Patil and Kumar provided a theoretical investigation into this topic in 2011 [34]. The authors proffered that different machine learning techniques can boost the accuracy and speed of leaf disease recognition systems. They argued that machine learning techniques are more efficient and scalable as compared to conventional image processing techniques.

According to Valliammai and Geethaiakshmi (2012), noise overpowering is significant for input images top to the extraction of reliable feature of plants [1]. The borders of the leaf vein are not apparent; as a result, the leaf image seems blurred when Gaussian noise is applied. Speckle noise has a substantial impact on the size, shape, and pattern of leaves. To restore noise-free leaf pictures for further processing, noise removal techniques (Gaussian and speckle) are required. These hybrid filter methods were created to reduce noise, increase image quality, and produce better results than other traditional filters.

The performance of multiple machine learning (ML) approaches for automatic analysis of plant disease was compared by Akhtar et al. (2013) [2]. This research shows that the best outcomes are achieved using features obtained from discrete cosine transform (DCT) and discrete wavelet transform (DWT) algorithms. These hybrid features were classified with SVM, and 94.45% accuracy was achieved.

Gawande and Gavhale (2014) examined and summarised image processing algorithms used to identify plant diseases in several plant species [3]. The most frequent algorithms for detecting plant illnesses were backpropagation neural networks (BPNNs), Support Vector Machines (SVMs), K-nearest neighbor (KNNs), and Spatial Gray-level Dependency Matrices (SGDM). These techniques can be used to analyze both sick and healthy plant leaves.

Xia and Li, in 2015, proposed an android architecture for an intelligent wheat disease diagnosis system [37]. People gather photographs of wheat disease using Android phones and send them over the network to a server for disease detection. The server performs picture segmentation after receiving disease images by transforming their color space from RGB to HSI. The grey level co-occurrence matrix and the color moment matrix will determine the disease's color and texture characteristics. After the preferred features are supplied into the support vector machine for recognition, the client receives the identification results.

Different techniques for classification such as Support Vector Machine (SVM), Genetic Algorithm, K-Nearest Neighbor Classifier (KNN), Principle Component Analysis, Probabilistic Neural Network(PNN), Fuzzy Logic and Artificial neural network(ANN) were explained by Dangre and Handore (2015) [4]. Using premature crop conditions and suitable management strategies can control the spread of leaf diseases.

Kaur and Singla (2016) studied pattern recognition approaches such as neural networks and backpropagation algorithms [5]. These methods provide reliable results, but not to the range that they could be considered for real-world agricultural diseases. As a solution, the authors demonstrated a nature-inspired technique with pre-processing for enhanced results. Noise reduction techniques were used for pre-processing.

Singh and Mishra (2016) used a genetic algorithm for plant disease detection. A genetic algorithm was used to develop an automatic segmentation and classification technique [6]. The primary focus was to achieve significant performance with less computational effort. Results show that the proposed approach can be used to detect leaf disease in the nascent stage. The detection performance can be enhanced using different algorithms like fuzzy logic, ANN, and hybrid algorithms.

In identifying leaf disease, Padmavathi and Thangadurai (2016) evaluated the outcomes of RGB and Grayscale images [7]. Color becomes a critical feature in spotting infected leaves and determining disease severity. To estimate the disease level, they used grayscale and RGB images and a median filter for image segmentation and enhancement to extract the sick region. The plant disease recognition algorithm based on leaf image categorization was built using convolutional networks. Thirteen different diseases have been identified using the ability to distinguish healthy leaves from their ambiances.

Using the dataset PlantVillage, Mohanty et al. 2016 investigated the performance of GoogLeNet and AlexNet CNN architectures for disease detection in plants [8]. They used 60 distinct combinations to conduct 60 different trials. They discovered that by using GoogLeNet with transfer learning performed the finest, by a 99.3% accuracy.

Misra and Singh (2017) have introduced an algorithm that assists in picture segmentation and categorization. The disease detection method is also aided by the genetic algorithm [6]. It is carried out in MATLAB. For classification, the K-mean clustering criterion and Minimum Distance Criterion were applied, with 86.54% accuracy.

Yang Lu et al. (2017) suggested a method for diagnosing illness in rice. This is accomplished through the use of a deep convolutional neural network (CNN) [9]. There are 500 photos in the dataset utilized for the analysis. For identifying purposes, ten different diseases of rice plants are used. 95.48% accuracy was attained during this process.

Jafari and Bakhshipour (2018) suggested a process for weed detection. Support Vector Machine and an artificial neural network (ANN) are used to determine each variety of plants [10]. ANN has a 92% accuracy rate, while SVM has a 96.67% accuracy rate.

With the use of CAE and SVM classifiers, Pardede et al. (Pardede et al., 2018) [11] created a system for automatic identification of disease for potato and corn plants. They used the PlantVillage dataset to extract leaf pictures of potato and corn plants. They detected diseases in potato and maize plants with 87.01% and 80.42% accuracy, respectively.

Park et al. (2018) successfully proposed a minimum redundancy and maximum relevance (MRMR) technique for feature selection in classification of fresh apple leaves [12]. A neural network framework is used to classify hyperspectral images. Apple leaves are employed as samples, and leaf samples are taken in six different conditions with five distinct spectral band lengths for analysis. This technique primarily aids in the shrinking of the band for improved leaf estimation. During this procedure, the complexity of computation is lowered.

Gan et al. (2018) have completed an important task of mapping citrus yield. The image processing technique is used to determine if the fruit is ripe or not [13]. For green fruit detection, a permutation of color and thermal images is used. Colour thermal combined property (CTCP) algorithm used to classify fruit images into ripe and unripe classes. Before the color is fused, the precision rate has grown from 78.1% to 90.4%. After the color was fused, the precision change from 86.6% to 95.5%.

Liang et al. (2019) used a severity estimation method to determine the severity of disease in the plant [14]. For diagnosing the disease, a method called PD2SE-Net. During identification step, the visualization and augmentation processes are carried out. As an auxiliary structure, the ResNet50 architecture is used. For disease severity, the accuracy is 0.99 and 0.98.

Kaya et al. (2015) discovered that manually classifying data has several key drawbacks: being costly, time-consuming, and requiring experts. A Deep Neural Network (NN) is used during the classification phase to solve such a problem [15]. The author looked at four different transfer learning methods used in a deep neural network to classify plants.

Too et al. (2019) developed a system for automatically detecting objects in an image, fast and precisely [16]. The primary objective of this research was to perform fine-tuning on CNN for achieving optimal results. The dataset consisted of 38 classes of images of pathological and healthy plant leaves. An architecture of CNN, such as inception V4, VGG, and ResNet series, dense net was examined. The accuracy of the testing achieved during this technique is 99.75%. Keras, CNMeM, CNN, and Theano are the libraries used.

Five different CNN architectures, namely Inception V3, Resnet18, Resnet50, Resnet 152, and VGG 16, were applied to detect leaf disease in banana plants; this research was conducted by Sanga et al. (2020) [17]. The authors concluded that ResNet152 is the best-performing model.

VGG19 and Inception V3 structures were demonstrated on the PlantVillage dataset by Chohan et al., 2020 [18]. Data augmentation was performed to augment the dataset synthetically. Experimental results demonstrate that the finest performing model is VGG19. The model demonstrated 98% training and 95% testing accuracy.

Another research work using multiple CNN architecture is presented by Tiwari et al., 2020 [19]. This research is based on the potato crop. The authors used CNN model such as VGG and Inception nets for extraction of features and later classified these with conventional classifiers such as ANN, SVM, logistic regression (LG), and K -nearest neighbors. This technique is helpful as it decreases time complexity to a great extent. With an accuracy rate of 97.8%, they discovered that VGG19 with logistic regression surpassed the competition.

It can be observed that most of the studies are based on a single crop only. Therefore, the presented comparisons between different techniques are not sufficient. Khamparia et al., 2020 [20] evaluated deep learning frameworks on three crops: maize, tomatoes, and potatoes. The dataset consisted of 900 images segregated into six categories, two (healthy and diseased) for each crop. The training accuracy obtained in this research is 100% but testing accuracy (86.78) is less than the above-mentioned approaches. This signifies the overfitting of data. This research proves that to develop a robust and scalable automated crop disease detection system, the technique must be trained and evaluated on different crops. Here, Table 1 represents the systematic view of previous papers.

Table 1. Overview of reviewed sources

Reference	Crop	Techniques	Accuracy
V. Singh et al. (2016) [6]	Banana, beans, lemon, rose	MDC(Minimum Distance Criterion) with k-mean	86.54%
		MDC	93.63%
		SVM	95.71%
Mohanty et al. (2016) [8]	Apple, grape, tomato, corn, cherry, blueberry, orange, bell pepper, potato, Soybean, strawberry	DCNN	99.35%
Lu, Jiang et al. (2017) [4]	Wheat	VGG-CNN-VD16	93.27%
		VGG-CNN-S	73%
A. Bakhshipour, A. Jafari (2018) [10]	Turnip beet, Pigweed Lambsquarters, Hare's-ear mustard, sugar weed	ANN &SVM (weed)	92.5% and 93.33%
		ANN & SVM (sugar beet)	93.33% and 96.67%
Akhtar et al. (2013) [2]	Rose	SVM	94.45%
Yang lu et al. (2017) [9]	Rice	CNN	95.48%
M. Chohan et al. (2020) [18]	Apple, grape, tomato, corn, cherry, raspberry, orange, Soybean, strawberry	CNN	98%
Gan et al. (2018) [13]	–	inceptionV3	95.50%
Liang et al	–	Resnet	0.99%
(2019) [14]			
Kaya et al. (2019) [15]	–	CNN + Transfer learning	99.60%
Too et al. (2019) [16]	–	CNN	99.75%
Sanga et al.(2020) [17]	Banana	ResNet – 152	99.20%
Tiwari et al. (2020) [19]	Potatoe	VGG	97.80%
Khamparia et al. (2020) [20]	Tomato, Potato, and maize	Deep Convolutional Encoder Network	86.78%
Yongquan Xia et al. (2015) [37]	Wheat	SVM	95%

3 Methodology

An automated leaf disease identification system consists of certain essential subroutines as shown in Fig. 1. The first step is to gather a standard dataset. The consequent step is preprocessing. Objective of this step is to improve quality of images for better classification results. Image resize, contrast enhancement, background elimination are some steps involved in preprocessing.

After preprocessing the acquired dataset is breaks into testing and training subsets. The dividing ratio can be 6:4 or 7:3 for training and testing respectively. Now, depending

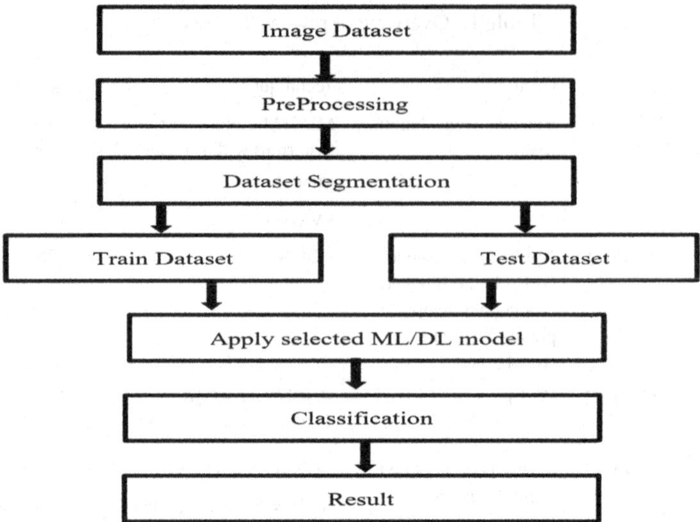

Fig. 1. Process flow of plant disease identification

on the problem an appropriate ML (machine learning) or DL (deep learning) model is selected. Number of images, classification levels and resource constraints are deciding factors for selection of a particular model. After training the developed model is tested on test data and results are observed.

Different ML or DL techniques is used for classification. Most widely used classification techniques are summarized in this section.

3.1 Decision Tree (DT)

The nodes, leaves, and branches of a decision tree (DT) form a hierarchical model [25]. A test on a characteristic is represented by each node in this model. The leaf represents a class label, and the branches represent groups of qualities that point to that class label as shown in Fig. 2. Classification policies are symbolised by the trail that leads from the root to the leaf. Logic identifies ways to split data sets according to decision trees. In each step of the tree-building process, information gain allows you to select which trait to divide. When both quantitative and categorical datasets are present, the decision tree method works well. The decision tree's key advantage is that it is resistant to outliers. The fundamental issue is that when some classes dominate, it might result in twisted learning trees. Because decision trees have a "if then other" structure, they can be altered in programming logic. They can also be utilised in categorization scenarios when attributes are thoroughly examined before a final category is defined.

3.2 Genetic Algorithm (GA)

Genetic algorithm imitates the process of natural selection. It draws its inspiration from the Charles Darwin theory of natural selection which is based on survival of the fittest [26]. The five important steps of genetic algorithm are shown in Fig. 3:

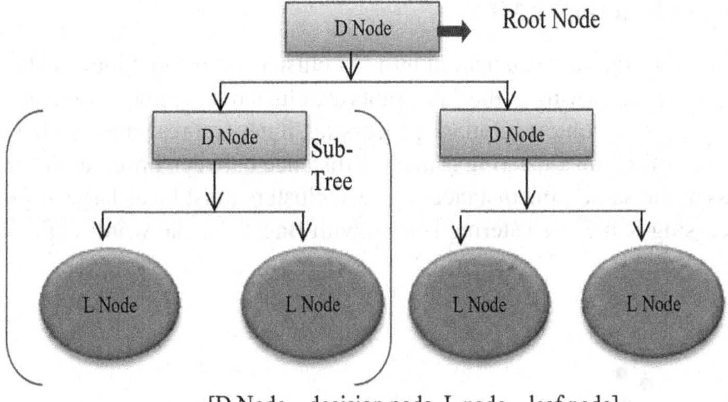

[D Node – decision node, L node – leaf node]

Fig. 2. Structure of decision tree

- Initial population generation
- Evaluation of fitness function
- Selection
- Crossover
- Mutation

The process starts with generation of initial set of chromosomes which represent initial population set. These chromosomes are evaluated on the basis of fitness function. The strongest chromosomes are selected. The selected chromosomes are crossed and mutated to generate off springs this process gets repeated until an optimal solution is reached.

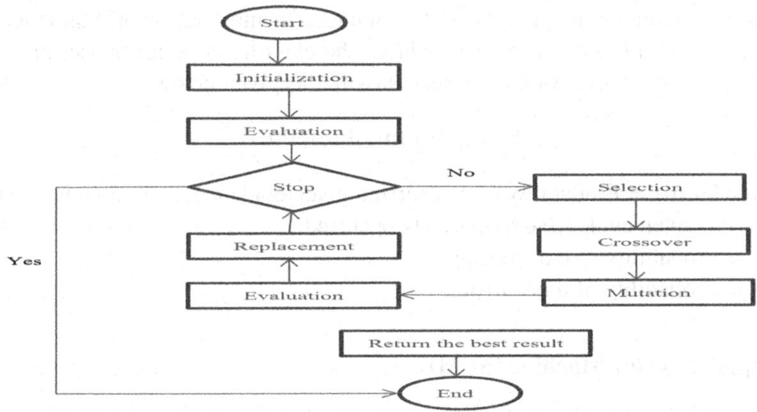

Fig. 3. Working principle of genetic algorithm

3.3 K Mean Clustering (KMC)

K means is used to split a given dataset into 'K' clusters or group. Figure 4 shows the K-mean clustering process where, the data points are clustered together based on a common characteristic [27]. In context of image processing, it may represent RGB channel value. The objective of K mean clustering is that the distance between observations in a cluster must be less at the same time distance between clusters must be as large as possible. In image processing K mean clustering is used with Bag of Visual words approach.

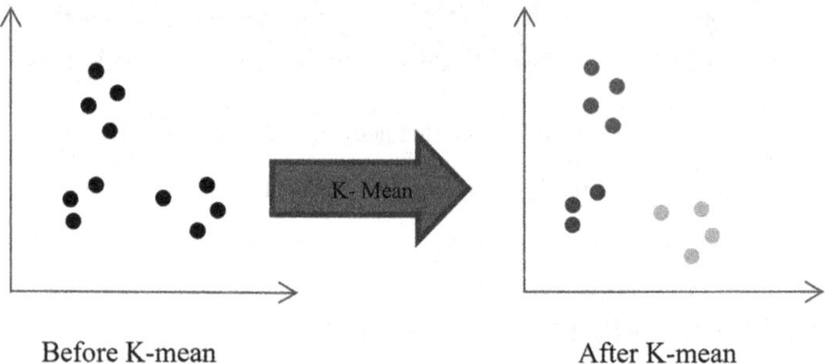

Fig. 4. K-means clustering process

3.4 Naive Bayes Classifier (NBC)

It is also called probabilistic machine learning model which is utilised for classification tasks [28]. The Bayes' theorem is used to create the classifier node. Some application of Naive bayes theorem are recommendation systems, spam filtering and sentiment analysis. They are quick and simple to implement, but their main disadvantage is that the predictor requirement must be self-contained. In most real-world situations, the predictors are interdependent, which reduces the classifier's performance. Probability distribution used in naïve Bayes to select classifier as given below.

$$P(c|d) = P(d|c)P(c)/P(d)$$

Where $P(c|d)$ = Probability of c occurring given evidence d has already occurred.
$P(d|c)$ = given evidence c has already occurred.
$P(c)$ = Probability of c occurring
$P(d)$ = Probability of d occurring.

3.5 Support Vector Machine (SVM)

It is most commonly used classification approach [29]. The objective of SVM is to divide the given observations into N classes on the basis of N dimensional hyperplane. It can be argued that there are infinite possibilities for selecting a hyperplane, however the hyperplane with greatest distance between two clusters is selected. Hyperplanes can be visualized as decision lines that aid in classification of data.

The number of entities determines the size of hyperplane, for instance for a problem involving two classes hyperplane is a row vector which shown in Fig. 5. Similarly, a two-dimensional hyperplane is required to classify three classes. It can be observed that the complexity of process increases with the number of classes.

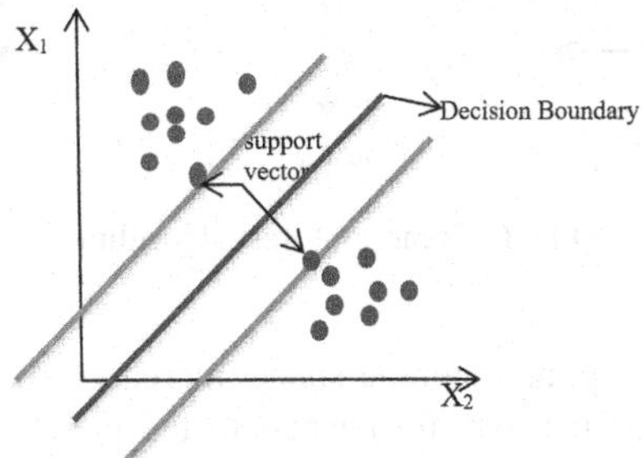

Fig. 5. Hyperplanes in SVM

3.6 Random Forest (RF)

RF is made up of several decision trees that work together to produce a single entity as shown in Fig. 6 [30]. These distinct trees divide the class prediction in a random forest, and the class with majority votes becomes the model prediction. The primary principle underlying the random forest is the combination of distinct separate trees functioning together for the forecasting process. Random Forest employs two ways to ensure that the behavior of each separate tree is not unduly connected among the behavior of any other trees in the model.

Bagging is the first way, and it takes advantage of the fact that decision trees are subtle to the data on which they are skilled, and even minimum variations to the training set can result in topologies like the first. Trees these are drastically different. The second method is feature randomness, in which every tree in a RF can only select after a subgroup of characteristics at random. As a result, the model's trees have a lot of variability, which leads to less co-relation and more diversification.

3.7 K-Nearest Neighbouring (KNN)

KNN is a well-known machine learning algorithm for solving classification and regression issues [31]. The distance function similarity measure is used to classify datasets into KNN. It's a non-parametric supervised learning technique used in data mining, image processing, intrusion detection, pattern recognition, and other domains. The classification of data in KNN is determined by the most appropriate closest neighbour votes. The number of neighbours is the most crucial decision factor in the classification process in KNN. In addition, during the test phase, the algorithm uses all of the training data.

As a result, they aren't required to create a training model. Data points those are unique. The amount of time and money required to examine all of the accessible data

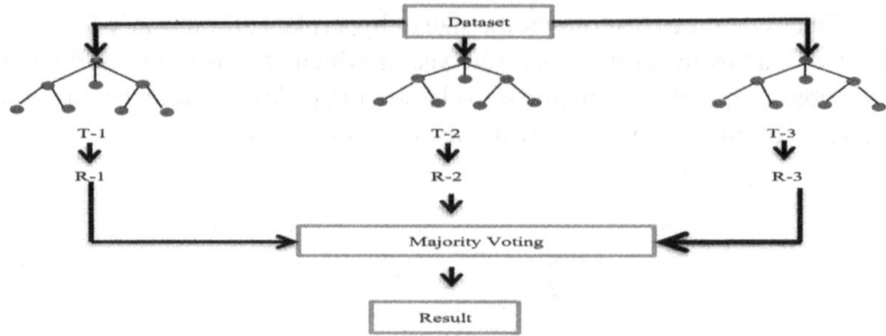

Fig. 6. Random forest algorithm

points, which necessitates greater storage space, increases. The KNN algorithm starts by selecting the k data points that are nearest to the points, then sorting the points based on the majority of votes for the k neighbours. The outcome is determined by which class receives the most votes from its objects. The distance functions Euclidean, Hammington, and Minkowski determine the distance between data points and find the closest neighbours. As the numeral of nodes in the KNN decreases, the performance improves. Figure 7 shows the assignment of a category to new data point using KNN.

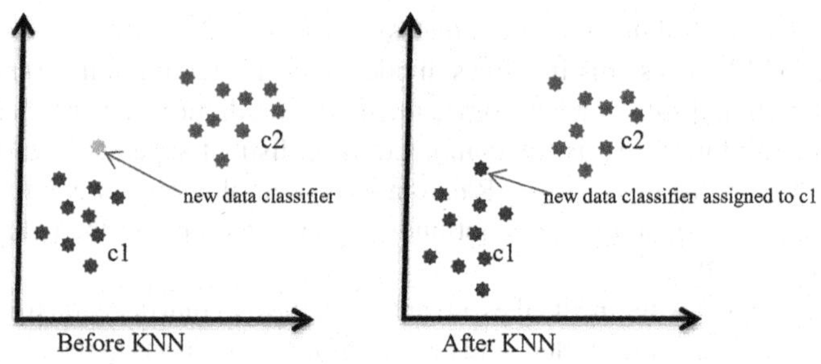

Fig. 7. KNN

3.8 Logistic Regression (LG)

Logistic Regression is a classification approach that leverages the concept of probability to assign observations to a specific set of classes [32]. Binary and multilinear logistic regression are the two forms of logistic regression (LG). LG is a type of linear regression that uses the "Sigmoid Function" as a cost function. Any value between 0 and 1 is transformed using this function. This function is used to link probabilities to forecasts. The optimization goal is represented by the cost function. This is accomplished by minimizing the cost function so that the error caused is as small as possible. Using gradient descent, the cost value is lowered.

3.9 Ensemble Learning (EL)

Ensemble learning is a technique for boosting the performance of single machine learning models by training a large number of them and making improvised predictions [33]. The predictors that have been taught to make predictions are referred to as "ensemble". Decision trees, which aid in the solution of quantitative challenges, are often used in ensemble learning. Set learning determines the final classification based on the set findings provided by the decision trees, rather than relying on predictive analysis and the conclusion of a single decision tree. Ensemble learning trains and presents the aggregated results of a range of machine learning (ML) models, such as logistic regression (LG), SVM classifier, random forest classifier, and others, to reach definite conclusions. There are two types of voting classifiers that are used to generate results: soft voting classifiers and hard voting classifiers.

3.10 Artificial Neural Network (ANN)

The Artificial Neural Network (ANN) is a deep learning (DL) system based on human brain's biological neural networks [34]. The creation of ANN was the outcome of an attempt to mimic the human brain's functions. ANN's operation is quite similar to that of biological neural networks, yet they are not identical.

Only numeric and structured data would be accepted as input by the ANN algorithm. CNN and recursive neural networks (RNN) are used to accept unstructured and non-digital data forms such as picture, text, and speech (RNN). Artificial Neural Networks are the only focus of this essay. The Fig. 8 shows the process of ANN.

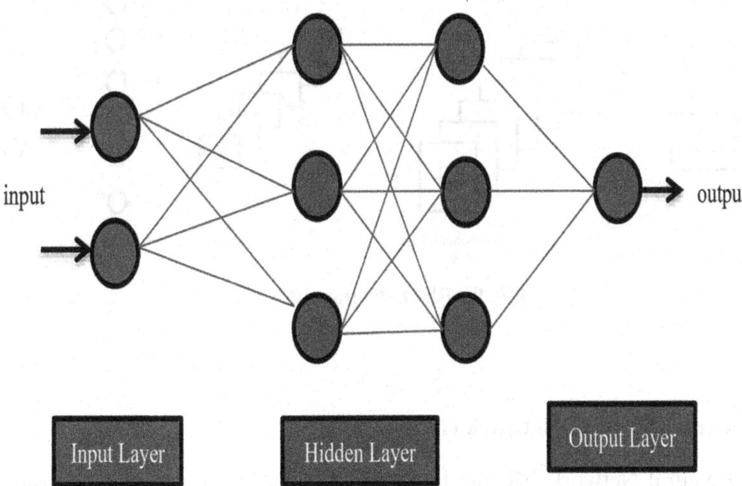

Fig. 8. Architecture of ANN

3.11 Convolution Neural Network (CNN)

In addition to assigning relevance (weight and prejudices of appetitatives) to various aspects/objects of an image, a Convolutional neural network is a deep learning (DL) system that can take image as an input and allot relevance (weight and prejudices of appetitatives) to various facets/things of the image [35]. Traditionally, the first ConvLayer is in charge of gathering low-level information like edges, colour, gradient direction, and so on. By adding layers to the design, we can handle high-end attributes while still having a system that contains the images in the dataset. The process yields two types of results: one in which the convoluted feature's dimensionality is lowered in comparison to the input, and another in which the dimensionality is raised or unchanged. Valid Padding is used in the first example, and the same padding is used in the second case.

- The level of grouping: The grouping level, like the convolutional level, is responsible for shrinking the convolved function's spatial size. The processing power required to process data is reduced by lowering dimensionality. Also effective for extracting rotation and dominant position invariant properties, which aids in the model training process.
- Max Pooling operates in the same way as Noise Suppression does. It denoises and reduces the dimensionality of the image by removing all noisy activations. Average pooling, on the other hand, is a noise reduction approach that merely reduces dimensionality. Figure 9 shows the basic structure of CNN layers.

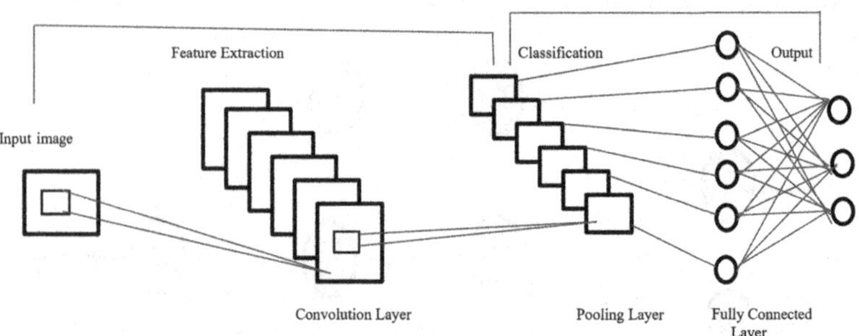

Fig. 9. Block diagram of CNN

3.12 Recurrent Neural Network (RNN)

Recurrent Neural Networks (RNN) is a part of ANN but it is applied on time series data [36]. In RNN the nodes are connected in a directed graph as shown in Fig. 10. This enables RNN to behave in a dynamic temporal manner. RNN are derived from feed forward neural network and can operate on large input sequence. Attention based RNN are often used to identify plant disease and predict the growth rate and remaining life. In conventional neural networks the consequent input data is independent of each other, but in RNN every new input depends upon its predecessor. This enables RNN to store the weight information of previous input in its memory and make consequent time-based predictions.

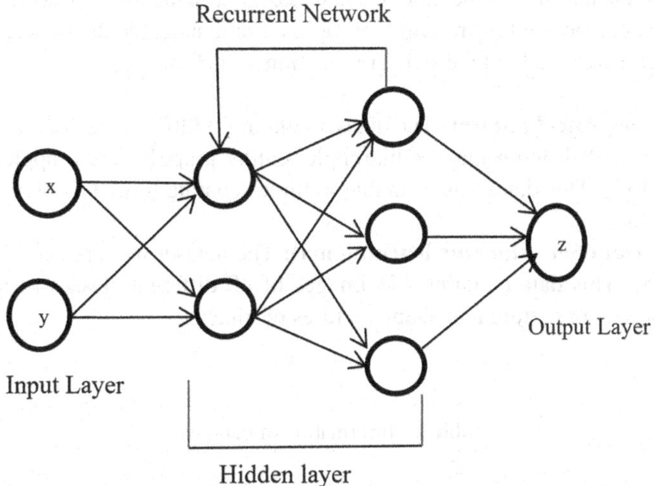

Fig. 10. Schematic representation of RNN

3.13 Visual Geometry Group (VGG)

VGG is a transfer learning model. Transfer learning models are applied on encoder layers of deep learning models to improve their performance [15]. Depth is the critical component of VGG. Its architecture is similar to CNN.

- An input image of 224 * 224 pixels is accepted.
- As a preprocessing step the mean RGB value is subtracted from each pixel.
- The consequent layers are convolution and spatial pooling.
- In the pretrained model 17 convolution layers with stride 1 and 3 * 3 * 3 padding is utilized. In spatial pooling layers the window size is 2 * 2 * 2 and stride is 2.
- In pretrained model 5 max pooling layers are present as shown in Fig. 11. Softmax and ReLU are used as fully connected layers.

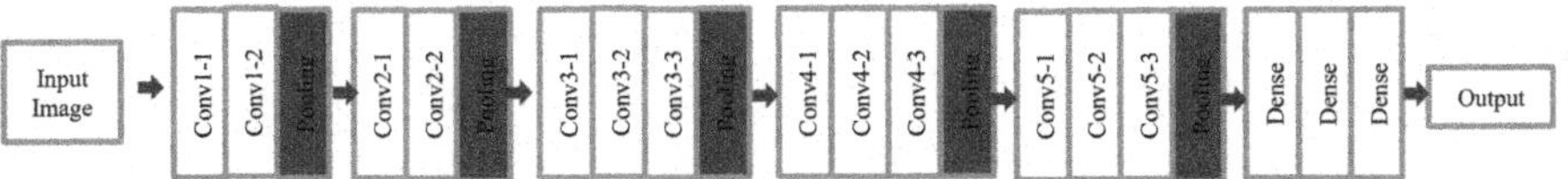

Fig. 11. Block diagram of VGG layers

4 Datasets

Selecting an appropriate dataset is the most crucial part of a self-acting system for identification of diseases in leaf. The quality and number of images of dataset direct influences the capabilities of the model. There are some standard datasets related to leaf disease which can be used for research purposes. Some noteable datasets are discussed below and information about datasets are mentioned in Table 2:

- **PlantVillage project dataset:** The dataset contains 54306 images of size 256 × 256 with labels for 38 different classes, like apple healthy, grape black rot, apple apple scab, and so on [34]. This dataset is available in three separate formats: colour, greyscale, and segmented.
- **UCI Leaf Data Set (Machine learning uci):** The dataset was created by sliva et al. (2013) [15]. This data contains 443 images of 40 different classes of plants. This dataset consists of texture and shape features of images.

Table 2. Information of datasets

Dataset	# of classes	# of images
PlantVillage dataset [34]	38	54306
UCI leaf dataset [15]	40	443

5 Limitations

From the literature reviewed it can be observed that most of the research work is based on an individual classifier. The advantages and limitations of individual classifiers are mentioned in Table 3.

In order to solve limitations of a single classifier, hybrid approaches can be used. It can be also seen that most of the studies are based on a single crop therefore, studies demonstrating the accuracy of a framework on different crops are required to comment upon the general usability.

Table 3. Advantages and limitations of classifiers used in literature

Technique	Advantages	Limitations
Decision tree [23]	Easy to interpret	Overfitting
Genetic algorithm [24]	Less computational efforts for optimum results	Complex structure
K-means clustering technique [25]	Fast convergence	Quality of convergence depends upon initial set of clusters
KNN [28]	Low learning cost	Complex interpretation
Naive bayes classifier [26]	Simple, accurate, agile	Requires strong assumptions on shape of data
Support vector machine [27]	Robust against bias	Low efficiency with high dimensional data
CNN [31]	High performance	Requires large amount of dataset
RNN [32]	Less training time on complex parameters	Training outcome can be non-deterministic in certain cases

6 Conclusion

Numerous machine learning-based initiatives to detect leaf diseases can be found in the literature. Several significant challenges, such as the utilization of real time information, automatically background removal, and uses of mobile devices for training and testing of CNN with number of parameters, are still significant research fields. All deep neural networks reviewed in this study used for feature extraction and classification purpose. Convolutional neural networks enable automated plant disease detection systems to identify and classify plant diseases by combining the expertise of plant pathologists with the capacity to extract symptomatic characteristics. The complexity of real-world problems and makes semantically categorizing data into representative datasets with large, labeled samples more difficult. This issue comes as a substantial and severe obstacle in the development of more practical machine learning systems.

Our analysis reveals considerable advancements in the application of CNN in the prediction of plant disease. Many approaches are increasingly giving new CNN architectures built upon the process of recognizing diseases of a plant. However, it is worth noting that most of the models discussed above employ many training parameters. With a high processing capacity, ample computation time is required to train a model with many training parameters. This prompted the researchers to experiment with transfer learning models to lower the complexity of time without compromising the accuracy of classification. It can be observed from the literature that transfer learning models demonstrate better disease recognition efficiency as compared to machine learning models, which are trained from scratch. It was also found that the research community is inclined towards some models; for instance, VGG16 is the most widely used model for leaf disease identification. Other transfer learning models such as Alex net, Resnet50, and GoogleNet are not frequently used. The biasness is also evident in crop selection as most of the research work on plant disease is based on crops such as rice, maize, and corn; fruits like apple, tomatoes, oranges are seldom used. We can work on these types of plant disease by using hybrid deep learning models and we will create a new dataset for this work.

References

1. Valliammai, N., Geethaiakshmi, S.N.: Multiple noise reduction using hybrid method for leaf recognition. In: 2012 International Conference on Devices, Circuits and Systems (ICDCS), 15–16 March 2012, Coimbatore, India (2012)
2. Akhtar, A., Khanum, A., Khan, S., Shaukat, A.: Automated plant disease analysis (APDA): performance comparison of machine learning techniques. In: Proceedings - 11th International Conference on Frontiers of Information Technology, FIT 2013, pp. 60–65 (2013). https://doi.org/10.1109/FIT.2013.19
3. Gavhale, K.R., Gawande, U.: An overview of the research on plant leaves disease detection using image processing techniques. IOSR J. Comput. Eng. **16**, 10–16 (2014)
4. Lu, J., et al.: An in-field automatic wheat disease diagnosis system. Comput. Electron. Agric. **142**, 369–379 (2017)
5. Kaur, P., Singla, S.: A review on plant leaf disease detection techniques. Int. J. Innov. Eng. Tech. **7**, 539–543 (2016)
6. Singh, V., Mishra, A.K.: Detection of plant leaf diseases using image segmentation and soft computing techniques. Inf. Process. Agric. (2016). http://dx.doi.org/10.1016/j.inpa.2016.10.005
7. Padmavathi, K., Thangadurai, K.: Implementation of RGB and grayscale images in plant leaves disease detection – comparative study. Indian J. of Sci. and Tech. **9**, 1–6 (2016)
8. Mohanty, S., Hughes, D., Salathe, M.: Using deep learning for image-based plant disease detection. Front. Plant Sci. **7**, 1419 (2016). https://doi.org/10.3389/fpls.2016.01419
9. Lu, Y., Yi, S., Zeng, N., Liu, Y., Zhang, Y.: Identification of rice diseases using deep convolutional neural networks. Neurocomputing **267**, 378–384 (2017)
10. Bakhshipour, A., Jafari, A.: Evaluation of support vector machine and artificial neural networks in weed detection using shape features. Comput. Electron. Agric. **145**, 153–160 (2018)
11. Pardede, H.F., Suryawati, E., Sustika, R., Zilvan, V.: Unsupervised convolutional autoencoder-based feature learning for automatic detection of plant diseases. In: 2018 International Conference on Computer, Control, Informatics and its Applications (IC3INA), pp. 158–162. IEEE, Tangerang (2018). https://doi.org/10.1109/IC3INA.2018.8629518
12. Park, K., Ki Hong, Y., Hwan Kim, G., Lee, J.: Classification of apple leaf conditions in hyper-spectral images for diagnosis of Marssonina blotch using mRMR and deep neural network. Comput. Electron. Agric. **148**, 179–187 (2018)
13. Gan, H., Lee, W.S., Alchanatis, V., Ehsani, R., Schueller, J.K.: Immature green citrus fruit detection using color and thermal images. Comput. Electron. Agric. **152**, 117–125 (2018)
14. Liang, Q., Xiang, S., Hu, Y., Coppola, G., Zhang, D., Sun, W.: PD2SE-net: computer assisted plant disease diagnosis and severity estimation network. Comput. Electron. Agric. **157**, 518–529 (2019)
15. Kaya, A., Keceli, A.S., Catal, C., Yalic, H.Y., Temucin, H., Tekinerdogan, B.: Analysis of transfer learning for deep neural network based plant classification models. Comput. Electron. Agric. **158**, 20–29 (2019)

16. Too, E.C., Yujian, L., Njuki, S., Yingchun, L.: A comparative study of fine-tuning deep learning models for plant disease identification. Comput. Electron. Agric. **161**, 272–279 (2019)
17. Sanga, S.L., Machuve, D., Jomanga, K.: Mobile-based deep learning models for Banana disease detection. Technol. Appl. Sci. Res. **10**(3), 5674–5677 (2020)
18. Chohan, M., Khan, A., Katper, S., Mahar, M.: Plant disease detection using deep learning. Int. J. Recent Technol. Eng. **9**(1), 909–914 (2020). https://doi.org/10.35940/ijrte.A2139.059120
19. Kumar, T., Mahrishi, M., Meena, G.: A comprehensive review of recent automatic speech summarization and keyword identification techniques. In: Fernandes, S.L., Sharma, T.K. (eds.) Artificial Intelligence in Industrial Applications. LAIS, vol. 25, pp. 111–126. Springer, Cham (2022). https://doi.org/10.1007/978-3-030-85383-9_8
20. Khamparia, A., Saini, G., Gupta, D., Khanna, A., Tiwari, S., de Albuquerque, V.H.C.: Seasonal crops disease prediction and classification using deep convolutional encoder network. Circ. Syst. Sign. Proc. **39**, 818–836 (2020). https://doi.org/10.1007/s00034-019-01041-0
21. Hasan, R., Palaniappan, S., Raziff, A.R.A., Mahmood, S., Sarker, K.U.: Student academic performance prediction by using decision tree algorithm. In: 2018 4th International Conference on Computer and Information Sciences (ICCOINS), pp. 1–5 (2018). https://doi.org/10.1109/ICCOINS.2018.8510600
22. Mirjalili, S.: Genetic algorithm. In: Mirjalili, S. (ed.) Evolutionary Algorithms and Neural Networks. SCI, vol. 780, pp. 43–55. Springer, Cham (2019). https://doi.org/10.1007/978-3-319-93025-1_4
23. Tete, T.N., Kamlu, S.: Detection of plant disease using threshold, k-mean cluster and ann algorithm. In: 2017 2nd International Conference for Convergence in Technology (I2CT), pp. 523–526 (2017). https://doi.org/10.1109/I2CT.2017.8226184
24. Islam, T., Sah, M., Baral, S., Roy Choudhury, R.: A faster technique on rice disease detection using image processing of affected area in agro-field. In: 2018 Second International Conference on Inventive Communication and Computational Technologies (ICICCT), pp. 62–66 (2018). https://doi.org/10.1109/ICICCT.2018.8473322
25. Mia, M.R., Roy, S., Das, S.K., Rahman, M.A.: Mango leaf disease recognition using neural network and support vector machine. Iran J. Comput. Sci. **3**(3), 185–193 (2020). https://doi.org/10.1007/s42044-020-00057-z
26. Guo, C., Zhang, J., Liu, Y., Xie, Y., Han, Z., Yu, J.: Recursion enhanced random forest with an improved linear model (RERF-ILM) for heart disease detection on the internet of medical things platform. IEEE Access **8**, 59247–59256 (2020). https://doi.org/10.1109/ACCESS.2020.2981159
27. Vaishnnave, M.P., Devi, K.S., Srinivasan, P., Jothi, G.A.P.: Detection and classification of groundnut leaf diseases using KNN classifier. In: 2019 IEEE International Conference on System, Computation, Automation and Networking (ICSCAN), pp. 1–5 (2019). https://doi.org/10.1109/ICSCAN.2019.8878733
28. Khan, M.U., Aziz, S., Bilal, M., Aamir, M.B.: Classification of EMG signals for assessment of neuromuscular disorder using empirical mode decomposition and logistic regression. In: 2019 International Conference on Applied and Engineering Mathematics (ICAEM), pp. 237–243 (2019). https://doi.org/10.1109/ICAEM.2019.8853684
29. Colbaugh, R., Glass, K., Rudolf, C., Tremblay, M.: Robust ensemble learning to identify rare disease patients from electronic health records. In: 2018 40th Annual International Conference of the IEEE Engineering in Medicine and Biology Society (EMBC), pp. 4085–4088 (2018). https://doi.org/10.1109/EMBC.2018.8513241
30. Kumari, C.U., Jeevan Prasad, S., Mounika, G.: Leaf disease detection: feature extraction with K-means clustering and classification with ANN. In: 2019 3rd International Conference on Computing Methodologies and Communication (ICCMC), pp. 1095–1098 (2019). https://doi.org/10.1109/ICCMC.2019.8819750

31. Ghosal, S., Sarkar, K.: Rice leaf diseases classification using CNN with transfer learning. In: 2020 IEEE Calcutta Conference (CALCON), pp. 230–236 (2020). https://doi.org/10.1109/CALCON49167.2020.9106423
32. Jayakumar, D., Elakkiya, A., Rajmohan, R., Ramkumar, M.O.: Automatic prediction and classification of diseases in melons using stacked RNN based deep learning model. In: 2020 International Conference on System, Computation, Automation and Networking (ICSCAN), pp. 1–5 (2020). https://doi.org/10.1109/ICSCAN49426.2020.9262414
33. Mahrishi, M., Hiran, K.K., Meena, G., Sharma, P. (eds.): Machine Learning and Deep Learning in Real-Time Applications. IGI Global (2020). https://doi.org/10.4018/978-1-7998-3095-5
34. Patil, K.J., Kumar, R.: Advances in image processing for detection of plant diseases. J. Adv. Bioinf. Appli. Res. **2**, 135–141 (2011)
35. Adedoja, A., Owolawi, P.A., Mapayi, T.: Deep learning based on NASNet for plant disease recognition using leave images. In: 2019 International Conference on Advances in Big Data, Computing and Data Communication Systems (icABCD), pp. 1–5 (2019). https://doi.org/10.1109/ICABCD.2019.8851029
36. Prajapati, H., Shah, J., Dabhi, V.: Detection and classification of rice plant diseases. Intell. Decis. Technol. **11**, 357–373 (2017)
37. Xia, Y., et al.: Intelligent diagnose system of wheat disease based on android phone. J. Inf. Comput. Sci. **12**(18), 6845–6852 (2015)

Detection of Network Intrusion Using Machine Learning Technique

L. K. Joshila Grace(✉), P. Asha, Mercy Paul Selvan, L. Sujihelen, and A. Christy

Department of Computer Science and Engineering, Sathyabama Institute of Science and Technology, Chennai, India
joshilagracejebin@gmail.com, {asha.cse,mercypaulselvan.cse, christy.cse}@sathyabama.ac.in

Abstract. The aim is to estimate a Windows device's risk of being corrupted by different malware families based on the machine's various properties. By merging movement and studies of threats obtained by Microsoft's endpoint security solution, Windows Defender, the simple data containing these properties and system infections was created. A Machine Identifier uniquely identifies each row throughout this dataset, which relates to a machine. The similarity measure has Detections, which means that Malware was found on the computer. You should estimate the value for Detections of each system in test.csv using the details and labels in train.csv. Using the Malware data collection, this analysis tests the output of the new proposed classifier algorithm. That proposed updated version of the Random forest classifiers provides improved outcomes in case of intrusion detection and false incidence, according to the empirical findings of our research.

Keywords: Security · Gradient boosting (GB) · Intrusion detection system (IDS) with feature selection · Malware data set · Unauthorized · Attacks

1 Introduction

For the past few years, the network has played a significant role in communication. The computer network allows the computing network devices to exchange information among different systems and individuals. Throughout computer networks, the resources of different organizations, corporations, schools, universities are accessed. This leads to massive growth in the networking field [12]. There has been a lot of interest among individuals in internet accessibility. In this context, the security of information has become a great challenge in this modern area. The information or data that we would like to send is supposed to be secured in such a way that a third party should not take control over them.

When we are talking about security, we have to keep three basic factors in our mind: Confidentiality, Integrity, and availability. Confidentiality means the privacy of information. It gives the formal users the right to access the system via the internet.

To recognize the authorized persons, this can be achieved properly along with monitoring programs. The second key factor is integrity. The integrity service means exactness

V. E. Balas et al. (Eds.): ICETCE 2022, CCIS 1591, pp. 373–382, 2022.
https://doi.org/10.1007/978-3-031-07012-9_32

of the information. It helps customers to include self-assurance that the data transmitted is acceptable and that an illegal person has still not altered it. An Intrusion Detection System (IDS) is used to watch malicious activities over the network [1].

The unfamiliar records can be sorted as normal or attack classes. First monitoring of the network traffic is done, and then the IDS sorts these network traffic records into either malicious class or regular class [9]. It acts as an alarm system that reports when illegal activity is detected. The exactness of the IDS depends upon the detection rate [8]. For the IDS, if the performance is better, so the identification accuracy is also greater. Some intrusion prevention devices appear to be able to detect and prevent attacks until they succeed. They are used to shield an association from attack. It is a relative concept that tries to identify a hacker when intrusion is attempted. Probably, when a successful attack is made, such a system will only alarm. The intrusion detection system is not an ideal solution for all types of attacks. The various goals that can be accomplished with an Intrusion Detection System are: The potential goals include the following: IDS detects threats, monitors user interaction from the point of entry, and issues warnings if appropriate. Detect design failures in the device [2, 5]. IDS can predict when the device is under attack and provides system protection without the need for non–expert personnel. Proof of intrusion is given.

2 Related Work

The Internet has just lately become an integral part of daily life. Accurate data processes dependent on the internet are vulnerable to a variety of attacks, which may result in significant losses as a result of multiple forms of harm. The value of information security is, therefore, evolving rapidly. With the rapid growth in the past two decades of information technology. Computer networks are commonly used in various fields of manufacturing, business and human life. Many researchers used the competition for intrusion detection to evaluate the use of machine learning for malware detection and find a data rate of around 91%, with false positive rates of just under 1%. One can also sanitize captured data by, e.g., removing or potentially sensitive information [4]. The important features for classification are selected using the feature selection process. A data collection containing only the most critical elements increases the model's acceptability and accuracy.

Domain and algorithm characteristics should be considered when selecting features [3]. In fact, the only way to know how well a model can generalize to new cases is to test it on new cases. Putting the model in development and tracking how well it works is one way to do that. Splitting the data into two sets is a safer option: the testing set and the evaluation set are two different types of sets. The training set is used to prepare your model, while the test set is used to test it.

The generalization error (or out-of-sample error) is related to the error value for cases registered, and you can approximate this error by testing your design on the test range. Machine learning can appear in many guises.We now address a variety of programs, the types of data they deal with, and eventually, in a much more stylized way, we formalize the issues [3].

For offline batch operation, many machine learning algorithms are better suited, and less so for settings involving real-time detection of low-latency. Non-machine learning detectors are also much simpler to incorporate even at high data rates in a streaming fashion [4]. Alternatively, much of the aim of machine learning is to eliminate a number of very diverse issues to a collection of pretty small models. The science of machine learning then focuses on solving certain problems and ensuring that the solution is valid. In their tests, both studies used large datasets and the findings revealed that Light BGM has strong advantages over conventional logistic regression models, such as high efficiency, reliability and usability. Also, the literature states that "Gradient Boosting Decision Tree (GBDT), due to its reliability and interoperability, is a commonly used machine learning algorithm".

Information management devices that are now dependent on the internet are insecure to a variety of attacks, resulting in a variety of damages and serious losses [10]. As a result, the value of information security is becoming exceedingly important [5].

The most fundamental objective of information protection would be to establish protective information technology that is protected against unauthorized access, usage, exposure, interference, alteration, or destruction [5].

In particular, data protection reduces the risks associated with 3 primary security principles of secrecy, integrity, and usability [2]. In the past, numerous systems were developed to detect and prevent Web attacks. Intrusion detection systems (IDS) are the most critical systems because they actively overcome malicious activity. Besides, IDS have a barrier of protection that amplifies Web attacks on personal computers. IDS could be used to identify various forms of threats where the conventional firewall does not work well on network communications and computer device use [5]. Detection of intrusions is based on the premise that the conduct of intruders differs from those of an authorized user [1]. Based on their identification methods, IDS can be Anomaly detection systems and misuse detection systems are divided into two groups [3]. Detection of abnormal breach decides whether it would be possible to flag deviations from the existing normal use behaviors as cyber attacks. Misinterpretation monitoring systems, but on the other side, accurately solve breaches of permits. Using these smart operatives and classification algorithm intrusion detection systems can be developed. Most IDS operate in two stages, which include the data pre-processing stage and the recognition stage of intrusion [3].

3 Detecting Intrusions in the Network

3.1 Dataset

The objective of the dataset, which is derived from the Malware dataset, is to calculate the probability of a device becoming corrupted with a variety of ransom ware based on the machine's various attributes. - The computer in this dataset has its own lines, with a unique "Machine Identifier" to identify it. The attribute "Has-Detections" indicates whether or not Malware has been detected on the device [11]. In addition, the dataset contains approximately 3 lakh rows and 83 parameters, including, among other things, hardware, device settings, and anti-virus security information. The data set is split into a test data set and a train data set in a 2:1 ratio in terms of picking a best standard.

3.2 Method

Data Cleaning: Since the data is large and takes a long time to load, the work started by reducing the amount of memory available and deploying the data to a lower form of data. Float 64 to float 32, for example. Memory use was then decreased by 4%.

Next, for each sector, this work rendered statistics on the Nan value. A field whose Nan percentage value was greater than 95% was deleted and the Nan value was replaced by an unoccupied integer value in the remaining fields.

3.2.1 Feature Selection

From the initial set of attributes, feature selection chooses a complex set of parameters [2]. This requires lesser training to understand the supervised learning and generates a broader classification, as it excludes redundant and irrelevant characteristics from its initial list.

Feature selection also enables visualization of data and interpretation of data [6]. Instead of measuring them precisely at the expense of deteriorated efficiency, time and space complexities, most real life issues certainly require an efficient and appropriate solution. It was therefore necessary for the study to be carried out using selected features.

The issue of choosing prominent variables from a Malware dataset for intrusion detection was too difficult to be expressed by a formula [1]. In addition, when all functions are used without feature filtering, measuring a solution precisely takes a long time. As a result, using a heuristic method is a viable option that efficiently performs the selection features.

Proposed Classifier:

Gradient boost is an approach in machine learning to resolve problems of regression and classification. As a result, a prediction model emerges from a number of poor prediction models in the Associate set [10]. One of the most importantlearning principles implemented in the last twenty years are boosting based algorithms. It has been developed to address problems with grouping, but can also be used to resolve problems with regression. A method which combines the results of a few "poor" classifiers generating a powerful "committee" has been used to improve gradients. It employs a poor classification algorithmic program to repeatedly alter versions of data, resulting in the creation of a "sequence of weak classifiers $G _ m(x)$, $m = 1, 2, …, M$".

The AdaBoost model recognizes vulnerabilities by using large data sets, while gradient boosting will do the same through using loss function gradients ($y = ax + b + e$, with e being especially important since it is the expression for error). The loss function is a calculation showing how well the coefficients of the model match the source data. a model F which tries to predict $y\hat{} = F(x)$ by minimizing the mean squared error(MSE). There is a certain poor model Fm at each stage of building an ensemble of M decision trees, where $1 \leq m \leq M$. The GBM algorithm builds the next tree $Fm + 1$ as $Fm + 1 = Fm(x) + h(x)$.

3.2.2 Confusion Matrix

A confusion matrix is a summary that shows the outcomes of classification problem estimation [7]. By counting values, the number of precise and incorrect predictions is compiled and broken down by class. This is the solution to the confusion matrix. When it makes predictions, the confusion matrix reveals the ways in which the classification model is confused. It gives one insight into not just the kinds of errors produced by a classifier, but also the types of errors made by a classifier.

Classification Rate/Accuracy:

$$\textbf{Accuracy} = \frac{TP + TN}{TP + TN + FP + FN}$$

Recall:

$$\textbf{\textit{Recall}} = \frac{TP}{TP + FN}$$

$$\textbf{\textit{Precision}} = \frac{TP}{TP + FP}$$

Precision:

Actual Values

Predicted Values	Positive (1)	Negative (0)
Positive (1)	TP	FP
Negative (0)	FN	TN

3.2.3 ROC Curve

Another common approach for binary classifiers appears to be the receiver operating characteristic (ROC) curve. The ROC curve is analogous to the accuracy/recall curve, except it contrasts the true positive rate (also called recall) with the false positive rate, instead of plotting accuracy against recall. The FPR is the ratio of negative cases that are wrongly considered positive. This is one minus the real negative rate, which is the percentage of correctly classified negative events. TNR is also known as precision. As a consequence, the ROC curve plots sensitivity (recall) rather than one precision. To start, use the roc curve() function to measure the TPR and FPR for various threshold values, to plot the ROC curve.

4 Results and Discussion

The two modals were trained using the given dataset. The ROC is seen in Fig. 3 with the KNN model. Though Fairuz Amalina Narudin's KNN model was also highly predictive of mobile malware, the highest KNN result of KNN was 0.56 and the Light GBM accuracy was 0.64. The research results are obviously unexpected, given that the dataset contains several nominal fields and the marks have no numerical value. KNN, On the other hand, since they rely on the distance from various samples of the data, they can't fit our dataset.

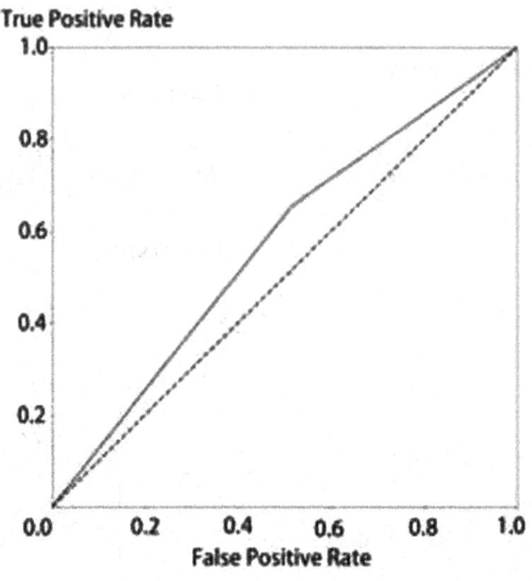

Fig. 1. ROC for KNN

Figure 1 displays the ROC based on the True Positive and False positive rate that narrates the specificity and sensitivity. Here, the ROC obtained for KNN classifier has the curves closest to the top-left corner, which acts as the best positive indication in terms of fit.

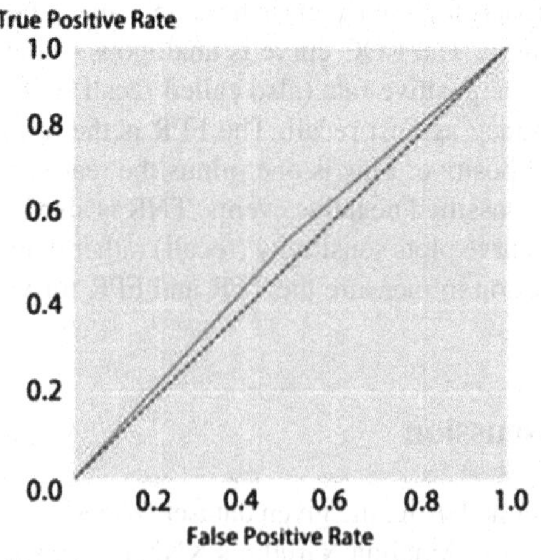

Fig. 2. ROC for light GBM

Figure 2 displays the ROC based on the True Positive and False positive rate for gradient boosting decision tree that narrates the specificity and sensitivity. This Classifier has produced a larger AUC (Area under Curve), which concludes that the proposed system has achieved good predictive accuracy.

```
Classification Report for Light GBM:
              precision    recall  f1-score   support

           0       0.64      0.65      0.65     49659
           1       0.64      0.64      0.64     49341

    accuracy                           0.64     99000
   macro avg       0.64      0.64      0.64     99000
weighted avg       0.64      0.64      0.64     99000

Confusion Matrix of Light GBM:
 [[32112 17547]
 [17764 31577]]
```

Fig. 3. Classification report and confusion matrix

Figure 3 displays the Confusion matrix obtained for gradient boosting decision tree in which the average accuracy rate is 64%, which is quite a better value.

```
In [23]: model.score(X_test,y_test)

Out[23]: 0.6428989898989899
```

Fig. 4. Accuracy of light GBM

When compared to KNN, GBM has gained better accuracy (Fig. 4).

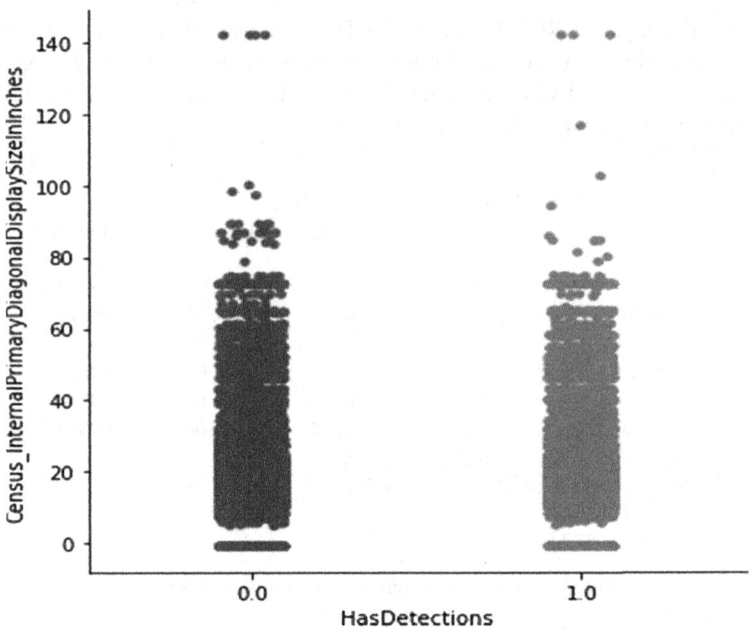

Fig. 5. "Censes_InternalPrimaryDiagnal" is associated with intrusion

Figure 5 depicts the amount of intrusion that has been experienced by the proposed system.

	feature	importance	normalized_importance	cumulative_importance
0	index	131677	0.069013	0.069013
1	AvSigVersion	130754	0.068529	0.137542
2	CityIdentifier	117879	0.061781	0.199324
3	Census_InternalPrimaryDiagonalDisplaySizeInInches	109288	0.057279	0.256603
4	Census_SystemVolumeTotalCapacity	105315	0.055197	0.311799
...	...	...	...	...
65	Census_IsAlwaysOnAlwaysConnectedCapable	158	0.000083	0.999948
66	OsVer	100	0.000052	1.000000
67	Census_DeviceFamily	0	0.000000	1.000000
68	Census_IsPortableOperatingSystem	0	0.000000	1.000000
69	SMode	0	0.000000	1.000000

70 rows × 4 columns

Fig. 6. Some of selected features

Figure 7 displays the intrusions experienced by the system based on few selected features (Fig. 6).

	MachineIdentifier	HasDetections
0	0000010489e3af074adeac69c53e555e	1
1	00000176ac758d54827acd545b6315a5	0
2	0000019dcefc128c2d4387c1273dae1d	0
3	0000055553dc51b1295785415f1a224d	0
4	00000574cefffeca83ec8adf9285b2bf	0

Fig. 7. Output whether machine is intruded or not

5 Conclusion

Our system for processing the "Malware Dataset" was presented in this report, along with various models that were tested and a focus on a few key fields. It was discovered that LightGBM had a higher accuracy when trained on a dataset with a range of features. However, owing to a variety of null data and unimportant functions, the AUC score is still less than 80%. The official research dataset should be used in the future and various methods could be used to compress the available values of those fields so that Chi-square is more valid in certain fields.

References

1. Kayack, H.G., Zincir-Heywood, A.N., Heywood, M.: Selecting features for intrusion detection: a feature relevance analysis on KDD 99 intrusion detection datasets. In: Third Annual Conference on Privacy, Security and Trust (2005)
2. Balakrishnan, S., Kannan, V.: Intrusion detection system using feature selection and classification technique. Int. J. Comput. Sci. Appl. **3**(4), 145–151 (2014)
3. Kok, S.H., Abdullah, A., Jhanjhi, N.Z., Supramaniam, M.: A review of intrusion detection system using nn and machine learning technique. Int. J. Eng. Res. Technol. **12**(1), 8–15 (2019)
4. Sommer, R., Paxson, V.: Outside the closed world: on using machine learning for network intrusion detection. In: IEEE Symposium on Security and Privacy, pp. 305–316 (2010)
5. Karatas, G., Sahingoz, O.K.: Neural network based intrusion detection systems with different training functions. In: 2018 6th International Symposium on Digital Forensic and Security (ISDFS), pp. 1–6 (2018)
6. Kabir, M.M., Islam, M.M., Murase, K.: Using a neural network, a new wrapper feature selection approach. Neuro Comput. **73**, 3273–3283 (2010)
7. Meena, G., et al.: performance comparison of network intrusion detection system based on different pre-processing methods and deep neural network. In: Proceedings of the International Conference on Data Science, Machine Learning and Artificial Intelligence, pp. 110–115 (2021)

8. Amiri, F., Yousefi, M.M.R., Lucas, C., Shakery, A., Yazdani, N.: Mutual information based feature selection for intrusion detection systems. J. Netw. Comput. Appl. **34**, 1184–1199 (2011)
9. Liu, H., Yu, L.: Towards combining grouping and clustering feature collection algorithms. IEEE Trans. Knowl. Data Eng. **17**, 491–502 (2005)
10. Scarfone, K., Mell, P.: Guide to intrusion detection and prevention systems (IDPS). National Institute Of Standards and Technology, February 2007
11. Shabu, S.L.J., Refonaa, J., Maran, S., Dhamodaran, S., Vedanarayanan: An integrated and dynamic wireless intrusion exposure solutions based on neural network. J. Phys.: Conf. Ser. **1770**(1), 012016 (2021)
12. Refonaa, J., Shabu, S.L.J., Dhamodaran, S., Grace, L.K.J., Vedanarayanan: To enhance security mechanism in ATM account using biometric system. J. Phys.: Conf. Ser. **1770**(1), 012017 (2021)

Scrutinization of Urdu Handwritten Text Recognition with Machine Learning Approach

Dhuha Rashid(✉) and Naveen Kumar Gondhi

Shri Mata Viashno Devi University, Katra, Jammu and Kashmir 182320, India
dhuha.rashid@gmail.com, naveen.gondhi@smvdu.ac.in

Abstract. Urdu Nastaliq script is a cursive and ligature-based script that is very difficult to recognize and detect using current computer algorithms dues to its complicated writing structure . Handwritten text imposes much more difficulty since all writers have their own writing styles. Many researchers have worked in this field on various aspects, from creating datasets to improving the recognition rate for its text. This paper is a literature survey on the work done for Urdu handwritten text recognition by optical character recognition using deep learning and machine learning techniques. It also compares the accuracy rates of respective algorithms over different datasets.

Keywords: Handwritten recognition · Machine learning · Urdu

1 Introduction

Urdu is a language that primarily belongs to the Indo-European category of languages. It is based on Nastaliq script which is a combination of two scripts viz; "Naskh" which is a script for writing Islamic calligraphy and "Taliq" which is also a script of Islamic calligraphy specifically for writing the Persian language.

The Urdu language is currently spoken by approximately 100 million people around the world with most of its native speakers from India, Pakistan, and other parts of Southeast Asia. It shares the phonology and grammar with Hindi language but the origin of the lexicons for both the languages are different i: e, Persian and Arabic for Urdu and Sanskrit for Hindi which makes them two separate languages. Its handwritten script is cursive and most of the characters form ligatures. It gets difficult to read this script and convert them to digital form because of these ligatures. For example, a letter 'چ' has four ligature forms as 'چ' and 'ﭼ'،'ﭽ'،' ﭻ'. They make it extremely difficult and complex to identify individual words in this language. Thus, there is a need for a good recognition system with high accuracy like KJv1 which can identify the script accurately. This identification can be very useful in document automation, preserving old Urdu literature, transliteration, and other aspects where Urdu language is used.

V. E. Balas et al. (Eds.): ICETCE 2022, CCIS 1591, pp. 383–394, 2022.
https://doi.org/10.1007/978-3-031-07012-9_33

2 Language and Dataset

2.1 Urdu Language and Script

Urdu language dates back to the 14th century, originating in Dakkan (Deccan). Its literary composition started at that time but was limited to the religious theme. Later its usage spread over to the north of India in the 15th century. A lot of Urdu Literature, manually written by people, is critically valuable but unarchived. Any OCR with high accuracy will dramatically help to preserve such handwritten Urdu literature.

Urdu script is an 'abjad' script where each symbol or glyph represents a consonant and it is up to the reader to derive or specify the appropriate vowel. Short consonants and long vowels are expressed by letters but short vowels and consonant length are not generally written. Also, it is cursive where the characters form different forms of ligatures on joining with other characters and multiple ligatures in combination to each other, form varieties of words. This property in linguistics is known as compound character.

In these modern times, attempts are being made to read all of the handwritten data computationally using various techniques like machine learning and deep learning. Tremendous improvements have been made in languages like English, Chinese, Korean, etc., and pushed their accuracies to around 99%. Their practical implications can be seen in daily lives from document automation to camera recognition of their letter patterns. Urdu language on the other hand is yet to achieve any such milestone. It is due to the above-stated complicacies in their writing and different ligature forms. Despite these complicacies, many researchers are working towards Urdu Character Recognition using various computational techniques.

2.2 Character Recognition System

Optical Character Recognition. Optical Character Recognition or OCR is a technology that recognizes characters, digitally takes handwritten letters as input, identifies them, and generates a computerized output form. The output produced is a format easily understandable by machines.

Reformulating the above statements, OCR is converting the image to its electronic or mechanical counterpart. From document automation to biometrics, OCRs show many practical functions and usages. Various offices and libraries have been digitized using this technology. Such great usage generates more demand for advanced OCRs in every field including one for Urdu Character recognition. Many old scripts and artifacts demand digitization and are waiting for improvements in Urdu OCRs.

Intelligent Character Recognition. Intelligent Character Recognition (ICR) is a software technology that can recognize the handwritten character of any style or font. It is much more versatile and advanced than OCRs. It specifically deals with handwritten recognitions. There are some issues in ICR such as a change in font, the slope of the line, different writing style even from a single writer, overlapping joining letters, missing placement of dots and diacritics aka secondary strokes, etc. that make the process of ICR more challenging than the recognition of the printed text [1].

Datasets. Various Datasets used for identification of Urdu text used are stated below. Their comparison is stated in Table 1.

UPTI. Urdu Printed Text Image (UPTI) database consists of 3645 classes and a large dataset of about 189003 ligatures. The total number of writers in this dataset is 250. To fetch the degraded images from the synthetic image, four degradation models have been used which manage jitter, elastic elongation, thresholding, and sensitivity. It was put forward by Nazly Sabbour and Faisal Shafait [2].

UNHD. Urdu Nastaliq Handwritten Dataset (UNHD) contains 312000 words written by 500 candidates. The number of words written by a single writer is approximately 624. It contains ligatures of up to five characters. It contains digits, nouns, pronouns, numbers, etc. of Urdu script. Noise removal, grayscale conversion, labeling, and annotation are some of the pre-processing steps done on this dataset. This dataset is publicly available and proposed by Salahuddin et al. [3].

CALAM. Cursive And Language Adaptive Methodologies (CALAM) is a multi-lingual database that includes the Urdu language [4]. Apart from Urdu, it includes a dataset for Hindi and English languages as well. This database is a full-length text sentence. This dataset consists of 46,664 words and 101,181 ligatures written by 725 writers.

HUCD. Handwritten Urdu Character Dataset (HUCD) consists of 106,120 samples of handwritten Urdu characters. This dataset was created by F. Mushtaq in which a sample was taken from 750 native writers of the Kashmir valley. They have divided the characters of Urdu into two groups, joiners, and non-joiners. 27 characters out of 36 have been categorized in joiners, 10 non-joiners and 1 character show no joining property at all. Also, no special characters have been used in this dataset [5].

Table 1. I Different datasets with their sizes

Dataset	Year	Reference	Size
UPTI	2013	[2]	189003 ligatures 3645 classes
UNHD	2017	[3]	312000 words
CALAM	2015	[4]	46,664 words 101,181 ligatures
HUCD	2021	[5]	106,120 samples

3 Problems In Recognition

While making attempts to ease the process of recognition of Urdu Handwritten characters, many difficulties occur in its recognition. This is because of many problems arising due to the complexities in writing the Urdu Script. One of the complexities arises by writing Urdu with a special pen with a slant-cut nib called ‘qat’ which gives a variety of width to the letter. Some of the intricacies are:

Table 2. Diagram illustrations

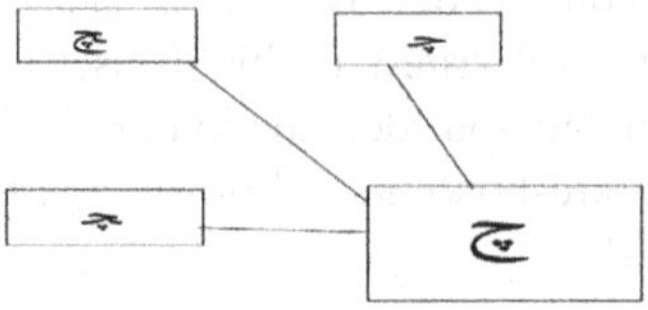

Diagram	Caption
	Fig. 1. Various Ligature Forms
 شاباشی	**Fig. 2.** Diagonality
ہندوستان ۱۴ اگست ۱۹۴۷ میں آزاد ہوا	**Fig. 3.** Bi-Directionality
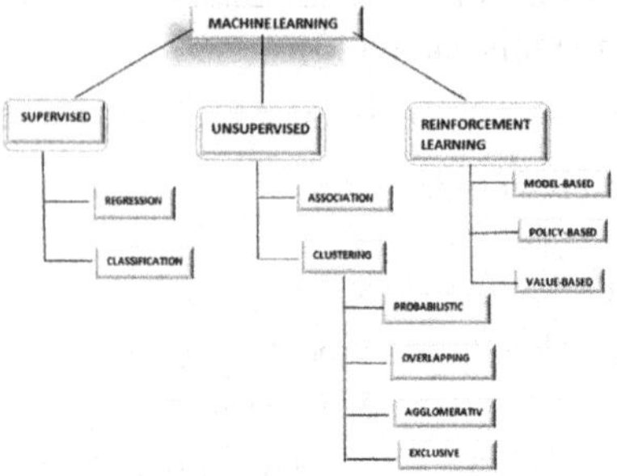 MACHINE LEARNING; SUPERVISED: REGRESSION, CLASSIFICATION; UNSUPERVISED: ASSOCIATION, CLUSTERING (PROBABILISTIC, OVERLAPPING, AGGLOMERATIV, EXCLUSIVE); REINFORCEMENT LEARNING: MODEL-BASED, POLICY-BASED, VALUE-BASED	**Fig. 4.** The Fig 4.1 Machine Learning and its types

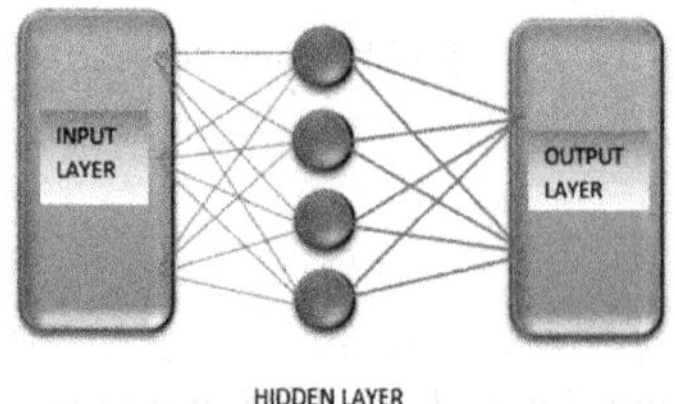

Fig. 5. Deep Learning Layers

3.1 Ligature Forms

When multiple forms of letters are combined to form a single glyph, it is a ligature. Urdu language comes with the property of forming ligatures. While forming the word, the letters are represented in different glyphs depending upon their position in the word.

For example, a letter 'چ' has four ligature forms as 'چ ' when written in the end 'چـ' when written in the beginning, 'ـچـ' when written in middle and 'چ' as an individual. It is depicted in Fig. 1.

3.2 Diagonality

When writing Urdu, some letters are written above the baseline while others go below it. This concept of writing gives it the diagonality i.e., the letters go diagonally from top right to bottom left while writing unlike English, Hindi in which the words are written along the baseline. Figure 2 shows the diagonality of the Nastaliq font.

3.3 Less Space

Nastaliq scripts are very less space-consuming. Each word is written very close to the following word creating grounds for confusion for segmentation and identification of each word individually. From Fig. 4, it can be seen that the words lack spaces in the sentence and create ambiguity in the identification.

3.4 Context Sensitivity

Writing Nastaliq font is context-sensitive. It means that the letters not only have many forms based on their position in the word but also depend on the letter it is preceded and followed by. For example, in both the words 'حج' and 'درج', the letter 'ج' is placed at the end but still takes up two different forms because of the letter that precedes it.

3.5 Diacritics

Writing Nastaliq font is context-sensitive. It means that the letters not only have many forms based on their position in the word but also depend on the letter it is preceded and followed by. For example, in both the words 'حج' and 'درج', the letter 'ج' is placed at the end but still takes up two different forms because of the letter that precedes it.

3.6 Bi-directionality

Being bi-directional means that it is two-directional. Nastaliq scripts have two directions for writing their script as represented in Fig. 3. The numerals are written from right to left while the words are written from left to right as mentioned earlier. It creates difficulties for machines to understand which way to go for the identification of the characters.

4 Learning Models for Recognition Systems

For recognition, various machine learning (ML) and deep learning (DL) models are used. Machine learning works on algorithms to parse data and learn from it to make future decisions. Deep learning on the other hand works on an artificially created neural network and makes intelligent decisions. The difference between the two is that ML still needs human interference and guidance but DL on the other hand once started completely works on its own with self-recovery in case of fault occurrences. Both ML and DL include various models and they are applicable for character recognition also. A researcher is supposed to choose a model according to its requirements.

4.1 Machine Learning

As specified earlier, machine learning is the learning of machines using a set of algorithms and code to be able to make independent decisions to some extent. The machines are first trained using some algorithms. An example for such a learning experience can be taken as an online music portal or YouTube platform which on the basis of one's viewed videos or music gives further similar recommendations using a set of algorithms.

Machine learning is classified into three groups based on the way it is trained (Fig. 4). These are: Supervised, Unsupervised and Reinforcement Learning.

In *Supervised learning*, machines are trained with 'labeled' data. A labeled data is one in which before the validation of data, some input data are associated with its correct output. That is, some data is labeled to correct output and using those data points it validates the data required. For example, we take a data set with images of pencils, erasers, sharpeners, and scales. The dataset will first be split into training and testing datasets. The images in the training dataset will be properly labeled to their respective labels. Then the image from the test dataset will be tested using this training dataset and the expected outcome is obtained.

Supervised learning is further divided into two categories viz, *classification* and *regression*. Classification algorithms are used in categorical training types of data. It means that if data can be presented only in two forms such as male or female, heads or tails, yes or no, classification algorithms are used. Some common algorithms of classification are Decision Tree, Random Forest, Logistic regression.

Support Vector Machine etc. On the other hand, Regression algorithms are used where the relationship between input and output variables is given e.g., weather predictions, food habits, and disease relationship, etc. Popular algorithms for regression are linear regression, Bayesians linear regression, polynomial regression, non-linear regression. Some algorithms can be used both for classification as well as regression such as k-NN, SVM, etc.

Unsupervised learning on the other hand works with 'unlabeled' data. That is the dataset is not labeled. It works without any supervision by finding some hidden patterns in the dataset. The input data does not have any corresponding output data, so the concept of regression and classification is not used directly. Instead, the idea of *clustering* and *association* is used. The data is clustered and associated with each other based on their similarities.

Take the examples of different kinds of tree leaves e.g., Populus, Chinar, Eucalyptus, and create the dataset from their image varieties without its prior labeling. Applying an unsupervised algorithm, will identify the similarities in the images and cluster them into groups. Some examples of unsupervised learning algorithms are K-mean clustering, k-NN, hierarchal clustering, NN, etc.

Reinforcement learning focuses on maximizing the reward. It learns from its previous experiences and is used by specific software to choose specific operations in particular situations which may yield the best output. It can be employed using value-based, policy-based, or model-based algorithms.

4.2 Deep Learning

Deep learning is a subset of machine learning and it learns exactly like the human brain. It creates a network called an artificial neural network which resembles a human neural network in its functioning. It uses neurons for the transfer of information just like the human brain. These neurons are also referred to as nodes sometimes. These artificial neural networks (ANN) form different mathematical models which solve complex problems using unstructured data. The neural network in deep learning is composed of three layers viz, input layer, hidden layer, and output layer (Fig. 5). Depending upon the type of data it uses for performing the calculations, there are various types of deep learning networks: Feedforward neural network, Radial basis function neural network, Long-short term, Convolution neural network, Autoencoder, Recurrent neural network, multi-layer perceptron.

5 Literature Review

Various researchers have worked in the field of handwritten Urdu text recognition. Multiple datasets were analyzed under different deep learning and machine learning algorithms and their accuracies are measured (Table 3).

5.1 Supervised Learning

Khalil Khan et al. in their paper [6] have used the decision tree model for Urdu text recognition. They have used J-48 which is a java implementation of the C4.5 decision tree. They have created a combination of handwritten and machine-written characters of Urdu to create their dataset of 441 images. In pre-processing, noise removal, binarization, normalization, skew correction, and edge detection has been used. They have used three different moments for feature extraction viz, hu moments, Zernike moments, and PCA. Further, classification is done using DT J-48. They achieved the best recognition of 92.06%.

Khan et al. [7] in their paper have given the idea of agglomerative hierarchal clustering for the creation of an OCR model for Urdu script. They have used a dataset of around 2430 ligatures and all these ligatures were used for training. In pre-processing, the images thresholding was done by applying global image thresholding. In the end, a bi-level image with a white background was produced. After this process, their different geometric features very extracted to which the clustering was applied. The clustering was divided into two levels. In the first levels, for every feature extracted, ligatures were bifurcated into 32 clusters. For the second level of clustering, the obtained output from first level clustering was clustered again. A total of 417 clusters were generated on which classification was done. Classification was done using decision tree, naïve Bayes, k-NN models. For testing, they have randomly selected 1215 ligatures.

For the decision tree, the recognition rate was found out to be 62%. Naïve Bayes and k-NN have an accuracy of about 73% and 100% respectively. One of the researches [8] recommended using an evolutionary method for the popularity of offline published Urdu script. In the preliminary stages, a textual content line picture is pre-processed. A holistic segmentation is used to section Urdu text into ligatures. After segmentation, 15 geometric and statistical functions are extracted from the segmented ligature images. These functions are then concatenated to shape the very last characteristic vector for every ligature picture. Later after characteristic extraction, intra-characteristic hierarchical clustering algorithm is used for clustering. Finally, Genetic Algorithm (GA) is used for optimization and popularity. The respective experiment was conducted on the UPTI dataset where the accuracy of the recognition rate came out to be around 96.72%.

5.2 Unsupervised Learning

Faisel Mushtaq et al. in their paper [5] proposed a deep neural network architecture to recognize handwritten Urdu characters and used the HUCD dataset. In pre-processing, they first did the segmentation where they put rectangular boxes around each image and then cropped each image. It is then followed by normalization. Each image further undergoes Otsu's method of global binarization. In the end, they have done the de-noising and smoothing before classifying the images to test and training sets. Also, they have used an explicit segmentation-based approach on their dataset. The training dataset of 74,285 samples was trained on a convolution neural network. Testing was done of 21,223 test samples. A total of 98.82% of recognition rate was observed. Shehryar Malik with other researchers worked on recognition with various deep learning techniques [9]. They used a new dataset and pre-processed the images. Techniques used in pre-processing were Otsu's binarization and normalization of the height of images to 64. Depending upon the position of the letter, each ligature changes its shape and based on these, different ids have been assigned to each of the forms. It implies that a segmentation-based approach has been used. The new dataset has 15,164 lines written by 490 writers. They have used CNN-RNN-CTC and encoder-decoder models over this dataset to check the recognition rate with various searches. It was observed that with greedy search algorithm in CNN-RNN-CTC and encoder, 88.50% & 89.52% of recognition rate was observed respectively. With beam search, the results were 88.75% and 90.07% respectively. Additionally, an n-gram language model has also been added. In a paper by Chhajro et al. [10] various models were implemented on a dataset that was created by collecting data from around 100 people. For each character, labeling was done. The total dataset was 4668 images with a size of 50 × 50. Multi-layer perceptron, Support Vector Machine, KNN, CNN, RNN, and random forest algorithms were implemented over this created dataset. The results showed the highest accuracy in CNN with 99% followed by multi-layer perceptron with

98%. Support vector machine and random forest both showed an accuracy of about 97%. The algorithms which had the least accuracies were RNN (80%) and KNN (38%).

A convolution neural network was used by Mujtaba Husnain et al. [11]. They proposed their novel dataset in which 800 images of each Urdu letter as well as numerals were taken. A total of 8000 number images and 30400-character images were taken segmented into a size of 28 × 28. In pre-processing, the noise was removed, images were converted to grayscale, and finally resized to lock the aspect ratio. They have also used the technique of feature extraction where geometrical aspects of the characters such as width, height, number of loops, and arcs are extracted. In their proposed CNN model, a 2D convolutional layer with a 5 × 5 kernel size was used as the first layer. Rectified linear unit (ReLU) activation function was used. A learning rate of 0.005 and batch size of 132 was chosen in this proposed model. It was observed that the model shows a 98.03% of accuracy rate.

ResNet is a type of artificial neural network which works on the idea of skip connection i.e., it skips some layers and uses shortcuts. This type of ANN was used by M. Kashif in his paper [12] where he used Resnet18 on a publicly available dataset UNHD. Features of the dataset such as shape, height, the width of the character, horizontal and vertical lines in the image were extracted. In his methodology, first, one-dimensional BLSTM was used. Since desired results could not be attained, ResNet18 was applied. It gave much more accurate results than the previously used networks. Stacked Denoising Autoencoder was also used by some researchers to experimentally check the recognition of the Urdu Handwritten datasets [13]. They used four different degradation techniques: jitter, elastic elongation, threshold, and sensitivity. UPTI dataset has been used with selectively choosing only 3732 ligatures and dropping others with any form of irregularities. Since UPTI is a sentence-based dataset, sentences from all the versions were first segmented into ligatures. Two versions of the dataset have been created for this purpose. One of 80 × 80 and 15 ×15 dimensions. Ten epochs were run for both dimensions. Accuracy ranged between 93% to 96% in the experimental results. In a paper [14] Urdu text detection is done using dataset and real-world images. Detection of 320 * 240 pixels is done by FasterRCNN and CNN. Regression Residual Neural Network (RRNN) is used for orientation of training and testing dataset and recognition was done using Two Stream Deep Neural Network (TSDNN). Dataset of 4.2 k and 51 k images were used. It was observed that TSDNN showed a recognition rate of 94.90% and 95.20% for character-based ligatures in the 4.2 k and 51 k datasets respectively.

Summary

Highest level of accuracy was seen in Khan et al. [7] research using k-NN model. It was observed that an accuracy of 100% was attained but the model is very small. In another paper [10], it showed lowest accuracy of 38% using k-NN which was lowest among all. This ambiguous outcome makes k-NN unfeasible for using it for recognition of Urdu Characters. UPTI is a good dataset and using Stacked Denoising autoencoder and Genetic algorithm considerably showed good recognition rates. Till date, the best results we observed was using CNN [11]. The recognition rate was 98% on an average dataset. Conclusively, we can say that the observed results in the literature survey is not sufficient to generalize the outputs for the recognition of Urdu Characters. In most of the research papers, the dataset used is not adequate. Thus, there is a need to improve the datasets and use the deep learning techniques on them to achieve much better recognition rate.

Table 3. Comparision of various datasets using different Machine Learning Algorithms

Author	Reference	Dataset	Method	Approach	Accuracy
Khalil Khan et al.	[6]	441 images	Decision Tree	Segmentation-free	92.06%
F.Mushtaq et al.	[5]	HUCD	CNN	Segmentation-based	98.82%
S.Malik et al.	[9]	New dataset (15,000 images)	CNN-RNN-CTC Attention-base encoder-decoder	Segmentation-based	88.75% 90.05%
Chhajro et al.	[10]	New dataset (4668)	MLP SVM k-NN CNN RNN Random Forest	Segmentation-based	98% 97% 38% 99% 80% 97%
Khan et al.	[7]	2430 images	Decision tree Naïve Bayes k-NN	Segmentation-based	62% 73% 100%
M.Husnain et al.	[11]	Novel dataset (38400 images)	CNN	Segmentation-based	98.03%
M.Kashif	[12]	UNHD	ResNet18	Segmentation-based	-
Ibrar Ahmad et al.	[13]	UPTI	Stacked Denoising Autoencoder	Segmentation-based	96%
Naila Khan et al.	[8]	UPTI	Genetic Algorithm	Holistic approach	96.72%
Syed Yasser	[14]	Synthetic dataset (4.1k & 51K images)	FasterRCNN and CNN	-	94.90% & 95.20%

6 Future Work and Conclusion

Among all the applied algorithms and techniques, it has been proved that Convolution Neural Network (CNN) till now has shown the best and most accurate recognition rate on the provided datasets. There is still a possibility of improvement when it comes to the dataset. Till now the publicly available dataset for Urdu Text is UNHD. Even this dataset has only kept 4,240 images public [9]. This creates a gap for the chance of improvement and the ability to create versatility in Urdu text recognition. Also, the only two datasets, UNHD and UPTI, have enough data points to make them promising datasets in Urdu Handwritten text.

Increasing the accuracy of the recognition rate is not enough. To make them practically useful, every task of natural language processing should apply to this language as well like it is for languages like English, Chinese, etc. Urdu on the other hand, is far behind achieving milestones like document automation, predictive text, email filters, etc. There is scope for further improvements in digitization of this language.

References

1. Husnain, M., Missen, M.M.S., Mumtaz, S., Coustaty, M., Luqman, M., Ogier, J.M.: Urdu handwritten text recognition: a survey ISSN 1751–9659. IET Image Process. **14**(11), 2291–2300 (2020). https://doi.org/10.1049/iet-ipr.2019.0401
2. Sabbour, N., Shafait, F.: A segmentation-free approach to Arabic and Urdu OCR. Doc. Recognit. Retr. XX **8658**, 86580N (2013). https://doi.org/10.1117/12.2003731
3. Ahmed, S.B., Naz, S., Swati, S., Razzak, M.I.: Handwritten Urdu character recognition using one-dimensional BLSTM classifier. Neural Comput. Appl. **31**(4), 1143–1151 (2017). https://doi.org/10.1007/s00521-017-3146-x
4. Jain, L.C., Behera, H.S., Mandal, J.K., Mohapatra, D.P.: Computational Intelligence in Data Mining – Volume 2: Proceedings of the International Conference on CIDM, 20-21 Dec 2014. Smart Innovation System Technologies, vol. 32, pp. 449–460 (2015).https://doi.org/10.1007/978-81-322-2208-8
5. Mushtaq, F., Misgar, M.M., Kumar, M., Khurana, S.S.: UrduDeepNet: offline handwritten Urdu character recognition using deep neural network. Neural Comput. Appl. **33**(22), 15229–15252 (2021). https://doi.org/10.1007/s00521-021-06144-x
6. Sharma, R., et al.: Index point detection for text summarization using cosine similarity in educational videos. IOP Conf. Ser. Mater. Sci. Eng. **1131**, 012001 (2021)
7. Khan, N.H., Adnan, A., Basar, S.: Urdu ligature recognition using multi-level agglomerative hierarchical clustering. Clust. Comput. **21**(1), 503–514 (2017). https://doi.org/10.1007/s10586-017-0916-2
8. Khan, N.H., Adnan, A., Waheed, A., Zareei, M., Aldosary, A., Mohamed, E.M.: Urdu ligature recognition system: an evolutionary approach. Comput. Mater. Contin. **66**(2), 1347–1367 (2020). https://doi.org/10.32604/cmc.2020.013715
9. Malik, S., Maqsood, M.N., Ali, A.R.: Urdu Handwriting Recognition using Deep Learning.
10. Chhajro, M.A.: Handwritten Urdu character recognition via images using different machine learning and deep learning techniques. Indian J. Sci. Technol. **13**(17), 1746–1754 (2020). https://doi.org/10.17485/ijst/v13i17.113
11. Husnain, M., et al.: Recognition of Urdu handwritten characters using convolutional neural network. Appl. Sci. **9**(13), 2758 (2019). https://doi.org/10.3390/APP9132758

12. Kashif, M.: Urdu Handwritten Text Recognition Using ResNet18 (2021) http://arxiv.org/abs/2103.05105
13. Ahmad, I., Wang, X., Li, R., Rasheed, S.: Offline Urdu Nastaleeq optical character recognition based on stacked denoising autoencoder. China Commun. **14**(1), 146–157 (2017). https://doi.org/10.1109/CC.2017.7839765
14. Arafat, S.Y., Iqbal, M.J.: Urdu-text detection and recognition in natural scene images using deep learning. IEEE Access **8**, 96787–96803 (2020). https://doi.org/10.1109/ACCESS.2020.2994214

Review on Analysis of Classifiers for Fake News Detection

Mayank Kumar Jain[1(✉)], Ritika Garg[1], Dinesh Gopalani[2], and Yogesh Kumar Meena[2]

[1] Swami Keshvanand Institute of Technology, Management and Gramothan, Jaipur 302017, Rajasthan, India
mayank.jain@skit.ac.in

[2] Malaviya National Institute of Technology, Jaipur 302017, Rajasthan, India
{dgopalani.cse,ymeena.cse}@mnit.ac.in

Abstract. The spread of false news on an online social media platform has been a major concern in recent years. Many sources, such as news stations, websites, and even newspaper websites, post news pieces on social media. Meanwhile, most of the new material on social media is suspect and, in some circumstances, deliberately misleading. Fake news is a term used to describe this type of information. Large volumes of bogus news on the internet have the potential to generate major societal issues. Accepting the stories and pretending that they are true is extremely harmful for our community. Many people believe that false news affected the 2016 presidential election in the United States. The term has since become commonplace as a result of the election. It has also attracted the interest of industry and academics, who are trying to figure out where it comes from, how it spreads, and what impacts it has. In this work, we looked at a number of different papers and compared all of the strategies for detecting false news.

Keywords: Natural language processing · Support vector machine · N-gram analysis · Machine learning

1 Introduction

The expansion of the Internet and recent technological advancements have had a huge influence on social relationships. People's use of social media to get information is becoming increasingly common. People also use various social media platforms to discuss their own activities, hobbies, and opinions. Many benefits of social media include simple access to information, minimal cost, and rapid dissemination of information. Because of these benefits, many individuals prefer to get their news through social media rather than common stories sources like television or newspapers [1]. As a result, social news is rapidly displacing traditional news sources. Although social platform has numerous advantages, news on the internet is not as reliable as traditional news sources.

However, the content of social media may occasionally be altered to serve other objectives. As a result, false news and rumors spread rapidly and widely. As a result of

V. E. Balas et al. (Eds.): ICETCE 2022, CCIS 1591, pp. 395–407, 2022.
https://doi.org/10.1007/978-3-031-07012-9_34

this circumstance, incorrect news stories are produced and disseminated. Furthermore, well-intentioned individuals distribute false news and disinformation without carefully vetting it [1, 3, 5]. Many websites exist on social media with the sole purpose of producing fake news.

After the mid-1990s, the way people communicate with one another altered. Multiple online social networks, such as Facebook, Twitter, make it easier for users to share real-time information with others on the same or other networks [9]. Online social networks have become a key means of communication and information sharing due to a variety of qualities such as simplicity of use, faster transformation, and lower cost. Almost all social network users nowadays get their news from internet sources [7]. However, as OSNs become more widespread, the Internet has become an excellent medium for communicating and distributing bogus news. False news is propagated through deceptive material, fake reviews, fake stories, advertising, false political statements and other means [5]. Fake news is currently spreading quicker on social media than in traditional media. Some information circulates on social media, causing people to be perplexed and distrustful. Fake news detection and identification on social media platforms is a difficult process. Fake news spreads quickly, affecting millions of people and their real-world surroundings. The spread of false news is not a new issue on social media platforms [5, 8]. Several firms and well-known individuals utilize a variety of social media networks to promote their products and build their reputation. Many users are influenced by these operations to share and like the news. Fake news propagated over the network as a result of this procedure. In terms of a topic, the content, style, and media platform of false news fluctuate with time, and fake news tries to distort linguistic form.

Furthermore, the user produced, shared, liked, and commented on a huge amount of material in an online social networking site. Many phone identities use repeated postings to propagate bogus news around the social media site [10]. Due to the large amount of data on the social network, it is impossible to discover all of these shared items. Fake news that spreads throughout the internet cannot be exposed by turning a blind eye. Some features of false news must be examined in order to identify it from genuine news. Numerous articles and blogs have been written as a result of public awareness see in Fig. 1.

Users' decisions have been influenced by user reviews, comments, and news posted on social media. The dissemination of low-quality news, particularly false news, has a detrimental impact on societal and individual attitudes [11]. Fake news is damaging to individuals, society, and companies, as well as governments. Fake news about the company, for example, spread by spam or malevolent individuals, might cause significant harm. As a result, detecting false news has become a major study topic.

Many websites, such as deversuardian.com and ABCnews.com, produced bogus news. False claims, fraudulent ads, conspiracy theories, satirical news, and fake news are just a few examples of fake news. These variations have an impact on people's life in every way. All of these stories have dominated public opinion, interest, and decision-making. Multiple authors and researchers have agreed on numerous criteria to detect false news pertaining to text, the reaction in the form of sharing, live, and the source from which it originated. The traditional technique of using a human editor and professional journalist to detect false news ignores the volume of content generated by

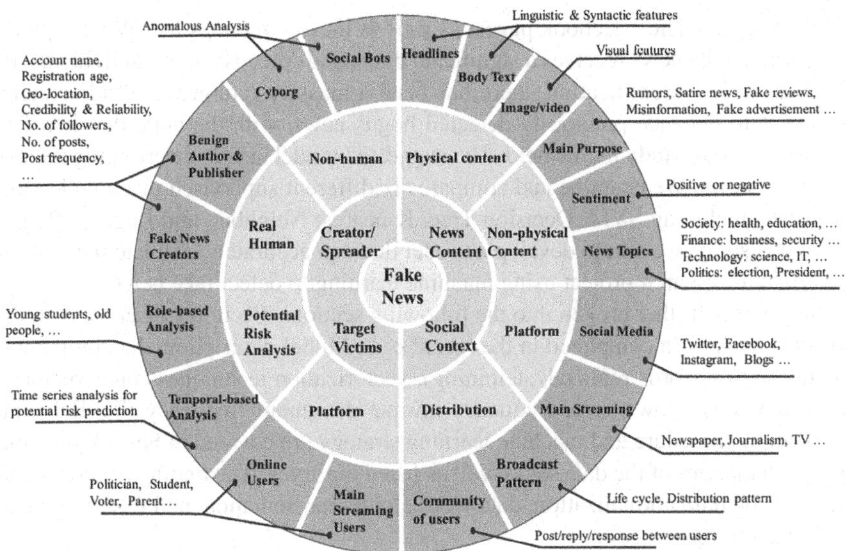

Fig. 1. Fake news and all that goes along with it [5]

social media platforms. To identify it in a timely manner, new computational approaches are necessary [14]. However, professional manual verification was required, as well as a computer technique to identify false news.

Many of the older approaches for detecting false news use a similar detection procedure. Preprocessing is applied to all datasets in order to eliminate noisy information. Even yet, using a big and false dataset, bogus stories on Twitter are discovered with a high level of accuracy. Linguistic features, linguistic cue approaches, tweet level characteristics, and NLP features are among the features retrieved from the news data [6]. Linguistic Analysis and Word Count (LIWC), Semantic Analysis, Probabilistic Context-Free Grammar, (TF-IDF) Term Frequency-Inverted Document Frequency, bag-of-words, n-grams, and Doc2Vec are some of the feature extraction approaches. Once the features have been recovered, they are fed into Machine Learning classifiers for training and testing the accuracy of fresh unlabeled news information prediction. After the features have been retrieved, Nave Bayes Classifier, Neural networks, Support Vector Machine (SVM), Random Forest, XGBoost, Decision Tree, Linear Regression, Logistic Regression, K Nearest Neighbors (KNN), AdaBoost Stochastic Gradient Descent (SGD) [6, 9], and Linear SVM are the majority of Machine Learning classifiers investigated in the previous fake news detection methods.

Our major goal is to distinguish between "true" and "false" news. We need to figure out whether there is an existing platform for detecting false news depending on context and content. Following this first identification, we must conduct large-scale experiments utilizing multiclass datasets and various machine learning algorithms. We'll employ a variety of machine learning methods in these exercises. Some fixed findings are accessible in prior models; therefore, we must compare these results to determine which news is true and which is false. This manifests as a result/outcome on social media sites such as Instagram and Facebook pages, as well as messaging apps like WhatsApp and Skype, where fake news receives a significant boost and spreads among individuals. For multi-class article identification, there has been a massive refulgence of adjustments. Our goal is to enhance previously detected bogus news, with the hope that it will be valuable for future study in this field. The suggested model aids in determining the genuineness of news. We examine and compare six different supervised machine learning approaches, including SVM, Decision Tree, K-nearest Neighbor, and Logistic Regression. In this study, we aim to develop a model that can accurately estimate the viability of a particular research project using machine learning to detect fake news.

The paper is further broken into the following sections. Section 2 discusses relevant works that have been completed in the recent past. Model description, data preparation techniques (stop words removal, stemming), vectorization techniques count vectorizer or bag of words (Bow), term frequency-inverse document frequency (TF-IDF), and classification procedure and machine learning strategy are covered in Sect. 3. Section 4 contains challenges of the data set as well as faced in survey. The findings of conclusion which include data normalization, data word cloud representation, and future scope are summarized in Last Section.

2 Related Work

Using machine learning, several academics have built models and algorithms for detecting bogus news. The challenge of detecting fake news has only lately been brought, but it has piqued the interest of those who have looked into it. Several methods have been developed to identify falsification in various forms of data.

Two stage strategies for identifying fakeness on social media platforms were discovered by Feyza et al. (2019) [1]. Several times pre-processing approach is performed to the entire data set in the first stage, resulting in the data set being changed from unstructured to structured. Document-Term matrix utilizing the TF weighting approach. In a secondary stage, data mining is used to apply the twenty-third supervised AI algorithm to the collection of data in order to transform it into structured format. The findings of this experiment are displayed in a variety of datasets, including: 1) BuzzFeed Political data set; accuracy in J48 is 0,655, whereas accuracy in F-measure is 1,000. 2) He obtained accuracy by using random political news data (1,000). 3) In the ISOT fakeness dataset, the best result is 1,000.

Somya et al. (2020) utilize an automated identification approach in the Chrome domain to detect manufactured tales on Facebook by analyzing some particular functions on fakebooks using deep learning [2]. The author employs user profile data functions, news data functions, and a combination of these two characteristics in this paper.

Pedro et al. (2020) [3] used five sets of data to categories and identify bogus articles using text attributes. They discriminate both texts and media stories in three distinct languages: Silvic, Germanic, and Latin. In addition, NLP tools such as Bags-of-words and Word2vec can be used. With a 92% accuracy rate.

Authors Julio C.S et al. (2019) offered a new set of features, compared their performance to prior data set features, and designed some new features such as lexical features, psycholinguistic features, semantic features, and subjectivity. And the actual positive rate is one, whereas the false positive rate is 0.4 [4].

The author of Xichen Zhang et al. (2019) [5] gives many sorts of characteristics for detecting bogus publications. And create a complete fake news detection ecosystem, which includes domains such as intervention, fact checking, alert systems, fake news detection, potential fake news prediction, suspiciousness analysis, and third-party verification to determine if the system is trustworthy.

M. Jain et al. (2020) introduced a new set of characteristics for ML classifiers that leverages two data sets on political articles. Extract linguistic/stylometric characteristics, a bag of words TF, and a BOW TF-IDF vector from dataset text fields, then use various ML approaches with bagging and boosting methods. This model has a stylometric accuracy of 87.26%, with Word vector having the greatest accuracy of 89.41% [6].

Mohamed K. et al. (2019) [7] propose a sequential survey for fake story detection. With certain predefined datasets, many types of data and various extracted characteristics are employed for detection.

Macro L et al. [8] develop a machine learning (ML) false tale detection system that blends news data with social narrative elements and outperforms current algorithms. Second, use a Messenger chat box to develop a technique and compare it to a real-world application. And gaining with the greatest accuracy (81.7%).

H. Liu et al. [9] established a system for detecting false news from several news sources called the Fake News Detector Based on Multi-Source Scoring (FNDMS). To assess the reliability of a single news source, content-based and author-based aspects are used. The Dempster-Shafer Theory (DST) model combined the veracity of many sources to arrive at a conclusion on the authenticity of a news occurrence. When compared to SVM, Logistic Regression, Random Forest, and AdaBoost, the framework's efficacy was confirmed. It would be preferable if this framework concentrated just on the source's purpose.

S. Helmstetter and H. Paulheim [10] used a big and noisy dataset to detect bogus news on Twitter. Features derived from a tweet include tweet level features, subject features, user-level features, emotion features, and text features. The suggested method was designed to detect false news tweets and sources. The source identified the user account from which the tweets were sent. Multiple classifiers were used to categories the tweets, including neural networks, Random Forest, SVM, Nave Bayes, and XGBoost classifiers. The XGBoost algorithm performed exceptionally well in classifying the tweets. The suggested technique has the disadvantage of focusing on only one source, namely Twitter. It also failed to target news pieces from a variety of sources.

H. Ahmed et al. [11] suggested an n-gram analysis technique for automatically detecting bogus news. The features were retrieved from the text using the TF and TF-IDF feature selection approaches. To verify the news's dependability, six different Machine Learning Classifiers and two approaches for choosing characteristics were examined. Six classifiers were compared: SVM, KNN, SGD, Linear SVM, and Decision Tree. Using the Unigram, TFIDF, and Linear SVM classifiers, the suggested model had the maximum accuracy. This approach does not consider the project's multi-source nature or legitimacy.

To Recognize fake news in news reports and social media platforms The writers have recommended a number of techniques. The developers use common language preparation processes to identify fake/phony news. Harita et al. [12] use stylometric highlights and word vectors highlights of the content of news items to create fake news with an accuracy of up to 95.49%.

Yang et al. [13] employed multimodal to detect false news in articles and used a convolution neural network to explicit and latent feature sets of text and picture data to recognize fake news. In Table 1 we show the limitation of some papers.

Table 1. Comparative analysis

Authors name	Dataset	Technology used	Results	Limitations
Hager et al. 2021 [24]	Dataset1, FakeNewsNet, FA-KES$_5$, The ISOT	Grid search, hyperopt optimization Techniques	Accuracy OPCNNFAKE = 97.84%, RNN = 86.76%	Not using Knowledge-base and fact-based approaches
Somya et al. 2020 [2]	More than 15 000 news contents from different Facebook users including both fake and real news	Fake news detection approach in chrome environment, KNN, SVM, Logistic Regression	Accuracy KNN = 99.3%, SVM = 99.3%, Logistic regression = 99.0%, Decision tree = 99.1%, LSTM = 99.4%	When compared to machine learning algorithms, deep learning algorithms take more time for testing and analysis
Abdullah et al. 2020 [25]	Fake news dataset from Kaggle	Hybrid CNN-LSTM Model	Accuracy 99.7%, 97.5% of training and testing	Doesn't use pre-processing
Feyza et al. 2019 [1]	BuzzFeed Political News Data set, Random Political News Data set, ISOT Fake News Data set	TF weighting method and Document-Term Matrix Twenty-three supervised artificial intelligence algorithms	Accuracy of 65.5% and 64.4%	They work on smaller data set of 1500 news articles
K. Shu et al. 2019 [22]	FakeNewsNet	Fake news data repository FakeNewsNet, Convolution neural network	Accuracy is 92% for Buzzfeed news and 93.6 for PolitiFact news	Accuracy is less due to a smaller number of features
J.C et al. 2019 [23]	None	Measure the prediction performance of proposed approaches to design auto detection system SVM, KNN	Accuracy RF = 85%, KNN = 80%, SVM = 79%	Accuracy for detecting fake account is very due to small dataset

3 Methodology

3.1 Data Pre-processing

Preparing raw data for use in a machine learning model is known as data pre-processing. It's the first and most crucial step in creating a machine learning model. Three phases make up data pre-processing: -

Data Cleaning - This is a method used to improve the quality of data. Normalizing data, reducing mistakes, smoothing noisy data, addressing missing data, finding unnecessary observations, and correcting errors are all part of this process.
Data Transformation - In this approach, raw data is converted into a format that allows data mining to collect strategic information quickly and efficiently. Because raw data is difficult to track, any information collected must first be pre-processed.
Data Reduction - Data reduction is a technique for shrinking the size of original data so that it may be represented in a lesser amount of space. Data reduction strategies maintain data integrity while lowering data (Fig. 2).

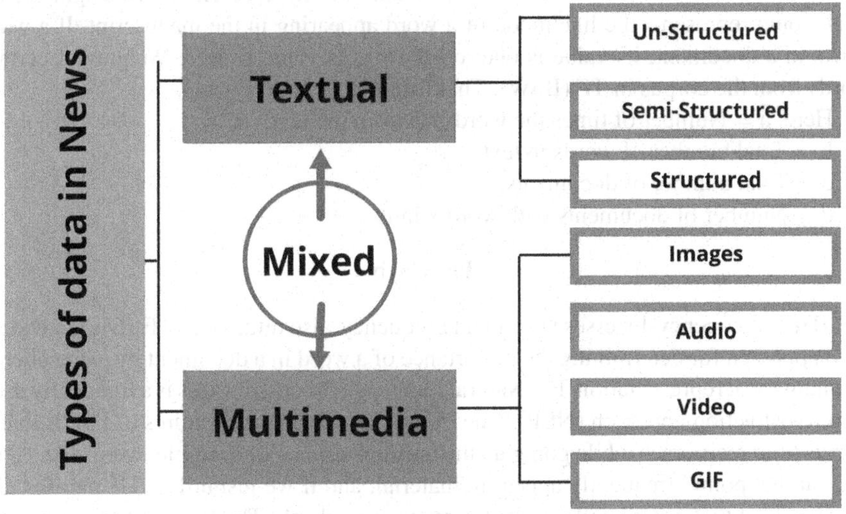

Fig. 2. Data types in the news

Stop Words Removal: Stop words have no significance until they are combined with other words. Stop words are considered noise in text classification when characteristics are retrieved from text. These are the terms that are utilized in article sentences to connect the concept and sentence structure. Stop words include prepositions, conjunctions, and articles. Words like by, for, from, how, of, on, that, the, too, was, what, when, about, and so on are examples. These terms are no longer in use. Total number of words in all articles and unique word count in the listed articles after preprocessing.

Stemming: The tokens are then converted into basic words as the next stage in text normalization. Stemming is a method of transforming a word into its proper form. Stemming is a technique for reducing the number of classes of words in data. Words' affixes are changed by stemming. We utilized the Porter Stemmer Algorithm to convert the term Python Ly to Python. Because it produces good stemming outcomes. In our data collection, words like extreme, very, government, transgender, and minister were replaced with extreme, govern, transgend, and minist, respectively.

3.2 Feature Extraction and Selection

Getting from high dimensional information is one of the challenges of text classifiers. There are a few concepts, phrases, and explanations in documents that contribute to the learning cycle's high computational weight. Furthermore, extraneous and excessive highlights might degrade the precision and execution of classifiers.

Term Frequency: It is commonly known as count vectorizer (CV), is a technique that assesses the correspondence of texts using the bag of words (BoW) technique. Each document comprises a collection of words that have been represented as a length vector. Each count represents the likelihood of a word appearing in the manuscript. If a word exists in a document, its value is one; otherwise, its value is zero. Weights of certain words from the corpus in TF (BoW). That formula given in Eq. (1).

Here, a = Number of times the words occur in the text.

b = Total number of words in text.

c = Total number of documents.

d = Number of documents with word x in it.

$$\mathbf{TF} = \mathrm{a/b} \tag{1}$$

TF-IDF: Frequency-Inverses Document Frequency, also referred as TF-IDF, is a widely used approach for determining the importance of a word in a document by using altered text numerical representation. For Natural Language Processing, this is a frequently used feature extraction approach (NLP). One of the most important features of IDF is that it affects term recurrence while going up the unusual terms. For example, words like "the" and "at that point" frequently appear in material, and if we just utilize TF, phrases like "the" and "at that point" will govern the recurrence check. The IDF, on the other hand, reduces the impact of these phrases. Below formula given in Eq. (2) and (3)

$$\mathbf{IDF} = \mathrm{Log}\,[\mathrm{c/d}] \tag{2}$$

$$\mathbf{TF{-}IDF} = \mathrm{TF} * \mathrm{IDF} \tag{3}$$

Content Based Feature: It is a ML approach that makes decisions based on feature similarity. It is frequently employed in recommender systems, which are algorithms that advertise or recommend items to people based on information gathered about them. Like-Number sentences, words, question mark, exclamation and capital letters, punctuation, negations (no, not), frequency used words.

Context-Based Feature: The contextual information aids in maximizing the consumers' understanding of the scene photographs on the web. Such details can help distinguish between ambiguous scene photos with intra-class variance and inter-class similarities, User_Name, User_Age, Registration Date, Registration Time, Follower, Following, Number of Posts.

3.3 Machine Learning Strategies for Detecting Fake News

Machine Learning is used to create models that make predictions based on past or historical data. It's a feature of AI; it learns from previous experiences and may anticipate our needs. Typically used to store and then analyses massive amounts of data. Nowadays, it is utilized to identify fraud. There are many other machine learning approaches, but we're focusing on supervised machine learning [15–22].

Naive Bayes: To forecast the class of the previous dataset, it is a basic, straightforward, and effective classification procedure. This machine learning classifier is used to make quick predictions. It forecasts data based on the likelihood of an object. It produces superior outcomes in multi-class predictions. This is mostly used for text categorization and consumes a significant quantity of data. When it employs NLP tasks like Sentimental analysis, it produces superior results.

It entails completing the following steps:

- First, make a frequency table based on the terms.
- Calculate the probability for each class based on the frequency table.
- Calculate the posterior probability for each class.
- The greatest posterior probability is the result of the prediction classifier.

K-Nearest Neighbours (KNN): It is based on a machine learning method that is supervised. KNN classification is mostly used to discover and identify abnormalities (false news). It is mostly used for classification problems, such as detecting bogus news; it is simple to use and takes little time, and it can readily reduce noise. It calculates the shortest distance using the Euclidean distance formula, which is K. Across all the input, this square root of the total of the square differences between a training data and a testing point.

Random Forest (RF): It is a supervised learning algorithm for text classification. It prevents overfitting by classifying the text and eliminating unpredictability. The "forest" suggestion is a collection of uncorrelated decision trees that are combined to eliminate variation and provide more accurate data predictions.

Extra Tree Classifier: It is a form of batch learning algorithm that outputs a classification result by aggregating several de-correlated decision trees collected in a "forest". It is quite similar to a Random Forest Classifier, and the only difference is that it informs the forest's decision trees. We didn't utilize it as a classifier in this case; instead, we used it as a feature selection approach to choose the most suited features, which we then used in the Classifiers to improve results and performance.

Support Vector Machine: This algorithm's purpose is to find the funniest line or decision boundary that can divide n-dimensional space into classes so that fresh data points can be readily placed in the correct category in the future. A hyperplane denotes the optimal choice boundary. Machine learning classification is shown in Fig. 3.

Logistic Regression: It's a method for predicting a categorical dependent variable from a set of independent variables. A categorical dependent variable's output is predicted using logistic regression. As a result, the result must be a discrete or categorical value. It can be Yes or No, 0 or 1, true or false, and so on, but instead of giving exact values like 0 and 1, it delivers probabilistic values that are somewhere between 0 and 1.

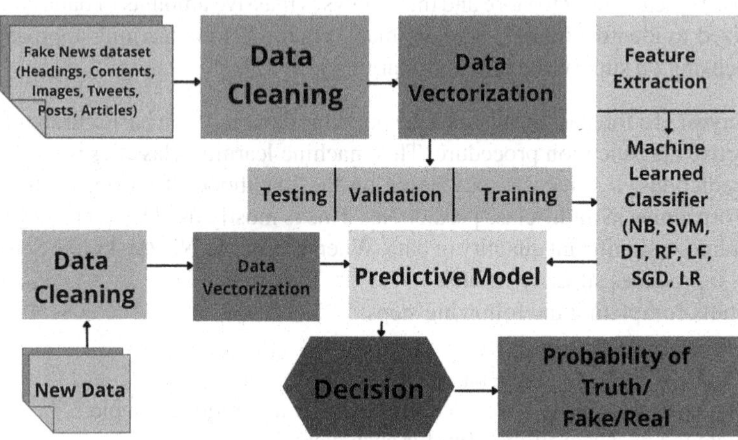

Fig. 3. Machine learning architecture

Convolutional Neural Networks: A CNN is a Deep Learning method that takes an image as input and assigns different weights and biases to different sections of the picture so that they can be distinguished. The Convolutional Neural Network Model can handle a variety of tasks in the Image Processing domain once they become differentiable, such as Image Recognition, Image Classification, Object and Face Detection, and so on.

Recurrent Neural Networks: Recurrent Neural Networks (RNNs) are neural networks that process temporal or consecutive data. To develop the best predictions, RNNs employ other data points in a series. They achieve this by taking in input and influencing the output by recycling the activations of previous nodes or later nodes in the sequence. Deep learning approach shown in Fig. 4.

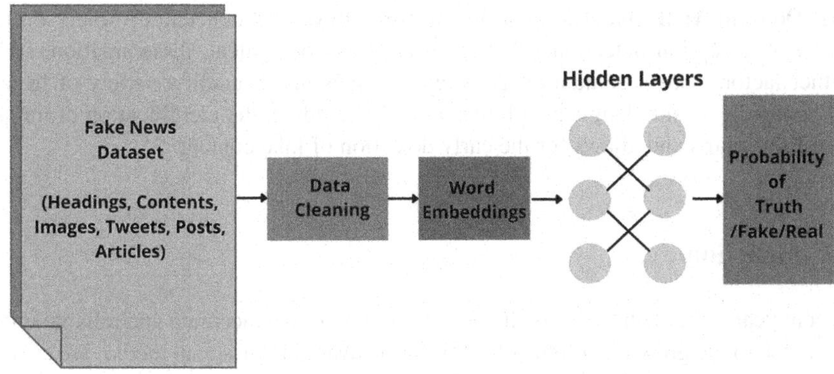

Fig. 4. Deep learning architecture

4 Challenges

Deep Learning: Deep learning technology can deal with any type of data, including text, photos, video, and audio. It may be adapted to a new type of problem and avoids feature engineering, which is the most time-consuming yet essential aspect of a machine-learning framework. Deep learning technologies, on the other hand, have the disadvantage of requiring a significant amount of time for model training with a relatively large amount of data and not providing interpretations of what the model has actually learned, making it almost a black box type of processing inside the model.

Multimedia False Information Detection: Data analytics, computer vision, and signal processing techniques are needed to generate fabricated and modified sounds, images, and movies. Machine learning and deep learning are essential for discovering signature traits of modified and produced multimedia.

Unsupervised Models: The majority of the current work is done utilizing supervised learning techniques. Due to the large amount of unlabelled data available on social media, unsupervised models must be constructed.

Datasets: Because most research is done on customised datasets, the production of persuasive gold standard datasets in this sector is critical. A benchmark comparison between multiple algorithms is impossible due to the absence of publicly available large-scale datasets.

Multilingual Platform: The majority of the research focuses on linguistic characteristics in English-language texts. Other popular and regional languages are not yet considered (multilingual platform for fake news identification).

Early Detection: Detecting fake news in its early phases, before it spreads widely, is a difficult effort that must be completed quickly in order to mitigate and intervene. It's nearly tough to change people's minds after bogus news has been widely accepted and trusted.

Cross Domain Analysis: The majority of present systems concentrate solely on the method of deception detection, whether in the form of content, dissemination, style, or other factors. Cross domain analysis, which takes into account a variety of factors such as topic, website, language, photos, and URL, aids in the identification of unique non-varying traits and allows for the early detection of fake content.

5 Conclusion

In recent years, people have found it more difficult to obtain accurate and reliable information due to the growing volume of information available on social media. We cover a variety of tools and methodologies for spotting fake news. In this work, we identified a number of machine learning approaches for detecting false news, including SVM, LR, NB, KNN, RF. On all types of datasets used by the authors, linear SVM performed better. To detect false news, several researchers employed deep learning models such as CNN, RNN, and Hybrid Models. In the future, we'll strive to construct a new real-time dataset for false news and use the graph convolution neural networks to detect false news.

References

1. Ozbay, F.A., Alatas, B.: Fake news detection within online social media using supervised artificial intelligence algorithms. Phys. A (2019). https://doi.org/10.1016/j.physa.2019.123174
2. Sahoo, S.R., Gupta, B.B.: Multiple features-based approach for automatic fake news detection on social networks using deep learning. Appl. Soft Comput. 106983 (2020). https://doi.org/10.1016/j.asoc.2020.106983
3. Arruda Faustini, P.H., Covões, T.F.: Fake news detection in multiple platforms and languages. Expert Syst. Appl. 113503 (2020). https://doi.org/10.1016/j.eswa.2020.113503
4. Reis, J.C.S., Correia, A., Murai, F., Veloso, A., Benevenuto, F., Cambria, E.: Supervised learning for fake news detection. IEEE Intell. Syst. **34**(2), 76–81 (2019). https://doi.org/10.1109/mis.2019.2899143
5. Zhang, X., Ghorbani, A.A.: An overview of online fake news: characterization, detection, and discussion. Inf. Process. Manage. **57**, 102025 (2019). https://doi.org/10.1016/j.ipm.2019.03.004
6. Kumar Jain, M., Gopalani, D., Kumar Meena, Y., Kumar, R.: Machine learning based fake news detection using linguistic features and word vector features. In: 2020 IEEE 7th Uttar Pradesh Section International Conference on Electrical, Electronics and Computer Engineering (UPCON), pp. 1–6 (2020). https://doi.org/10.1109/UPCON50219.2020.9376576
7. Elhadad, M.K., Fun Li, K., Gebali, F.: Fake news detection on social media: a systematic survey. In: 2019 IEEE Pacific Rim Conference on Communications, Computers and Signal Processing (PACRIM) (2019). https://doi.org/10.1109/pacrim47961.2019.8985
8. Della Vedova, M.L., Tacchini, E., Moret, S., Ballarin, G., DiPierro, M., de Alfaro, L.: Automatic online fake news detection combining content and social signals. In: 2018 22nd Conference of Open Innovations Association (FRUCT) (2018). https://doi.org/10.23919/fruct.2018.846830

9. Liu, H., Wang, L., Han, X., Zhang, W., He, X.: Detecting fake news on social media: a multi-source scoring framework. In: 2020 IEEE 5th International Conference on Cloud Computing and Big Data Analytics (ICCCBDA), pp. 524–531. IEEE, April 2020
10. Helmstetter, S., Paulheim, H.: Weakly supervised learning for fake news detection on Twitter. In: 2018 IEEE/ACM International Conference on Advances in Social Networks Analysis and Mining (ASONAM) pp. 274–277. IEEE (2018)
11. Ahmed, H., Traore, I., Saad, S.: Detection of online fake news using n-gram analysis and machine learning techniques. In: Traore, I., Woungang, I., Awad, A. (eds.) ISDDC 2017. LNCS, vol. 10618, pp. 127–138. Springer, Cham (2017). https://doi.org/10.1007/978-3-319-69155-8_9
12. Reddy, H., Raj, N., Gala, M., Basava, A.: Textmining-based fake news detection using ensemble methods. Int. J. Autom. Comput. 1–12 (2020)
13. Yang, Y., Zheng, L., Zhang, J., Cui, Q., Li, Z., Yu, P.S.: TI-CNN: convolutional neural networks for fake news detection. arXiv preprint arXiv:1806.00749 (2018)
14. Meel, P., Vishwakarma, D.K.: Fake news, rumor, information pollution in social media and web: a contemporary survey of state-of-the-arts, challenges and opportunities. Expert Syst. Appl. **153**, 112986 (2019)
15. Perez-Rosas, V., Kleinberg, B., Lefevre, A., Mihalcea, R.: Automatic detection of fake news. arXiv preprint arXiv:1708.07104 (2017)
16. Kumar, T., Mahrishi, M., Meena, G.: A comprehensive review of recent automatic speech summarization and keyword identification techniques. In: Fernandes, S.L., Sharma, T.K. (eds.) Artificial Intelligence in Industrial Applications. LAIS, vol. 25, pp. 111–126. Springer, Cham (2022). https://doi.org/10.1007/978-3-030-85383-9_8
17. Loper, E., Bird, S.: NLTK: the natural language toolkit. In: Proceedings of the ACL-02 Workshop on Effective Tools and Methodologies for Teaching Natural Language Processing and Computational Linguistics, ETMTNLP 2002, vol. 1, pp. 63–70. Association for Computational Linguistics, Stroudsburg (2002). https://doi.org/10.3115/1118108.1118117
18. Rubin, V.L., Chen, Y., Conroy, N.J.: Deception detection for news: three types of fakes. Proc. Assoc. Inf. Sci. Technol. **52**(1), 1–4 (2015). https://doi.org/10.1002/pra2.2015.145052010083
19. Shu, K., Sliva, A., Wang, S., Tang, J., Liu, H.: Fake news detection on social media: A data mining perspective. ACM SIGKDD Explor. Newsl. **19**(1), 22–36 (2017a). Shu, K., Wang, S., Liu, H.: Exploiting tri-relationship for fake news detection. arXiv:1712.07709 (2017b)
20. Tausczik, Y.R., Pennebaker, J.W.: The psychological meaning of words: LIWC and computerized text analysis methods. J. Lang. Soc. Psychol. **29**(1), 24–54 (2010). https://doi.org/10.1177/0261927X09351676
21. Mahrishi, M., et al.: Video index point detection and extraction framework using custom YoloV4 darknet object detection model. IEEE Access (2021). https://doi.org/10.1109/ACCESS.2021.3118048
22. Shu, K., Mahudeswaran, D., Wang, S., Lee, D., Liu, H.: Fake newsnet: a data repository with news content, social context and dynamic information for studying fake news on social media. arXiv preprint arXiv:1809.01286 (2018)
23. Reis, J.C., Correia, A., Murai, F., Veloso, A., Benevenuto, F., Cambria, E.: Supervised learning for fake news detection. IEEE Intell. Syst. **34**(2), 76–81 (2019)
24. Saleh, H., Alharbi, A., Alsamhi, S.H.: OPCNN-FAKE: optimized convolutional neural network for fake news detection. IEEE Access **9**, 129471–129489 (2021). https://doi.org/10.1109/ACCESS.2021.3112806
25. Abdullah, A., Awan, M., Shehzad, M., Ashraf, M.: Fake news classification bimodal using convolutional neural network and long short-term memory. Int. J. Emerg. Technol. Learn **11**, 209–212 (2020)

A Machine Learning Approach for Multiclass Sentiment Analysis of Twitter Data: A Review

Bhagyashree B. Chougule and Ajit S. Patil(✉)

Kolhapur Institute of Technology's College of Engineering (Autonomous), Shivaji University, Kolhapur 416234, India
patil.ajit@kitcoek.in

Abstract. Sentiment analysis or opinion mining is a prominent and most demanding research topic in today's world. The main idea behind this research topic is to recognize the user's opinions and emotions towards the aspect of service or product via a text basis. Sentiment analysis involves mining text, lexicon construction, extracting features and finally finding polarity of text. Even though numerous amounts of research work were conducted in this field through different methods, opinion mining is still considered a challenging field for research.

Most of the prior research concentrated on the binary or ternary classification of sentiments such as positive, negative, neutral. Some studies have done an analysis of Twitter sentiment based on ordinal regression, but by turning the problem of ordinal regression into a problem of binary classification. The aim of this study is to review the multiclass sentiment analysis of Twitter text data using an automated i.e., machine learning approach. This review paper intends to focus on existing work for Twitter sentiment analysis with multiple polarity categorization and explore gaps with future scope in the said research area.

Keywords: Machine learning · Twitter sentiment analysis · Opinion mining · Ordinal regression

1 Introduction

In recent years the use of social networks and blogging websites has been improved vigorously. People tend to spend several hours from their daily routine on social media, which makes the social network a rich source of information. Microblogs are a good source of online information-sharing platforms with the growth of social media. People prefer to use microblogging websites to express their thoughts, opinions regarding a variety of contexts e.g., products, movies, social issues, events, organizations, etc. The nature of microblogs, the ability to post real-time messages became the reason behind the growth in the usage of social media.

Millions of people use social networking sites like Facebook, Twitter, Tumbler to share their opinions or sentiments regarding products, current issues, complaints, etc. In between them, Twitter is the most popular platform for people to express their sentiments with varying users from common man to politicians or presidents of different nations.

V. E. Balas et al. (Eds.): ICETCE 2022, CCIS 1591, pp. 408–416, 2022.
https://doi.org/10.1007/978-3-031-07012-9_35

Analyzing Twitter data is a critical task because of its informal language of writing, short length Tweets, use of special characters like hashtags (Elbagir et al. 2018).

The lexicon-based approach and the machine learning approach are two methods of doing Twitter sentiment analysis. This study mainly focuses on the machine learning approach for Twitter sentiment analysis as it gives the most prominent results.

Along with this, the study focuses on the ordinal regression problem. Ordinal regression problems are considered in-between classification and regression which has received considerable attention in the research area (Elbagir et al. 2019).

The contribution of this paper can be depicted as follows:

- To undergo a review of sentiment analysis on Twitter data.
- To review different machine learning approaches used for sentiment analysis along with their polarization classes and evaluation measures.
- To find out research gaps in the existing study of sentiment analysis.

2 Literature Review

(Peng et al. 2016), proposed a computation pipeline for drug-related classification and sentiment analysis on Twitter to extract Adverse Drug Event (ADE). To do this author followed a five-step pipeline: Capturing English Tweets, Data pre-processing (cleaning, tokenization, stemming), Drug classification, Tweets sentiment classification, and ADE extraction. Tweets were classified in positive and negative feelings using the Support Vector Machine (SVM) and Naïve Bayes (NB) classifier. Experiments showed that SVM was superior to Naïve Bayes in accuracy.

(Pagolu et al. 2016), analyzed the correlation between Tweet sentiments and stock market movements and elaborated that positive opinion in Tweet will encourage to invest in a company's stock indirectly increasing the stock price of a company. For this, Microsoft stock price Tweets are pre-processed in three stages of filtering: tokenization, stop-word removal, and regular expression matching for special character removal. To perform correlation, the problem was turned out to be a classification (0 & 1) problem. Sentiment analysis was done with the use of Random Forest, Logistic Regression, Sequential Minimal Optimization (SMO) sentiment analyzers.

(Deshwal et al. 2016), stated that no previous work gave a conclusive way of features extraction and classification algorithms selection that is perfect to develop efficient sentiment classification systems. The author proposed a system with six classifiers-Naïve Bayes, Discriminative Multinomial Naïve Bayes, Bayes Network, Hyperpipes, Random Forest, and Sequential Minimal Optimization. To design an appropriate sentiment categorization system combination of feature extraction techniques was used with unigram as a feature. Tweets are categorized in ternary polarities (positive, negative, neutral) by the execution of all classifiers on extracted features.

(Juneja et al. 2017), used seven different machine learning classifiers (Stochastic Gradient Descent, Multinomial Naïve Bayes, Naïve Bayes, Bernoulli Naïve Bayes, Logistic regression, NuSVC, and Linear SVC) to classify tweets as positive, negative sentiments, and predicted Delhi corporation election result. A slightly different approach used in the proposed system was like cross-domain sentiment analysis, ML classifiers are trained with movie reviews data and then ted with political parties pre-processed Tweets.

(Aziz et al. 2017), achieved successful results in cross-domain sentiment analysis using machine learning methods like Random Forest, Multinomial Bayes Classifier, Stochastic Gradient Descent, SVM-RBF (Radial Basis Function), and SVM-LK. To enhance model performance author replaced the frequency matrix approach with TF-IDF for feature selection.

(Bouazizi et al. 2017), introduced a novel approach for the classification of Tweets in multiple sentiment classes (hate, sadness, anger, love, happiness, fun, and neutral) rather than binary and ternary. The author used a user-friendly SENTA tool for the collection of features along with unigram and applied one of the machine learning techniques, Random Forest for experimentation. In this work, the author mentioned that this work can be extended for the classification of text into multiple classes and is scalable. Also, quantifying sentiments were not covered (scores for sentiment subclasses) which is a challenging task.

(Nagarajan et al. 2018) proposed a hybrid model for Tweets ternary classification as positive, negative, and neutral classes. One machine learning classifier-Decision Tree along with two optimization algorithms-Genetic Algorithm and Practical Swarm Optimization were preferred for hybridization by the author, which introduced a new model with improved accuracy for sentiment classification. Accuracy, recall precision, and F-measure metrics were used for performance evaluation. Compared to other ML classifiers and optimization methods, the experiment achieved over 90% accuracy.

(Fouad et al. 2018) introduced an efficient ML model for binary classification of Tweets. The author used information gain as a feature extraction technique to minimize the feature vector dimension because the large dimension of a feature vector gives rise to complex calculation issues. After pre-processing feature extraction is done with Bag-of-words containing unigrams and bigrams. A classifier ensemble was used by imposing a majority voting ensemble with Support Vector Machine, Naïve Bayes, Logistic Regression as the base classifiers to take advantage of individual classifiers.

(Wang et al. 2018) performed the binary classification of sentences by using a modified chi-square-based feature clustering and weighting scheme. With the POS tagging (Adverb, Adjective, Verb), the author considered distinguishability as well as the dependency of words. To retrieve features unigram feature extractor is used and Multinomial Naïve Bayes as a text classifier. Precision, recall, and F1-measure were used as performance evaluation metrics for a proposed scheme which showed improvement in the performance of a system as compared to other schemes. (El_Rahman et al. 2019) suggested a model using a combination of both unsupervised and supervised machine learning algorithms for real Twitter data ternary classification. After pre-processing, an unsupervised learning algorithm with a lexicon-based approach was used to classify Tweets as positive, negative, and neutral. Finally, for training, supervised ML algorithms (NB, SVM, ME, DT, RF, Bagging) are used. To show which one from McDonald and KFC is popular, the results were tested in terms of cross-validation, and F-score.

(Dritsas et al. 2019) developed a pre-processing tool that processes natural language data and text to remove noise, which helps to get an efficient and optimum way of sentiment analysis. Pre-processing includes three data representations-unigram, bigram, trigram, and the “Apply features” method for positive and negative labeling for features. To differentiate the proposed work from others, the author also taken into account emojis in pre-processing. SVM, NB classifiers used with different train-test values to measure performance depending on the percentage of accuracy.

(Aziz et al. 2019) focused on detecting degradation in the performance of the sentiment analysis model as all previous work emphasized only improving the result of sentiment analysis and sentiment analysis does not guarantee the same performance of the model against real-time data. The author suggested a Contextual Analysis method to construct a Hierarchical Knowledge Tree (HKT) to discover discrepancies and similarities between the train and the actual dataset, also to cluster words as positive and negative automatically. To measure accuracy, experiments are done with RF and MNB classifiers and they proved change in dataset breaks down the performance of the ML model.

(Kausar et al. 2019) proposed a system with five polarity classes as strongly negative, negative, neutral, positive, strongly positive with three polarity features. Authors used different combinations of adverb features based on their senses in review categories. Performance of the system was evaluated with six classifiers such as Naïve Bayes, Random Forest, Decision Tree, Gradient Boosting, Sequence to Sequence Model, Support Vector Machine, and three evaluation measures -Precision, Recall, F-measure.

(Elbagir et al. 2019), the authors' goal is to carry out a sentiment analysis of Twitter data based on ordinal regression with methods of machine learning, SoftMax, Support vector regression, Random Forest, Decision Tree. The author performed five sentiment polarities classes a highly negative, moderate negative, neutral, moderate positive, highly positive. Accuracy measured by 10-fold cross-validation and results shows that Decision tree, Random Forest have better performance than Support vector regression.

3 General Workflow for Twitter Data Sentiment Analysis

The workflow of sentiment analysis typically followed by all previous researchers with slight differences is shown in Fig. 1. After doing a systematic review of previous work it is observed that Twitter sentiment analysis consist of five major modules: Collection of tweets from Twitter API, Data Cleaning by tokenization, stemming, stop words removal, etc., Extracting Features using Bag of Words or Term Frequency-Inverse Document Frequency techniques along with N-gram approach, Applying Machine Learning Classifiers and finally visualization of results.

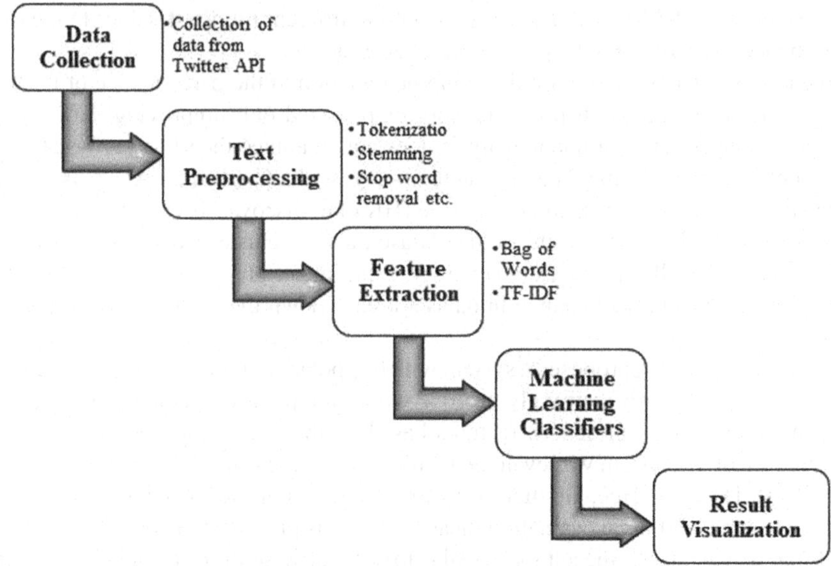

Fig. 1. Sentiment analysis workflow

4 Various Machine Learning Algorithms Along with Number of Polarization Classes and Performance Measures

Table 1 shows various machine learning algorithms used to perform Twitter data sentiment analysis along with performance measures. The main focus of this summary is on polarization classes along with grams used during feature extraction. From the Table 1, it is observed that previous researchers proposed lots of methods while performing sentiment analysis. The common way followed by most of the researchers is classifying text (Tweets or reviews) in positive, negative, and neutral classes. Only a few researchers tried to classify text into more than three classes (multiclass polarity). Also, while doing multiclass classification only unigram features are considered which raises the possibility of a lack of understanding of the meaning of words resulting in wrong polarity classification.

Table 1. Analysis of polarization classes performed along with gram features for twitter sentiment analysis

S no.	Author & year	Publisher	Dataset	Polarization classes	Machine learning techniques	Accuracy	
1	Peng et al. 2016	IEEE/ACM	Tweets with five drug names	Ternary	SVM, Naive Bayes	63% 60%	
2	Pagolu et al. 2016	IEEE	Microsoft tweets from Twitter API	Binary with n-gram	Random Forest, Logistic Regression, SMO	70.49% 57.14% 65.84%	
3	Deshwal et al. 2016	IEEE	Sanders twitter dataset	Ternary with unigram	Naïve Bayes, Bayes Net DMNB, SMO, Hyperpipes, Random Forest	F-score 53% 63% 76% 75% 67% 64%	
4	Juneja et al. 2017	IEEE	Twitter tweets from Twitter API, NLTK Library data	Binary	Original NB, MultinomialNB, BernoulliNB Linear Reg, SGDC, Linear SVC, NuSVC	72.0% 80.0% 75.0% 76.0% 69.0% 71.0% 78.0%	
5	Aziz et al. 2017	IEEE	Amazon and IMDB dataset	Binary for cross domain	Multinomial Naive Bayes, Random Forest Classifier, Stochastic Gradient Descent, Support Vector Machine-Radial Kernel, Support Vector Machine-Linear Kernel	–	
6	Bouazizi et al. 2017	IEEE	Tweeter tweets	7 classes with unigrams	Random Forest "love", "happiness" "fun", "neutral", "hate", "sadness" "anger"	75% 54% 40% 67% 90% 52% 62%	

(continued)

Table 1. (*continued*)

S no.	Author & year	Publisher	Dataset	Polarization classes	Machine learning techniques	Accuracy	
7	Nagarajan et al. 2018	Springer	Twitter API	Ternary	Genetic Algorithm, Particle Swarm Optimization (Optimization algorithms), Decision Tree (ML technique)	90%	
8	Fouad et al. 2018	Springer	Tweets from-Stanford Twitter Sentiment Corpus, Sanders Dataset, Health Care Reform	Binary with unigram and bigrams	Support Vector Machine, Linear Regression, Naïve Bayes, Majority Voting Ensemble	For Sanders data 92.71% 90.11% 91.91% 93.94%	
9	Wang et al. 2018	Springer	Tweets	Binary with unigrams	Multinomial Naïve Bayes	71%	
10	El_Rahman et al. 2019	IEEE	McDonalds and KFC tweets	Ternary with word cloud	Maximum entropy, Naive Bayes, support vector machine, random forest, decision tree, bagging	F-score McD 31% 51% 40% 16% 57% 40%	KFC 67% 55% 67% 73% 60% 74%
11	Dritsas et al. 2019	Springer	Self-driving cars dataset, Airline's dataset	Ternary with unigram, bigram, and trigram	Naive Bayes-unigram bigram Trigram Support Vector machine unigram, bigram trigram	Car 78.66% 79.92% 79.29% 73.04% 76.59% 75.93%	Airlines 85.38% 79.99% 74.26% 77.26% 69.32% 65.25%
12	Kausar et al. 2019	IEEE	Office Product and musical DVD reviews (**not tweets**)	Five classes with adverb features as unigram and bigrams	Support Vector Machine, Decision Tree, Naive Bayes, Random Forest, Gradient Boosting, Sequence to Sequence model	94% 95% 91% 81% 95% 91%	
13	Elbagir et al. 2019	IEEE	Twitter dataset	Five classes with unigram features	Decision Tree, Support Vector Regression, Multinomial Logistic Regression, Random Forest	91.81% 81.95% 67.2% 83.2%	

5 Research Gaps and Challenges

Sentiment analysis is a subtask of natural language processing. It is compared with text classification. Classifying text as positive and negative is a very simple analysis, one of the challenges is how to detect a more in-depth emotion. Sometimes, a given text to analyze will exhibit multipolarity. In these cases, having only the total result of the analysis can be misleading and can sometimes hide valuable information. Here, the total sentiment polarity will be missing key information. Some other challenges can be listed as:

A. Border Between Two Sentiments: The task of multiclass sentiment analysis presents many challenges, from which to state at the beginning, it is difficult even for humans to make difference between two sentiments that are very close to each other. As an example, positive and very positive make it harder for the classifier to define the border between the two.
B. Negation: It is harder to tell that a negative word present in a tweet expresses the opposite sentiment of the negated text.
C. Irony or Sarcasm: Several words in natural language have multiple meanings depending on their context. Their meaning could be similar or totally opposite.
D. Poly-Sentiments: Though tweets are short in length, they may contain multiple sentiments. It is hard to detect all sentiments present within a tweet. This leads to misclassification of tweets, as the classifier unable to determine the dominant sentiment.

6 Conclusions

In this paper, different Twitter sentiment analysis papers have reviewed, concerning polarity classes and feature extraction used. Earlier contributions employed different machine learning algorithms and various implementations for sentiment analysis. It is observed that a little amount of work was done for multiclass sentiment analysis. Also, the present review explores some research gaps and challenges in sentiment analysis which are useful for doing further research in the field.

Acknowledgments. I would like to thank Mr. A. S. Patil, my guide, who guided me to finalize my research topic. Also like to thank my family who helped me and provided me with a thorough understanding of the research.

Referencess

Aziz, A.A., Starkey, A., Bannerman, M.C.: Evaluating cross domain sentiment analysis using supervised machine learning techniques. In: Intelligent Systems Conference, London, UK, 7–8 September 2017. IEEE (2017). 978-1-5090-6435-9/17/$31.00

Aziz, A.A., Starkey, A.: Predicting supervise machine learning performances for sentiment analysis using contextual-based approaches. IEEE Access **8**, 17722–17733 (2019). https://doi.org/10.1109/ACCESS.2019.2958702

Bouazizi, M., Ohtsuki, T.: A pattern-based approach for multi-class sentiment analysis in twitter. IEEE Access **5**, 20617–20639 (2017). https://doi.org/10.1109/ACCESS.2017.2740982

Deshwal, A., Sharma, S.K.: Twitter sentiment analysis using various classification algorithms. In: 5th International Conference on Reliability, Infocom Technologies and Optimization (ICRITO), Uttar Pradesh, Noida. IEEE (2016). 978-1-5090-14897/16/$31.00

Dritsas, E., et al.: Pre-processing framework for twitter sentiment classification. In: MacIntyre, J., Maglogiannis, I., Iliadis, L., Pimenidis, E. (eds.) AIAI 2019. IAICT, vol. 560, pp. 138–149. Springer, Cham (2019). https://doi.org/10.1007/978-3-030-19909-8_12

Elbagir, S., Yang, J.: Sentiment analysis of twitter data based on ordinal classification. ACAI. Sanya, China. ACM (2018). ISBN 978-1-4503-6625-0/18/12…$15.00. https://doi.org/10.1145/3302425.3302488

Elbagir, S., Yang, J.: Twitter sentiment analysis based on ordinal regression. IEEE Access **7**, 163677–163685 (2019). https://doi.org/10.1109/ACCESS.2019.2952127

Fouad, M.M., Gharib, T.F., Mashat, A.S.: Efficient twitter sentiment analysis system with feature selection and classifier ensemble. In: Hassanien, A.E., Tolba, M.F., Elhoseny, M., Mostafa, M. (eds.) AMLTA 2018. AISC, vol. 723, pp. 516–527. Springer, Cham (2018). https://doi.org/10.1007/978-3-319-74690-6_51

Juneja, P., Ojha, U.: Casting online votes: to predict offline results using sentiment analysis by machine learning classifiers. In: 8th ICCCNT, IIT Delhi, Delhi, India, 3–5 July 2017. IEEE (2017)

Kausar, S., Huahu, X., Ahmad, W., Shabir, M.Y., Ahmad, W.: A sentiment polarity categorization technique for online product reviews. IEEE Access **8**, 3594–3605 (2019). https://doi.org/10.1109/ACCESS.2019.2963020

Nagarajan, S.M., Gandhi, U.D.: Classifying streaming of Twitter data based on sentiment analysis using hybridization. Neural Comput. Appl. (2018). https://doi.org/10.1007/s00521-018-3476-3

Pagolu, V.S., Reddy Challa, K.N., Panda, G., Majhi, B.: Sentiment analysis of twitter data for predicting stock market movements. In: International Conference on Signal Processing, Communication, Power and Embedded System (SCOPES). IEEE (2016). 978-1-5090-4620–1/16/$31.00

Peng, Y., Moh, M., Moh, T.: Efficient adverse drug event extraction using twitter sentiment analysis. In: IEEE/ACM International Conference on Advances in Social Networks Analysis and Mining (ASONAM). IEEE, San Francisco (2016). 978-1-5090-28467/16/$31.00

El_Rahman, S.A., Feddah A.A., AlShehri, W.A.: Sentiment analysis of twitter data. IEEE (2019). 978-1-5386-8125-1/19/$31.00

Wang, Y., Kim, K., Lee, B., Youn, H.Y.: Word clustering based on POS feature for efficient twitter sentiment analysis. Human Centric Comput. Inf. Sci. (2018). https://doi.org/10.1186/s13673-018-0140-y

Soft Computing

Efficacy of Online Event Detection with Contextual and Structural Biases

Manisha Samanta(✉), Yogesh K. Meena, and Arka P. Mazumdar

Department of CSE, Malaviya National Institute of Technology Jaipur, Jaipur, Rajasthan, India
{2015rcp9524,ymeena.cse,apmazumdar.cse}@mnit.ac.in

Abstract. Social media streams have brought about a new way of information dissemination in today's world. People use social media streams such as Twitter, Facebook and so on, and see it as an important means of communication. However, huge amount of data generated real time is difficult to manage. One such challenge is the efficient clustering of data to check for a potential event. To tackle this problem, we applied different approaches to find the optimal clustering for event detection using Twitter as our primary source of data. Also, we proposed a theoretical model for purging of dead events. Although the work is experimental, it holds great potential to be further established as an efficient and effective event detection approach.

Keywords: Online event detection · Context · Structural · Similarity · Social media

1 Introduction

Event detection is a new paradigm gaining popularity with technological advancements and research. An event is any occurrence in a determinable space and time, online or offline, with or without the participation of human agents. It may be a part of a group of similar incidents occurring earlier or later in the timeline. The general idea of online event detection (OED) is to effectively and efficiently indicate possible events occurring online or offline anywhere globally by analyzing the user posts on social media. Traditionally, events were detected through RSS feeds, news articles, and classified columns. However, due to the advent of social media streams such as Twitter, Facebook, Instagram, phenomenal growth, popularity, and their use as a communication medium, a massive amount of real-time data is available at our disposal. The textual contents obtained through any social media platform give insights into why particular events are popular with most of the population.

Among the OED approaches available in the literature, it can be observed that either they solely use context similarity among the documents, or they additionally use structural information for grouping these documents. Here, context denotes the text posted by a user, and structural information is any relevant

V. E. Balas et al. (Eds.): ICETCE 2022, CCIS 1591, pp. 419–430, 2022.
https://doi.org/10.1007/978-3-031-07012-9_36

information except the context, such as sender, receiver, location, followers, and others. The primary contribution of this work is to analyze the effect of each of these two classes of parameters in detecting events. This study also aims to analyze their balance and the effect of the threshold for clustering to achieve improved outcomes. The experimental results show that the event detection system performs better in the presence of both of the parameters; however, it also depends significantly on the selection of their ratio and the clustering threshold value.

The rest of the article is arranged as follows. Section 2 presents a brief discussion on the related works in literature. The computational model for both context and structural similarity as considered here is presented in Sect. 3. The next section presents the experimental results and detailed discussions on the same. Finally, the article is concluded in Sect. 5.

2 Related Work

Initial event detection-related works focused primarily on news documents (i.e., news wires, radio broadcast). The basic assumption by these works is that all the documents carry some meaningful event information [2,18,25] as they used retrospective data sources, such as new channel broadcast data, emails, blogs, etc. The scenario has greatly shifted recently, as social media streams are mostly used as the source of information for the NED (New Event Detection) technique. These social media posts are full of jabbers and highly unstructured. To process such data streams, unsupervised machine learning techniques, such as clustering, are introduced [2,6,11,18,20,24,25]. Before clustering, preprocessing is performed with the help of natural language processing tools, such as tokenization, stemming and stop-word removal, pos tagging, NER, etc. [1,3,11,20]. Then contextual similarity of incoming social streams are compared with the existing event clusters and put into the most similar cluster only if it satisfies a certain threshold, else forms a new cluster by itself [1,3].

The weights of each of the terms in these documents are measured using a similarity measurement technique. One of the seminal techniques for the purpose is TF-IDF [15,16] and was proposed by (Johns *et al.*). To date, it is a highly accepted metric in the field of unsupervised document clustering. To improve the efficacy of TF-IDF, Deisy *et al.* [7] introduced Modified Inverse Document Frequency (MIDF) scheme for text categorization. But some specific terms which carry different meanings or features based on different contexts, TF-IDF fails to measure such complicated situational terms; therefore, Lertnattee [17] introduced the concept of Inverse Class Frequency (ICF). Along with this class function, Ren *et al.* [21] proposed two contrasts of TF-IDF, those are, TF-IDF-ICF (Term frequency-inverse document frequency) and TF-IDF-ICSDF (Term frequency-inverse class space density frequency). The second one showed better results with a much higher positive rate of term differentiation for all the terms.

To measure the distinguishing power of a term in the field of text categorization, TF-IGM (term frequency-inverse gravity moment) is proposed by

Chen et al. [5]. Further, enhanced variants of TF-IGM are introduced, such as SQRTTF-IGM-imp and TF-IGM-imp [9]. The authors conducted an experimental study on the well-known datasets, Reuters-21578, and mini-20-news-groups, to prove the performance of the proposed metric. The article then proposes another set of metrics, TF-MONO and SRTF-MONO [10], by considering the maximum-occurrence and non-occurrence of terms.

The direct or indirect human intervention of an event can be described as the structural property of the event [23]. In [1], the authors proposed a structural similarity metric by considering the information of involved people (sender and receiver) of the post. In Nguyen *et al.* [19], the real-time data flow of social media is compared with the behavior of users. Therefore, to estimate the score of a keyword, they include various information related to users. To formulate twitter-based information-theoretic metric [12], the authors proposed modeling of user credibility, such as follower list, retweet count, etc., being essential aspects. The proposed work in [26] suggests the inclusion of structural properties, such as information related to participants, organizations, and location. They proposed a deep clustering method to indicate possible event clusters with this information. Cao et al. [4] proposed a Knowledge-Preserving Incremental Heterogeneous Graph Neural Network (KBGNN) model to detect social events by incrementally expanding its knowledges from SNS texts. Further, to design the social streams, the authors proposed using the user information of the sender and any other user name mentioned in the text. In work presented by Girolamo *et al.* [8], we observed that during the prepossessing stage of tweet data, user information is taken into account to compute structural similarity.

3 Structural and Context Similarities in Online Event Detection

The literature shows that two classes of approaches exist for online event detection based on the information used for the purpose. While a group of researchers focuses on contextual similarities, others introduce structural similarities and context similarities together. Here, context similarity is performed through the text of the tweets, and structural similarity is through the network structure and the meta-data of the messages. The structural information also conveys the relationships of the senders and the receivers. Since such similarity measurements require information other than the textual and temporal information about any document, defining the information model of the social media posts for such approaches is of utmost importance. Let a streaming object in social media denote an exchange of information between actors present in the media. Each of such streaming objects contains at least the following three information:

- *Text document:* It consists of the textual content of the social stream message between two entities as posted or communicated by the sender. This is primarily used to compute textual similarity.

- *Sender:* The sender is the originator of the text document. The meta-data of each document on social media generally contain the unique identity of the sender.
- *Receiver:* The receiver here is the set of zero or more actors who receive the message posted by the sender. The meta-data of each document generally contains the unique identity of all the receivers.

Here, the *sender* and the *receiver* fields are structural information. Though most social networks, such as Twitter, Facebook, and Instagram, offer a larger variety of structural information, these two are common fields and are provided across all platforms. This section, next, explores both of these approaches and discusses their effects in the process.

3.1 Structure and Context Based Event Detection

Many of the proposed works in the literature advocate inclusion of relationships [8,12,20] among the actors in the social network to identify probable events as the messages exchanged between closely related groups of actors can indicate events of common interest. Therefore, these online real-time event detection approaches have two primary tasks: (a) the ability to use contextual similarity and (b) the ability to combine structural similarity with contextual similarity for calculating the total similarity of an incoming tweet with the existing clusters.

One such seminal contribution towards this direction is proposed by Aggarwal *et al.* [1], and later followed by [8,12,20]. Among many available similarity metrics, TF-IDF is the most widely used due to its simplicity and effectiveness. TF-IDF is a statistical measure that intends to reflect how important a word is to a document in a collection of documents. The TF value increases proportionally to the number of times a word appears in the document. However, this value is offset by the frequency of this word in the rest of the corpus, represented by IDF. The latter property helps adjust the word's importance measurement because some words appear more frequently in general and do not convey much information. To cluster similar documents that may indicate a new event, several articles in the literature used cosine similarity [22]. With the help of a similarity measure technique, TF-IDF in this case, cosine similarity then computes the context similarity between an incoming object's text document and the existing clusters as shown in Eq. 1.

$$SimC(D_i, C_r) = \frac{\sum_{k=1}^{t}(T_{ik} \cdot T_{rk})}{\sqrt{\sum_{k=1}^{t}(T_{ik})^2 \cdot \sum_{k=1}^{t}(T_{rk})^2}} \tag{1}$$

where, D_i is the incoming tweet document, C_r is the candidate cluster. All the terms in a cluster and the incoming document are kept in a term vectors of size t, where the values of kth term in this vectors, are given by T_{ik} andT_{rk} respectively, represent the term frequencies of the same. It is to be noted that, both the term vectors are modified to same size by adding entries of unavailable words from each other.

Structural similarity represents the similarity in the set of senders and receivers between a cluster and the incoming stream object. It into the picture based on the observation that a closely related group of people are more likely to discuss related social events than a sparsely connected group of people [1]. This information may help identify the similarity in the network structure of the incoming social stream object and the cluster in question. The structural similarity ($SimS$) can be computed shown in Eq. 2

$$SimS(S_i, C_r) = \frac{\sum_{t=1}^{S_r} b_t \cdot v_{rt}}{\sqrt{||R_i \bigcup \{q_i\}||\sum_{t=1}^{S_r} v_{rt}}} \tag{2}$$

Here, S_i is the incoming streaming object, C_r in the cluster being compared, v_{rt} denotes the frequency of tth node in C_r, R_i represents the set of recipients of the ith stream object, and q_i is the sender of the object. Moreover, b_t represents the binary information for t^{th} node in $R_i \bigcup \{q_i\}$. In Eq. 2, the numerator represents the summation of the number of common nodes between the cluster and the incoming object. Here, the common nodes can be from either the sender or the receiver. The denominator shows the summation of the frequency of the sender and the receivers present in the cluster, multiplied by the *L1 norm* [14] of the node list of the incoming social stream object. L1 norm indicates the portion of nodes that are similar between the cluster and the incoming object concerning the total number of node occurrences in the cluster.

After the structural and the context similarity are calculated, they can be combined using a linear equation to calculate the total similarity as given in Eq. 3

$$Sim = \lambda \cdot SimC(D_i, C_r) + (1 - \lambda) \cdot SimS(S_i, C_r) \tag{3}$$

In the above equation, the parameter *lambda*'s value is between 0 to 1 and denote the weights of both $SimC$ and $SimS$. The λ can be varied to attain different weights of $SimC$ and $SimS$ to indicate their importance in the total similarity computation. The value of this parameter is to be specified by the designers based on their requirements.

The similarity value must be greater than an assumed threshold value to cluster a particular document. If no matching cluster is found, a new cluster is created with this document. The threshold value can be calculated as shown in Eq. 4, where α denotes the computed threshold value, and μ and σ are computed as given in Eqs. 5 and 6, respectively.

$$\alpha = \mu - (3 * \sigma) \tag{4}$$

$$\mu = \frac{M_1}{M_0} \tag{5}$$

$$\sigma = \sqrt{\frac{M_2}{M_0} - (\frac{M_1}{M_0})^2} \tag{6}$$

In the above equations, M_0, M_1, and M_2 are zero, first and second-order moments [14] of the closest similarity values maintained continuously and can be updated additive after every cluster update.

3.2 Context Based Event Detection

Most of the approaches for online event detection in the literature [2,3,24] do not consider any structural property of the streaming objects. The model assumed by this class of works is more straightforward than the model described in Sect. 3.1. Here, each of the streaming objects contains one critical piece of information, the *Text document*. Therefore, the *sender* and *receiver* fields are of no importance to these approaches.

One crucial work in this category, which is also similar to the one previously discussed in Sect. 3.1, is presented by Becker *et al.* [3], and later extended by the works in [12,13]. This class of work aims to identify any real-world events that are being discussed over social media. The proposed approaches are generally clustering-based methods, where the incoming tweets are first compared with already formed clusters based on the computed similarity values. Otherwise, a new cluster is formed to accommodate the tweet. The work here proposes using an incremental clustering algorithm with a threshold tuned empirically during a training phase. For scalability, a centroid representation of each cluster is formed that consists of a statistical summary of all the documents in the cluster. Each document is represented as a TF-IDF weight vector of its textual content and uses the cosine similarity metric to calculate the similarity value, which is similar to the Eq. 1 in Sect. 3.1.

4 Comparison of Implemented Works

In this section, we present a detailed analysis of both the two approaches as discussed in Sect. 3. We first discuss the dataset and the evaluation metrics to be used for the experiments, respectively. Finally, we present and discuss the experimental results in detail based on these parameters.

4.1 Data Set

As the data from Twitter can be freely accessed, this article aims to use tweet data set for the purpose of experiments. However, due to a very high amount of tweets generated over Twitter, it is challenging to evaluate these algorithms in real-time. Therefore, we first accumulate and prepare a Twitter social streaming dataset crawled from *Twitter* through streaming API. Further, as the structural information, as mentioned in Sect. 3, between the actors is required for measuring structural similarity, the list of receiver IDs for all the tweets is also downloaded.

The evaluation of the algorithms requires labeling ground truths manually to each downloaded twitter document, where each possible event is given a unique integer number. One of these numbers is assigned to each of these documents, acting as the ground truth.

4.2 Evaluation Metric

As this article aims to analyze the effects of structural and context similarities on event detection techniques, we chose to compare the two approaches based on the two parameters: homogeneity and cluster purity [1].

Event detection only makes sense if a cluster depicts a meaningful event. To represent a meaningful event, such clusters should contain only those tweets that are about the same or similar event, thereby providing information toward it. If a cluster contains tweets from other possible events or some of its member documents are unrelated to most of the tweets in the cluster, it reduces the quality of the cluster. Homogeneity is a metric that assigns scores to each cluster representing the percentage of tweets that correlate to the event that the cluster indicates. This metric also penalizes the clusters that contain a higher number of tweets unrelated to the topic of the corresponding cluster. The next metric, purity, or the cluster purity denotes the ratio between the number of homogeneous clusters that denote an event and the total number of clusters as shown in Eq. 7.

$$Purity = \frac{(No.of HomogeneousClusters)}{(TotalClusters)} \tag{7}$$

Here, we deviate from the existing definition of the metric such that, as the system is online streaming-based, the total number of documents can not be used here.

4.3 Experimental Results

We present the experimental results in four phases. First, we try to analyze the effect of structural similarity in event detection process by comparing the two techniques discussed in Sect. 3. Next, we try to analyze the effect of selecting different values for λ to indicate the effect of context and structural similarities

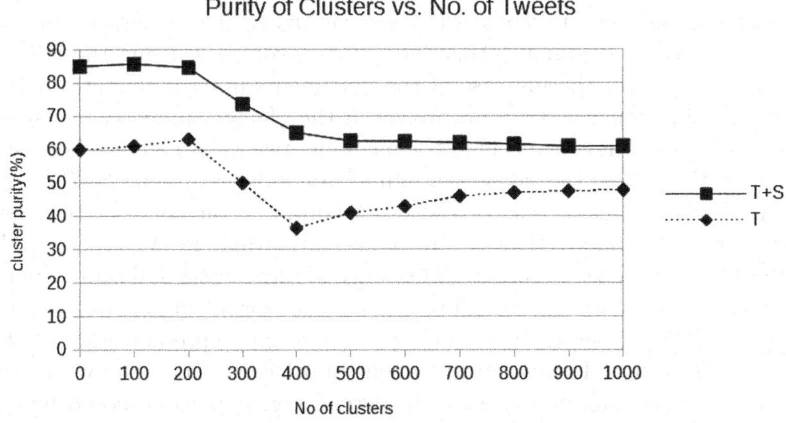

Fig. 1. Purity of clusters with respect to the number of tweets

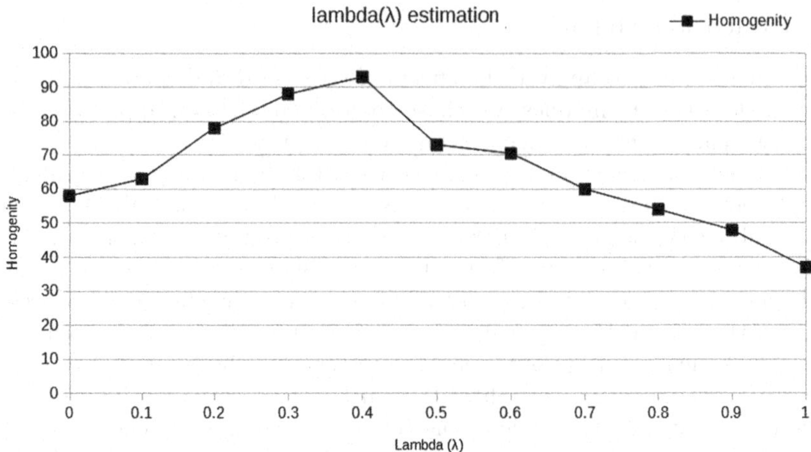

Fig. 2. Effect of different Λ on average homogeneity

in the event detection efficacy. Similarly, we present the effect of the threshold value (α) in overall purity of clustering. Finally, we compare both the approaches with their suitable values for λ and α from the previous two analyses. All these experimental results are presented here. To run the experiments, the dataset is first divided into three sub sets. We calculated homogeneity and purity of clusters as discussed above on each part of data using both the approaches and the final results are present by taking mean of these experiments.

Figure 1 exhibits that the purity of clusters as the number of the cluster increases for both context similarity, shown as T, and structural similarity, shown as $T + S$. The figure shows that when the number of clusters is small, the purity for structural similarity is 25% higher than the context similarity-based approach. It can also be observed that, though the performance of both the approaches increases slightly, as the number of classes grows, the cluster purity eventually falls heavily and then stabilizes eventually. As the new tweets come, many of them carry common terms and actors involved, causing the fall in the purity. However, with the increase of the documents in each cluster, important words eventually replace the weak words in the cluster summary, causing the reduction in the wrong clustering of the incoming tweets. It can also be observed that, in all the cases, the structural similarity induces about 20–25% better cluster purity just by using only two structural information alongside the context.

Next, we try to analyze the effect of choosing the value for λ as shown in Eq. 3 that indicates the linear combination between the structural and contextual similarity. In most of the works, including [1], the value for λ (ranging between 0 and 1) is assumed to be user-defined. In Fig. 2, the x-axis represents the weightage of λ, varied from 0 to 1, with an increment of 0.1 each time. The y axis represents the corresponding homogeneity in percentage. It is to be noted here that the $\lambda = 0$ represents context similarity only. From the graph, it is clear that the

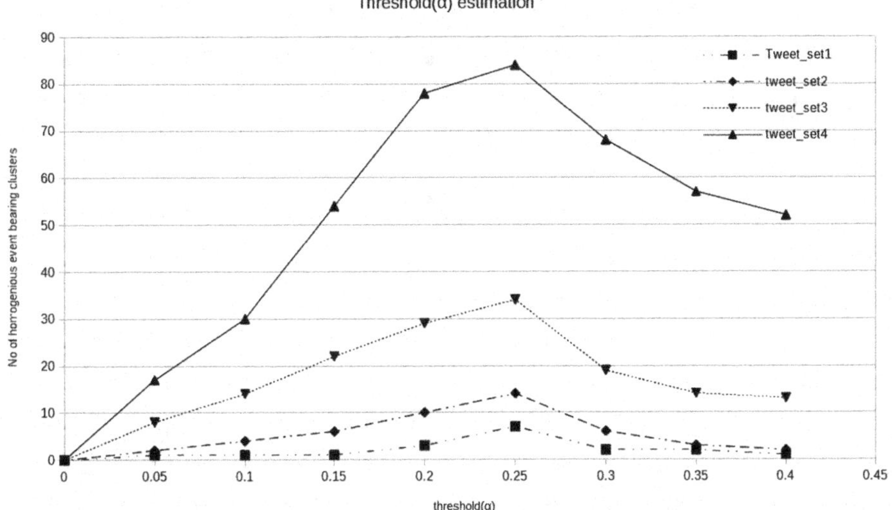

Fig. 3. Number of homogeneous clusters with respect to α

homogeneity also increases with the increment of λ. However, to a certain extent, the homogeneity increases until the λ up to 0.4 in our case. The homogeneity starts decreasing when it is increased further. This implies that the inclusion of structural similarities in event detection may increase the homogeneity of the clusters as compared to the absence of either of them, denoted by $\lambda = 0$ and $\lambda = 0$ respectively.

The parameter α denotes the threshold value that is dynamically calculated by the Eq. 4 as discussed previously. If a high α value is assumed, tweets about similar events may be clustered differently, thereby reducing the purity of the clusters as the number of clusters will increase unnecessarily. On the other hand, if we keep a lower α value, it will eventually cluster dissimilar tweets together, resulting in reduced homogeneity of that cluster. Basically, the alpha value tells us the minimum similarity required for an incoming tweet to be clustered inside an existing cluster. If these minimum total similarity criteria are not met, then the incoming tweet is used to make a new cluster with just this particular tweet. If we keep a higher alpha value, it would require an extremely high total similarity value for a tweet to get clustered inside a cluster. Since the value of alpha can be dynamically determined, the user does not control setting its value. However, this can lead to the selection of non-optimal α as there can be a high number of irrelevant terms in a cluster, as shown in Fig. 2. Therefore, this article next analyses the effects of the different values of α, as depicted in Fig. 3. The figure presents the number of homogeneous event-bearing clusters with respect to different threshold values for all four subsets of our data sets. It can be observed that the homogeneous clusters increase with higher values of α, though after a point (for the value 0.25 in our case), it starts decreasing, for the reason as discussed previously.

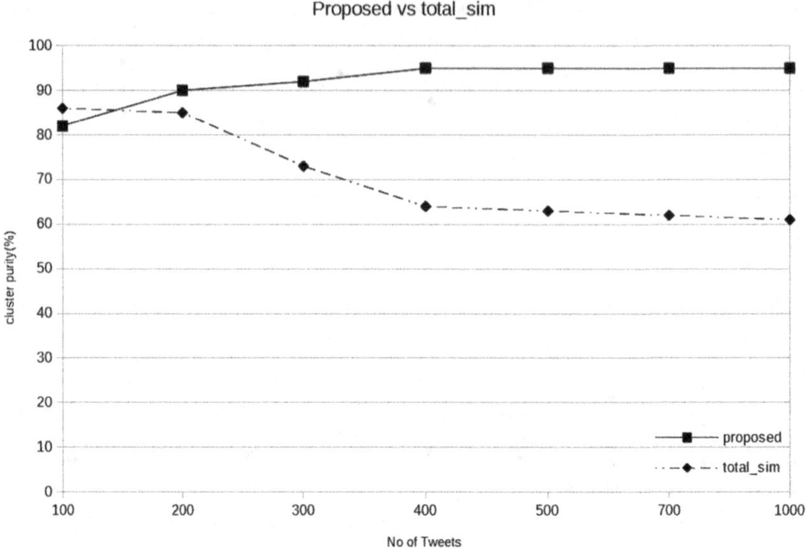

Fig. 4. Cluster purity with respect to computed α vs. analysed α

Finally, we aim to analyze the quality of the α estimation, as shown in the previous section. Figure 4 depicts the comparison of estimated α, computed as 0.4 in this case, and the proposed value 0.25, as observed in Fig. 3. It presents the comparison in terms of purity with respect to the number of tweets. From the above graph, we can infer that the value of α calculated independently yielded better purity as the number of tweets increases than the selection using Eq. 4, indicating that the selection induces more error in the presence of a larger number of documents.

5 Conclusion

Event detection from social media streams is a trending topic in today's world of social media craze. Due to the popularity of social media streams, event detection on social media has gained attention from the research community. Literature indicates two classes of approaches, where either only context similarity is used, or structural similarity is also considered alongside. This article aims to analyze the effect of both approaches in detecting the open event and found that a balance between both is desired to achieve better cluster purity. Moreover, it also observed that selection of the threshold value also impacts greatly in attaining homogeneity and purity of the system, where either purity or both of may reduce based on the selected value. Though this article is able to point out the considerations to attain an effective event detection system, it assumes a simple model of computation with few limitations, such as limited structural information, limited corpus size, presence of sarcasm in the documents, erroneous data in the dataset. We wish to pursue and continue our work in these directions in the future.

References

1. Aggarwal, C.C., Subbian, K.: Event detection in social streams. In: Proceedings of the 2012 SIAM International Conference on Data Mining, pp. 624–635. SIAM (2012)
2. Allan, J., Papka, R., Lavrenko, V.: On-line new event detection and tracking. In: Proceedings of the 21st Annual International ACM SIGIR Conference on Research and Development in Information Retrieval, pp. 37–45 (1998)
3. Becker, H., Naaman, M., Gravano, L.: Beyond trending topics: real-world event identification on twitter. In: Proceedings of the International AAAI Conference on Web and Social Media, vol. 5 (2011)
4. Cao, Y., Peng, H., Wu, J., Dou, Y., Li, J., Yu, P.S.: Knowledge-preserving incremental social event detection via heterogeneous gnns. In: Proceedings of the Web Conference 2021, pp. 3383–3395 (2021)
5. Chen, K., Zhang, Z., Long, J., Zhang, H.: Turning from TF-IDF to TF-IGM for term weighting in text classification. Expert Syst. Appl. **66**, 245–260 (2016)
6. Chen, L., Roy, A.: Event detection from flickr data through wavelet-based spatial analysis. In: Proceedings of the 18th ACM Conference on Information and Knowledge Management, pp. 523–532 (2009)
7. Deisy, C., Gowri, M., Baskar, S., Kalaiarasi, S.M.A., Ramraj, N.: A novel term weighting scheme midf for text categorization. J. Eng. Sci. Technol. **5**(1), 94–107 (2010)
8. Di Girolamo, R., Esposito, C., Moscato, V., Sperlí, G.: Evolutionary game theoretical on-line event detection over tweet streams. Knowl.-Based Syst. **211**, 106563 (2021)
9. Dogan, T., Uysal, A.K.: Improved inverse gravity moment term weighting for text classification. Expert Syst. Appl. **130**, 45–59 (2019)
10. Dogan, T., Uysal, A.K.: A novel term weighting scheme for text classification: TF-MONO. J. Informetrics **14**(4), 101076 (2020)
11. Dong, X., Mavroeidis, D., Calabrese, F., Frossard, P.: Multiscale event detection in social media. Data Min. Knowl. Dis. **29**(5), 1374–1405 (2015). https://doi.org/10.1007/s10618-015-0421-2
12. Doulamis, N.D., Doulamis, A.D., Kokkinos, P., Varvarigos, E.M.: Event detection in twitter microblogging. IEEE trans. cybern. **46**(12), 2810–2824 (2015)
13. Hasan, M., Orgun, M.A., Schwitter, R.: TwitterNews+: a framework for real time event detection from the twitter data stream. In: Spiro, E., Ahn, Y.-Y. (eds.) SocInfo 2016. LNCS, vol. 10046, pp. 224–239. Springer, Cham (2016). https://doi.org/10.1007/978-3-319-47880-7_14
14. Horn, R.A., Johnson, C.R.: Norms for vectors and matrices. Matrix analysis, pp. 313–386 (1990)
15. Jones, K.S.: A statistical interpretation of term specificity and its application in retrieval. J. Documentation (1972)
16. Jones, K.S.: A statistical interpretation of term specificity and its application in retrieval. J. Documentation (2004)

17. Lertnattee, V., Theeramunkong, T.: Analysis of inverse class frequency in centroid-based text classification. In: IEEE International Symposium on Communications and Information Technology, 2004. ISCIT 2004, vol. 2, pp. 1171–1176. IEEE (2004)
18. Li, Z., Wang, B., Li, M., Ma, W.Y.: A probabilistic model for retrospective news event detection. In: Proceedings of the 28th Annual International ACM SIGIR Conference on Research and Development in Information Retrieval, pp. 106–113 (2005)
19. Nguyen, D.T., Jung, J.E.: Real-time event detection for online behavioral analysis of big social data. Future Gener. Comput. Syst. **66**, 137–145 (2017)
20. Nguyen, D.T., Jung, J.J.: Real-time event detection on social data stream. Mobile Networks Appl. **20**(4), 475–486 (2014). https://doi.org/10.1007/s11036-014-0557-0
21. Ren, F., Sohrab, M.G.: Class-indexing-based term weighting for automatic text classification. Inf. Sci. **236**, 109–125 (2013)
22. Salton, G., McGill, M.J.: Introduction to Modern Information Retrieval. mcgraw-hill, New York (1983)
23. Samanta, M., Meena, Y.K., Mazumdar, A.P., Govil, M.C.: EDOF: an open framework for event detection systems. In: 2017 8th International Conference on Computing, Communication and Networking Technologies (ICCCNT), pp. 1–7. IEEE (2017)
24. Sayyadi, H., Hurst, M., Maykov, A.: Event detection and tracking in social streams. In: Third international AAAI Conference on Weblogs and Social Media (2009)
25. Yang, Y., Pierce, T., Carbonell, J.: A study of retrospective and on-line event detection. In: Proceedings of the 21st Annual International ACM SIGIR Conference on Research and Development in Information Retrieval, pp. 28–36 (1998)
26. Zhang, J.: Robust social event detection via deep clustering. In: 2021 IEEE International Conference on Parallel and Distributed Processing with Applications, Big Data and Cloud Computing, Sustainable Computing and Communications, Social Computing and Networking (ISPA/BDCloud/SocialCom/SustainCom), pp. 814–819. IEEE (2021)

Popularity Bias in Recommender Systems - A Review

Abdul Basit Ahanger, Syed Wajid Aalam, Muzafar Rasool Bhat(✉), and Assif Assad

Islamic University of Science and Technology, Awantipora, India
muzafar.rasool@islamicuniversity.edu.in

Abstract. With the advancement in recommendation techniques, focus is diverted from just making them more accurate to making them fairer and diverse, thus catering to the set of less-popular items (the long tail) that often get neglected due to inherent biases in recommender systems. Popularity bias has been recently acknowledged as a major bias of critical concern in the field of recommender systems. Although research on popularity bias has gained pace from the last couple of years, this field is believed to be still in its infancy. To advance research in this area, this paper thoroughly investigates current state of the art and could have a very positive impact on further research in popularity bias. Besides the mitigation techniques discussed in this paper, allied evaluation metrics that were used in measuring popularity bias have also been discussed.

Keywords: Popularity bias · Algorithmic bias · Recommender systems

1 Introduction

1.1 Recommender Systems

Recommender Systems (RS) have evolved, since their development in these three decades, from information filtering systems designed to tackle with the information overload of humongous data available online to accurate online prediction systems. The main goal of recommendation systems is to suggest the most relevant items to the users as accurately as possible to the best satisfaction of the user through analysis of their preferences by observing their interactions (clicks, ratings, likes etc.) with the system in question. The evaluation of the accuracy of the recommender system has improved from error based metrics like Mean Absolute Error to accuracy based metrics like precision, recall and nDCG [10].

As the prediction accuracy of recommender systems rose, research focus shifted to making recommendations more fair, diverse, novel and serendipitous by removing biases inherent in the data and the bias amplification that arose from the algorithms themselves.

Popularity bias is one of the biases found in recommender systems. Research has found that user satisfaction in popular trend suggestions is of the same

V. E. Balas et al. (Eds.): ICETCE 2022, CCIS 1591, pp. 431–444, 2022.
https://doi.org/10.1007/978-3-031-07012-9_37

order of magnitude as the suggestions made by state of the art recommender systems with lower cost of development and maintenance [15] which raises the question whether trying to mitigate the popularity bias is of any significance. However there are many ethical issues that arise from such biases in addition to the research showing considerable interest of users in the non popular category of items.

Recommending popular trends is a good option for non-personalized systems or when the user-item interactions are sparse, however in personalized systems user satisfaction can be increased with the introduction of novel and diverse choices that would help increase the serendipity in the systems.

The effect of popularity bias in recommender systems has been first studied by [16] to show the effect of popularity in music recommendations. And a factor was naturally found dominant in collaborative filtering recommender systems over content based recommender systems. The paper also laid the groundwork in measuring popularity by using the long tail model.

Interpreting popularity is one of the first tasks in trying to establish the existence of popularity bias, measuring it, analysing its effects on recommendation quality and trying to mitigate it (if needed) to the extent necessary.

1.2 Popularity

To understand popularity bias, it is important to settle on a formal definition of popularity and how it is used in the literature, particularly in terms of recommender systems.

Abdollapuri, 2020 [1] defines popularity of an item i in a distribution of user-item-rating list as ratio of users who rated the item to the total number of users in the list with K as the set of users and I as the set of items:

$$popularity(i) = \frac{\sum_{k \in K} \mathbb{I}(i \in P_k)}{|K|} \tag{1}$$

where P_k is the user profile of user k in

Popularity is also calculated by [26] as the ratio between the number of ratings an item i gets to the number of ratings of the most rated item.

$$popularity(i) = \frac{|ratio(i)|}{\max_{i \in I} ratio(i)} \tag{2}$$

The above definition is categorized as *total popularity* by [15] and is most commonly seen in literature [16,23,26].

This notion, however, treats every interaction with the system as a positive interaction. There are instances where a user negatively interacts with the system or their interaction is neutral i.e. they are neither pleased or displeased with the system although the system accounts for these interactions in favour of the item. When only positive interactions of a user (either implicit or explict) are taken into account it is called *relevant popularity* [15].

2 Popularity Bias Evaluation and Mitigation

The error based metrics like RMSE initially used to evaluate Recommender Systems were shown to be insufficient for proper evaluation. Information Retrieval Metrics like MAP and NDCG were found better suited for Recommender System Evaluation as they preferred ranking over prediction errors. The evaluation and mitigation of popularity bias has been studied over the course of time with each proposed mitigation technique usually having their own evaluation metric. Some recent and promising mitigation techniques along with their evaluation metrics are studied in this paper in addition to few studies that only proposed evaluation methods.

2.1 Statistical Treatment to Training Data

The popularity bias mitigation techniques can be carried out through pre-processing the data itself. The idea is to remove the bias in the training data which would in turn let the RS avoid such biases in their recommendations. [10,23]

Two such methods are proposed by [10]. In the first method *Percentile Based Approach (P1R)*, the data is divided into popular and non-popular items and fed separately to the RS. The difference between the predicted rating of the popular set and the non-popular set is not much since the RS was trained on statistically fairer datasets.

In second approach *Uniform Item Test Profile (UAR)* [10] treats the dataset such that each item has same amount of test rating which would remove the statistical advantage that the popular items might have otherwise.

In both cases the size of the training set gets considerably reduced causing more sparsity in user-item interaction matrix, which negatively impacts the overall accuracy of the recommender system.

2.2 Model Based and Reranking Techniques

The two most widely used techniques in literature are model based and re-ranking techniques [1]. In *Model Based* technique, the objective function is modified to adjust for item popularity in final rating prediction. The *re-ranking* technique is a post processing method where the recommendations produced by a RS are reordered to minimize the effect of popularity bias.

The various mitigation techniques and evaluation methods found in literature are discussed from here on.

2.3 Using Novelty Score

A Novelty Based Approach (MCFNR) is proposed by [9]. They introduce novelty score to tackle with the popularity bias. They categorize users into persistent and occasional by calculating their login count and comparing it to a threshold.

Normal Collaborative Filtering is used for occasional users as popularity bias is less likely to impact the satisfaction level of occasional users. The proposed (MCFNR) is used on persistent users whose satisfaction level may be improved by recommending items that are novel yet relevant to them by introducing Novelty Score metric on the items distributed in the long tail. After identifying the user as persistent and predicting unseen but relevant items for the user, they introduce a novelty score as a product of item frequency and log of total users to users who are interested in an item i. The Novelty Score (NS) of an item i is given as :

$$NS(i) = K(i) \times L(i) \tag{3}$$

K, *Item Frequency* and L *Inverse User Frequency* are in-turn computed as:

$$K(i) = \frac{ratings\ of\ item\ i}{max\ ratings\ received\ by\ any\ item} \tag{4}$$

$$L(i) = \log\left\{\frac{total\_users}{users\ who\ rated\ i}\right\} \tag{5}$$

The novelty score is used to introduce new items to the user from the long tail that were still relevant but less likely to show up in the recommendation list due to popularity bias.

They used Precision, Recall and F-Measure for evaluation of recommendation accuracy however this study suggested no metric to evaluate how much popularity bias was mitigated.

2.4 ReGularization

Popularity bias mitigation can be treated synonymous to long tail coverage of items [3]. They use regularization based techniques (techniques used to avoid over-fitting) in learning to rank algorithms (Alternating Least Square in this case). Like Bedi, et al. [9], they also didn't experiment on occasional users, thus removing users who have given fewer ratings. They used a modified version of RankALS [31] by introducing regularization in it to control popularity bias.

In order to mitigate popularity bias in their main objective function

$$\min_{K,Q} aob(P, Q) + \lambda rg(K, Q), \tag{6}$$

where K is the user matrix and Q, the item matrix, $aob()$ is the objective function, $rg()$ is the regularizer and λ is the balancing factor. They identify and try to minimize the regularization component of the equation by assuming that a fair recommendation distribution would have half of its recommendations from short head and half from the medium tail. Dissimilarity matrix D is used to compute the regularizer. It contains pairwise dissimilarities between the items, in such a way that the recommendation list thus produced tries to create a balance between accuracy and intra-list diversity (ILD).

Average Percentage of Tail Items (APT) metric is used by [3]. This metric (also used in [5]), calculates the proportion of the long tail items in the user's recommendation list. The metric is defined as :

$$APT = \frac{1}{|K|} \sum_{k \in K} \frac{|i, i \in L_k \cap \Phi|}{|L_k|} \tag{7}$$

where L_k is the recommendation profile of user k from the set of users K. Φ is the set of long tail items.

2.5 Personalized Long-Tail Promotion

The limitation of using regularizer in [3] is that the regulzarizer has to be encoded within the algorithm itself and its performance highly depends on the type of algorithm that is being used. The same authors also propose a method [5] which takes personalization into account to boost long tail. The methods is discussed below.

For a user k in a recommendation list R, we can re-rank R to produce a new list S ($|S| < R$), iteratively using the following equation :

$$P(v|k) + \lambda P(v, S'|k) \tag{8}$$

where $P(v|k)$ is the probability that $k \in K$ is interested in $v \in V$ (K and V being the user-list and the item list) and $P(v, S'|k)$ is the probability that k is interested in v and v is currently not in S. In order to diversify the recommendation from both short head Γ and long tail Γ', $P(v, S'|k)$ is computed as :

$$P(v, S'|k) = \sum_{d \in \{\Gamma, \Gamma'\}} P(d|k)P(v|d) \prod_{i \in S} (1 - P(i|d, S)) \tag{9}$$

where $P(v|d)$ is 1 if $v \in d$ or else 0.

In addition to the APT referred in Eq. 7, they used *Average Recommendation Popularity (ARP)* taken from [36]. ARP computes the average popularity of items in a recommendation list by computing the sum of the number of items an item appears for each user in the list of users L and averaging it.

$$ARP = \frac{1}{|K|} \sum_{k \in K} \frac{\sum_{i \in L_k} \phi(i)}{|L_k|} \tag{10}$$

where L_k is the recommendation list of user k, $\phi(i)$ is the rating frequency of item i in the training set (i.e. how popular the item is) and K is the number of users. The higher values of ARP are desired for more balanced recommendation. The metric is susceptible to outliers and can show a low ARP only because a few extremely unpopular items popped up in the recommendation list.

They also used *Average Coverage of Long Tail Items (ACLT)* [5]. ACLT evaluates the extent to which long-tail items have been covered by the recommendation list. This removes the problem of ACT which would give higher values irrespective of the ratio of long-tail items in the data. ACLT is computed as:

$$ACLT = \frac{1}{|K_t|} \sum_{k \in K_t} \sum_{i \in L_k} \mathbb{1}(i \in \Gamma) \tag{11}$$

where $\mathbb{1}(i \in \Gamma)$ is 1 when i is a long tail item.

2.6 Bias Aware Variational Encoders

After showing significant improvement in recommendation accuracy through Variational Autoencoders(VAC) [25], VAC were proven to suffer from popularity bias [12] like other Collaborative Filtering RS. To mitigate popularity bias [13] gives penalty to popular item recommendations of Collaborative Filtering Recommendation Systems based on Variational Autoencoders by controlling a weighing parameter of VAC to find a perfect balance between accuracy and bias mitigation. Their experiments showed 8% reduction in bias while affecting the recommendation accuracy by 3%.

The authors introduced a weighing variable λ in the decoder part of the VAC to penalize the popular items.

Variational Autoencoders work by minimizing the Evidence Lower Bound function (ELB), where ELB is computed as:

$$ELB = \mathbb{E}_z[\log d(x|z)] - \beta \cdot D_{KL}[e(z|x||d(z)] \tag{12}$$

where $e(z|x)$ is the encoder part to estimate the z space from input x and $d(x|z)$ is the decoder part to estimate the original data x from the latent space. This algorithm works by introducing a regularizing parameter λ in calculating $d(x|z)$ that would penalize the popular items.

Given an item set $i \in I$, user set $u \in U$ and a probability distribution $\pi(z_u)$ for latent representation of user profile z_u, the decoder part is given as :

$$\log d(x_u|z_u) = \sum_{i=0}^{N} X_{ui} \log \pi_i(z_u) \cdot \lambda \tag{13}$$

$$\lambda = \left(1 - \frac{inter(x_i) - min(inter(\mathbf{x}))}{max(inter(\mathbf{x})) - min(inter(\mathbf{x}))}\right) \tag{14}$$

where $inter(\mathbf{x})$ is total interactions of item x. λ supresses the most popular items to prefer the less popular items. The authors introduce a compensation metric $POPB$, inspired from NDGC which is computed as

$$POPB = \frac{DCP}{IDCP} \tag{15}$$

DCP is the discounted cumulative popularity and is given as :

$$DCP@K = \sum_{j=1}^{|K|} \frac{p(x_j)}{\log_2(j+1)} \tag{16}$$

K is the recommendation list of user u, sorted as per relevance of items.

$IDCP$ is the ideal discounted cumulative popularity which measures discounted cumulative popularity when the list is ordered by popularity REL_k and is given as:

$$IDCP@K = \sum_{i=1}^{|REL_K|} fracp(x_i)\log_2(i+1) \tag{17}$$

2.7 NISER: Normalized Item and Session Representations

This method given by Gupta et al. [19] works on session based recommender systems which use past interactions in a user session to recommend items. The authors used a variation of Graph Neural Network(GNN) for recommendations. They work in both offline as well as online settings where introduction of novel and thus non-popular items is very frequent.

The proposed model *NISER* Normalized Item and Session Representations deals with popularity bias in session based recommendations. In DNN based models, close embeddings of a target item are computed through softmax function.

The paper uses Average Recommendation Popularity metric *ARP* described above to measure the popularity bias. Their model $NISER^+$, a variation of NISER in which position embeddings are added to base model and dropout is applied to input item embeddings show significantly lower values on ARP model as compared to $GNN+$

2.8 FairBoost

Inspired from fair boosting algorithm on classification [18] proposed a model based mitigation algorithm $FairBoost$ works which like [9] by weighing up the non-popular items. Their model is based on *AdaBoost* [28] which combines several weak learners into a single powerful learner. This modular approach makes solving popularity bias easier.

They measure popularity bias as the error gap between non-popular and popular items and is given as :

$$PB(\hat{A}, \tau) = \frac{\sum_{(u,m)} : m \in \mathcal{NPS}\delta_{u,m}(A_{um}, \hat{A}_{um})B_{um}}{\sum_{(u,m)} : m \in \mathcal{NPS}B_{um}} - \tag{18}$$

$$\frac{\sum_{(u,m)} : m \in \mathcal{PS}\delta_{u,m}(A_{um}, \hat{A}_{um})B_{um}}{\sum_{(u,m)} : m \in \mathcal{PS}B_{um}} \tag{19}$$

$\hat{A}$ is the prediction matrix, A is the user-item interaction matrix, τ is the threshold to differentiate between popular and non-popular items by counting the rating of items, B_um is 1 if a rating exist between user u and item m otherwise 0. $\mathcal{NPS}$ is the set of non-popular items and $\mathcal{PS}$ is the set of popular items, δ can either be MAE or MSE between A and $\hat{A}$.

2.9 Popularity Bias in Group Recommendations

The group based recommender systems try to combine the individual recommendations to provide a unified recommendation to a particular group [24]. In group based recommendations, [35] proposes two model based mitigation techniques to counter popularity bias. The first method *AdaptedVaR* or Adapted Value Aware Ranking is based on Value aware Ranking by [4]. The idea behind AdaptiveVaR is to give more weightage to long tail items. In AdaptiveVaR, the item weights are calculated for each group separately. The item weight for item i in a group g is calculated as

$$w_{ci} = \frac{1}{\log_2(pop_c(i))} \tag{20}$$

where $pop_c(i)$ is the popularity of an item i in group c. The group rating R_{ci} is determined using the aggregation process and then normalized in the interval $[0, 1]$ before adding the item weightage w_{gi} as shown below:

$$Y_{ci} = \alpha w_{ci} + (1 - \alpha)\overline{R_{ci}} \tag{21}$$

where Y_{ci} is the ranking score of group g and $\overline{R_{ci}}$ is the normalized group rating. The α is tuned randomly which makes it difficult to optimize alpha for a good balance between accuracy and quality and a single item rating by a user complicates the ranking process. Therefore the authors propose *Enhanced Reranking procedure* that tries to integrate group rating and item weights either by multiplicative approach as:

$$Y_{ci} = w_{ci} \times \overline{R_{ci}} \tag{22}$$

or through Augmentative approach as :

$$Y_{ci} = \overline{R_{ci}} + (\overline{R_{ci}} \times (w_{ci}) \tag{23}$$

ERP initially works similar to AdaptiveVaR. However, instead of favouring the long tail items, ERP penalizes the popular items instead. Therefore, the weigh parameter becomes

$$w_{ci} = 1 - pop(i) \tag{24}$$

The author uses Group Average Popularity *GAP* metric [6] to measure popularity bias. Their results show that using ERP improves the recommendation quality without much loss in ranking accuracy.

GAP measures Popularity Bias of the recommendation list as a whole, divides the user list into groups based on their preferences and then calculates the popularity bias of the individual groups. GAP is measured for the items in a user's profile and the items in the user's recommendation list. Their normalized relative difference is treated as the metric for measuring popularity bias in that group.

$$\Delta GAP(g) = \frac{GAP(g)_\rho - GAP(g)_r}{GAP(g)_\rho} \tag{25}$$

Where $GAP(g)_\rho$ is the GAP value of the user's item profile and $GAP(g)_r$ is the GAP value of user's recommendation profile. GAP value can be calculated using the below equation.

$$GAP(g) = \frac{\sum_{u \in g} \frac{\sum_{i \in \rho_u} \psi(i)}{|\rho_u|}}{|g|} \tag{26}$$

where $\psi(i)$ is the popularity ratio of calculated as number of rated items over the number of total users.

2.10 Calibrated Popularity

This method unlike above methods uses user-oriented approach where different recommendations are given based on the taste of different users. Inspired from Steck [30] and based on their previous work in [6], Abdollahpuri et al., divide the users according to the preferences of users in their historical data and accordingly recommend items to them based on their preferences.

Similar to [3], Calibrated Popularity [7] derives its list by re-ranking an initial recommendation list R_u for user u and measuring the distributional differences between the recommendation list R and user's historical profile in each item category $C = c_1, c_2, ..., c_k$ taken as $C = H, M, T$ (H is the set of most popular items, M is set of items with medium popularity and T is the set of items with very low popularity) in the paper. To find the re-ranked recommendation list S, sum of relevance and calibration is maximized in the following objective function:

$$S_r = \arg\max_{R_u} (1 - w) \cdot Rl(R) - \lambda \mathfrak{J}(P, Q(R_u)) \tag{27}$$

w controls the relevance and calibration, $\mathfrak{J}()$ is *jenson-shanon divergence* between the items in the recommendation profile and recommendation list.

The author proposes a metric that takes into account the users preferences with respect to popular items. UPD or User Popularity Deviation compares the ratio of item groups given in a user's historical profile with the ratio of user group in their recommendation list. $UPD = 0$ means the ratio is exactly the same. UPD for a user group U_g is measured by computing the JensenShanon divergence $\mathfrak{J}$ between the historical items Q and recommended items R.

$$UPD(c) = \frac{\sum_{u \in c} \mathfrak{J}(Q(\rho_u), R(L_u))}{|g|} \tag{28}$$

where p_c is the historical item profile of user u and L_u is the recommendation profile of user u.

The probability of an item i in an item group c for user u is calculated by taking the ratio of the sum of ratings of items of users u's profile in item group c to the sum of ratings of user u's profile in item groups other than c i.e. c'.

$$q(c|u) = \frac{\sum_{i \in k_u} rating(u,i) if(i \in c)}{\sum_{c' \in C} \sum_{i \in k_u} rating(u,i) if(i \in c')} \tag{29}$$

where $rating(u,i)$ indicates that user u has rated the item i and k_u is the user profile of user u.

Similarly the probability of an item in an item group in the users recommendation list R is computed by taking the ratio of the sum of recommended items in item group c to the sum of recommended items in groups other than c i.e. c'.

$$r(c|u) = \frac{\sum_{i \in R_u} (i \in c)}{\sum_{c' \in C} \sum_{i \in R_u} (i \in c')} \tag{30}$$

The total UPD is calculated by finding the average UPD of individual groups. The lower values of UPD are desirable.

$$UPD = \frac{\sum_{c \in C} UPD(c)}{|C|} \tag{31}$$

2.11 Reducing Popularity Bias over Time

Time aware recommender systems have been studied in the literature [14][21]. Abdollahpouri [2] takes time as a factor to look at long tail coverage in regular intervals of time. Explicit Query Aspect Diversification (xQuAD) [27] algorithm is used in information retrieval to improve the diversity. They propose a variant to xQuAD, *Time Smooth xQuAD* that would factor in personalization past interests of user in long tail items. The method is a re-ranking mitigation algorithm. Instead of observing a recommendation list S at a given time, all the previous historical recommendations are observed upto the current time.

They use Average Recommendation Popularity (ARP), Long-tail coverage Ratio (LCR) and Cumulative LCR (CLCR) to measure the popularity bias. Their results show more exposure to unique long-tail items over the baseline algorithms.

2.12 Other Evaluation Metrics

Other significant evaluation metrics that have been used to measure popularity bias are *Gini Index* and *Aggregate Diversity.*

Gini Index for recommender systems computes the inequality in recommended item distribution. The higher values of Gini Index represent that less items are in the long tail [17].

For a recommendation list S for users K and Items I. The Gini Index is given as:

Table 1. Summary of Mitigation Techniques

Mitigation Technique	Evaluation Metrics used	Datasets used
Percentile Based Approach (P1R) [10]	nDCG	MovieLens [20], Last.fm [29]
Uniform Item Test Profile (UAR) [10]	nDCG	MovieLens [20], Last.fm [29]
Modified Collaborative Filtering Approach for Novel Recommendations (MCFNR) [9]	Precision, Recall, F-Measure	Book dataset from [33]
ReGularization [3]	Average Percentage of Tail Items (APT) [3]	Movielens 1M [20], Epinions [32]
Personalized Long-tail Promotion [5]	Average Percentage of Tail Items (APT) [3], Average Recommendation Popularity (ARP) [5], Average Coverage of Long Tail items (ACLT) [5]	Movielens 1M [20], Epinions [32]
Bias Aware Variational Encoders [13]	Compensation Metric POPB [13]	Movielens 20M [20]
NISER: Normalized Item and Session Representations [19]	Average Recommendation Popularity ARP [3]	
FairBoost [18]	Popularity Biasedness (PB) [18]	Netflix [11], Yahoo [34], Amazon [22], Movielens [20]
Popularity Bias in Group Recommendations [35]	Group Average Popularity (GAP) [6]	Movielens 100K [20], Movielens 1M [20], Ciao20
Calibrated Popularity (CP) [7]	User Popularity Deviation [7]	MovieLens 1M [20], Last.fm [29]

$$Gini(S) = \frac{1}{1-|I|}\sum_{j=1}^{|I|}(2j-|I|-1)pop(i_j|S) \tag{32}$$

where $pop(i)$ is the ratio of items to the total no of recommended items $|S|$ and is sorted in increasing order.

$$p(i|S) = \frac{|u \in U : i \in S_u|}{\sum_{u \in U}|S_u|} \tag{33}$$

Aggregate Diversity Agg-Div is the ratio between item count a recommendation list to the total available items. This values shows the extent to which of long tail items have been covered by the recommendation algorithm. However this metric doesn't take into account the recommendation frequency of an item [8]

$$AggDiv = \frac{|\bigcup_{u \in U} L_u|}{|I|} \tag{34}$$

Conclusion

The mitigation techniques reviewed above, precisely summarized in **Table** 1, show the state of art in popularity bias research of Recommendation Systems. The proposed mitigation techniques are mostly either model based techniques with mitigation algorithms added to the objective function as a regularizer or re-ranking techniques where mitigation is implemented over the recommendation generated by a Recommender system.

Most of the proposed mitigation methods accompany with them a separate evaluation metric specific to their problems which reveal a lack of a standard benchmark metric and therefore could be the focus of new research studies.

There also seems to be a need to focus on bias amplifications arising in deep models which are very likely to dominate the future recommendation systems.

Acknowledgement. This research is funded by IUST Kashmir, J&K under grant number IUST/Acad/RP_APP/18/99.

References

1. Abdollahpouri, H.: Popularity bias in recommendation: a multi-stakeholder perspective. Ph.D. thesis (2020)
2. Abdollahpouri, H., Burke, R.: Reducing popularity bias in recommendation over time. arXiv (2019)
3. Abdollahpouri, H., Burke, R., Mobasher, B.: Controlling popularity bias in learning-to-rank recommendation. In: RecSys 2017 - Proceedings of the 11th ACM Conference on Recommender Systems, pp. 42–46 (2017)
4. Abdollahpouri, H., Burke, R., Mobasher, B.: Popularity-aware item weighting for long-tail recommendation (2018)
5. Abdollahpouri, H., Burke, R., Mobasher, B.: Managing popularity bias in recommender systems with personalized re-ranking. arXiv, (Anderson 2006), pp. 413–418 (2019)
6. Abdollahpouri, H., Mansoury, M., Burke, R., Mobasher, B.: The unfairness of popularity bias in recommendation. arXiv (2019)
7. Abdollahpouri, H., Mansoury, M., Burke, R., Mobasher, B., Malthouse, E.: User-centered evaluation of popularity bias in recommender systems, vol. 1. Association for Computing Machinery (2021)
8. Adomavicius, G., Kwon, Y.: Improving aggregate recommendation diversity using ranking-based techniques. IEEE Trans. Knowl. Data Eng. **24**(5), 896–911 (2012)
9. Bedi, P., Gautam, A., Sharma, C.: Using novelty score of unseen items to handle popularity bias in recommender systems. In: Proceedings of 2014 International Conference on Contemporary Computing and Informatics, IC3I 2014, pp. 934–939 (2014)

10. Bellogín, A., Castells, P., Cantador, I.: Statistical biases in Information Retrieval metrics for recommender systems. Inf. Retrieval J. **20**(6), 606–634 (2017). https://doi.org/10.1007/s10791-017-9312-z
11. Bennett, J., Lanning, S.: The netflix prize. In: KDD Cup and Workshop, pp. 3–6 (2007)
12. Borges, R., Stefanidis, K.: On measuring popularity bias in collaborative filtering data. In: CEUR Workshop Proceedings, p. 2578 (2020)
13. Borges, R., Stefanidis, K.: On mitigating popularity bias in recommendations via variational autoencoders. In: Proceedings of the ACM Symposium on Applied Computing, pp. 1383–1389 (2021)
14. Campos, P.G., Díez, F., Cantador, I.: Time-aware recommender systems: a comprehensive survey and analysis of existing evaluation protocols. User Model. User-Adap. Inter. **24**(1), 67–119 (2013). https://doi.org/10.1007/s11257-012-9136-x
15. Cañamares, R.: An analysis of popularity biases in recommender system evaluation and algorithms. Ph.D. thesis, Universidad Autonoma de Madrid (2019)
16. Celma, O., Cano, P.: From hits to niches? Or how popular artists can bias music recommendation and discovery. In: Proceedings of the 2nd KDD Workshop on Large-Scale Recommender Systems and the Netflix Prize Competition, NETFLIX 2008 (2008)
17. Eskandanian, F., Mobasher, B.: Using stable matching to optimize the balance between accuracy and diversity in recommendation. In: UMAP 2020 - Proceedings of the 28th ACM Conference on User Modeling, Adaptation and Personalization, pp. 71–79 (2020)
18. Gangwar, A., Jain, S.: An adaptive boosting technique to mitigate popularity bias in recommender system, vol. 1. Association for Computing Machinery (2021)
19. Gupta, P., Garg, D., Malhotra, P., Vig, L., Shroff, G.: Normalized item and session representations to handle popularity bias, NISER (2019)
20. Harper, F.M., Konstan, J.A.: The movielens datasets. ACM Trans. Interact. Intell. Syst. **5**(4), 1–19 (2016)
21. Harshvardhan, G.M., Gourisaria, M.K., Rautaray, S.S., Pandey, M.: UBMTR: unsupervised Boltzmann machine-based time-aware recommendation system. J. King Saud Univ. - Comput. Inf. Sci. (2021)
22. He, R., McAuley, J.: Ups and downs: modeling the visual evolution of fashion trends with one-class collaborative filtering. In: 25th International World Wide Web Conference, WWW 2016, pp. 507–517 (2016)
23. Jannach, D., Lerche, L., Kamehkhosh, I., Jugovac, M.: What recommenders recommend: an analysis of recommendation biases and possible countermeasures. User Model. User-Adap. Inter. **25**(5), 427–491 (2015). https://doi.org/10.1007/s11257-015-9165-3
24. Joy, J., Pillai, R.V.G.: Review and classification of content recommenders in E-learning environment. Journal of King Saud University - Computer and Information Sciences (2021)
25. Liang, D., Krishnan, R.G., Hoffman, M.D., Jebara, T. Variational Autoencoders for Collaborative Filtering. In: Proceedings of the 2018 World Wide Web Conference on World Wide Web - WWW 2018, vol. abs/1808.0, pp. 689–698. ACM Press, New York (2018)
26. Niemann, K., Wolpers, M.: A new collaborative filtering approach for increasing the aggregate diversity of recommender systems. In: Proceedings of the ACM SIGKDD International Conference on Knowledge Discovery and Data Mining, Part F1288, pp. 955–963 (2013)

27. Santos, R.L.T., Macdonald, C., Ounis, I.: Exploiting query reformulations for web search result diversification. In: Proceedings of the 19th International Conference on World Wide Web, WWW 2010, pp. 881–890 (2010)
28. Schapire, R.E.: A Brief Introduction to Boosting Generalization error. In: Ijcai, vol. 99, pp. 1401–1406 (1999). T Labs, Park Avenue, A Room, and Florham Park
29. Schedl, M.: The LFM-1b dataset for music retrieval and recommendation. In: Proceedings of the 2016 ACM on International Conference on Multimedia Retrieval, pp. 103–110. ACM, New York (2016)
30. Steck, H.: Calibrated recommendations. In: RecSys 2018–12th ACM Conference on Recommender Systems, pp. 154–162 (2018)
31. Takács, G., Tikk, D.: Alternating least squares for personalized ranking. In: RecSys 2012 - Proceedings of the 6th ACM Conference on Recommender Systems, pp. 83–90 (2012)
32. Tang, J., Gao, H., Liu, H., Das Sarma, A.: eTrust: understanding trust evolution in an online world. In: Proceedings of the ACM SIGKDD International Conference on Knowledge Discovery and Data Mining, pp. 253–261 (2012)
33. Vashisth, P., Chandoliya, D., Yadav, B.K., Bedi, P.: Trust enabled argumentation based recommender system. In: International Conference on Intelligent Systems Design and Applications, ISDA, pp. 137–142 (2012)
34. Yahoo. R2 - Yahoo! Music User Ratings of Songs with Artist, Album, and Genre Meta Information (2006)
35. Yalcin, E., Bilge, A.: Investigating and counteracting popularity bias in group recommendations. Inf. Process. Manag. **58**(5), 102608 (2021)
36. Yin, H., Cui, B., Li, J., Yao, J., Chen, C.: Challenging the long tail recommendation. Proc. VLDB Endowment **5**(9), 896–907 (2012)

Analysis of Text Classification Using Machine Learning and Deep Learning

Chitra Desai(✉)

Department of Computer Science, National Defence Academy, Pune, India
chitragdesai@gmail.com

Abstract. This paper analyzes the performance of text classification using machine learning and deep learning models. The data is preprocessed, followed by text preprocessing. Natural language processing requires the conversion of text data to numerical vectors before they are passed to machine learning or deep learning models. Bag of words and Term Frequency and Inverse Document Frequency (TFIDF) techniques are used for converting text to a numeric vector. The machine learning models demonstrated are Naïve Bayes Classifier and Logistic Regression. Here, the Bag of words is used with Naïve Bayes Classifier, and TFIDF is used with the logistic regression model. The deep learning models demonstrated here are Long Short-Term Memory (LSTM) and Gated Recurrent Unit (GRU). The data set used consists of 1786 instances. Train test ratio used is 80:20. The performance of text classification models depends upon the dataset and how well its features are preprocessed as well as the selection of train test data. In this paper, four different algorithms are applied to the same dataset to analyze the performance of text classification.

Keywords: Text classification · Bag of words · Term Frequency and Inverse Document Frequency (TFIDF) · Logistic regression · Naïve Bayes Classifier · Long Short-Term Memory (LSTM) · Gated Recurrent Unit (GRU)

1 Introduction

Natural Language Processing (NLP) [1], can be referred to as an artificial intelligent method of communication with the intelligent systems using natural language used by humans to communicate with each other. Humans communicating in common language can easily find the context of words in their communication and infer their understanding in spite of ambiguity and impreciseness in natural language, but the same becomes a huge challenge for machines to implement.

The NLP task can be classified at a low and high level [2]. At the low level, it includes tokenization [3], pos tagging [4], morphological decomposition, problem-specific segmentation, to list a few. At a high level, it includes spelling and grammatical error identification and recovery and Named Entity Recognition (NER).

In natural language processing (NLP), the text data needs to be converted to a numeric vector so that it is understood by the machine. Techniques like a bag of word [5] and

V. E. Balas et al. (Eds.): ICETCE 2022, CCIS 1591, pp. 445–455, 2022.
https://doi.org/10.1007/978-3-031-07012-9_38

Term Frequency and Inverse Document Frequency (TF-IDF) [6] are used in NLP to transform raw text data to numeric vectors for the purpose of classification, summarization, clustering, etc. Tokens are another important aspect that needs to be dealt with while working in NLP. Tokenization can be done at the character level, word level or sentence level. The unique tokens obtained from the raw text forms the vocabulary for the corpus. Count vectorizer and TF-IDF use vocabulary as features.

When the features are available, text classification can be performed using a statistical or machine learning approach. The significance of machine learning algorithms comes in with the necessity of a large volume of data to be processed and automation. Depending upon the learning criteria, learning algorithms can be supervised, unsupervised or semi-supervised. The supervised learning algorithms are further classified as parametric classifiers and non-parametric classifiers [7]. In a parametric classifier, the algorithm is applied to training data to estimate unknown function f, whereas non-parametric algorithms make explicit postulation about the function form of f. In parametric models' data is summarized based on underlying parameters. Logistic regression and Naïve Bayes algorithms are examples of parametric classifiers. SVM [8], decision tree [9], KNN, etc. are examples of non-parametric algorithms.

This paper analyzes the performance of text classification using machine learning and deep learning models. The data is preprocessed, followed by text preprocessing. Natural language processing requires the conversion of text data to numerical vectors before they are passed to machine learning or deep learning models. Bag of words and Term Frequency and Inverse Document Frequency (TFIDF) techniques are used for converting text to a numeric vector. As the present study is limited to parametric classifiers, the machine learning models demonstrated are Naïve Bayes Classifier [10] and Logistic Regression. Here, Bag of words is used with Naïve Bayes Classifier, and TFIDF is used with a logistic regression model. The deep learning models demonstrated here are Long Short-Term Memory (LSTM) and Gated Recurrent Unit (GRU).

2 Methodology

This paper presents text classification using Machine Learning and Deep learning algorithms and analyzes its performance. The methodology adopted here is – the raw data is preprocessed followed by text preprocessing. There are two labeled columns in the data set. One column is the feature (text) and the other is target (label). The text classification to a particular class is performed using four different algorithms. The data is split into train test data in the ratio of 80:20. Then train and test features are transformed into numeric vector using TFIDF. After, transformation logistic regression algorithm is used for predicting text classification. The same train test features are again transformed using count vectorizer into numeric vector and Naïve Bayes algorithm for prediction is applied. In case the machine learning algorithms (logistic regression and Naïve Bayes), the target variable, which consists of four classes was encoded. In the next step of the experiment, before the application of deep learning algorithms, the target variable was transformed to categorical and then the train test split was performed. Tokenization was done and post padding sequences were added. Using LSTM and GRU models are generated for the prediction of text classification.

3 Data and Text Pre-processing

Given any data set [11] for text classification, it may consist of various columns which may not be relevant for the task under consideration. Therefore, it becomes essential to first do feature selection to select appropriate columns that are relevant for text classification. Also, before we proceed for text preprocessing, it is required to perform basic data preprocessing which may help to gain insight into the data. Data and text preprocessing is equally important to model building. In this section, we will gain primary insight into the data and then proceed with text preprocessing after selecting essential features for text classification. Figure 1 shows the first five sample records of raw data.

	text	label
0	Trial Balance and Rectification of Errors 199...	accounts
1	Issue and Redemption of Debentures 1 03 perc...	accounts
2	Trial Balance and Rectification of Errors 21 ...	accounts
3	Accounting for Share Capital 1 9 In practice...	accounts
4	80 Accountancy : Company Accounts and Analysi...	accounts

Fig. 1. First five sample records of subject text classification dataset

3.1 Data Preprocessing

There are no null values in the dataset. There are 90 duplicate values in the dataset as shown in Fig. 2. The duplicate values are dropped and the number of instances is reduced to 1696 instances. The dataset as mentioned earlier consists of 2 columns – text and label. In this dataset, the label is our target variable. The target variable consists of four unique classes – 'accounts', 'biology', 'physics' and 'geography'. The value count across each class is - accounts are 284, biology is 595, geography is 93, physics is 724 as shown in Fig. 3. It is observed the data is highly imbalanced.

The column label which is our target variable has data values that are non-numeric. The non-numeric values are transformed to numeric values by applying label encoding. This transforms accounts to 0, biology to 1, physics to 2, geography to 3.

The length of text for each of the instances in the text column is computed. Table 1 provides statistical insight into the text column. The mean text length is observed to be 2204. The minimum, maximum and interquartile range values indicate the presence of outliers.

	text	label
316	BIOLOGY Mammals from colder climates general...	biology
400	TRANSPORT IN PLANTS the loss of water in the...	biology
416	ORGANISMS AND POPULATIONS Figure 13.7 Mut...	biology
424	ORGANISMS AND POPULATIONS (iii) Migrate: The...	biology
453	TRANSPORT IN PLANTS examples of imbibition a...	biology
...	...	...
1698	(l'rmx'vrsit'rll . ' (":mum'ntul Rm "3111' Ha...	geography
1718	EURASIAN PLATE NORTH AMERICAN 60° PLATE V M...	geography
1766	HOSZCHH CHAPTER ou have already learnt that...	geography
1771	98 (ii) (iii) (iV) FUNDAMENTALS OF PHYSIC...	geography
1775	96 of the earth. These are extremely dense a...	geography

90 rows × 2 columns

Fig. 2. Duplicate values in dataset

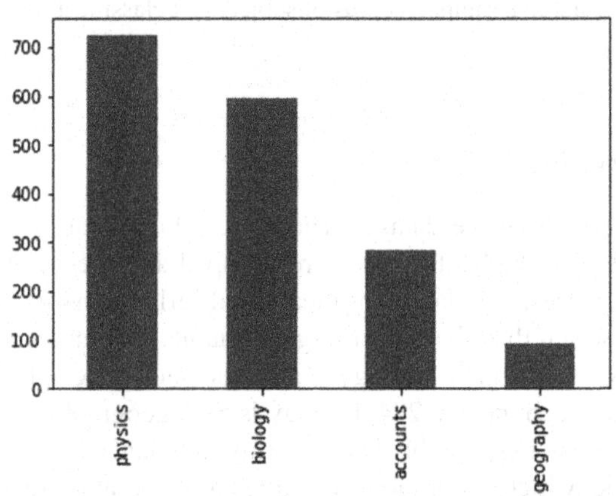

Fig. 3. Value count across each class of target variable

3.2 Text Preprocessing

After the primary data preprocessing step, text preprocessing [12] is followed. Initially, all the text is converted to lower case. The contracted words in the text is de-contracted, for example, can't will be de-contracted to cannot, 've to have and so. The text is further cleaned by removing stopwords [13], punctuations, numbers, html and xml markup, white spaces. Figure 4 shows first five records after text preprocessing.

Table 1. Text length statistics before text preprocessing

	Text length
Count	1696.00000
Mean	2204.846108
Std	772.701475
Min	278.00000
25%	1607.750000
50%	2246.00000
75%	2740.75000
Max	5012.00000

	text	label
0	trial balance rectification error trial balanc...	0
1	issue redemption debenture percentage usually ...	0
2	trial balance rectification error illustration...	0
3	accounting share capital practice entry amount...	0
4	accountancy company account analysis financial...	0

Fig. 4. Results after text cleaning

The text length after text preprocessing is again computed and the result is as shown in Table 2. Minimum has dropped from 278 to 8, maximum from 5012 to 3367 and mean text length has dropped by 36.12%.

Table 2. Text length statistics after text preprocessing

	Text length
Count	1696.00000
Mean	1408.629717
Std	513.246585
Min	8.00000
25%	1019.00000
50%	1424.00000
75%	1783.25000
Max	3367.00000

The statistical counts in Table 2 shows that there are outliers in the data. To check those outliers, instances with text length greater than 3000 and less than 100 are identified as shown in Figs. 5 and 6 respectively. There are 3 instances with text length greater than 3000 and 4 instances less than 100. To impact the mean, three instances above 3000 can be dropped. However, for further experiment, these records are retained here. The unique words found in 1696 instances is 18632. The data after text preprocessing is split into 80% training data and 20% of testing data.

	text	label	TextLen
938	duol nature radiation every body temperature e...	3	3175
1265	g ravitati india leap space india started spac...	3	3367
1688	geography discipline ocean geographical factor...	2	3048

Fig. 5. Instances with text length greater than 3000

	text	label	TextLen
1172	l physic	3	8
1254	simulate propagation electromagnetic wave e	3	43
1316	physic hysterisis magnetic material f	3	37
1660	l physic	3	8

Fig. 6. Text length less than 100

Figure 7 shows distribution of text length for each of the four classes in the target variable. The X-axis represents the text length and the Y-axis represents the number of instances.

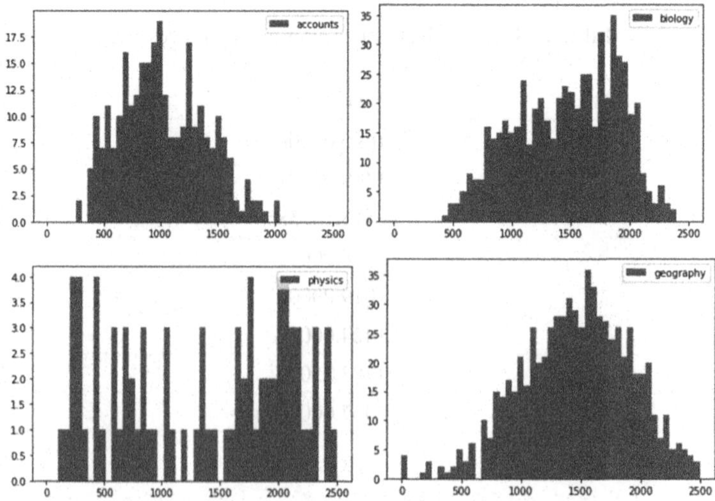

Fig. 7. Distribution of text length for each class in target variable

4 Feature Extraction and Machine Learning Model

In this section, TF_IDF and bag of words technique is used for feature extraction and text classification is performed using machine learning algorithms - logistic regression and Naïve Bayes. The machine learning algorithm infers from training data and make prediction for new data. In the process of text classification, the text in column *text* is categorized to one of the four classes based on its content. The logistic regression model is trained using features extracted by TF-IDF technique. The dimensions of train and test features obtained using TF-IDF is as shown in Table 3. It is observed that the dimensions after TFIDF transformation is high, where, the number of features exceed the number of observations. As linear classifier works well for large and high dimension data set, logistic regression is used. The logistic regression model is fit to the training vectors obtained using TFIDF. The accuracy score obtained using logistic regression is 96.47%.

Table 3. Dimension of feature vector for train test using TF-IDF

	Dimensions
X_train_vectors_tfidf	(1356, 17010)
X_test_vectors_tfidf	(340, 17010)

The Naïve Bayes model is trained using features extracted by the Bag of words technique. The Naïve Bayes model is trained using features extracted by the Bag of words technique. For creating a bag of words model, the maximum features of the count vectorizer are set to 1500. Table 4 shows train and test dimensions for the feature vector. The reason to use the Naïve Bayes algorithm is that it has proved useful in most of the natural processing tasks and in particular, the strong conditional independence assumption makes it more suitable for classification.

Table 4. Dimension of feature vector for train test using count vectorizer

	Dimensions
X_train_vectors_cv	(1356, 1500)
X_test_vectors_cv	(340, 1500)

The Naïve Bayes model is fit to the training vectors obtained using count vectorizer. The accuracy score obtained using Naïve Bayes is 97.05%.

For the given data set it is observed that using TFIDF and count vectorizer the number of instances is equal but there is huge reduction in the number of columns. That is in case of count vectorizer, only 8.8% of TFIDF columns, we have obtained better accuracy using Naïve Bayes classifier.

5 Deep Learning

The accuracy obtained by above two machine learning algorithms is further compared to deep learning models. The models used are Long Short-Term Memory (LSTM) and Gated Recurrent Unit (GRU). Before applying LSTM and GRU, following basic steps are observed:

1. The text column data is stored in array X and labels are stored in array y.
2. As mentioned above the label consist of four classes. These four classes are encoded. Figure 8 shows sample output of categorical label class after encoding. Let us assume after encoding the y is stored in categorical labels.

```
categorical_labels

array([[1., 0., 0., 0.],
       [1., 0., 0., 0.],
       [1., 0., 0., 0.],
       ...,
       [0., 0., 1., 0.],
       [0., 0., 1., 0.],
       [0., 0., 1., 0.]], dtype=float32)
```

Fig. 8. Categorical labels after encoding

3. The data is split into train test data in the ratio of 80% of training data and 20% of testing data. Let us assume the data after splitting is stored in X_train, X_test, y_train, y_test. It is important here to note the dimensions of train and test data. Refer Table 5.

Table 5. Train test data for deep learning algorithms

	Dimensions
X_train	(1356)
X_test	(340)

While we worked with TFIDF and count vectorizer, the y, that is the target variable was one dimensional, whereas, as shown in Fig. 10, the target variable is two-dimensional y_train = (1356, 4) and y_test = (340, 4).

4. The average length of the text after text pre-processing is approximately 1400. So, the max length is set to 1400. The train and test data are tokenized and using max length = 1400, post padding sequences are generated. Following is the sample output for X_train.

$$\text{array}([87, 11, 275, \ldots, 0, 0, 0], \text{dtype} = \text{int32})$$

5. The number of words is 18632 as stated above, so the word index length is 18632 and therefore the vocabulary size is set to 18632.
6. The embedding dimension is set to 16.
7. To apply LSTM, the following are the steps:

   a) A sequential model is initialized.
   b) Embedding layer with input dimension = vocabulary size and output dimension = embedding dimension is added.
   c) LSTM layer with embedding dimension = 16, dropout = 0.2 and return sequence = True is added.
   d) A layer to flatten the input tensor is added.
   e) Dense layer with 'relu' activation followed by 'softmax' activation is added. Figure 9 presents the model summary.
   f) The model is compiled using RMSProp.
   g) There are four classes in the target variable so the loss is computed using categorical cross-entropy.
   h) The model accuracy and model loss is as shown in Fig. 10.
   i) Using LSTM the text classification is performed. With epochs 15 and batch of 1024, the training accuracy is 0.9801, and validation accuracy is 0.9441.

```
Model: "sequential_6"
_________________________________________________________________
Layer (type)                 Output Shape              Param #
=================================================================
embedding_6 (Embedding)      (None, 1400, 16)          298128
_________________________________________________________________
lstm_6 (LSTM)                (None, 1400, 16)          2112
_________________________________________________________________
module_wrapper_12 (ModuleWra (None, 22400)             0
_________________________________________________________________
module_wrapper_13 (ModuleWra (None, 512)               11469312
_________________________________________________________________
dense_6 (Dense)              (None, 4)                 2052
=================================================================
Total params: 11,771,604
Trainable params: 11,771,604
Non-trainable params: 0
_________________________________________________________________
```

Fig. 9. Model summary for LSTM

8. To apply GRU, the following are the steps:

   a) A sequential model is initialized.
   b) Embedding layer with input dimension = vocabulary size and out put dimension = embedding dimension is added.
   c) GRU layer with embedding dimension = 16, dropout = 0.2 and return sequence = True is added.

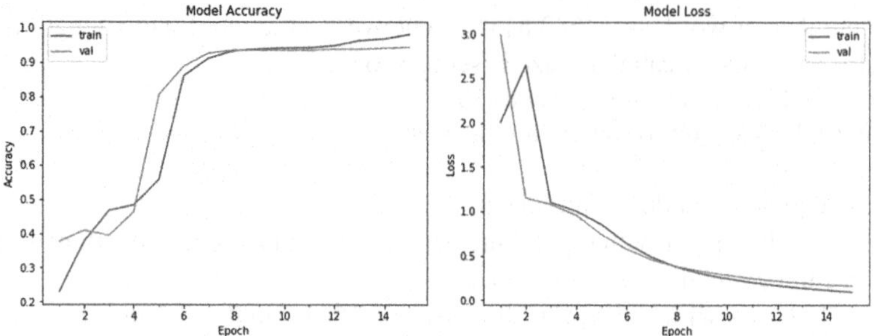

Fig. 10. Model accuracy and model loss for LSTM

d) A layer to flatten the input tensor is added.
e) Dense layer with 'relu' activation followed by 'softmax' activation is added.
f) The model is compiled using RMSProp.
g) There are four classes in target variable so the loss is computed using categorical cross entropy.
h) The model accuracy and model loss is as shown in Fig. 11.
i) Using GRU the text classification is performed. With epochs 15 and batch of 1024, the train accuracy is 0.9874 and validation accuracy is 0.9676.

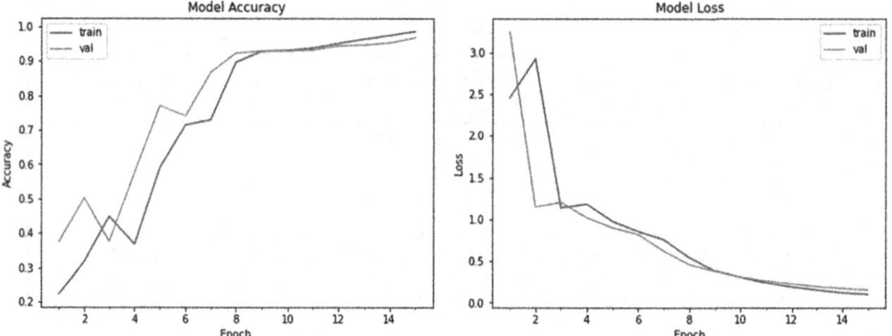

Fig. 11. Model accuracy and model loss for GRU

6 Conclusion

This paper analyzes the performance of text classification for multiclass target variables using machine learning and deep learning models. The data set consists of two columns - text and labels. There are a total of 4 classes - accounts, physics, biology, and geography. Text processing is applied to the data to clean the text. Initially, TFIDF is used for transformation and then Logistic Regression is applied for classification. Using Logistic Regression, the Accuracy score obtained is 0.9647. Then the cleaned data set is again transformed using a Bag of words (Count vectorizer) and then Naive Bayes Algorithm is applied for text classification. Using the Naive Bayes Algorithm, the Accuracy score obtained is 0.9705. Using LSTM, the text classification is performed. With epochs 15 and batch of 1024, the training accuracy is 0.9801, and validation accuracy is 0.9441. Using GRU the text classification is performed. With epochs 15 and batch of 1024, the training accuracy is 0.9874, and validation accuracy is 0.9676. Thus, it is observed that for the subject text classification data set, GRU gives better accuracy compared to Logistic Regression, Naïve Bayes algorithm, and LSTM.

References

1. Chowdhary, K.R.: Natural language processing. In: Fundamentals of Artificial Intelligence, pp. 603–649. Springer, New Delhi (2020). https://doi.org/10.1007/978-81-322-3972-7_19
2. Nadkarni, P.M., Ohno-Machado, L., Chapman, W.W.: Natural language processing: an introduction. J. Am. Med. Inform. Assoc. **18**(5), 544–551 (2011)
3. Rai, A., Borah, S.: Study of various methods for tokenization. In: Mandal, J.K., Mukhopadhyay, S., Roy, A. (eds.) Applications of Internet of Things. LNNS, vol. 137, pp. 193–200. Springer, Singapore (2021). https://doi.org/10.1007/978-981-15-6198-6_18
4. Bulusu, A., Sucharita, V.: Research on machine learning techniques for POS tagging in NLP. Int. J. Recent Technol. Eng. (IJRTE) 8(1S4), 897-900 (2019). ISSN: 2277-3878
5. Brownlee, J.: A Gentle Introduction to the Bag-Of-Words Model (2020). https://Machinelearningmastery.Com/Gentle-Introduction-Bag-Words-Model/
6. Uther, W., et al.: TF–IDF. In: Sammut, C., Webb, G.I. (eds.) Encyclopedia of Machine Learning, pp. 986–987. Springer, Boston (2010). https://doi.org/10.1007/978-0-387-30164-8_832
7. Burkov, A.: The Hundred-Page Machine Learning Book. Andriy Burkov, Quebec City (2019)
8. Cortes, C., Vapnik, V.: Support-vector networks. Mach. Learn. **20**(3), 273–297 (1995)
9. Quinlan, J.R.: C4.5: Programs for Machine Learning. Elsevier, Amsterdam (2014)
10. Malpani, P., Bassi, P., Mahrishi, M., Jain, V.: A novel framework for extracting geospatial information using SPARQL query and multiple header extraction sources. In: Afzalpulkar, N., Srivastava, V., Singh, G., Bhatnagar, D. (eds.) Proceedings of the International Conference on Recent Cognizance in Wireless Communication & Image Processing, pp. 489–499. Springer, New Delhi (2016). https://doi.org/10.1007/978-81-322-2638-3_56
11. Dataset. https://Drive.Google.Com/Drive/U/0/Folders/1d1h9honqjtss7lsxdp2557w-Hcreifry. Accessed 20 Oct 2021
12. Méndez, J.R., Iglesias, E.L., Fdez-Riverola, F., Díaz, F., Corchado, J.M.: Tokenising, stemming and stopword removal on anti-spam filtering domain. In: Marín, R., Onaindía, E., Bugarín, A., Santos, J. (eds.) CAEPIA 2005. LNCS (LNAI), vol. 4177, pp. 449–458. Springer, Heidelberg (2006). https://doi.org/10.1007/11881216_47
13. Jusoh, S., Alfawareh, H.M.: Techniques, applications and challenging issues in text mining. Int. J. Comput. Sci. Issues **9**(6), 431 (2012). ISSN (Online): 1694-0814

Evaluation of Fairness in Recommender Systems: A Review

Syed Wajid Aalam, Abdul Basit Ahanger, Muzafar Rasool Bhat(✉), and Assif Assad

Islamic University of Science and Technology, Awantipora, India
muzafar.rasool@islamicuniversity.edu.in

Abstract. Recent advances in Recommender Systems have shifted the attention of researchers towards fair and unbiased recommendations. A growing number of users use Recommender Systems for decision making and information retrieval and in turn have a significant impact on public opinion. Therefore, it is of pivotal importance to address the unfairness issue in recommender systems for fair inclusion of disadvantaged groups. The study of fairness in recommender systems is a relatively new field with a vast scope for further research and improvement. This study presents a thorough investigation of existing metrics in fairness evaluation from different contexts like user fairness, item fairness, group fairness, individual fairness, multi-sided fairness, etc.

Keywords: Fairness · Recommender systems · Multi-sided fairness

1 Introduction

Recommender systems are part of our lives as these help in decision making, information retrieval, listening to preferred music, and a lot more. Recommender systems mostly rely on historical interactions, so accuracy is increased as users interact with the system [37]. However, with newly added interactions the inherent bias of users like cognitive bias, cultural bias, gender bias, etc. [4] is transferred into machines as well. A recommender system should be fair enough in recommending relevant items to users and it should help users to find what they need even if they are unaware of their needs. The items of advantaged groups are more likely to get recommended irrespective of their quality or user's preferences [1] which makes fairness a critical issue in recommender systems.

The majority of existing works focused on the algorithmic aspect, leaving the non-algorithmic aspects behind. There is a need to consider different approaches in making the recommendations neutral by considering both algorithmic, as well as non-algorithmic approaches [11]. Recommendations could generate unfair results as these are data-driven systems that rely on huge amounts of data and algorithms which may be biased [22]. Non-algorithmic aspects mentioned in [5] has been used by [35] in his work.

V. E. Balas et al. (Eds.): ICETCE 2022, CCIS 1591, pp. 456–465, 2022.
https://doi.org/10.1007/978-3-031-07012-9_39

Quantifying fairness can get complicated when we try to see fairness as more than just a skewed distribution by adding more perspectives to it [10]. Algorithmic decisions have overcome human decision-making in some areas. However recent studies claim that algorithmic decisions are not fair towards certain social groups (e.g. women, black) by focusing on one of the notions of un(fairness) [34]. In an attempt to overcome the unfairness, the researchers have come up with the following notions of fairness:

- *disparate treatment*: This is intentional discrimination or direct discrimination towards an individual or group of people when a decision changes with a change in sensitive information.
- *disparate impact*: This is unintentional discrimination towards certain sensitive value feature groups (blacks, women). This is also referred to as statistical parity.
- *disparate mistreatment*: A decision-making process suffers from this disparity if it computes the accuracy of the feature group differently than others.

All three notions are very important and should be a consideration at the same time, but considering only one will not give a fair system.

2 Related Work

Recommender systems can be evaluated through error-based metrics like Mean Absolute Error, Root Mean Squared Error, Mean Squared Error, and Normalized Mean Absolute Error. Which used to be popular in measuring the accuracy of recommender systems. These metrics show the deviation of predicted values from actual values and how accurate our predictions are. To predict the accuracy

Predicted Values		**Actual Values**	
		Negative	Positive
	Negative	True negative	False negative
	Positive	False positive	True positive

Fig. 1. Confusion matrix

of any Collaborative Filtering recommender system, [9] has used Mean Absolute Error as the prediction evaluation metric. MAE simply measures the arithmetic average between absolute errors and is defined as:

$$|\overline{E}| = \frac{\sum_{i=1}^{N} |p_i - r_i|}{N} \tag{1}$$

RMSE, MSE, and NMAE are variations of MAE. RMSE shows how far the predictions are from measured true values and is popular for the evaluation of the accuracy of predicted ratings [13].

Recommender systems can also be evaluated by observing how the user interacts with the system. There are four different ways how a user interacts with the system [9,24,33] and constitute the components of *confusion* matrix which is used to measure model performance where actual values are on one axis and predicted values are on another axis.

- *TP: True Positive is the item count of recommended items with positive interaction by the user.*
- *FP: False positive is the item count of the recommended items with no interaction by the user.*
- *FN: False Negative is the item count with which a user interacted positively but wasn't recommended by the system.*
- *TN: True Negative is the item count which a user didn't interact with and was also not recommended by the system either.*
  Various recommendation techniques are based on these interactions and numerous metrics are available to measure recommendation accuracy [6,18].

Confusion matrix is also used to approximate classification and misclassification correctly. Accuracy, Sensitivity [33] and Specificity [28] are calculated from the components of the confusion matrix.

Accuracy measures the correct predictions (True value) across the entire dataset.

$$Accuracy = \frac{TP + TN}{TP + TN + FP + FN} \tag{2}$$

Sensitivity also called as True Positive Rate (TPR) refers to probability of actual positives.

$$Sensitivity\ (TPR) = \frac{TP}{TP + FN} \tag{3}$$

Specificity also called as True Negative Rate (TNR) refers to probability of actual negatives.

$$Specificity\ (TNR) = \frac{TN}{TN + FP} \tag{4}$$

Precision, recall, and nDCG are the accuracy evaluation metrics used to measure the quality of recommendations [6,9]. Recommendation accuracy metrics are Precision (P@K), Recall (R@K) [2] and nDCG (nDCG@K) [10] where K is the cutoff. *Precision* and *Recall* have been mentioned as information retrieval evaluation metrics by [16] and recommendation quality metrics by [36] and are calculated through *confusion matrix.*

Precision is the successful recommendations out of total recommendations.

$$Precision = \frac{|tp|}{|tp| + |fp|} \tag{5}$$

Recall on the other hand measures successful recommendations concerning items the user liked.

$$Recall = \frac{|tp|}{|tp| + |fn|} \tag{6}$$

nDCG (normalized cumulative gain): DCG (Discounted Cumulative Gain) is a metric that is used to measure the rank but the length of the recommendation list may vary, so DCG needs to be normalized, which is defined as:

$$nDCG_u@K = \frac{1}{IDCG@K} \sum_{i=1}^{K} \frac{2^{r}_{ui} - 1}{log_2(1+i)} \tag{7}$$

where i holds the rank position in the list and IDCG@K as the score obtained in the recommendation list by an ideal ranking.

The metrics discussed above are used to measure the accuracy of recommender systems. Besides these metrics, several metrics have been proposed to measure the fairness of any recommender system which are discussed below.

MAD (Mean Absolute Difference), and *KS* (Kolmogorov-Smirnov statistics) are recommendation fairness metrics for different groups (dominant and protected) used by [36] and are computed as:

$$MAD = \left| \frac{\sum R^0}{|R^0|} - \frac{\sum R^1}{|R^1|} \right| \tag{8}$$

where R^0 and R^1 are the predicted ratings of different groups and greater the value of MAD is depicted as unfairness.

$$KS = \left| \sum_{j=1}^{N} l \times \frac{G(R^0, j)}{|R^0|} - \sum_{j=1}^{N} l \times \frac{G(R^1, j)}{|R^1|} \right| \tag{9}$$

where N, $G(R^0, j)$ are the number of intervals and size of each interval respectively. Distributions are more alike if the value of KS is lower.

Table 1. Comperative analysis

Metric	Dataset used	Algorithm	Perspective
MAD [36] KS [36]	MovieLens [14] User-expert-topic Twitter data [36]	FATR [36]	Group perspective
Value unfairness [32] Absolute unfairness [32] Underestimation unfairness [32] Overestimation unfairness [32]	Movielens [14] Real movie rating data [14]	Matrix factorization [19]	Group perspective
UGF [21]	Amazon 5-core dataset [http://jmcauley.ucsd.edu/data/amazon]	BiasedMF [19] PMF [27] NeuMF [15] STAMP [23]	Group perspective
GRU [12]	Amazon item e-commerce dataset	-	Group perspective
IRU [12]	Amazon item e-commerce dataset	Heteroembed [12] PGPR [12] KGAT [12]	Individual perspective
R_{indv} [25]	MovieLens [14]	Matrix factorization [25]	Individual perspective
$rND(\tau)$	ProPublica German credit	-	Item sided perspect ive
Score disparity [20] Recommendation disparity [20]	MovieLens [www.grouplens.org] -	Normalized matrix factorization [20] K-nearest neighbour [20]	User sided perspective
$(GF)_{consumer}$ [30] $(GF)_{producer}$ [30]	MovieLens [14] LastFM [www.last.fm]	BPRMF [26] WRMF [17] NGCF [29]	Multi sided perspective

Different types of metrics for rating prediction of recommender systems that measures disparity among binary groups (disadvantaged users and advantaged users) [32] were introduced using Movielens Million Dataset [14]. The first metric is *value unfairness* which measures the inconsistency between actual and predicted ratings of binary groups (men and women). *Value unfairness* occurs when one group is preferred over the other by giving it a higher prediction and is defined as:

$$U_{val} = \frac{1}{m}\sum_{i=1}^{m} |(E_l[y]_i - E_l[r]_i) - (E_{\sim l}[y]_i - E_{\sim l}[r]_i)| \tag{10}$$

where $E_l[y]_i$ and $E_l[r]_i$ are average predicted score and average rating of disadvantaged groups for ith item, $E_{\sim l}[y]_i$ and $E_{\sim l}[r]_i$ are average predicted score and average rating of advantaged groups for ith item respectively, $\sim$ sign represents negation. If the value of *value unfairness* is low that means overestimation and underestimation of groups are evenly balanced. If advantaged groups are overestimated and disadvantaged groups are underestimated then the value of *value unfairness* is high, which shows that the direction and magnitude of error are the same for both groups.

The next one is *absolute unfairness* which is unsigned and the direction of the error is not pondered, *absolute unfairness* takes the only absolute difference of average predicted score and an average rating of each group. e.g. if the disadvantaged group is given preference below their true preference by 0.5 and advantaged group by 0.5 above their true preference, then their *absolute unfairness* will be zero [3].

The inconsistency in underestimating and overestimating the true ratings are measured in the next two metrics. The next metric is *underestimation unfairness* which is defined as:

$$U_{under} = \frac{1}{m}\sum_{i=1}^{m} |max\{0, E_l[r]_i - E_l[y]_i\} - max\{0, E_{\sim l}[r]_i - E_{\sim l}[y]_i\}| \quad (11)$$

The next metric is converse to *underestimation unfairness* giving *overestimation* to recommendations and is defined as:

$$U_{over} = \frac{1}{m}\sum_{i=1}^{m} |max\{0, E_l[y]_i - E_l[r]_i\} - max\{0, E_{\sim l}[y]_i - E_{\sim l}[r]_i\}| \quad (12)$$

In both the cases of over and under, users don't get what they want and what is fruitful for them. In the first case, users may be provided with too many recommendations and in the latter case, users may not be recommended with proper items according to their preferences.

Different perspectives of fairness are taken into consideration by [22] like group fairness & individual fairness, single fairness & multi-sided fairness, static fairness & dynamic fairness, associative fairness & causal fairness, etc. Among these perspectives, some are used to calculate user-sided fairness and some for item-sided fairness. Multiple users are often part of the recommendation system at the same time like watching TV, listening to Music, etc. so, it is important to consider many people's interests at the same time to make things better.

- *Group fairness & Individual Fairness*: These are user-sided fairness metrics. In group fairness, the two groups i.e. protected as well as advantaged groups should be treated as equal while individual fairness demands similar treatment for every individual item [7,24]. Group fairness discussed by [21] is more user-oriented and is defined as:

$$UGF(G_1, G_2, W) = \left| \frac{1}{G_1}\sum_{i \in G_1} \mathcal{M}(W_i) - \frac{1}{G_2}\sum_{i \in G_2} \mathcal{M}(W_i) \right| \leq \varepsilon \quad (13)$$

where $\leq \varepsilon$ represents strictness in fairness, $\mathcal{M}(W_i)$ represents the recommendation quality, and G_1, G_2 are two different groups having ϕ in common and W is the vector of top k recommendations.
Recommendation quality has been evaluated by [12] for user sided group and individual perspectives. For group fairness, GRU (Group Recommendation Unfairness) metric is used and is defined as:

$$GRU(G1, G2, Q) = \left| \frac{1}{|G1|} \sum_{i \in G1} \mathcal{M}(W_i) - \frac{1}{|G2|} \sum_{i \in G2} \mathcal{M}(W_i) \right| \tag{14}$$

For individual fairness IRU (Individual Recommendation Fairness) metric is computed as:

$$IRU(W) = \frac{\sum_{W_x, W_y} |\mathcal{M}(W_x) - \mathcal{M}(W_y)|}{2m \sum_{i=1}^{m} \mathcal{M}(W_i)} \tag{15}$$

Individual fairness mentioned by [25] need some improvement in minimization and is computed as:

$$R_{indv}(W, \hat{W}) = \frac{1}{m^2} \sum_{j=1}^{m} \sum_{l>j} (\mathcal{L}_j - \mathcal{L}_l)^2 \tag{16}$$

where $\mathcal{L}_j$ & $\mathcal{L}_l$ are mean squared error of user j and l over known ratings.

- *User vs Item fairness*: For the recommender system to be fair, both item-sided and user-sided perspectives need to be considered. Users must get fairer recommendations of items and items should be fairly recommended to users. For item sided fairness [31] proposed a metric(rND) which is defined as:

$$rND(\tau) = \frac{1}{X} \sum_{j=20,30}^{M} \frac{1}{log_2 j} \left| \frac{|S^+_{1...j}|}{j} - \frac{|S^+|}{M} \right| \tag{17}$$

where X is the normalizer, M is the number of items and S^+ is the size of the protected group.
For user sided fairness [20] has proposed two metrics, *Score Disparity* and *Recommendation Disparity.* Score disparity is defined as:

$$\mathcal{D}_s = \frac{\sum_{v1, v2 \in U} |\mathcal{A}(v_1) - \mathcal{A}(v_2)|}{2n \sum_{v \in U} \mathcal{A}(v)} \tag{18}$$

where $\mathcal{A}(v)$ is user satisfaction and is calculated as a ratio of the sum between preference score for recommended items and preference score of top-k items.
Recommendation Disparity is defined as:

$$\mathcal{D}_R = \frac{\sum_{v_1, v_2 \in U} |sim(v_1) - sim(v_2)|}{2n \sum_{v \in U} sim(v)} \tag{19}$$

where sim(v) is calculated as ratio of similarity between recommended items and top k items.

– *Multi-Sided Fairness*: Multi-sided fairness covers both consumer side fairness (c-fairness) where the disparate impact of recommendations is viewed for the consumer side and producer side fairness (p-fairness) where fairness is preserved from provider side only [8]. A recent study [30] has focused on two-sided markets as there is a need to contemplate both the consumer side and producer side objectives. They have proposed two metrics which are mentioned below.

$$(GF)_{consumer} = \frac{1}{\frac{m}{2}} \sum_{j=1}^{m} \sum_{k=1}^{j} ||\overline{s}_j - \overline{s}_k||_2^2 \tag{20}$$

where combination of pairs is calculated by $\frac{m}{2}$, $\overline{s}_j$ & $\overline{s}_k$ represents average satisfaction between users of jth and k^{th} groups.
The second metric is defined as:

$$(GF)_{producer} = || \in - \in^* ||_2^2 \tag{21}$$

where $\in$ & $\in^*$ are the vectors for item exposure dispersion and target exposure dispersion respectively. $\in^*$ is normally depicted as flat distribution.

Conclusion

In this paper, we surveyed evaluation metrics for fairness on recommender systems. We observed different users or groups are not fairly recommended. Several works have focused on algorithmic aspects only and left behind the non-algorithmic aspects.

Fifteen fairness evaluation metrics were comparatively analyzed (Table 1) and it was found that most studies used Movielens dataset and variants of Matrix Factorization. During the survey, we observed that the Individual, multi-sided perspective of fairness was least explored and needs more attention. The study of unfairness in recommender systems is quite novel and lacks consensus on standardized evaluation metrics for measuring fairness or common characteristics for in-depth comparative analysis.

Acknowledgements. This research is funded by IUST Kashmir, J & K under grant number IUST/Acad/RP_APP/18/99.

References

1. Abdollahpouri, H., Burke, R.: Multi-stakeholder recommendation and its connection to multi-sided fairness (2019)
2. Anwar, T., Uma, V.: CD-SPM: cross-domain book recommendation using sequential pattern mining and rule mining (2019)
3. Ashokan, A., Haas, C.: Fairness metrics and bias mitigation strategies for rating predictions. Inf. Process. Manage. **58**(5), 1–18 (2021)
4. Baeza-Yates, R.: Bias on the web. Commun. ACM **61**(6), 54–61 (2018)

5. Barocas, S., Selbst, A.D.: Big data's disparate impact. Calif. L. Rev. **104**, 671 (2016)
6. Kouki, A.B.: Recommender system performance evaluation and prediction: an information retrieval perspective. 272 (2012)
7. Biega, J.: Enhancing privacy and fairness in search systems (2018)
8. Burke, R.: Multisided fairness for recommendation (2017)
9. Chung, Y., Kim, N., Park, C., Lee, J.-H.: Improved neighborhood search for collaborative filtering. Int. J. Fuzzy Log. Intell. Syst. **18**(1), 29–40 (2018)
10. Deldjoo, Y., Anelli, V.W., Zamani, H., Bellogín, A., Di Noia, T.: A flexible framework for evaluating user and item fairness in recommender systems. User Model. User-Adapt. Interact. **31**(3), 457–511 (2021). https://doi.org/10.1007/s11257-020-09285-1
11. Elahi, M., Abdollahpouri, H., Mansoury, M., Torkamaan, H.: Beyond algorithmic fairness in recommender systems, pp. 41–46 (2021)
12. Fu, Z., et al.: Fairness-aware explainable recommendation over knowledge graphs. In: Proceedings of the 43rd International ACM SIGIR Conference on Research and Development in Information Retrieval, pp. 69–78 (2020)
13. Shani, G., Gunawardana, A.: Evaluating recommendation systems. In: Ricci, F., Rokach, L., Shapira, B., Kantor, P.B. (eds.) Recommender Systems Handbook, pp. 257–297. Springer, Boston (2011). https://doi.org/10.1007/978-0-387-85820-3_8
14. Harper, F.M., Konstan, J.A.: The MovieLens datasets: history and context. ACM Trans. Interact. Intell. Syst. **5**(4), 1–19 (2015)
15. He, X., Liao, L., Zhang, H., Nie, L., Hu, X., Chua, T.S.: Neural collaborative filtering. In: Proceedings of the 26th International Conference on World Wide Web, pp. 173–182 (2017)
16. Herlocker, J.L., Konstan, J.A., Terveen, L.G., Riedl, J.T.: Evaluating collaborative filtering recommender systems. ACM Trans. Inform. Syst. **22**(1), 5–53 (2004)
17. Hu, Y., Koren, Y., Volinsky, C.: Collaborative filtering for implicit feedback datasets. Proceedings - IEEE International Conference on Data Mining, ICDM, pp. 263–272 (2008)
18. Joy, J., Pillai, R.V.G.: Review and classification of content recommenders in E-learning environment. J. King Saud Univ. - Comput. Inf. Sci. (2021)
19. Koren, Y., Bell, R., Volinsky, C.: Matrix factorization techniques for recommender systems. Computer **42**(8), 30–37 (2009)
20. Leonhardt, J., Anand, A., Khosla, M.: User fairness in recommender systems. In: The Web Conference 2018 - Companion of the World Wide Web Conference, WWW 2018, pp. 101–102 (2018)
21. Li, Y., Chen, H., Fu, Z., Ge, Y., Zhang, Y.: User-oriented fairness in recommendation. In: The Web Conference 2021 - Proceedings of the World Wide Web Conference, WWW 2021, pp. 624–632 (2021)
22. Li, Y., Ge, Y., Zhang, Y.: Tutorial on fairness of machine learning in recommender systems. In Proceedings of the 44th International ACM SIGIR Conference on Research and Development in Information Retrieval, pp. 2654–2657 (2021)
23. Liu, Q., Zeng, Y., Mokhosi, R., Zhang, H.: STAMP: short-term attention/memory priority model for session-based recommendation. In: Proceedings of the 24th ACM SIGKDD International Conference on Knowledge Discovery & Data Mining, pp. 1831–1839 (2018)
24. Bozzoni, A., Cammareri, C.: Which fairness measures for group recommendations? (2020)

25. Rastegarpanah, B., Gummadi, K.P., Crovella, M.: Fighting fire with fire: using antidote data to improve polarization and fairness of recommender systems. In: Proceedings of the Twelfth ACM International Conference on Web Search and Data Mining, pp. 231–239 (2019)
26. Rendle, S., Freudenthaler, C., Gantner, Z., Schmidt-Thieme, L.: BPR: Bayesian personalized ranking from implicit feedback. In: Proceedings of the 25th Conference on Uncertainty in Artificial Intelligence, UAI 2009, pp. 452–461 (2009)
27. Mnih, A., Salakhutdinov, R.R.: Probabilistic matrix factorization (2009)
28. Sokolova, M., Japkowicz, N., Szpakowicz, S.: Beyond accuracy, F-score and ROC: a family of discriminant measures for performance evaluation. In: Sattar, A., Kang, B. (eds.) AI 2006. LNCS (LNAI), vol. 4304, pp. 1015–1021. Springer, Heidelberg (2006). https://doi.org/10.1007/11941439_114
29. Wang, X., He, X., Wang, M., Feng, F., Chua, T.S.: Neural graph collaborative filtering. In: SIGIR 2019 - Proceedings of the 42nd International ACM SIGIR Conference on Research and Development in Information Retrieval, pp. 165–174 (2019)
30. Wu, H., Ma, C., Mitra, B., Diaz, F., Liu, X.: A multi-objective optimization method for achieving two-sided fairness in e-commerce recommendation (2021)
31. Yang, K., Stoyanovich, J.: Measuring fairness in ranked outputs. In: Proceedings of the 29th International Conference on Scientific and Statistical Database Management, pp. 1–6 (2017)
32. Yao, S., Huang, B.: Beyond parity: fairness objectives for collaborative filtering. Adv. Neural. Inf. Process. Syst. **30**, 2922–2931 (2017)
33. Yedder, H.B., Zakia, U., Ahmed, A., Trajković, L.: Modeling prediction in recommender systems using restricted Boltzmann machine. In: 2017 IEEE International Conference on Systems, Man, and Cybernetics (SMC), pp. 2063–2068. IEEE (2017)
34. Zafar, M.B., Valera, I., Gomez-Rodriguez, M., Gummadi, K.P.: Fairness constraints: a flexible approach for fair classification. J. Mach. Learn. Res. **20**(1), 2737–2778 (2019)
35. Zafar, M.B., Valera, I., Gomez Rodriguez, M., Gummadi, K.P.: Fairness beyond disparate treatment and disparate impact: learning classification without disparate mistreatment. In: 26th International World Wide Web Conference, WWW 2017, pp. 1171–1180 (2017)
36. Zhu, Z., Hu, X., Caverlee, J.: Fairness-aware tensor-based recommendation. In: Proceedings of the 27th ACM International Conference on Information and Knowledge Management, pp. 1153–1162 (2018)
37. Zhu, Z., Kim, J., Nguyen, T., Fenton, A., Caverlee, J.: Fairness among New Items in Cold Start Recommender Systems. In: SIGIR 2021 - Proceedings of the 44th International ACM SIGIR Conference on Research and Development in Information Retrieval, pp. 767–776 (2021)

Association Rule Chains (ARC): A Novel Data Mining Technique for Profiling and Analysis of Terrorist Attacks

Saurabh Ranjan Srivastava(✉), Yogesh Kumar Meena, and Girdhari Singh

Department of Computer Science and Engineering, Malaviya National Institute of Technology, Jaipur, Jaipur, Rajasthan, India
saurabhranjansrivastava@gmail.com, {ymeena.cse, gsingh.cse}@mnit.ac.in

Abstract. In data mining, association rule mining algorithms executing over massive datasets, generate vast amounts of rules. These rules pose a gruesome strain of knowledge post-processing on the user. This compels the user to dig through the rules to find relevant knowledge for decision making. For simplification of this practice, we propose the concept of association rule chains. Contrary to conventional association rules, the association rule chains map the values from dataset according to a user defined hierarchy of features, irrespective of the value constraints of interestingness measures. This mapping of values into associations rule chains relies on feature-wise hierarchical clustering of data items with the feature at the root of hierarchy termed as profile. The discovered association rule chains allow the user to assemble the domain relevant information from limited profile based subsets of the search space into a comprehensive feature specific knowledge base. In this paper, we demonstrate our approach by implementing it on a section of global terrorism database (GTD) of terrorist attacks for 4 regions. Our approach facilitates modeling of event profiles for specific classes of terrorist attack events by generating the relevant association rule chains for preemptive event analysis. Finally, we evaluate our approach against occurrences of terrorist attacks from the test dataset.

Keywords: Association rule · Interestingness · Data mining · Profiling · Terrorist attack

1 Introduction

Data mining is a paradigm that utilizes a wide range of techniques to explore relationships and patterns existing in available data and generate useful and valid forecasts [1]. Currently, various data mining techniques are being devised and employed to harness the patterns of data into emerging trends, and process these trends into meaningful forecasts. The data patterns of events can be utilized to forecast the specifics and impacts of the upcoming similar events in future. On close inspection of events, it can be established that significant geospatial events of a type [2], share a common set of feature values

V. E. Balas et al. (Eds.): ICETCE 2022, CCIS 1591, pp. 466–478, 2022.
https://doi.org/10.1007/978-3-031-07012-9_40

and thus can be related to a unique class of events or event profile. Extensive research literature has been published on several interestingness measures for association rules [3] and algorithms for deriving them. Contrary to this, minimal literature is available on analysis of results achieved from association rule mining. Our motivation for the work proposed in this paper is based on the idea of mapping associations emerging from a common class of events into a specialized profile for comprehensive analysis of its event specific information.

In this paper, we propose a novel concept of discovering association rules chains from the itemsets of the mapped feature values of given dataset. We model profiles for specific classes of itemsets by mapping the relevant association rule chains for anticipatory event analysis. Later, we employ our approach for mapping the feature values of terrorist attack events performed by Al-Qaeda terrorist organization into event profiles of 4 regions: Abyan, Shabawh, Gao and Timbuktu and evaluate it by forecasting the geospatial details [2] of terrorist attacks at various locations of the 4 regions from the test dataset. Our approach considers both frequent as well as rare itemsets for profiling of terrorist attack events. To implement our proposed approach, we utilize the subset of Global Terrorism Database (GTD) [4, 5] for a time span of year 1990 to 2017. We also compare the performance of the explored association rule chains against Apriori association rule mining algorithm [6]. In Sect. 2 of this paper, we discuss various terms and concepts associated with the subject under literature review. Then we present the proposed approach in Sect. 3 and discuss the results with case studies in Sect. 4. The findings are concluded with directions of future research in Sect. 5.

2 Literature Review

2.1 Association Rule Mining

The research of mining association rules originated from market basket analysis and today is being employed in various decision making scenarios. It is a prominent data mining technique for exploring patterns or informal structures and associations among them from information repositories and transactional databases [1]. Syntactically, association rules represent a derived conclusion (e.g. purchase of a specific item) associated with a set of conditions (e.g. purchase of another specific item) accompanied by a value of occurrence probability.

For example, the rule `Milk, Bread → Eggs (284, 41.3%, 0.87)` implies that eggs are often bought by customers whenever milk and bread are also bought together. This purchase observation or 'rule' is observed in 284 records, applicable to 41.3% of the transactional data of the store and is 87% consistent.

The 2 fundamental measures widely utilized to judge the quality or interestingness of an association rule are support and confidence [1, 7]. Support is an objective parameter that represents the percentage of data items that satisfy a specific association rule, while another objective parameter confidence, evaluates the degree of certainty of an identified rule. To understand the limitations of conventional association rule mining, a short summarization of these 2 measures is necessary. The formal definition of support and confidence for items A and B are as follows:

$$\text{sup}\ (A \rightarrow B) = \text{sup}\ (B \rightarrow A) = P\ (A \text{ and } B)$$

Support can be considered as a frequency based constraint that suffers with the limitation of rare item pruning tendency, that eliminates the infrequently occurring items capable of generating valuable associations.

`conf (A → B) = P (B | A) = P (A and B) / P (A) = sup (A → B) / sup (A)`

Similar to support, confidence also filters out rules exceeding a minimum threshold value. It can be considered as conditional probability of item B against item A, such that Bs with higher support values also generate higher confidence values even without any association among items. Other interestingness measures such as lift and conviction also employ support and/or confidence for evaluation of association rules [3].

2.2 Issues with Association Rules and Interest Measures

In the domain of association rule mining, most of the work is focused on methods of discovering rules that satisfy the constraints of minimal support and confidence values for a dataset [6]. Parallel to this, reduction in processing time and size of result set is also a subject of research [8]. Contrary to this development, minimal work has been proposed to improvise user interpretation of association rules. The existing association rule mining techniques generate rules on a 'generate and test' basis within the complete search space of the available dataset [9] without considering any specific section. Here, we discuss the major issues of association rule mining [10] that limit its utility for massive datasets.

Selection of Appropriate Values of Interest Measures: Association rule mining algorithms need values of parameters as support, confidence and lift to be configured before execution. For appropriate judgement of these values, the user is expected to acquire adequate experience to achieve an adequate number of best suitable association rules.

Generation of Large Volume of Rules: For a provided set of interest measure values, conventional association rule mining algorithms generate a large number of rules without any guarantee of their relevance to user requirements. The pruning of rules is performed on the basis of threshold values of parameters irrespective of the context or quality of the rules achieved.

Poor Comprehensibility of Discovered Rules: The subjective understandability of discovered rules is usually ignored by traditional association rule mining techniques. A major cause for this ignorance is the dependence of subjectivity over the domain expertise and knowledge of the user. Contrary to this, accuracy of the technique is guided by the threshold values of interest measures. In this case, a tradeoff between understandability and accuracy of the technique is evident.

In this paper, we propose an approach to enhance the comprehensibility of association patterns among items of dataset by incorporating context knowledge of the records.

2.3 Event Profiling

The profile of a subject can be defined as a set of data values and their correlations that reveal the characteristics or traits of the subject evaluated or ranked against a scale [11].

The purpose of profiling is to assemble the knowledge of the characteristics relevant to the subject. Profiling can be also used as an effective tool for forecasting about a subject of interest. In case of event forecasting, most of the existing techniques, concentrate on temporal traits of events (e.g. sports, elections) while ignoring the geospatial features of the events and their correlations.

Few techniques of profile representation are semantic networks, weighted concepts, weighted keywords and association rules [12]. Among these, association rules are the most elaborate and comprehensive technique of profile representation [13].

3 Proposed Work

In this section, we present our proposed approach of modeling and analysis of event values (itemsets). Our approach works by clustering these event values or itemsets as systematic structures called event profiles, and then exploring patterns from them in form of relevant chain of rules. Later, these chains are utilized to forecast the geospatial details of upcoming events having profiles similar to their parent past events. Now we discuss our approach with the help of an example and algorithmic pseudocode for it.

3.1 Association Rule Chains

The proposed association rule chains can be viewed as an improvisation of traditional association rules discovered by classical association rule mining techniques [6]. Contrary to traditional association rules, where the pattern of occurrence is split between antecedent and consequent sections of features (e.g. A → B), association rule chains are composed by user defined hierarchy of features according to his preference. Here, the values of each feature are clustered for a common value of upper feature in the hierarchy termed as a profile.

The topmost feature of the hierarchy denoted as profile becomes the root from which every association rule chain originates. Basically, profile defines the selective subset of the search space or dataset from which association patterns of chains will be explored or mined. Selection of a feature to form profile can be based on user defined criteria or random preference. Similarly, the hierarchy of features to be followed in association rule chains can be also guided by a user defined criteria.

To formally define association rule chains, let us consider a group of features f_x, f_1, f_2, …, f_n in a given dataset with random values. On selecting f_x as the profile for the hierarchy of features f_1, f_2, …, f_n; the proposed structure of association rule chain will be:

$$f_x : f_1 \rightarrow f_1 \ldots \rightarrow \ldots f_n$$

The chain can attain a maximum length of count of all features included in the hierarchy, while a minimum of 2 features will be required for defining the smallest association rule chain. In other words, the chain can be as long as per the required number of features defined in the hierarchy.

Here, the hierarchy of features plays a significant role in modeling the profile of feature f_x discussed as follows:

1. The hierarchy eliminates the need of conventional structure of association rules for representing the patterns among feature values or itemsets, making the analysis of patterns more compact and therefore comprehensible.
2. It also eliminates the constraint of specifying the minimum threshold values of support and confidence measures, thereby automatically also ending the necessity of pruning of rules. This results in an in-depth and broader scanning of patterns with maximum to minimum number of occurrences of itemsets inclusive in a profile.
3. Due to the coverage of all features in the chain, the count of generated rules is reduced drastically (more than 70% in this paper), resulting into a compact rule base for implementation. We further elaborate our proposed concept with the help of an example ahead.

Example. For a given dataset with a random number of records and features X, A, B, C and D, let feature X define the profile domain to be analyzed. With 400 records of feature value type X_1; A_1, B_1, (C_1, C_2), (D_1, D_2, D_3) are the values for features A, B, C and D respectively, that together compose itemsets like (A_1 B_1 C_1 D_1), (A_1 B_1 C_1 D_2), … (A_1 B_1 C_2 D_3).

Here we suppose that feature value X_1 forms the profile to be analyzed. Now, for clustering of feature values falling inside the profile of X_1, we compute average support of the 4 features. If the maximum value of support or occurrence of a feature value is 1, then dividing it by the number of distinct types of feature values generates the average support (σ_{avg}) for the feature as follows:

$\sigma_{avgA} = 1$ (for 1 item, A_1), $\sigma_{avgB} = 1$ (for 1 item, B_1), $\sigma_{avgC} = 0.5$ (for 2 items, C_1 and C_2), $\sigma_{avgD} = 0.33$ (for 3 items D_1, D_2 and D_3)

Now on hierarchically rearranging records by decreasing values of their average support and clustering them by common value of upper feature, we achieve following observations for provided number of occurrences.

Out of given 400 records of profile X_1, A_1 appears 100 times, achieving a support value of 0.25. Similarly, from 100 occurrences of A_1, B_1 co-appears with it 15 times. From 15 occurrences of A_1 B_1; C_1 co-appears 7 and C_2 co-appears 8 times. Under A_1 B_1 C_1, co-occurrence of D_1 is 2, D_2 is 3 and D_3 is 2 times. While under A_1 B_1 C_2, co-occurrence of D_1 is 2, D_2 is 4 and D_3 is observed 2 times.

Now we generate association rule chains based on profile of feature value X_1, along with their corresponding confidence values from the achieved itemsets in previous stage. We further compute the profile confidence ($conf_{Pr}$) by computing product of confidence values of sub-chains that represent the probability of individual occurrence of a specific itemet belonging to a profile.

Here, for instance, the confidence (0.285) of chain $A_1 \rightarrow B_1 \rightarrow C_1 \rightarrow D_3$ presents the 2 co-occurances of item D_3 among 7 occurances of $A_1B_1C_1$. But, the profile confidence of $A_1 \rightarrow B_1 \rightarrow C_1 \rightarrow D_3$ illustrates the 2 co-occurances of these 4 feature values under 400 values of profile feature value X_1.

X_1: $A_1 \rightarrow B_1$, conf (15/100 = 0.15), $conf_{Pr}$ (15/400 = 0.375)
X_1: $A_1 \rightarrow B_1 \rightarrow C_1$, conf (7/15 = 0.466), $conf_{Pr}$ (7/400 = 0.0175)
X_1: $A_1 \rightarrow B_1 \rightarrow C_2$, conf (8/15 = 0.533), $conf_{Pr}$ (8/400 = 0.02)
X_1: $A_1 \rightarrow B_1 \rightarrow C_1 \rightarrow D_1$, conf (2/7 = 0.285), $conf_{Pr}$ (2/400 = 0.005)
X_1: $A_1 \rightarrow B_1 \rightarrow C_1 \rightarrow D_2$, conf (3/7 = 0.428), $conf_{Pr}$ (3/400 = 0.0075)

X_1: $A_1 \rightarrow B_1 \rightarrow C_1 \rightarrow D_3$, conf (2/7 = 0.285), $conf_{Pr}$ (2/400 = 0.005)
X_1: $A_1 \rightarrow B_1 \rightarrow C_2 \rightarrow D_1$, conf (2/8 = 0.25), $conf_{Pr}$ (2/400 = 0.005)
X_1: $A_1 \rightarrow B_1 \rightarrow C_2 \rightarrow D_2$, conf (4/8 = 0.5), $conf_{Pr}$ (4/400 = 0.01)
X_1: $A_1 \rightarrow B_1 \rightarrow C_2 \rightarrow D_3$, conf (2/8 = 0.25), $conf_{Pr}$ (2/400 = 0.005)

The visual interpretation of the explored itemsets or association rule chains can be generated as follows in Fig. 1.

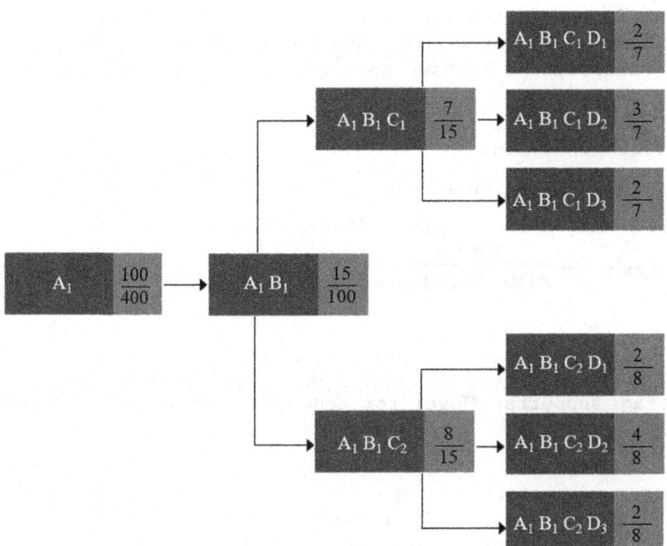

Fig. 1. Visual hierarchy of the explored association rule chains provided in example.

The organization of the proposed approach as an algorithmic pseudocode is discussed ahead.

3.2 Association Rule Chain (ARC) Algorithm

Here we present the proposed approach of event profiing by generation of association rule chains in form of algorithmic pseudocode.

```
Association Rule Chain (ARC)
{
   Input: Dataset features (fx, f1, f2, …, fn)
   Output: Association rule chains of format
        fx: f1 → f1 ... →… fn

   // Data preprocessing      {
      Prepare Dataset: (fx, f1, f2, …, fn)
       a. Selecting features to form hierarchy for chain.
       b. Defining fx feature to form the profile of hierarchy.
       c. Merging, error & duplicate removal, tagging values.
       d. Partitioning dataset into training & testing subsets.
   }

   // Profile processing      {
       Average support(σavg): {
       Computing average support for each feature as:
```

$$\sigma avg_{\text{feature}} = \frac{\sum \sigma_{feature}}{\#\,(itemtype)_{feature}}$$

```
      }
      Hierarchy formation:    {
       Rearrange features f1, f2, …, fn by decreasing values of their
       average supports (σavg) for top-down flow of profile for feature
       fx.
      }
      Cluster generation:     {
       Recursively cluster the records of dataset to:
       a. Generate clusters of feature values falling under a common
feature value of an upper feature in the hierarchy.
       b. Continue until
          (i). all features are covered in itemset generation
          (ii). all records become part of a unique itemset (event
        profile)
      }
      Chain generation:       {
       Recursively increase the association rule chain:
       a. Start from the feature at the bottom of hierarchy
       b. Count and store occurrence of feature values in a cluster
          co-occurring with its associated upper level feature value
       c. Continue until
          (i). all features are covered in chain generation
          (ii). all records become part of a unique chain (event
profile)
   }

   // Interest computation    {
       For each discovered association rule chain and its subsets
       a. Confidence (conf): no. of co-occurrences of feature value &
itemset/ no. of occurrences of itemset in chain
```

```
        b. Profile confidence (confPr): no. of co-occurrences of all
    feature values in the largest chain / no. of
    occurrences of profile feature value.
    }  }
```

The proposed algorithm is partitioned into 3 sections. The data preprocessing section selects and defines features for the association rule chain to be generated. Merging, error and duplicate removal and tagging of feature values as well as division of training and test data is also handled in this section. The profile processing section performs computation of average support for features, hierarchy and cluster formation as well as chain generation. Finally, the quantitative evaluation of generated association rule chains is performed by computing confidence (conf) and profile confidence ($conf_{Pr}$) in interest computation section. The implementation and results of the approach proposed in this algorithm on an event dataset is discussed in next section.

4 Results and Discussions

Now we discuss the application of our proposed event profiling approach on a segment of the Global Terrorism Database (GTD) developed by National Consortium for the study of terrorism and responses of terrorism (START) at university of Maryland [4, 5]. From GTD, we have fetched data of terrorist attacks by Al-Qaeda terrorist organization in 2 terror inflicted regions of Mali and Yemen each: Gao, Timbuktu, Abyan and Shabwah for the time span of year 1990 to 2017.

Every terrorist attack attains a set of attributes or features that inherently relate to specific patterns. On treating these various features of a terrorist attack like target, location and attack type as itemsets, a profile of the respective attack can be modeled by discovering the association rule chains. Later, the modeled profile can be utilized for mining patterns about the details of the possible attacks in upcoming future.

The basic reason for selecting data of attacks by Al-Qaeda is that factors such as selection of tactics, location, attack types and targets remain crucial to success of a terrorist attack. Despite of painstaking efforts of terrorist organizations to invoke the element of surprise in attack, these factors remain common in many events and leave pattern trails that can be utilized to analyze the future behavior of the terrorist group. Also, to investigate this element of surprise in attacks, we deliberately focus on mapping attack profiles with minimum confidence scores. This implies that after training our system on this data, the repetition of attack profiles with minimum occurrences will be investigated among forecasted attacks by Al-Qaeda that occurred in year 2018. Referring to the attack attributes of GTD with highest weightage [14], we have employed ‘target’, ‘location’ and ‘attack type’ features to showcase our approach.

The terrorist attacks executed by Al-Qaeda in these 4 regions have been mapped under their respective profiles as association rule chains. Confidence value for each chain is then computed to record its interest measures. Subsequently, attacks profiles with minimum confidence scores are selected to forecast similar attacks in year 2018.

4.1 Case Study: GAO Profile

The implementation of our approach for profile of GAO region has been presented here. GAO is a terror inflicted region of country Mali in sub-Saharan Africa. For the considered time span, a total of 15 terrorist attacks were reportedly executed in this region by Al-Qaeda. Out of these 15 attacks, civilians were targeted in 2, a French woman was kidnapped in 1 attack and military was targeted in remaining 12 attacks resulting into a support score of 0.8. Out of the 12 attacks on military, 7 have been reported at locations such as military bases and check posts scoring to a confidence value of 0.583. Similarly, 4 out of 7 attacks at military locations were performed by use of gun firing by terrorists generating a confidence score of 0.571. The value of profile confidence ($conf_{Pr}$) for the GAO: MIL → CKP → FRG association rule chain results into 0.266 as the highest type of attack in profile of GAO region. Interestingly, it is also the highest attack type in all 4 regions. The snapshot of selective association rule chains for profiles of all 4 regions along with their confidence values and profile confidence is presented in Table 1. Further, the computation of all confidence and profile confidence values for GAO profile region are presented in Fig. 2 as a case study of our approach.

4.2 Comparisons

To establish the efficiency of our approach, now in Table 2, we compare the number of association rules generated against the number of association rule chains explored for each profile. The count of association rules chains generated by our proposed profiling approach is reduced by 70% than the number of association rules explored by Apriori algorithm running over Weka data mining tool [15]. Here, a maximum reduction of 77.32% is visible for number of 172 rules, while the lowest reduction of 74.39% is achieved for 110 rules for GAO profile. From this observation, it can be deduced that our proposed system improvises its performance with increase in items and subsequent escalation in number of association rules.

4.3 Forecast Evaluations

Now we evaluate the forecast efficiency of our approach by mapping the confidence of association rule chains from the profiles of 4 regions against the test case studies of terrorist attacks executed by Al-Qaeda during year 2018. For this, we investigate the recurrence of attacks with minimum confidence values that occurred during 1990 to 2017. The evaluations reveal that to maintain the element of surprise in their attacks, Al-Qaeda repeated the attacks of profiles that were executed by them for least number of times in past. A summary of terrorist attacks performed by Al-Qaeda in past and repeated in year 2018 has been provided in Table 3.

We illustrate this repetition of attacks of rare class by the instances of suicide car bombing attacks from the profile of Shabwah region in Yemen. On September 20, 2013, an explosive-laden vehicle was driven into the military barracks of Al-Mayfaa town in Shabwah by a suicide bomber. Despite being a premature explosion, the blast claimed lives of 38 Yemeni soldiers [16].

Table 1. Selective association rule chains for profiles of 4 regions with their confidence values

Association rules for region based profile	Conf.	Association rule chains for region based profile	Conf.	$Conf_{Pr}$
ABYAN				
GOVT → OFFICE	1.0	GOVT → OFFICE → BOMBING	0.5	0.055
OFFICE → BOMBING	0.5			
CIVILIAN → PUBLIC	1.0	CIVILIAN → PUBLIC → SUICIDE	1.000	0.055
PUBLIC → SUICIDE	1.0			
MILITARY → OPEN	0.33	MILITARY → OPEN → AMBUSH	0.066	0.055
OPEN → AMBUSH	0.2			
MILITARY → TRANSPORT	0.266	MILITARY → TRANSPORT → FIRING	0.5	0.111
TRANSPORT → FIRING	0.5			
SHABWAH				
CIVILIAN → OFFICE	1.0	CIVILIAN → OFFICE → FIRING	0.5	0.0741
OFFICE → FIRING	0.5			
MILITARY → CHECKPOST	0.5	MILITARY → CHECKPOST → FIRING	0.4	0.1428
CHECKPOST → FIRING	0.4			
MILITARY → PUBLIC	0.3	MILITARY → PUBLIC → SUICIDE	0.33	0.0741
PUBLIC → SUICIDE	0.3			
POLICE → OFFICE	1.0	POLICE → OFFICE → ASSAULT	0.5	0.0741
OFFICE → ASSAULT	0.5			
TIMBUKTU				
CIVILIAN → TRANSPORT	0.25	CIVILIAN → TRANSPORT → AMBUSH	1.0	0.0741
TRANSPORT → AMBUSH	1.0			
MILITARY → CHECKPOST	0.375	MILITARY → CHECKPOST → FIRING	0.66	0.1428
CHECKPOST → FIRING	0.66			
MILITARY → OPEN	0.25	MILITARY → OPEN → ASSASINATION	0.5	0.0741
OPEN → ASSASINATION	0.5			
POLICE → CHECKPOST	1.0	POLICE → CHECKPOST → HOSTAGE	0.5	0.0741
CHECKPOST → HOSTAGE	0.5			
GAO				
MILITARY → CHECKPOST	0.583	MILITARY → CHECKPOST → FIRING	0.5714	0.266
CHECKPOST → FIRING	0.5714			
MILITARY → PUBLIC	0.166	MILITARY → PUBLIC → BOMBING	0.5	0.066
PUBLIC → BOMBING	0.5			
CIVILIAN → CHECKPOST	1.0	CIVILIAN → CHECKPOST → BOMBING	0.5	0.066
CHECKPOST → BOMBING	0.5			
F-WORKER → OPEN	1.0	F-WORKER → OPEN → KIDNAP	1.0	0.066
OPEN → KIDNAP	1.0			

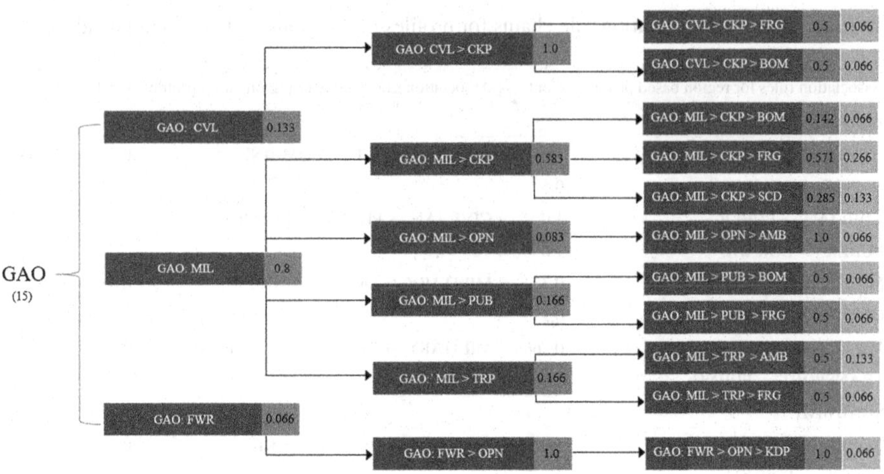

Fig. 2 Confidence and profile confidence values for GAO profile region

Table 2. Number of association rules generated against the number of association rule chains explored for profile of each region.

Profile	Number of association rules	Number of association rule chains	Percentage reduction of rules
ABYAN	172	39	77.325%
SHABWAH	126	31	75.396%
GAO	110	28	74.545%
TIMBUKTU	138	33	76.086%

Table 3. Mapping of association rule chains against real terrorist attacks by Al-Qaeda

Profile	Association rule chains	Conf_{Pr}	Observed on	Repeated on
ABYAN	MILITARY → PUBLIC → BOMBING	0.066	10-JAN-2017	16-FEB-2018
	MILITARY → CHECKPOST → FIRING	0.066	16-JAN-2017	18-JUL-2018
SHABWAH	MILITARY → CHECKPOST → SUICIDE	0.0741	20-SEP-2013	30-JAN-2018
GAO	MILITARY → TRANSPORT → FIRING	0.066	11-SEP-2017	06-APR-2018
TIMBUKTU	MILITARY → CHECKPOST → FIRING	0.1428	05-FEB-2016, 17-JUN-2017	14-APR-2018

After almost 4 years, this rare of its type attack was repeated on January 30, 2018, when another suicide bomber killed 15 newly recruited Yemeni soldiers by crashing his explosive-laden car into a military checkpoint located in the Nokhan area of Shabwah province [17]. In both attacks, Yemenis military was on target and hit at locations under military occupation i.e. barracks and check post. Also, suicide car bombing was employed as the method of attack common in both events. The SHABWAH: MILITARY → CHECKPOST → SUICIDE association rule chain depicts this attack with the of lowest confidence score of 0.1 and profile confidence of 0.0741.

5 Conclusions and Future Work

In this paper, we have demonstrated that by treating features of a dataset as a combination of constituent itemsets, they can be modeled into profiles capable of revealing useful patterns. Contrary to traditional association rule mining techniques [18], we have also proposed a new approach of mapping associations among feature values or itemsets related to a profile of a subject domain in form of association rule chains. These chains efficiently present association patterns of feature values with significantly reduced volume of results discovered from the search space. To prove our approach, we implemented it on data of terrorist attacks performed by Al-Qaeda in Abyan and Shabwah of Yemen, and Gao and Timbuktu regions of Mali nations fetched from Global Terrorism Database (GTD).

It can be concluded that support (occurrence) of an individual item (feature value) contributes to the ultimate itemset confidence of the itemset associations in a profile. However, the frequency of support and confidence can alter the impact of the profile according to the nature of the problem. As a proof of this fact, in the domain of terrorism, we observed that an attack of low confidence score can lay a high impact; while a rarely sold item of low support value in a shopping basket application leaves negligible effect on sales data.

It is also evident that attacks with lower confidence scores indicate a higher surprise rate of occurrence and can generate a high impact on their repetition. As a part of future work, synonymous to temporal association rules, we aim to enhance the coverage of profiles to improve their forecast efficiency for a user defined timeline. Moreover, introduction of modeling mechanisms like directed graphs, supervised-unsupervised learning, possibility logic and uncertainty theory can enhance the precision and impact of this technique as an effective forecasting solution.

Acknowledgements. This work is supported by the project 'Forecasting Significant Social Events by Predictive Analytics Over Streaming Open Source Data', hosted by Malaviya National Institute of Technology, Jaipur and funded by the Science and Engineering Research Board (SERB) under Department of Science & Technology, Government of India.

References

1. Han, J., Pei, J., Kamber, M.: Data mining: concepts and techniques. Elsevier (2011)

2. Butkiewicz, T., Dou, W., Wartell, Z., Ribarsky, W., Chang, R.: Multi-focused geospatial analysis using probes. IEEE Trans. Vis. Comput. Graph. **14**(6), 1165–1172 (2008)
3. McGarry, K.: A survey of interestingness measures for knowledge discovery. Knowl. Eng. Rev. **20**(1), 39–61 (2005)
4. LaFree, G., Dugan, L., Fogg, H.V., Scott, J.: Building a Global Terrorism Database (2006)
5. Start.umd.edu: (2019) Global Terrorism Database [Online]. Available: https://www.start.umd.edu/gtd/. Accessed 16 Nov. 2021
6. Agrawal, R., Srikant, R.: Fast algorithms for mining association rules. In: 20th International Conference on Very Large Data Bases, VLDB, vol. 1215, pp. 487–499 (1994)
7. Agrawal, R., Imielinski, T., Swami, A.: Mining association rules between sets of items in large databases. ACM Sigmod Rec. **22**(2), 207–216 (1993)
8. Woon, Y.K., Ng, W.K., Das, A.: Fast online dynamic association rule mining. In: Second International Conference on Web Information Systems Engineering 2001. vol. 1, pp. 278–287. IEEE Press (2001)
9. Weng, C.-H., Huang, T.-K.: Observation of sales trends by mining emerging patterns in dynamic markets. Appl. Intell. **48**(11), 4515–4529 (2018). https://doi.org/10.1007/s10489-018-1231-1
10. Mahrishi, M., Sharma, G., Morwal, S., Jain, V., Kalla, M.: Data model recommendations for real-time machine learning applications: a suggestive approach. In: Machine Learning for Sustainable Development, pp. 115–128. De Gruyter, Boston (2021)
11. Nadee, W.: Modelling user profiles for recommender systems. Doctoral dissertation, Queensland University of Technology (2016)
12. Gauch, S., Speretta, M., Chandramouli, A., Micarelli, A.: User profiles for personalized information access. In: Brusilovsky, P., Kobsa, A., Nejdl, W. (eds) The Adaptive Web. LNCS, vol. 4321. Springer, Heidelberg (2007). https://doi.org/10.1007/978-3-540-72079-9_2
13. Mobasher, B.: Data mining for web personalization. In: Brusilovsky, P., Kobsa, A., Nejdl, W. (eds) The Adaptive Web. LNCS, vol. 4321. Springer, Heidelberg (2007). https://doi.org/10.1007/978-3-540-72079-9_3
14. Tutun, S., Khasawneh, M.T., Zhuang, J.: New framework that uses patterns and relations to understand terrorist behaviors. Expert Syst. Appl. **78**, 358–375 (2017)
15. Weka 3: Machine Learning Software in Java (n.d.). Retrieved from https://www.cs.waikato.ac.nz/ml/weka/. Accessed 16 Nov. 2021
16. CBC News, Al-Qaeda militants kill 38 troops in Yemen attacks [Online]. Available: https://www.cbc.ca/news/world/al-qaeda-militants-kill-38-troops-in-yemen-attacks-1.1861592. Accessed 16 Nov. 2021
17. Xinhua | English.news.cn: 15 soldiers killed in suicide car bombing at checkpoint in Yemen [Online]. Available: https://www.xinhuanet.com/english/2018-01/30/c_136936300.htm, 30 Jan. 2018. Accessed 16 Nov. 2021
18. Huang, M.-J., Sung, H.-S., Hsieh, T.-J., Wu, M.-C., Chung, S.-H.: Applying data-mining techniques for discovering association rules. Soft. Comput. **24**(11), 8069–8075 (2019). https://doi.org/10.1007/s00500-019-04163-4

Classification of Homogeneous and Non Homogeneous Single Image Dehazing Techniques

Pushpa Koranga(✉), Sumitra Singar, and Sandeep Gupta

Bhartiya Skill Development University, Jaipur, India
Pushpa.Koranga@ruj-bsdu.in

Abstract. Due to the suspended particles in the atmosphere formed by hazy or foggy weather conditions produces a degraded image which causes change of color, reduced contrast and poor visibility. Practically haze cannot always be uniform in nature sometime non uniform haze can be formed. On the basis of haze formation single image dehazing can be categorized into two type's i.e. homogeneous and non homogeneous haze. In this paper we have discuss different methods of single image dehazing used for homogeneous and non homogeneous haze. As qualitative parameter such as Peak Signal to Noise ratio (PSNR) and Structure Similarity Index Measures (SSIM) is used to check the quality of output dehazed image. On the basis of result obtained we have compared different techniques of single image dehazing. Average PSNR and SSIM value also calculated for 30 set of samples for both homogeneous and non homogeneous hazy images. Result is also analysis with plotted graph of average PSNR and SSIM value.

Keywords: Single image dehazing · Homogeneous haze · Non homogeneous haze · Transmission map · Atmospheric light

1 Introduction

Due to the suspended particles in the environment in the form of smoke, fog or other floating particles may degrade the quality of image. Image dehazing techniques try to enhance the quality of image by removing haze amount in hazy image which occurs due to various environmental conditions. Image processing and computer vision tasks require haze removal from hazy images so that it becomes feasible for different algorithms of computer vision to analyze it. Otherwise it would affect security of video surveillance, purpose system, traffic surveillance camera etc. The haze formation model is given by [1]-

$$I(x) = J(x)t(x) + A(1 - t(x)) \tag{1}$$

where *I(x)* is observed hazy image value, *J(x)* is clear scene radiance, *A* is global atmospheric light value, and *t(x)* is scene transmission. For homogeneous haze, *t(x)* can be given by-

$$t(x) = e^{-\beta d(x)} \tag{2}$$

V. E. Balas et al. (Eds.): ICETCE 2022, CCIS 1591, pp. 479–493, 2022.
https://doi.org/10.1007/978-3-031-07012-9_41

where $t(x)$ is scene transmission map, β is scattering coefficient and $d(x)$ is scene depth. Techniques of image dehazing are categorized into two types i.e. Single image dehazing and multiple image Dehazing techniques. In Single Image Dehazing only Single Image is needed for dehazing techniques while in Multiple Image Dehazing more than one image of the same scene is needed. Various image dehazing techniques have been adopted till now for single input image. In recent time many techniques were used to remove uniform haze from hazy images due to the hypothesis made that haze formation is homogeneous in nature, but violated in real life due to non uniform distribution of haze in the real world [2]. Single image dehazing can be further categorized into two types i.e. homogeneous image dehazing techniques and non homogeneous image dehazing techniques. In homogeneous haze, uniform haze is captured throughout the scene while in non homogeneous haze; non-uniform hazy image is captured throughout the scene (Fig. 1).

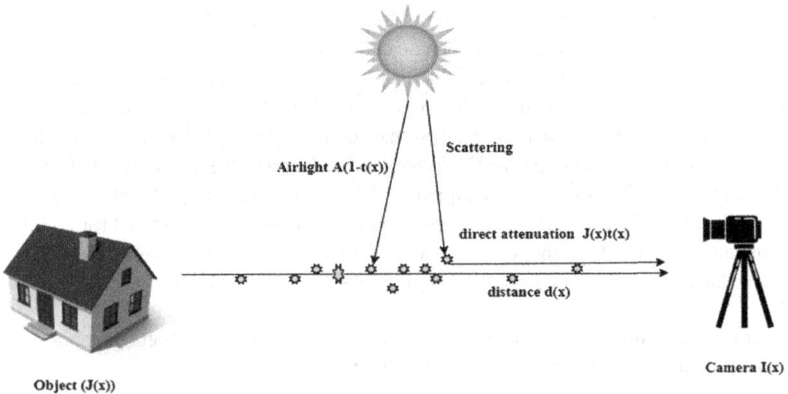

Fig. 1. Haze formation due to scattering and attenuation of light

2 Method

Image dehazing is classified into two types single image dehazing and multiple image dehazing. Single image dehazing techniques involves only single input for image dehazing. Single image dehazing on the basis of formation of haze can be described into two types i.e. homogeneous single image dehazing and non homogeneous single image dehazing [3]. It can be seen that homogeneous haze has uniform haze throughout the scene. Therefore calculation of transmission map and atmospheric value is kept constant for certain image dehazing techniques. While non homogeneous haze has non uniform haze throughout the scene. Different method has been adopted to calculate transmission map and atmospheric value which is discussed below (Fig. 2).

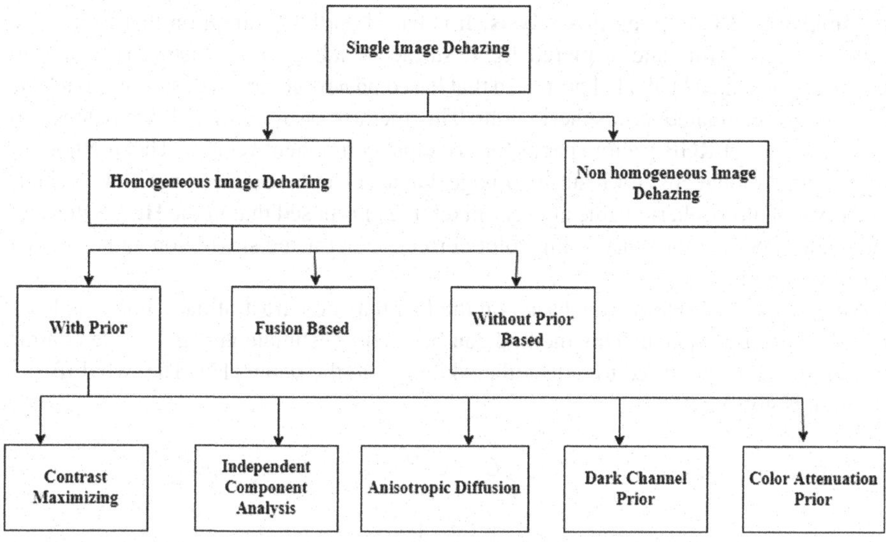

Fig. 2. Classification of single image Dehazing

3 Homogeneous Single Image Dehazing

Homogeneous Single Image Dehazing: Generally homogeneous image dehazing techniques for single image can be categorized into three types:

3.1 With Prior Based

With prior based method is based on some previous assumptions and knowledge for estimating dehazed image. There are several method based on with priors discussed below-

a) Contrast Maximizing: In contrast maximizing assumption made that contrast of haze free image is more than hazy image. Haze from hazy image is improved by optimizing the contrast, degradation of gradient and finest detail constrained to minimum level [3]. Estimation of Atmospheric light is done by using quad tree decomposition and transmission map is estimated by objective function measured by image entropy and information fidelity of image [4]. Variational Contrast enhancement method is used to minimize image energy [5]. As in other proposed method atmospheric light was considered constant but in the color line model assumption made that atmospheric light color is constant but intensity may vary in the image [6]. Multi scale gradient domain contrast enhancement approaches do not depend on estimated atmospheric light value as it does not rely much on transmission map of scene. This method is useful to improve color and contrast of hazy images [7]. Detail of haze free image is estimated by laplacian template based on multiple scattering models [8]. In [9] for scene priors two objectives are proposed to derive transmission map based on both theoretically and heuristics.

b) Independent Component Analysis: It is based on the assumption that to increase scene visibility eliminate scattered light, surface shading and transmission map are locally uncorrelated [10]. [11] proposed that based on new geometry assumption in RGB vector space estimated atmospheric light. This method assumes that all the patches are through origin of RGB vector spaces for principal component vectors. This method has high computational cost for resulting image due to selection of image patch with candidate pixel of atmospheric value at every pixel. [12] Proposed that in the Hue, Saturation and value (HSV) color space using color correction extracted sky region and estimated transmission map.

c) Anisotropic Diffusion: It is based on the fact that a contrast of hazy image is lower than the haze free image. This method can be used to estimate air light while contrast of hazy image is increased to reduce the whiteness in the scene [13]. The two diffusion equation is given by-

$$C = e^{-D^2} \tag{3}$$

$$C = \frac{1}{1 + D^2} \tag{4}$$

$D =$ information of edges, $C =$ conduction coefficient of the diffusing process.

d) Dark Channel Prior: It is based on fact that hazy image formed by dark pixel and one color channel of RGB has low intensity. In [14] uses dark channel prior and wavelet transform for image dehazing. This method uses guided filters instead of soft matting to estimate the scene depth of the image. The dark channel of image is given by-

$$J^{dark}(x) = \frac{min}{y\epsilon\Omega(x)}\left(\frac{min}{c\epsilon(r, g, b)}\right)J^c(y) \tag{5}$$

where $J^c =$ intensity of color channel $\boldsymbol{\Omega}(x) =$ local patch centered at x c $\in$ *(r, g, b)* of RGB color image. In Fast dehazing, soft matting replaces with bilateral filter which is fast and efficient [15]. It uses guided filter to reduce time complexity the rough transmission map and guidance image uses nearest neighbor interpolation to reduce processed time [16]. Block pixel method is used for dark channel evaluation in which firstly the block level of the dark channel after that pixel level is calculated [17]. MODCM is proposed to prevent halo artifacts and preserve scene structure detail [18].

e) Color Attenuation prior: This technique is also based on the prior or assumption that depth scene is estimated by calculating attenuation of different color channels and their difference [19].

The equation is given by-

$$d(X) = \theta_0 + \theta_1 V(x) + \theta_2 S(x) + \epsilon(x) \tag{6}$$

To overcome the drawback of multiple scattering techniques which makes blur image change of detail prior is used [20]. In non local dehazing techniques some assumptions are made that with a few hundred color images can be represented and this method is based on pixels which make it faster and robust. The main disadvantage of this technique is that it cannot be used for such scenes where air light is brighter [21].

3.2 Fusion Based

The fusion method takes input from two types of version adapted from the original image such that it gives a haze free image [22]. In this method brightness is improved and visibility enhancement is improved using light, image and object chromaticity to recover low visibility and change information of color [23]. Fusion of two methods i.e. Contrast enhancement and white balancing is done to improve visibility. Laplacian pyramid representation is used to minimize artifacts [24]. In this technique propagating deconvolution is used to convert smog into fog then dark channel prior carried out to enhance the detail and contrast [25]. In this method quad tree subdivision is used to estimate atmospheric light value where objective function is maximized which consists of information fidelity as well as image entropy is used to estimate transmission map. Finally the weighted least square method is used to refine transmission map [26]. Image quality assessment of dehazed image is done by building a database then subjective study carried out [27]. Wavelet transforms and Guided filter is employed to estimate scene depth map and to increase contrast of hazy image contrast enhancement is done [28]. Dark channel prior is used in optimized design method to evaluate atmospheric light then white balancing is done. At last contrast stitching is used to improve visual effect of image [29].

3.3 Without Prior Based

This method is based on learning features not on some previous assumption used for dehazing. It have mainly three steps i.e. 1) estimating transmission map value 2) estimating Atmospheric value A 3) Estimating haze free scene through computing. Basically to estimate accurate transmission map convolutional neural network method was proposed which avoid inaccurate estimation of transmission map. It uses encoder and decoder where encoder is used to find transmission map and decoder uses residual function so that all hidden layer is fully connected to all other neuron in second layer. The advantages of this it converge training data set model and improves learning rate. It also established non linear relationship between ground truth image and hazy image. Residual loss functions and Mean square error plays important role in training of dataset of image [30]. Convolutional Neural Network consist of six layer which is given by-

First layer consist of convolution layer, to extract features input of convolution layer connected to upper receptive field.

Equation for convolution layer is given by-

$$Conv = f(X * W + b) \tag{7}$$

where X represent image matrix, W represent convolution kernel, b represent offset value, f represent ReLU function and * represent convolution operator. The second layer consist of slice layer, which slice input on the basis of dimensions as 25 * 14 * 14 feature map is sliced to convert in size of 5 * 5 * 14 * 14. The third layer consists of Eltwise layer. There are three operation in it i.e. sum or subtract (add, minus), product (dot multiply) and max (take large value). The fourth layer consists of multi scale mapping to increase the accuracy of extraction of feature. The fifth layer consist of 5 * 5 convolution kernel so that to perform maximum extraction from pixel level neighbor. The last layer have convolution layer to convert with a dimension of 1 * 1 input data into a feature map. At last transmission map is estimated by reconstructed transmission map and corresponding ground truth map [31].

Learning based algorithms based on robust dehazing found dark channels as most useful feature [32]. Stochastic enhancement method uses stochastic iterative algorithm and visibility map is estimated [33]. DehazeGan is an unsupervised technique which explicitly learns atmospheric light and transmission map of hazy image by using adversarial composition network [34]. Robust color sparse representation uses haze density information for color and luminance channels to determine transmission map [35]. Statistical color ellipsoid used in RGB space fit in haze pixel cluster. Ellipsoid surfaces have a minimum color component which can increase the contrast at any noise level or at any haze [36]. In ImgsensingNet proposed to check air quality by taking haze images using unmanned aerial vehicle (UAV) and ground based wireless sensor for air quality index. This method uses 3 dimensional CNN by adding prior feature map in dimensions [37] (Fig. 3).

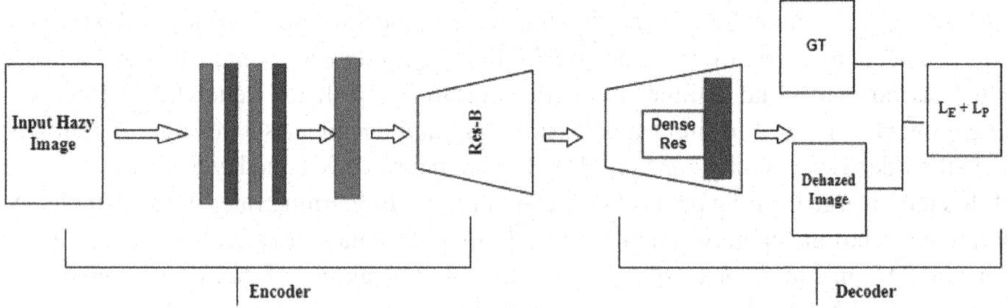

Fig. 3. CNN model using Encoder and Decoder

4 Non Homogeneous Single Image Dehazing

4.1 Non Homogeneous Single Image Dehazing

For homogeneous atmosphere, attenuation coefficient is considered constant while for non homogeneous atmosphere, attenuation coefficient reduced exponentially with increase in height of atmosphere [38]. As coarse transmission map value are further refined by guided filter also adjusted by comparing global atmospheric light with detected intensity of sky regions however fail in dense haze. In this method image quality assessment measures a human visual system with different dynamic range capable of comparing image pair. In [39] visibility map is estimated of the haze and stochastic enhancement for non homogeneous haze. This method has good results for visibility, colorfulness and contrast. AOD network uses end to end convolution neural networks to obtain haze free images. AOD net techniques can be used on both natural and synthetic images [40]. For non-homogeneous underwater image three approach is used firstly segmentation using K-means clustering secondly depth map using morphological operation finally post processing produces better results for color fidelity, preserving natural color but fail in medium light [41]. Adaptive patch based CNN method uses quad tree decomposition method to produce multiple patches such that homogeneous haze regions are divided into large patch sizes while non homogeneous patches are divided into small adaptive patch sizes in which local detail is preserved [42]. Back projected pyramid network (BPPnets) method gave better results for NTIRE 2018 dataset for homogeneous haze, NTIRE 2019 dataset for dense haze and NITIRE 2020 dataset for non homogeneous dataset. BPPnets provide training architecture at end to end point and provide good result for indoor image while poor result for outdoor image [43]. Transmission refining and inflection point estimation used to estimate visibility distance for non homogeneous images [44]. Skyline methods as well as atmospheric scattering model accurately estimate atmospheric light. Transmission map is optimized by a least square filtering method to highlight edge detail and reduce halo effect [45]. A hierarchical network which is fast and deep is used for non homogeneous haze and also very fast method use for real time application [46]. AtJwD which is based on deep learning used to estimate physical parameters of hazy images. Different features map generated by channel attention scheme while missing features generated at decoder using non local features [47]. Ensemble dehazing network (EDN), Ensemble dehazing U net architecture (ENU) and EDN-AT is effective and gives good results for non homogeneous haze. EDN and EDU result in rich hierarchical networks with different types of modeling capabilities while EDN - AT effective for physical model incorporation into framework of deep learning [48]. Knowledge transfer techniques uses teacher network to utilize clean image for training with robust and strong prior for image. For more attention, attention mechanism is introduced combining pixel attention with channel attention [2]. Patch map selection (PMS) Network based on automatic selection of patch size, CNN is used to generate patch maps from input image [49]. Trident dehazing network constructed from three sub networks firstly, Encoder decoder reconstruct coarse haze free image secondly coarse feature maps is refined with high frequency detail at last haze density map generation without any supervision reconstruct haze density map [50].

5 Experimental Result and Discussion

We compared different techniques of single image dehazing for homogeneous and non homogeneous haze. The dataset of 30 set of sample of image is taken from NTIRE 2018 for homogeneous haze which contains different outdoor scenes along with ground truth image [51]. While the dataset of 30 set of sample of image for non homogeneous haze is taken from NH-HAZE which contains different outdoor scenes along with ground truth image [52].

PSNR of image can be calculated by-

$$PSNR = 10\ log_{10}\left[\frac{MAX^2 I(X)}{MSE}\right] \tag{7}$$

where I(x) is haze free scene. MSE is mean square error between hazy and haze free scene.

Structural Similarity Index Measure (SSIM) measures similarity between hazy and haze free Images based on three parameter such as luminance, contrast and structure of image. SSIM is calculated by [53]-

$$SSIM = F(L(i), C(i), S(i)) \tag{8}$$

where L(i) is luminance, C(i) is Contrast and S(i) is Structure of an Image. Different single image dehazing techniques is implemented using python. Average PSNR and SSIM value is calculated for dehazed image with the reference of ground truth image for 30 set of sample of homogeneous and non homogeneous hazy image. Result is compared on the basis of graph plotted for average PSNR and SSIM value. As qualitatively and quantitatively result of image dehazing directly depend on PSNR and SSIM value (Figs. 6, 7 and Tables 1, 2).

a) Hazy Image (Zhu) b) Ground Truth Image c) CAP

d) ImgSensingNet (He) (Rashid) e) Dark Channel (Yang) f) CNN

Fig. 4. Homogeneous image Dehazing using different techniques

a) Hazy Image (Zhu) b) Ground Truth Image c) CAP

d) ImgSensingNet (He) (Rashid) e) Dark Channel (yang) f) CNN

Fig. 5. Non homogeneous image Dehazing using different techniques

Table 1. PSNR and SSIM value for homogeneous and non homogeneous image using different techniques (Fig. 4 and Fig. 5)

Single Image Dehazing techniques	Homogeneous		Non-Homogeneous	
	PSNR	SSIM	PSNR	SSIM
CAP (Zhu et al.)	27.77	0.59	27.90	0.59
Dark Channel (He et al.)	27.74	0.62	28.11	0.57
ImgSensingNet (Yang et al.)	27.75	0.57	27.93	0.57
CNN (Rashid et al.)	28.24	0.67	27.93	0.67

Table 2. Average PSNR and SSIM value using different techniques for 30 set of sample hazy image

Different single image dehazing techniques	Homogeneous Haze		Non Homogeneous haze	
	PSNR	SSIM	PSNR	SSIM
CAP (Zhu et al.)	27.97	0.56	28.01	0.50
Dark channel Prior (He et al.)	28.00	0.63	27.93	0.52
ImgsensingNet (yang et al.)	27.84	0.58	27.94	0.49
CNN (Rashid et al.)	28.05	0.61	27.88	0.50

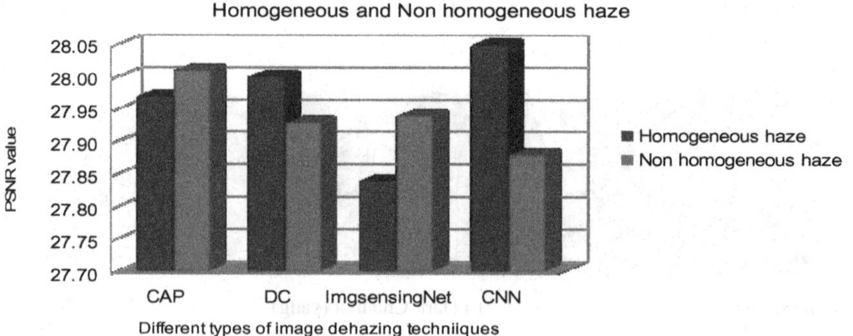

Fig. 6. Comparison of average PSNR value for Homogeneous and Non Homogeneous Haze

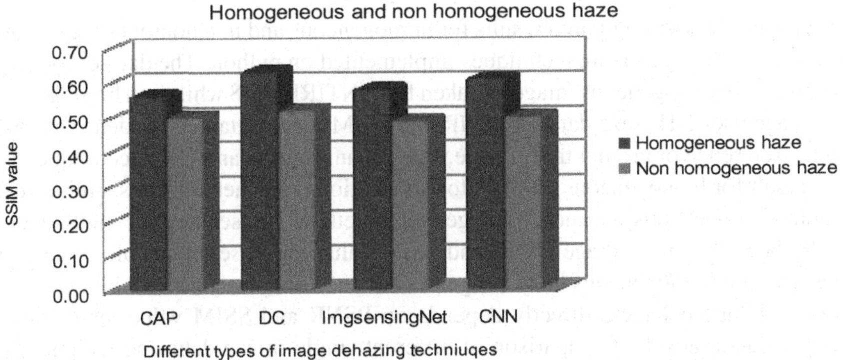

Fig. 7. Comparison of average SSIM value for Homogeneous and Non Homogeneous Haze

Table 3. Comparison of Homogeneous and Non Homogeneous haze using different Image Dehazing techniques

Techniques	Advantages	Disadvantages
CAP (Zhu et al.)	• Better PSNR value for non homogeneous haze	• It causes color distortion • Constant atmospheric value is calculated
Dark channel Prior (He et al.)	• It is fast and easy to implement	• Based on prior knowledge or assumption • Do not work well for sky region • As constant atmospheric value is calculated where it vary for non homogeneous haze
ImgsensingNet (yang et al.)	• It use refined transmission map value • Visually better result	• It causes color distortion • Poor SSIM value for homogeneous and non homogeneous haze
CNN (Rashid et al.)	• Do not based on prior assumption or knowledge • Unlike DCP it does not have constant atmospheric value as it is determined at the time of training • Giving better result for both homogeneous as well as non homogeneous haze	• Visually poor quality image for non homogeneous haze • Based on training of dataset • High Computational time • PSNR and SSIM value need to be improved

6 Conclusion

In this paper we have compared results for homogeneous and non homogeneous images using single image dehazing techniques implemented on python. The dataset for 30 set of samples for homogeneous images is taken from NTIRE 2018 while non homogeneous hazy image uses NH Haze dataset. PSNR and SSIM value obtained for dehazed image with the reference of ground truth image. Dark channel prior and CNN technique gives better result for homogeneous haze. Color attenuation prior method gives similar result for both homogeneous and non homogeneous method. ImgsensingNet method gives visually better result. Average PSNR and SSIM value for 30 set of sample of homogeneous and non homogeneous hazy image are obtained and analysis with plotted graph. Result of image dehazing directly depends on PSNR and SSIM value quantitatively as well as qualitatively. Comparison of different single image dehazing techniques is discussed in Table 3. In future more focus required for non homogeneous hazy images. Also further improvement required for night time haze as mostly techniques focused for day time hazy images.

References

1. Narasimhan, S.G., Nayar, S.K.: Contrast restoration of weather degraded images. IEEE Trans. Pattern Anal. Mach. Intell **25**(6), 713–724 (2003)
2. Wu, H., Liu, J., Xie, Y., Qu, Y., Ma, L.: Knowledge transfer dehazing network for nonhomogeneous dehazing. In: Proceedings of the IEEE/CVF Conference on Computer Vision and Pattern Recognition Workshops, pp. 478–479 (2020)
3. Ancuti, C., Ancuti, C.O.: Effective contrast-based dehazing for robust image matching. IEEE Geosci. Remote Sens. Lett. **11**(11), 1871–1875 (2014). https://doi.org/10.1109/LGRS.2014.2312314
4. Park, D., Park, H., Han, D.K., Ko, H.: Single image dehazing with image entropy and information fidelity. In: 2014 IEEE International Conference on Image Processing (ICIP), Paris, pp. 4037–4041 (2014). https://doi.org/10.1109/ICIP.2014.7025820
5. Galdran, A., Vazquez-Corral, J., Pardo, D., Bertalmío, M.: A variational framework for single image dehazing. In: Agapito, L., Bronstein, M., Rother, C. (eds.) Computer Vision–ECCV 2014 Workshops. ECCV 2014. LNCS, vol. 8927, pp. 259–270. Springer, Cham (2015). https://doi.org/10.1007/978-3-319-16199-0_18
6. Santra, S., Chanda, B.: Single image dehazing with varying atmospheric light intensity. In: Fifth National Conference on Computer Vision, Pattern Recognition, Image Processing and Graphics (NCVPRIPG), Patna, 2015, pp. 1–4 (2015). https://doi.org/10.1109/NCVPRIPG.2015.7490015
7. Mi, Z., Zhou, H., Zheng, Y., Wang, M.: Single image dehazing via multi-scale gradient domain contrast enhancement. IET Image Process. **10**, 206–214 (2016)
8. Zhang, H., Li, J., Li, L., Li, Y., Zhao, Q., You, Y.: Single image dehazing based on detail loss compensation and degradation. In: 2011 4th International Congress on Image and Signal Processing, Shanghai, pp. 807–811 (2011). https://doi.org/10.1109/CISP.2011.6100341
9. Lai, Y., Chen, Y., Chiou, C., Hsu, C.: Single-image dehazing via optimal transmission map under scene priors. IEEE Trans. Circuits Syst. Video Technol. **25**(1), 1–14 (2015). https://doi.org/10.1109/TCSVT.2014.2329381
10. Fattal, R.: Single image dehazing. ACM Trans. Graph. (TOG) **27**(3), 1–9 (2008)

11. Sulami, M., Glatzer, I., Fattal, R., Werman, M.: Automatic recovery of the atmospheric light in hazy images. In: 2014 IEEE International Conference on Computational Photography (ICCP), pp. 1–11. IEEE (2014)
12. Yoon, I., Kim, S., Kim, D., Hayes, M.H., Paik, J.: Adaptive defogging with color correction in the HSV color space for consumer surveillance system. IEEE Trans. Consum. Electron. **58**(1), 111–116 (2012)
13. Tripathi, A.K., Mukhopadhyay, S.: Single image fog removal using anisotropic diffusion. IET Image Process. **6**(7), 966–975 (2012)
14. Carlevaris-Bianco, N., Mohan, A., Eustice, R.M.: Initial results in underwater single image dehazing. In: OCEANS 2010 MTS/IEEE SEATTLE, Seattle, WA, pp. 1–8 (2010). https://doi.org/10.1109/OCEANS.2010.5664428
15. Xu, H., Guo, J., Liu, Q., Ye, L.: Fast image dehazing using improved dark channel prior. In: Proceedings of International Conference on Information Science and Technology (2012). https://doi.org/10.1109/ICIST.2012.6221729
16. Zhang, Q., Li, X.: Fast image dehazing using guided filter. In: IEEE 16th International Conference on Communication Technology (ICCT), Hangzhou, pp. 182–185 (2015). https://doi.org/10.1109/ICCT.2015.7399820
17. Yu, T., Riaz, I., Piao, J., Shin, H.: Real-time single image dehazing using block-to-pixel interpolation and adaptive dark channel prior. IET Image Process. **9**(9), 725–34 (2015)
18. Liu, Y., Li, H., Wang, M.: Single image dehazing via large sky region segmentation and multiscale opening dark channel model. IEEE Access **5**, 8890–8903 (2017)
19. Xie, B., Guo, F., Cai, Z.: Improved single image dehazing using dark channel prior and multi-scale Retinex. In: International Conference on Intelligent System Design and Engineering Application, Changsha, pp. 848–851 (2010).https://doi.org/10.1109/ISDEA.2010.141
20. Li, J., et al.: Single image dehazing using the change of detail prior. Neurocomputing **156**, 1–11 (2015)
21. Berman, D., Treibitz, T., Avidan, S.: Non-local image dehazing. In: Proceedings of the IEEE Conference on Computer Vision and Pattern Recognition (2016)
22. Ancuti, C.O., Ancuti, C., Bekaert, P.: Effective single image dehazing by fusion. In: 2010 IEEE International Conference on Image Processing. IEEE (2010)
23. Lien, C., Yang, F., Huang, C.: An efficient image dehazing method. In: 2012 Sixth International Conference on Genetic and Evolutionary Computing, Kitakushu, pp. 348–351 (2012). https://doi.org/10.1109/ICGEC.2012.47
24. Ancuti, C.O., Ancuti, C.: Single image dehazing by multi-scale fusion. IEEE Trans. Image Process. **22**(8), 3271–3282 (2013). https://doi.org/10.1109/TIP.2013.2262284Aug
25. Wang, R., Wang, G.: Single smog image dehazing method. In: 2016 3rd International Conference on Information Science and Control Engineering (ICISCE), pp. 621–625. IEEE (2016)
26. Park, D., Park, H., Han, D.K., Ko, H.: Single image dehazing with image entropy and information fidelity. In: 2014 IEEE International Conference on Image Processing (ICIP), pp. 4037–4041. IEEE (2014)
27. Ma, K., Liu, W., Wang, Z.: Perceptual evaluation of single image dehazing algorithms. In: IEEE International Conference on Image Processing (ICIP), Quebec City, QC, 2015, pp. 3600–3604 (2015). https://doi.org/10.1109/ICIP.2015.7351475
28. Fu, Z., Yang, Y., Shu, C., Li, Y., Wu, H., Xu, J.: Improved single image dehazing using dark channel prior. J. Syst. Eng. Electron. **26**(5), 1070–1079 (2015)
29. Mahrishi, M., Morwal, S., Muzaffar, A.W., Bhatia, S., Dadheech, P., Rahmani, M.K.I.: Video index point detection and extraction framework using custom YoloV4 Darknet object detection model. IEEE Access **9**, 143378–143391 (2021). https://doi.org/10.1109/ACCESS.2021.3118048

30. Rashid, H., Zafar, N., Iqbal, M.J., Dawood, H., Dawood, H.: Single image dehazing using CNN. Procedia Comput. Sci. **147**, 124–130 (2019)
31. Li, J., Li, G., Fan, H.: Image dehazing using residual-based deep CNN. IEEE Access **6**, 26831–26842 (2018)
32. Tang, K., Yang, J., Wang, J.: Investigating haze-relevant features in a learning framework for image dehazing. In: 2014 IEEE Conference on Computer Vision and Pattern Recognition, Columbus, OH, 2014, pp. 2995–3002 (2014). https://doi.org/10.1109/CVPR.2014.383
33. Bhattacharya, S., Gupta, S., Venkatesh, K.S.: Dehazing of color image using stochastic enhancement. In: 2016 IEEE International Conference on Image Processing (ICIP). IEEE (2016)
34. Zhu, H., Peng, X., Chandrasekhar, V., Li, L., Lim, J.H.: DehazeGAN: when image dehazing meets differential programming. In: IJCAI, pp. 1234–1240 (2018)
35. Huang, S., Donglei, W., Yang, Y., Zhu, H.: Image dehazing based on robust sparse representation. IEEE Access **6**, 53907–53917 (2018)
36. Bui, T.M., Kim, W.: Single image dehazing using color ellipsoid prior. IEEE Trans. Image Process. **27**(2), 999–1009 (2017)
37. Yang, Y., Hu, Z., Bian, K., Song, L.: ImgSensingNet: UAV vision guided aerial-ground air quality sensing system. In: IEEE INFOCOM - IEEE Conference on Computer Communications, Paris, France, 2019, pp. 1207–1215 (2019). https://doi.org/10.1109/INFOCOM.2019.8737374
38. Zhenwei, S., Long, J., Tang, W., Zhang, C.: Single image dehazing in inhomogeneous atmosphere. Optik **125**(15), 3868–3875 (2014)
39. Bhattacharya, S., Gupta, S., Venkatesh, K.S.: Dehazing of color image using stochastic enhancement. In: 2016 IEEE International Conference on Image Processing (ICIP), pp. 2251–2255. IEEE (2016)
40. Li, B., Peng, X., Wang, Z., Xu, J., Feng, D.: AOD-NET: all-in-one dehazing network. In: Proceedings of the IEEE International Conference on Computer Vision, pp. 4770–4778 (2017)
41. Borkar, S.B., Bonde, S.V.: Oceanic image dehazing based on red color priority using segmentation approach. Int. J. Ocean. Oceanogr. **11**(1), 105–119 (2017)
42. Kim, G., Ha, S., Kwon, J.: Adaptive patch based convolutional neural network for robust dehazing. In: 2018 25th IEEE International Conference on Image Processing (ICIP), pp. 2845–2849. IEEE (2018)
43. Singh, A., Bhave, A., Prasad, D.K.: Single image dehazing for a variety of haze scenarios using back projected pyramid network. arXiv preprint arXiv: 2008.06713 (2020)
44. Nankani, H., Mahrishi, M., Morwal, S., Hiran, K.K.: A formal study of shot boundary detection approaches—comparative analysis. In: Sharma, T.K., Ahn, C.W., Verma, O.P., Panigrahi, B.K. (eds.) Soft Computing: Theories and Applications. AISC, vol. 1380, pp. 311–320. Springer, Singapore (2022). https://doi.org/10.1007/978-981-16-1740-9_26
45. Fu, H., Bin, W., Shao, Y., Zhang, H.: Scene-awareness based single image dehazing technique via automatic estimation of sky area. IEEE Access **7**, 1829–1839 (2018)
46. Das, S.D., Dutta, S.: Fast deep multi-patch hierarchical network for nonhomogeneous image dehazing. In: Proceedings of the IEEE/CVF Conference on Computer Vision and Pattern Recognition Workshops (2020)
47. Metwaly, K., Li, X., Guo, T., Monga, V.: Nonlocal channel attention for nonhomogeneous image dehazing. In: Proceedings of the IEEE/CVF Conference on Computer Vision and Pattern Recognition Workshops, pp. 452–453 (2020)
48. Yu, M., Cherukuri, V., Guo, T., Monga, V.: Ensemble Dehazing Networks for Non-Homogeneous Haze. In: Proceedings of the IEEE/CVF Conference on Computer Vision and Pattern Recognition Workshops, pp. 450–451 (2020)

49. Chen, W.T., Ding, J.J. and Kuo, S.Y.: PMS-net: robust haze removal based on patch map for single images. In: Proceedings of the IEEE Conference on Computer Vision and Pattern Recognition, pp. 11681–11689 (2019)
50. Liu, J., Wu, H., Xie, Y., Qu, Y., Ma, L.: Trident dehazing network. In: Proceedings of the IEEE/CVF Conference on Computer Vision and Pattern Recognition Workshops, pp. 430–431 (2020)
51. Ancuti, C., Ancuti, C.O., Timofte, R.: Ntire 2018 Challenge on image dehazing: methods and results. In: Proceedings of the IEEE Conference on Computer Vision and Pattern Recognition Workshops, pp. 891–901 (2018)
52. Ancuti, C.O., Ancuti, C., Timofte, R.: NH-HAZE: an image dehazing benchmark with non-homogeneous hazy and haze-free images. In: Proceedings of the IEEE/CVF Conference on Computer Vision and Pattern Recognition Workshops, pp. 444–445 (2020)
53. Yousaf, R.M., Habib, H.A., Mehmood, Z., Banjar, A., Alharbey, R., Aboulola, O.: Single image dehazing and edge preservation based on the dark channel probability-weighted moments. Math. Probl. Eng. 2019 (2019)

Comparative Study to Analyse the Effect of Speaker's Voice on the Compressive Sensing Based Speech Compression

Narendra Kumar Swami(✉) and Avinash Sharma

Jaipur Institute of Technology Group of Institutions, Jaipur, Rajasthan, India
er.nkswami@gmail.com

Abstract. It is essential to store speech and audio data in various speech processing and detection applications. Sometimes this data is too much bulkier and requires ample storage. So, there is a simple method to compress the signal and then again decompress. But this method is too much old and requires a higher sampling rate as per Nyquist's theorem. Proper reconstruction can also be done with the least values of samples, and the Compressive Sensing Theory proves it. This work will elaborate on how efficiently this technique can reconstruct speech and audio signals using very few measurements. We will also represent the effect of the male and female voices on the compression process. The effectiveness of the reconstruction is measured by using MSE & PSNR values. The time-frequency response for the actual and reconstructed signal is also represented in this paper by using that we can easily understand the compressive sensing phenomena.

Keywords: Compressive sensing · Time-Frequency (T-F) analysis · Speech processing · Peak Signal to Noise Ratio (PSNR) · Mean Square Error (MSE)

1 Introduction

The Nyquist theory is used to acquire and reformulate signals from compressed data. Compressive Sensing is the technique that is used in medical, satellite communication, mathematics, and other technical domains. This technique adopts a novel approach that will not completely agree with traditional sampling theory. This section consists of the introductory part of the Compressive Sensing approach. Sparsity and compressibility are the backbone structure for Compressive Sensing. Then we will move ahead with the most valuable question, how to get back an actual signal with the help of fewer samples. Sparsity is the answer to this question; if the signal is compressible and sparse in nature, we can easily reconstruct the signal.

Compressed Sensing is how to compact the signal before storing the signal [1, 2]. The demand for high operational speed receives a high sampling rate. To minimize the overload of the system, Compressed Sensing is the only solution. We can use this methodology to sample the signal below Nyquist's sampling rate [3, 4]. This technique uses few samples instead of large samples and can efficiently recover the original signal

V. E. Balas et al. (Eds.): ICETCE 2022, CCIS 1591, pp. 494–504, 2022.
https://doi.org/10.1007/978-3-031-07012-9_42

using these few samples. Thus this approach improves the operating speed and the put incapacity of the system [5].

Compressive Sensing is the technology that can convert analog data into digital data, which is already compressed in a very economical manner. This frugal word shows that it required fewer data storage with very few samples for reconstruction, and the noise is also significantly less in this type of processing. For example, recent transform coders like JPEG2000 employ the aspect that various signals have a significant representation in sparse domain with a fixed basis. This means that we can only select some of the adaptively found coefficients using the transform coding from the whole number of samples. The algorithm consists of a method that picks up the coefficients that are useful for the proper reconstruction, or we can say that which have large values and discard all other coefficients with less magnitude. So this type of process, which is used in Compressive Sensing, can overcome a large amount of data storage in traditional coding. In the next part of the paper, we will elaborate on work done in a similar domain by various researchers.

2 Literature Survey

The compressive Sensing Technique is a well-known approach and used by various researchers in different applications. Fereshteh Fakhar Firouzeh et al., in their work, discuss the improvement for the reconstruction approach [6]. In their work, the SNR improvement is shown. Compression factor versus SNR difference plots is mentioned in work. In their work, a comparative study between traditional coding and compressive sensing methods is done by Abdelkader Boukhobza et al. [7]. For the comparison purpose, speech coding techniques like Code-excited linear prediction and Lloyd-Max quantization were compared with the compressive sensing approach. In their work, Karima Mahdjane and Fatiha Merazka considered multi-frequency audio signals and then applied a compressive sensing approach to reconstruct the signal by using different reconstruction algorithms [8]. The authors calculate Computational Complexity and Signal to Noise ratio for original and reconstructed signals to evaluate the performance. Fabrice Katzberg et al., in their work, focus on the measurement of sound when the microphones are non-stationary [9]. Dynamic measurement of sound can be possibly affected by spatial sampling issues. So, a Compressive Sensing Based Framework is proposed by the authors that can create a robust model for the moving microphones. Two components are measured to check the effectiveness of the method one is the energy of the residual signal in dB, and another is mean normalized system misalignment. The main objective of the authors is on Discrete Cosine Transform with context to sparsity. Then they use the Compressive Sensing framework and methods to compress the audio signals [10] further.

As we know that the frequency components in the male and female voice are different, so the computation complexity for both cases are different. In this work, we are going to analyze the impact of this frequency component change on the effectiveness of the Compressive Sensing reconstruction strategy. The following section consists of some reconstruction techniques for compressed signals.

3 Methodology

In compressive Sensing, we require some reconstruction algorithm for the desired data to recover the data from the small number of measurement samples. Figure 1 classifies the popular techniques in compressive sensing recovery algorithms. The sparse recovery depends upon the following factors:

- Signal recovery timing from the measurement vectors.
- Measurement of samples calculates the storage requirement.
- Significantly less complexity of the implementation.
- For the smooth execution need to possible portability of the hardware section.
- Fidelity property of the signal recovery.

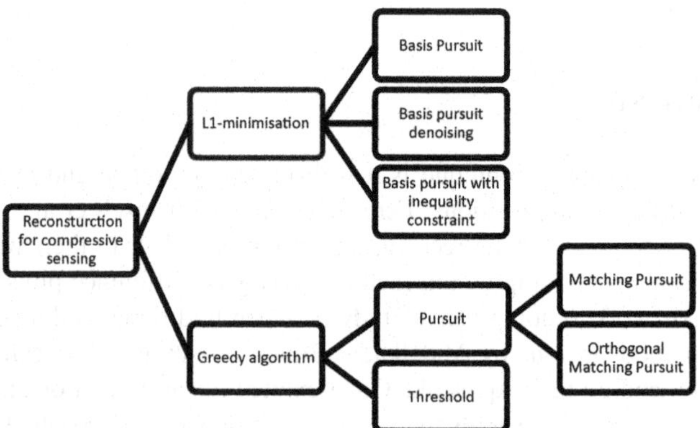

Fig. 1. Various compressive sensing reconstruction algorithms

3.1 Orthogonal Matching Pursuit (OMP)

OMP is used over the matching pursuit. It selects the selected coefficients at ith iteration steps; it never repeats the similar coefficients like the matching pursuit. OMP technique is more complicated schemes for a single iteration cost, and storage is higher.

Steps for Orthogonal Matching Pursuit:

Input: E, y, k
Output: k-sparse approximation $\hat{A}$ of the test signal A.

Initialization:

1. Set $\Upsilon^0 = 0$
2. Set $r^0 = y$

Iteration: During iteration l, do.

1. Find $\lambda^l = \text{argmax}_{j=1,\ldots,N} \left|\langle r^{l-1}, b_h\rangle\right|$ where b_h is the hth column of E
2. Augment $\Upsilon^l = \Upsilon^{l-1} \cup \{\lambda^l\}$
3. Solve $\tilde{A}^l = \arg\max_A \left\|y - E_{\Upsilon^l} A\right\|_2$

Calculate new approximation of data as $\tilde{y}^l = E_{\Upsilon^l}\tilde{A}^l$, and new residual as.

All the critical components of Orthogonal Matching pursuit are mentioned above. We are using the minimum l_1 reconstruction algorithm that entirely depends on Basis Pursuit with some basic features of the Orthogonal Matching Pursuit.

4 Result and Analysis

It is not so simple to differentiate the quality of the reconstruction algorithm for male and female voices. So, to check the quality of reconstruction, we are using Quality Assessment Parameters like Mean Square Error and the Signal to Noise Ratio. These Parameters are able to represent the actual picture of the Compressive Sensing Based Reconstruction process. Compression Ratio, Signal to Noise Ratio, and Mean Square Error is discussed further.

4.1 Compression Ratio

Compression Ratio is a mathematical quantity that indicates the level of compression. It is the ratio of the number of audio samples considered for reconstruction to the total audio samples [11].

$$\text{Compression Ratio}\, CR = k/n \tag{1}$$

Here, k = Samples used for reconstruction. n = Overall samples in the audio signal.

4.2 Signal to Noise Ratio (SNR)

SNR is a well-known parameter to measure the quality of various types of signals as audio and video. Evaluation of SNR is too easy. To calculate the SNR in dB following expression is used [12, 13].

$$SNR = 10 * \log_{10} \frac{\left[\sum_n (Q_i^2)\right]}{\sum_n [(Q_i - S_i)^2]} \tag{2}$$

4.3 Mean Square Error (MSE)

MSE is simply the mathematical parameter that will evaluate the similarity between the actual signal and the recovered signal [14].

$$MSE(Q, S) = \frac{1}{n}\sum_{1}^{n}((Q_i - S_i)^2) \tag{3}$$

where Q = Actual voice signal, S = reconstructed voice signal, n = total count of observations.

We will elaborate on the effect of male and female voices when speaking the same word and further process it using a Compressive Sensing approach.

Table 1. Compression ratio v/s PSNR

CR	PSNR (dB)									
	Boot		Bough		Boy		Bud		Buy	
	Male	Female	Male	Female	Male	Female	Male	Female	Male	Female
0.2	14.27	3.37	5.85	9.10	13.18	15.92	8.37	9.04	8.39	10.05
0.3	17.98	5.19	11.16	16.11	20.07	20.81	13.73	14.46	16.01	15.93
0.4	21.76	6.77	17.83	20.81	24.96	23.62	19.44	18.84	23.62	20.41
0.5	24.40	9.37	24.49	25.97	28.87	26.05	25.31	22.06	30.73	23.96
0.6	27.35	12.79	29.86	29.83	33.20	29.15	28.80	24.93	34.97	27.12
0.7	30.45	15.60	33.68	32.23	36.77	32.98	33.95	28.94	39.33	31.47
0.8	34.24	20.25	39.20	36.58	40.49	37.41	37.63	32.76	42.37	35.88

Table 2. Compression ratio v/s PSNR

CR	PSNR (dB)									
	Cease		Church		Deed		Fife		For	
	Male	Female	Male	Female	Male	Female	Male	Female	Male	Female
0.2	4.75	1.67	7.35	3.95	5.53	8.26	5.53	7.07	12.64	12.92
0.3	7.43	3.68	12.19	6.03	9.50	10.75	10.10	9.97	19.08	14.63
0.4	10.02	5.95	16.12	7.74	13.57	13.18	15.06	12.27	23.67	15.76
0.5	12.49	8.79	19.29	9.73	16.37	15.21	19.49	14.24	27.05	17.08
0.6	15.05	11.85	22.59	12.30	19.21	18.01	23.37	16.04	29.99	18.79
0.7	18.53	15.12	26.13	15.30	22.50	20.65	25.80	18.28	32.58	20.67
0.8	22.36	18.68	29.52	19.30	26.03	24.26	28.56	21.22	36.16	23.02

Table 3. Compression ratio v/s PSNR

CR	PSNR (dB)									
	Gag		Hand		Judge		Kick		Loyal	
	Male	Female	Male	Female	Male	Female	Male	Female	Male	Female
0.2	8.67	8.97	5.00	5.36	6.32	7.80	4.57	3.78	10.69	13.85
0.3	13.16	12.25	8.76	9.02	10.24	9.85	7.68	6.02	15.67	20.19
0.4	17.34	16.66	14.58	12.74	15.78	11.69	13.50	8.21	21.23	23.95
0.5	21.48	19.61	23.50	15.79	19.71	14.14	18.63	10.21	25.37	26.84
0.6	24.60	22.51	27.24	19.79	23.03	16.75	22.77	12.24	29.85	30.02
0.7	28.61	26.38	31.52	24.08	27.33	20.02	26.94	15.01	34.15	33.43
0.8	33.34	29.50	36.08	29.95	31.44	23.69	31.08	18.68	38.88	37.17

Table 4. Compression ratio v/s MSE

CR	Mean Square Error							
	Boot		Bough		Boy		Bud	
	Male	Female	Male	Female	Male	Female	Male	Female
0.2	4.17E–04	0.0013	0.001522	0.001642	7.13E–04	4.14E–04	0.001382	7.51E–04
0.3	1.78E–04	9.2E–04	4.49E–04	3.26E–04	1.82E–04	1.34E–04	4.03E–04	2.16E–04
0.4	7.44E–05	6.39E–04	9.66E–05	1.11E–04	5.91E–05	7.03E–05	1.08E–04	7.86E–05
0.5	4.04E–05	3.52E–04	2.08E–05	3.38E–05	2.41E–05	4.01E–05	2.80E–05	3.75E–05
0.6	2.05E–05	1.60E–04	6.05E–06	1.39E–05	8.87E–06	1.97E–05	1.25E–05	1.94E–05
0.7	1.00E–05	8.37E–05	2.51E–06	7.98E–06	3.90E–06	8.14E–06	3.83E–06	7.68E–06
0.8	4.20E–06	2.87E–05	7.03E–07	2.93E–06	1.66E–06	2.94E–06	1.64E–06	3.19E–06

Table 5. Compression ratio v/s MSE

CR	Mean Square Error							
	Buy		Cease		Church		Deed	
	Male	Female	Male	Female	Male	Female	Male	Female
0.2	9.72E–04	0.0011	0.0017	0.0082	0.002595	0.00559	0.001312	8.49E–04
0.3	1.68E–04	2.87E–04	9.20E–04	0.0051	8.52E–04	0.00347	5.26E–04	4.78E–04
0.4	2.91E–05	1.02E–04	5.07E–04	0.0030	3.45E–04	0.00233	2.06E–04	2.74E–04
0.5	5.67E–06	4.51E–05	2.87E–04	0.0015	1.66E–04	0.001480	1.08E–04	1.72E–04
0.6	2.14E–06	2.18E–05	1.59E–04	7.92E–04	7.77E–05	8.18E–04	5.62E–05	9.00E–05
0.7	7.83E–07	8.00E–06	7.14E–05	3.72E–04	3.44E–05	4.10E–04	2.64E–05	4.89E–05
0.8	3.89E–07	2.90E–06	2.96E–05	1.64E–04	1.58E–05	1.63E–04	1.17E–05	2.13E–05

Table 6. Compression ratio v/s MSE

CR	Mean Square Error							
	Fife		For		Gag		Hand	
	Male	Female	Male	Female	Male	Female	Male	Female
0.2	5.5360	0.0019	0.00118	6.07E–04	0.00184	7.77E–04	0.00653	0.00442
0.3	0.0010	0.00102	2.70E–04	4.10E–04	6.57E–04	3.65E–04	0.00275	0.00190
0.4	3.34E–04	6.04E–04	9.37E–05	3.16E–04	2.46E–04	1.32E–04	7.21E–04	8.09E–04
0.5	1.20E–04	3.83E–04	4.30E–05	2.33E–04	9.67E–05	6.70E–05	9.24E–05	4.01E–04
0.6	4.94E–05	2.54E–04	2.19E–05	1.57E–04	4.71E–05	3.44E–05	3.91E–05	1.60E–04
0.7	2.81E–05	1.51E–04	1.20E–05	1.02E–04	1.87E–05	1.41E–05	1.46E–05	5.94E–05
0.8	1.49E–05‘	7.68E–05	5.28E–06	5.93E–05	6.30E–06	6.87E–06	4.5E–06	1.54E–05

Table 7. Compression ratio v/s MSE

CR	Mean Square Error							
	Judge		Kick		Loyal		Measure	
	Male	Female	Male	Female	Male	Female	Male	Female
0.2	0.002	9.32E–04	0.00201	5.78E–04	0.00196	0.00136	0.003134	0.001469
0.3	0.0010	5.82E–04	9.85E–04	3.45E–04	6.24E–04	3.18E–04	0.001530	7.21E–04
0.4	2.87E–04	3.80E–04	2.58E–04	2.08E–04	1.74E–04	1.34E–04	6.17E–04	4.02E–04
0.5	1.16E–04	2.17E–04	7.90E–05	1.32E–04	6.68E–05	6.87E–05	1.78E–04	2.37E–04
0.6	5.41E–05	1.19E–04	3.05E–05	8.25E–05	2.38E–05	3.30E–05	6.47E–05	1.15E–04
0.7	2.01E–05	5.60E–05	1.17E–05	4.35E–05	8.85E–06	1.51E–05	2.12E–05	4.77E–05
0.8	7.80E–06	2.40E–05	4.50E–06	1.87E–05	2.98E–06	6.36E–06	6.37E–06	1.91E–05

Table 1, 2 and 3 is wholly based on the computation of compression ratio v/s Signal to Noise ratio for 16 different words spoken by males and females. So, this computation will elaborate regarding the impact of voice on the value of SNR that is directly proportional to the quality of reconstruction. Further, Mean square error analysis is also mentioned, which can provide us with the exact value of change in the quality of the reconstructed and original signals. Table 4, 5, 6 and 7 show the precise values of Compression Ratios v/s Mean Square Error. The mean square error value for the female voice content is higher than the male voice content.

Another critical analysis that is conducted here is the Time-Frequency analysis that is given below. Time-Frequency analysis is very much essential if we want to identify the variation in the signal with respect to time and frequency. To see the effect of compressive sensing technique on the actual male and female voice signal it is very much important to understand it's time frequency analysis. In this work we have considered only one word to do the experimental time-frequency analysis.

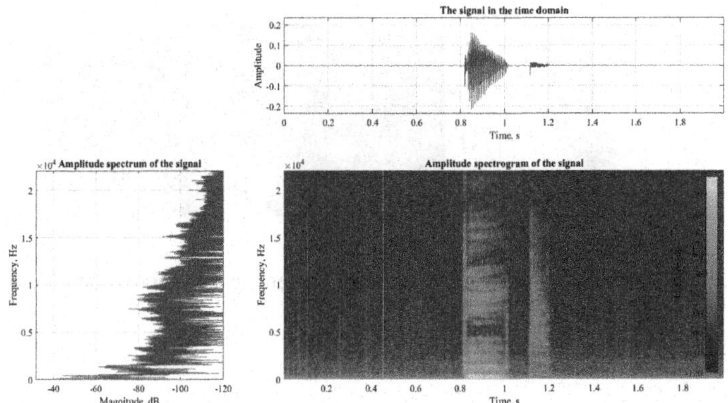

Fig. 2. T-F response of actual signal (boot) for male

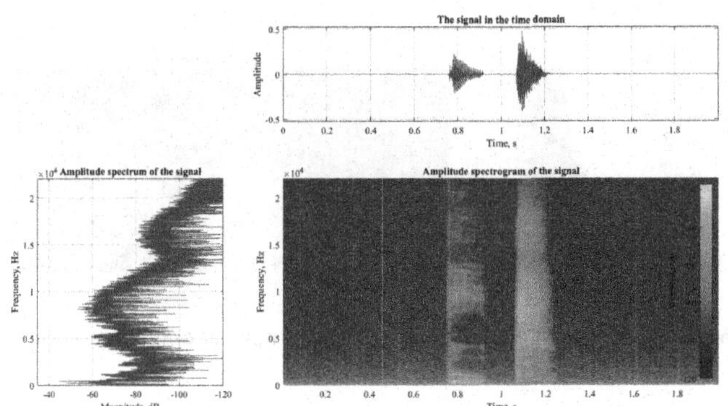

Fig. 3. T-F response of actual signal (boot) for female

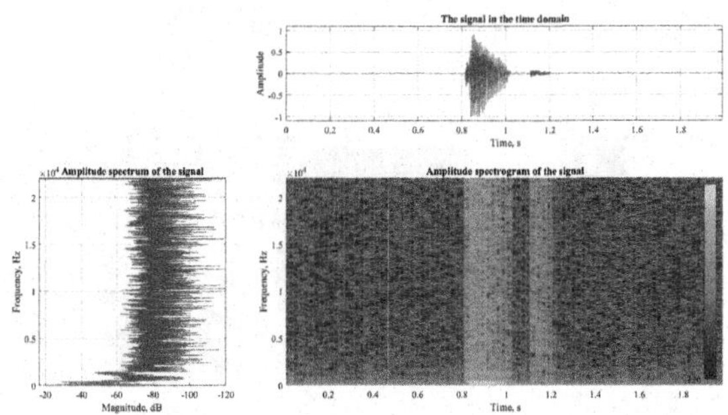

Fig. 4. T-F response of actual signal (boot) for male at compression ratio 0.20

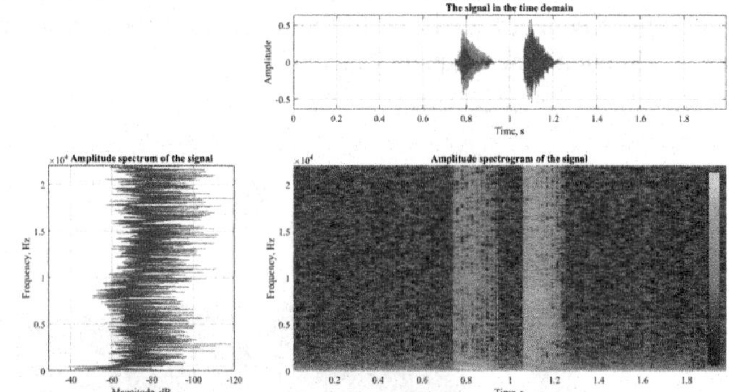

Fig. 5. T-F response of actual signal (boot) for female at compression ratio 0.20

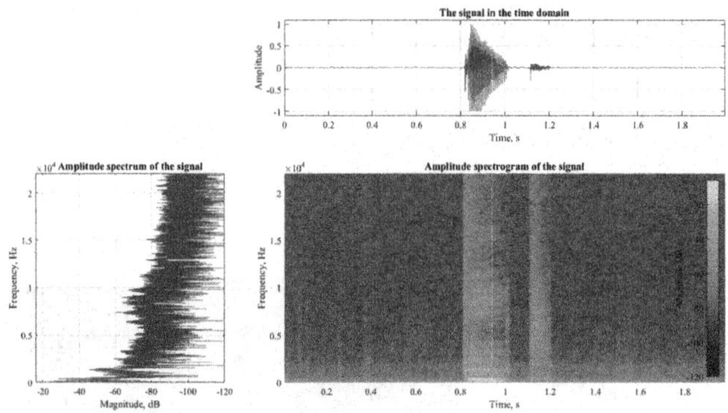

Fig. 6. T-F response of actual signal (boot) for male at compression ratio 0.80

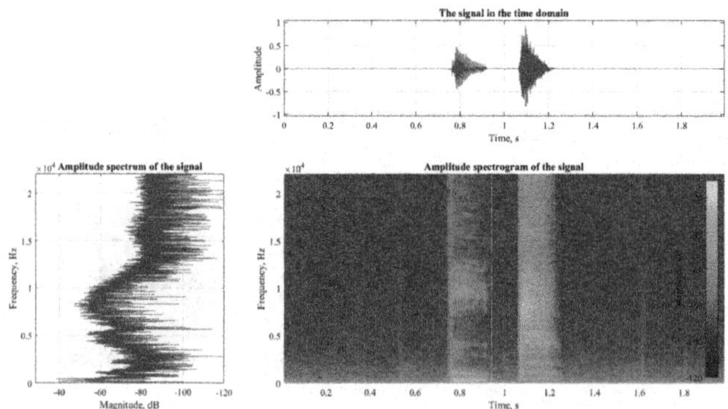

Fig. 7. T-F response of actual signal (boot) for female at compression ratio 0.80

Figure 2, 3, 4, 5, 6 and 7 is exactly representing the time-frequency response for the Actual signal, both male and female. We also describe the time-frequency response on the most minor and highest value of the Compression Ratio. The following section consists of the conclusion.

5 Conclusion

The experimental analysis that is conducted above has some outcomes or interesting concluding remarks. Our primary motive is to compress the signal and to reconstruct it effectively. For that purpose, Compressive Sensing is used. Theoretical and mathematical aspects are shown in the previous section. After that, the results which we got through the analysis are discussed. As shown in the results, 16 words are considered for which the male and female voice signal is compressed and then recovered using the Compressive Sensing approach. The Concluding remarks are as follows.

- SNR value for the male voice is higher than the female voice for the same word. This means either the voice component of the female signal is low; else, it is not much easier to recover the female voice components for a particular word.
- MSE value is most negligible for the male speaker and high for the female speaker.
- The time-frequency analysis represents the signal in the time domain, Magnitude v/s frequency curve, & Time v/s frequency curve.
- If we further examine the time-frequency curve, we can understand that the male speaker signal in the time domain is more structured than the female speaker signal for the same word.
- The compressive Sensing approach represents highly efficient results at low compression ratios too.

References

1. Candès, E.J., Romberg, J., Tao, T.: Robust uncertainty principles: exact signal reconstruction from highly incomplete frequency information. IEEE Trans. Inform. Theory **52**(2), 489–509 (2006)
2. Donoho, D.: Compressed sensing. IEEE Trans. Inform. Theory **52**(4), 1289–1306 (2006)
3. Baraniuk, R.: Compressive sensing. IEEE Signal Process. Mag. **24**(4):118–120, 124 (2007)
4. Shukla, U.P., Patel, N.B., Joshi, A.M.: A survey on recent advances in speech compressive sensing. In: 2013 IEEE International Multi-Conference on Automation, Computing, Communication, Control and Compressed Sensing (iMac4s), pp. 276–280 (2013)
5. DeVore, R.A.: Nonlinear approximation. Acta Numer. **7**, 51–150 (1998)
6. Firouzeh, F.F., Abdelazez, M., Salsabili, S., Rajan, S.: Improved recovery of compressive sensed speech. In: 2020 IEEE International Instrumentation and Measurement Technology Conference (I2MTC), pp. 1–6. IEEE (2020)
7. Boukhobza, A., Hettiri, M., Taleb-Ahmed, A., Bounoua, A.: A comparative study between compressive sensing and conventional speech conding methods. In: 2020 1st International Conference on Communications, Control Systems and Signal Processing (CCSSP), pp. 215–218. IEEE (2020)

8. Mahdjane, K., Merazka, F.: Performance evaluation of compressive sensing for multi-frequency audio signals with various reconstructing algorithms. In: 2019 6th International Conference on Image and Signal Processing and their Applications (ISPA), pp. 1–4. IEEE (2019)
9. Katzberg, F., Mazur, R., Maass, M., Koch, P., Mertins, A.: Compressive sampling of sound fields using moving microphones. In: 2018 IEEE International Conference on Acoustics, Speech and Signal Processing (ICASSP), pp. 181–185. IEEE (2018)
10. Stanković, L., Brajović, M.: Analysis of the reconstruction of sparse signals in the DCT domain applied to audio signals. IEEE/ACM Trans. Audio Speech Lang. Process. **26**(7), 1220–1235 (2018)
11. Joshi, A.M., Upadhyaya, V.: Analysis of compressive sensing for non stationary music signal. In: 2016 International Conference on Advances in Computing, Communications and Informatics (ICACCI), pp. 1172–1176. IEEE (2016)
12. Upadhyaya, V., Salim, M.: Compressive sensing: methods, techniques, and applications. In: IOP Conference Series: Materials Science and Engineering, vol. 1099, no. 1, p. 012012. IOP Publishing (2021)
13. Upadhyaya, V., Salim, M.: Basis & sensing matrix as key effecting parameters for compressive sensing. In: 2018 International Conference on Advanced Computation and Telecommunication (ICACAT), pp. 1–6. IEEE (2018)
14. Upadhyaya, V., Sharma, G., Kumar, A., Vyas, S., Salim, M.: Quality parameter index estimation for compressive sensing based sparse audio signal reconstruction. In: IOP Conference Series: Materials Science and Engineering, vol. 1119, no. 1, p. 012005. IOP Publishing (2021)

Employing an Improved Loss Sensitivity Factor Approach for Optimal DG Allocation at Different Penetration Level Using ETAP

Gunjan Sharma and Sarfaraz Nawaz(✉)

Department of Electrical Engineering, Swami Keshvanand Institute of Engineering, Management and Gramothan, Jaipur, India
sarfaraz@skit.ac.in

Abstract. The potential of distributed generation to deliver cost-effective, ecologically sustainable, rising, and more reliable solutions has resulted in an increase in its use for electricity generation globally. DGs and capacitors are chosen over centralized power generation to bring electricity demand closer to load centres. The proper positioning and capability of DG is vital in solving common power system concerns such as power network loss reduction, stability, and voltage profile enhancement. An Analytical method is used in this paper to determine optimal size and site for placement of DG units. Type-I and type-II DGs are integrated in the system to achieve the objective. Analytical method is tested on IEEE 69-bus system at varying penetration levels (30–70%). The results are compared to those obtained by other latest approaches and found superior.

Keywords: Improved loss sensitivity factor · Distribution power loss · Distributed generation · Capacitor

1 Introduction

The load on power system utilities has increased as global demand for electricity has increased. Since traditional fuels used for electricity generation have a negative impact on the environment, renewable energy sources are preferred for generation to meet this demand. The ability of DG to deliver both reactive and active power aids in the enhancement of power quality. Along with providing power, DG improves the voltage profile of the system, making it more reliable. The size and location of distributed generation have a huge impact on its performance. To select the ideal location for DG installation, several methodologies have been utilized. Using a hybrid BPSO-SLFA algorithm, Abdurrahman Shuaibu Hassan et al. [1] found the best position for DG. Waseem Haider et al. [2] optimised the placement of DG using PSO to reduce system power losses and improve voltage profile. Oscar Andrew Zongo et colleagues [3] used a hybrid PSO and NRPF technique to install DG and improve voltage stability while reducing actual and reactive power losses in the system. For optimal DG sizing and seating, Mohamed A. Tolba et al. [4] employed hybrid PSOGSA. By incorporating DG, MCV Suresh et al.

V. E. Balas et al. (Eds.): ICETCE 2022, CCIS 1591, pp. 505–515, 2022.
https://doi.org/10.1007/978-3-031-07012-9_43

[5] employed a hybrid GOA-CS approach to reduce power losses in the system. Tuba Gözel et al. [6] proposed an analytical method for DG placement and sizing based on a loss sensitivity factor. Other goals for DG installation are also considered in addition to power loss mitigation and voltage profile enhancement. Srinivasa Rao Gampa et al. [7], for example, properly placed and sized DG by taking average hourly load fluctuation into account. The EHO approach was used by C. Hari Prasad et al. [8] to do a cost-benefit analysis for proper DG allocation in a distribution system. The level of penetration is another crucial component in DG integration. Several studies have discovered that integrating DG beyond a certain point is harmful to the system. Minh Quan Duong et al. [9] integrated a wind and photovoltaic system into the grid and discovered that when the penetration of these generators is less than 30%, the system functions satisfactorily. K. Balamurugan et al. [10] investigated the effects of DG at various penetration levels and in different parts of the system. The effects of high DG penetration levels on distribution line thermal constraints and voltage rise were investigated by Desmond Okwabi Ampofo et al. [11].. The types of DGs are as follows.

- Type I: which produces active power
- Type II: which produces reactive power
- Type III: which produces active and reactive power
- Type IV: which generates real power and absorbs reactive power [12]

Each type has a different impact on the electricity system. A.M. Abd-rabou et al. studied the influence of various DG types on the electricity system [13].

In this paper, improved loss sensitivity factor (ILSF) is used to determine the ideal position for DG placement in the IEEE 69-bus test system. On the best location, the type I and type II DGs are integrated at various levels. The simulation is done using ETAP software. The size of DG in the test system is small enough to achieve minimum losses. The results are then compared to those of other studies.

2 Problem Formulation

The purpose of this study is to reduce the reactive power losses in the test system. At node i the total real and reactive power can be calculated as follows (Fig. 1):

$$P = P_{\mathrm{i}} + R_{\mathrm{ik}} \frac{P_{\mathrm{i}}^2 + jQ_{\mathrm{i}}^2}{V_{\mathrm{i}}^2} \tag{1}$$

$$Q = Q_{\mathrm{i}} + X_{\mathrm{ik}} \frac{P_{\mathrm{i}}^2 + jQ_{\mathrm{i}}^2}{V_{\mathrm{i}}^2} \tag{2}$$

The system's power losses are given by,

$$Pl_{(\mathrm{ik})} = R_{\mathrm{ik}} \frac{P_{\mathrm{i}}^2 + jQ_{\mathrm{i}}^2}{V_{\mathrm{i}}^2} \tag{3}$$

$$Ql_{\mathrm{ik}} = X_{\mathrm{ik}} \frac{P_{\mathrm{i}}^2 + jQ_{\mathrm{i}}^2}{V_{\mathrm{i}}^2} \tag{4}$$

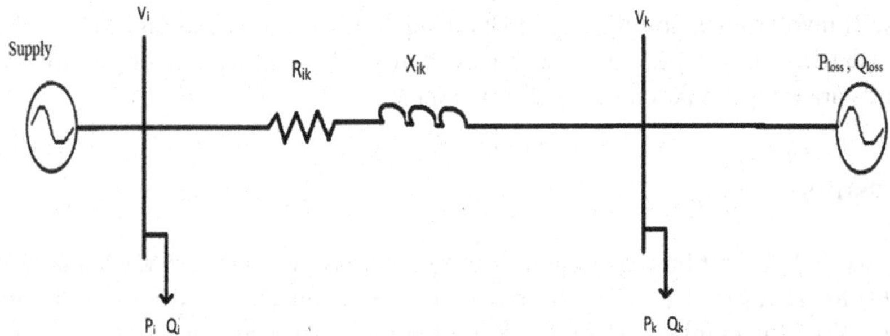

Fig. 1. Radial Distribution system

A distribution system with buses i and k is depicted in the diagram above. The voltages on these buses are V_i and Vk, respectively, with a line impedance of R_{ik} + j X_{ik}.

3 Analytical Method

The improved loss sensitivity factor technique, which aids in selecting the bus with the highest reduction in loss when DG is applied, is used to discover the best location for DG. The shift in losses as a result of the compensation given by placing the DGs is referred to as loss sensitivity [14] The ILSF is given by,

$$\frac{\partial Ql}{\partial Q_{net}} = \frac{2 \times Q_k \times X_{ik}}{V_k^2} \tag{5}$$

Equation (5) is used to calculate the ILSF. The optimal locations for DG deployment are chosen based on loss sensitivity factors and voltage magnitude. The ILSF aids in determining the priority order, and the magnitude of the voltage aids in evaluating the requirement for correction. The step for obtaining results:

- Obtain load and line data, then run the load flow
- Calculate real and reactive power losses
- Obtain ILSF using Eq. (5)
- Store ILSF values in descending order
- Choose the top four buses for DG allocation
- Check the losses after placing DG at all the locations obtained
- Starting from 30% Penetration level, increase the DG size (by + 10 kW and $_+$10 kVar) up to 70%
- Choose the values with maximum loss reduction as the ideal size and site for DG placement

Three distinct penetration levels are used to penetrate the DG: 30%, 50%, and 70%. The different cases considered are.

- In case I, type I is pierced at various penetration level at all of the locations.

- Case II involves penetrating type II DG using at various penetration levels.
- In case III, both type I & II DGs, using every feasible combination are employed. The results are then compared with other methods.

4 Results

In this work, IEEE 69-bus test system is used. The base voltage and MVA are 12.66kV and 100 MVA respectively. The 69-bus system'Ss connected load is 3802.1 kW and 2694.6 kVar. Bus numbers 61,11,17,18 are the ideal positions for DG placement. For all cases, results are derived using single DG, two DG and three DG at 30%, 50% and 70% penetration level. After exploring all possible combinations, the best results are compared to those obtained using alternative approaches (Fig. 2).

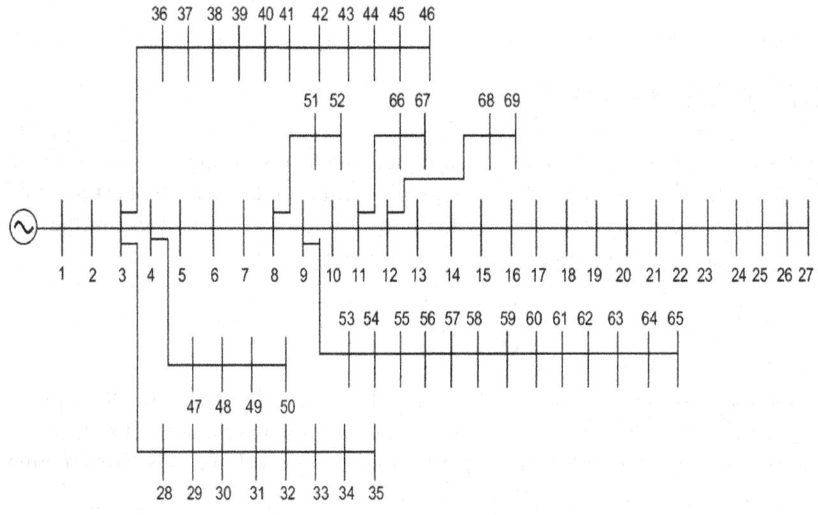

Fig. 2. IEEE 69-bus as single line diagram

Case I

For single DG integration, it is observed that when type I DG was integrated, the minimum losses obtained were 42 KVar and 85 KW at 50% penetration level. The minimum losses obtained with two DG are 73 KW and 37 KVar at 70% penetration level. The minimum losses obtained for three DGs are 71 KW and 36 KVar at 70% penetration.

Simulation results using proposed method for different cases at different penetration levels are shown in table below (Table 1).

Table 1. Results obtained using type I DG

Penetration level	DG size (kW)	Bus no. - DG size (kW)	Losses		% V_{min} (Bus no.)
			Active power (kW)	Reactive power (kVar)	
1 DG					
30%	1140	61	111	53	0.9483 (65)
50%	1830	61	85	42	0.9668 (27)
70%	2640	61	91	43	0.9710 (27)
2 DG					
30%	1140	17–316 61–828	119	57	0.9397 (65)
50%	1900	17–527 61–1376	84	41	0.9588 (65)
70%	2310	17–532 61–1775	73	37	0.9714 (65)
3 DG					
30%	1140	11–243 17–316 61–587	134	63	0.9329 (65)
50%	1910	11–264 17–403 61–1242	87	43	0.9552 (65)
70%	2590	11–522 17–360 61–1710	71	36	0.9712 (65)

The table below shows the comparison of results obtained using proposed method with other latest techniques, with type I DG integration (Table 2).

Case II

For single DG integration, the minimum losses obtained when type II DG is integrated in the system are 148 kW and 69 KVar at 50%. The minimum losses obtained with two DG are 148 KW and 69 kVAr at approximate 50% penetration level. The minimum losses obtained with three DGs are 146 kW and 68 KVar at 70% penetration level. The table below compares the outcomes obtained using different methodologies (Table 3).

Table 2. Simulation results comparison for type I DG combined with the 69-bus test system.

Reference	Technique	Size of DG (kW)/Location	Size of DG (kW)	Results obtained		
				Overall active power loss (kW)	Overall reactive power loss (kVar)	V_{min}
1 DG						
	Proposed	1830 (61)	1827	85	42	0.9712
[15]	KHA	1864.6446(61)	1864.6446	85	42	0.9670
[15]	SKHA	1864.6376(61)	1864.6376	85	42	0.9670
[14]	Dragonfly algorithm	1872.7(61)	1872.7	85	41	0.9671
[16]	HHO	1872.7 (61)	1872.7	85	41	0.9671
[16]	TLBO	1872.7 (61)	1872.7	85	41	0.9671
2 DG						
	Proposed	530(17), 1770(61)	2307	73	37	0.9714
[17]	HGWO	531(17), 1781(61)	2312	73	37	0.9715
[16]	HHO	532.1422(17), 1780.7918(61)	2312.934	73	37	0.9715
[16]	TLBO	531.8405(17), 1781.4909(61)	2313.3344	73	37	0.9715
[15]	KHA	972.1609 (51), 1726.9813(61)	2699.1422	80	39	0.9711
3 DG						
	Proposed	520(11), 360(17), 1710(61)	2592	71	36	0.9712
[15]	KHA	549.1027(15), 1013.969(49), 1768.4752(61)	3331.547	72	33	0.9713
[15]	SKHA	370.880(11), 527.1736(17), 1719.0677(61)	2617.125	71	36	0.9715
[16]	HHO	542.1997(11), 372.6548(17), 1715.8598(61)	2630.7503	71	36	0.9716

(*continued*)

Table 2. (*continued*)

Reference	Technique	Size of DG (kW)/Location	Size of DG (kW)	Results obtained		
				Overall active power loss (kW)	Overall reactive power loss (kVar)	V_{min}
[16]	TLBO	528.3797(11), 380.1178(17), 1725.5912(61)	2634.0887	71	36	0.9715

Table 3. Results obtained using type II DG

Penetration level (approx.)	DG size (kVar)	Bus no. - DG size (kVar)	Losses		% V_{min} (Bus no.)
			Active power (kW)	Reactive power (kVar)	
1 DG					
30%	810	61	169	78	0.9208 (65)
50%	1310	61	148	69	0.9277 (65)
70%	1850	61	156	72	0.9355 (65)
2 DG					
30%	810	17–243 61–564	174	80	0.9182 (65)
50%	1560	17–339 61–1221	148	69	0.9277 (65)
70%	1880	17–570 61–1314	148	69	0.9297 (65)
3 DG					
30%	810	11–243 17–150 61–414	181	83	0.9165 (65)
50%	1350	11–150 17–243 61–954	154	72	0.9242 (65)
70%	1790	11–306 17–243 61–1242	146	68	0.9287 (65)

The table below shows the comparison of results obtained using proposed method with other latest techniques, with type II DG integration (Table 4).

Table 4. Simulation results comparison for type II DG combined with the 69-bus test system.

Reference	Technique	Size of DG (kVAr)/Location	Total DG size (kVAr)	Results obtained		
				Overall active power loss (kW)	Overall reactive power loss (kVar)	V_{min}
1 DG						
	Proposed	1310(61)	1314	153	71	0.9279
[16]	HHO	1328.2988(61)	1328.2988	153	71	0.9281
[16]	TLBO	1329.951(61)	1329.951	153	71	0.9281
[17]	HGWO	1330(61)	1330	153	71	0.9281
2 DG						
	Proposed	340(17), 1221(61)	1560	148	69	0.9277
[16]	HHO	353.3152(17), 1280.4326(61)	1633.7478	148	69	0.9285
[16]	TLBO	360.311(17), 1274.4034(61)	1634.7149	148	69	0.9285
[17]	HGWO	364(17), 1227(61)	1641	148	69	0.9285
3 DG						
	Proposed	310(11), 240(17), 1240(61)	1791	146	68	0.9287
[16]	HHO	318.6356(11), 258.1434(17), 1242.1248(61)	1845.9038	146	68	0.9289
[16]	TLBO	413.3228(11), 230.5589(17), 1232.5096(61)	1876.3913	146	68	0.9288
[18]	FPA	450(11), 150(22), 1350(61)	1950	146	68	0.9303

Case III

The integration of DG along with capacitor provide a better result as compared to DG and capacitors alone. The table below provides a comparative analysis of results obtained using ILSF. The study shows that minimum losses obtained are 5.5 kW and 7.5 kVar when three DGs are placed with three capacitors in the test system (Table 5).

Table 5. Simulation results comparison for type I & II DG combined with the 69 bus test system.

References	DG size		Location	Active power loss (kw)	Reactive power loss (kvar)	Vmin
	Type I (kw)	Type II (kvar)				
Proposed [19]	1830	1310	61	25	15	0.9713
	1200(60)	1222.5(61)		33.45	–	0.9712
Proposed	530 1775	340 1221	17 61	8	9	0.9907
Proposed	400 280 1580	360 250 1150	11 17 61	5.5	7.5	0.99
[20]	479(12) 365(19) 1680(61)	300(12) 269(19) 1202(61)	12 19 61	6	8	0.9899
[21]	481 (11) 359 (19) 1678 (61)	289 (11) 278 (18) 1182 (61)		6	8	0.9943

5 Conclusion

In this study, an analytical technique is used to find the best position and size for DG. Type I and type II DG are integrated at these location at different penetration levels. It is observed that minimum power losses are obtained when both type I and II DGs are integrated together in the system at 70% penetration level. The minimum losses for each case are obtained at 70% penetration level with three DGs. Also, it is observed that the losses acquired using the suggested technique are lower than those obtained using other methods. DG integration can reduce system losses, although the penetration level differs depending on the type of DG and the number of DG employed in the system. Increasing penetration levels above 50% leads in a modest change in percentage loss reduction.

References

1. Hassan, A.S., Sun, Y., Wang, Z.: Multi-objective for optimal placement and sizing DG units in reducing loss of power and enhancing voltage profile using BPSO-SLFA. Energy Rep. **6**, 1581–1589 (2020). https://doi.org/10.1016/j.egyr.2020.06.013
2. Haider, W., Ul Hassan, S.J., Mehdi, A., Hussain, A., Adjayeng, G.O.M., Kim, C.H.: Voltage profile enhancement and loss minimization using optimal placement and sizing of distributed generation in reconfigured network. Machines **9**(1), 1–16 (2021). https://doi.org/10.3390/machines9010020

3. Zongo, O.A., Oonsivilai, A.: Optimal placement of distributed generator for power loss minimization and voltage stability improvement. Energy Procedia **138**, 134–139 (2017). https://doi.org/10.1016/j.egypro.2017.10.080
4. Tolba, M.A., Tulsky, V.N., Diab, A.A.Z.: Optimal sitting and sizing of renewable distributed generations in distribution networks using a hybrid PSOGSA optimization algorithm. In: 2017 IEEE International Conference on Environment and Electrical Engineering and 2017 IEEE Industrial and Commercial Power Systems Europe (EEEIC/I&CPS Europe) (2017). https://doi.org/10.1109/EEEIC.2017.7977441
5. Suresh, M.C.V., Edward, J.B.: A hybrid algorithm based optimal placement of DG units for loss reduction in the distribution system. Appl. Soft Comput. J. **91**, 106191 (2020). https://doi.org/10.1016/j.asoc.2020.106191
6. Jiyani, A., et al.: NAM: a nearest acquaintance modeling approach for VM allocation using R-Tree. Int. J. Comput. Appl. **43**(3), 218–225 (2021)
7. Gampa, S.R., Das, D.: Optimum placement and sizing of DGs considering average hourly variations of load. Int. J. Electr. Power Energy Syst. **66**, 25–40 (2015). https://doi.org/10.1016/j.ijepes.2014.10.047
8. Prasad, C.H., Subbaramaiah, K., Sujatha, P.: Cost–benefit analysis for optimal DG placement in distribution systems by using elephant herding optimization algorithm. Renew. Wind Water Solar **6**(1), 1–12 (2019). https://doi.org/10.1186/s40807-019-0056-9
9. Duong, M.Q., Tran, N.T.N., Sava, G.N., Scripcariu, M.: The impacts of distributed generation penetration into the power system. In: 2017 11th International Conference on Electromechanical Power System, SIELMEN 2017 - Proceedings, vol. 2017-Janua, pp. 295–301 (2017). https://doi.org/10.1109/SIELMEN.2017.8123336
10. Malpani, P., Bassi, P., et al.: A novel framework for extracting geospatial information using SPARQL query and multiple header extraction sources. In: Afzalpulkar, N., Srivastava, V., Singh, G., Bhatnagar, D. (eds) Proceedings of the International Conference on Recent Cognizance in Wireless Communication & Image Processing. Springer, New Delhi (2016). https://doi.org/10.1007/978-81-322-2638-3_56
11. Ampofo, D.O., Otchere, I.K., Frimpong, E.A.: An investigative study on penetration limits of distributed generation on distribution networks. In: Proceedings - 2017 IEEE PES-IAS PowerAfrica Conference Harnessing Energy, Information Communication Technology Affordable Electrification, Africa, PowerAfrica 2017, pp. 573–576 (2017). https://doi.org/10.1109/PowerAfrica.2017.7991289
12. Dinakara, P., Veera, V.C., Gowri, M.: Optimal renewable resources placement in distribution networks by combined power loss index and whale optimization algorithms. J. Electr. Syst. Inf. Technol. **5**(2), 175–191 (2018). https://doi.org/10.1016/j.jesit.2017.05.006
13. Abd-rabou, A.M., Soliman, A.M., Mokhtar, A.S.: Impact of DG different types on the grid performance. J. Electr. Syst. Inf. Technol. **2**(2), 149–160 (2015). https://doi.org/10.1016/j.jesit.2015.04.001
14. Suresh, M.C.V., Belwin, E.J.: Optimal DG placement for benefit maximization in distribution networks by using Dragonfly algorithm. Renew. Wind, Water, and Solar **5**(1), 1–8 (2018). https://doi.org/10.1186/s40807-018-0050-7
15. ChithraDevi, S.A., Lakshminarasimman, L., Balamurugan, R.: Stud Krill herd Algorithm for multiple DG placement and sizing in a radial distribution system. Eng. Sci. Technol. an Int. J. **20**(2), 748–759 (2017). https://doi.org/10.1016/j.jestch.2016.11.009
16. Ponnam, V., Babu, K.: Optimal integration of different types of DGs in radial distribution system by using Harris hawk optimization algorithm. Cogent Eng. **7**(1), 1823156 (2020). https://doi.org/10.1080/23311916.2020.1823156
17. Sanjay, R., Jayabarathi, T., Raghunathan, T., Ramesh, V., Mithulananthan, N.: Optimal allocation of distributed generation using hybrid grey Wolf optimizer. IEEE Access **5**, 14807–14818 (2017). https://doi.org/10.1109/ACCESS.2017.2726586

18. Tamilselvan, V., Jayabarathi, T., Raghunathan, T., Yang, X.S.: Optimal capacitor placement in radial distribution systems using flower pollination algorithm. Alexandria Eng. J. **57**(4), 2775–2786 (2018). https://doi.org/10.1016/j.aej.2018.01.004
19. Mahdad, B., Srairi, K.: Adaptive differential search algorithm for optimal location of distributed generation in the presence of SVC for power loss reduction in distribution system. Eng. Sci. Technol. an Int. J. **19**(3), 1266–1282 (2016). https://doi.org/10.1016/j.jestch.2016.03.002
20. Yuvaraj, T., et al.: Optimal integration of capacitor and distributed generation in distribution system considering load variation using bat optimization algorithm. Energies **14**(12), 3548 (2021). https://doi.org/10.3390/en14123548
21. Thangaraj, Y., Kuppan, R.: Multi-objective simultaneous placement of DG and DSTATCOM using novel lightning search algorithm. J. Appl. Res. Technol. **15**(5), 477–491 (2017). https://doi.org/10.1016/j.jart.2017.05.008

Empirical Review on Just in Time Defect Prediction

Kavya Goel(✉), Sonam Gupta, and Lipika Goel

Ajay Kumar Garg Engineering College, Ghaziabad, India
{Kavya2010004m,guptasonam}@akgec.ac.in

Abstract. Just In Time Defect Prediction abbreviated as JITDP refers to a software that helps to detect whether any change made to that software leads to defect or not and that too immediately in no time. That is why the name of the model is just in time defect prediction. It helps to develop a defect free software. Various researchers have proposed their models to showcase their accuracy, efficiency and also the scenario in which they perform the best. A systematic review is performed on the proposed approaches which have the following conclusions:1. Random forest algorithm proves to be the most efficient algorithm and outperforms all the other proposed models. 2. Random Forest is observed to be a generalized algorithm which can be used from various perspectives. 3. Apart from random forest, deep learning fusion technique, AB+LMT and DeepJIT are also efficient algorithms with an average accuracy of 0.80. 4. CBS+ is the most efficient algorithm in case of long term approach followed by random forest.

Keywords: Just in time defect prediction · Defect-prone · Recall

1 Introduction

Software has become an integral and indispensable part in each and every field be it healthcare, banking, research, agriculture etc. For efficient working we need to have efficient software which are not only fast but also defect-prone and reliable. The demand for software is increasing with every passing day and more complex software having size in many KLOC are being developed. Not only it is time consuming and involves high testing costs, it is also highly impractical to test each and every possible test case. Moreover many developers nowadays take help of online code repositories like GitHub to develop the software, which implies that any code can be used at any stage of its evolution or lifecycle because of which there is no guarantee of the code to be defect-free. But because of this, the quality of software cannot be tampered at any cost. So here comes the need of Just in time defect prediction system which helps both developers and testers to anticipate which changes in the software might lead to defects. Also it helps them to identify the amount of risk associated with that defect (high/medium/nominal).

V. E. Balas et al. (Eds.): ICETCE 2022, CCIS 1591, pp. 516–528, 2022.
https://doi.org/10.1007/978-3-031-07012-9_44

In this paper, a detailed analysis is done about the various approaches proposed by different researchers. All the research work done on the topic helps us to dive into the knowledge pool of the traditional and modern techniques of JITDP to have ample of knowledge and get to know the strengths and limitations of each and every approach and also be able to decide which approach is most feasible and efficient and in which scenario. The aim of this paper is to identify the answers to the following research questions.

RQ1 Which algorithm proves to have maximum efficiency in JITDP?

RQ2 Which algorithm can be used in the generalized perspective for JITDP?

This paper is divided into five sections. The need and importance of JITDP is discussed in the introduction in Sect. 1. The related work and proposed approaches are described in Sect. 2 under the head literature survey. A case study involving various datasets and approaches are presented in Sect. 3 discussion, conclusion in Sect. 4, following them is Sect. 5 named as research gaps where the limitations and future scope is discussed.

2 Literature Review

There are two ways in which defect prediction was evaluated (i) confusion matrix based metrics and (ii) effort aware based metrics. A confusion matrix is a technique for summarizing the performance of a classification algorithm. They are useful because they give direct comparisons of values like True Positives, False Positives, True Negatives and False Negatives. Confusion matrix based metrics is used for analyzing predictive performance. But one drawback of confusion matrix is that it only focuses on the positive parameters and neglect the negative ones which make the analysis partial or biased. Second method is effort aware based metrics. Some examples of effort aware based metrics are accuracy and Popt. This method sets a code churn inspection coverage limitation and is used to check defect inspection efficiency. Both these methods focus on prediction accuracy and defect inspection efficiency but one of the most important feature i.e. software reliability is lacking. We evaluate performance measures from both the methods to ensure which method serves the purpose in which scenario. People often understand software reliability and software quality as the same terms, but they have different meanings. Software quality refers to, how well the software complies with, or conform to a given design, based on functional requirements or specifications. Software reliability pertains to the probability of failure-free software operation for a specified period of time in a particular environment.

The proposed approaches on just in time defect prediction are as follows:

i. Time period based approach
ii. Supervised machine learning
iii. Unsupervised machine learning
iv. Local and global model
v. Deep Learning approach

Time Period Based Approach: Time period based approach is broadly classified into two categories namely short term and long term and was proposed by Yuli Tian and Ning Li [1]. Short term approach aims to detect whether the exposed defect is an early exposed defect or not to prevent defects which are identified at early stages. In order to classify the defects, a threshold value (θ) is taken, latency value lesser than the threshold value will classify the defect as early exposed defect.

$$\theta = \min(4\text{weeks},\ 1\% \times (\tau - \tau 0)),$$

where, τ0 is the start time and τ is the end time of the dataset. Firstly software changes and software failures are collected from open source repositories available online. Then these two variables are mapped and the available dataset is divided. This process consists of two phases namely training phase and application phase. So out of the 100% dataset available, it is divided into parts of 25% and 75% for training phase and application phase respectively as shown in Fig. 1. In the training phase, the system is trained to predict whether the code is defect-prone or not. Then application phase is used to test whether the system is performing as per the training or not and also aims at continuous improvement in accuracy. Then confusion matrix and effort aware based matrix are calculated to evaluate the performance of the system.

Long term based approach aims at software quality and reliability for long term perspective. One major attribute of software is durability which indicates how long the software is able to meet the requirements of the client. This approach comprises of three phases which are training phase, observation phase and the application phase. In this case the data is divided into three parts i.e. 25%, 25% and 50% for training phase, observation phase and the application phase respectively as described in Fig. 2. The working of the training and application phase is similar to that of short term based approach. The process includes division of dataset into three parts. Then after training and application phase, the results are simulated and observed in the observation phase. Thereafter, software reliability growth model techniques like SS and GO are applied to assess software reliability.

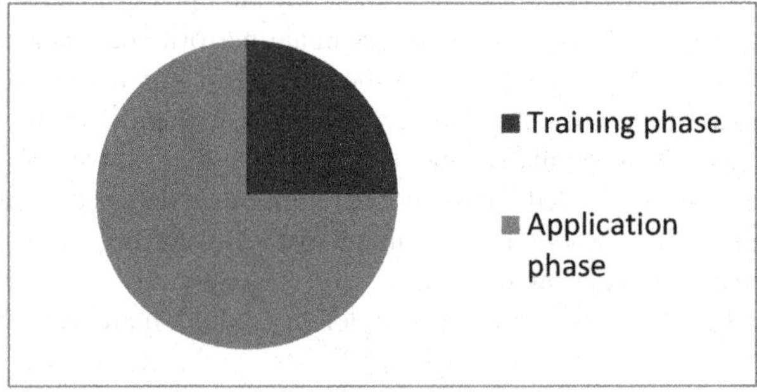

Fig. 1. Short term data division

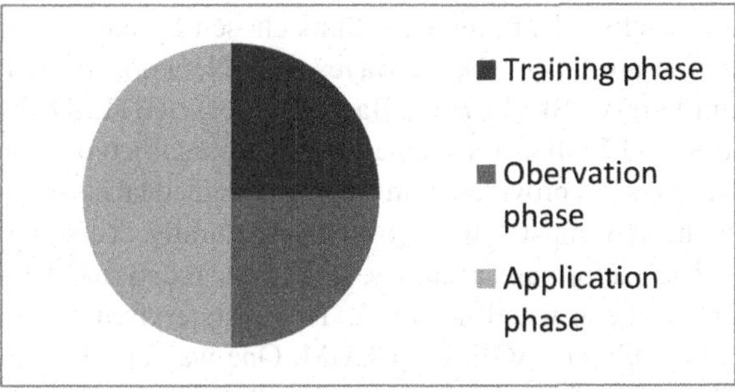

Fig. 2. Long term data division

Researchers have applied different techniques and different algorithms on sample data set as well as original dataset to analyze which algorithm works best under which circumstances. So working on different techniques and evaluating their accuracy in prediction, their recall value, P opt value and several other factors, they found and concluded that for short term, Random Forest serves as the best option while CBS+ is most suitable when one opts for long term software.

Machine Learning Approach: Supervised ML is a task of learning a function that maps an input to an output based on example input-output pairs. Researchers have taken several factors into consideration for evaluating the performance of each algorithm. Those factors are accuracy, precision, recall and F1 score. The formulae for all the above listed factors are listed below in Fig. 3 where TP, FP, TN and FN refer to true positive, false positive, true negative and false negative values respectively.

$$
\begin{aligned}
precision &= \frac{TP}{TP + FP} \\
recall &= \frac{TP}{TP + FN} \\
F1 &= \frac{2 \times precision \times recall}{precision + recall} \\
accuracy &= \frac{TP + TN}{TP + FN + TN + FP} \\
specificity &= \frac{TN}{TN + FP}
\end{aligned}
$$

Fig. 3. Evaluation metrics: precision, recall, F1, accuracy and specificity

The supervised machine learning algorithms chosen by the researchers for just in time defect prediction are CBS+, Naïve Bayes, RBFNetwork (Radial Basis Function Network), Random Forest, IBk (Instance Based learner), AB+LMT, LR, SVM, Linear SVM, J48, Adaboost and XGboost. Unsupervised learning is a type of machine learning in which the algorithm is not provided with any pre-assigned labels or scores for training the data. As a result, they must self discover any naturally occurring patterns in that training data set. The factors are accuracy, precision, recall and F1 score. All these formulae have already been listed in Fig. 3. The unsupervised algorithms taken into consideration in our study are AGE, LT, CCUM, One way and K-means. Xiang Chen [10] in his research paper indicated that unsupervised models outperforms the supervised models based on his evaluation on different algorithms. JITLine is a machine learning based approach which was proposed by Chanathip Pornprasit [3]. According to him this model can predict defect introducing commits and also identify associated defective lines for that particular commit. Some of the key advantages of this model are: fast, cost effective and fine grained.

Local and Global Model Approach: According to Xingguang Yang [2], local models involve grouping of clusters based on their extent of similarity or we can say, homogeneity uses k-mediods and WHERE algorithm. But there is no such case in global model. Global model uses whole dataset as a single unit and uses it for training and prediction. The functionality of local models and global models can be easily understood by Figs. 4 and 5 respectively. On researches and analysis made on local and global models, research scholars observed their performance in three scenarios: cross validation, cross project validation and time-cross validation. Some observed on the basis of their cost effectiveness. The results for classification and effort aware logistic regression are mentioned below in Table 1.

Table 1. Local model and global models.

Process name	Cross validation	Cross project validation	Timewise cross validation
Classification process	GLOBAL > LOCAL	GLOBAL > LOCAL	GLOBAL > LOCAL
EALR	LOCAL > GLOBAL	LOCAL > GLOBAL	GLOBAL > LOCAL

Deep Learning

Deep learning is a quite fascinating topic and various algorithms of deep learning have been tried and tested by various researches with different objectives. T. Hoang [5] used deep learning along with CNN (Convulution Neural Network) model to generate features automatically and named it as DeepJIT. It does not focus on the hierarchical structure of commits. It uses two CNNs and then concatenates them to finally calculate the probability of the defect introducing commit. CC2Vec [4] model uses commit message and the code changes as inputs and then produce vector format in two CNN forms. DeepJIT does not focus on the hierarchical structure of commits. To improvise this model H. J. Kang

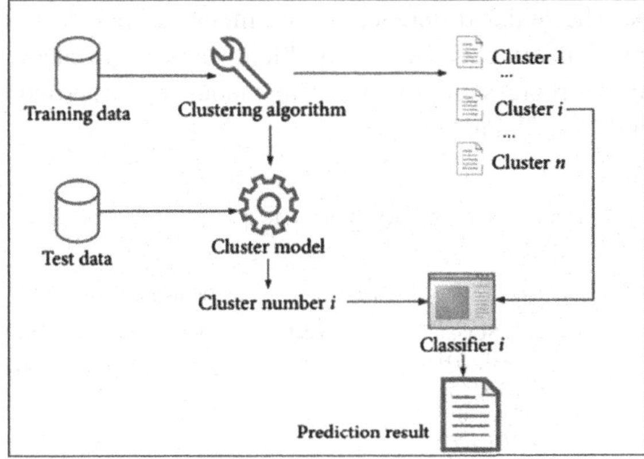

Fig. 4. Local model [9]

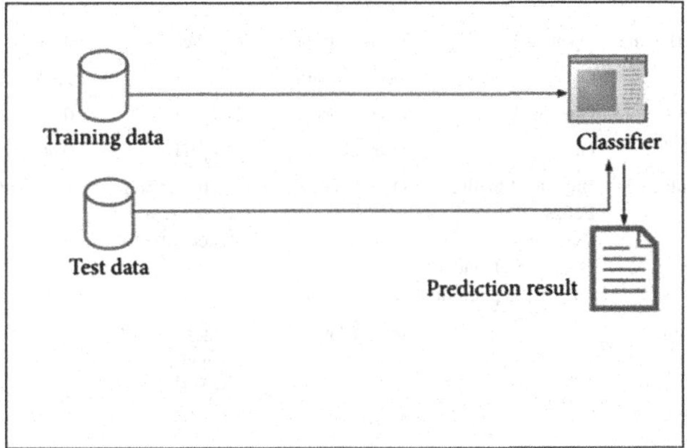

Fig. 5. Global model [9]

proposed new method with a fusion of deep learning and CC2Vec that focuses on the hierarchical representation of code commits with the help of HAN (Hierarchical Attention Network).It learns the relationship between the semantics and the code changes in the software. For training this model both training as well as unlabelled testing dataset are required. This method outperformed the above mentioned DeepJIT model.A model was proposed by Saleh Albahli [7] having a fusion of deep learning, random forest, XGBoost and Rainbow based reinforcement algorithm. The input is passed through the fusion of all the three methods namely random forest, XGBoost and rainbow reinforcement algorithm. The output is presented in terms of accuracy and false alert. The optimization of the model is done with the help of ADAM (Adaptive Moment Estimation) algorithm. The model keeps improving its functioning by reward and punishment points. Deep learning and neural network together helps to extract the important and essential features automatically which makes it time saving and fast. Neural Network based Regression model is also able to calculate complex computations with the help of neural network. The major difference in this model is that in this case the defect prediction is taken as binary classification problem (ranking problem) while all others have considered it as a regression problem. This model was proposed by Lei Qiao and Yan Wang. Lei Qiao, Yan Wang [8].

Table 2. Analysis

Year	Author	Dataset	Approach	Alorithm	F1	Acc	Recall
2019	Hindawi	Bugzilla (BUG), Eclipse JDT (JDT), Mozilla (MOZ), COL, PLA and PostgreSQL (POS)	Local model global model	Local model	0.629	0.493	-
				global model	0.678	0.314	-
2021	Chanathip Pornprasit	Openstack	Machine Learning	JITLine	0.33	0.83	0.26
		Qt	Machine Learning	JITLine	0.24	0.82	0.17
2020	T. Hoang, HJ Kang, D. Lo	Openstack	Deep Learning	CC2Vec	0.24	0.77	0.99
		Qt	Deep Learning	CC2Vec	0.19	0.81	0.96
2019	T. Hoang, HK Dam, Y Kamei	Openstack	Deep Learning	DeepJIT	0	0.75	0
		Qt	Deep Learning	DeepJIT	0.06	0.76	0.3
2019	Salen Albahli	Bugzilla, Mozilla, Eclipse JDT, Columba, PostgresSQL and the Eclipse platform	Deep Learning	Deep Learning + XGBoost + Random forest		0.82	-
2019	Lei Qiao, Yan Wang		Deep Learning	Deep Learning + Neural Network	-	-	0.69
2021	Yang, Y.; Zhou, Y.; Liu, J.; Zhao, Y.; Lu, H.; Xu, L.; Leung, H		Machine learning	LR	0.671	0.655	
				SVM	0.688	0.655	
				LINEAR SVM	0.019	0.473	
				J48	0.671	0.6600	
				ADABOOST	0.663	0.6900	
				XGBOOST	0.698	0.70000	
				DNN	0.682	0.6275	
				K-MEANS	0.009	0.4775	
2020	Yuli Tian, Ning Li, Jeff Tian and Wei Zheng	18 large open-source projects: ActiveMQ, Ant, Camel, Derby, Geronimo, Hadoop, HBase, IVY, JCR, Jmeter, LOG4J2, LUCENE, Mahout, OpenJPA, Pig, POI, VELOCITY and XERCESC	Supervised machine learning approach	EALR (Effort aware logistic regression)	0.15	0.79	0.48
				Random Forest	0.30	0.86	0.76

(continued)

Table 2. (*continued*)

Year	Author	Dataset	Approach	Alorithm	F1	Acc	Recall
				CBS+	0.13	0.78	0.45
				Naïve Bayes	0.14	0.78	0.46
				RBF Network	0.02	0.70	0.07
				IBK	0.15	0.79	0.49
				AB+LMT	0.28	0.85	0.73
			Unsupervised machine learning approach	AGE	0.00	0.68	-
				LT	0.01	0.70	-
				CCUM	0.01	0.70	-
				One way	0.01	0.69	-

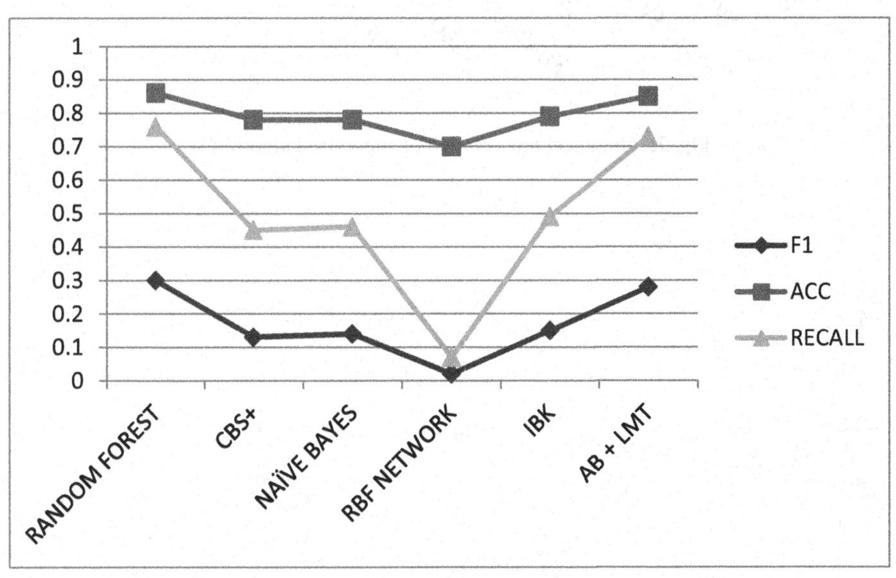

Fig. 6. Graphical presentation of supervised algorithms

3 Discussion

This section includes a systematic review of the proposed approaches by comparing them on the three parameters F1 score, accuracy and recall value. The datasets used in the approaches are not same, many open source datasets, Openstack, Qt etc. are used so the results may vary if these algorithms are implemented on the same dataset. And that is why average values are taken into consideration [8].

Table 2 gives a systematic analysis of various algorithms on the basis of their F1 score, Accuracy and Recall value in order to analyze their performance. Table 3 gives an

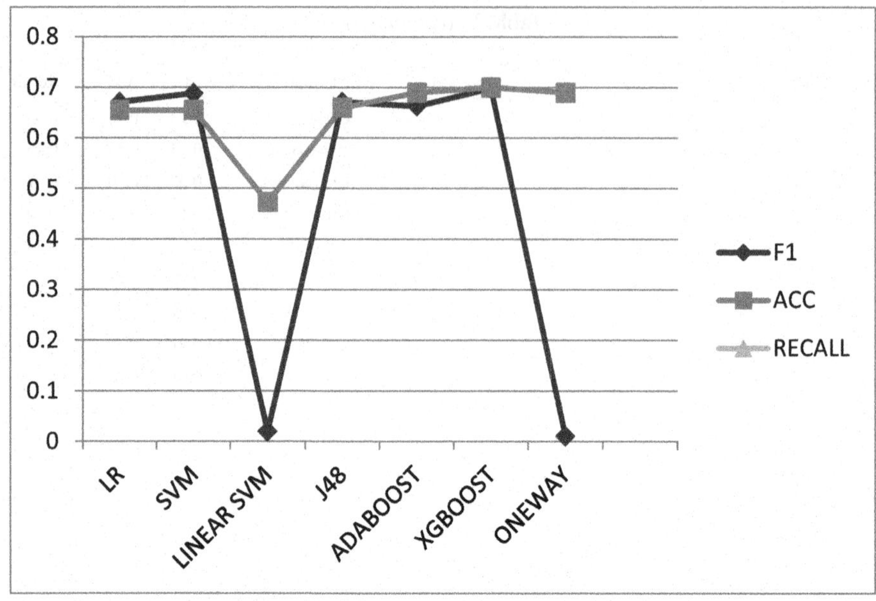

Fig. 7. Graphical presentation of supervised algorithms

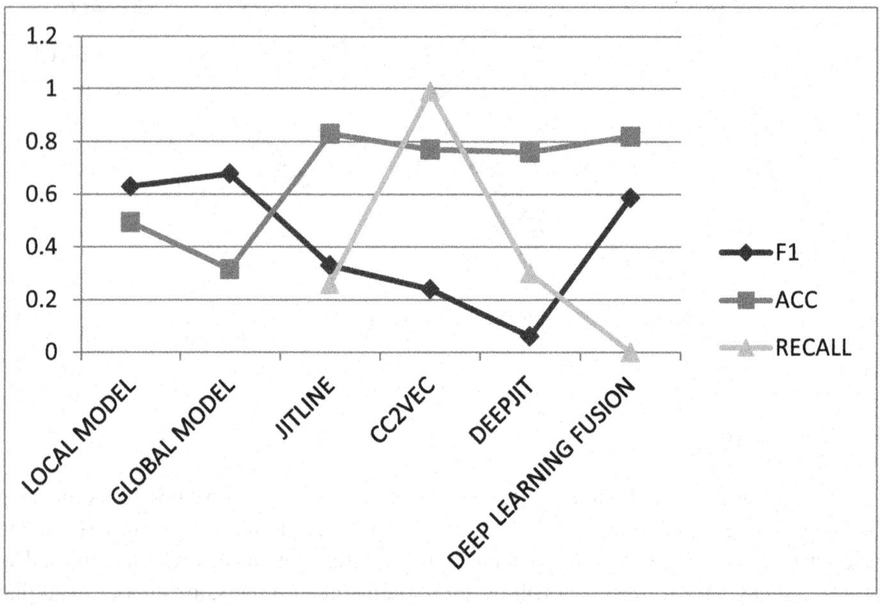

Fig. 8. Graphical presentation of various algorithms

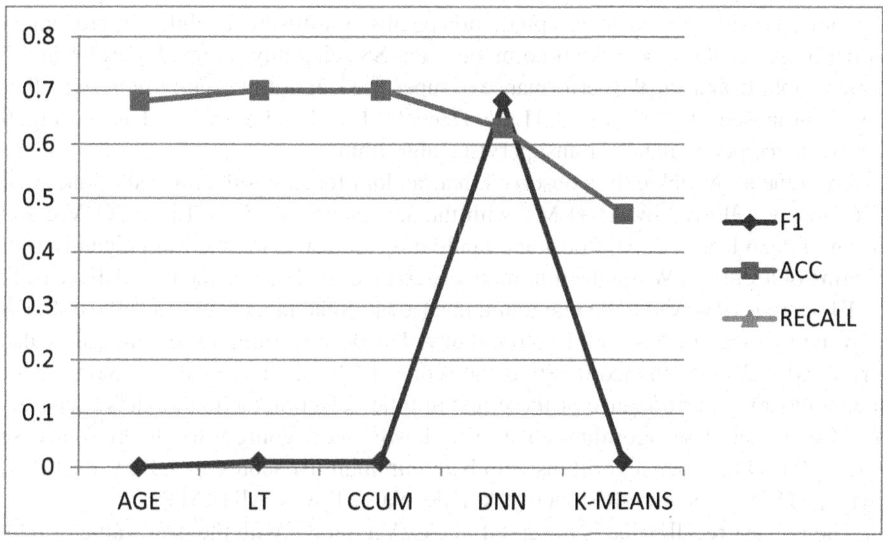

Fig. 9. Graphical presentation of unsupervised algorithms

Table 3. Early exposed defects prediction results on software reliability improvement

Algorithm	GO	SS
	Reliability improvement	Reliability improvement
EALR	206.30%	189.82
CBS+	**216.33**	**200.20**
Naïve Bayes	202.84	187.25
RBF network	91.67	81.24
Random forest	**207.72**	**192.18**
IBK	171.37	156.89
AB+LMT	182.31	168.23
AGE	78.11	68.76
LT	76.35	67.02
CCUM	155.89	141.66
One way	155.89	141.66

overview of early exposed defects prediction results on software reliability improvement on the basis of GO (Goel Okumoto model) and SS reliability scores. Figures 6 and 7 depict graphs to analyze the performance of supervised algorithms, Fig. 8 represent other models proposed like CC2Vec, JITLine, DeepJIT, Local and global models and Fig. 9 represents the performance of unsupervised algorithms.

From the above table, it is observed that random forest has the highest accuracy of 0.86 closely followed by AB+LMT with the accuracy of 0.85. JITLine, CC2Vec and fusion of deep learning, XGBoost and random forest also performed significantly well in terms of accuracy. While the minimum accuracy of 0.70 is obtained in RBFnetwork.

F1 score is observed to be maximum in case of global models with a value of 0.678 followed by local models with F1score 0.629. The deep learning fusion algorithm also performed well and attained 0.586 as the score of F1. F1 score is also a major factor to determine the performance of these just in time defect prediction models because a large fraction of these algorithms have a very low F1 score going as low to 0(0.6) in case of DeepJIT. Other such algorithms with bare minimum F1 scores are CBS+, AGE, LT, EALR, CCUM, One way, RBF network, IBK, Naïve Bayes, AB+LMT.

The highest recall value is observed in CC2Vec model with the peak value of 0.99 followed by random forest, AB+LMT and fusion of deep learning algorithm. The lowest values have been observed in DeepJIT and RBF Network with 0.3 and 0.07 recall values respectively.

From the above discussion we will try to answer the above mentioned research questions which are as follows.

RQ1 Which algorithm proves to have maximum efficiency in JITDP?

From Table 2, it is clearly evident that the maximum efficiency in just in time defect prediction is showcased by random forest algorithm as it has the maximum accuracy compared to all other algorithms with the value of 0.86 and F1score and recall value as 0.3 and 0.76 respectively.

RQ2 Which algorithm can be used in the generalized perspective for JITDP?

This paper focuses on finding a generalized algorithm for just in time defect prediction which can be used in every perspective with optimum efficiency. So based on the analysis of various algorithms, we observe that random forest serves as the generalized algorithm. Mostly researchers either use machine learning or deep learning for JITDP and random forest proves to be the machine learning algorithm with maximum accuracy. Also it is observed that it is ranked second in detecting the early exposed defects which proves it is also efficient in time period based approach i.e. short term. Other approaches like CC2Vec does not comply with detecting defects immediately and DeepJIT and JITLine does not have high recall, precision, popt and accuracy so they cannot be the generalized model.

4 Conclusion

The Just in Time Defect Prediction system aims to predict the defects immediately so that appropriate steps can be taken in order to avoid them. Some approaches have attained high accuracy but lack in other crucial parameters. Some models are fast and cost effective but do not have accuracy. So each model has its own advantages and disadvantages. Different scenario leads to different conclusions on different datasets. On comparing supervised and unsupervised machine learning approaches it is observed that supervised algorithms are much more effective when there is labeled dataset while on the contrary unsupervised algorithms perform better in case of unlabelled dataset [6] [8]. In supervised algorithm, random forest is seen to outperform other algorithms. In time based approach it is observed that short term algorithms are analyzed upon the ability of early exposed defects while long term approach is analyzed on the basis of overall reliability improvement. So as per the observations in Table 3 we can observe CBS+ proves to be the most efficient algorithm with the highest values followed by random forest.

The following conclusions are made from this paper, which are as follows:

1. Random forest algorithm proves to be the most efficient algorithm and outperforms all the other proposed models.
2. Random Forest is observed to be a generalized algorithm which can be used from various perspectives.
3. Apart from random forest, deep learning fusion technique, AB+LMT and DeepJIT are also efficient algorithms with an average accuracy of 0.80.
4. CBS+ is the most efficient algorithm in case of long term approach followed by random forest.

5 Research Gaps

This section consists of all the gaps which are present in this model till date. All those research gaps are enlisted below.

1. Although much research work has been done on this topic but there is no generalization for the JITDP model. It is true that many approaches have been proposed but there is no generalized model that can work in all scenarios with optimum efficiency.
2. CC2vec does not comply with the basic guidelines of JITDP. It does not yield the defects instantaneously and this is the major limitation of CC2Vec. But still it is considered in studying the JITDP model.
3. Another problem faced by the researchers is the unavailability of proper datasets which makes it difficult for the researchers to focus and analyze their implementations.
4. Local model works efficiently only when the value of k is set to 2 in k-mediods. Another problem associated with the model is that it lags in classification performance o it does not guarantee success in every scenario.

References

1. Tian, Y., Li, N., Tian, J., Zheng, W.: "How well just-in-time defect prediction techniques enhance software reliability?." In: Proceedings of the IEEE International Conference on Software Quality, Reliability and Security (QRS) (2020)
2. Yang, X., Yu, H., Fan, G., Shi, K., Chen, L.: "Local versus global models for just-in-time software defect prediction" (2019)

3. Pornprasit, C., Tantithamthavorn, C.K.: JITLine: a simpler, better, faster, finer-grained just-in-time defect prediction (2021)
4. Hoang, T., Kang, H.J., Lo, D., Lawall, J.: "CC2Vec: distributed representations of code changes." In: Proceedings of the International Conference on Software Engineering (ICSE), pp. 518–529 (2020)
5. Hoang, T., Dam, H.K., Kamei, Y., Lo, D., Ubayashi, N.: "DeepJIT: an end-to-end deep learning framework for just-in-time defect prediction." In: Proceedings of the International Conference on Mining Software Repositories (MSR), pp. 34–45 (2019)
6. Qiao, L., Wang, Y.: Effort-aware and just-in-time defect prediction with neural network. PLoS One **14**(2), e0211359 (2019). https://doi.org/10.1371/journal.pone.0211359
7. Mahrishi, M., Hiran, K.K., Meena, G., Sharma, P. (eds.): Machine Learning and Deep Learning in Real-Time Applications. IGI Global, Beijing (2020). https://doi.org/10.4018/978-1-7998-3095-5
8. Li, N., Shepperd, M., Guo, Y.: A systematic review of unsupervised learning techniques for software defect prediction. Inf. Softw. Technol. **122**, 106287 (2020)
9. Yang, Y., et al.: "Effort-aware just-in-time defect prediction: simple unsupervised models could be better than supervised models." In: Proceedings of the 2016 24th ACM SIGSOFT International Symposium on Foundations of Software Engineering. Seattle, WA, USA 13–18, pp. 157–168, November 2016
10. Chen, X., Zhao, Y., Wang, Q., Yuan, Z.: "MULTI: Multi-objective effort-aware just-in- time software defect prediction." (2017)
11. Mende, T., Koschke, R.: Effort-aware defect prediction models. In: European Conference on Software Maintenance and Reengineering (2010). https://doi.org/10.1109/CSMR.2010.18
12. Huang, Q., Xia, X., Lo, D.: Revisiting supervised and unsupervised models for effort-aware just-in-time defect prediction. Empir. Softw. Eng., 1–40 (2018). Research Collection School Of Information Systems
13. Yan, M., Fang, Y., Lo, D., Xia, X., Zhang, X.: File-level defect prediction: unsupervised vs. Supervised Models, pp. 344–353 (2017). https://doi.org/10.1109/ESEM.2017.48
14. MSR 2014. In: Proceedings of the 11th Working Conference on Mining Software Repositories, pp. 172–181, May 2014. https://doi.org/10.1145/2597073.2597075
15. Kamei, Y., Fukushima, T., McIntosh, S., Yamashita, K., Ubayashi, N., Hassan, A.E.: Studying just-in-time defect prediction using cross-project models. Empir. Softw. Eng. **21**(5), 2072–2106 (2015). https://doi.org/10.1007/s10664-015-9400-x
16. Yang, X., Lo, D., Xia, X., Zhang, Y., Sun, J.: "Deep learning for just-in-time defect prediction." In: 2015 IEEE International Conference on Software Quality, Reliability and Security, pp. 17–26 (2015). https://doi.org/10.1109/QRS.2015.14

Query-Based Image Retrieval Using SVM

Neha Janu[1(✉)], Sunita Gupta[2], Meenakshi Nawal[1], and Pooja Choudhary[3,4]

[1] Department of Computer Science and Engineering, Swami Keshvanand Institute of Technology Management and Gramothan, Jaipur, India
{nehajanu,meenakshi.nawal}@skit.ac.in

[2] Department of Information Technology, Swami Keshvanand Institute of Technology Management and Gramothan, Jaipur, India
sunita@skit.ac.in

[3] Department of Electronics and Communication Engineering, MNIT, Jaipur, India
pooja.choudhary@skit.ac.in

[4] Department of Electronics and Communication Engineering, Swami Keshvanand Institute of Technology Management and Gramothan, Jaipur, India

Abstract. In today's world, images are used to extract information about objects in a variety of industries. To retrieve images, many traditional methods have been used. It determines the user's query interactively by asking the user whether the image is relevant or not. Graphics have become an important part of information processing in this day and age. The image is used in image registration processing to extract information about an item in a variety of fields such as tourism, medical and geological, and weather systems calling. There are numerous approaches that people use to recover images. It determines an individual's query interactively by asking users whether the image is relevant (similar) or not. In a content-based image retrieval (CBIR) system, effective management of this image database is used to improve the procedure's performance. The study of the content-based image retrieval (CBIR) technique has grown in importance. As individuals, we have studied and investigated various features in this manner or in combinations. We discovered that image Registration Processing (IRP) is a critical area in the aforementioned industries. Several research papers examining color feature and texture feature extraction concluded that point cloud data structure is best for image registration using the Iterative Closest Point (ICP) algorithm.

Keywords: CBIR · SVM · Feature extraction · Point cloud

1 Introduction

Today's image processing plays an important role in image registration. In the 1990s, a new research field emerged: Content-based Image Retrieval, which aims to index and retrieve images based on their visual contents. It's also known as Query by Image Content (QBIC), and it introduces technologies that allow you to organize digital pictures based on their visual characteristics. They are built around the image retrieval problem in data bases. CBIR includes retrieving images from a graphics database to a query image. Similarity comparison is one of CBIR's tasks.

V. E. Balas et al. (Eds.): ICETCE 2022, CCIS 1591, pp. 529–539, 2022.
https://doi.org/10.1007/978-3-031-07012-9_45

1.1 Image Registration Process

Image registration is a fundamental task in image processing that is used for pattern matching of two or more images taken at different time intervals from different sensors and from different perspectives. Almost all large industries that value images require image registration as a steadfastly associated operation. Identical for objective reorganization real time images are targeted in literal guesses of anomalies where image registration is a major component.

On 3-D datasets, two types of image registration are performed: manually and automatically. Human operators are responsible for all of the process corresponding features of the images that are to be registered when manually registering. To obtain logically good registration results, a user must select a significantly large number of feature pairs across the entire image set. Manual processes are not only monotonous and exhausting, but they are also prone to inconsistency and limited accuracy. As a result, there is a high demand for developing automatic registration techniques that require less time or no human operator control [1].

In general, there are four steps for image registration for supporters, such as:

Detection of Features: It is a necessary step in the image registration process. It detects features such as closed-boundary fields, ranges, edges, outlines, crossing line points, angles, and so on.

Feature Matching: In this step, the process of matching the features extracted by Database Images and Query Images that result in a visually similar result is performed.

Estimation of the Transform Model: The sort and parameters of the so-called mapping purposes, uses, aligning the sensed image with the statement, and direction image are all assigned a value. The parameters of the mapping purposes and uses are worked out using the made specific point letters.

Image Re-sampling and Transformation: The sensed image is significantly altered as a result of the mapping purposes and uses. Image values in non-integer orders are calculated using the correct interpolation expert method of art, and so on.

1.2 Fundamentals of CBIR System

In typical CBIR systems, the visual content of the pictures is extracted from the database (Fig. 1). And it described by multi-dimensional feature vectors. These features vector of the images in the database forms a feature database. To retrieve the images, users offer the retrieval system with example images (Fig. 1).

There are three main components used in this CBIR system such as:

1. User's Query, 2. Database Unit, 3. Retrieval system.

Query Unit: The query unit then extracts three features from the picture and saves them as a feature vector. There are three distinct characteristics: texture, color, and form. The color moment might be included with a feature. Images may be translated and partitioned into grids by extracting their color features. A few seconds have been shaved off the end result. Wavelet transforms with pyramidal and tree-structured structures are employed to extract texture information. It is converted to grayscale and the Daubechies wavelet is used. The image's feature vector is used to aggregate all of the image's features.

Database Unit: The graphics' feature vectors are gathered and stored in the database unit in the same way that a feature database is (FDB). In order to do a comparison query, both pictures and database images must have the same capabilities. As a result, features are shown.

Retrieval Unit: You may use this unit to store your database feature vector and query picture. An SVM classifier is tinkered with in this game. It is then sorted into one of these categories. To get the 20 most similar photos to the one being searched for, it compares the image being searched for to all of the others in the class.

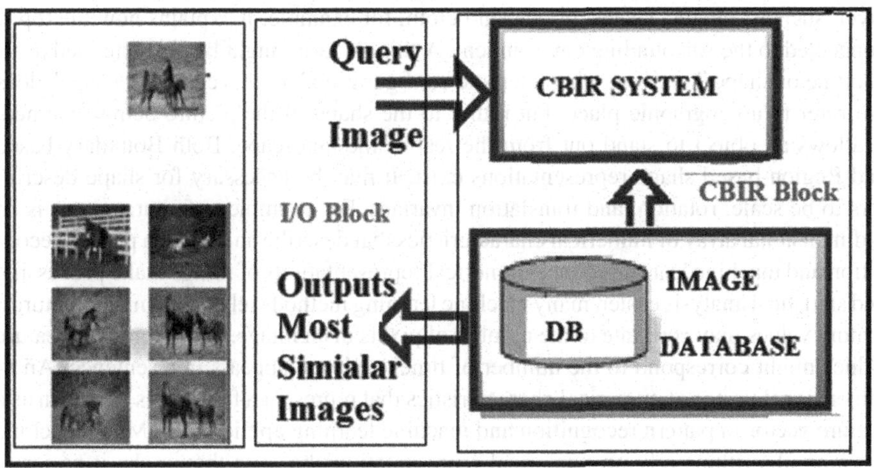

Fig. 1. Block diagram of CBIR

1.3 CBIR Applications

There are many applications for CBIR technology as below:

1. To identification of defect and fault in industrial automation.
2. It is used as face recognition & copyright on the Internet.
3. In medical plays a very important role in Tumors detection, Improve MRI and CT scan.
4. For the weather forecast, satellite images.
5. To map making from photographs.
6. For crime detection using fingerprints matching.
7. In the defense used for detection of targets.

1.4 Feature Extraction

When extracting the features, feature extraction techniques are applied. It has properties such as the colour, the texture, the form, and the ability to show vectors (Fig. 2). It doesn't matter what the size of a picture is, colour is employed extensively to convey an image's meaning. Color is a feature of the eye and the mind's ability to absorb information. Color helps us communicate the distinctions between different locations, things, and epochs. Colors are often specified in a colour space. Similarity dimension key components and colour space, colour quantization are used in colour feature extraction. RGB (red, green, and blue), HSV (hue, saturation, and value), or HSB (hue, saturation, and brightness) can be used (Hue, Saturation, and Brightness). These colour histogram moments are referred to as cases. The independent and picture representation of the magnitude of an image is commonly utilised for colour [2].

A surface's texture specifies the visual patterns it creates, and each pattern has a certain degree of uniformity. It contains crucial information on the structure of this exterior layer, such as; clouds, leaves, bricks, and cloth, for example. It explains how the top is connected to the surrounding environment. A thing's form might be characterised as its "outline or shape," or as its "characteristic configuration." In this context, "shape" does not refer to a geographic place, but rather to the shape of the picture being searched. It allows an object to stand out from the rest of the landscape. Both Boundary-based and Region-based shape representations exist. It may be necessary for shape descriptors to be scale, rotation, and translation invariant. For example, a feature vector is an n-dimensional array of numerical characteristics that describe an object in pattern recognition and machine learning. Since numerical representations of things make processing and statistical analysis easier, many machine learning methods rely on them. For pictures, feature values may correlate to the number of pixels in an image, while for texts, feature values might correspond to the number of times a phrase appears in a sentence. An n-dimensional vector of numerical characteristics that represent a few things is known as a feature vector in pattern recognition and machine learning applications. Many machine learning algorithms rely on numerical representations because they make it easier to conduct statistical analysis and processing. Textual representations may use pixels to indicate the value of a feature, while pictures may use feature values corresponding to the pixels.

1.4.1 Basic Architecture of Feature Extraction System

1.4.2 Feature Extraction Techniques

There are following techniques used as follows:

In terms of Color-Based Feature Extraction, there are two main options: Color Moments and Color Auto-Correlation. An image's colour distribution may be measured using a concept known as "colour moments." Image similarity scores are calculated for each comparison, with lower scores indicating a greater degree of resemblance between pictures. It is the primary use of colour moments that is colour indexing (CI). It is possible to index images such that the computed colour moments may be retrieved from the index. Any colour model may be used to compute colour moments. Per channel, three different colour moments are calculated (9 moments if the colour model is RGB and 12

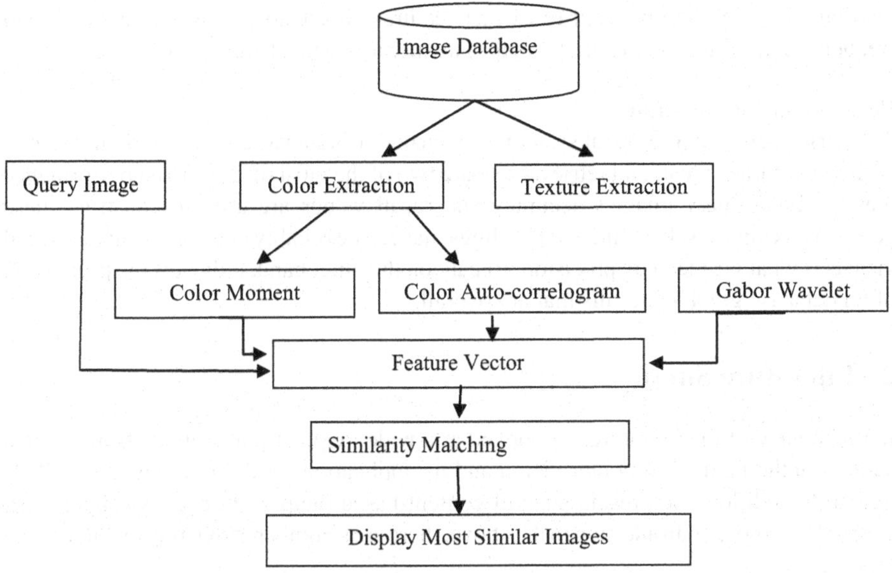

Fig. 2. Feature extraction system

moments if the colour model is CMYK). Using colour pairings (i,j) to index the table, a colour auto-correlogram may be stored as a correlogram, with each row indicating how often it is to be found between two adjacent pixels (i,j).

As opposed to storing an auto-correlogram in the form of a table where each dth item represents the likelihood of finding a pixel I from the same one at the given distance. Auto-correlograms therefore only display the spatial correlation between hues that are the same [2]. In this Texture-based approach, there are a number of Texture attributes, such as Coarseness, Directionality, Line likeness, Regularity and Roughness, that may be extracted from the image. Techniques like Wavelet Transform and Gabor Wavelet can be used to categories textures. You may extract information from any kind of data using a wavelet: this includes audio signals and pictures. Analysis allows us to use very large time intervals in which we are looking for specific information. Other techniques of signal analysis omit elements like self-similarity, breakdown points, discontinuities in higher derivatives, and patterns that this approach is capable of exposing about the data. When applied to a signal, wavelet analysis de-noises or compresses it without causing significant deterioration [3]. A filter named after Dennis Gabor; the Gabor filter is a wavelet. Also known as a Gabor wavelet filter, the Gabor filter. The extraction of texture information from pictures for image retrieval is generally accepted. Gabor filters may be thought of as a collection of wavelets, each of which directs energy in a certain direction and at a specific frequency. Like the human body, the Gabor filter has frequency and orientation representations. Extraction and harmonic functions are achieved via Gabor filters using Fourier transforms of the function's Fourier transforms. This set of energy distributions may then be used to extract features. The Gabor filter's orientation and frequency adjustable characteristic makes texture assessment simple [4].

Performance Evaluation

The performance of a retrieval system is evaluated using a variety of criteria in system. Average accuracy recall and retrieval rate are two of the most often utilized performance metrics. Each query image's accuracy and recall values are used to construct these performance metrics. Percentage of findings that can be reliably deduced from recovered visuals is what we mean by precision. Recall, on the other hand, indicates the proportion of all database graphics results that are overall.

2 Literature Survey

In this review of the literature, we looked at a wide range of publications from diverse authors in the field of computer vision and its applications to discover the latest CBIR trends. In addition, the present state of content-based picture retrieval systems is also covered in this paper. In addition to the above, there are a number of existing technological aspects.

2.1 Support Vector Machine (SVM)

It was proposed by V. Vapnik in the mid-1990s that support vector machine (SVM) be implemented. Since the beginning of the millennium, it has been the most common machine learning algorithm. As a sophisticated machine learning technique, SVM is currently commonly employed in data mining projects like CRM. Computer vision, pattern recognition, information retrieval, and data mining all rely on it these days. A binary labelled dataset's linear separation hyperplane may be found using this technique. It's a classifier with a hyperplane for dividing classes (Fig. 3).

The SVM model depicts the Examples as objects in space, with a visible gap as broad as possible between the various kinds. Non-linear classification and classification are routine for it. High-dimensional feature spaces can be mapped from their inputs. The classifier accomplishes this by focusing on the linear combination of the characteristics while reaching a classification decision. SVM is a type of binary decision-making option. Classification process that uses labelled data from two groups as input and generates sparks in a file to maximize the number of new data points possible. Analyzing and training are the two main phases. In order to train an SVM, one must provide the SVM with a finite training set of previously known statistics and decision values. A two-class classification problem can be found as input data is mapped. Using the RBF kernel and hyper-plane linear interpolation Vectors closest to the decision border are placed in this space by the classifier.

Let m-dimensional inputs = xi (i = 1,2,3… ………M) belong to Class – "*" or Class– " +" and the associated labels be yi = 1 for Class A and – 1 for Class B.

Here Class -A is Red Star and Class -B is Green Plus.

Decision function for SVM is

$$D(x) = w T x + b; \qquad \text{(i)}$$

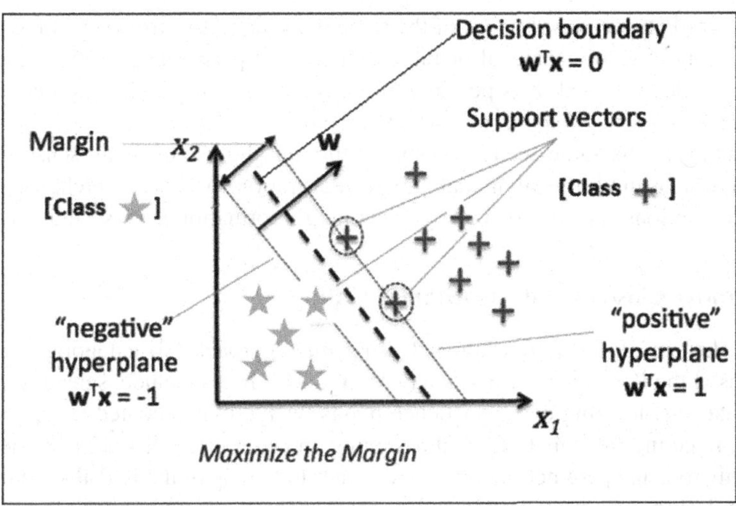

Fig. 3. SVM maximize the Hyperplane and Margin two different class Red Star and Green plus (under 2D)

where w is an m-dimensional vector, b is a scalar, and yi

$$D\ (xi) \geq 1 \ \text{for}\ i = 1,\ 2,\ 3\ldots,\ M. \tag{ii}$$

The distance between the separating hyper plane D(x) = 0 and the training datum nearest to the hyper plane is called the margin. The hyper plane D(x) = 0 with the maximum margin is called the optimal separating hyper plane.

2.2 Surface Based Image Registration

For image registration in the medical business, the three-dimensional edge surface of an anatomical item or composition is intrinsic and provides substantial geometrical feature data. By combining various registration approaches, it is possible to create equivalent surfaces in many pictures and then compute the transformation between the images. On the surface, there is just one object that can be collected. An implicit or parametric surface, such as a B-spline surface, is revealed by a facial view. Skin and bone may be easily extracted from CT and MR images of the head [5].

2.2.1 Applications of Surface-Based Image Registration

Many fields, including remote sensing, medical imaging, computer vision, and others, make use of image registration in some capacity or another. Surface-based image registration may be broken down into a number of subcategories, some of which include:

1. Image acquisition is done at various times for the same view (Multiview registration). Changes in perspective can be found and estimated by using the mean.

2. Different Perspectives: Images of the same scene are taken from several angles. To get, a huge two-dimensional or three-dimensional perspective of the scene being photographed is used. It is possible to use remote sensing and computer vision to create mosaics of photos from a scanned region.
3. A picture of the same scene is obtained through the use of many sensors, which is known as multimodal modal image registration. More comprehensive scene representations can be achieved by integrating information from several sources.

2.3 Iterative Closest Point Algorithm (I.C.P)

Medical data in 3D may be evaluated using this approach. To categories the many variations of the ICP, several criteria may be used. Examples include: selecting portions of 3D data sets, locating correspondence points, weighting estimated correspondence pairings, rejecting false matches and assigning an error score. If true, then another is necessarily true and parametric curves and comes to the top of the ICP algorithm which is a general-purpose, representation-independent shape-based number on a list algorithm that can be used with a range of geometrical early people including point puts, line part puts, triangle puts (much-sided comes to the top).

$$\sqrt{\sum_{j=1}^{N} W_j^2 d^2(T(xj),\ y} = \sqrt{\sum_{j=1}^{N} W_j^2 ||T(xj) - yj||^2} \tag{iii}$$

$$\text{Where } \mathrm{y_j} = \mathrm{C(T(xj),\ y)} \tag{iv}$$

2.4 Academic Proposals

Understanding, extracting, and obtaining knowledge about a certain field of a topic is critical to doing a literature study. In this work, a number of existing approaches to picture retrieval are evaluated, including:

ElAlami, [6] who stated that the 3D colour histogram was a better option than the 2D colour histogram. Furthermore, the Gabor filter technique is capable of describing the image's attributes in a matter of seconds. Color coherence vectors and wavelets were used to improve recovery speed in a new version. Preliminary and reduction processes are used to extract the features from the collection. After then, the method is used to decrease the search area and the time spent retrieving information.

In this paper, [7] Darshana Mistry, et al. Detectors such as references and feel graphics can be used to compare two or more images of the same scene taken at different times, from different perspectives, or at different times of day. According on location and feature, they are grouped into categories. The picture registration process has four stages.

1. Feature detection
2. Feature matching
3. Alter model estimation
4. Image Re-sampling and transformation.

Data sets utilising point-cloud data organisation were used by Bohra, et al. [8] to minimise errors and time in surface-based image registration methods. To store CT, MRI, and Tumor pictures, as well as to construct 3D models of sets of these images. ICP method that registers two 3D data collections and finds the closest points into data collections based on the given tolerance distance for each data collection.

Enhances the overall efficacy of picture retrieval, according to N. Ali, et al. [9]. Images are divided into two halves by using the rule of thirds to identify focal points in the cortical lines that make up this grid. Color and texture are two features that provide the human visual system a suitable response.

Gabor feature extractions are more accurate than other wavelet transforms and discrete cosine transforms, according to Pratistha Mathuretl, Neha Janu, [10]. That Gabor's analytical features reveal that Gabor is a better method for extracting edge or shape features than DCT and DWT feature extraction, as shown in the findings. A percentage of the features are lost in the low-frequency feature sub-band (LL) and other frequency groups during the feature extraction process in DWT and DCT. This results in lesser accuracy as compared to Gabor. Gabor with a scale projection was more accurate than Gabor without a scale.

As a result of the incorporation of two new ways, such as graphic feel and spatial importance of pairs of colour within feature, G.S. Somnugpong, Kanokwan Khiewwan provided the high image that is shifting more robustly. In order to improve the precision of averaging, new strategies must be developed. It's possible to use colour correlograms to process information, whereas EDH gives geometric information when it comes to the image. Better than any one feature is a combination of low-end characteristics. Dimensioning can benefit from using Euclidean distance [11].

Shape descriptors, shape representations, and texture aspects were all introduced in this article. Using a CBIR method, information may be accessed both locally and globally. In order to combine colour and texture data, they proposed a novel CBIR approach. Extracting colour information may be done using the Color Histogram (CH). Discrete Wavelet Transform and Edge Histogram Descriptor are used to extract features. A lot of work has been done on the features. They've integrated a few elements to get greater results than a single feature could have on its own. As a result, the human visual system is well-served by texture and colour. Characteristic Extraction technique was introduced by Neha Janu, Pratistha Mathuretl., [12]. They include Gabor filter, DCT and DWT. J. Cook, et al. [13] developed a novel method of 3D face identification using the Iterative Closest Point (ICP) algorithm in this newspaper. Face and face modelling is effective for recording stiff parts when contrast is present in a certain area. This method employs 3D registration methods. It is used to correct for the surface's characteristics and build a correlation between target and assessment using the ICP algorithm.

BiaoLi [14] and YinghuiGao Wang [14] contributed the data. Measures and representations of attribute similarity. When used in conjunction with a convolution neural network (CNN), the SVM is used in this study in order to develop an image-discriminating hyperplane that can distinguish between visually similar and different picture pairings at a large level. The approach has been shown to improve the overall efficacy of CBIR in tests. SVM is used to learn the similarity measures, while CNN is used to extract the feature representations in this research work.

3 Conclusions and Research Motivation

In this study, a quick introduction to image retrieval and its structure are given. Researchers are using feature extraction algorithms to extract visuals from their training database. Textures, forms, and a variety of unique visuals are included in each of these databases. In order to retrieve pictures based on colour, Gabor filters and colour histograms (RGB and HSV) have been determined to be the most effective techniques for performing texture-based extraction. SVM classifiers and surface-based image registration approaches are frequently used in 3-D data sets image enrollment by a wide range of organisations. Finally. I.C.P (Iterative Closest Point) is the most common and efficient algorithm for surface-centered picture registration, and it is beneficial for improved image registration. A service vector system may be able to make this work easier and more efficient, according to a number of studies. The ICP algorithm provides excellent results and selects the images that best suit the needs of the user. ICP and SVM may be used in the future to classify images in datasets.

References

1. Mistry, D., Banerjee, A.: Review: image registration. Int. J. Graph. Image Process. **2**(1) 2012
2. Yu, J., Qin, Z., Wan, T., Zhang, X.: Feature integration analysis of bag-of-features model for image retrieval. Neurocomputing **120**, 355–364 (2013)
3. Hassan, H.A., Tahir, N.M., Yassin, I., Yahaya, C.H.C., Shafie, S.M.: International Conference on Computer Vision and Image Analysis Applications. IEEE (2015)
4. Yue, J., Li, Z., Liu, L., Fu, Z.: Content-based image retrieval using color and texture fused features. Math. Comput. Model. **54**(34), 1121–1127 (2011)
5. Amberg, B., Romdhani, S., Vetter, T.: Optimal Step Nonrigid ICP Algorithms for Surface Registration. This work was supported in part by Microsoft Research through the European PhD Scholarship Programme
6. Neha, J., Mathur, P.: Performance analysis of frequency domain based feature extraction techniques for facial expression recognition. In: 7th International Conference on Cloud Computing, Data Science & Engineering-Confluence, pp. 591–594. IEEE (2017)
7. Elalami, M.E.: A novel image retrieval model based on the most relevant features. Knowl.-Based Syst. **24**(1), 23–32 (2011)
8. Bohra, B., Gupta, D., Gupta, S.: An Efficient Approach of Image Registration Using Point Cloud Datasets
9. Ali, N., Bajwa, K.B., Sablatnig, R., Mehmood, Z.: Image retrieval by addition of spatial information based on histograms of triangular regions
10. Jost, T., Hügli, H.: Fast ICP algorithms for shape registration. In: Van Gool, L. (ed.) DAGM 2002. LNCS, vol. 2449, pp. 91–99. Springer, Heidelberg (2002). https://doi.org/10.1007/3-540-45783-6_12
11. Nazir, A., Ashraf, R., Hamdani, T., Ali, N.: Content based image retrieval system by using HSV color histogram, discrete wavelet transform and edge histogram descriptor. In: International Conference on Computing, Mathematics and Engineering Technologies (iCoMET), Azad Kashmir, pp. 1–6 (2018)
12. Somnugpong, S., Khiewwan, K.: Content based image retrieval using a combination of color correlograms and edge direction histogram. In: 13th International Joint Conference on Computer Science and Software Engineering. IEEE (2016). 10.1109

13. Janu, N., Mathur, P.: Performance analysis of feature extraction techniques for facial expression recognition. Int. J. Comput. Appl. **166**(1) (2017). ISSN No. 0975-8887
14. Cook, J., Chandran, V., Sridharan, S., Fookes, C.: Face recognition from 3D data using iterative closest point algorithm and Gaussian mixture models, Greece, pp. 502–509 (2004)
15. Fu, R., Li, B., Gao, Y., Wang, P.: Content-based image retrieval based on CNN and SVM. In: 2016 2nd IEEE International Conference on Computer and Communications, pp. 638–642
16. Kostelec, P.J., Periaswamy, S.: Image registration for MRI. In: Modern Signal Processing, vol. 46. MSRI Publications (2003)
17. Won, C.S., Park, D.K., Jeon, Y.S.: An efficient use of MPEG-7 color layout and edge histogram descriptors. In: Proceedings of the ACM Workshop on Multimedia, pp. 51–54 (2000)
18. Kato, T.: Database architecture for content-based image retrieval. In: Image Storage and Retrieval Systems, Proceedings SPIE 1662, pp. 112–123 (1992)
19. Elseberg, J., Borrmann, D., Nüchter, A.: One billion points in the cloud – an octree for efficient processing of 3D laser scans. Proc. ISPRS J. Photogram. Remote Sens. **76**, 76–88 (2013)
20. Kumar, A., Sinha, M.: Overview on vehicular ad hoc network and its security issues. In: International Conference on Computing for Sustainable Global Development (INDIACom), pp. 792–797 (2014). https://doi.org/10.1109/IndiaCom.2014.6828071
21. Nankani, H., Mahrishi, M., Morwal, S., Hiran, K.K.: A Formal study of shot boundary detection approaches—Comparative analysis. In: Sharma, T.K., Ahn, C.W., Verma, O.P., Panigrahi, B.K. (eds.) Soft Computing: Theories and Applications. AISC, vol. 1380, pp. 311–320. Springer, Singapore (2022). https://doi.org/10.1007/978-981-16-1740-9_26
22. Dadheech, P., Goyal, D., Srivastava, S., Kumar, A.: A scalable data processing using Hadoop & MapReduce for big data. J. Adv. Res. Dyn. Control Syst. **10**(02-Special Issue), 2099–2109 (2018). ISSN 1943-023X

Data Science and Big Data Analytics

A Method for Data Compression and Personal Information Suppressing in Columnar Databases

Gaurav Meena[1(✉)], Mehul Mahrishi[2], and Ravi Raj Choudhary[1]

[1] Central University of Rajasthan, Ajmer, India
gaurav.meena@curaj.ac.in

[2] Swami Keshvanand Institute of Technology, Management and Gramothan, Jaipur, India
mehul@skit.ac.in

Abstract. Data privacy and security is the need of the hour. The importance elevates further when we submit the data over a network. It is very easy to extract personnel's history like illness, vulnerability etc. through the information posted. Through this research, a technique for data compression and abstraction, particularly in columnar databases is proposed. It provides domain compression at the attribute level to both row and columnar databases using Attribute Domain Compression (ADC). Large datasets may be stored more efficiently by using this approach to minimize their size. Because we want the second process to be as flexible as possible, we give it the value (n) so that it can find all n − 1 more tuples.

Keywords: Anonymization · Columnar database · Compression · Data · Privacy · Security

1 Introduction

There has been a lot of study done on data compression in the scientific world. The most apparent reason to consider database compression is to free up disc space. The study explores whether query processing times can be reduced by applying compression techniques to reduce the amount of data that has to be read from the disc. There has recently been increasing interest in using compression techniques to speed up database performance [14]. Data compression is currently accessible in all major database engines, with distinct approaches employed by each. A subset of the queries was run with the following setups to evaluate the performance speedup gained with the Compression:

- No compression
- Proposed Compression
- compression of categories and compression of descriptions [2]

V. E. Balas et al. (Eds.): ICETCE 2022, CCIS 1591, pp. 543–559, 2022.
https://doi.org/10.1007/978-3-031-07012-9_46

Next, the paper looks at the two primary compression techniques used in row-oriented databases: n-anonymization and binary compression domain encoding. Finally, the research introduces two complex algorithms combined to build a novel optimum domain compression technique. The article also includes practical examples that are run on a columnar-oriented platform called 'Infobright.'

2 Literature

2.1 Query Intensive Applications

In the mid-1990s, a new era of data management emerged, characterized by query-specific data management and massive, complicated data volumes. OLAP and data mining are two examples of query-specific DBMS [15].

Online Analytical Processing. This tool summarizes data from big data sets and visualizes the response by converting the query into results using 2-D or 3-D visuals. "Give the percent comparison between the marks of all students in B. Tech and M. Tech," says the OLAP query. The response to this question will most likely take the shape of a graph or chart. "Data Cubes" are three-dimensional and two-dimensional data visualizations [10].

Data Mining. Data mining is now the more demanding application of databases. It is also known as 'Repeated OLAP.' Data mining aims to locate the subgroups that require some mean values or statistical data analysis to get the result [8]. The typical example of a data mining query is "Find the dangerous drivers from a car insurance customer database." It is left to the data mining tool to determine the characteristics of those dangerous customers group. This is typically done by combining statistical analysis and automated search techniques like artificial intelligence.

2.2 The Rise of Columnar Database

Column-store At first, transposed files were looked at as table attribute grouping in the 1970s, when DBMSs started. A completely deconstructed storage model (DSM, an early predecessor of column stores) has several benefits over NSM (normal row-based storage) as early as the mid-1980s [5].

Today's relational databases are dominated by OLTP (online transactional processing) applications. To a transaction (like buying a laptop from an online store), each row in a relational database is equivalent. RDBMS systems have always been designed on a row-by-row basis, and this remains the case today. With this strategy, incoming data entry for transactional-based systems may be effectively managed [11].

A recent study found that the size of data warehouses is increasing every three years, which is common practise in major corporations. Aside from that, these warehouses have a large hourly workload, encountering about 20 lakh SQL queries per hour [7].

The privacy of individuals is jeopardised by any breach or illicit publication of data in warehouses. An OLTP database design may not be the ideal choice for highly read-demanding and selective applications. Business intelligence and analytical software frequently conduct queries on certain database properties. The columnar approach's low cost can be attributed to its simplicity and effectiveness [16].

The columnar database, sometimes referred to as a "columnar database", transforms the way databases store data. Since neighbouring records are more likely to be kept on disc when data is stored this way, it opens the potential of compressing By using a row-based approach to add and remove transactional data, this design introduces a new paradigm. A columnar strategy, on the other hand, focuses on only a small subset of the table's columns [3].

Row-oriented databases provide for capabilities like indexing, global views, horizontal partitioning, and more. When it comes to performing queries, they're more efficient, but there are some drawbacks. Tables are either over-indexed (causing load and maintenance difficulties) or incorrectly indexed since it's difficult to predict which columns would need indexing in business intelligence/analytic situations. This causes many queries to execute much slower than planned [13].

2.3 Row Oriented Execution

After going through a few different implementation options, we'll look at some commercial row-oriented database management systems that support column-database architectures [13].

Vertical Partitioning. Column-store replication is made simple by dividing a row store vertically into all of the relations. Each logical column in the schema has its own physical table. Column I of the logical schema and the corresponding value of the position column are contained in the ith table's column I, accordingly. Queries are changed to conduct joins on the position property instead of receiving several columns from the same relation [12].

Index-Only Plans. There are two limitations to using vertical partitioning. A big header must be kept on each tuple when using row-stores since the position attribute must be stored in each column. This wastes both space and disc bandwidth. As a result, we adopt a new method known as Index only strategies to deal with these issues. There is no need to employ a distributed B+Tree index on top of a typical row-oriented design because the important relations are preserved [4].

Materialized Views. Finally, we'll look at viewpoints that are manifested. By employing this method, we can create the optimal number of materialized views for each query flight. In the optimum view, just the columns required to answer the questions in that flight's queries are shown. In these views, no columns from various tables are pre-joined [6].

2.4 Column Oriented Execution

In this part, we'll look at three popular column-oriented database system efficiency enhancements.

Compression. Data compression employing column-oriented compression techniques has been proven to enhance query performance by up to an order of magnitude when kept in this compressed format while being acted on. Save information in columns so that it may be sorted by anything like name or phone number. There's no denying that phone numbers are more comparable than other types of text, such as e-mail addresses or names. By sorting the data by that column, the data becomes even more compressed [1].

After a lengthy hiatus in database compression, there is a resurgence in the field to enhance database quality and performance. Data compression is now available in the primary database engines, with different methodologies being used in each of them. Column stores are thought to provide improved compression because of the increased similarity and redundancy of data inside columns, using less storage hardware and performing quicker since, among other things, they read fewer data from the disc. Furthermore, because the items in the columns are similar, the compression ratio in a columnar database is more excellent [3]. The statistical distribution of the frequency of symbols appearing in the data is used in Huffman and Arithmetic encoding. An ordinary character gets a shorter compression code in Huffman coding, whereas a rare sign gets longer. If there are four symbols a, b, c, and d, each having a probability of 1 3/16, 1/16, 1/16, and 1/16, each character requires two bits to represent without compression. A possible Huffman coding is the following: a = 0, b = 10, c = 110, d = 111. As a result, the average length of a compressed symbol equals: $1 \cdot 13/16 + 2 \cdot 1/16 + 3 \cdot 1/16 + 3 \cdot 1/16 = 1.3$ bits. Arithmetic encoding is like Huffman encoding, except it assigns an interval to the whole input string based on the statistical distribution.

Invisible Joins. Queries over data warehouses, particularly over data warehouses, are frequently structured as follows:

- Limit the number of tuples in the fact table using selection predicates on one (or more) dimension tables.
- This is followed by aggregation on the confined data set using additional dimension table properties to identify groups.

Each time a predicate or aggregate grouping is used, the fact table and dimension tables need to be joined together [9].

We propose an invisible join approach for foreign-key/primary-key joins in column-oriented databases as an alternative to standard query strategies. Predicates are created by turning joins on the fact table's foreign key fields to predicates. A hash lookup (which simulates a hash join) or more elaborate methods beyond the scope of our study can be used to test these predicates.

2.5 Anonymization

As a result, breaches and unauthorized information releases in warehouses pose a severe threat to people's privacy. N-Anonymity is a well-liked technique for collecting anonymous data. The strategy's purpose is to determine the value of a tuple, say n, so that n may be used to find other n – 1 tuples or, at the very least maximal tuples [15].

As the number of n grows, so does the level of protection. One technique to create an identical tuple inside the identifying attributes is to generalize values within the features, such as in an address property, by deleting city and street information [10].

Identifying data may be done in a variety of ways, but generalization is the most efficient. Multidimensional recoding generalization, global generalization, and local generalization are examples of generalization approaches. The process of generalizing an attribute at the global level is known as global recording. People's ages are displayed in 10-year intervals, for instance [13].

2.6 Conventional Compression

Database compression strategies are used to increase the speed of a database by reducing its size and improving its input and output active/query performance. Compression's primary premise is that it delimits storage and maintains data next to each other, reducing the size and number of transfers. This section demonstrates the two different classes of compression in databases:

1. Domain Compression
2. Attribute Compression

The classes can be implemented in either a column or row-based database. Queries done on compressed data are more efficient than queries executed on a decompressed database [7]. We'll go through each of the sections above in-depth in the areas below.

Domain Compression. We'll look at three types of compression algorithms in this section: numeric Compression using NULL values, string compression, and dictionary-based Compression. Because all three compression algorithms suit domain compression, we'll continue with domain compression for the characteristics.

Numeric Compression in the Presence of NULL Values
This approach compresses numeric characteristics, such as integers, and has some NULL values in their domain. The essential notion is that a tuple's consecutive zeros or blanks are eliminated, and a description of how many and where they appeared is supplied at the end. It is occasionally advised to encode the data bitwise to remove the variation in size of the attribute due to null values, i.e., an integer of 4 bytes is substituted by 4 bits [12].

For example:

Bit value for 1 = 0001
Bit value for 2 = 0011
Bit value for 3 = 0111
Bit value for 4 = 1111
And all '0s' for the value 0

- *String Compression*: In a database, a string is represented by the char data type, and SQL has previously suggested and implemented compression by providing the varchar data type. This approach offers an extension of traditional string compression. The recommendation is that after changing the char type to varchar, it is compressed in the second step using any of the compression algorithms available, such as Huffman coding, LZW algorithm, etc. [8]
- *Dictionary Encoding*: Instead of using an ordinary data structure, this one uses a "Dictionary". Only a few variables in the database are used repeatedly. Therefore this is a benefit. It is estimated using dictionary encoding that a single column property requires X bits of coding (which can be calculated directly from the number of unique values of the attribute). There are a few ways to do this. The first is to determine how many X-bit encoded items may be stored in one byte. One value cannot be encoded in a single byte if the property includes 32 values: three in two bytes, four in three bytes, or six in four bytes [4].

Attribute Compression. As we all know, all compression algorithms were created with data warehouses in mind, where a large quantity of data is stored, which is often made up of numerous textual properties with low cardinality. However, this part will illustrate strategies used in traditional databases such as MYSQL, SQL SERVER, and others [8].

Dimension tables with numerous rows take up a lot of space, therefore this technology's primary purpose is to reduce that space while also improving speed.

2.7 Layout of Compressed Tuples

The entire layout of a compressed tuple is shown in Fig. 1. A tuple can be made up of up to five pieces, as shown in the diagram: 1. Dictionary-based compression (or any other fixed-length compression approach) compresses all fields in the first half of a tuple [5–7]. 2. The second component uses a variable-length compression technique to keep track of all compressed fields' encoded length information. This is like the numerical compression approaches described above. 3. Third-section data includes (uncompressed) integer, double, and CHAR values, excluding VARCHARs and CHARs compressed into VARCHARs. 4. Integers, doubles, and dates are all stored as variable-length compressed values in the fourth section. The compressed size of a VARCHAR field might be included in the fourth section. The third component of a tuple has an uncompressed integer value when the size information for a VARCHAR field is not compressed. 5. Finally, the VARCHAR field's string values are saved in the tuple's fifth component (compressed or uncompressed). While the division into five discrete components may appear to be difficult, it is relatively intuitive. To begin, splitting tuples into fixed and variable sizes is a logical choice that is used by most database systems today. Because the first three tuple elements have set sizes across all tables, their sizes will be consistent. There are no additional address computations required because this data contains compression information and the value of a field. Integer, double, and date fields that aren't compressed can be read regardless of whether the rest of the data is [9].

Finally, because tiny fields' length information may be encoded in less than a byte, we distinguish between short variable-length (compressed) fields and possibly large variable-length string fields. Even yet, the length information for extensive areas is encoded in two phases. These five components are not present in every database tuple. Tuples with no compressed fields, for example, have just the third and maybe the fifth part. Remember that because all tuples in the same database are compressed using the same techniques, they all have the same structure and components.

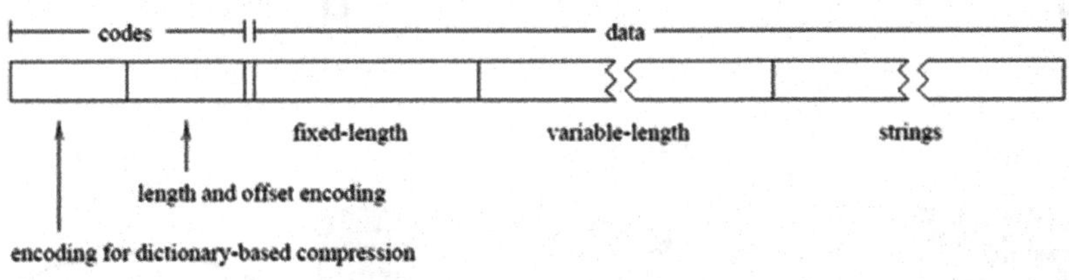

Length	NOT NULL	NULL allowed
0	—	000
1	00	001
2	01	010
3	10	011
4	11	100

Table 1: Length Encoding for Integers

Length	NOT NULL	NULL allowed
0	—	00
4	0	01
8	1	10

Table 2: Length Encoding for Doubles

Fig. 1. Layout of compressed tuple

3 Methodology

As discussed in literature section, queries are conducted on a platform that performs query rewriting and data decompression as needed by using compression techniques. The query execution is on a tiny scale that gives far better results than uncompressed queries on the same platform. This section displays the various compression methods used on the tables before comparing the results graphically and in tabular format. Figure 2 represents an OLTP process in which two queries, insert and lookup, are executed on a student table. Figure 3 depicts the OLAP access pattern, which necessitates processing a few characteristics and access to a large amount of data. In comparison to OLTP, the number of queries executed per second are less.

It's worth noting that WHERE clause queries need to be updated since selection and projection procedures don't necessitate looking for a tuple with a specific attribute.

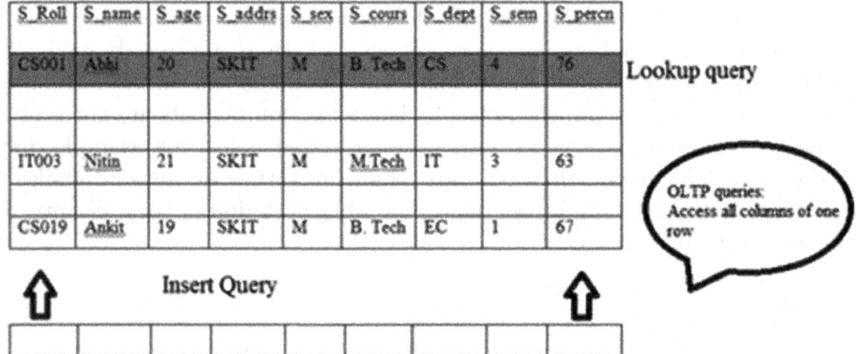

Fig. 2. OLTP access

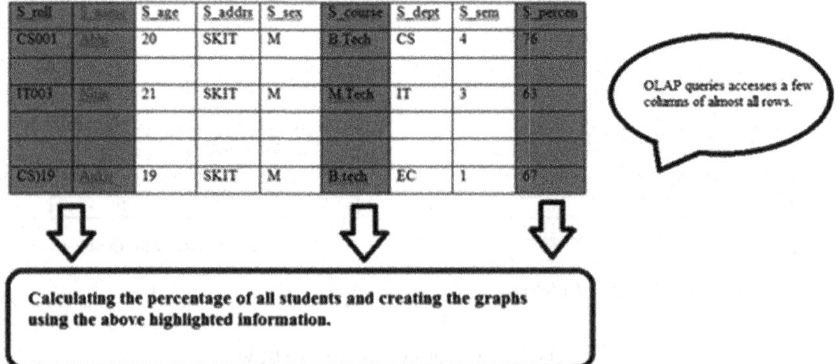

Fig. 3. OLAP access

Even though data storage technology has progressed, disc access technology has not. The speed of RAM and CPUs, on the other hand, has improved. This advancement in technology led to the introduction of data compression, which reduces the amount of data stored in exchange for a small amount of processing overhead (to compress and decompress data).

While reading data from a disc or running queries, data gets compressed. Static and dynamic implementations of compression algorithms are possible. Although compression and decompression increase execution time, the smaller amount of data that must be read/stored on discs in databases and warehouses offsets this [2].

3.1 Compression Scheme

Compression is done through the following steps:

1. Attributes are analyzed, and a frequency histogram is built.
2. Most common values are encoded as a one-byte code, whereas the least common values are encoded as a two-byte code. One byte is usually enough, however in rare circumstances two bytes aren't.
3. The codes table, as well as any relevant information, is stored in the database.
4. The property is updated, with the matching codes replacing the original values (the compressed values).

The below example of an employee table represents the compression technique:

Table 1. Employee table with type and cardinality

Attribute name	Attribute type	Cardinality
SSN	TEXT	1000000
EMP_NAME	VARCHAR(20)	500
EMP_ADD	TEXT	200
EMP_SEX	CHAR	2
EMP_SAL	INTEGER	5000
EMP_DOB	DATE	50
EMP_CITY	TEXT	95000
EMP_REMARKS	TEXT	600

Table 1 shows some of the most common characteristics of a client dimension in a data warehouse, which may be a significant dimension in many enterprises (e.g., e-business). For example, various properties, such as EMP NAME, EMP ADD, EMP SEX, EMP SAL, EMP DOB, EMP CITY, and EMP REMARKS, are suitable for coding.

Table 2 shows an example of potential codes for the EMP CITY property if we wish to code it. Binaries express the codes to be more accurate with the

Table 2. Code table example

City name	City postal code	Code
DELHI	011	00000010
MUMBAI	022	00000100
KOLKATA	033	00000110
CHENNAI	044	00001000
BANGALORE	080	00001000 00001000
JAIPUR	0141	00000110 00000110
COIMBATORE	0422	00001000 00001000 00001000
COCHIN	0484	00010000 00010000 00010000

notion. Rather than use two-byte codes to represent the 256 least frequent values (such as Delhi and Mumbai), we'll use one-byte codes to represent the 256 most common values (like Delhi and Mumbai) (e.g., Jaipur and Bangalore). The values listed in Table 2 (represented in binary) are the ones kept in the database, not the more significant integers. Instead of holding "Jaipur," 6 ASCII characters long, we just record one byte with the binary cone "00000110 00000110."

3.2 Query Execution

It is necessary to rewrite queries when the WHERE clause makes use of coded attributes for filtering. The coded values in these searches must be used to replace the values used to filter the results. Examples of query rewriting in its simplest form are shown below. The value 'JAIPUR' is replaced with the corresponding code obtained from codes Table 3.

Table 3. Query execution

Actual query	Coded query
Select EMPLOYEE.NAME	Select EMPLOYEE.NAME
From EMPLOYEETABLE	From EMPLOYEETABLE
Where EMPLOYEE_LOC = 'JAIPUR'	Where EMPLOYEE_LOC = '00000110 00000110'

3.3 Decompression

To decode the attributes, they must be selected from the query select list. There are times when it is necessary to decompress features with compressed values after the query has been run. There will be little decompression time compared to the total query execution time because data warehousing queries typically return a limited number of results.

3.4 Prerequisites

The experiments' goal is to empirically quantify the advantages in storage and performance obtained using the proposed technique. The examinations were divided into two halves. In the first phase, just category compression was used. In the second phase, we used categories compression in conjunction with descriptions compression.

4 Results and Discussions

Since the start of the research, our goal has been to make every tuple in a table like n – 1 other tuples. Identity-related traits can be used to identify people at a table. An elderly guy living in a distant region with the postcode 312345 may have his Oral cancer concern disclosed if Table 4 is released. This record has a unique postcode. Because of this, we may generalize the value of the Gender and Postcode attributes so that each tuple in the Gender, Age, and Postcode attribute sets only appears twice.

Table 4. Table contains the actual data of patients in a particular city

No.	Gender	Age	Postcode	Problem
01	Female	Old	312345	Oral Cancer
02	Male	Old	312245	Asthma
03	Female	Teenage	312783	Obesity
04	Female	Young	312434	Breast Cancer
05	Male	Child	312656	Polio

Table 5 shows the result of this generalization. Because different nations use different postal schemes, we utilize a simplified postcode scheme in which the hierarchy 312345, 312***, 31****, 3***** refers to rural, city, region, state, and unknown, respectively.

Table 5. Table contains the globally recorded data of patients in a particular city

No.	Gender	Age	Postcode	Problem
01	*	Old	312*	Oral Cancer
02	*	Old	312*	Asthma
03	*	Teenage	312*	Obesity
04	*	Young	312*	Breast Cancer
05	*	Child	312*	Polio

4.1 Identifier Attribute Set

A set of identifier attributes is a collection of qualities that may identify persons in a database. In Table 4, for example, the identifier attribute set Gender, Age, Postcode is an identifier attribute set.

4.2 Equivalent Set ϵ

It is referred to as an equivalent set of a table if all the tuples have the same value for the attribute set. A similar collection of qualities may be seen in Table 4 for the characteristics of gender and age and postcode and problem. Because two characteristics are utilized to identify the table contains the globally recorded data of patients in a particular city. Table 5 is the 2-Anonymity view of Table 4.

The results simply illustrate that as the comparable set grows, compression becomes simpler, and as a result, the cost of Anonymization increases as the equivalent set grows. We can deduce Eq. 1 from this hypothesis.

$$CAVG = \frac{\Sigma RECORDS}{\Sigma \epsilon} \tag{1}$$

4.3 Domain Compression Through Binary Conversion

We can deduct from the study that the larger the equivalent set, the easier it is to compress. The cost of Anonymization is a factor of the equivalent set.

Encoding of Distinct Values. This compression strategy assumes that the table we've published has a small number of different domains of attributes and that these values recur across the database's massive number of tuples. As a result, binary encoding of each attribute's other matters, then representing the tuple values in each column of the relation with the corresponding encoded values, would compress the entire connection [1].

We'll count how many different values are in each column and encode the data in bits accordingly. Consider the example below, which depicts the two primary characteristics of a connection Patients. We may now translate the existing realm of attributes to a broader domain by using the N-Anonymization with global recording approach. For example, as illustrated in Table 7, age may be translated into a 10-year interval.

Table 6. Patient table instance

Age group	Common problems
10	Polio
20	Depression
30	Obesity
50	Cancer

Assume age is of integer type and has 5 unique values, as shown in Table 6, to investigate the compression gains obtained by this strategy. If there are 50 patients, the Age property will require total storage of 50 * size of (int) = 50 * 4 = 200 bytes. We discovered that there are 9 unique values for age using our compression approach. Thus, we'll need the upper bound of log (9), i.e., 4 bits, to represent each data value in the Age field. It's simple math to figure out that we'd need 50 * 4 (bits) = 200 bits = 25 bytes, which is a lot less.

Table 7. Stage 1 compression for patient table

Age group	Common problems
10 to 19	Polio
20–29	Depression
30–49	Obesity
50 above	Cancer

This is the first stage of compression, and it simply converts one column into bits as per Table 8. The outcome will be considered if we apply this compression to all the table's columns.

Table 8. Binary compression for patient table

Age group	Common problems
00	Polio
01	Depression
10	Obesity
11	Cancer

4.4 Paired Encoding

The following example shows that, in addition to reducing redundancy (repetition values), the encoding mentioned above approach also helps to reduce the memory needs of the relations. That is, in addition to the distinct values of column 1 or column 2, it is likely that there are a few values of even (column 1, column 2) joined together. After that, the two columns are combined, and the pair values are translated in accordance with the encoding result.

Using the bit-encoded database from Stage 1 as input, compression proceeds in Stage 2, where columns are coupled in pairs of two using the distinct-pairs technique. Combining the 'Age' and 'Problem' columns to see how much more compression is feasible. Our upper bound is around log (5) = 2 bits since there are five possible pairings in Table 9 (10, Polio), (20, Depression), (30, Obesity), (50,

Cancer). Table 9 shows the results of stage 2 compression. After the feature has been compressed, the characteristics are paired or coupled, as shown in Table 10. Similarly, all the columns are linked in two-by-two pairings. It's easy if the table has an even number of columns. Even if the columns aren't aligned properly, we can pick all of them to be uncompressed intelligently. We can simply compute the needed amount of space after applying this compression strategy as below:

Before compression: 5 * (4) + 4 * (4) = 36 bytes

After Compression and coupling: 4 * 2 = 8 bits.

Table 9. Stage 2 compression for Patient table

Age group	Common problems
00	000
01	001
10	010
11	011

Table 10. Coupling

Age group- Common problems
00000
01001
10010
11011

4.5 Add-ons to Compression

Some of the findings were produced by altering the coupling of characteristics after successful compression over relation and domains. The following points demonstrate some of these options.

Functional Dependencies. Functional dependencies exist between attributes and states that: "If and only if each value of X is connected with at most one value of Y, a set of characteristics Y in R is said to be functionally reliant on another attribute X." This means that the value of characteristics in set X may be determined by the value of attributes in set Y.

We discover that grouping columns with relationships like functional dependencies produce better compression results by rearranging the characteristics. Table 11 shows an example of functional dependencies-based compression. The amount of compression was checked using two different test situations. As illustrated in Table 12, the test case links the characteristics (name, age) (Gender, problem) and then checks individual and linked distinct values. Coupling is done with the supplied features (name, gender) (Age, Problem) in test scenario 2.

Table 11. Representing functional dependency-based coupling

Name	Gender	Age	Problem
Harshit	M	10	Oral Cancer
Naman	M	20	Asthma
Aman	M	30	Obesity
Rajiv	M	50	Breast Cancer

Table 12. Represents the number of distinct values in each column

Column name	Distinct values
Name	19
Gender	2
Age	19
Problem	19

Few Distinct Tuples. Occasionally, a database will include columns with only a few specific values. For example, the domain of the Gender property will always be either male or female. As a result, it is advised that such traits be combined with those with many different values. Consider the four qualities name, gender, age, and problem, where name is 200, gender is 2, age is 200, and the issue is 20.

Consider the coupling, gender, name and age, problem. The result would be
200 * 2 + 200 * 20 = 4400 distinct tuples.

Whereas coupling gender, problem and name, age. The result would be
2 * 20 + 200 * 200 = 40040 distinct tuples.

5 Conclusion

In this paper, we look at ways to increase database speed by using compression techniques. We also present an approach for compressing columnar databases after comparing the results. We looked at the following study topics:

- Compression of different domains of databases: We looked at how different database domains like varchar, int, and NULL values may be dealt with while compressing a database. Unlike other compression methods, our method considers the diverse nature of string characteristics and employs a thorough strategy to select the most efficient encoding level for each string property. Our tests demonstrate that utilizing HDE techniques leads to a higher compression ratio than using a single current method and the optimal mix of I/O savings and decompression overhead.
- Compression-aware query optimization: When it comes to query speed, we've found that knowing when to decompress string properties is critical. Traditional optimizers may not always offer effective plans when combined with

a cost model that incorporates compression's input/output advantages and the CPU expense of decompression. The research findings show that query performance depends on a mix of effective compression technologies and compression-aware query optimization. Our compression and optimization methods outperform those of our competitors by a factor of ten. These improved results suggest that a compressed database's query optimizer must be tuned for better performance.
- Compressing query results: We advocated that domain information about the query increases the compression effect on query results. We used a variety of compression approaches in our approach, which we expressed using an algebraic framework.

6 Future Work

This research has a lot of potential applications in the future.

- Compression-aware query optimization: Start by investigating how intermediate (decompressed) results might be stored to reduce the costs of transitory decompression. This is important. Second, we intend to analyze how our compression approaches will cope with upgrades. Third, we'll investigate how hash join affects our query optimization efforts.
- Result compression: We intend to look at the problem of query and compression plan optimization together. Currently, the query plan given by query optimization is used for compression optimization. The total cost of a query plan plus a compression plan, on the other hand, is not the same as the cost of the query plan alone. For example, a more costly query plan may sort the results so that the sorted-normalization approach may be used, resulting in a reduced total cost.

References

1. Liu, Y., et al.: DCODE: a distributed column-oriented database engine for big data analytics. In: Khalil, I., Neuhold, E., Tjoa, A.M., Da Xu, L., You, I. (eds.) ICT-EurAsia 2015. LNCS, vol. 9357, pp. 289–299. Springer, Cham (2015). https://doi.org/10.1007/978-3-319-24315-3_30
2. Abadi, D.J., Madden, S.R., Hachem, N.: Column-stores vs. row-stores: how different are they really? In: Proceedings of the 2008 ACM SIGMOD International Conference on Management of Data, SIGMOD 2008, pp. 967–980. Association for Computing Machinery, New York (2008). https://doi.org/10.1145/1376616.1376712
3. Bhagat, V., Gopal, A.: Comparative study of row and column oriented database. In: 2012 Fifth International Conference on Emerging Trends in Engineering and Technology, pp. 196–201 (2012). https://doi.org/10.1109/ICETET.2012.56
4. Chernishev, G.: The design of an adaptive column-store system. J. Big Data **4**(1), 1–21 (2017). https://doi.org/10.1186/s40537-017-0069-4
5. Davis, K.C.: Teaching database querying in the cloud. In: 2019 IEEE Frontiers in Education Conference (FIE), pp. 1–7 (2019). https://doi.org/10.1109/FIE43999.2019.9028440

6. Dwivedi, A.K., Lamba, C.S., Shukla, S.: Performance analysis of column oriented database versus row oriented database. Int. J. Comput. Appl. **50**(14), 0975-8887 (2012)
7. Ghane, K.: Big data pipeline with ML-based and crowd sourced dynamically created and maintained columnar data warehouse for structured and unstructured big data. In: 2020 3rd International Conference on Information and Computer Technologies (ICICT), pp. 60–67 (2020). https://doi.org/10.1109/ICICT50521.2020.00018
8. Mademlis, I., Tefas, A., Pitas, I.: A salient dictionary learning framework for activity video summarization via key-frame extraction. Inf. Sci. **432**, 319–331 (2018)
9. Mahrishi, M., Morwal, S.: Index point detection and semantic indexing of videos—a comparative review. In: Pant, M., Kumar Sharma, T., Arya, R., Sahana, B.C., Zolfagharinia, H. (eds.) Soft Computing: Theories and Applications. AISC, vol. 1154, pp. 1059–1070. Springer, Singapore (2020). https://doi.org/10.1007/978-981-15-4032-5_94
10. Mahrishi, M., Shrotriya, A., Sharma, D.K.: Globally recorded binary encoded domain compression algorithm in column oriented databases. Glob. J. Comput. Sci. Technol. (2012). https://computerresearch.org/index.php/computer/article/view/417
11. de Moura Rezende dos, F., Holanda, M.: Performance analysis of financial institution operations in a NoSQL columnar database. In: 2020 15th Iberian Conference on Information Systems and Technologies (CISTI), pp. 1–6 (2020). https://doi.org/10.23919/CISTI49556.2020.9140981
12. Mehra, R., Puram, B., Lodhi, N., Babu, R.: Column based NoSQL database, scope and future. Int. J. Res. Anal. Rev. **2**, 105–113 (2015)
13. Stonebraker, M., et al.: C-store: a column-oriented DBMS. In: Proceedings of the 31st International Conference on Very Large Data Bases, VLDB 2005, pp. 553–564. VLDB Endowment (2005)
14. Vieira, J., Bernardino, J., Madeira, H.: Efficient compression of text attributes of data warehouse dimensions. In: Tjoa, A.M., Trujillo, J. (eds.) DaWaK 2005. LNCS, vol. 3589, pp. 356–367. Springer, Heidelberg (2005). https://doi.org/10.1007/11546849_35
15. Westmann, T., Kossmann, D., Helmer, S., Moerkotte, G.: The implementation and performance of compressed databases. SIGMOD Rec. **29**(3), 55–67 (2000). https://doi.org/10.1145/362084.362137
16. Zhou, X., Ordonez, C.: Computing complex graph properties with SQL queries. In: 2019 IEEE International Conference on Big Data (Big Data), pp. 4808–4816 (2019). https://doi.org/10.1109/BigData47090.2019.9006312

The Impact and Challenges of Covid-19 Pandemic on E-Learning

Devanshu Kumar[1], Khushboo Mishra[2], Farheen Islam[3], Md. Alimul Haque[1](✉), Kailash Kumar[4], and Binay Kumar Mishra[2]

[1] Department of Computer Science, Veer Kunwar Singh University, Ara 802301, India
shadvksu@gmail.com
[2] Department of Physics, Veer Kunwar Singh University, Ara 802301, India
[3] Department of Education, Patna Women's College, Patna, India
[4] College of Computing and Informatics, Saudi Electronic University, Riyadh, Kingdom of Saudi Arabia

Abstract. COVID-19 has caused havoc on educational systems around the globe, impacting over 2 billion learners in over 200 countries. University, college, and other institutional facility cutbacks have impacted more than 94% of the world's largest population of students. As a result, enormous changes have occurred in every aspect of human life. Social distance and mobility restrictions have significantly altered conventional educational procedures. Because numerous new standards and procedures have been adopted, restarting classrooms once the restrictions have been withdrawn seems to be another issue. Many scientists have published their findings on teaching and learning in various ways in the aftermath of the COVID-19 epidemic. Face-to-face instruction has been phased out in a number of schools, colleges, and universities. There is concern that the 2021 academic year, or perhaps more in the future, will be lost. Innovation and implementation of alternative educational systems and evaluation techniques are urgently needed. The COVID-19 epidemic has given us the chance to lay the groundwork for digital learning. The goal of this article is to provide a complete assessment of the COVID-19 pandemic's impact on e-learning and learning multiple papers, as well as to suggest a course of action. This study also emphasises the importance of digital transformation in e-learning as well as the significant challenges it confronts, such as technological accessibility, poor internet connectivity, and challenging study environments.

Keywords: COVID–19 · Online education · e-Learning experience · Technological challenges · Smart classes · University and College students

1 Introduction

The universal COVID-19 pandemic has spread throughout the world, hitting nearly all nations and regions. The epidemic was initially found in Wuhan, China, in December 2019. Countries all over the world have issued warnings to their citizens to be cautious. Lockdown and stay-at-home strategies have been implemented as the necessary steps

V. E. Balas et al. (Eds.): ICETCE 2022, CCIS 1591, pp. 560–572, 2022.
https://doi.org/10.1007/978-3-031-07012-9_47

to flatten the curve and reduce the spread of the virus [32]. In the interim, mobility was permitted, offices reopened, and schools and colleges reopened for chosen levels while continuing with online classes for others. The school and colleges shutdown have impacted around 10 lakhs students in India. The effect has been far-reaching, disrupting education throughout the academic session and possibly even more in the subsequent months. Face-to-face instruction has been phased out in a number of schools, colleges, and universities. Substitute instructional and assessment techniques must be developed and implemented quickly. The COVID-19 epidemic has given us the chance to lay the groundwork for ICT learning [9]. ICT can provide the better solutions to reduce the challenges of conventional approaches like Online classes, distance learning and virtual campuses for universities. We shall experience a digital transformation in schools and college education as a result of the COVID-19 pandemic, which will be enforced via video conferencing, online classes, online examinations, and communications in digital worlds [13, 33]. Since 1960s, European universities have been using e-learning technologies. Digital education is a process that allows professors to deliver educational content to their students by using online media, web services, or other data communications media. E-learning is the technological evolution of the traditional classroom setting and material [29].

1.1 Advantages and Limitations of E-Learning

A. Advantages of E-Learning
Compared with traditional learning, e-Learning offers the following advantages:

Accessibility
E-learning-based skills are developed according to learning progression, offering learning wherever and anytime and facilitating networking connections. The network manager has no trouble managing a session with a high percentage of students.

Price and Choice
There has not been a high cost for a session. To fulfill the expanding demands for learning of all societies, learning can also be chosen according to individual requirements and ambitions.

Flexibility
Students may not have to study all contents while taking a new learning course. Therefore in this situation, learning development can be speeded up. Learning programs may also be routinely and rapidly updated.

B. Limitations of E-Learning
Although e-Learning has many advantages, this mode of learning has some drawbacks, as follows:

- E-learning might demand that students work alone and be conscious of themselves. Furthermore, learners want to present their capacity to cooperate effectively with professors and other Fellows and discuss things through the network.
- Students must also understand how to create a solution that outlines their specific requirements and how to apply what they have planned.

- E-learning could be less effective for students whose technical abilities are inadequate.
- In other words, a significant influence on advancement and learning outcomes is being made by technical facilities (web, speed, price…).
- E-learning could also be misled by the potential to copy and paste, pirated and duplicated, which are susceptible to weak selection skills.
- E-learning as an educational approach allows students to be solitary, distant and lacking in connection and relationships. It consequently takes strong inspiration and time management abilities to reduce these consequences.

1.2 Types of E-Learning

Online Learning
This is a process of teaching in which a program is completely carried out using a Learning Management System in the communication network. Although, E-learning uses the benefits of e-learning exclusively and ignores the virtues of classroom learning. Online learning is classified into two modes, i.e. Synchronous Learning and Asynchronous learning.

In the Synchronous learning mode, the instructor and learner, both involved in the Learning Management System. On the other hand, in Asynchronous Learning, students and instructors involved in the Learning Management System at any time.

Blended/Hybrid Learning
This is a mode of learning in which a course is delivered using both online and physical learning. E-learning helps the teaching-learning process and offers only the most relevant material and concepts. The rest of the content is still supplied individually and to the maximum extent possible. These two forms of learning are pretty well planned, with strong and complementary connections to ensure the goal of ensuring the education standards.

Below are the two basic categories of blended/hybrid courses.

- Blended Classroom Course: A traditional classroom is used for a slightly larger percentage of the course.
- Blended Online Course: A more significant component of the course is delivered via the Internet.

According to the study, several problems exist, such as a lack of virtual learning apparatus, a lack of experience for educators to online teaching, a communication, learners haven't smart home with IoT and AI based E – Learning equipment, equality, and poor academic performance in university system [12, 14]. This paper explained the worldwide effect of the COVID-19 pandemic on the academic activities. During the COVID-19 pandemic, the difficulties and potential of E - Learning and continuing education are reviewed, and a path ahead is offered.

This paper is classified as follow. Section 2 will describe the theoretical background. Section 3 shows the impact of COVID – 19 on education. Section 4 presents the challenges and Sect. 5 highlights opportunities of teaching and learning during pandemic. Section 7 concludes with discussion.

2 Literature Review

Last two to three decades, scientists have analyzed the study on the use of ICT in education in both developed and developing countries.

The article [30] explained that underdeveloped countries do not have complete access to digital technology. It also has to deal with physical and behavioral issues. Author described that support the use of ICT in E – Learning because it enhances and expands learning possibilities for learners [21]. This article also explained the need of providing instructors with a basic knowledge of technology as a teaching skill. According to author [22], there is sufficient proof that ICT improves e-learning outcomes. If interactive learning methods are followed, it's achievable. ICT approaches in e-learning, according to author [20] act as a hindrance to professionally developing educators, and teacher training courses should be set up to tackle this. According to author [26], utilizing ICT in classrooms motivates children to study, resulting in better learning results. According to the report, technology helps pupils stay current and adapt quickly. Author [8] highlights the need for academic institutions throughout the world to reconsider and update their strategic plan in order to adapt to advances in technology. To overcome this difference, i.e., the gap between conventional and advanced working techniques, a significant amount of time and effort would be necessary.

According to author [34], countries that employ technological tools in education have an edge over those that do not. According to [3], learners like and seek for a rich digital world. According to the study, university students were angered by the lack of access to laptops and other tools, whereas science students were provided with all these facilities. According to author [6, 7], organizational change and leadership transformation occur as a result of the adoption of digitization in the concerned industry. According to [19], digital learning and education programs increase the pace and educational standards. It states that digital teaching programmes assist to speed up the process of knowledge dissemination by making it more organized and adaptable, as well as providing pupils with a higher cognitive understanding. According to [19, 25], study is being done by various paper, E-Learning system enhanced with the help of various ML algorithms like end semester evaluation or assessment.

According to [11], there is still a massive disparity in the integration of digital technology and resources in Europe's university environment. As described in [4], governments should work with numerous stakeholders, corporations, and individuals that agree in implementing digital into many areas, especially education, so that everyone advantages. The influence of digitalization on the education industry is discussed in the study paper [5]. Based on secondary data, the research employs a descriptive study technique. According to the report, instilling digitalization in India's education system will bring it up to speed with global institutions. As author described in [18], higher education institutions must strive quickly to adhere to the adjustments necessary, particularly in the present Covid-19 scenario, while maintaining educational standards. Any resistance to change will have an impact on the current and future education system and will be damaging to academia. According to [15], learners with hearing impairments have problems when studying online. The authors of the article [23, 27, 31] expressed concern that the epidemic is already causing worry among students. It has long been recognized that the availability of ICT has always boosted learning efficacy. In addition, teaching method based on such techniques promotes practicality and adaptability, resulting in a richer educational environment [1, 2, 35].

In light of previous studies, it is critical to determine how this epidemic has driven us to embrace the notion of e - learning and to identify the issues that have arisen as a result. In addition, the study will assist us in determining whether or not these arrangements should be extended in the near future.

3 Deterioration and Impact on Educational System Due to COVID – 19 Pandemic

Understanding the impact of e-learning on academic performance as well as the societal implications of continuing to provide such a type of education is essential. Many scholars have studied the impact of online learning on academia and determined that it has a number of benefits, including ensuring educational continuation, ensuring ongoing learning, and reducing the extra expenditures associated with traditional education. Since the educator and the student were in various locations, restrictions such as instructional techniques, schedule, and timing have remained.

The effect didn't seem to stop with the education sector; it also had an effect on learners' learning experiences when it came to obtaining research and study materials. In Turkey, Hebebci carried out an investigation to see what lecturers and learners felt of the COVID-19 pandemic's online classes technologies [16]. According to the survey, 42.9% of those expressed that online educational students are struggling with team projects owing to a lack of on-campus interaction. He thoroughly expressed the benefits and drawbacks of E-learning; he stated that digital classes can save money and time and provide flexibility. But some demerit like greater risk of interruption, the use of sophisticated technology, the lack of social connection, the challenge in remaining in connection with teachers, and the fact that online certificates are not accepted by employment international market.

Figure 1 presented the critical challenges and factors of E-learning system during COVID-19 pandemic. UNICEF data show as in Fig. 2 that COVID -19 impact on education very badly. Approximately 60% of governments implemented digital E-Learning strategies for pre-primary schooling, while 90% of nations had done e-learning setup earlier. Internationally, three out of every four pupils who may not be accessible by e - learning programs seem to be from remote regions and/or originate from the poorest families.

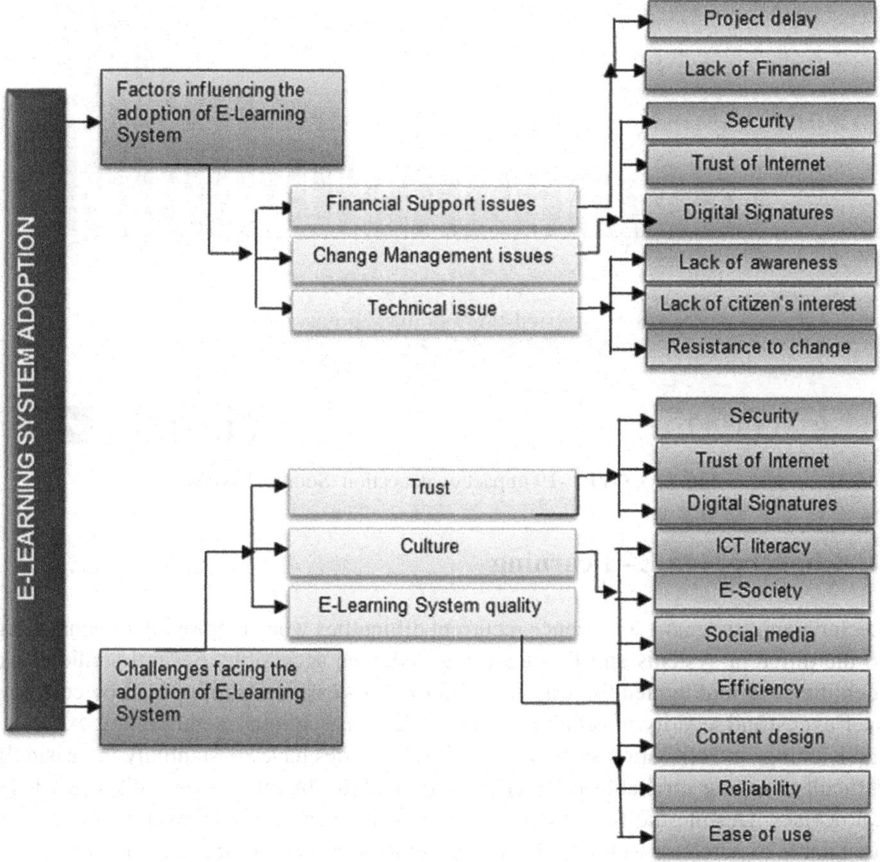

Fig. 1. The critical challenges and factors of E-learning system during COVID-19 pandemic

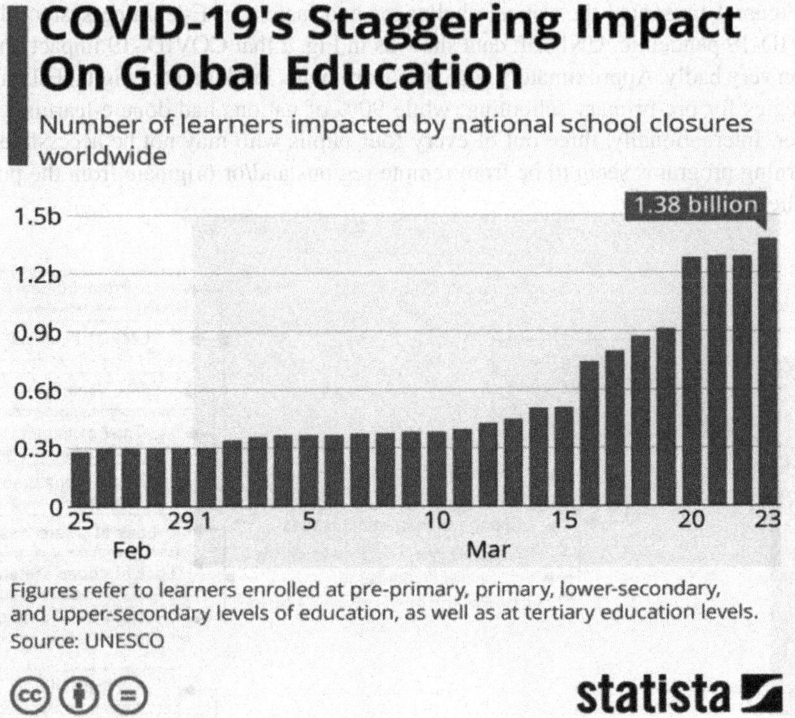

Fig. 2. COVID -19 impact on education. Source: UNESCO

4 Challenges in E - Learning

Lecturer and students—experience recurrent difficulties when employing or connecting to the different systems and E – Learning materials accessible. Several studies have identified and highlighted the various challenges: E-learning is influenced by concerns such as cost and security, affordability, adaptability, instructional methodology, experiential learning, and education system [10]. Most countries have substantially increasingly difficulty keeping a reliable Internet connection and obtaining pace with technological devices. Despite the fact that so many poor learners in underdeveloped countries are unable to purchase web-based learning equipment. Face-to-face activities and self-exploratory learning have become increasingly big challenges for learners as a result. Especially, those students whose performance is overall good in the classroom, face several obstacles during online education due to money. Real workspaces that are suitable to different types of learners have various challenges. Essentially motivated students have fewer problems in the classroom because they require minimum supervision and direction, whereas students who struggle with learning face difficulties. Due to the parent's financial condition, many highly qualified sstudents have not attended or access online courses in the pandemic. As a result, students' academic performance is deteriorating day by day, and its effect on internal and as well as semester examination [32]. Most of the time, students are assessed for plagiarism and other issues during their exams, which can be challenging and confusing. The various methods used for organizing exams vary depending on the expertise of the teachers. Due to pandemic schools and colleges has been shut down and it is not only affected internal examinations but also affected the main public qualifications examination in the entire country. Depending on how long the lockdown lasts, postponing or canceling the entire examination evaluation is a realistic prospect. (United Nations, 2020). Due to the outbreak of the COVID-19 epidemic, various examinations, including the state-level board examinations, were canceled across India. Several entrance test suspended or delayed due to this pandemic. The current crisis has had a significant influence on the quality education in the country. According to a research conducted in France, the abandonment of conventional examination processes in 1968 as a result of student riots had favorable long-term labor market implications for the afflicted group [24, 28].

Aside from being enjoyable for the youngsters, school time facilitates positive knowledge and competencies. While children are away from the real academic calendar, they face financial, emotional, and psychological effects. Most of these students are now taking online classes and spending a lot of time on web sites, putting them at risk of online abuse. Children have been associated with potential hazardous and aggressive information, as well as an increased chance of cyberbullying, as a result of excessive and unsupervised time spent on social networking sites. So much parents are depend on digital and technological alternatives to keep their young kids involved in the learning process, entertained, and associated from the outside world, and not all students have the right information, abilities, and funds to support themselves secure online. In India, the number of online learners come from remote areas where their families are relatively uneducated farmers. Pupils support their families with agricultural tasks such as farming, livestock care, and home duties. Numerous pupils even asked that their exams be moved to the afternoon since they wanted to work on the farms in the early morning. Several students mentioned that they were expected to attend to their sick relatives, grandmother, or relatives, including transporting them to clinics. When they go home at night, this became tough for them to stay up with the lessons. In addition to inadequate Internet access, the vast majority of children don't have such connection to mobile phones or computers at residence. Due to the closure of businesses and workplaces, a large segment of the population has no or little earnings. The data plan (charges) are quite expensive in comparison to the low income earned, and uninterrupted Internet connection is a pricey enterprise for the agrarian society. Most people prefer online face-to-face lessons (video); nevertheless, some students (particularly those from low-income families) have complained that the face-to-face online class requires additional data plans. Instructors are confused as to who to listen to and which tools to use.

5 Opportunities for Educators and Learners

The COVID-19 epidemic presented many individuals with the opportunity to develop a learning system. It has strengthened the bond between teachers and parents. For the first time, online platforms such as Google Classroom, Zoom, virtual learning environments, and social networking sites, as well as other community forums such as WhatsApp, Facebook, Viber, and WeChat, are being researched and tested for education - learning in order to maintain education. Even once face-to-face teaching begins, this may be studied more, and these channels can give more materials and guidance to the students. Professors must design innovative efforts to help overcome the constraints of digital education. Educators are actively cooperating with each other at the local level to enhance online teaching approaches. As instructors, parents, and students have similar experiences, there are unmatched chances for collaboration, innovative solutions, and readiness to learn from others and try new methods. [17]. Due to the increasing popularity of e-learning, various educational institutions have made their technology available for free. This allows teachers and students to experience and learn in a more dynamic and exciting manner.

6 Issues that May Cause E – Learning to Failing

There are various issues that may causes e-learning to underperform or fail such as "Technological Constraint as shows in Fig. 3", "Distractions as shown in Fig. 4", "Instructor's Incompetency" as presented in Fig. 5, "Learner's Inefficacy as shown in Fig. 6" and "Health Issues as shown in Fig. 7" in E - Learning.

Themes	Criteria	Example
Digital divide	Gap between the learners who have ready access to internet and computers and those who don't	"not all rural students have the privilege of internet and laptop"
Data limit	Insufficient data pack to access the material/to attend the class	"online classes consume large amount of data which is difficult to afford"
Poor connectivity	Interrupted internet supply that makes the learners difficult to learn	"dragging of classes due to network problem"
Issues with the device	Lack of device or device incompatibility to the applications used for online classes	"Unavailability of gadgets with some of the students"
Non Recordable videos	Online classes that cannot be downloaded or recorded for future	learning "videos that cannot be recorded or watched later pose difficulty in learning"
Technical issues	Low quality audio or video; low bandwidth	"lack of voice clarity and poor signal strength"
Virtual presence only	No face to face interaction between the learners and teachers	"only one-way communication and no scope for interaction"

Fig. 3. Technological Constraint

Themes	Criteria	Example
Poor learning environment	Lack of congenial learning environment	"home environment is not suitable for learning as it leads to lot of disturbances from children and relatives"
Noise	Distractions that deviate learner from learning	"two-way communication is loathsome as the voices from all the sides are raising"

Fig. 4. Distractions

Themes	Criteria	Example
Technophobia	Instructors fear of handling ICT's	"Lack of technical expertise of the teacher"
Poor teaching skills	Inability of instructor to render the subject matter	"prolonged monotonous lectures with improper explanations"
Unstructured content	Curriculum which is designed improperly lacking a clarity in course objective	"poorly designed content where classes are held for name sake"
No follow up	Instructor doesn't take the feedback/online exams nor addresses their queries	"doubts cannot be cleared effectively compared to classroom environment"

Fig. 5. Instructor's incompetency

Themes	Criteria	Example
Indiscipline	Irresponsible and unorganised behaviour of learner	"lack of discipline as no one is there to control the students"
Student attritions	Reduction in the number of learner's attending the classes	"Poor attendance and students miss the classes without any reason"
Unmotivated	Learner does not possess interest or enthusiasm in learning	"Lack of interest and laziness among the students to learn during vacations"

Fig. 6. Learner's inefficacy

Themes	Criteria	Example
Strain	Difficulty in concentration due to the harmful radiations from device used for online classes	"prolonged usage of mobile phone for classes causes headaches due to harmful rays"
Worsening of existing health issues	Deteriorating the already existing health issues due to prolonged online classes	"continuous classes lead to worsening of migraine and backache"

Fig. 7. Health issues

7 Discussion and Conclusion

Although numerous investigations have been conducted on the impact on the impact of the COVID-19 pandemic on E-learning over all worldwide. The lockdown of COVID-19 has raised the major challenges to school, colleges and university academic activities. Although many learners use digital learning tools, several experience significant E-learning obstacles like Internet access problems, specialized study area, mobile phone for joining classes online, and the feeling of anxiety. According to this study, the great number of students had never taken an online course prior to the epidemic. Surprisingly, over half of the learners have spent less time s than they did before the epidemic. In this current situation more research and study required for effective pedagogy for online teaching and learning.

The cost and scalability of teaching resources for all students from all socioeconomic levels has been recognized as a hurdle, prompting instructional tool developers to place an emphasis on customization. Involvement from the administration is also essential. Given the current situation, educational standards throughout the globe, including India, must spend in E-Learning, particularly in Information and Communication Technology and successful pedagogy. Developing E – learning tools more attractive, creative and learner's friendly is also the new and interesting another research area. It would help and preparing the educational system for future uncertainty. The COVID-19 epidemic has shown us that instructors and students/learners should be educated on how to use various online educational technologies. Teachers and students will return to normal lessons when the COVID-19 epidemic has passed. E-Learning has the potential to convert remote educational experiences into more creative and successful learning environments. Finally, this study explore the experience of teachers and students in the context of E- learning. And also focuses on the challenges, e-learning failing factors and future prospects of education with useful recommendation for the successful e-learning technology.

References

1. Al-Rahmi, A.M., et al.: The influence of information system success and technology acceptance model on social media factors in education. Sustainability **13**(14), 7770 (2021)
2. Al-Rahmi, W.M., et al.: Big data adoption and knowledge management sharing: an empirical investigation on their adoption and sustainability as a purpose of education. IEEE Access **7**(2019), 47245–47258 (2019)
3. Beauchamp, G., Parkinson, J.: Pupils' attitudes towards school science as they transfer from an ICT-rich primary school to a secondary school with fewer ICT resources: does ICT matter? Educ. Inf. Technol. **13**(2), 103–118 (2008)
4. Bejinaru, R.: Assessing students' entrepreneurial skills needed in the knowledge economy. Manag. Mark. **13**(3), 1119–1132 (2018)
5. Bloomberg, J.: Digitization, digitalization, and digital transformation: confuse them at your peril. Forbes (2018). Accessed 28 Aug 2019
6. Bratianu, C.: A new perspective of the intellectual capital dynamics in organizations. In: Identifying, Measuring, and Valuing Knowledge-Based Intangible Assets: New Perspectives, pp. 1–21. IGI Global (2011)

7. Brătianu, C., Anagnoste, S.: The role of transformational leadership in mergers and acquisitions in emergent economies. Manag. Mark. **6**(2) (2011)
8. Brătianu, C.: Foreword. The knowledge economy: the present future. Manag. Dyn. Knowl. Econ. **5**(4), 477–479 (2017)
9. Dhawan, S.: Online learning: a panacea in the time of COVID-19 crisis. J. Educ. Technol. Syst. **49**(1), 5–22 (2020)
10. Godber, K.A., Atkins, D.R.: COVID-19 impacts on teaching and learning: a collaborative autoethnography by two higher education lecturers. Front. Educ. 291 (2021)
11. Mahrishi, M., Sharma, G., Morwal, S., Jain, V., Kalla, M.: Data model recommendations for real-time machine learning applications: a suggestive approach (chap. 7). In: Kant Hiran, K., Khazanchi, D., Kumar Vyas, A., Padmanaban, S. (ed.) Machine Learning for Sustainable Development, pp. 115–128. De Gruyter, Berlin (2021). https://doi.org/10.1515/9783110702514-007
12. Haque, Md.A., Haque, S., Sonal, D., Kumar, K., Shakeb, E.: Security enhancement for IoT enabled agriculture. Mater. Today Proc. (2021). https://doi.org/10.1016/j.matpr.2020.12.452
13. Haque, M.A., Sonal, D., Haque, S., Nezami, M.M., Kumar, K.: An IoT-based model for defending against the novel coronavirus (COVID-19) outbreak. Solid State Technol. **2020**, 592–600 (2020)
14. Haque, S., Zeba, S., Haque, Md.A., Kumar, K., Basha, M.P.A.: An IoT model for securing examinations from malpractices. Mater. Today Proc. (2021). https://doi.org/10.1016/j.matpr.2021.03.413
15. Harsha, R., Bai, T.: Covid-19 lockdown: challenges to higher education. Cape Comorin **2**(4), 26–28 (2020)
16. Hebebci, M.T., Bertiz, Y., Alan, S.: Investigation of views of students and teachers on distance education practices during the coronavirus (COVID-19) pandemic. Int. J. Technol. Educ. Sci. **4**(4), 267–282 (2020)
17. Hollweck, T., Doucet, A.: Pracademics in the pandemic: pedagogies and professionalism. J. Prof. Cap. Commun. (2020)
18. Jadhav, S.J., Raghuwanshi, S.V.: A study on digitalization in education sector. Int. J. Trend Sci. Res. Dev. 43–44 (2018). https://doi.org/10.31142/ijtsrd18667. Special Is, Special Issue-ICDEBI 2018
19. Jaleel, S.: A study on the metacognitive awareness of secondary school students. Univers. J. Educ. Res. **4**(1), 165–172 (2016)
20. Jung, I.: ICT-pedagogy integration in teacher training: application cases worldwide. J. Educ. Technol. Soc. **8**(2), 94–101 (2005)
21. Kelly, M.G., McAnear, A.: National Educational Technology Standards for Teachers: Preparing Teachers to Use Technology. ERIC (2002)
22. Laird, T.F.N., Kuh, G.D.: Student experiences with information technology and their relationship to other aspects of student engagement. Res. High. Educ. **46**(2), 211–233 (2005)
23. Manzoor, A., Ramzan, Q.: Online teaching and challenges of COVID-19 for inclusion of persons with disabilities in higher education. Dly. Times (2020)
24. Mahrishi, M., Sharma, P., et al. (eds.): Machine Learning and Deep Learning in Real-Time Applications. IGI Global (2020). https://doi.org/10.4018/978-1-7998-3095-5
25. Kumar, K., Singh, N.K., Haque, Md.A., Haque, S.: Digital Transformation and Challenges to Data Security and Privacy. IGI Global (2021). https://doi.org/10.4018/978-1-7998-4201-9
26. Mishra, R.C.: Women Education. APH Publishing (2005)
27. Mahrishi, M., Morwal, S., Muzaffar, A.W., Bhatia, S., Dadheech, P., Rahmani, M.K.I.: Video index point detection and extraction framework using custom YoloV4 darknet object detection model. IEEE Access **9**, 143378–143391 (2021). https://doi.org/10.1109/ACCESS.2021.3118048

28. Murgatroyd, S.: The precarious futures for online learning
29. Pokhrel, S., Chhetri, R.: A literature review on impact of COVID-19 pandemic on teaching and learning. High. Educ. Futur. **8**(1), 133–141 (2021)
30. Sahoo, B.P., Gulati, A., Haq, I.U.: Covid 19 and challenges in higher education: an empirical analysis. Int. J. Emerg. Technol. Learn. **16**(15) (2021)
31. Sayaf, A.M., Alamri, M.M., Alqahtani, M.A., Al-Rahmi, W.M.: Information and communications technology used in higher education: an empirical study on digital learning as sustainability. Sustainability **13**(13), 7074 (2021)
32. Sintema, E.J.: Effect of COVID-19 on the performance of grade 12 students: implications for STEM education. Eurasia J. Math. Sci. Technol. Educ. **16**(7), em1851 (2020)
33. Sonal, D., Pandit, D.N., Haque, Md.A.: An IoT based model to defend Covid-19 outbreak. Int. J. Innov. Technol. Explor. Eng. **10**(7), 152–157 (2021). https://doi.org/10.35940/ijitee.G9052.0510721
34. Stensaker, B., Maassen, P., Borgan, M., Oftebro, M., Karseth, B.: Use, updating and integration of ICT in higher education: linking purpose, people and pedagogy. High. Educ. **54**(3), 417–433 (2007)
35. Ullah, N., Al-Rahmi, W.M., Alzahrani, A.I., Alfarraj, O., Alblehai, F.M.: Blockchain technology adoption in smart learning environments. Sustainability **13**(4), 1801 (2021)

Analysing the Nutritional Facts in Mc. Donald's Menu Items Using Exploratory Data Analysis in R

K. Vignesh(✉) and P. Nagaraj

Department of Computer Science and Engineering, Kalasalingam Academy of Research and Education, Krishnankoil, Virudhunagar, India
{Vignesh.k,Nagaraj.p}@klu.ac.in

Abstract. A quantitative data analytic tradition is the Exploratory Data Analysis (EDA) based on the original work of John Tukey. In Statistics, EDA is the process of cleaning data and using appropriate visualizations to extract insights and summarize the data in the given dataset. The computational and core conceptual tools of EDA cover the use of interactive data display and graphics, diagnosing, evaluating, an emphasis on model building, and addressing the issues of the fundamental measurements that are consorted with several distributions and also undertaking some of the procedures that are resistant to mislead or flawed results because of the unpredictable change of real world data. EDA can be further classified as Graphical or non-graphical and Univariate or multivariate data. EDA can be helpful for expressing what does the data refers before the modeling task. It is not easy to see at a large dataset or a whole spreadsheet and to decide important characteristics of the data. It may be typical to obtain insights by seeing at plain numbers. So, EDA techniques have been evolved as an assist in this type of situation. In this article we are going to take a nutritional dataset and visualize it using EDA. This visualization contains Bar chart, Histogram, Boxplot, Scatterplot, Barcode and Violin plot.

Keywords: Exploratory data analysis · Boxplot · Histogram · Scatterplot · Bar chart · Barcode · Univariate and Multivariate analysis

1 Introduction

Exploratory Data Analysis is an analytical process of performing initial inquiry on data to originate models, to certain hypotheses, and to see hypothesis with the help of review of graphical representation and statistics. EDA which is also known as Visual Analytics, is a heuristic search technique which is used in large datasets for finding significant relationships between variables. It is generally the first technique when approaching unstructured data [8]. We apply Exploratory data analysis techniques to the selected subsets of data to gain insight into the sources and distribution. These exploratory methods include graphical data analysis, multivariate compositional data analysis and geochemical modelling [7]. Enormousdata is produced by healthcare industries, called big data

V. E. Balas et al. (Eds.): ICETCE 2022, CCIS 1591, pp. 573–583, 2022.
https://doi.org/10.1007/978-3-031-07012-9_48

that accommodates hidden knowledge or model for decision making. Enormous volume of data is used to make decision which is more accurate than perception. Exploratory Data Analysis detects mistakes, finds appropriate data and also verifies assumptions and determines the correlation between exploratory variables [2]. The research present in this article is an exploratory case study of nutritional facts. The aim of this study is to discover elements that had an impact on people's health who takes food in Mc. Donald's menu, and to understand the need of exploratory analysis. We used a variety of tools for visualization and prediction [12, 14]. In this article, we are going to do visualizations with bar charts, Boxplot, Histogram, Scatterplot, Bar chart, and Barcode on a dataset using EDA analysis in R.

2 Related Work

Nathalie Lyzwinski et al., research advise that when a person experience stress, then the relationship with the food–body reward pathway is changed and high-calorie food with a low nutritional value (junk food) creates an addictive system. Students are at risk for having an unhealthy lifestyle under stress, desire for junk food, binge eating before exams, and consuming fast food [1]. Padman et al., stated exploratory study, to improve the nutrition of children we analyze game telemetry through virtual reality-based entranced mobile gaming which uses AI(Artificial Intelligence) to achieve personalized behavior support to understand the user interactions from playing Fooya, an iOS/Android based mobile app [3]. Nedyalkova et al., analyzed the application of EDA to classify, model and interpret clinical data of DMT2 patients who have many facetsand to predict DMT2 among huge group of patients by sorting methods, to model the trajectories of the disease by interpretation of specific indicators, to study diabetic problems, to identify metabolic and genetic biomarkers in patients with DMT2 and factors associated with cardio-vascular by chemometric methods [4]. Lunterova et al., aim of this research was to interconnect efficiently a large data of over 8400 food data points with a trans-disciplinary method. There were two main focus points- the usability of the visualization and communicativeness [5]. Markazi-Moghaddam et al., described an E-clusters and classes are discovered by using Clustering analysis and Discretization of time intervals from the data. The partitioning algorithm used was K-means clustering to subdivide time intervals into clusters [6]. Arora et al., discovered the graphical representation for the better visualization of outcomes using OriginPro 2016 software is the data collected, organized, and processed [9].

Taboada et al., identified EDA usually consists of six steps namely: (i) categorize attributes; (ii) the data is characterize using univariate data analysis; (iii) By performing bivariate and multivariate analysis the interactions among attributes are detected; (iv) detecting and minimizing the influence of missing and aberrant values; (v) detecting outliers; (vi) feature engineering, where we can transform the features or combine to generate the new features [10].

Petimar et al., done an experiment which was conducted among adults, adolescents, and children. In this natural experiment, we found that McDonald's voluntary calorie labeling was not associated with large changes in calorie content of meals that are purchased when compared to the unlabeled fast-food restaurant meals. Exploratory analysis discovered that calorie labeling was associated with less calories which are purchased among black as well as white adolescents with obesity [12].

Vega et al., described exploratory analysis of experimental data are administered by box plots, ANOVA, display methods and unsupervised outline recognition in an effort to favor sources of disparity [13]. Jayman, stated Chefs adopting a school (CAAS), is an evidence-based food education program which is implemented in schools in UK (United Kingdom) which helps children to know the importance of nutritional literacy and also encourages to take healthy food. This qualitative study examined the underlying components of the CAAS program are donated to improve children's healthy diet [15].

3 Design Methodology

In R Language, we are going to perform exploratory data analysis on a nutritional dataset. This dataset is taken from the Kaggle website. This dataset has 260 observations of 24 variables. By using this dataset, we are going to perform

1. Preliminary data Transformation and cleaning.
2. Exploring the distribution of calories across different menu items, food categories and serving sizes with the help of visualization methods such as bar chart, histogram, to estimate smooth density, box plot, and scatter plot.
3. To explore the distribution of nutritional values across different types of food items and quantity which have been served with the help of visualization methods such as histogram, barcode, box plot etc.
4. Finding average Nutrition in different categories using violin plot.
5. Finding food items having lowest and highest calorific values, high cholesterol containing items and which items are good to have complete day food at McDonald's.
6. A framework of EDA is demonstrated in Fig. 1.

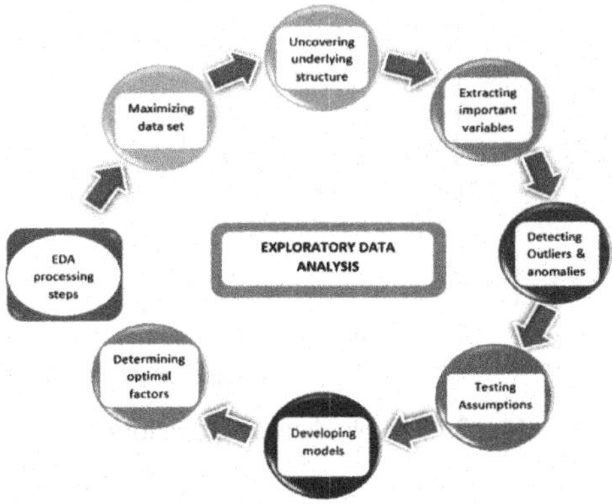

Fig. 1. Exploratory data analysis.

4 Experimental Result

An EDA process and visualization are described in the Fig. 2, 3, 4, 5, 6, 7, 8, 9, 10, 11 and 12.

Step 1. Preliminary data transformation and cleaning.

For this dataset implementation, we require packages such as ggplot2, readr, tidyverse, ggthemes, string, plotly, data.table packages.

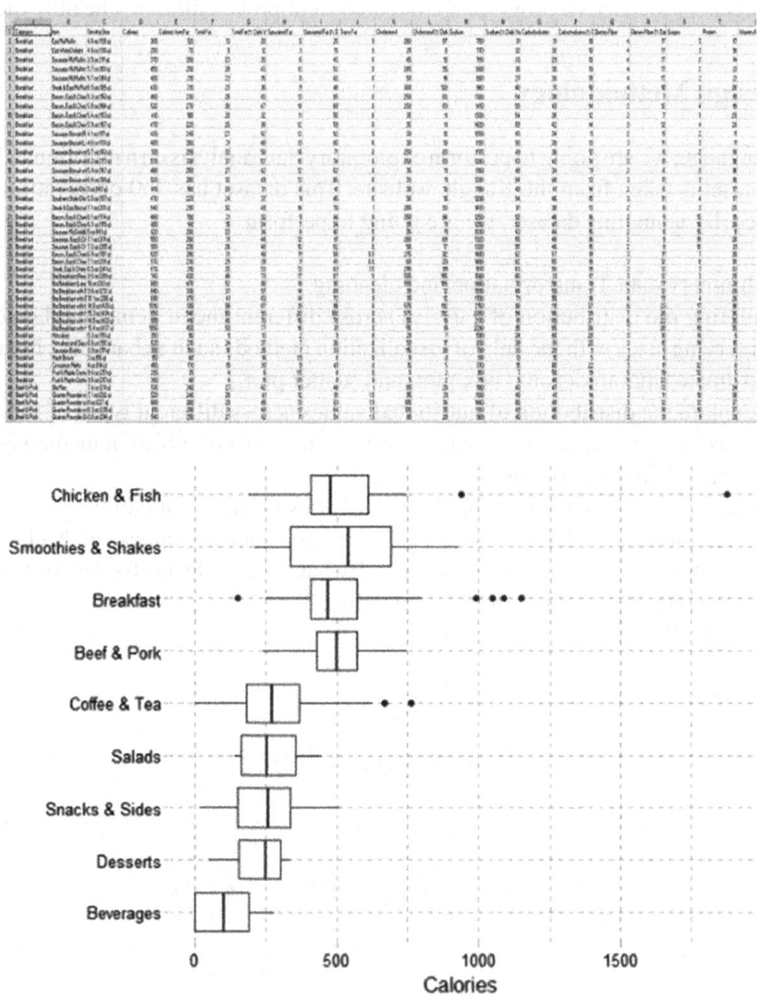

Fig. 2. Sample dataset and calories

This is the simple data set. From this dataset we can observe that serving size is recorded in an unfriendly manner, so we are going to mix ounces and grams in a single column. The serving size variable is encoded as an element in McDonald's data set. We are going to convert it into a single numeric variable represented by grams, and milliliters.

Step 2. Exploring the distribution of calories across different kinds of menu items, food categories and serving sizes.

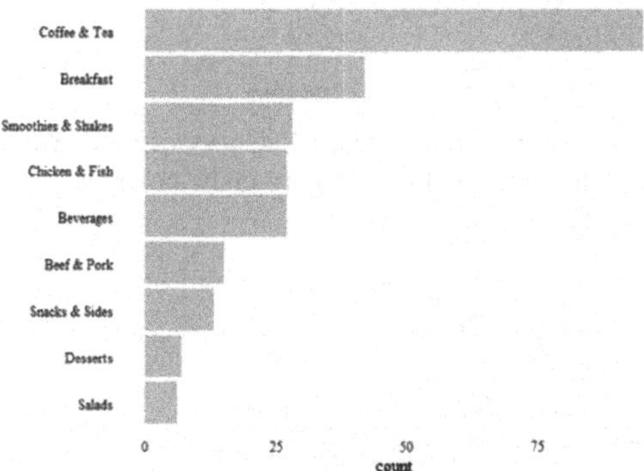

Fig. 3. Bar chart contains the distribution of food categories in the menu. Coffee & Tea is a foremost category in the menu.

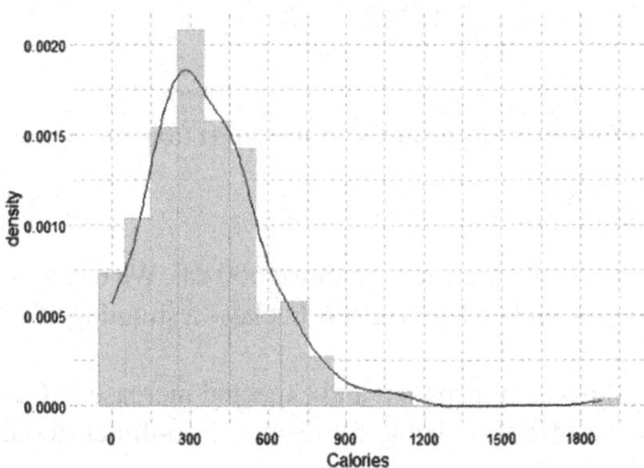

Fig. 4. Histogram with density line shows distribution of Caloric values. Density of distribution of caloric values of peaks around 300.

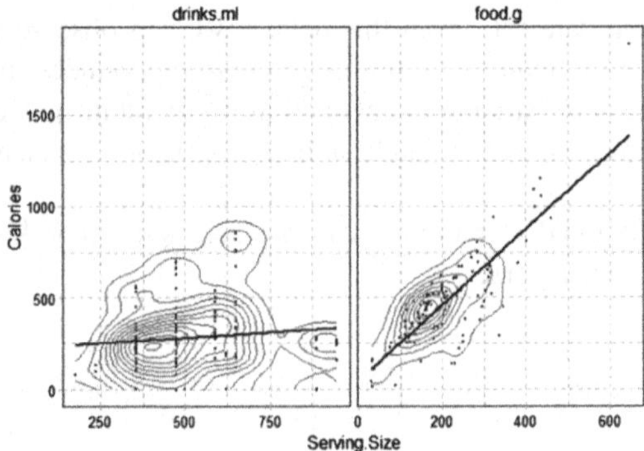

Fig. 5. Caloric distribution of **Box plot** in each Category of food. We are now going to see a monster-caloric outlier in the Chicken, Fish category in the box plot.

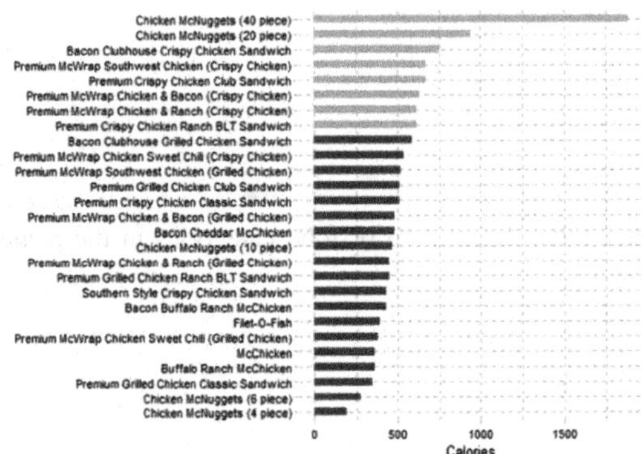

Fig. 6. Caloric values for specific menu Items for **Bar chart** from Chicken and Fish food category. (Color figure online)

The red color bars shown are equal or above 600 cal. We can now see that a massive outlier. It is a 40 pieces Chicken Mc Nuggets bucket - a stunning of 1880 cal (1060 from fat).

Regression line shows a transparent and expected increase in Calories with Serving Size, but no such clear effect is visible for drinks. Two-dimensional density estimates clearly a larger variance of serving size values.

Note: remember we're looking a rather measure scale for drinks (ml) and food (grams).

Step 3. Exploring distributions of nutritional value across different food categories and serving sizes.

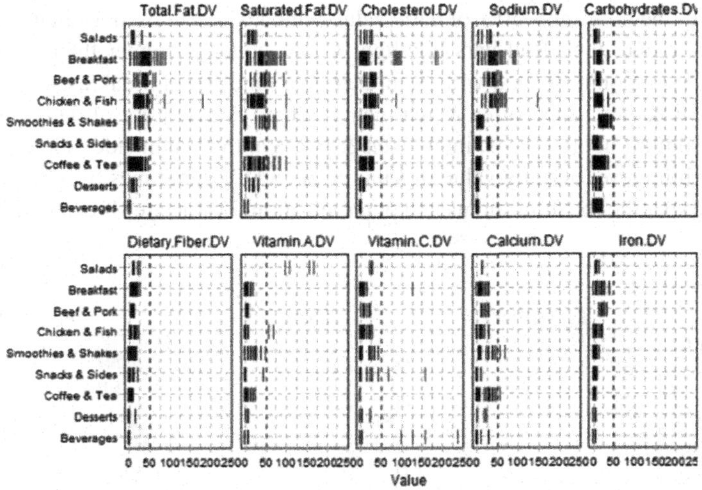

Fig. 7. Scatter plots of Calories and Serving Size, aspect by type of menu food.

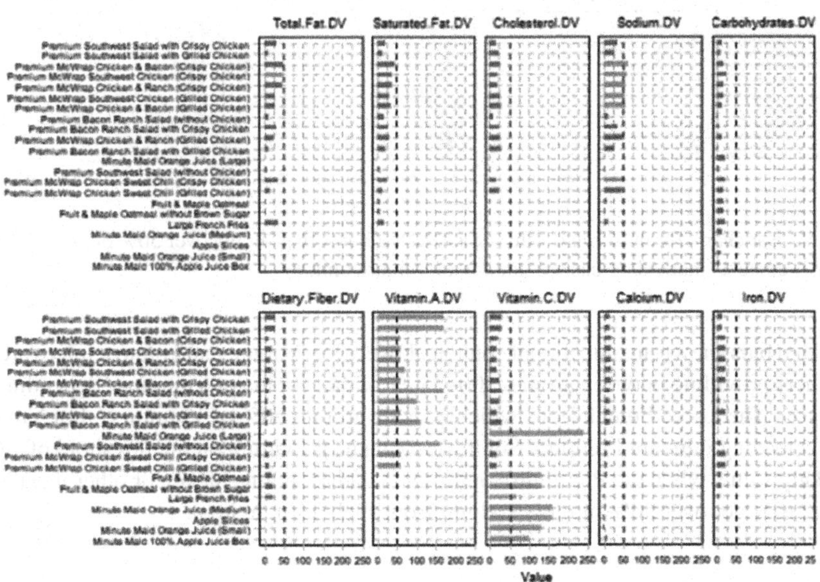

Fig. 8. Barcode plots showing distribution of specific nutritional values in different food categories.

Dashed line represents 50% of daily value-which is sort of high when it comes from only a one menu item. Red bars highlight values equal or higher than 50% of daily values. A big amount of food with Saturated Fat comes from breakfasts, smoothies & shakes, beef & pork and coffee & tea menus. There is also a lot of daily and routine value for sodium in breakfasts, in chicken & fish. On the bright side- some items have extremely high level of daily value in Vitamin A and C, sometimes over 150% of daily value. The specific 'healthy' items and other nutritional value are shown in this barcode plot.

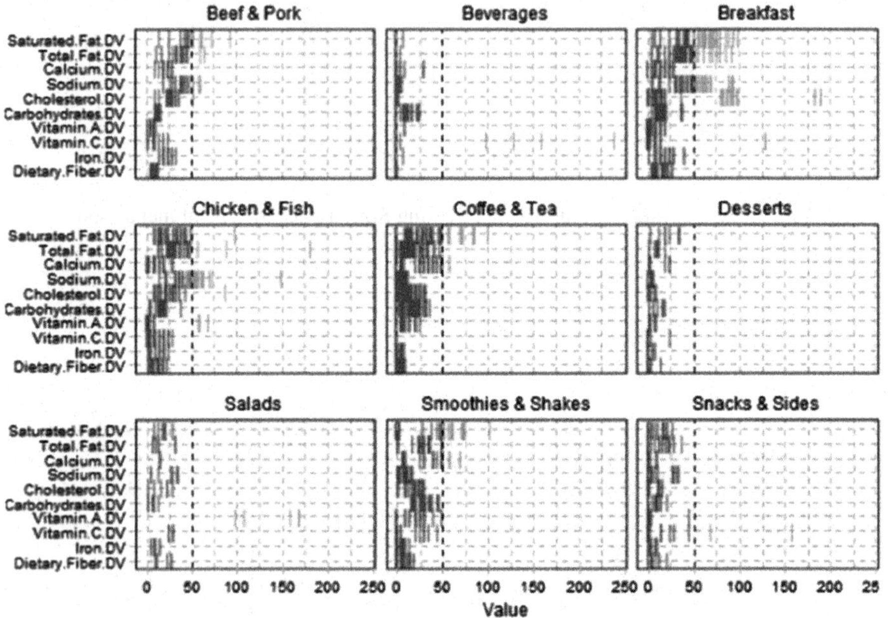

Fig. 9. Bar charts for nutritional values for selected items that supply over 50% of daily Vitamin C or D content (marked by dashed line). (Color figure online)

Red bars shown are nutritional content above 50%. Minute maid orange juices are the winners providing between around 100–250% of daily value of Vitamin C depending on the size. The Large French Fries which provide around 60% of Vitamin C. For Vitamin A the leading item are Premium Southwest Salads, providing on the threshold of 170% of daily value, when the Premium Southwest Salads contain chicken or other meat, then you easily get over 50% of daily sodium from such a single item.

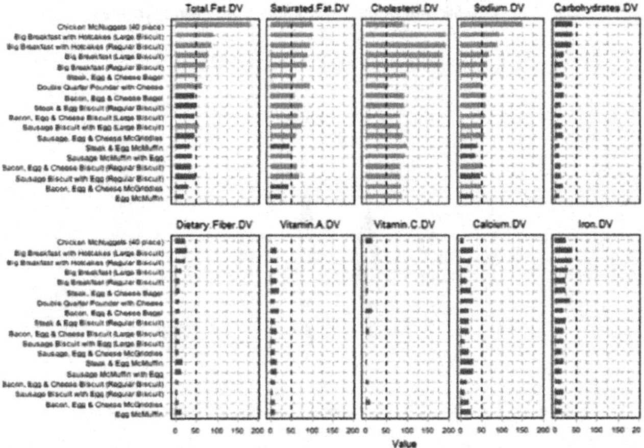

Fig. 10. Bar charts for nutritional values for selected items that supply over 50% of daily Cholesterol content. (Color figure online)

Heavy Breakfast are the killers - from each of them you get almost 200% of daily amount of cholesterol.

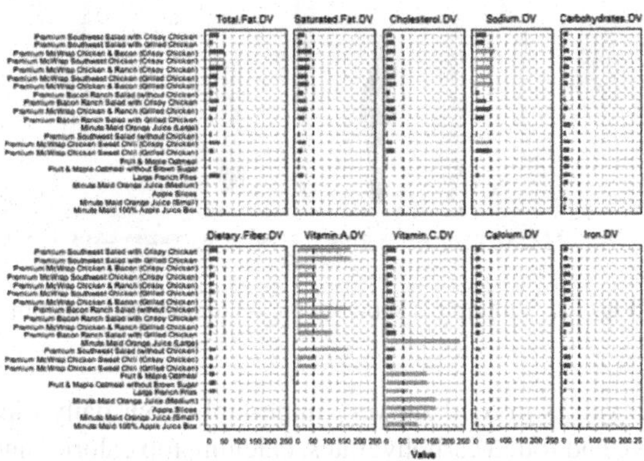

Fig. 11. Barcode plots showing distribution of specific nutritional values, faceted by different food categories. (Color figure online)

Red Bars highlight values above 50%. Confirms very large content of Saturated Fat, Total Fat, Sodium and Cholesterol in breakfast category - possibly the foremost on this menu if we consider health threat of high saturated fat, cholesterol and sodium consumption.

Step 4. Finding average Nutrition in different categories using violin plot.

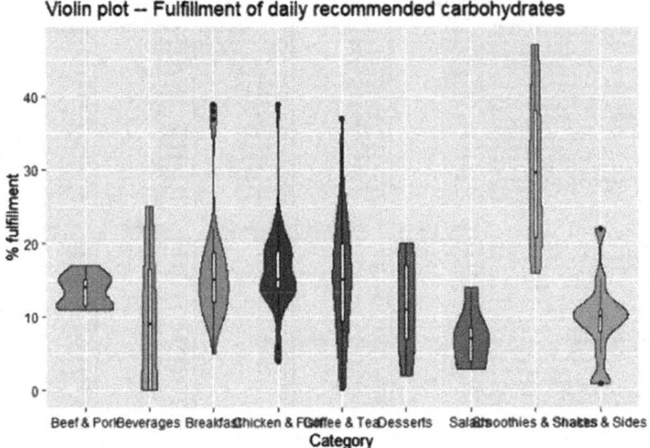

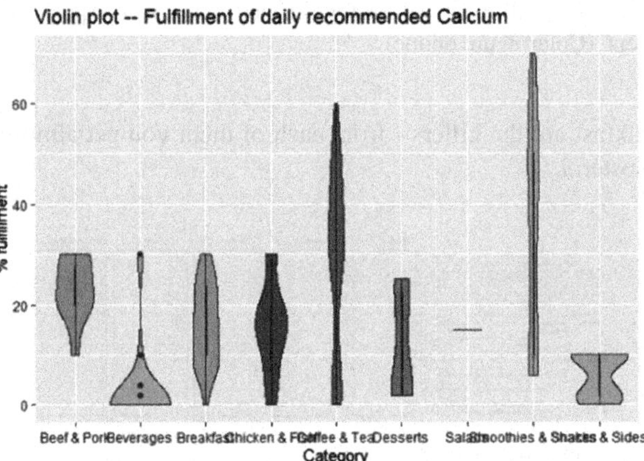

Fig. 12. Violinplots showing the recommended calcium

The nutritional values in food are very important in our daily life. From the above implementation; we had found carbohydrates, calcium, top calories and low calories per gram that contain in Mc Donald's menu. We also implemented the plots of different types to study and to analyze the dataset to know the nutritional value in Mc Donald's menu and which food item is better to take from the menu to maintain good health when we are taking full day meals at Mc Donald's.

5 Conclusion

We conclude that Exploratory Data Analysis is the process of examining a dataset to comprehend the shape of the data, signals in the data that may be useful for building predictive models in future enhancement, and correlations between features. It is helpful to be ready to complete this task in both scripting language and a Query language.

References

1. Nathalie Lyzwinski, L., Caffery, L., Bambling, M., Edirippulige, S.: University students' perspectives on mindfulness and mHealth: a qualitative exploratory study. Am. J. Health Educ. **49**(6), 341–353 (2018)
2. Indrakumari, R., Poongodi, T., Jena, S.R.: Heart disease prediction using exploratory data analysis. Proc. Comput. Sci. **173**, 130–139 (2020)
3. Padman, R., Gupta, D., Prakash, B.S., Krishnan, C., Panchatcharam, K.: An exploratory analysis of game telemetry from a pediatric mhealth intervention. In: MEDINFO 2017: Precision Healthcare Through Informatics: Proceedings of the 16th World Congress on Medical and Health Informatics, vol. 245, p. 183. IOS Press, January 2018
4. Nedyalkova, M., et al.: Diabetes mellitus type 2: exploratory data analysis based on clinical reading. Open Chem. **18**(1), 1041–1053 (2020)
5. Lunterova, A., Spetko, O., Palamas, G.: Explorative visualization of food data to raise awareness of nutritional value. In: Stephanidis, C. (ed.) HCII 2019. LNCS, vol. 11786, pp. 180–191. Springer, Cham (2019). https://doi.org/10.1007/978-3-030-30033-3_14
6. Markazi-Moghaddam, N., Jame, S.Z.B., Tofighi, E.: Evaluating patient flow in the operating theater: an exploratory data analysis of length of stay components. Inform. Med. Unlocked **19**, 100354 (2020)
7. Bondu, R., Cloutier, V., Rosa, E., Roy, M.: An exploratory data analysis approach for assessing the sources and distribution of naturally occurring contaminants (F, Ba, Mn, As) in groundwater from southern Quebec (Canada). Appl. Geochem. **114**, 104500 (2020)
8. Taboada, G.L., Han, L.: Exploratory data analysis and data envelopment analysis of urban rail transit. Electronics **9**(8), 1270 (2020)
9. Arora, A.S., Rajput, H., Changotra, R.: Current perspective of COVID-19 spread across South Korea: exploratory data analysis and containment of the pandemic. Environ. Dev. Sustain. **23**, 6553–6563 (2021). https://doi.org/10.1007/s10668-020-00883-y
10. Taboada, G.L., Seruca, I., Sousa, C., Pereira, Á.: Exploratory data analysis and data envelopment analysis of construction and demolition waste management in the European Economic Area. Sustainability **12**(12), 4995 (2020)
11. Sharma, R., et al.: Index point detection for text summarization using cosine similarity in educational videos. IOP Conf. Ser.: Mater. Sci. Eng. **1131**, 012001 (2021)
12. Petimar, J., et al.: Evaluation of the impact of calorie labeling on McDonald's restaurant menus: a natural experiment. Int. J. Behav. Nutr. Phys. Act. **16**(1), 99 (2019)
13. Vega, M., Pardo, R., Barrado, E., Debán, L.: Assessment of seasonal and polluting effects on the quality of river water by exploratory data analysis. Water Res. **32**(12), 3581–3592 (1998)
14. Chris, N.: Exploratory Data Analysis and Point Process Modeling of Amateur Radio Spots (2020)
15. Jayman, M.: An exploratory case study of a food education program in the United Kingdom: chefs adopt a school. J. Child Nutr. Manag. **43**(1) (2019)

Applied Machine Tool Data Condition to Predictive Smart Maintenance by Using Artificial Intelligence

Chaitanya Singh[1], M. S. Srinivasa Rao[2], Y. M. Mahaboobjohn[3], Bonthu Kotaiah[4], and T. Rajasanthosh Kumar[5](✉)

[1] Computer Engineering Department, Swarrnim Startup and Innovation University, Gandhinagar, India
[2] Department of Mechanical Engineering, VNRVJIET, Hyderabad 500090, Telangana, India
[3] ECE, Mahendra College of Engineering, Salem, India
[4] Maulana Azad National Urdu Central University, Gachibowli, Hyderabad, Telangana, India
[5] Department of Mechanical Engineering, Oriental Institute of Science and Technology, Bhopal, India

trsk.mech@gmail.com

Abstract. We describe how to integrate data-driven predictive maintenance (PdM) in machine decision-making and data collection and processing. A brief overview of maintenance methods and procedures is provided. In this article is a solution for a real-world machining issue. The answer requires many stages, which are clearly described. The outcomes demonstrate that Preventive Maintenance (PM) might be a PdM method in an actual machining process. To offer a Visual examination of the cutting tool's RUL. This study proves shown in one procedure but reproducible for most of the series piece productions.

Keywords: PdM · RUL · Machining process · Concept drift · Real application

1 Introduction

Professionals refer to this as "Industry 4.0" or "The Fourth Industrial Revolution". (I4.0) Industry 4.0 is concerned with integrating physical and digital manufacturing systems [1]. I4.0 has arrived. PHM is an inevitable trend in the moreover, it provides a solid foundation for big industrial data and smart manufacturing industrial equipment health management solution I4.0 and its core technologies include the automated production of industrial systems [2, 3]. Machine/component data gathering Type machine based on gathered data Automated defect identification and diagnosis using learning algorithms are possible. But it's a to choose suitable ML methods, data types, data sizes, and hardware Industrial systems using ML. Choosing the wrong predictive maintenance method Infeasible maintenance schedule due to dataset and data size. Thus, It will assist academics and practitioners choose suitable ML methods, data quantity, and data quality by reviewing the current research. A viable ML application. It can detect the

V. E. Balas et al. (Eds.): ICETCE 2022, CCIS 1591, pp. 584–596, 2022.
https://doi.org/10.1007/978-3-031-07012-9_49

performance deterioration of industrial equipment. Hidden risks, failures, pollution, and near-zero. Manufacturing process accidents [4]. These massive quantities of ML data provide essential information. information that may enhance overall manufacturing process and system productivity in many domains, including condition-based decision assistance regular maintenance and wellness checkups Thanks to recent technological advancements, data Collecting vast amounts of information via computerized control and communication networks. multiple equipment's operating and process data streams in As a source of data for automated FDD [6]. Those data gathered may be used to improve intelligent preventative [7] PdM stands for preventative maintenance. In addition to saving money on maintenance, ML applications may also help prevent damage to Improvements in spare component life and inventory reductions and improved operator safety. A rise in total earnings and many more. A vast and significant link exists between these benefits and maintenance methods. Aside from that, defect detection is an essential part of predictive maintenance. Industry to find problems early. Maintenance techniques may be used to classifications:

1. Correctional or unscheduled maintenance is also known as Run 2 Failure (R2F). It's one of the easiest maintenance methods is only done when the Failed. It may cause equipment downtime and subsequent failures, resulting in costly repairs. Increase production defects by a significant margin. Preventive Maintenance (PvM): also known as planned or time-based maintenance (TBM). PvM is planned scheduled maintenance.
2. To predict failures, It may lead to unneeded maintenance, which increases costs. We want to enhance equipment efficiency by reducing production hiccups [18]. Condition-based Maintenance (CBM): that can only be carried out when they are a must When the maintenance activities on the process is degraded. As a rule, CBM arranged ahead of time.
3. Statistical-based maintenance: maintenance plans are only taken when Required. A lot like CBM, it relies on constant monitoring of the machine. To determine when such maintenance activities are required, it uses prediction techniques. Maintenance may be planned. It also enables early failure detection using. Using machine learning techniques to forecast the past data. Statistical inferences, visual characteristics (such as colour, wear) engineering methods and methodologies Any maintenance plan should aim to reduce equipment failure. Should enhance equipment quality, extend equipment life, and decrease Repairs & upkeep Fig. 1 depicts an overview of maintenance categories.

PdM swung among other maintenance methods, one of the most promising. The approach has lately been used in several areas to achieve such qualities [19]. Researches It has been used in I4.0 because it attracts industries' attention. To optimize asset usage and management. Has recently emerged as one of the most potent weapons available. Used to create intelligent prediction algorithms in many applications. It is a creation. Spanning decades into a variety of fields. ML is a technique that allows us. Using previous input data, a model can predict outcomes. And its production ML, according to Samuel, A.L. non-programmed ability to solve problems ML methods are known to they can handle multivariate, high-dimensional data and Can find hidden connections in data in chaotic, dynamic, and complex settings. The performance and benefits may vary depending on the ML method used. To date, machine learning (ML) methods are extensively used in manufacturing (including (upkeep, optimization, troubleshooting, and control) As a result, demonstrate the latest ML advances in PdM. This extensive study mainly utilizes Scopus to find and acquire papers. Used. Globally, this article seeks to identify and classify ML method, ML category, data acquisition device, applied data datatype and data size.

2 Predictive Maintenance

Even though all machines eventually fail, the failure modes and effects vary widely. If a machine breaks down unexpectedly, people may be left in the dark for hours, resulting in the loss. Businesses need a well-defined maintenance strategy. To avoid costly outages and reduce the damage caused by failures. Predictive analytics is the most sophisticated method, and it requires significant resources to execute.

After-the-fact maintenance often referred to as reactive or breakdown maintenance occurs after the equipment has already failed. This method conserves both time and resources by eliminating the need for further preparation and assistance. It's possible while dealing with non-critical, easily repairable, and redundant equipment. Say, for example, that light bulbs are only changed after they are completely depleted.

When you include overtime pay, reduced usable life of assets, reputational harm, and safety concerns, corrective maintenance may be very costly over the long term, even if there are no upfront expenses. According to Marshall Institute estimates, reactive maintenance costs businesses up to five times as proactive maintenance methods.

Regular equipment inspections are part of preventive maintenance, and they help to minimise deterioration and failure. Assets' usable lives and efficiency are extended by pre-planned operations like lubrication and filter replacements. All of this equates to savings in monetary terms. According to research, proactive maintenance saves between 12 and 18% compared to reactive maintenance.

Preventive methods, on the other hand, can't completely prevent catastrophic failures from occurring. This technique requires preparation and extra personnel. Over or under-checking is common when it comes to dependability checks.

3 Literature Review

There hasn't been a review of the literature on intelligent sensors for intelligent maintenance. There are presently reviews in the literature that deal with each of these topics individually. When it comes to economic and human losses, [4], look to intelligent sensors to keep tabs on rock bolts' condition and integrity. Multifunctional sensors suited for industrial production, [11] describes sensors for intelligent gas sensing in the literature review, and [6] illustrates smart parking sensors substituting ultrasonic sensors with machine learning in engineering. After that, sensors for health monitoring are defined. A significant component of the 4.0 industry idea, intelligent sensors and smart factories, are often seen in literature reviews [13]. Smart sensors are used in an intelligent factory according to assess and troubleshoot specific equipment. Discuss how big, intelligent factories are being transitioned and implemented in real life. [14, 15] equally concerned in putting the ideas into practice. There are several studies in the area of intelligent maintenance with the resounding concept of predictive maintenance. Introduces machine learning and AI techniques, which they see as a potential tool for predictive maintenance. Predictive maintenance is taking the role of reactive maintenance, as seen in the context of Industry 4.0. Thermal power plants focus.

4 Data Acquisition

Data acquisition is the process of gathering and storing data from a physical process in a system. A predictive maintenance software collects two kinds of data: event data and condition data. Tracking data While the event data includes information on the asset's fate, The condition monitoring data related to the physical asset's health measures created a notification system resources for manufacture Signals include tremors, acoustics, and oil. Temperature, humidity, and climate. Many sensors gather this data. have been created, including ultrasonic sensors and rain sensors. Many Sensors and computers are being improved, implying a more straightforward way to store data.

5 Data Processing

Acquired data may include missing, inconsistent, or noisy values. Data quality has a significant effect on data mining outcomes. Pre-processing techniques may enhance these findings. Pre-processing data is a crucial step. This prepares and transforms the original dataset. Pre-processing data It May be classified into three types:

5.1 Data Cleaning

Raw data, particularly event data, are typically incomplete, noisy, or inconsistent. Data mistakes may be produced by human or sensor errors, and identifying and eliminating them improves data quality. Challenging to read data. There is no easy method to clean during mining or in general. Techniques rely on human examination aided by a graphical tool. Median is used to pad uncertain numbers with zeros. Wei et al. missing data recovery for metabolomics data (zero, half minimum, QRILC (Quantile Regression In-Line Components) Censored Data Imputation)). Aside from missing data, noisy values are a concern for data erasure. [15] suggested using clustering techniques. To identify the noise, Clustering methods may identify data outliers when comparable values are grouped. Outliers are values that are outside the clusters. Using a one-class SVM technique, Jin et al. would indicate a system deterioration. Maronna et al. three-sigma edit rule used by Jimenez Cortadi et al. turning method.

5.2 Data Transformation

Data transformation is to get a more suitable data representation for one step. Modelling progress Standardization, when data are scaled to a tiny signal's range, and comparability Smoothing is also used to segregate data. And the loudness. Smoothers may be forward or backward for a particular dataset. Looking. We shall only examine smoothers that replace one observation with another. a mix of comments before and after. A brief review of several smoothing techniques. Symbolic Aggregate Approximation found it superior to the criteria for time series categorization. Smoothing regressions and clustering.

5.3 Data Reduction

The high computational cost of having a large amount of data may make machine decision making problematic. Hardware time is increased by the quantity of data processed. When data is in plenty, computing costs increase. Thus systems have grown and changed through time. It is the primary element. Feature reduction may be an inverse version of feature generation, which takes input with many features and reduces it to fewer (yet independent) ones (SOM).

6 Maintenance of Predictive Machine Tool Systems

6.1 Predictive Maintenance of Machine Tool Systems

A cutting tool that revolves is used to remove material and shape the workpiece. Tool shape variability occurs owing to the interaction between the device and the workpiece as the machining progresses. In the creation of heat and stress, a tool is worn out. Surface quality is negatively impacted because of the cutting tool's performance. A product's surface finish indicates its overall quality. A quality product is achieved by ensuring that the condition of the cutting tool is always well-maintained. A lousy product may be produced if devices cannot be monitored and discarded because of quality issues. The tools' practical, helpful life (EUL) may be estimated using a precision tool in the machining process. The wear on the cutting tool is shown in Fig. 1, and it is evident from that wear that the device is in a specific condition. In this simulation, a mechanism is described as being in one of three states: fresh, worn, or overworn, depending on the duration of flank wear (standard, warning, and failure).

6.2 Predictive Maintenance of the Spindle Motor

Spindles have a direct effect on quality and production. Therefore their design is an important factor. The spinning shafts that power machine tools also constantly expose rolling components to static and dynamic forces. For example, continuously applied forces wear out bearings, rotors, and shafts, and this wear and tear results in mechanical failure. Repairs such as changing damaged spindles and calibrating such tool runout are difficult projects. Components are kept for planned maintenance since they may be used for replacement. Estimating the life of a component is difficult, and making an accurate assessment is much more complicated in a worn condition. The practice of using a PdM

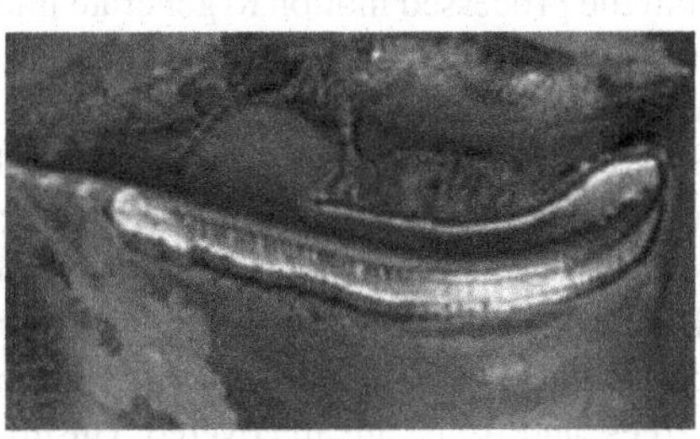

Fig. 1. Cutting tool wear

to help minimise future maintenance efforts while increasing productivity and quality is therefore beneficial to operations. Spindle damage comes from bearing failure. Piezoelectric force sensors and accelerometers detect variations in the race diameter, which cause vibration. Rolling element bearing geometry may be used to predict characteristic frequencies of local defects in a ball bearing (Fig. 2). Defects include either a bad inner or outer race, a bad rolling element, or a bad bearing cage. It is possible to utilise these four frequencies to examine spindle conditions. To help identify flaws in bearing spindles, ANN (Artificial Neural Network), fuzzy logic, and Bayesian classification, in addition to time and frequency domain investigations, were used.

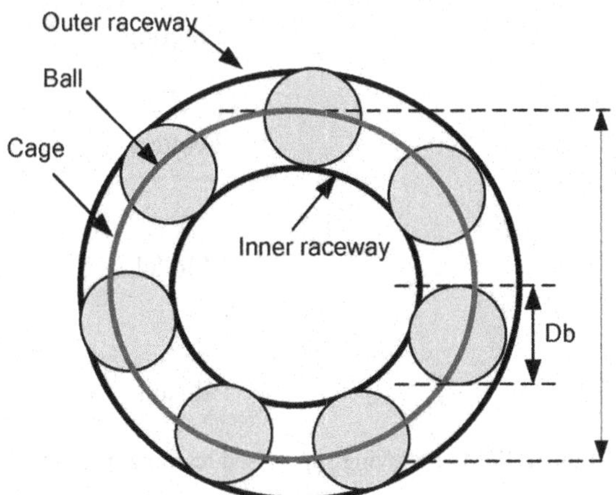

Fig. 2. Bearing rolling element geometry

7 Methodology for Condition Monitoring

Sensor signals from outside the machine may translate the circumstances of the tool (e.g., accelerometer, microphone, dynamometer, and thermometer). The initial step to processing raw analogue data eliminates any unwanted frequency spectrums to access helpful information. Next, extract qualities from the processed motion to generate information about the environment and then compress it. Features are retrieved in the time or frequency domain depending on the technology (for example, the Fourier transform of vibration signal from a spindle indicates a local defect of ball bearings clearly than the signal in the time domain). This will aid in training AI systems. Under a myriad of learning schemes, such as supervised, unsupervised, and reinforcement learning. Extracted features are trained using labels in supervised learning. In other words, supervised learning algorithms may help analyse spindle monitoring to determine whether it's normal or faulty. Regression models, SVM, decision trees, and ANN, are all covered. Unsupervised learning is known to create estimate models. However, the estimate models are also missing labels. The model learns from reward and penalty information and uses the gathered information to create the best policy for the goal. An AI model's results may be estimated after training. For example, a prediction model called the spindle health model might be built with the help of the processed accelerometer data and information about the spindle state (labels). The standard method for supervised learning techniques, which is how PdM usually utilises PdM (Fig. 3).

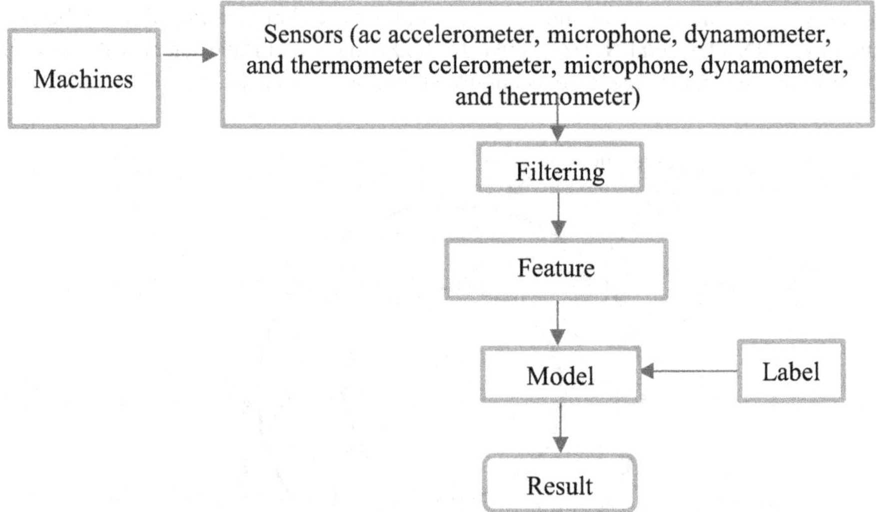

Fig. 3. PdM employing supervised learning methods.

8 Manufacturing Data

The PdM systems were constructed utilising the datasets from the milling and bearing experiments. Studies utilise sensors and get data in real-time to collect information on in-service equipment and then run until the device stops working. Failure incidents are used to evaluate post-training monitoring systems.

8.1 Milling Dataset

The cast iron workpiece, measuring 483 mm × 178 mm × 51 mm, was utilised in the experiment. The flank wear measurements were taken at a feed of 0.25 mm/rev and a speed of 200 m/min (826 rev/min) (VB). Six signals were gathered while the operator ran: (1) DC spindle motor current (2) AC spindle motor current (3) table vibration (4) spindle vibration, and (5) two table vibration frequencies (VTwavetable) and (6) machine orientation (VBspindle) (5) emitting from the table (6) emitting from the spindle (AEspinlde). Every sensor was sampled at 250 Hz, and 3500 data points were collected between the two cuts.

8.2 Dataset

An apparatus was utilised to evaluate the bearings' performance. A shaft was set at 2000 RPM by an AC motor attached via a rub belt. To push the directions and the post, the mechanism used a radial force of 6000 lb. A 20 kHz accelerometer that is sensitive to vibrations was placed in each bearing housing. The project used a fail-first method where ideas were tried until they stopped working. Three separate experiments were done (test1, test2, and test3) (Fig. 4).

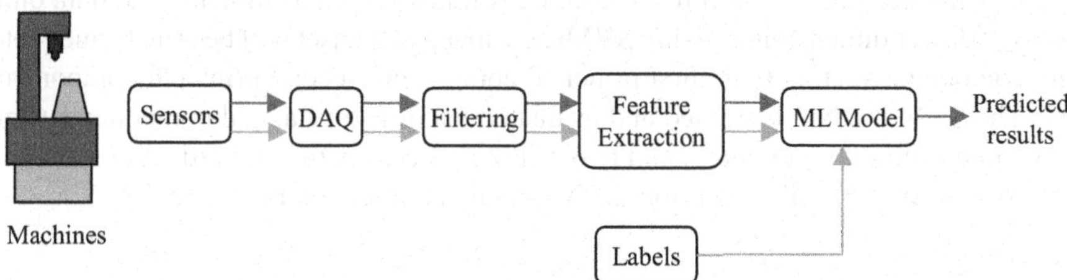

Fig. 4. SVM-based condition classification tool

9 Simulation Results

Two experimental datasets (one involving milling and the other involving bearings) are used to build and test a prediction model for a PdM system. Cutting tool and bearing conditions are monitored and forecasted using the design and testing of support vector machines and artificial neural networks, which are subtypes of artificial neural networks: recurrent neural networks and convolutional neural networks. To evaluate classification performance, confusion matrices are used. A confusion matrix is a useful approach to convey the results of a classification method. Actual and projected values are shown in each row and column, with errors and accuracy highlighted. By putting accuracy and errors together, the matrices can tell which classifier is confused.

9.1 Monitoring the Conditions of the Cutting Tool

To guarantee that the cutting tool is replaced as soon as possible, it is essential to examine its condition. SVMs are used to classify the tool's situations. SVM is a multi-class classification machine learning method [20]. The software first generates hyperplanes. In this case, the features (support vectors) indicate the best hyperplanes for separating classes. Linear quadratic optimization is used to accomplish this. Functions (kernel functions) are used to non-linear signals to transform the original input space into a higher dimensional feature space that can better classify the gestures. Data sets that are not linearly separable may be separated in a higher-dimensional space using Vapnik-Chervonenkis (VC) statistical learning theory [20]. Linear and polynomial Kernel Functions are evaluated for their performance. The method is explained. As previously indicated, the condition of the tool's flanks is used to categorize it. Warning and failure are the three tool wear states. Several raw signal characteristics To build the SVM input dataset must be gathered. The collected dataset's central tendency or dispersion may better describe the tool's context. To sum it up: There are 14 pieces to the machining dataset: arithmetic mean and 50th percentile (as well as the trimmed and interquartile range) as well as mean absolute deviation and content as well as standard deviation and kurtosis. The dataset's dimensionality is too high to use as an input dataset. To this end, PCA is used in the feature dataset to derive new practical features (also known as principal components). While keeping global information, PCA translates multi-dimensional data onto new axes (lower dimensions). Using SVMs, the mapped dataset will be transformed into main component vectors (e.g., first principal component, second principal component, etc.). The SVM method is trained and evaluated using two principal components (first and second primary ingredients) and two milling process parameters (depth of cut and feed) to classify the tool's condition. SVM-based classification is seen in Fig. 5.

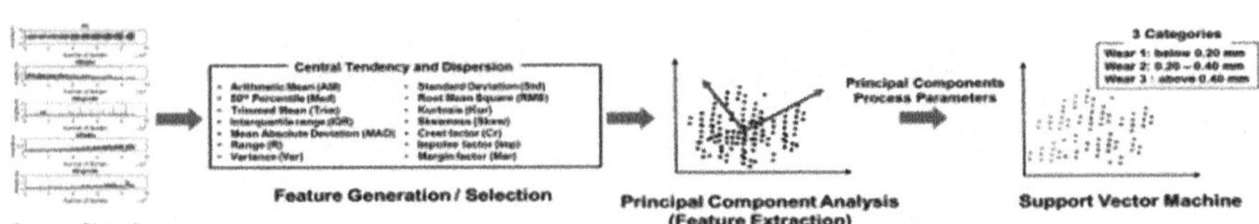

Fig. 5. Diagram of the SVM-based condition classification tool.

The polynomial kernel with d = 3 (Cubic SVM) has the most excellent average accuracy of 87%, as shown in Fig. 6.

9.2 Monitoring the Conditions of Bearing

The bearing's state is monitored using two AI methods, the recurrent neural network (RNN) and the convolutional neural network (CNN). After the raw signals are processed, algorithms do the processing. Bearing conditions are expected, exemplary, and failing, respectively translating to the life span of up to 100% (acceptable), 66.7% (not bad), and 33.4% (poor). RNNs are a connectionist model capable of tracking data dynamics

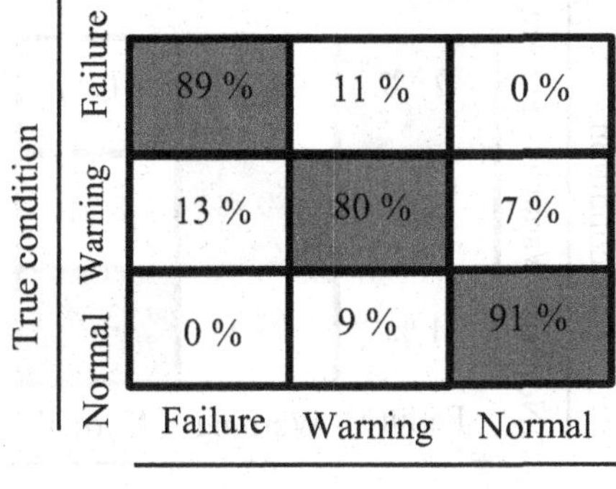

Fig. 6. SVM confusion matrix for cutting tool condition classification.

in sequences. RNNs, unlike feedforward neural networks, may use the internal state to process input sequences (memory). This is important for time-sequential data in PdM monitoring systems. To build a time series, feature datasets were created by chopping 50% overlapping windows from raw accelerometer signals. Figure 7 has the specifics.

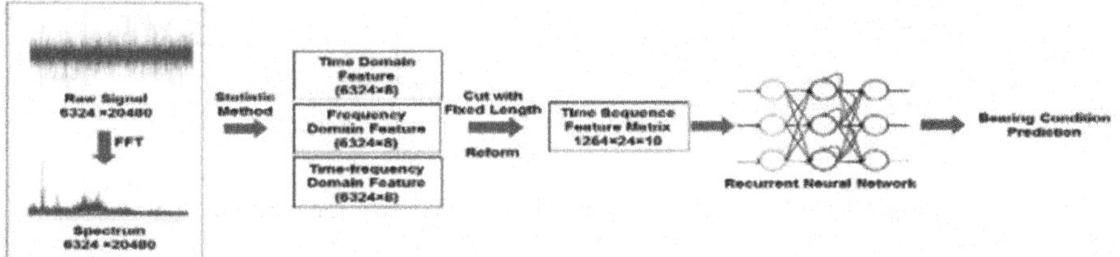

Fig. 7. RNN-based bearing condition classification.

A random dataset is divided into a training group and a testing group. Testing results are shown in the confusion matrix (Fig. 8). (average accuracy of 93%). The model is far more accurate in the critical stage (failure) despite some confusion at first. The CNN was built to sort images (2D). To detect single-point motion, the technique for normalised image data is not required to have feature extraction capabilities. This allows the CNN to receive normalised signals in the time and frequency domains.

Several research papers use spectrum analysis to transform encapsulated time-domain vibration data into the frequency domain in order to measure bearing performance. This model monitored the frequencies in these experiments associated with providing physical attributes (e.g., rotational speed and pitch and ball diameter). Bearing defect frequency may be different for each type of fault. The bearing fault results in amplitude variations that increase harmonious movement at a defined frequency. Bearing information, such as timing and frequencies, is input to a particular algorithm to see whether it can accurately forecast bearing conditions. Signals are sorted according to whether they're for training or testing after pre-processing. Figure 9 illustrates the simulation's structure. Figure 10 shows the simulation results. Time- and frequency-domain

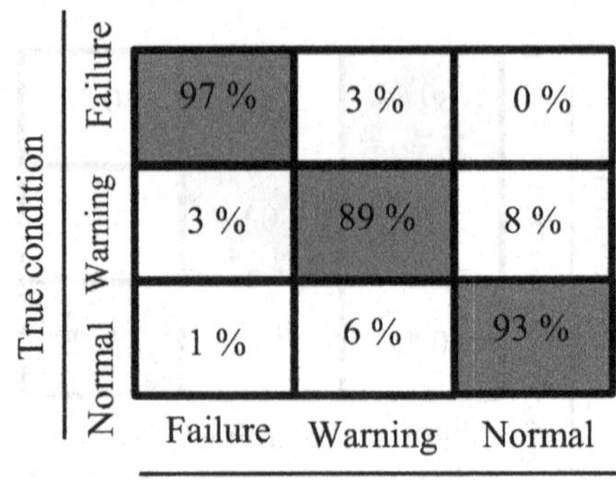

Fig. 8. Confusion matrix for RNN bearing classification.

signals each provide a model accuracy of around 84 and 98%, respectively. Bearing conditions may be predicted better by using frequency-domain data rather than time-domain inputs.

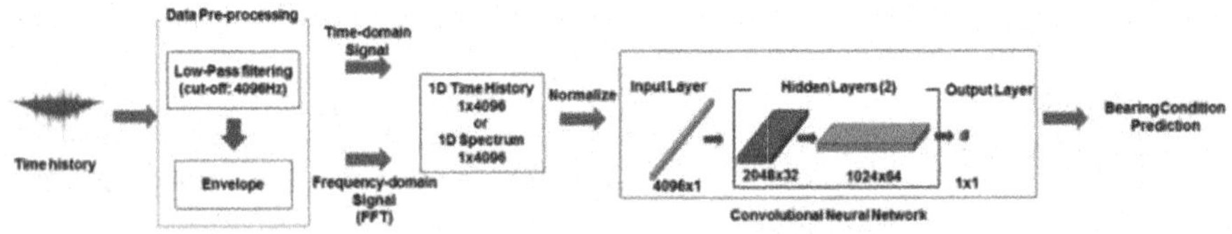

Fig. 9. CNN-based bearing condition categorization diagram.

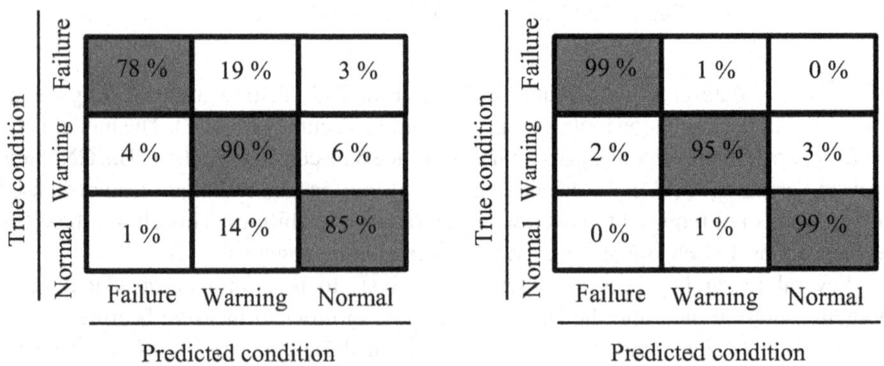

Fig. 10. Time history data (left) and spectrum data (right) for CNN categorization of bearing situations (right).

10 Conclusion

The PdM systems are designed using AI algorithms, and the PdM systems use the cutting tool and the spindle. The bearing's states (normal, warning, and failure) are tracked by its RUL and the wear it exhibits. Classification of tool conditions is done using the SVM and ANN (RNN and CNN) algorithms. The experiments' data was studied to discover the accuracy of a tool's wear and tear diagnosis. The algorithm's frequency domain properties are better in assessing bearing illness than its time-domain qualities. Raw manufacturing data must be cleaned and processed to obtain information that helps AI systems identify relevant attributes. Simply, decreasing machine downtime and improving component RUL may be accomplished via the use of PdM in machine tool systems. Optimizing maintenance work, supplier changes, and machine safety ensure consistent production with appropriate machines.

References

1. Vathoopan, M., Johny, M., Zoitl, A., Knoll, A.: Modular fault ascription and corrective maintenance using a digital twin. IFAC-PapersOnLine **51**, 1041–1046 (2018)
2. Swanson, E.B.: The dimensions of maintenance. In: Proceedings of the 2nd International Conference on Software Engineering, Wuhan, China, 13–15 January 1976, pp. 492–497 (1976)
3. Aramesh, M., Attia, M., Kishawy, H., Balazinski, M.: Estimating the remaining useful tool life of worn tools under different cutting parameters: a survival life analysis during turning of titanium metal matrix composites (Ti-MMCs). CIRP J. Manuf. Sci. Technol. **12**, 35–43 (2016)
4. Wu, R., Huang, L., Zhou, H.: RHKV: an RDMA and HTM friendly key-value store for data-intensive computing. Future Gener. Comput. Syst. **92**, 162–177 (2019)
5. Zhang, W., Yang, D., Wang, H.: Data-driven methods for predictive maintenance of industrial equipment: a survey. IEEE Syst. J. **13**, 2213–2227 (2019)
6. Lee, J., Lapira, E., Bagheri, B., Kao, H.A.: Recent advances and trends in predictive manufacturing systems in big data environment. Manuf. Lett. **1**, 38–41 (2013)
7. Biswal, S., Sabareesh, G.R.: Design and development of a wind turbine test rig for condition monitoring studies. In: Proceedings of the 2015 International Conference on Industrial Instrumentation and Control (ICIC), Pune, India, 28–30 May 2015, pp. 891–896 (2015)
8. Amruthnath, N., Gupta, T.: A research study on unsupervised machine learning algorithms for early fault detection in predictive maintenance. In: Proceedings of the 2018 5th International Conference on Industrial Engineering and Applications, ICIEA, Singapore, 26–28 April 2018 (2018)
9. Wuest, T., Weimer, D., Irgens, C., Thoben, K.-D.: Machine learning in manufacturing: advantages, challenges, and applications. Prod. Manuf. Res. **4**, 23–45 (2016). https://doi.org/10.1080/21693277.2016.1192517
10. Cao, H., Zhang, X., Chen, X.: The concept and progress of intelligent spindles: a review. Int. J. Mach. Tools Manuf. **112**, 21–52 (2017). https://doi.org/10.1016/j.ijmachtools.2016.10.005
11. Qiu, M., Chen, L., Li, Y., Yan, J.: Bearing Tribology: Principles and Applications. Springer, Heidelberg (2016). https://doi.org/10.1007/978-3-662-53097-9
12. Andersen, B., Xu, L.: Hybrid HVDC system for power transmission to island networks. IEEE Trans. Power Deliv. **19**, 1884–1890 (2004)

13. Aten, M., Shanahan, R., Mosallat, F., Wijesinghe, S.: Dynamic simulations of a black starting offshore wind farm using grid forming converters. In: 18th Wind Integration Workshop, Energynautics GmbH, Dublin (2019)
14. Blasco-Gimenez, R., Añó-Villalba, S., Rodríguez-D'Derlée, J., Morant, F., Bernal-Perez, S.: Distributed voltage and frequency control of offshore wind farms connected with a diode-based HVdc link. IEEE Trans. Power Electron. **25**, 3095–3105, 2010
15. Gogino, A., Goebel, K.: Milling Data Set (2007). http://ti.arc.nasa.gov/project/prognostic-data-repository/
16. Mahrishi, M., et al.: Data model recommendations for real-time machine learning applications: a suggestive approach. In: Machine Learning for Sustainable Development, pp. 115–128. De Gruyter (2021)
17. Suhr, D.: Principal component analysis vs. exploratory factor analysis. In: SUGI 30 Proceedings 2005, 203-30 (2005). https://doi.org/10.1177/096228029200100105
18. Eren, L.: Bearing fault detection by one-dimensional convolutional neural networks. Math. Probl. Eng. **2017**, 1–9 (2017). https://doi.org/10.1155/2017/8617315
19. Goel, A., Bhujade, R.K.: A functional review, analysis and comparison of position permutation-based image encryption techniques. Int. J. Emerg. Technol. Adv. Eng. **10**(7), 97–99 (2020)
20. Peng, Y., Zheng, Z.: Spectral clustering and transductive-SVM based hyperspectral image classification. Int. J. Emerg. Technol. Adv. Eng. **10**(4), 72–77 (2020)
21. Alanazi, B.S., Rekab, K.: Fully sequential sampling design for estimating software reliability. Int. J. Emerg. Technol. Adv. Eng. **10**(7), 84–91 (2020)
22. Gujarati, Y., Thamma, R.: A novel compact multi-axis force sensor. Int. J. Emerg. Technol. Adv. Eng. **10**(12), 1–12 (2020)
23. Laber, J., Thamma, R.: MATLAB simulation for trajectory/path efficiency comparison between robotic manipulators. Int. J. Emerg. Technol. Adv. Eng. **10**(11), 74–88 (2020)
24. Da Silva, J.: Use of smartphone applications in vibration analysis: critical review. Int. J. Emerg. Technol. Adv. Eng. **10**(8), 27–31 (2020)
25. Manikyam, S., Sampath Kumar, S., Sai Pavan, K.V., Rajasanthosh Kumar, T.: Laser heat treatment was performed to improve the bending property of laser welded joints of low-alloy ultrahigh-strength steel with minimized strength loss **83**, 2659–2673 (2019)
26. Kumar, S.S., Somasekhar, T., Reddy, P.N.K.: Duplex stainless steel welding microstructures have been engineered for thermal welding cycles & nitrogen (N) gas protection. Mater. Today: Proc. (2021). https://doi.org/10.1016/j.matpr.2020.11.091
27. Kumar, T., Mahrishi, M., Meena, G.: A comprehensive review of recent automatic speech summarization and keyword identification techniques. In: Fernandes, S.L., Sharma, T.K. (eds.) Artificial Intelligence in Industrial Applications. LAIS, vol. 25, pp. 111–126. Springer, Cham (2022). https://doi.org/10.1007/978-3-030-85383-9_8
28. Dhotre, V.A., Mohammad, H., Pathak, P.K., Shrivastava, A., Kumar, T.R.: Big data analytics using MapReduce for education system. Linguistica Antverpiensia 3130–3138 (2021)
29. Gurugubelli, S., Chekuri, R.B.R.: The method combining laser welding and induction heating at high temperatures was performed. Des. Eng. 592–602 (2021)
30. Pavan, K.V.S., Deepthi, K., Saravanan, G., Kumar, T.R., Vinay, A.: Improvement of delamination spread model to gauge a dynamic disappointment of interlaminar in covered composite materials & to forecast of material debasement. PalArch's J. Archaeol. Egypt/Egyptol. **17**(9), 6551–6562 (2020)

Detection of Web User Clusters Through KPCA Dimensionality Reduction in Web Usage Mining

J. Serin[1], J. Satheesh Kumar[2(✉)], and T. Amudha[2]

[1] Research and Development Center, Bharathiar University, Coimbatore, India
[2] Department of Computer Applications, Bharathiar University, Coimbatore, India
{j.satheesh,amudhaswamynathan}@buc.edu.in

Abstract. Dimensionality Reduction is a technique that performs feature extraction to enhance the evaluation of the model which uses data mining techniques. When the number of features in the model increases, the number of samples taken for tested also increases proportionally. The samples should also be selected to include all possible combination of the features. Web Mining plays a significant role in understanding the user's browsing behavioral pattern on their web usage data. Web Usage mining records the user's visited web page in their website. Each user's visited umpteen number of pages as well as each website also contains more number of web pages. These pages help to build the model which finds the user's behavioral pattern leads to a complex model which may have lot of inconsistent or redundant features in the data there by increase the computation time. When a greater number of features are included, there is also a high possibility of overfitting. Thus, to address the issue of overfitting and increase in computation time dimensionality reduction technique is introduced. Feature extraction technique helps to find the smaller set of new pages with the combination of the given independent web pages. Comparing three dimensionality reduction techniques such as PCA, Isomap and KPCA in which non-linear reduction technique of KPCA is used. Furthermore, unsupervised clustering technique of fuzzy Clustering technique is applied to identify the similar user behavioral pattern. Empirical study proved that the clustering accuracy is improved with the non-linear dimensionality reduction technique of KPCA is used with CTI dataset of DePaul University and MSNBC dataset is taken from the Server logs for msnbc.com

Keywords: Dimensionality reduction · PCA · Isomap · KPCA and fuzzy clustering

1 Introduction

Dimensionality Reduction plays a major role to avoid over fitting [1]. This technique filters only significant features required for training. Feature extracted data leads to less computation time and it requires less storage also. It removes redundant features, noise and also reduces the dimensions which makes the data applicable for many algorithms that are unfit for large number of dimensions. It works only with the highly correlated variables. With the higher dimensions of data, it is difficult to visualize the data which is

V. E. Balas et al. (Eds.): ICETCE 2022, CCIS 1591, pp. 597–610, 2022.
https://doi.org/10.1007/978-3-031-07012-9_50

overcome by dimensionality reduction. Dimensionality Reduction can be implemented in two different ways.

i) Feature Selection
ii) Feature Extraction
iii) Feature Selection

It is the process of choosing more relevant variables from our sample. The main advantages of these methods include easiest and maintaining interpretability of our variables and at the same time we gain no information from the eliminated variables.

Feature Extraction is the process of finding the smaller set of new variables with the combination of the given independent variables. It will keep the most important variable to predict the value and drop the least important independent variable. The following are the most well-known linear dimensionality reduction methods.

2 Literature Survey

Haakon et al. [2] introduced new methodology to find traffic anomalies using PCA. This research work consist of four subcomponents (i) First component: Data collected from Abilene with 11 – node research backbone that connects all the research labs across continental United States and Geant with 23-node network that connects national research and education network in European countries. Data flow statistics are taken from these two networks using Juniper's J-Flow tool to effectively detect the anomalies across the networks by PCA are evaluated.

Telgaonkar Archana et al. [3] performed classification techniques through PCA and LDA and on highly dimensional set such as UMIST (University of Manchester Institute of Science & Technology), COIL (Columbia Object Image Library) and YALE. These datasets consists of object images and human faces in which PCA is applied to reduce the dimensionality of huge data by retaining some useful information in the original dataset. Feature Extraction are done by using PCA and then classification are applied on feature vectors. The results of the dimensionality reduction techniques PCA and LDA are compared and the accuracy is evaluated with mean average and correlation.

Zhong et al. [1] processed data mining techniques for the prediction of the stock market return by using ANN (Artificial Neural Network) classification with the dimensionality reduction techniques. Three different dimensionality reduction techniques such as principal component analysis (PCA), fuzzy robust principal component analysis (FRPCA) and kernel – based principal component analysis are used to select the principal components among the 60 different financial economic features to forecast the stock returns. These principal components are applied to ANN classification to forecast their daily stock returns. Results proved that ANN with PCA got higher accuracy when compared the other two.

Amir Najafi et al. [4] introduced a new methodology named Path-Based Isomap a nonlinear dimensionality reduction method to identify pattern recognition and image classification. Significant improvement in execution time with less memory requirements and low dimensional embedding was achieved through path mapping algorithm with geodesic distances. They used stochastic algorithm to find the shortest paths which is called Stochastic Shortest Path Covering (SPSS).

Guihua Wen et al. [5] used improved Isomap with the graph algebra to optimize good neighborhood structure for visualization and time complexity. The neighborhood graphs were built within the dimension of the projected space using geodesic distances between two points which gave lower residual variance and robustness. This optimization reduced the time complexity because of the good neighborhood graph speed up the subsequent optimization process. This approach was tested and validated with benchmark datasets.

Paul Honeine [6] proposed a recursive kernel based algorithm to extract linear principal component with a reduced-order model. The classical kernel principal component analysis and Iterative kernel principal component analysis techniques are compared and evaluated on both synthetic datasets and images of hand written digits.

In this research work, different techniques of dimensional reduction have been examined to extract the important principal components taken for our further study. Related work stated that dimensionality reduction technique has been chosen based on the type of dataset and then applied on any data mining techniques. The experimental results proved that the accuracy of the result improved better after implementation of dimensionality reduction techniques.

The research work of the paper is structured as follows: Sect. 3 discusses about the three different feature extraction techniques in data mining. Section 4 discusses about the grouping technique as fuzzy c-means clustering. Section 5 shows the experimental results by applying three different unsupervised dimensionality techniques as PCA, KPCA and Isomap with fuzzy clustering and compare their accuracies. Section 6 concludes this paper with a review of the better dimensionality reduction technique.

3 Techniques in Dimensionality Reduction

The process of reducing the random variables for considering further process in machine learning, information theory and statistics is called dimensionality reduction [7]. Machine learning techniques involve both feature selection and feature extraction steps by considering both linear and non-linear dimensionality reduction models.

3.1 Principal Component Analysis (PCA)

To visualize the relational distance between the data points for linear data sets is measured by the statistical tool called PCA [8, 9]. It acts an essential tool for data analysis to make predictive models. It is a technique which finds a new set of variables with the existing huge number of variables in the given dataset and the new set of independent variables is called principal components. The linear combination of these variables called principal components which is extracted in such way that the principal components describe maximum variance in the dataset. It enables us to identify correlations and patterns in the dataset. After applying PCA, all variables become an independent to one another. It is benefitted under the assumptions of a linear model.

The three major steps involved in PCA are as follows

i) Covariance matrix for the given dataset is constructed
ii) Calculation of eigen values and vector of the covariance matrix
iii) Transform our data for reduced dimensionality reduction by using eigen values and vectors.

3.2 Kernel PCA (KPCA)

PCA works well for the linear datasets and it can be applied to linear dataset which are linearly separable. If the dataset is non-linear, optimal solution might not be attained during dimensionality reduction [10, 11]. The non-linear function is much needed to reduce the dimensionality of dataset for linearly inseparable data into linearly separable data by projecting them into high dimensional space. Mapping function is required to trait the data from a lower dimensional space into higher dimensional space to make the dataset into linearly separable. The function which is used to overcome the problem of linearly inseparable data is kernel function and it used in principal component analysis. PCA will be done in high dimensional space to reduce the dimensions. The following steps helps to perform the KPCA:

i) Choose the kernel function as
   β (Xi, Xj) and let T be any transformation to a higher dimension
ii) Construct the kernel matrix which applies kernel function to all pairs of data
iii) The new kernel matrix is defined
iv) Sort our eigen vectors based on the calculated eigen values in decreasing order
v) Choose the value of m and find the product of that matrix with our data, where m be the number of dimensions for our reduced dataset.

3.3 Isometric Mapping (Isomap)

Dimensionality Reduction methods use Euclidean distance measure to find the principal components of data and it gives better approximation for linear data [12, 13]. It fails to discover more complex structures of data and it reduces the geometry of data. Isomap algorithm belongs to the part of non-linear dimensionality reduction family and it main focus is to reduce the dimensions. It finds low dimensional embedding of the dataset through eigenvalue decomposition of the geodesic distance matrix. Euclidean distance metric holds good if the neighborhood points are linear. Isomap creates similarity matrix by using eigen value decomposition. It uses the local information to create a global similarity matrix and uses Euclidean metrics to find the neighborhood graph. It measures the shortest path between two points by approximating geodesic distance measure.

4 Fuzzy Clustering

Traditional grouping of data in unsupervised algorithms such as k-means, k-mediods helps to cluster the data points into one cluster where the data belong to only one cluster. In order to predict the user behavioral pattern, the users may be interested in more than one page and obviously the user may belong to more than one cluster. To identify such type of clusters, Fuzzy clustering was first introduced by Dunn [14] and improved by Bezdek [15] in which every object is a member to every other cluster to a certain degree. The data object which lies closer to the center of the cluster will have a high degree of belonging or the data objects that lies far away from the center will have a lower degree of pertaining or membership to that cluster. The objective function used for minimizing the dissimilarity by Fuzzy C Means Clustering (FCM) is given by equation:

$$\mathrm{Jm} = \sum\nolimits_{i=1}^{c} \sum\nolimits_{j=1}^{m} Q_{ij}^{n} d_{ij}^{2},\ 1 \le \mathrm{n} < \infty \tag{1}$$

$$\text{where } \mathrm{d_{ij}} = \left|\left|\mathrm{x_j} - \mathrm{c_j}\right|\right|^2$$

where $\mathrm{Q_{ij}}$ is the degree of membership function matrix, C is the mentioned number of clusters, m is the total number of data records, n is any real number greater than 1, $\mathrm{d_{ij}}$ is the distance from $\mathrm{x_j}$ to $\mathrm{v_{ij}}$. $\mathrm{v_j}$ denotes the cluster center of the $\mathrm{j^{th}}$ cluster to the $\mathrm{i^{th}}$ iteration.

5 Proposed Work

The preprocessed logfile has been taken as input for three different kinds of unsupervised feature extraction techniques used as Principal Component Analysis, Isometric mapping

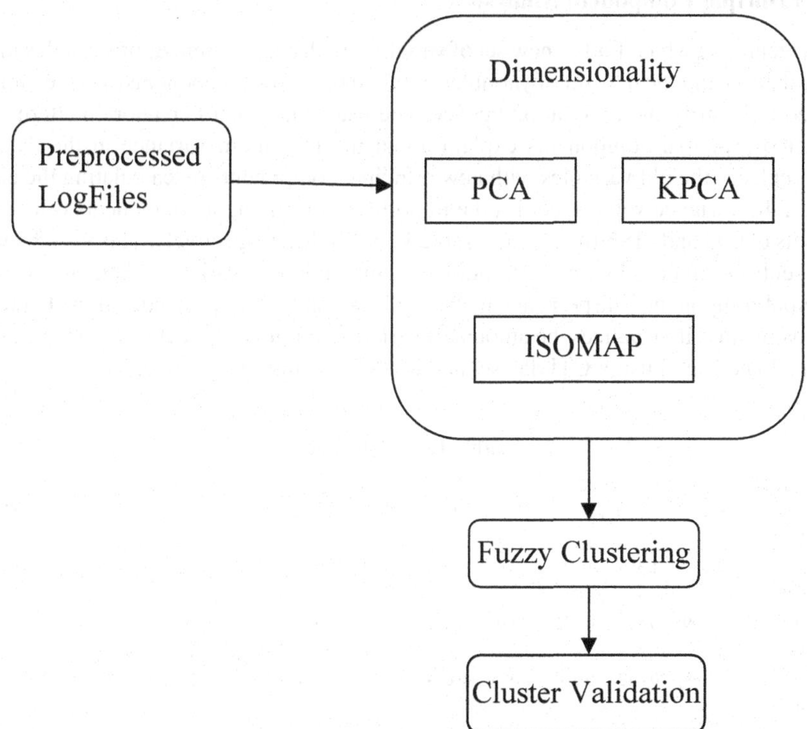

Fig. 1. Methodology of the work flow

and Kernel Principal Component Analysis. The important principal components are generated and taken as an input for fuzzy clustering. The goodness of the clustering results is validated through their Silhouttee Coefficient and Dunn index by comparing their results. The diagrammatical representation of the proposed work is shown in Fig. 1.

6 Experimental Evaluation

The three different unsupervised dimensionality reduction techniques are implemented in R programming by giving the preprocessed logfiles. The input has been taken from MSNBC anonymous web dataset and also the dataset taken from DePaul University (https://cds.cdm.depaul.edu/resources/datasets/). The raw log files have been taken from this site has performed the following preprocessing techniques implemented in the log files as Data Cleaning, User Identification, Session Identification and Path Completion [14].

The preprocessed data has been transformed into Weighted Session Page view matrix to reduce the complexity to apply clustering techniques for finding similar web users. In this paper, fuzzy clustering technique is applied with the new principal components which is identified by the feature extraction techniques in order to get more accurate results.

6.1 Principal Component Analysis

It is a technique which finds a new set of variables with a large number of variables in the given dataset and the new set of variables is called principal components. This technique is used to identify the correlation between the user's and their behavioral pattern. The extracted principal components explain about the maximum variance in the dataset. PCA replaces the old variables with new principal components by calculating the eigen values, the variance value specifies the importance of principal components of the two datasets of CTI and MSNBC given in Table 1 and Table 2 respectively. This visualization represents the eigen values corresponding to their amount of variation explained by each principal component. The point at which the elbow bends shows the indication of optimal dimensionality. It visualizes the amount of variation captured by each component shown in Fig. 2 and Fig. 3 using CTI dataset and MSNBC dataset respectively.

Table 1. CTI dataset

Importance of Components	PC1	PC2	PC3	PC4	PC5	PC6	PC7	PC8	PC9	PC10	PC11	PC12
Standard Deviation	1.531	1.356	1.227	1.153	1.035	0.972	0.857	0.786	0.749	0.655	0.638	0.452
Proportion of Variance	0.199	0.153	0.125	0.110	0.089	0.078	0.061	0.051	0.046	0.035	0.033	0.017
Cumulative Proportion	0.199	0.348	0.474	0.585	0.674	0.753	0.814	0.866	0.913	0.948	0.982	1.000

Table 2. MSNBC dataset

Importance of Components	PC1	PC2	PC3	PC4	PC5	PC6	PC7	PC8	PC9	PC10	PC11	PC12	PC13	PC14	PC15	PC16
Standard Deviation	2.197	1.374	1.228	1.022	1.005	0.989	0.985	0.899	0.860	0.830	0.805	0.753	0.698	0.678	0.630	0.566
Proportion of Variance	0.268	0.104	0.083	0.058	0.056	0.056	0.054	0.053	0.044	0.041	0.038	0.036	0.031	0.271	0.025	0.022
Cumulative Proportion	0,268	0.373	0.457	0.515	0.571	0.625	0.679	0.724	0.765	0.804	0.840	0.871	0.923	0.966	0.984	1.000

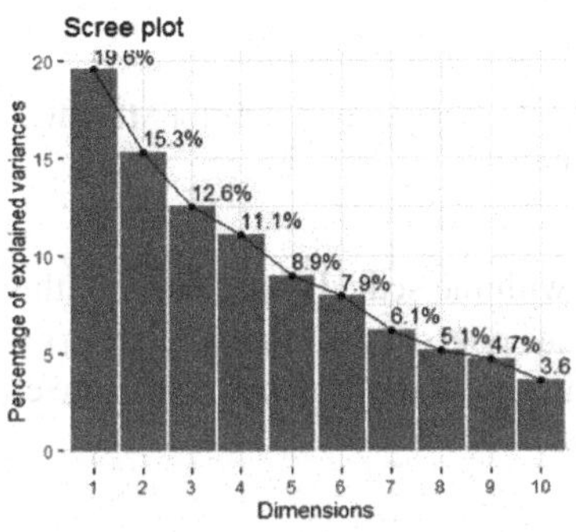

Fig. 2. Variations in CTI dataset

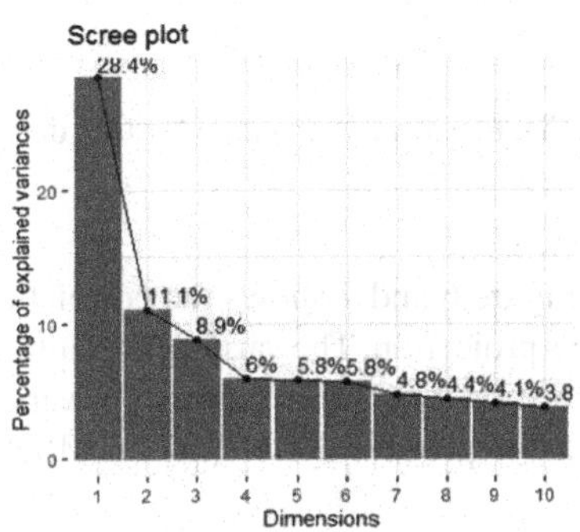

Fig. 3. Variations in MSNBC dataset

From the above scree plot it is inferred that the five principal components used for CTI dataset and first three principal components used for MSNBC dataset. After that there is a steady decrease for the remaining components. A graphical display of correlation matrix shows the positive and negative correlation.

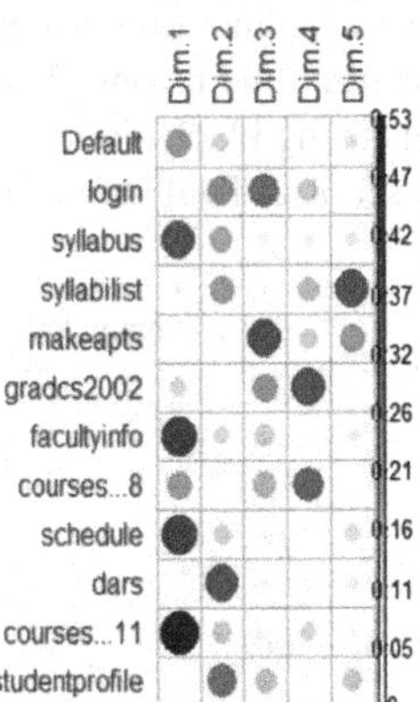

Fig. 4. Correlation graph of CTI Dataset

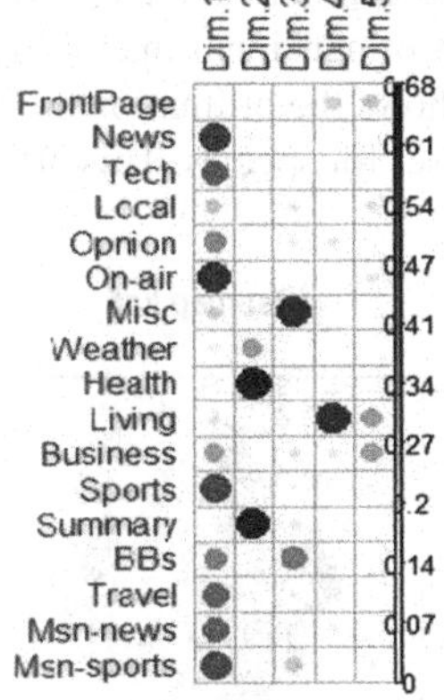

Fig. 5. Correlation graph of MSNBC dataset

Positive correlation which represents that the both variables move in the same direction and the negative correlation represents that the one variable decreases the other variable increases and vice-versa. In this Fig. 4 and Fig. 5, indicated that the bigger blue dot represents positive correlation and the smaller one represents negative correlation for CTI and MSNBC dataset.

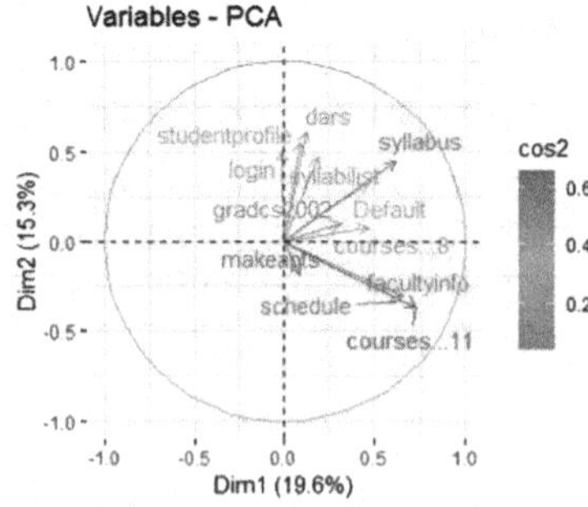

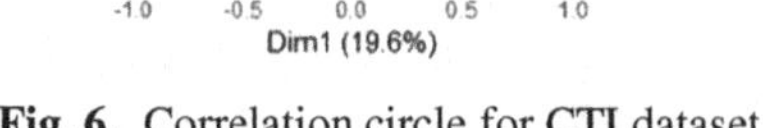

Fig. 6. Correlation circle for CTI dataset

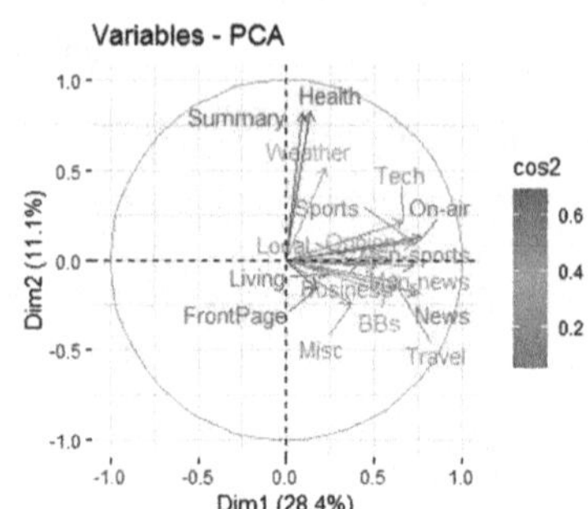

Fig. 7. Correlation Circle for MSNBC Dataset

Figure 6 and 7 shows the correlation circle with the scaled coordinates of the variable's projection. The variable Default, syllabus and course 8 are closely relate to positive correlation for CTI dataset and the variables Sports, msn-sports, On-air, Tech are closely related to positive correlation for MSNBC dataset.

6.2 Isometric Mapping

Isomap is a non-linear dimensionality reduction and geodesic distance calculation is calculated between the data points in the neighborhood graph. The first step is to determine the neighborhood points to retain the shortest dissimilarities among objects and the second step is to construct the neighborhood graph by estimating the dissimilarities as shortest path distances. The next step is to find the shortest path between the data points. By performing multi-dimensional scaling, the higher dimensional data is reduced into lower dimensional data. Non-linear mapping between our higher dimensional data and our lower dimensional manifold is done to reduce feature dimensions. This Fig. 8 (a) and (b) shows the ordination diagram for both data sets using black circles as pages and blue crosses for species. This plot graph shows the subset of DePaul University CTI data

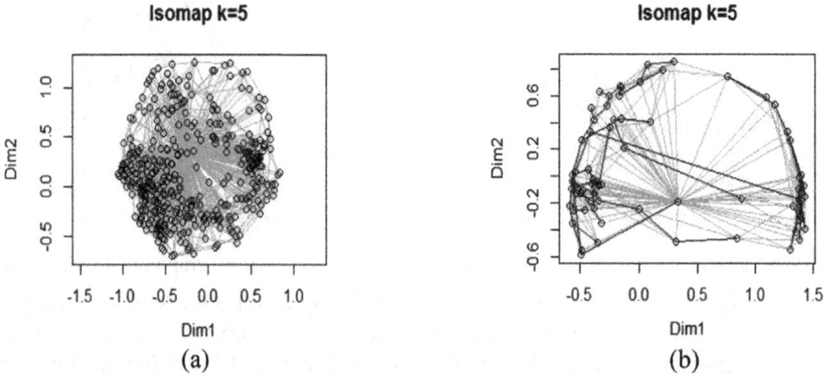

Fig. 8. (a) Ordination diagram for CTI and (b) MSNBC dataset

and MSNBC dataset after isomap dimension reduction. If the manifold is not sampled well and also contains holes these technique can't be performed well.

6.3 Kernel Principal Component Analysis

Kernel Principal Component Analysis is a nonlinear dimensionality reduction which effectively used to compute principal components in high dimensional feature spaces. The data passed in the kernel function and the Gaussian value used to map the high dimensional feature space and thereby principal components are calculated. This Fig. 9 (a) and (b) shows the plot graph of the principal components data as two dimensional data for CTI and MSNBC dataset.

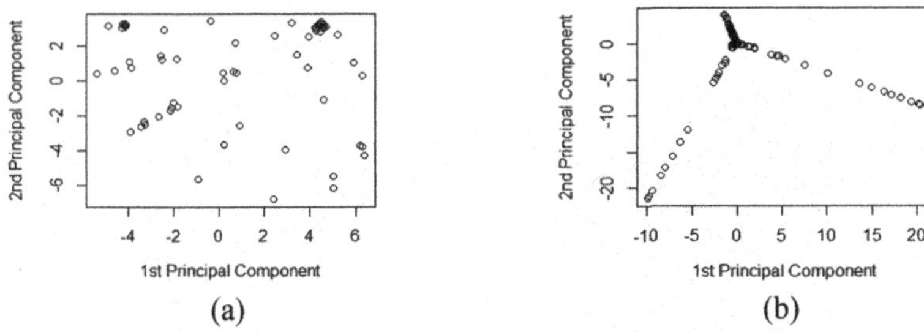

Fig. 9. (a) Principal components for CTI b) MSNBC dataset

6.4 Fuzzy Clustering

It is a clustering method where the data points is included in more than one cluster. It classify the data points into various groups based on similarities between the items. A data point with a membership grade is in between 1 and 0 where 0 represents that the data point is farthest from the cluster center and 1 represents that the data point is closest to the cluster center. Data points in the cluster should be similar as possible to

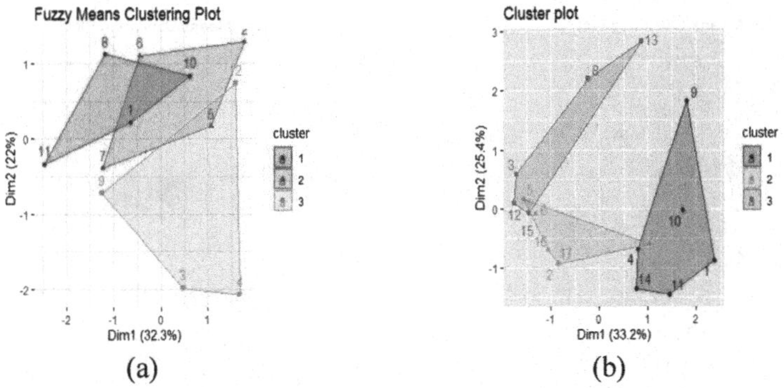

Fig. 10. (a) Fuzzy Clustering using CTI (b) Fuzzy Clustering using MSNBC using PCA

each other and as dissimilar to other points in the cluster. The Fig. 10 (a) and (b) shows fuzzy c-means clustering form three clusters after applying PCA with DePaul University and MSNBC dataset. The diagram shows that same data point belongs to more than one cluster. Obviously, the same user may be interested in more than one page. They might be interested sports, health, news and so on. It will be useful to identify the user's behavioral pattern also.

Isomap stands for Isometric mapping which is an unsupervised dimensionality reduction technique based on the spectral theory. It preserves the geodesic distances in the lower dimension. Fit a manifold with five nearest neighbors and reduce the dimension into two components and fuzzy clustering is applied with their isomap coordinates. The Fig. 11 (a) and (b) shows the fuzzy cmeans with three and four clusters done on CTI dataset.

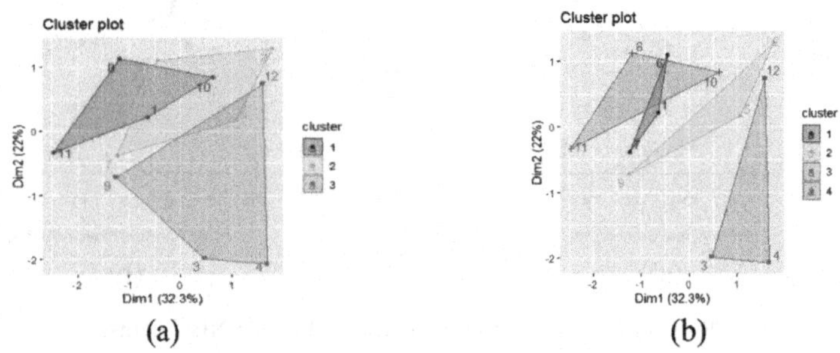

(a) (b)

Fig. 11. (a): Isomap with Fuzzy Clustering (m = 3) b) Isomap with Fuzzy Clustering (m = 4)

Kernel Principal Component Analysis is a non-linear version of the PCA which is capable of apprehend higher order statistics to represent the data in a reduced format [16]. To capture higher order statistics the data of the input space is mapped into another space, called feature space [17]. Fuzzy Clustering is applied to the principal components of KPCA and the clustering result in shown in Fig. 12 (a) and (b) with the value of m 3 & 4 clusters respectively. Three clusters are formed in CTI dataset and the four clusters are formed using MSNBC dataset and the dimensionality variance of the principal

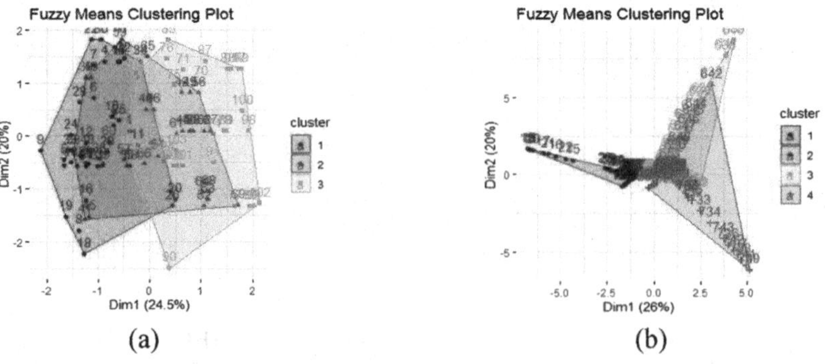

(a) (b)

Fig. 12. (a) KPCA for Fuzzy Clustering with m = 3 b) KPCA for Fuzzy Clustering with m = 4

component is also shown in the figure. Overlapping clusters shows the user's browsing interesting is involved in more than cluster.

6.5 Analysis of Results

The procedure for evaluating the goodness of the clustering algorithm is represented by the term cluster validation. Internal cluster validation is one of the categorical classes for clustering validation. This validation technique uses the internal information to evaluate the goodness of the structure without referring any external information. The silhouette coefficient measures the average distance between the clusters and the silhouette plot displays how close each point in one cluster is to point the neighboring clusters.

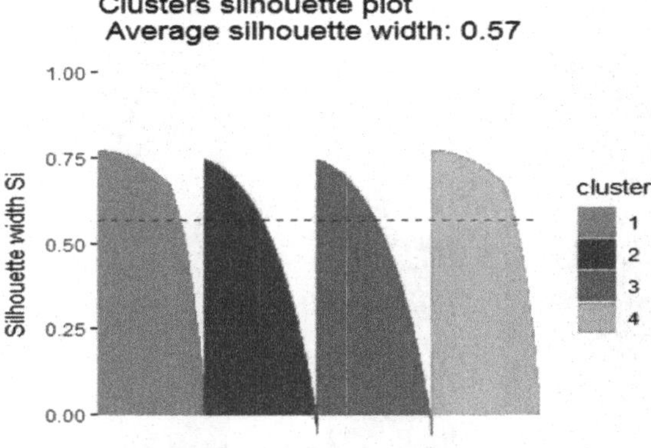

Fig. 13. Silhouette width for KPCA

The Fig. 13 represents average silhouette width taken from four clusters for CTI dataset and each cluster silhouette width is represented as Table 3.

The silhouette average width for cluster formation is shown in Fig. 14 by using three different unsupervised dimensionality reduction techniques for the two different datasets. The silhouette average width closes to 1 represents that the observation is well matched to the assigned cluster. By using this measurement statistics, KPCA gets high value of silhouette width which indicates a good clustering when compare to other two techniques.

Table 3. Silhouette width for each cluster

Cluster no	Sil. width
1	0.75
2	0.67
3	0.47
4	0.72

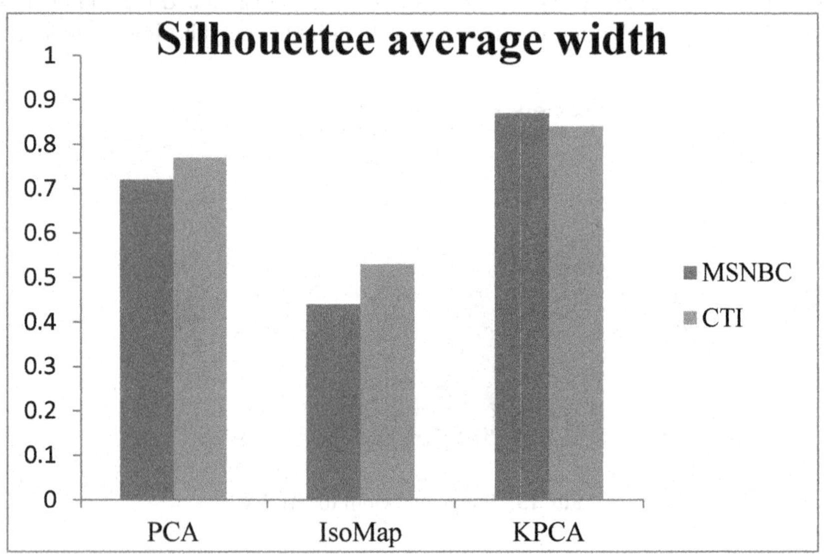

Fig. 14. Silhouette Average Width

The dunn index is another internal clustering measure which can be computed by using the formula

$$D = \frac{min.seperation}{max.diameter} \tag{2}$$

The min. denotes the minimum value of inter-cluster separation and the max. denotes the maximal value of intra cluster distance.

This Fig. 15 represents the Dunn Index measurement which evaluate the quality of the clusters. If the data points involved in the clusters are compact and well separated indicated that the diameter of the cluster is expected to be small and the distance between the clusters is expected to be large. Therefore, the value of the Dunn Index should be maximized. By comparing the dunn values among these three dimensionality reduction techniques, KPCA dunn value is maximized when compared to other two techniques.

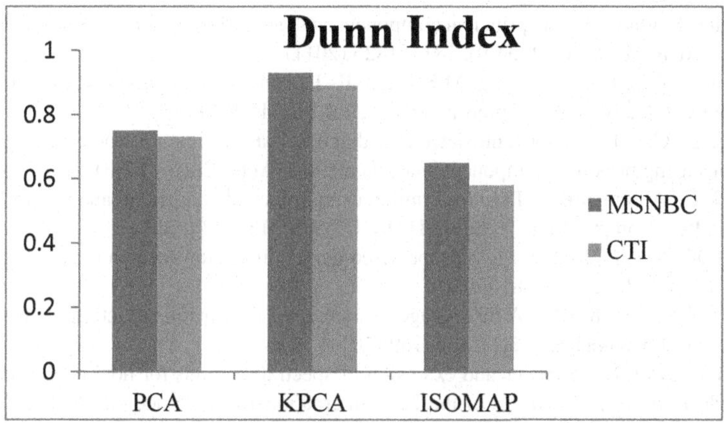

Fig. 15. Dunn Index values

7 Conclusion

This paper described and implemented three different unsupervised dimensionality techniques. They are Principal Component Analysis (PCA), manifold learning as Isometric mapping (ISOMAP) and Kernel based Principal Component Analysis (KPCA). The choice of these three different approaches is based on the dataset and the mining techniques going to be applied. PCA extracted their principal components by finding a new set of variables with the existing huge number of variables and it is a well-known statistical tool for linear datasets which is used to predict models also. Isomap is a non-linear dimensionality reduction method which preserves true relationship between the data points. It is computational is very expensive and it fails for manifold with holes. KPCA is a nonlinear dimensionality reduction which effectively used to compute principal components in high dimensional feature spaces. To handle non-linear data with its effective computationally time KPCA is used for dimension reduction.

References

1. Zhong, X., Enke, D.: Forecasting daily stock market return using dimensionality reduction. Expert Syst. Appl. **67**, 126–139 (2017)
2. Ringberg, H., Soule, A., Rexford, J., Diot, C.: Sensitivity of PCA for traffic anomaly detection. In: Proceedings of the 2007 ACM SIGMETRICS International Conference on Measurement and Modeling of Computer Systems, pp. 109–120, June 2007
3. Sachin, D.: Dimensionality reduction and classification through PCA and LDA. Int. J. Comput. Appl. **122**(17), 1–5 (2015)
4. Tiwari, A., Sharma, V., Mahrishi, M.: Service adaptive broking mechanism using MROSP algorithm. In: Kumar Kundu, M., Mohapatra, D.P., Konar, A., Chakraborty, A. (eds.) Advanced Computing, Networking and Informatics- Volume 2. SIST, vol. 28, pp. 383–391. Springer, Cham (2014). https://doi.org/10.1007/978-3-319-07350-7_43
5. Wen, G., Jiang, L., Shadbolt, N. R.: Using graph algebra to optimize neighborhood for isometric mapping. In: IJCAI, pp. 2398–2403, January 2007

6. Honeine, P.: Online kernel principal component analysis: a reduced-order model. IEEE Trans. Pattern Anal. Mach. Intell. **34**(9), 1814–1826 (2011)
7. Jorgensen, P.E., Kang, S., Song, M.S., Tian, F.: Dimension Reduction and Kernel Principal Component Analysis. arXiv preprint arXiv:1906.06451 (2019)
8. Wang, S., Cui, J.: Sensor-fault detection, diagnosis and estimation for centrifugal chiller systems using principal-component analysis method. Appl. Energy **82**(3), 197–213 (2005)
9. Paul, L.C., Al Sumam, A.: Face recognition using principal component analysis method. Int. J. Adv. Res. Comput. Eng. Technol. (IJARCET) **1**(9), 135–139 (2012)
10. Wang, M., Sha, F., Jordan, M.: Unsupervised kernel dimension reduction. Adv. Neural. Inf. Process. Syst. **23**, 2379–2387 (2010)
11. Bharti, K.K., Singh, P.K.: A three-stage unsupervised dimension reduction method for text clustering. J. Comput. Sci. **5**(2), 156–169 (2014)
12. Sha, F., Saul, L.K.: Analysis and extension of spectral methods for nonlinear dimensionality reduction. In: Proceedings of the 22nd International Conference on Machine Learning, pp. 784–791, August 2005
13. Najafi, A., Joudaki, A., Fatemizadeh, E.: Nonlinear dimensionality reduction via path-based isometric mapping. IEEE Trans. Pattern Anal. Mach. Intell. **38**(7), 1452–1464 (2015)
14. Dunn, J.C.: A fuzzy relative of the ISODATA process and its use in detecting compact well-separated clusters (1973)
15. Bezdek, J.C.: Pattern Recognition with Fuzzy Objective Function Algorithms: Springer (2013)
16. Schölkopf, B., Smola, A., Müller, K.R.: Nonlinear component analysis as a kernel eigenvalue problem. Neural Comput. **10**(5), 1299–1319 (1998)
17. Datta, A., Ghosh, S., Ghosh, A.: PCA, kernel PCA and dimensionality reduction in hyperspectral images. In: Naik, G.R. (ed.) Advances in Principal Component Analysis, pp. 19–46. Springer, Singapore (2018). https://doi.org/10.1007/978-981-10-6704-4_2

Integrating Machine Learning Approaches in SDN for Effective Traffic Prediction Using Correlation Analysis

Bhuvaneswari Balachander[1(✉)], Manivel Kandasamy[2], Venkata Harshavardhan Reddy Dornadula[3], Mahesh Nirmal[4], and Joel Alanya-Beltran[5]

[1] Saveetha School of Engineering, Saveetha Institute of Medical and Technical Sciences, Kanchipuram, India
bhuvaneswari@saveetha.com

[2] Unitedworld School of Computational Intelligence, Karnavati University, Gandhinagar, Gujarat, India
manivel.k@karnavatiuniversity.edu.in

[3] IT & ITES, Startups Mentoring Society, Tirupati, India

[4] PREC, Loni, India

[5] Electronic Department, Universidad Tecnológica del Perú, Lima, Peru
C18121@utp.edu.pe

Abstract. The study shows that numerous academic researchers are utilizing machine learning and artificial intelligence approaches to regulate, administer, and run networks, as a result of the recent explosion in interest in these fields. In contrast to the scattered and hardware-centric traditional network, Software Defined Networks (SDN) are a linked and adaptive network that offers a full solution for controlling the network quickly and productively. The SDN-provided network-wide information may be used to improve the efficiency of traffic routing in a network environment. Using machine learning techniques to identify the fewest overloaded path for routing traffic in an SDN-enabled network, we investigate and demonstrate their application in this study. These years have seen an increase in the number of researchers working on traffic congestion prediction, particularly in the field of machine learning and artificial intelligence (AI). This study topic has grown significantly in recent years on account of the introduction of large amounts of information from stationary sensors or probing traffic information, as well as the creation of new artificial intelligence models. It is possible to anticipate traffic congestion, and particularly short-term traffic congestion, by analyzing a number of various traffic parameter values. When it comes to anticipating traffic congestion, the majority of the studies rely on historical information. Only a few publications, on the other hand, predicted real-time congestion in traffic. This study presents a comprehensive summary of the current research that has been undertaken using a variety of artificial intelligence approaches, most importantly distinct machine learning methods.

Keywords: Software Defined Network (SDN) · Machine Learning (ML) · Artificial intelligence · Traffic · Quality of Service (QoS) · Information · Datasets

V. E. Balas et al. (Eds.): ICETCE 2022, CCIS 1591, pp. 611–622, 2022.
https://doi.org/10.1007/978-3-031-07012-9_51

1 Introduction

Conventional networks are naturally dispersed, and they route or forward signals using hop-based routing strategies. The path with the fewest hops between any pair of nodes is picked, and all traffic is routed down that path. This method of transferring data worked well when networks were first created, but as they gotten bigger and utilization, issues like as latency began to emerge, strangling the lines and requires the interaction inefficient. Due to the substantial control overhead imposed by communicating modified network information in a distributed fashion, routing techniques such as hop-based approaches do not include the current amount of network problems into their route computations. However, this often results in wasteful network resource consumption, since signals may be routed along a little long corridor that is less crowded than the congested least steps path, resulting in improved traffic stress sharing in the networks [1].

SDN (Software Defined Networking) is a revolutionary networking architecture that allows for centralized management of network assets by combining control and information planes [2]. It acts as a customization link between network elements and the centralized administrator. The centralized controller offers a worldwide view of the whole network, allowing for more versatility in regulating and running the network to effectively fulfill the required Quality of Service (QoS) need. Conventional networks lack such a coherent worldwide perspective of the network. Machine Learning is now widely employed in a wide range of applications & industries. Using labeled samples, supervised machine learning techniques are used to anticipate future occurrences [3]. The method infers a suitable function to produce effective findings by studying a known experimental dataset. Unsupervised machine learning methods, on the other hand, are utilized when no previous knowledge is able to instruct the algorithms.

It attempts to derive assumptions from the datasets in attempt to characterize underlying commonalities and traits in the unorganized information. Semi-supervised machine learning methods train using a mix of labeled and unlabeled input. This ensemble may be used to significantly increase acquisition accuracy. Machine learning techniques that use reinforcement learning create actions and alter them in response to punishments or incentives. This technology enables machines & software programs to autonomously find the optimal behavior in a given scenario in order to optimize their efficiency. Due to their dispersed and dumb nature, machine learning methods are difficult to integrate and implement in conventional networks to govern and run networks. SDN opens up new avenues for incorporating intelligence into networks [4].

2 Objective

The research aimed to fulfill the following objectives:

1. To study the Machine Learning
2. To study the Software Defined Networking (SDN)
3. To study the Traffic Prediction
4. To study the Traffic Classification
5. To study Optimized Routing on the Basis of Traffic Classification

3 Methodology

Machine learning & artificial intelligence to regulate networks. SDNs transform traditional networks into linked, complex networks that provide a full network management platform. SDN's network-wide data may help enhance traffic routing. For traffic monitoring in an SDN network, we examine and show the application of machine learning approaches. In current history, traffic congestion prediction has become a major topic in Artificial Intelligence (AI) and machine learning (ML). New AI models and visualization of information from permanent sensors and investigation devices have expanded this research area. Different traffic parameters are assessed to predict short-term traffic congestion.

4 Machine Learning

Machine Learning is defined as the branch of research that investigates how computers may learn without even being specifically programmed in a certain way. This might be viewed of as a more formalized version of an earlier casual judgment. Machine learning is defined as: When a computer programme learns from experiences E with regard to a group of tasks T as well as a performance measure P, it is said to have learned from experience E if its performance at tasks in T, as measured by P, increases over time. It is possible to describe machine learning (ML) as the research journal of techniques & statistical modelling that computerized networks use to do a certain job without utilizing detailed instructions, rather depending on trends & assumptions [5] as a subfield of machine learning.

Table 1 shows the results of the study. Machine Learning can be divided into 4 main categories: supervised, unsupervised, semi-supervised, & reinforcement learning. Supervised learning is the most common kind of machine learning. Detailed descriptions of various algorithms are provided in the next section [6] (Fig. 1).

Table 1. The results of the study

Learning method	Description
Supervised machine learning	To build a supervised learning model, you provide the system with learning algorithm, that is, inputs and anticipated outcomes/outputs. With fresh data, it can forecast using the learned connection. The training data is labelled. Supervised learning issues fall into two categories: regression & categorization. While in regression issues, the theory predicts continuously outcomes, classification models generate discontinuous outcomes. Many supervised learning methods exist [7]
Unsupervised machine learning	Machine learning without supervision is the absolute opposite of supervised learning. The system is provided with a set of inputs information without their matching outputs in this paradigm. The algorithm groups the information regarding the relationships between the information's "features" [8]. k-means & self-organizing mappings are 2 main instances of such techniques
Semi-supervised machine learning	Semi-supervised learning is a kind of learning that falls between supervised and unsupervised. In this paradigm, a portion of the input information has associated outputs (labelled), whereas the rest does not (unlabeled). 'Transductive' SVM & graph-based approaches are examples of such techniques [9]
Reinforcement learning	It is the challenge of determining how individuals should conduct in a given environment in order to maximize progressive rewards [10]. An agent interacts with its surroundings in order to learn the appropriate behaviors to do in order to maximize long-term benefit

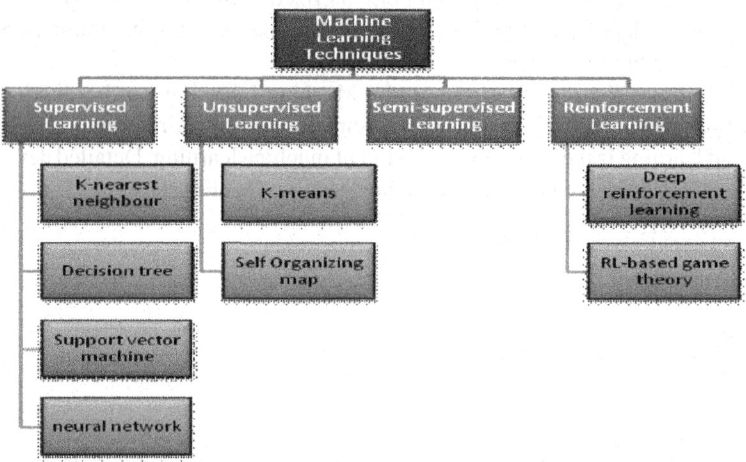

Fig. 1. Techniques for machine learning

5 Software Defined Networking (SDN)

The internet as we understand it is mostly built of the connectivity of old or heritage networking, which is what we call the backbone of the internet. Traditional networks are comprised of specialized network equipment such as networking equipment, which is liable for both processes of moving choices and transmitting data throughout the network. These network devices route traffic based on just a limited understanding of the network's configuration. The network structure of the Legacy system is scattered. Network devices are composed of many thousands of code lines that are exclusive and do not allow for any degree of customization. The intricacy of conventional networks [11] is one of their most significant drawbacks. The demands placed on networks have changed throughout the years. In response to these evolving needs, the more complicated the architecture of network devices becomes, as well as the increased cost of this equipment. It is also difficult to apply new regulations in old networks because of their established infrastructure. Adding an extra device or introducing a new service in an already big network would need a thorough setup that would include numerous network devices being used in tandem. This will take a significant amount of time & financial assets (Fig. 2).

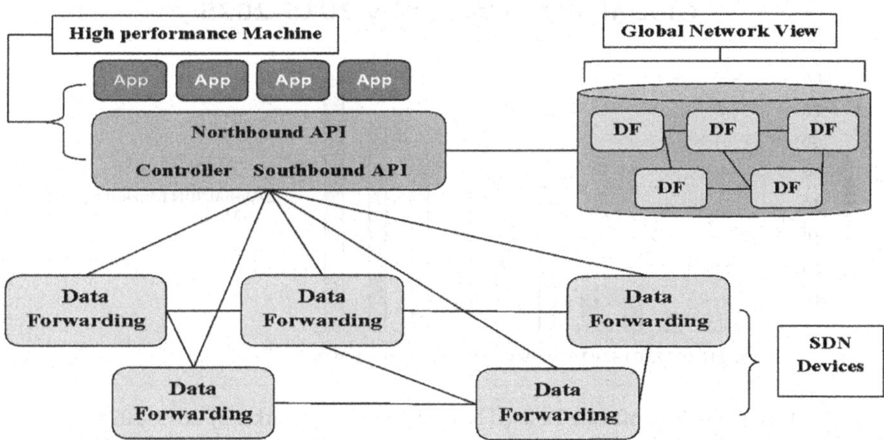

Fig. 2. Design of Software Defined Networking (SDN)

Because the usage of Internet of Things devices is growing at such a fast pace, conventional networks are plainly falling short of the mark to satisfying today's network needs. In current history, there has been a significant increase in interest in software-defined networking. Many studies have been conducted in recent years to determine how we might leverage the flexibility provided by SDN to control traffic and enhance computer connectivity, particularly in light of the increasing usage of Internet of Things (IoT) devices. SDN is essentially the splitting of the network switch into stages [12] and is not complicated. This centralized component, which manages networks gadgets such as switches, shields the network's intellect away from the rest of the network. When compared to the restrictions of conventional networks, SDN provides more network flexibility.

In classical networks, a router or switch is considered to be a singular object that comprises both the control plane (the brain) that makes decisions and the information plane or transmitting plane that is accountable for the forwarding of packets of information. With the introduction of SDN, there seems to be a decoupling between the control plane & the information planes. In other words, there is a specialized central controller that is independent of the rest of the system and that manages the transmitting switches. The transmitting mechanisms on the information plane are completely incompetent. Their intellect might be limited or non-existent if they are built in this way. There are several benefits to this concept [1]. It provides network managers with the ability to programme networks in order to meet more specialized or customized demands. In addition, as compared to traditional networking, where the switches only get a limited view of the network, the central unit has a worldwide perspective of the network, which is advantageous (Fig. 3).

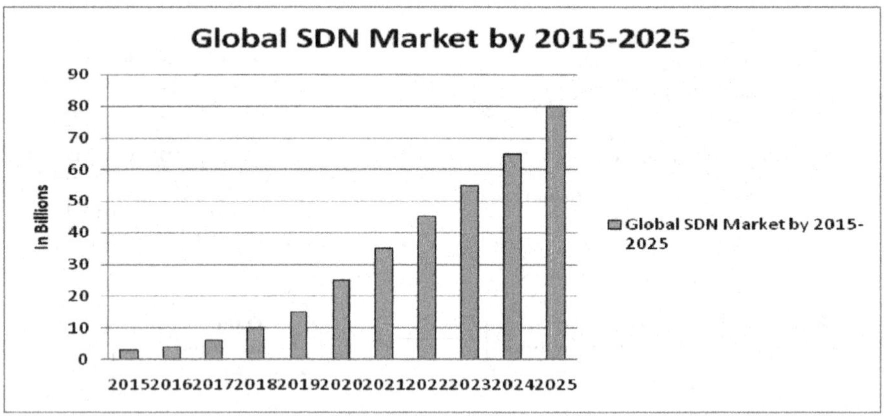

Fig. 3. Global Software Defined Networking (SDN) market by 2015–2025

6 Traffic Prediction

Network traffic prediction is critical in the management & administration of today's more sophisticated and diversified networks, and it is becoming more important. It includes predicting future traffic volumes, and has typically been handled via the use of time series forecasting (TSF). The goal of TSF is to develop a regression framework that is suitable of drawing an adequate link amongst projected volume of traffic and recently known volume of traffic [13]. Current TSF models for traffic prediction may be divided into two categories: statistically analytical methods and supervised machine learning (ML) models. Advanced auto - regressive integrating movement averages (ARIMA) methods are often used in statistical methods, but supervised neural networks (supervised NNs) are used to train the vast number of traffic prediction algorithms in practice.

Generalized auto-regressive (AR) or movement averages (MA) methods are used in combination to conduct auto-regression on differences or "stationarized" information in the ARIMA model, which is a common strategy for TSF. As a result of the fast expansion of networks and the rising sophistication of network traffic, classic TSF models appear to be undermined, resulting in the development of more powerful machine learning models [14]. Other than traffic volume, attempts have been made in latest days to minimize inefficiency and/or increase correctness in traffic predictions by including information from flows, which are distinct from traffic volume, into the forecast process. Numerous traffic prediction algorithms that make use of machine learning are described in the following subsections, and their results are summarized in Table 2.

Table 2. Numerous traffic prediction algorithms

ML technique	Application	Dataset	Features	Output
Supervised:- SVR	Prediction of link load in ISP networks (TSF)	Internet traffic gathered at an ISP network's POP (N/A)	Link stress measured on a timeframe	Link load prediction
Supervised:- MLP-NN with different training algorithms (GD, CG, SS, LM, Rp)	Network traffic prediction (TSF)	1000 points datasets (N/A)	Previous observations	Traffic volume that is expected
Supervised:- NNE trained with Rp	Predictions of link stress and traffic flow in ISP networks (TSF)	Traffic on a transatlantic connection aggregating traffic in the ISP backbone (N/A) SNMP traffic information from two ISP networks	Estimated traffic volume in the last few minutes multiple days	Traffic volume that is expected
Supervised:- MLP-NN	Prediction of route bandwidth availability from beginning to finish (TSF)	NSF Turgid datasets (N/A)	Load maximum, minimum, and average values measured in the last 10 s to 30 s	The amount of capacity that will be available on an end-to-end link in a future era
Supervised:- KBR-LSTM-RNN	Using historical internet flow data to predict future traffic volume (regression)	Over a 24-week timeframe, network traffic volumes and flowing count data were recorded every 5 min (public)	Flow count	Traffic volume that is expected

7 Traffic Classification

Traffic categorization is crucial for network administrators in order to carry out a broad variety of network administration and managerial functions efficiently and effectively. Capacity management, security & detection techniques, quality of service & service segmentation, efficiency measurement, & resource management are just a few of the capabilities available. For instance, a corporate network operator may also want to prioritize traffic for business-critical apps, recognize an unknowable traffic for outlier detection, or undertake workload characteristics in required to design efficient material managerial strategies that meet the efficiency and resource specifications of a variety of apps [15].

It is necessary to be able to properly correlate network traffic with pre-defined classes of concern in order to perform traffic classification. It is possible that these categories of interests will be classified as classes of applications (e.g., e-mail), classes of apps (e.g., Skype, YouTube, or Netflix), or classes of services [16]. An application or class of apps that have the same QoS criteria are grouped together as a class of service, for example, based on Quality of Service (QoS). As a result, it is feasible that apps that seem to function differently from one another are really part of the same level of services [17].

In generally, network traffic categorization algorithms may be divided into four main groups based on specific port, packet content, host behavior, and flow characteristics [18]. The traditional technique to traffic categorization simply links applications with port numbers that have been published with the Internet Assigned Numbers Authority (IANA). But, because it is no longer the de facto standard, and because it does not offer itself to learning owing to the need for minimal lookup, it is not included in the scope of this study. Furthermore, it has been evidenced that simply relying on dock numbers is ineffectual, owing in large part to the use of vibrant port bargaining, tunneling, and the mishandling of port assigning numbers to the well apps for the purpose of distorting traffic and ignoring firewalls, among other factors. Although several classifiers use port numbers in combination with other strategies [19] to increase the effectiveness of the traffic classifiers, this is not always the case. Following that, we will explore the different traffic categorization algorithms that make use of machine learning, which will be summarized in Table 3.

Table 3. Different traffic categorization algorithms

ML technique	Datasets	Characteristics	Classes	Evaluation	
				Settings	Result
Supervised RF, Boost, Boost	Exclusive: business network	Packet dimension (from 1 to N packets), packet timestamp (from 1 to N packets), inter-arrival duration (from 1 to N packets), source/destination MAC, source/destination IP, source/destination port, flow length, packet count and byte count are all variables to consider	YouTube, Skype, Face book, Web Browsing	N = 5	RF: Accuracy 73.6–96.0% SGBoost: Accuracy 71.2–93.6% XGBoost: Accuracy 73.6–95.2%
Semi-supervised Laplacian-SVM	Univ. network is a unique	unpredictability of packet length, packet transmission duration, due to impaired, destination port, packages to react, min package length from destinations to sources, packet inaction degrees from transmitting data	Voice/video conferencing, streaming, large-scale data transmission, and more services are available	N = 20, Laplacian-SVM parameters $\lambda = 0.00001-0.0001$, $\sigma = 0.21-0.23$	Accuracy > 90%
Supervised k-NN, Linear-SVM, Radial-SVM, DT, RF, Extended Tree, AdaBoost, Gradient-AdaBoost, NB, MLP	KDD	Connection numbers, connection percentages, login status, error rate	Attack types	Dynamic classification and features collection selection	Accuracy = 95.6%

8 Optimized Routing on the Basis of Traffic Classification

In this part, we will look at works that have utilized machine learning to categorize traffic and then gone on to identify the ideal route based on traffic class that has been identified. [20] makes use of machine learning and software-defined networking to enhance scheduling & control in information centers. There were two stages to the framework. The initial phase entailed applying Machine Learning to categories the traffic at the network's edges in order to identify patterns. Using the categorization results in combination with its worldwide perspective of the network, the centralized SDN controller is able to deploy efficient routing options. This study did not go into detail on which methods and processes would be utilized for categorization & routing, nor did it specify which methods and procedures will be used for classification and routing. Another piece of work, which combines traffic categorization with route optimization, is provided in this publication. Within a software-defined network (SDN) context, they presented a Deep Neural Network (DNN) classifiers for network traffic categorization [21].

A virtual network function (VNF) is incorporated into the design in order to lessen the strain on the SDN. When a flow comes, the SDN switches routes the flow in accordance with the flow rules that have been configured. If a package does not have a flow rule associated with it, it makes a query to the controllers. Depending on the network architecture, the controllers determine the best path to take for the flow to be routed and then installs these patterns in the SDN switches in addition, the controller creates a flow rule that directs a duplicate of the packet to a port where the VNF will be listening for it. Traffic is captured and characteristics are extracted for use in traffic categorization using the trained deep neural network (DNN). The SDN controller receives the specified class data, which is marked in the DSCP field and transmitted to it. A route that meets the QoS criteria of the defined traffic class is now sought by the SDN controller, which then assigns major value to the resulting flows in all switches along the route, giving them preference over the flows that were originally installed [22]. This paper proposes a strategy for improving connection speeds, however it does not suggest any optimization techniques that may be employed by the SDN controller after classifications in order to find the best routing for the network traffic. In order to route the recognized packet, the controllers simply utilize its standard routing method, which is based on the worldwide perspective.

9 Conclusion

In this study, we have suggested a machine learning approach for congestion conscious traffic prediction in software defined networks. The suggested model learns updated network information at various periods of time and integrates that knowledge for picking least overloaded channel for traffic management. The study and predictions of traffic has been a focus of continuing research in several sub-fields of computer networks. Countless group of experts have been developed an efficient network traffic technique for the analysis and the predictions of network traffic and also, we evaluated the optimum routing on the basis of traffic categorization.

References

1. Kreutz, D., Ramos, F.M., Verissimo, P.E., Rothenberg, C.E., Azodolmolky, S., Uhlig, S.: Software-defined networking: a comprehensive survey. Proc. IEEE **103**(1), 14–76 (2015). https://doi.org/10.1109/jproc.2014.2371999
2. Crucianu, M., Boujemaa, N.: Active semi-supervised fuzzy clustering. Pattern Recogn. **41**(5), 1834–1844 (2008). https://doi.org/10.1016/j.patcog.2007.10.004

3. Panwar, V., Sharma, D.K., Kumar, K.V.P., Jain, A., Thakar, C.: Experimental investigations and optimization of surface roughness in turning of en 36 alloy steel using response surface methodology and genetic algorithm. Mater. Today: Proc. (2021). https://doi.org/10.1016/j.matpr.2021.03.642
4. Abar, T., Ben Letaifa, A., El Asmi, S.: QoE enhancement over DASH in SDN networks. Wirel. Pers. Commun. **114**(4), 2975–3001 (2020). https://doi.org/10.1007/s11277-020-07513-w
5. Rimal, Y.: Machine learning prediction of Wikipedia. Int. J. Mach. Learn. Netw. Collab. Eng. **3**(2), 83–92 (2019). https://doi.org/10.30991/ijmlnce.2019v03i02.002
6. Yu, F., Huang, T., Xie, R., Liu, J.: A survey of machine learning techniques applied to software defined networking (SDN): research issues and challenges. IEEE Commun. Surv. Tutor. **21**(1), 393–430 (2019). https://doi.org/10.1109/comst.2018.2866942
7. Mahrishi, M., Morwal, S., Muzaffar, A.W., Bhatia, S., Dadheech, P., Rahmani, M.K.I.: Video index point detection and extraction framework using custom YoloV4 Darknet object detection model. IEEE Access **9**, 143378–143391 (2021). https://doi.org/10.1109/ACCESS.2021.3118048
8. Holzinger, A.: Introduction to machine learning and knowledge extraction. Mach. Learn. Knowl. Extr. **1**(1), 1–20 (2017). https://doi.org/10.3390/make1010001
9. Jain, A., Pandey, A.K.: Multiple quality optimizations in electrical discharge drilling of mild steel sheet. Mater. Today Proc. **4**(8), 7252–7261 (2017). https://doi.org/10.1016/j.matpr.2017.07.054
10. Henderson, P., Bellemare, M., Pineau, J.: An introduction to deep reinforcement learning. Found. Trends Mach. Learn. **11**(3–4), 219–354 (2018). https://doi.org/10.1561/2200000071
11. Meena, G., et al.: Traffic prediction for intelligent transportation system using machine learning. In: 2020 3rd International Conference on Emerging Technologies in Computer Engineering: Machine Learning and Internet of Things (ICETCE), pp. 145–148 (2020). https://doi.org/10.1109/ICETCE48199.2020.9091758
12. Kapre, P., Shreshthi, R., Kalgane, M., Shekatkar, K., Hande, Y.: Software defined networking based intrusion detection system. Int. J. Recent Trends Eng. Res. **3**(5), 475–480 (2017). https://doi.org/10.23883/ijrter.2017.3252.m8yed
13. Jain, A., Pandey, A.K.: Modeling And optimizing of different quality characteristics in electrical discharge drilling of titanium alloy (grade-5) sheet. Mater. Today: Proc. **18**, 182–191 (2019). https://doi.org/10.1016/j.matpr.2019.06.292
14. Troshin, A.V.: Machine learning for LTE network traffic prediction. Infokommunikacionnye tehnologii 400–407 (2019). https://doi.org/10.18469/ikt.2019.17.4.06
15. Zúquete, A.: Traffic classification for managing applications' networking profiles. Secur. Commun. Netw. **9**(14), 2557–2575 (2016). https://doi.org/10.1002/sec.1516
16. Yu, C., Lan, J., Xie, J., Hu, Y.: QoS-aware traffic classification architecture using machine learning and deep packet inspection in SDNs. Procedia Comput. Sci. **131**, 1209–1216 (2018). https://doi.org/10.1016/j.procs.2018.04.331
17. Padmanabhan, V., Seshan, S., Katz, R.: A comparison of mechanisms for improving TCP performance over wireless links. ACM SIGCOMM Comput. Commun. Rev. **26**(4), 256–269 (1996). https://doi.org/10.1145/248157.248179
18. Akodkenou, I., Soule, A., Salamatian, K.: Traffic classification on the fly. ACM SIGCOMM Comput. Commun. Rev. **36**(2), 23–26 (2006). https://doi.org/10.1145/1129582.1129589
19. Jain, A., Yadav, A.K., Shrivastava, Y.: Modelling and optimization of different quality characteristics in electric discharge drilling of titanium alloy sheet. Mater. Today Proc. **21**, 1680–1684 (2020). https://doi.org/10.1016/j.matpr.2019.12.010
20. Cost efficient resource management framework for hybrid job scheduling under geo distributed data centers. Int. J. Mod. Trends Eng. Res. 4(9), 77–84 (2017). https://doi.org/10.21884/ijmter.2017.4282.yq9ne

21. Zhang, C., Wang, X., Li, F., He, Q., Huang, M.: Deep learning-based network application classification for SDN. Trans. Emerg. Telecommun. Technol. **29**(5), e3302 (2018). https://doi.org/10.1002/ett.3302
22. Deebalakshmi, R., Jyothi, V.L.: Smart routing based on network traffic classification techniques and considerations. Int. J. Public Sect. Perform. Manag. **5**(1), 1 (2019). https://doi.org/10.1504/ijpspm.2019.10016149

UBDM: Utility-Based Potential Pattern Mining over Uncertain Data Using Spark Framework

Sunil Kumar(✉) and Krishna Kumar Mohbey

Department of Computer Science, Central University of Rajasthan, Ajmer, India
2sunil.cs@gmail.com, kmohbey@gmail.com

Abstract. In practical scenarios, an entity presence can depend on existence probability instead of binary situations of present or absent. This is certainly relevant for information taken in an experimental setting or with instruments, devices, and faulty methods. High-utility patterns mining (HUPM) is a collection of approaches for detecting patterns in transaction records that take into account both object count and profitability. HUPM algorithms, on the other hand, can only handle accurate data, despite the fact that extensive data obtained in real-world applications via experimental observations or sensors are frequently uncertain. To uncover interesting patterns in an inherent uncertain collection, potential high-utility pattern mining (PHUPM) is developed. This paper proposes a Spark-based potential interesting pattern mining solution to work with large amounts of uncertain data. The suggested technique effectively discovers patterns using the probability-utility-list structure. One of our highest priorities is to improve execution time while increasing parallelization and distribution of all workloads. In-depth test findings on both real and simulated databases reveal that the proposed method performs well in a Spark framework with large data collections.

Keywords: Potential utility mining · Big data · Uncertainty · Spark · Data mining

1 Introduction

IoT applications create enormous volume of data, many of which are entities of measurements, occurrences, log records, and other data. IoT information has a diversity of integrated metadata, such as quantity and variability, as well as variable degrees of importance (e.g., usefulness, cost, threat, or significance) [10,26]. Frequent pattern mining cannot be utilized to uncover necessary insights for decision-taking process when dealing with missing or inaccurate data. UApriori [7], a compressed uncertain frequent pattern (CUFP)-tree [15], and Uncertain frequent pattern (UFP) growth method [14] have been used to deal with uncertain data. By adopting a strategy based on generation and testing and a breadth-first search technique to extract knowledge, UApriori was a pioneer in

V. E. Balas et al. (Eds.): ICETCE 2022, CCIS 1591, pp. 623–631, 2022.
https://doi.org/10.1007/978-3-031-07012-9_52

this field. CUFP-Tree is an extension of UP-growth that uses a tree structure to discover rules. It also uses an advance data structure for storing information. Based on stored information, advance mining can be performed. Lin et al. [18] introduced an efficient method to generate high utility patterns. It can handle the massive amount of uncertain data.

In order to find the most useful HUIs, this strategy looked at both their usefulness and their degree of uncertainty at the same time. Zhang et al. [27] also proposed an approach for handling uncertain data. This approach can produce high utility-probability patterns. Ahmed et al. [4] incorporate uncertainty and utility as two parameters to predict utility-based patterns. They also used an evolutionary computing approach for their work. Shrivastava et al. [22] proposed an effective approach to predict patterns from uncertain datasets.

Retailers may utilize uncertainty to learn more about their customers' buying habits and make better judgments about the quality of their customers. A Frequent-based Pattern Mining (FPM) method, on the other hand, ignores this information and may uncover valueless knowledge, such as objects with lesser usefulness [8]. To put it another way, standard FPM does not work well when handling quantitative datasets or extracting useful patterns for data analytics. High-Utility Pattern Mining (HUPM) [9,11,21] is a novel utility-driven data analytics paradigm that addresses the primary shortcomings of FPM. There are several ways in which the notion of utility may be used, and it is crucial to note that it can be used to represent numerous participants' happiness, utility, revenue, risk, or weight [6]. When data collected from many sources, an item's existence in real-life scenarios maybe depends on different factors. The probability is also included with it.

A product's "utility" is usually defined as the product's weight, costs, risk; benefit; or value to end-users. "utilities" may be taken into account most effectively in implementations when the database is uncertain. Uncertain databases represent inaccurate values in transaction data. There are various reasons for the uncertainty and each record may be connected with a probability within a transaction record. Suppose in a transaction like "X:7 Y:3 Z:9 A:1, 90%", different items "X Y Z A" purchased with "7 3 9 1" quantity and its approximate probability maybe 90%.

In the subject of risk prediction, an event's risk may also be referred to its likelihood. It comprises three "ACF" items with occurrence frequencies of "3 2 4" and an existence probability of 80%, as shown by the "A:3, C:2, F:4, 80%". The "utility" component may be applied to various circumstances and domains, including significance, weight, interest, and profit. Therefore, the utility idea can be used in a wide range of situations and disciplines.

To discover new knowledge, it is required to integrate the utility with the uncertainty in various disciplines. Lin et al. [17] introduced PHUI-UP and PHUI-List approach for finding HUIs in uncertain data. This approach is based on an Apriori-like algorithm. Lin et al. [18] then provided many pruning algorithms to minimize the search area for HUIM in uncertain datasets to improve mining performance. For knowledge discovery, Ahmed et al. [4] uses evolutionary computing to look for answers that are not dominant in terms of utility and uncertainty. It's difficult to deal with large datasets in many sectors using the preceding methods, even when they combine utility with unknown aspects.

Because of the inherent inaccuracies and uncertainties, traditional pattern identification algorithms are impossible to use them straightforward to mine the necessary knowledge or information in these types of contexts. Since "utility" and "uncertainty" are two separate elements, we cannot compare them. Instead, we should see utility as a semantic measure of how much a pattern is valued based on its users' experience, aim, and understanding. Although utility-based techniques have been researched to deal with exact information, incapable of dealing with uncertainty [17]. Mining information with a low likelihood may provide meaningless or misleading results, if the uncertain component is not considered.

We have developed an effective approach for detecting PHUIs in this study. It uses utility and uncertainty based patterns. Also, the Spark framework is used for implementation. The abbreviations used in the paper are shown in Table 1. The major contribution is as follows:

1. The proposed UBDM approach discovers PHUIs from big uncertain data.
2. The UBDM can handle an extensive dataset.
3. The proposed UBDM is a spark framework-based algorithm, and the spark framework works faster than Hadoop or other distributed frameworks by up to hundred times.

Table 1. List of acronyms

HUPM	High Utility Pattern Mining
PHUPM	Potential High Utility Pattern Mining
HTWUI	High Transaction-Weighted Utility Itemset
1-HTWUI	First-level High Transaction-Weighted Utility Itemset
k-itemset	An itemset with k number of items
FPM	Frequent Pattern Mining
HDFS	Hadoop Distributed File System
TU	Total Utility
TWU	Transaction-Weighted Utility
HUI	High Utility Itemset
M-U	Minimum Utility
M-P	Minimum Probability
UD	Uncertain Database
PD	Partition Database
UFP	Uncertain Frequent Pattern

The remaining paper is organized as subsequent: The past research and preliminary efforts relevant to the planned work are described in Sect. 2. The suggested approach is described in depth in Sect. 3. The results and implications are presented in Sect. 4. Finally, Sect. 5 explains the conclusion.

2 Related Work

Binary data describes the presence or absence of any item. This dataset is mostly used in association rules to discover frequent items. A large quantity of obtained data is erroneous or incomplete in real-world applications [1]. This problem has surfaced in recent years and is becoming more relevant. Frequent itemset can also be discovered from uncertain datasets by employing the expected support model [14] and probabilistic frequency model [12].

UApriori was designed by Chui et al. [7] using an Apriori-based level-wise technique, which was subsequently built upon by other researchers to efficiently mine frequent itemsets in uncertain databases after it was first developed. The anticipated support threshold for frequent mining of patterns in an uncertain database was thus computed for the patterns that occur often. UFP-growth approach is an extension of FP-tree used to mine UFIs [17]. This approach uses the BFS technique's divide and conquer strategy to discover most FP without candidate generation. This method was proven to outperform the UApriori algorithm in terms of performance. A recursive version of H-Mine was UHmine that also uses divide and conquer with BFS to generate UFIs [2]. Afterward, Wang et al. [25] published the MBP method, which looks at a narrower pool of potential candidates than the UApriori approach. A CUFP-tree structure was later created by Lin et al. [15] for effective mining of UFIs.

To discover new knowledge, it is required to integrate the utility with the uncertainty in various disciplines. Lin et al. [17] presents PHUI-UP and PHUI-List approach for finding HUIs in uncertain data. This approach is based on an Apriori-like algorithm. Lin et al. [18] provided many pruning algorithms to minimize the search area for HUPM in imprecise datasets to improve mining performance. For knowledge discovery, evolutionary computing looks for answers that aren't dominant in utility and uncertainty [4]. It's impossible to deal with large datasets in many sectors using the preceding methods, even when they combine utility with unknown aspects.

Bernecker et al. [5] introduced the frequent probabilistic model, which differs significantly from the preceding "anticipated support-based" approach. Furthermore, Sun et al. [23] suggested two efficient methods (p-Apriori and TODIS) for discovering common patterns from the bottom up or the top-down. SUF-growth and UF-streaming [13] are used to discover useful items from streaming-based imprecise data. Tong et al. [24] developed an approach for discovery of frequent patterns in uncertain datasets.

3 Proposed Method

To mine PHUIs more efficiently, a novel approach called UBDM (Uncertain Big Data Mining) is presented in this section. The suggested technique uses the PU-list [17] layout. The PU-list is a particularly efficient vertical data format. It keeps a list of the TIDs of records in which the itemset exists, as well as properties for each itemset. The purpose of the PU-list is to store and condense all necessary details from an uncertain dataset in order to explore the search space for a given itemset. The proposed algorithm UBDM, partition data among workers and evaluates it separately on each node. It incorporates a divide-and-conquer mechanism, during which the job is broken down into smaller chunks and handled separately on each worker. To begin, calculates the TWU of 1-itemset and order each elements according to TWU in ascending sequence and eliminates non-PHUI 1-itemsets and filter transactions for these itemsets. A sorted 1-itemset divides the search space evenly among the nodes. According to the search space, records are processed locally on workers is generated. Each node builds its own PHUI using the allotted partition data. In the suggested frameworks, there are two separate algorithms that are accountable for varying purposes, as stated below.

Algorithm 1. UBDM (Uncertain Big Data Mining)

Input: UD: Uncertain Database, M-U: Minimum Utility, M-P: Minimum Probability
Output: PHUI's

Compute total utility (TU) and total probability (Tpro) of transaction dataset
Calculate TWU of each item
for all record r $\in$ UD **do**
  Remove unpromising items if(TWU(item) $\leq$ M-U)
end for
Generate revised dataset by removing empty records from UD
Partition revised dataset (par-data) among cluster worker nodes
for each worker node **do**
  Mining(par-data, itemset, M-U, M-P)
end for

Algorithm 2. Mining

Input: PD: Partition Database, M-U:Minimum Utility, M-P:Minimum Probabilty
Output: PHUI's at a worker node

Create PU-list for each item
for all item in PU-list **do**
  if(Utility(item)$\geq$=M-U and Pro(item)$\geq$M-P)
  PHUI $\leftarrow$ item
end for
Generate projected dataset
Mining(itemset, PU-list, projected dataset, M-U, M-P)

4 Experiment Results

4.1 Dataset Detail

The experiments used both real-world (retail[1]) and synthetic dataset T10I4D100K (with reference footnote 1) to test our approach. A total of 88,162 records from a super market are included in the retail dataset. It has 16,470 unique items, with a maximum transaction size of 76 items and an average transaction size of 10.3. The IBM Quest Synthetic Dataset Generator [3] was used to create the synthetic dataset T10I4D100K. It has 100k unique elements, 870 records, with an average record size of 10.1 and a maximum record size of 29. A simulation approach developed in earlier research [19,20] was used to attribute both frequency (internal) and benefit (external) to the element in the datasets. Furthermore, because of the uncertainty attribute, each transaction in these datasets was assigned a distinct probabilistic estimate in the limit (0.5, 1.0).

4.2 Experimental Setting

Extensive experiments were conducted in this section to evaluate the performance of the suggested algorithm. The results of the proposed strategy are compared to PHUI-List [17], and HEWI-Tree [16] methods for mining PHUIs. Experiments were carried out by altering the minimal utility threshold while keeping the minimum potential probability threshold constant for the other parameters. The experiment is done on a four-node Spark cluster with one master and three workers. Nodes are configured with an Ubuntu 18.04 Linux system with RAM (8 GB), an Intel(R) Core i7 processor operating at 3.20 GHz, a 1TB HDD, Spark 2.3.0, Scala 2.13, and Java 8.

4.3 Experimental Outcomes

In Fig. 1, and 2, the execution time of the analyzed methods for different M-U and a static M-P is compared and illustrated. On all datasets, the proposed approach is faster than the PHUI-List and HEWI-Tree algorithms. The findings are presented in Tables 2 and 3 with a varying minimum utility which is in percent of the total utility of dataset.

Table 2. Execution time (ms) comparison on Retail dataset

Algorithm	M-U (minimum utility)				
	0.04%	0.06%	0.08%	0.10%	0.12%
HEWI-Tree	130000	180000	200000	220000	260000
PHUI-List	80000	81000	82000	83000	85000
UBDM	**53902**	**57000**	**63000**	**67000**	**74000**

[1] http://www.philippe-fournier-viger.com/spmf/.

Table 3. Execution time (ms) comparison on T10I4D100K dataset

Algorithm	M-U (minimum utility)				
	0.05%	0.04%	0.03%	0.02%	0.01%
HEWI-Tree	84000	88000	91000	94000	96000
PHUI-List	58000	60000	65000	70000	75000
UBDM	**17000**	**19000**	**21000**	**24000**	**27000**

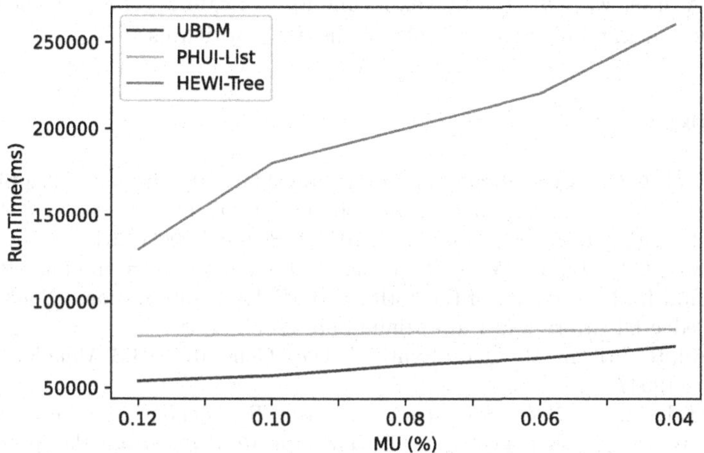

Fig. 1. Running time comparison for retail dataset

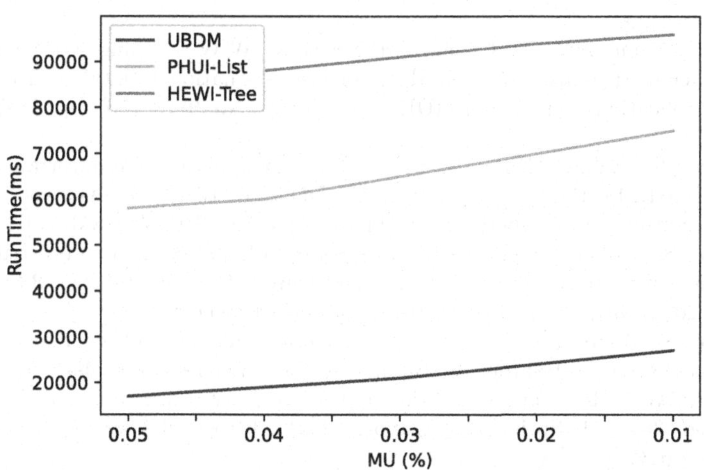

Fig. 2. Running time comparison for T40I10D100K dataset

5 Conclusion

With the fast expansion of information, an effective mining method to uncover knowledge for decision-making is required. Conventional mining techniques were restricted in their capacity to manage the massive dataset. In this work, we proposed a big data approach, UBDM, which efficiently discover PHUIs from uncertain massive dataset. UBDM implemented on Spark framework. To check the efficiency, we have used both real dataset (retail) and synthetic dataset (T10I4D100K). Experiments showed that the developed UBDM has better performance than the state-of-the-art methods such as PHUI-List and HEWI-Tree. In future works, we will conduct experiments to evaluate the proposed work's scalability and memory against state-of-the-art algorithms.

References

1. Aggarwal, C.C.: An introduction to uncertain data algorithms and applications. In: Aggarwal, C. (ed.) Managing and Mining Uncertain Data, pp. 1–8. Springer, Heidelberg (2009). https://doi.org/10.1007/978-0-387-09690-2_1
2. Aggarwal, C.C., Li, Y., Wang, J., Wang, J.: Frequent pattern mining with uncertain data. In: Proceedings of the 15th ACM SIGKDD International Conference on Knowledge Discovery and Data Mining, pp. 29–38 (2009)
3. Agrawal, R., Srikant, R.: Quest Synthetic Data Generator. IBM Almaden Research Center (1994)
4. Ahmed, U., Lin, J.C.W., Srivastava, G., Yasin, R., Djenouri, Y.: An evolutionary model to mine high expected utility patterns from uncertain databases. IEEE Trans. Emerg. Top. Comput. Intell. **5**(1), 19–28 (2020)
5. Bernecker, T., Kriegel, H.P., Renz, M., Verhein, F., Zuefle, A.: Probabilistic frequent itemset mining in uncertain databases. In: Proceedings of the 15th ACM SIGKDD International Conference on Knowledge Discovery and Data Mining, pp. 119–128 (2009)
6. Cai, C.H., Fu, A.W.C., Cheng, C.H., Kwong, W.W.: Mining association rules with weighted items. In: Proceedings of the International Database Engineering and Applications Symposium, IDEAS 1998 (Cat. No. 98EX156), pp. 68–77. IEEE (1998)
7. Chui, C.-K., Kao, B., Hung, E.: Mining frequent itemsets from uncertain data. In: Zhou, Z.-H., Li, H., Yang, Q. (eds.) PAKDD 2007. LNCS (LNAI), vol. 4426, pp. 47–58. Springer, Heidelberg (2007). https://doi.org/10.1007/978-3-540-71701-0_8
8. Kumar, S., Mohbey, K.K.: A review on big data based parallel and distributedapproaches of pattern mining. J. King Saud Univ. - Comput. Inf. Sci. **34**(5), 1639–1662 (2022). https://doi.org/10.1016/j.jksuci.2019.09.006
9. Kumar, S., Mohbey, K.K.: High utility pattern mining distributed algorithm based on spark RDD. In: Bhateja, V., Satapathy, S.C., Travieso-Gonzalez, C.M., Flores-Fuentes, W. (eds.) Computer Communication, Networking and IoT. LNNS, vol. 197, pp. 367–374. Springer, Singapore (2021). https://doi.org/10.1007/978-981-16-0980-0_34
10. Mohbey, K.K., Kumar, S.: The impact of big data in predictive analytics towards technological development in cloud computing. Int. J. Eng. Syst. Model. Simul. **13**(1), 61–75 (2022). https://doi.org/10.1504/IJESMS.2022.122732

11. Kumar, S., Mohbey, K.K.: Memory-optimized distributed utility mining for big data. J. King Saud Univ. - Comput. Inf. Sci. (2021). https://doi.org/10.1016/j.jksuci.2021.04.017
12. Lehrack, S., Schmitt, I.: A probabilistic interpretation for a geometric similarity measure. In: Liu, W. (ed.) ECSQARU 2011. LNCS (LNAI), vol. 6717, pp. 749–760. Springer, Heidelberg (2011). https://doi.org/10.1007/978-3-642-22152-1_63
13. Leung, C.K.S., Hao, B.: Mining of frequent itemsets from streams of uncertain data. In: 2009 IEEE 25th International Conference on Data Engineering, pp. 1663–1670. IEEE (2009)
14. Leung, C.K.-S., Mateo, M.A.F., Brajczuk, D.A.: A tree-based approach for frequent pattern mining from uncertain data. In: Washio, T., Suzuki, E., Ting, K.M., Inokuchi, A. (eds.) PAKDD 2008. LNCS (LNAI), vol. 5012, pp. 653–661. Springer, Heidelberg (2008). https://doi.org/10.1007/978-3-540-68125-0_61
15. Lin, C.W., Hong, T.P.: A new mining approach for uncertain databases using CUFP trees. Expert Syst. Appl. **39**(4), 4084–4093 (2012)
16. Lin, J.C.W., Gan, W., Fournier-Viger, P., Hong, T.P., Chao, H.C.: Mining weighted frequent itemsets without candidate generation in uncertain databases. Int. J. Inf. Technol. Decis. Making **16**(06), 1549–1579 (2017)
17. Lin, J.C.W., Gan, W., Fournier-Viger, P., Hong, T.P., Tseng, V.S.: Efficient algorithms for mining high-utility itemsets in uncertain databases. Knowl.-Based Syst. **96**, 171–187 (2016)
18. Lin, J.C.-W., Gan, W., Fournier-Viger, P., Hong, T.-P., Tseng, V.S.: Efficiently mining uncertain high-utility itemsets. Soft. Comput. **21**(11), 2801–2820 (2016). https://doi.org/10.1007/s00500-016-2159-1
19. Liu, M., Qu, J.: Mining high utility itemsets without candidate generation. In: Proceedings of the 21st ACM International Conference on Information and Knowledge Management, pp. 55–64 (2012)
20. Liu, Y., Liao, W., Choudhary, A.: A two-phase algorithm for fast discovery of high utility itemsets. In: Ho, T.B., Cheung, D., Liu, H. (eds.) PAKDD 2005. LNCS (LNAI), vol. 3518, pp. 689–695. Springer, Heidelberg (2005). https://doi.org/10.1007/11430919_79
21. Mohbey, K.K., Kumar, S.: A parallel approach for high utility-based frequent pattern mining in a big data environment. Iran J. Comput. Sci. **4**, 195–200 (2021)
22. Srivastava, G., Lin, J.C.W., Jolfaei, A., Li, Y., Djenouri, Y.: Uncertain-driven analytics of sequence data in IoCV environments. IEEE Trans. Intell. Transp. Syst. **22**, 5403–5414 (2020)
23. Sun, L., Cheng, R., Cheung, D.W., Cheng, J.: Mining uncertain data with probabilistic guarantees. In: Proceedings of the 16th ACM SIGKDD International Conference on Knowledge Discovery and Data Mining, pp. 273–282 (2010)
24. Tong, Y., Chen, L., Cheng, Y., Yu, P.S.: Mining frequent itemsets over uncertain databases. arXiv preprint arXiv:1208.0292 (2012)
25. Wang, L., Cheng, R., Lee, S.D., Cheung, D.: Accelerating probabilistic frequent itemset mining: a model-based approach. In: Proceedings of the 19th ACM International Conference on Information and Knowledge Management, pp. 429–438 (2010)
26. Wu, J.M.T., et al.: Mining of high-utility patterns in big IoT-based databases. Mob. Netw. Appl. **26**(1), 216–233 (2021)
27. Zhang, B., Lin, J.C.W., Fournier-Viger, P., Li, T.: Mining of high utility-probability sequential patterns from uncertain databases. PLoS ONE **12**(7), e0180931 (2017)

Blockchain and Cyber Security

Penetration Testing Framework for Energy Efficient Clustering Techniques in Wireless Sensor Networks

Sukhwinder Sharma[1](✉), Puneet Mittal[1], Raj Kumar Goel[2], T. Shreekumar[1], and Ram Krishan[3]

[1] Mangalore Institute of Technology and Engineering, Mangalore, India
sukhwinder.sharma83@gmail.com
[2] Noida Institute of Engineering and Technology, Greater Noida, India
[3] Mata Sundri University Girls College, Mansa, India

Abstract. Limited node energy has been always remained key challenge for the designers of clustering techniques in Wireless Sensor Networks (WSNs). From homogeneous to heterogeneous WSNs, there is no best fit clustering technique; therefore plenty of different techniques have been proposed by researchers from time to time. This paper aims to develop a generalized deterministic framework, Penetration Testing Framework (PTF), for researchers to validate their proposed techniques on a common platform giving best fit techniques for homogeneous as well as heterogeneous WSNs. The PTF consists of a set of network scenarios having different network parameters and performance parameters mixes. Some of the most relevant as well as popular clustering techniques designed for homogeneous and/or heterogeneous WSNs have been tested under PTF to validate its applicability and generality. A rigorous in-depth analysis has been made through development of different network scenarios and performance parameters mixes. The results attained are quite interesting, having useful directions for the designers of clustering techniques for modern WSNs.

Keywords: Clustering · Energy · Performance · Stability period · WSN

1 Introduction

Wireless Sensor Networks (WSNs) have been a popular choice for monitoring and control applications [1–3] in hard to reach ecosystems like forests, oceans and wetlands. It provides sensor nodes having in-built or replaceable batteries that can survive from months to years for providing useful information in an economical and reliable manner to sink node(s), which further transfer(s) it to some remote location for possible monitoring and decision making. Limited node energy has been always remained key challenge [4] for the designers of clustering techniques in WSNs. Attempts are being made by researchers to develop energy efficient techniques to increase the stability period and lifetime of these networks through reduced energy consumption [5]. The development of energy-efficient clustering-based techniques is one of the most promising areas in this

V. E. Balas et al. (Eds.): ICETCE 2022, CCIS 1591, pp. 635–653, 2022.
https://doi.org/10.1007/978-3-031-07012-9_53

direction [6, 7]. From homogeneous to heterogeneous WSNs, there is no best fit clustering technique while plenty of different techniques have been proposed by researchers from time to time. This paper aims to develop a generalized deterministic framework, Penetration Testing Framework (PTF), for researchers to validate their proposed techniques on a common platform giving best fit techniques for WSNs. The PTF consists of a set of network scenarios having different network parameters and performance parameters mixes. Some of the most relevant as well as popular clustering techniques designed for homogeneous and/or heterogeneous wireless sensor networks have been tested under PTF to validate its applicability and generality. A rigorous in-depth analysis has been made through development of different network scenarios and performance parameters mixes. The results attained are quite interesting, having useful directions for the designers of clustering techniques for modern wireless sensor networks.

Section 1 gives introduction, followed by the details of proposed PTF in Sect. 2. The clustering concept and working of clustering techniques tested under PTF are discussed in Sect. 3, while the performance analysis of chosen clustering techniques is discussed in the Sect. 4. The final conclusions of this work are showcased in Sect. 5.

2 Penetration Testing Framework (PTF)

After designing a new clustering technique, the researcher(s) look for some generalized environment having set of input and output parameters as well as some existing well known techniques to validate the performance of proposed technique and attained improvements over other techniques. The PTF consists of various network parameters that include a) radio model considered for deciding how the energy dissipation will be calculated based on the node deployment and data transmission in the network, b) network model indicating the physical parameters about the network size, number of nodes and types, their placement along with that of the sink node, c) performance metrics, and d) simulation setup.

2.1 Assumptions

To restrict the applicability of clustering techniques within a deterministic framework, following assumptions are made while implementing this work, as is done by other researchers working in this area [8–12]:

Deployment of all the sensor nodes is in random uniform manner.

a) Sensor nodes are static, and hence no change in their positions once deployed.
b) All sensor nodes have similar processing and communication capabilities.
c) There is only one sink positioned at the center of the sensing field having fixed location.
d) Sensor nodes always have data for communication to the sink.
e) Cluster heads directly forward data to the sink.

2.2 Radio Energy Consumption Model

The 1st-order radio energy dissipation model is used in this work [10]. It has been frequently used by researchers in this field for network performance evaluation and analysis [8–13].As per this radio energy-consumption model (see Fig. 1), during data transmission, node energy is dissipated to run its power amplifier and transmit electronics. Some energy is also dissipated by the receiver during data reception for running the receive electronics. Based upon the transmitter receiver distance, two transmitter amplifier models: i) free space channel model, and ii) multi path fading model are used.

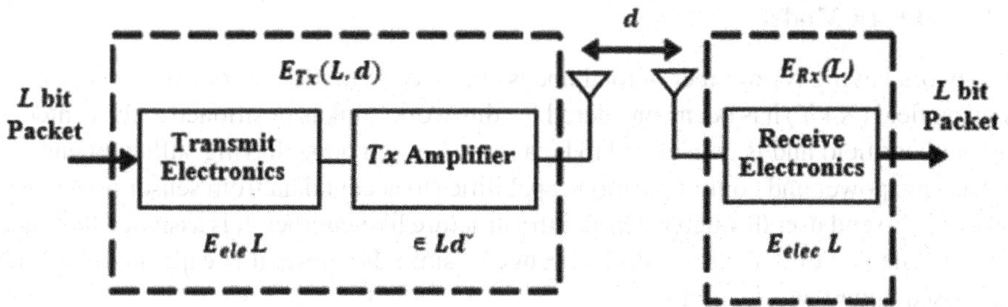

Fig. 1. Radio energy-consumption model [10].

According to the 1st-order radio energy-consumption model, the transmission of one L-bit packet over distance d, the energy consumed by the radio (such that the Signal-to-Noise Ratio (SNR) is least) is represented by:

$$E_{Tx}(L, d) = \begin{cases} L * E_{elec} + L * E_{fs} * d^2 \;\; if \;\; d < d_0 \\ L * E_{elec} + L * E_{mp} * d^4 \; if \; d \geq d_0 \end{cases} \tag{1}$$

$$E_{Rx}(L) = L * E_{elec} \tag{2}$$

where E_{elec} is the quantity of energy expended per bit to run the transmitter/receiver circuit E_{fs} and E_{mp} rely upon the free space or multi path transmitter-amplifier model used. If transmission distance between sender and receiver node d is less than a threshold value d_0, free space model is used, else multipath model is used.

Two components of energy consumption in communication operations are- electronics energy and amplifier energy; while, energy consumed in processing operations is represented by data aggregation. Electronics energy E_{elec} depends on different factors like digital coding, modulation, filtering, and spreading of the underlying signal; whereas, the amplifier energy ($= E_{fs} * d^2$ or $E_{mp} * d^4$) mainly depends upon the transmission distance and acceptable bit-error rate. Table 1 represents the typical radio characteristics for calculating the energy consumed by various communications and processing operations.

Table 1. Characteristics of radio [10].

Operation(s)	Energy dissipation
Transmitter or receiver electronics	$E_{elec} = 50\text{nJ/bit}$
Transmit amplifier, if ($d \geq d_0$)	$E_{mp} = 0.0013\text{pJ/bit/m}^4$
Transmit amplifier, if ($d < d_0$)	$E_{fs} = 10\text{pJ/bit/m}^2$
Data aggregation	$E_{DA} = 5\text{nJ/bit/signal}$

2.3 Network Model

A network having N_T number of total nodes deployed in random uniform manner over a sensing field ($X * Y$) has been considered for this work. Sink is positioned at the center of the network field and is considered to be a resource rich node, having sufficient energy, processing power and communication capabilities to accept data from sensor nodes, and forward it to end user through network infrastructure like internet. It is assumed that each sensor node is within the transmission range of sink; therefore, it is capable to become CH based upon the selection.

Two types of WSNs- homogeneous WSN and heterogeneous WSN are considered under this work. Homogeneous WSN is designed under the assumption that all N_T sensor nodes are having equal initial energy, computational power, link levels and memories. Heterogeneous networks comprise of N_n number of normal sensor nodes and N_a number of advanced sensor nodes ($N_T = N_n + N_a$). Two energy heterogeneity parameters- fraction of advanced sensor nodes (m), and the additional energy factor b/w normal and advanced sensor nodes (α) are used to calculate the number of normal and advanced node(s) in the WSN. Advanced sensor nodes are allocated with α times additional initial energy than the normal sensor nodes, while carrying an equal amount of computational power, memory and link levels. Following equations represent the calculation of number of normal and advanced sensor nodes in the underlying network:

$$N_a = m * N_T \tag{3}$$

$$N_n = (1 - m) * N_T \tag{4}$$

2.4 Performance Metrics

To evaluate these clustering techniques, performance parameters used are as follows-

- Stability Period: It is the period from the start of network operation to the death of the first sensor node. This period is also called the "stable region". Higher values of stability period are desirable to ensure reliable communications.

- Network Lifetime: It is the period from the beginning of network operation to the death of the last living sensor node. Higher values of network lifetime give data communication for longer duration, whether reliable or non-reliable.
- Throughput: It gives the quantitative measure of data communicated by the sensor nodes. It is evaluated as the total no. of packets received by all CHs and by the sink in all rounds.
- Energy Dissipation: The quantitative measure of energy consumed per round by the network is called energy dissipation. An efficient network requires energy dissipation at lower rate to keep the network alive for longer duration. For reliable communication, higher energy dissipation in stable region is desirable to provide increased stability period.
- Advanced Nodes and Normal Nodes Alive per Round: It represents the number of advanced nodes and normal nodes that have not fully dissipated their energy. It gives the quantitative measure of alive nodes in the underlying network. It indicates ability of clustering technique to utilize energy of advanced nodes to shield normal nodes from earlier dying.
- No. of CHs per Round: It is the no. of sensor nodes elected as CHs per round. It represents ability of clustering technique to maintain an average no. of CHs per round equal to the pre-decided optimal no. of cluster heads.

These performance metrics are used by many researchers in this field [8–12].

2.5 Simulation Setup

Introduction of additional energy nodes, as in heterogeneous networks, leads to increasing the overall energy of underlying network, which may automatically increase the lifetime of underlying network. So, to have a critical analysis of these techniques, it is desirable to examine whether the performance of a clustering technique is improved owing to increased network energy or due to the judiciousness of technique itself.

So, for the purpose, three system scenarios are designed and implemented in this work (Table 2). In scenario 1 and 2, network is considered to be purely homogeneous with 0.50 J and 0.80 J respectively as the initial energy of each node leading to total network energy as 50 J and 80 J respectively.

The Scenario 3 represents a heterogeneous network with a fraction (20% of total nodes) of normal nodes replaced by advanced nodes having three time extra additional energy than that of the normal nodes. Here, 80 normal nodes are equipped with 0.50 J as initial node energy, and 20 advanced nodes are equipped with 2 J of individual initial energy. It results in same total network energy as that in Scenario 2 as 80 J. The purpose of Scenario 2 and 3 is to examine the impact of network energy on the clustering techniques in a more critical manner, as in Scenario 2 the additional network energy is equally distributed among all the network nodes rather than localized to few additional advance nodes.

Table 2. Simulation parameters.

Simulation parameters	Homogeneous WSN		Heterogeneous WSN
	Scenario 1	Scenario 2	Scenario 3
Total no. of nodes (N_T)	100	100	100
Field size	100 m × 100 m	100 m × 100 m	100 m × 100 m
Energy of normal node - initial (E_0)	0.50 J	0.80 J	0.50 J
Total network energy (E_{total})	50 J	80 J	80 J
No. of rounds (R)	5,000	5,000	5,000
Fraction of special or advanced nodes (m)	—	–	0.2
Advanced and normal nodes additional energy factor (α)	—	–	3
Message size (L)	4,000 bits	4,000 bits	4,000 bits
Optimal CH selection probability, (p)	0.1	0.1	0.1

3 Working of Clustering Techniques

WSN nodes communicate within their respective clusters with other nodes or their respective CH. Nodes can sense, store and communicate their own data or can receive and forward data of other nodes in the cluster. The CHs perform sensing, and communication of their own data as well as data received from associated nodes in the cluster. Clustering techniques decide number of clusters in the network and form clusters through nomination of CHs based upon certain network parameters and their thresholds. The primary aim of these techniques is to provide efficient communication of information with effective resources utilization.

Clustering enhances bandwidth utilization and diminishes separate transmission overhead of each node. The network lifetime can be prolonged by wisely distributing the tasks of becoming CH among different network nodes. Amid dynamic restructuring of network clusters, there is dynamic exchange of responsibilities from simple node to cluster head and vice versa. Clustering also reduces energy consumption through conversion of long distance communications to short distance multi-hop communications resulting less expenses on inter- and intra-cluster communication costs.

The clustering techniques considered in this part of the work are briefly overviewed in this section, so as to adjudge their salient features and limitations.

3.1 Working of LEACH

LEACH works on the principle of equal opportunity for all sensor nodes in the deployed network to become CH in one of the network clusters for sensing and communication.

It initially decides the optimal no. of network clusters k which gives the required no. of CHs on average per round of communication. The value of k is calculated as [14]:

$$k = \sqrt{\frac{N_T}{2\pi}} * \frac{2}{0.765} \tag{5}$$

Then, the values of total no. of network nodes and optimal no. of network clusters are used to calculate the optimal probability, p of a sensor node to be designated as a CH in the current round. The value of p is calculated as:

$$p = \frac{k}{N_T} \tag{6}$$

LEACH technique ensures an epoch of $1/p$ rounds for each network node to be designated as CH. Non-designated sensor nodes will be placed in set G with increased CH election probability after each round. Nodes from the set G choose a random no. in range 0 to 1 at the beginning of each round independently. Thereafter the CH selection threshold $T(s)$ is computed for the present round as:

$$T(s) = \begin{cases} \frac{p}{\left(1-p*\left(r*mod\frac{1}{p}\right)\right)} & if\ s \in G \\ 0 & otherwise \end{cases} \tag{7}$$

A sensor node is designated as CH for a particular round whenever the chosen random no. value is less than the threshold $T(s)$. Dynamic selection of CHs provides balanced load distribution among all the nodes, thereby improves the network lifetime up to certain extent.

3.2 Working of LEACH-C

Due to distributed CH selection process, LEACH gives uneven no. of CHs in each round, thereby degrades reliability and efficiency of the underlying network. Another technique, LEACH-C provides centralized selection of optimal number of CHs per round. It authorizes the energy-rich sink node to select the CHs out of capable sensor nodes.

3.3 Working of ALEACH

ALEACH follows distributed cluster formation through autonomous decision making without centralized control or without network global information requirements. It removes long distance communications with the sink thereby reduces the energy consumption. Similar to LEACH, ALEACH divides the network into clusters, where CHs forward data, collected from associated nodes in cluster, to the sink node. While LEACH considers only current state probability, ALEACH uses general probability as well as current state probability to decide the energy threshold. In ALEACH, the threshold equation, Eq. 7 has been improved with the introduction of two new terms: the General probability, G_p, and the Current State probability, CS_p. The threshold value $T(s)$ for the current round can be calculated as:

$$T(s) = G_p + CS_p \tag{8}$$

Here,

$$G_p = \frac{k}{N_T - k * \left(r * mod \frac{N_T}{k}\right)} \tag{9}$$

and

$$CS_p = \frac{E_{current}}{E_{max}} * \frac{k}{N_T} \tag{10}$$

Where $E_{current}$ represents the current (residual) energy of the node, and E_{max} represents the network energy assigned at the time of node deployment called as initial network energy.

So, the final value of threshold with present values of G_p and CS_p in the Eq. 8 will be:

$$T(s) = \frac{k}{N_T - k * \left(r * mod \frac{N_T}{k}\right)} + \frac{E_{current}}{E_{max}} * \frac{k}{N_T} \tag{11}$$

3.4 Working of SEP

In SEP technique, higher CH selection probability p_{adv} is assigned to the nodes with higher initial energy, advanced nodes, in comparison to the CH selection probability p_{nrm} for the normal nodes with lower initial energy. It results in frequent selection of advanced nodes as CH than the normal nodes. This technique assures that the each advanced node will be designated as CH every $1/p_{adv}$ rounds, while each normal node will be designated as CH every $1/p_{nrm}$ rounds as

$$p_{adv} = \frac{p * (1 + \alpha)}{(1 + \alpha * m)} \tag{12}$$

$$p_{nrm} = \frac{p}{(1 + \alpha * m)} \tag{13}$$

where p is node's optimal probability for CH selection when nodes in the network carry equal probability for selection.

Normal nodes will have current round threshold in the present round given as $T(s_n)$:

$$, T(s_n) = \begin{cases} \frac{p_{nrm}}{\left(1 - p_{nrm} * \left(r * mod \frac{1}{p_{nrm}}\right)\right)} & if \;\; s_n \in G' \\ 0 & otherwise \end{cases} \tag{14}$$

where G' is set of those normal nodes those were not declared CHs in the last $1/p_{nrm}$ rounds.

Similarly, threshold for advanced nodes is given by $T(s_a)$:

$$T(s_a) = \begin{cases} \frac{p_{adv}}{\left(1-p_{adv} * \left(r * mod \frac{1}{p_{adv}}\right)\right)} & if \;\; s_a \in G'' \\ 0 & otherwise \end{cases} \tag{15}$$

where, G'' is the set of those advanced nodes those were not declared CHs in the last $1/p_{adv}$ rounds.

3.5 Working of DEEC

SEP based clustering technique considers sensor node's initial energy to decide the cluster heads. It results in higher penalization of the advanced nodes carrying additional energy initially than the normal nodes. It results in faster energy depletion and frequent dying of these advanced nodes due to increased energy consumption being CHs. So, advanced nodes at long distance from sink may die out earlier than normal nodes. DEEC protocol, on the other hand, uses residual energy of nodes at individual level as well as average network energy for deciding the CHs in a particular round. In this technique, the ratio of residual node energy and average network energy is used to elect the CHs in a round. It ensures higher probability of becoming CH for the nodes with higher amount of initial as well as residual energy.

Average energy of network, E_{avg} at round r is calculated as:

$$E_{avg}(r) = \frac{1}{N_T} * \sum_{i=1}^{N_T} E_i(r) \tag{16}$$

where $E_i(r)$ represents the i th node's residual energy in current round, r.

In DEEC, average probability for CH selection of i th node in round r is given as:

$$p_i = p\left[1 - \frac{E_{avg}(r) - E_i(r)}{E_{avg}(r)}\right] = p * \frac{E_i(r)}{E_{avg}(r)} \tag{17}$$

Here, p represents reference value of the avg. probability, p_i. In homogeneous WSNs, all sensor nodes are equipped with equal amount of initial node energy, therefore, p is used as reference energy for the probability p_i. This value of p varies based upon the initial node energy in the heterogeneous WSNs. The calculation for value of p in a heterogeneous WSN having two energy levels is given under Eq. 12 and 13, terming the two types of nodes as normal nodes and special or advanced nodes. Average probability of becoming CH in DEEC technique for the normal and the special or advanced nodes is represented as:

$$p_i = \begin{cases} \frac{p * E_i(r)}{(1+\alpha * m) * E_{avg}(r)} & if \; s_i \; is \; the \; normal \; node \\ \frac{p * (1+\alpha) * E_i(r)}{(1+\alpha * m) * E_{avg}(r)} & if \; s_i \; is \; the \; advanced \; node \end{cases} \tag{18}$$

The threshold for these normal and special or advanced nodes, $T(s_i)$ is given by:

$$T(s_i) = \begin{cases} \frac{p_i}{1-p_i * \left(r * mod \frac{1}{p_i}\right)} & if \;\; s_i \in G \\ 0 & otherwise \end{cases} \tag{19}$$

Where the set of nodes, G are eligible for election of CH in a round, r.

3.6 Working of DDEEC

DEEC does not completely solve the penalization problem of advanced nodes due to consideration of node initial energy in CH election average probability calculation. A node carrying higher amount of residual energy in a particular round,r will become a CH more frequently than the normal nodes. As the nodes keep depleting their energy in each round, the residual energy of advanced nodes will become almost equal to normal nodes after some rounds. Still, the DEEC technique continues penalizing the advanced nodes by making them CHs more frequently than normal nodes carrying equal or higher residual energy at this stage. It results in earlier death of these overburdened advanced nodes. It makes DEEC an underperforming technique due to its poor CH selection. This unbalanced situation of node residual energy is taken into consideration by another technique, DDEEC which takes into account the node residual energy as well as average network energy while selecting the CHs and introduces a new threshold residual energy parameter as, Th_{Res} given by:

$$Th_{Res} = E_o * \left(1 + \frac{a * E_{disNN}}{E_{disNN} - E_{disAN}}\right) \tag{20}$$

where, E_{disNN} is the amount of energy a normal node dissipates, and E_{disAN} is the amount of energy an advanced node dissipates. It makes CH selection more balanced and efficient. The CH selection probability of normal nodes becomes equal to advanced nodes, when both types of nodes attain residual energy threshold limit due to higher energy consumption in advanced nodes so far. The CH selection in DDEEC is done based upon average probability p_i as follows:

$$p_i = \begin{cases} \frac{p * E_i(r)}{(1+a*m) * E_{avg}(r)} & for\ Normal\ nodes,\ E_i(r) > Th_{res} \\ \frac{(1+a) * p * E_i(r)}{(1+a*m) * E_{avg}(r)} & for\ Advanced\ nodes,\ E_i(r) > Th_{res} \\ \frac{(1+a) * p * E_i(r)}{(1+a*m) * E_{avg}(r)} & for\ Advanced,\ Normal\ nodes,\ E_i(r) \leq Th_{res} \end{cases} \tag{21}$$

4 Simulation Results and Analysis

The above clustering techniques, considered for performance evaluation and analysis, are implemented under MATLAB environment. A network model consisting 100 sensor nodes, which are deployed over the square sensing field (size 100 m * 100 m) in a uniform random manner is considered. The location of sink is at the sensing field center, and is considered to have sufficient resources to perform until the full lifetime of network.

LEACH, LEACH-C, and ALEACH, are evaluated being homogeneous WSNs clustering techniques for all the three scenarios. The purpose is to see the impact of equal probability assignment for CH selection, as is done by these homogeneous techniques, to heterogeneous nodes, as in scenario 3, on the overall performance of the network. Further, SEP, DEEC, and DDEEC clustering techniques, which are specially designed for the heterogeneous WSNs, are also evaluated for scenario 3 for comparative analysis and to adjudge their goodness over one another.

4.1 Performance Comparison of Homogeneous Clustering Techniques

The performance comparison of LEACH, LEACH-C and ALEACH for parameters- stability period, network throughput, energy dissipation, and the advanced nodes alive per round parameters for above three scenarios is shown below, where technique name is followed by scenario- (1) represents the scenario 1, (2) represents the scenario 2, and (3) represents the scenario 3.

Stability Period Comparison

The stability period for the three techniques under three network scenarios is shown in Fig. 2. For scenario 1 and 2, it is seen that the homogeneous techniques work well for networks with pure homogeneous nodes due to their principal of equal treatment for all the nodes.

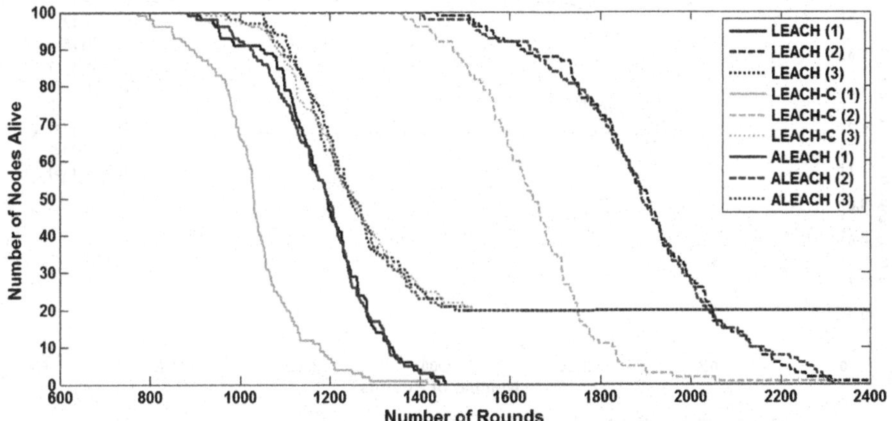

Fig. 2. Stability period comparisons of homogeneous clustering techniques.

The values of stability period get substantially improved with increase in network energy. For example with total network energy as 50 J (scenario 1) the stability of LEACH, LEACH-C and ALEACH is around 906, 771, and 886 rounds, while it jumps to 1438, 1365, and 1412 respectively for the second scenario with total network energy as 80 J. The total network energy for heterogeneous network scenario 3 is kept deliberately same as that of scenario 2 and is 80 J.

Here, it is seen that for LEACH, LEACH-C and ALEACH techniques, stability period of the network falls down to 1053, 923, and 974 rounds, around 27%, 32% and 31% reduction respectively, as compared to scenario 2. It indicates that clustering techniques meant for homogeneous networks failed to utilize the energy heterogeneity in an effective way, as beyond the first node failure, the energy of advanced nodes remain almost unutilized.

Throughput Comparison

The throughput for the three techniques under three network scenarios is shown in Fig. 3. It is seen that throughput for these three techniques under scenario 2 is much higher than scenario 1 and 3. It is expected as these techniques are meant for homogeneous clustering and work well when all the nodes have higher initial energies as in scenario 2. For scenario 3, although the total network energy is same as that of scenario 2, it is unequally distributed among advanced and normal nodes. Homogeneous techniques do not fit well as they assign equal probability to all nodes for CH selection. Accordingly, although the throughput for scenario 3 is higher than scenario 1, it does not match up with that of scenario 2. Further, it can be observed that LEACH and ALEACH techniques have higher throughput than LEACH-C technique for all the three scenarios.

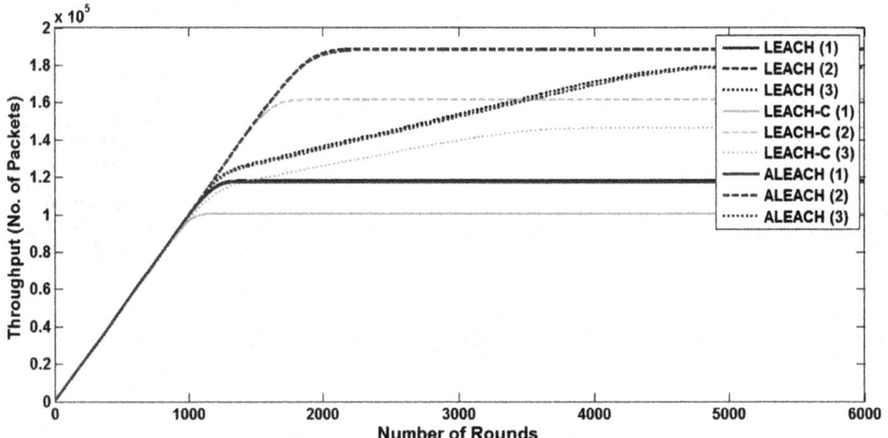

Fig. 3. Throughput comparisons of homogeneous clustering techniques.

Energy Dissipation Comparison

The energy dissipation for the three techniques under three network scenarios is shown in Fig. 4. It can be observed that LEACH and ALEACH techniques have better energy dissipation than LEACH-C technique for all the three scenarios. Further, energy dissipation for all three techniques under scenario 2 is better than scenario 3. Although in scenario 3, energy exists beyond that of the other two scenarios, it is not available in the stable region period, and hence the network reliability suffers.

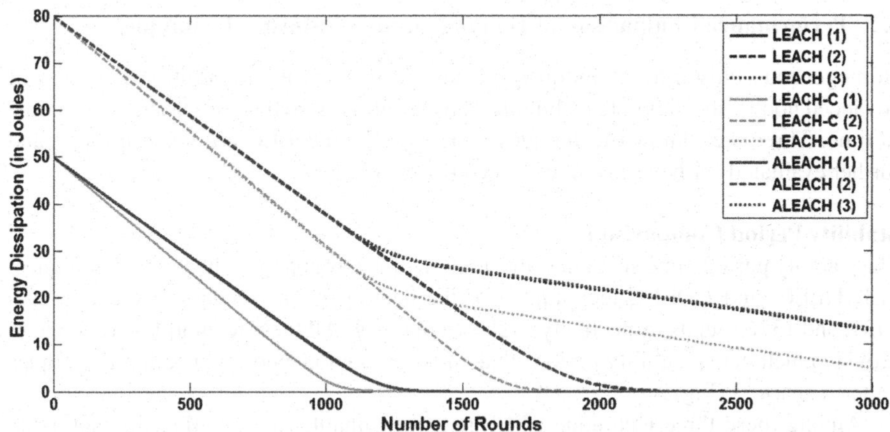

Fig. 4. Energy dissipation comparisons of homogeneous clustering techniques.

Advanced Nodes Alive per Round Comparison

The advanced nodes alive per round for the three techniques under network scenario 3 are shown in Fig. 5. As advanced nodes are the additional energy nodes introduced only in scenario 3, the first advanced node dies at 3804, 2652 and 3843 rounds respectively for LEACH, LEACH-C and ALEACH techniques.

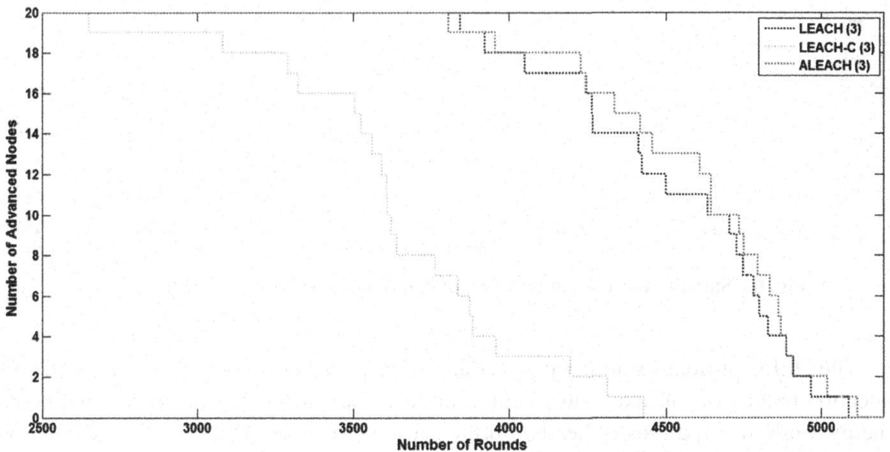

Fig. 5. Advanced nodes alive comparisons of homogeneous clustering techniques.

Due to under utilization of additional energy of advanced nodes, as homogeneous clustering techniques give no special treatment to heterogeneous network nodes, normal nodes start dying quite earlier while advanced nodes start dying at more than 2600 rounds.

4.2 Performance Comparison of Heterogeneous Clustering Techniques

Homogeneous clustering techniques are unable to exploit the additional energy of advanced nodes, so, elaborate techniques have been suggested by researchers from time to time. Three well known techniques - SEP, DEEC, DDEEC are taken up here and further evaluated for heterogeneous network scenario 3.

Stability Period Comparison

The stability period, time spent in stable region, of heterogeneous clustering techniques-SEP, DEEC, and DDEEC techniques for the heterogeneous scenario 3 equals 1292, 1431, and 1376 rounds respectively as shown in Fig. 6. All these techniques are found to provide much better stability periods than those by homogeneous clustering techniques under scenario 3 (Fig. 2).

Among these three techniques, SEP remained unable in exploiting the additional initial energy of the advanced nodes in a judicious manner, as normal nodes die earlier due to their participation in CH selection process.

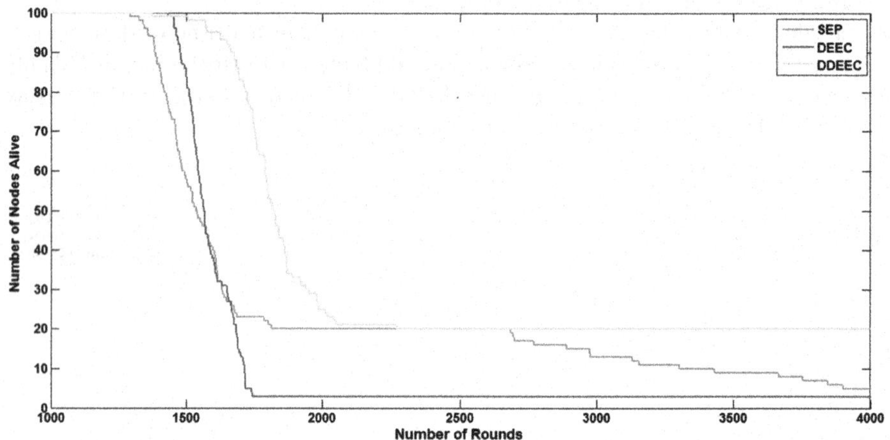

Fig. 6. Stability period comparisons of heterogeneous clustering techniques.

The DEEC prolongs stability period due to better load distribution and balanced CH selection process, as it takes into account initial node energy as well as residual node energy while making cluster heads. Further, it is pointed out that the residual energy threshold concept as used by DDEEC to improve upon DEEC does not help much in improving stability period.

Throughput Comparison

Throughput results for the three techniques, under scenario 3, are shown in Fig. 7. It is seen that overall throughput of DDEEC is highest among all three techniques.

Individual values of packets to the sink and the CHs reveal that DDEEC undergoes unbalanced CH selection process after few rounds, thereby creates very large number of CHs resulting increased throughput due to increased direct communications in unstable

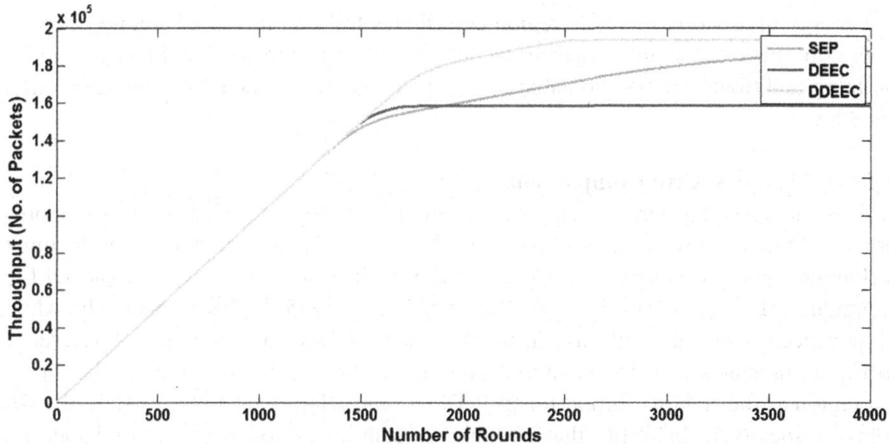

Fig. 7. Throughput comparisons of heterogeneous clustering techniques.

region. In the stable region of the underlying network, however, DEEC and DDEEC performed equally well. As the stability period of SEP is lowest among all the three so is its throughput.

Energy Dissipation Comparison

Energy dissipation for the three techniques under scenario 3 is displayed in Fig. 8. It represents the quantity of total network energy consumed by network nodes for their operations, which is decreasing in each round. It is observed that SEP, DEEC and DDEEC techniques dissipate 68%, 76%, and 68% of total network energy in stable region, while LEACH, LEACH-C and ALEACH techniques dissipate 55%, 57%, and 51% in stable region for scenario 3.

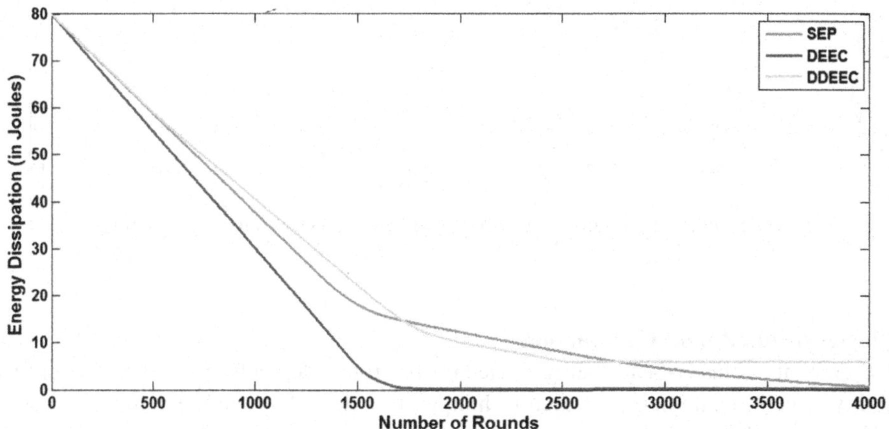

Fig. 8. Energy dissipation comparisons of heterogeneous clustering techniques.

Energy dissipation in stable region contributes towards increased stability period. Therefore, heterogeneous clustering techniques provide improved stability period by utilizing additional energy in stable region as compared to homogeneous clustering techniques.

Advanced Nodes Alive Comparison

Advanced nodes play a major role in enhancing the lifetime of a network if used appropriately. In this sub-section, numbers of advanced nodes that remain alive for the three techniques under scenario 3 are shown in Fig. 9. It is seen that for SEP, and DEEC techniques, the first advanced node dies at 2685 and 1455 rounds respectively, while all advanced nodes are still alive at 6000 rounds for DDEEC technique. A reference to Fig. 6, indicates that first advanced node in all three techniques dies in the unstable region as the stability period for SEP, DEEC and DDEEC is 1292, 1431, and 1376 rounds respectively. In DEEC, that gives highest stability period, first advanced node die near stable region and much earlier than other techniques. This is due to its better load distribution among advanced and normal nodes.

To prolong the overall lifetime of underlying network, the desirable thing is that the advanced nodes contribute to their maximum in the stable period. So, going by the analysis carried out it is felt that these techniques do not fully exploit the additional energy of advanced nodes.

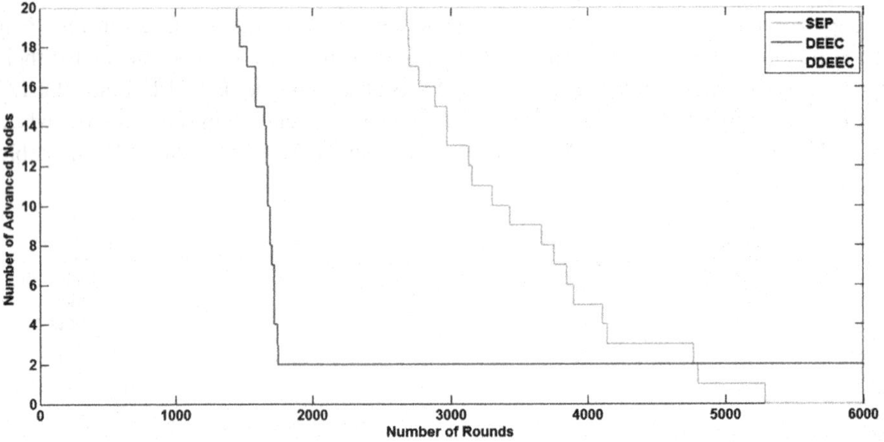

Fig. 9. Advanced nodes alive comparisons of heterogeneous clustering techniques.

Cluster-heads/Round Comparison

The amount of cluster-heads being created in each round depends upon respective probability selection parameter for each technique. It is desirable that the count of chosen CHs per round remain close to the optimal value that is around 10% of the total nodes. As the performance of a network gets affected by this number, so in this sub-section, this parameter is analyzed in detail and results shown in Fig. 10.

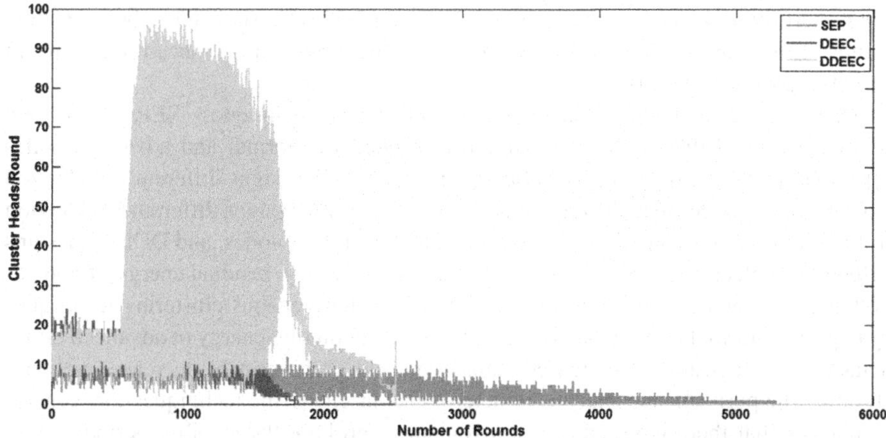

Fig. 10. CHs per round comparisons of heterogeneous clustering techniques.

It is gathered that SEP and DEEC techniques choose uniform average number count for CHs in each round within the desirable 10% limit, whereas for the DDEEC, the average number of CHs drastically increases up to 90% after about 500 rounds. This is owing to the complex process of optimal probabilities calculation for CH selection in DDEEC. It results in more direct transmissions to sink by all the CHs, thereby resulting in unbalanced clustering formation and working of DDEEC.

5 Conclusions

A penetration testing framework for energy efficient clustering techniques is proposed in this paper, where three different network scenarios are provided for performance comparison of well known homogeneous and heterogeneous clustering techniques. Scenario 1 and 2 represent homogeneous network having normal nodes with equal initial energy. Scenario 3 represents a heterogeneous network having normal as well as the advanced nodes carrying higher initial energy level than the normal nodes. Scenario 2 and 3 are chosen for evaluation and analysis of performance for underlying clustering techniques in homogeneous and heterogeneous networks having equal total network energy.

Based upon performance analysis of homogeneous clustering techniques- LEACH, LEACH-C and ALEACH, it is observed that these techniques give equal treatment to all the network nodes. These perform well when the overall energy of network is increased with uniform allocation of additional energy to all the network nodes. So, homogeneous clustering techniques are suitable for homogeneous WSNs. In heterogeneous WSNs, the overall energy of network is increased through replacement of some network nodes with advanced nodes carrying higher energy at initial level. The performance analysis for scenario 3 reveals that homogeneous clustering techniques failed to perform well, and their performance sharply degraded by almost 30% as compared to scenario 2 having same total network energy. This is due to their equal treatment for all nodes, which result in earlier death of equally burdened energy deficient normal nodes. Normal nodes start dying at very early stage, while advanced nodes start dying beyond 2600 rounds. This reflects the need for special kind of treatment for advanced nodes meant for heterogeneous networks.

Performance analysis of heterogeneous clustering techniques- SEP, DEEC and DDEEC revealed that different treatment is needed for normal and advanced nodes to prolong their stability period or effective lifetime. SEP assigns different CH election probabilities to nodes with different initial energies, DEEC assigns different CH election probabilities based upon initial as well as residual energy of nodes, and DDEEC assigns different CH election probabilities based upon initial energy, residual energy of node as well as predefined energy threshold. In this way, heterogeneous clustering techniques try to protect normal nodes from dying earlier while depleting energy of advanced nodes at faster rate. It results in increased stability period at the cost of decreased network lifetime. The performance analysis of heterogeneous clustering techniques for scenario 3 indicates that there is scope of improvement to prolong the stability period of heterogeneous network by efficient utilization of additional energy of advanced nodes. As design complexity increases from SEP to DDEEC, and CH selection process becomes unbalanced, the proposed technique should be capable to prolong stability period with balanced CH selection process at minimal complexity. So, the techniques designed for heterogeneous WSN need further improvisation to prolong the stability period.

Finally, it can be observed that testing a proposed clustering technique under PTF will definitely help researchers to ascertain quality of their technique individually as well as compare its performance with other relevant techniques.

References

1. Kandris, D., et al.: Applications of wireless sensor networks: an up-to-date survey. Appl. Syst. Innov. **3**(14), 1–24 (2020)
2. Ramson, S.R.J., et al.: Applications of wireless sensor networks—a survey. In: International Conference on Innovations in Electrical, Electronics, pp. 325–329. Instrumentation and Media Technology, Coimbatore (2017)
3. Xu, G., et al.: Applications of wireless sensor networks in marine environment monitoring: a survey. Sensors **14**, 16932–16954 (2014)
4. Sharma, S., et al.: Issues and challenges in wireless sensor networks. In: International Conference on Machine Intelligence and Research Advancement, Katra, India, pp. 58–62 (2013)
5. Anastasi, G., et al.: Energy conservation in wireless sensor networks: a survey. Ad Hoc Netw. **7**(3), 537–568 (2009)
6. Sharma, S., et al.: Energy efficient clustering techniques for heterogeneous wireless sensor networks: a survey. J. Mob. Comput. Commun. Mob. Netw. **3**(1), 1–9 (2016)
7. Ambekari, J.S., et al.: Comparative study of optimal clustering techniques in wireless sensor network: a survey. In: ACM Symposium on Women in Research, New York, USA, pp. 38–44 (2016)
8. Elbhiri, B., et al.: Developed Distributed Energy-Efficient Clustering (DDEEC) for heterogeneous wireless sensor networks. In: 5th IEEE International Symposium on I/V Communications and Mobile Network (ISVC), Rabat, Morocco, pp. 1–4 (2010)
9. Heinzelman, W.B., et al.: An application-specific protocol architecture for wireless microsensor networks. IEEE Trans. Wirel. Commun. **1**(4), 660–670 (2002)

10. Jain, P., et al.: Impact analysis and detection method of malicious node misbehavior over mobile ad hoc networks (IJCSIT). Int. J. Comput. Sci. Inf. Technol. **5**(6), 7467–7470 (2014)
11. Qing, L., et al.: Design of a distributed energy-efficient clustering algorithm for heterogeneous wireless sensor networks. Comput. Commun. **29**(12), 2230–2237 (2006)
12. Smaragdakis, G., et al.: SEP: a stable election protocol for clustered heterogeneous wireless sensor networks. In: 2nd International Workshop on Sensor and Actor Network Protocols and Applications, Boston, Massachusetts, pp. 251–261 (2004)
13. Ali, M.S., et al.: ALEACH: advanced LEACH routing protocol or wireless microsensor networks. In: International Conference on Electrical and Computer Engineering, Dhaka, Bangladesh, pp. 909–914 (2008)
14. Bandyopadhyay, S., et al.: Minimizing communication costs in hierarchically clustered networks of wireless sensors. Comput. Netw. **44**(1), 1–16 (2004)

Audit and Analysis of Docker Tools for Vulnerability Detection and Tasks Execution in Secure Environment

Vipin Jain[1,2](✉), Baldev Singh[1], and Nilam Choudhary[2]

[1] Vivekananda Global University, Jaipur 303012, India
ervipin.skit@gmail.com

[2] Swami Keshvanand Institute of Technology, Management and Gramothan, Jaipur 302017, India

Abstract. Containers have existed for many years, Dockers have not invented anything in this sense or almost nothing, but it should not be taken away from them. Every timeless monolithic application is developed and based on modules, allowing a more agile, fast, and portable development. Well-known companies like Netflix, Spotify, Google, and countless startups use microservices-based architectures in many of the services they offer. Limit and control the resources that the application accesses in the container, they generally use their file system like UnionFS or variants like aufs, btrfs, vfs, overlayfs, or device mapper which are layered file systems. The way to control the resources and capabilities it inherits from the host is through Linux namespaces and cgroups. Those Linux options aren't new at all, but Docker makes it easy, and the ecosystem around it has made it so widely used. In this paper, different Docker security levels are applied and tested over the different platforms and measure the security attribute to access the legitimate resource. Additionally, the flexibility, comfort, and resource savings of a container provided by a virtual machine, or a physical server are also analyzed.

Keywords: Virtualization · Container · Docker · Security · Access control · Variability · Security practices

1 Introduction

Docker has gained a lot of popularity and usability in business and educational environments; due to the great virtues they possess in terms of deployment facilities and the apparent security options it brings. Docker is a secure and lightweight software container that allows you to wrap pieces of software in a file system along with everything you need to run and run one or more applications independently, ensuring that they always run as usual. Same way, regardless of environment or operating system, docker allows accelerating the development process, eliminates inconsistencies in environments, improves code distribution, and allows much more effective collaboration between developers (Fig. 1).

V. E. Balas et al. (Eds.): ICETCE 2022, CCIS 1591, pp. 654–665, 2022.
https://doi.org/10.1007/978-3-031-07012-9_54

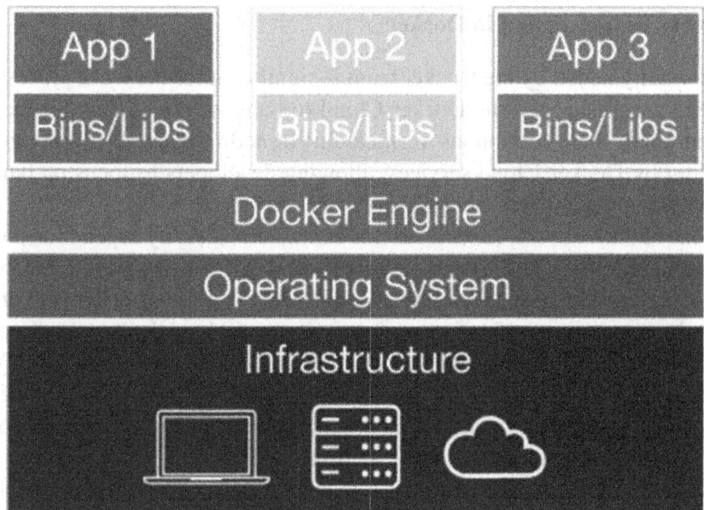

Fig. 1. Docker architecture

1.1 Docker Security Service (DSS)

The DSS service is triggered when a user uploads an image to a Docker Cloud repository: the service receives the image and breaks it down into its respective components. It then compares it to the Known Vulnerabilities databases (CVE databases) and finally verifies all the components at a binary level to validate their content. Once the scan result is ready, it is sent to the Docker Cloud service graphical interface [1]. When a new vulnerability is reported to the CVE databases, the DSS service activates and checks if there are any images or components with this vulnerability in the repository and sends a report via email to the administrators of the repository. The information presented both in the graphical interface and in the emails contains data about the vulnerabilities and the images that present them (Fig. 2).

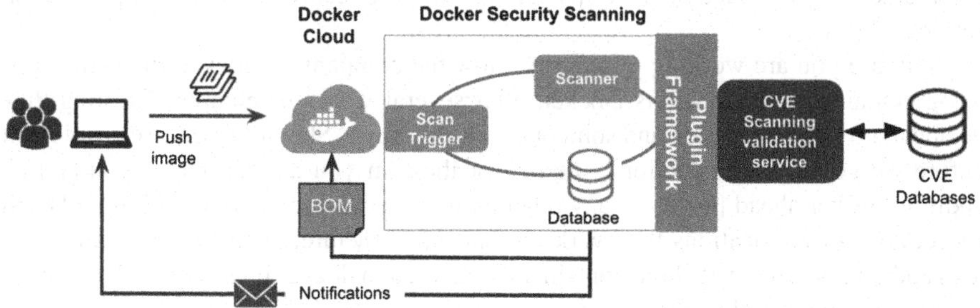

Fig. 2. Docker security service operation flow chart

1.2 Security Enhancement in Docker

This functionality added by the Docker team is significant for the work processes of the development team that uses the Docker Cloud to increase the security of the released images and allow their continuous verification. In addition, the DSS service is automatically run by Docker Cloud services, allowing developers to focus on building the product and having a much more agile and effective workflow [2].

Here, one of the main emphases is that the container's security depends on the security of the host and its kernel. These weaknesses exposed in the investigations are intrinsic to the containers, but very little taken into account when deploying systems in production or even in the integration by host operating systems, such as the one reported in CVE 2018-8115This takes advantage of the validation of input parameters of a Windows system library just when loading a Docker image. In Docker, administrators need to use tools to validate some of the risks, mainly validating the relationship between the host and the container, even before executing an image [3].

Tools like Dockscan allow validating that relationship between the host and the containers, and like every application that is enabled, it can generate risks. In the above case, the tool shows the lack of user limits and free traffic between the network of the host and that of the containers. When running a container and exposing some service, new gaps arise that must be taken into account, where the lack of access controls is seen in the container that is marked as a high-risk vulnerability [4].

There are many Open Source and other paid tools that allow validation of the Docker daemon configuration or the exposed services, such as:

1. Clair Scanner
2. Docker Bench Test
3. OpenScap Docker Notary
4. StackRox

The integration of these or other tools within containerized enterprise DevOps systems such as Docker, incorporate a necessary layer of security for the production implementation of this technology. This layer is increasingly important, generating new tools and methodologies called SecDevOps that seek to have security schemes for production services [5].

Suppose you are working in security. Now the company decides to run some apps using containers, the choice is Docker, after several weeks or months of testing they decide to go into production and someone suddenly says "Should we do a security audit before we go out to production", the rest of the film you already know, you and an audit of Docker ahead [6]. As in any other audit, you can use your arsenal of tools and procedures for applications that work on that infrastructure, analyze file permissions, logs and others. But what about containers, images, and docker files or even the security of orchestration and clustering tools.

Some Specific Considerations for This Audit:

1. Images and packages should be up-to-date, as well as bug-free.

2. Automate the audit: we should be able to automate any task. It will save us a lot of time, and we will be able to run such an audit as many times as necessary, forgetting manual audits unless you are learning [7].
3. Container links and volume: when using the "docker diff" command, it will be simpler to discover errors if the underlying filesystem is read-only.
4. There will be greater difficulty in auditing larger images. As much as feasible, try to reduce the images.
5. The host system kernel, which is regularly updated, is accessible by all containers running within the same server.

2 Security Review in Docker Architecture

Using Docker, an application's container configuration may be changed dynamically without affecting the system's overall performance. As a key component, Docker Engine is a lightweight application that includes built-in functionality for orchestration; scheduling; networking; and security. The Docker Engine may then be used to run containers [8]. A layered file system ensures that necessary components to operate the application are included in each Docker image [9]. Registries like Docker Cloud and Docker Trusted Registry are used by teams that create and deploy apps to manage and share images. Containers for running applications are scheduled, coordinated, and controlled by the Docker Universal Control Plane during runtime. Docker Datacenter, an integrated platform for contemporary application environments based on container technology, provides access to these technologies.

In this paper, we'll use the following Docker Security taxonomy:

Subjects are objects that accomplish tasks. We can associate a subject with a user easily, but users can only act through an agent or a process. A Subject could be viewed as an abstract of a process.

Objects are accessible components or resources of a file system. Virtual storage or a network interface are two different types of resources that might be associated with an item. Processes are typically made available for use by other processes.

The *access matrix* is a table of subject and object. The mode of accusations of the subject and matching object is indicated in each cell for particular rows and columns. Because most cells are empty, the matrix isn't built as a two-dimensional array. Store access modes by rows, registering the list of the objects and access modes allowed for each subject, or store access modes by columns, registering a list of the subjects who have access to the object. The modes of access are authorized for each subject for each object. Here, we're talking about a system based on capability lists in the first scenario, and an access control list-based system in the second (ACLs Access Control Lists). Among the most commonly used operating systems, only FreeBSD provides an implementation of a *capability* list. In Linux, the control mechanism of access described in capabilities, which we will refer to later, does not save a relationship with the *capabilities* just described.

The most common access control model is the so-called discretionary (DAC, Discretionary Access Control). This model allows the owner of an object to randomly modify the object's access mode. Normally, the pro-Presario of an object is the subject that created it. ACL-based systems such as Windows, UNIX, Linux, or VMS use this model. A specific problem of Unix systems, and which Linux inherited is that, regardless of limitations in DAC access control, this type of access control only applies to systematically form objects of a file system. It is not possible, for example, to control access to a network interface by applying an ACL.

Mandatory Access Control (MAC) designates a set of access control models in which the permissions of a subject about an object are determined by an access control policy that the user cannot modify (although later we will clarify this aspect). Thus, for example, a certain MAC policy may prevent the owner of an object grants writing permissions to other subjects there are several MAC models. In the context of this work, those that are relevant SELinux applies, mainly TE (Type Enforcement), and especially when Applies in a container environment, MCS (Multi-Category Security).

Multilevel Security (MLS) implementations supported the MLS model by allocating subjects and objects a tag with two attributes. There may be hierarchical classifications for the main, but this isn't the case here. Multi-valued attributes are possible for the second attribute. To set the access modes, a pre-defined policy was used that compared the topic's and the object's labels. Any shape that corresponds to the source can be used to label the whole model of MLS. After this point in time, we'll be requesting this label using the term "security level" instead of the more specific term "security sensitivity". "The MLS paradigm is implemented as an SELinux configuration in Linux. The level of security just described corresponds to the fourth field or attribute of SELinux context (we will define context later). In this work, we have more references to MLS models. We mention it only because it is the basis on which a policy designed specifically for management is built containers, MCS (Multi-category Security), which we describe further TE, and that reuses the "security level" field of SELinux context in a different way.

TE (Type Enforcement) in these models' objects and subjects are labeled; subject access modes by an object, and from subject to subject are specified in two matrices. Subjects, objects, and types are all allocated tags as part of the denominating domains. Subjects and objects are both assigned a type in SELinux's TE implementation, which controls a single database of permissible access modes. A second matrix of access (MAC) is therefore required in addition to the one that regulates DAC access: each process has a unique UID, GID, and security identifier in addition to the standard UIDs and GIDs (SID).

Multi-Category Security (MCS) unlike historical models, MCS is a model based on an implementation-specific to SELinux and is only described informally in some articles online by authors involved in their development. According to (Ibrahim et al. 2020): "MCS is, in fact, an MLS modification". The MLS label field, MLS kernel code, MLS strategy constructs, labeled printing, and label encoding is all reused in SE Linux. MCS is a policy modification, accompanied by a few user hacks to hide some of the undesirable MLS things, from a technological perspective. There are a few kernel changes, but they all pertain to making it easier to switch to MCS or MLS without requiring a full file system upgrade.

2.1 Virtualization System Security

The technique of virtualization paves the way for better usage of IT and infrastructure resources. With so many options, some are more popular than others when it comes to virtualization Hypervisors are an example of a technology that has elevated virtualization from a novelty to a useful tool. On the physical level, virtual machines virtualize the hard drive, CPU, and RAM. This implies that the operating system (OS) may access a virtualized version of the actual device via the hypervisor. Although Docker containers and virtual servers have many similarities, Docker produces virtual environments that share the OS's kernel, which is referred to as the "core" in Linux parlance. As a result, containers are now virtualized in software rather than on physical devices [7].

2.2 Containers Security

Containers are used to provide isolation and low overhead in a Linux environment. The management interface or the container engine interacts with the kernel components and provides tools for creating and managing the containers. Since containers are lightweight, we can achieve higher density and better performance as compared to virtual machines. A small, lightweight virtual machine is created on non-Linux-based OS like Mac or Windows. The hardware virtualization feature should be enabled on the host machine to support the spinning up of a virtual machine. Once a small Linux OS is running, containers are created using the virtual machine's kernel.

2.3 Hypervisors Security

For a long time, hypervisors were the preferred means of hosting servers, making virtualization accessible to the general public as well as large corporations. The user situations of these two categories of stakeholders are vastly different, yet the essential description and technology of hypervisors are the same. Hypervisors have established themselves as a critical component of virtualization technologies as a result of significant virtualization adoption and market share.

2.4 Security in Hardware Level and Virtualization

Hardware Level Virtualization has widely been used over the last decade for implementing virtualization. Hypervisors create a layer between the software that runs on top of them and the underlying hardware. A hypervisor is a device that may be used to monitor software that creates virtual resources which run directly on the hardware isolated from the underlying host operating system [12].

2.5 System Level Virtualization

Virtualization has gained traction in last few years because of its low overhead [11]. Hardware level virtualization is considered as heavy weighted because it relies on hardware emulation. Alternatively, in OS-level virtualization, there is a container engine that uses kernel features like cgroups, namespaces, chroot, etc. for creating isolated user-space instances known as containers, on top of the host machine. The containers share the host machine's kernel with the help of container engine instead of running a full operating system as it does in hypervisor-based virtualization, thereby reducing the overall overhead [8] (Fig. 3).

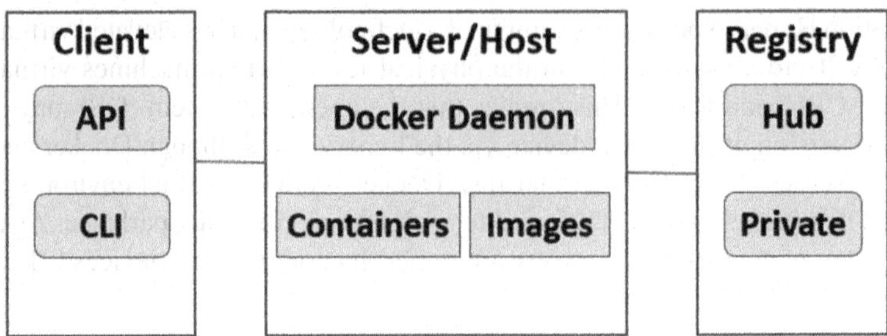

Fig. 3. Container architecture in Docker

The inability to deploy a new paradigm of systems administration and the difficulty in mitigating threats, according to the Portworx study [13] researchers in 2017, was a key impediment to container adoption. They attribute this to persistent storage. It is clear that the methods used in Linux to create, deploy, interconnect and manage application containers do not fit individually very well to traditional practices in systems administration. Regarding the security issue, the isolation of workloads executed in a container with respect to other containers or with respect to Host system (host) is considered, quite widely, lower than that provided by hypervisors [14].

3 Security Analysis in Docker

Security is an increasingly important factor in the day-to-day of companies which required different good practice to be proposed to increase the security in containers once deployed.

3.1 Docker Environment Settings

Within the Docker world, different good practices are recommended to carry out when you want to deploy a service with Docker.

Host. The Docker environment's most critical component might be considered the Host machine, given that it is where all the infrastructure is supported and where the containers are. It is therefore highly recommended and at the same time very important to follow some *Good Practice* guidelines when configuring the machine where the different containers [9].

Docker Specific Partition. The first point to keep in mind when you want to install Docker is partitioning of the hard disk because Docker needs to store the different images that the user has been using (previously downloaded). It is essential that Docker is installed in the */var/lib/docker* partition. This partition must be mounted on the root directory "*/*" or on the directory "*/var*". To verify the correct installation, in the first instance, it must be verified if you have installed Docker in the specified directory.

The second check to be performed, if the directory "/var/lib/docker", located inside. "/Dev/sda1" is mounted where it belongs. Otherwise, you should manage with the Linux Logical Volume Manager (LVM) tool to be able to create the corresponding partition for docker [10].

User Permits. Once Docker is installed in the correct directory, you need to limit which users can control the Docker daemon. By default, the daemon can only be controlled by the *root* user of the system. Next, we run the basic *Hello-World* container for testing that everything works correctly. If it returns an error as the user you do not have sufficient permissions to control the daemon. When Docker is installed on a machine, by default, a *docker* group is created, where only those who are integrated into this group will permit to control the daemon [11]. At the same time, Docker can also be controlled by all those users who are in the *sudo* group. Therefore, it is necessary to verify that users have permission to control the daemon and add those as necessary that can execute actions on Docker [10]. If we execute the same command but with *superuser* since the user is in the sudo group.

To speed up this whole process, it is necessary to add the user in question to the *docker* group [5]. At this point, it is advisable to carry out good custody of the credentials of the user to be added, because they will have permission to perform any action with the Docker daemon.

3.2 Files and Directories Audit

Since the Docker daemon operates as root, it's imperative that all folders and files be inspected on a regular basis to keep tabs on what's going on. To properly audit all events, the Linux Audit Daemon *framework* [11] should be used. With Linux Audit Daemon it allows:

1. Audit file accesses and modifications.
2. Monitor system calls
3. Detect intrusions
4. Record commands by users

The file "/etc./audit/rules.d/audit.rules" has to be updated to include new rules for the audit daemon in order to function properly. To be able to audit directories, we will need to add the relevant rules (Table 1):

Table 1. Docker commands to add rules for security

## Rules	Commands
## R-0	-w/usr/bin/docker -k docker
## R-1	-w/var/lib/docker -k docker
## R-2	-w/etc./docker -k docker
## R-3	-w/usr/lib/systemd/system/docker.service -k docker
## R-4	-w/usr/lib/systemd/system/docker.socket -k docker
## R-5	-w/usr/lib/systemd/system/docker.socket -k docker
## R-6	-w/etc./docker/daemon.json -k docker
## R-7	-w/usr/bin/docker-containerd -k docker
## R-8	-w/usr/bin/docker.runc -k docker

The audit daemon will need to be restarted when the new rules have been added. Lastly, you may see the audit logs under the Audit Logs section:

/var/log/audit/audit.log

3.2.1 Docker Daemon

The Docker daemon is the most crucial component for controlling the lifecycle of the many images on the host machine. We have made the "*securitization*" for managing the resources, either to provide or to limit.

3.2.2 Traffic Limitation Between Containers

Once you are already working with deployed containers, by default, Docker allows traffic between containers deployed on the same *Host*, therefore you can allow each container to have the ability to view the traffic of the other containers since all of them are in the same network. To limit the traffic between the different containers deployed and that, therefore, visibility is available between them within the same *Host* machine, there is a command to disable this option. With the *docker* command, you are accessing the options offered by Docker to make modifications to the Docker daemon.

$ docker COMMAND
Therefore, to limit traffic, the following command must be entered:
$ docker –icc = false

3.2.3 CLI Authorization

Authorization for the *Command Line Interface* is based on docker groups. *User permissions*, add users control permissions who need access to the Docker to deploy the necessary services.

3.2.4 Log Management

If we are interested in keeping tabs on what's going on in the containers at all times, log management is essential. A centralized log management system is required since several containers might be operating on the same Host at the same time, each creating its own logs. There are several commands for monitoring the logs, such as:

$ docker logs <container_id>

$ docker service logs <SERVICE | TASK>

The above commands are for cases where the conventional output (STDOUT and STDERR) is configured correctly. In some cases, using the above commands will not be sufficient since the displayed information will not be possible to exploit it in an ideal way. In those cases, the following steps are required:

- In the case of a program to treat the *logs*, it would not be advisable to use the instructions "*docker logs*" since it will not show the required information.
- You must do log redirection if a non-interactive process (such as a web server) within the container has a service that is already delivering logs to some file; as a result, the traditional outputs are not found enabled. You'll find the choices for redirecting and formatting logs below, so you can take use of them to their full potential. Command and several alternatives will be available:

  $ docker COMMAND

3.2.5 Driver Description

none: The *docker logs* command will not display any output since the *logs were* not available.
json-file: It is Default driver. The *logs* will be formatted as JSON.
syslog: The *logs* will be formatted as a syslog. The syslog daemon the system should be running.
journaled: The logs will be formatted as journaled. The demon of system journaled should be running.
gelf: Write *logs* to an endpoint Graylog Extended Log Format (GELF).
fluentd: The *logs* will be formatted as fluentd.
awslogs: Write the *logs* in Amazon CloudWatch Logs.
splunk: Write the *logs* in splunk.
gcplogs: Write the *logs* in Google Cloud Platform (GCP) Logging.

$ docker —log-driver syslog

3.2.6 Access Demon Configuration Files

The Docker daemon operates with root privileges. If you haven't already done so, we'll now offer you some advice on how you can allow all other users access to the daemon's folders and files. Next, The default permissions for each of these files (there is only one user with permissions to control):

3.2.7 File/Directory User: Group Permissions

docker. service root: root 644 (rw-r - r--)
docker. socket root: root 644 (rw-r - r--)
Docker socket root: docker 660 (rw-rw ----)
/etc./docker root: root 755 (rwxr-xr-x)
daemon. json root: root 644 (rw-r - r--)
/etc./default /docker root: root 644 (rw-r - r--)

Conclusion. In this analysis, there is a lot of room for improvement and adapting these or other tools to make them really useful at the level of business security analysis. It is not a bad start, at least to test what we have regardless of the state of our environments. It remains to be seen what great players can contribute on this topic and containers (Google, MS, AWS, etc.) in favor of all of them. We concluded that it is quite difficult to keep up with Docker since new versions are published with many changes and news almost every week. It seems that is the price to pay when we work with emerging technologies. In the future, orchestration systems like Kubernetes or even other Docker will be deployed like AWS ECS for security purposes.

References

1. Blial, O., Ben Mamoun, M.: DockFlex: a Docker-based SDN distributed control plane architecture. Int. J. Future Gener. Commun. Netw. **10**(12), 35–46 (2017)
2. D'Urso, F., Santoro, C., Santoro, F.F.: Wale: a solution to share libraries in Docker containers. Future Gener. Comput. Syst. Int. J. Esci. **100**, 513–522 (2019)
3. De Benedictis, M., Lioy, A.: Integrity verification of Docker containers for a lightweight cloud environment. Future Gener. Comput. Syst. Int. J Esci. **97**, 236–246 (2019)
4. Farshteindiker, A., Puzis, R.: Leadership hijacking in Docker swarm and its consequences. Entropy **23**(7), 914 (2021)
5. Ibrahim, M.H., Sayagh, M., Hassan, A.E.: Too many images on DockerHub! How different are images for the same system? Empir. Softw. Eng. **25**(5), 4250–4281 (2020). https://doi.org/10.1007/s10664-020-09873-0
6. Ibrahim, M.H., Sayagh, M., Hassan, A.E.: A study of how Docker compose is used to compose multi-component systems. Empir. Softw. Eng. **26**(6), 1–27 (2021). https://doi.org/10.1007/s10664-021-10025-1
7. Kwon, S., Lee, J.H.: DIVDS: Docker image vulnerability diagnostic System. IEEE ACCESS **8**, 42666–42673 (2020)
8. Martin, A., Raponi, S., Combe, T., Di Pietro, R.: Docker ecosystem - vulnerability analysis. Comput. Commun. **122**, 30–43 (2018)
9. Mondal, P.K., Sanchez, L.P.A., Benedetto, E., Shen, Y., Guo, M.Y.: A dynamic network traffic classifier using supervised ML for a Docker-based SDN network. Connect. Sci. **33**(3), 693–718 (2021)
10. Shu, J.W., Wu, Y.L.: Method of access control model establishment for marine information cloud platforms based on Docker virtualization technology. J. Coast. Res. **82**, 99–105 (2018)
11. Jain, P., et al.: Impact analysis and detection method of malicious node misbehavior over mobile ad hoc networks. Int. J. Comput. Sci. Inf. Technol. (IJCSIT) **5**(6), 7467–7470 (2014)

12. Tien, C.W., Huang, T.Y., Tien, C.W., Huang, T.C., Kuo, S.Y.: KubAnomaly: anomaly detection for the Docker orchestration platform with neural network approaches. Eng. Rep. **1**(5), e12080 (2019)
13. Zerouali, A., Mens, T., Decan, A., Gonzalez-Barahona, J., Robles, G.: A multi-dimensional analysis of technical lag in Debian-based Docker images. Empir. Sofw. Eng. **26**(2), 1–45 (2021). https://doi.org/10.1007/s10664-020-09908-6
14. Zhu, H., Gehrmann, C.: Lic-Sec: an enhanced AppArmor Docker security profile generator. J. Inf. Secur. Appl. **61**, 102924 (2021)

Analysis of Hatchetman Attack in RPL Based IoT Networks

Girish Sharma[1,2], Jyoti Grover[1(✉)], Abhishek Verma[1], Rajat Kumar[1], and Rahul Lahre[1]

[1] Malaviya National Institute of Technology, Jaipur 302017, India
{2020rcp9012,jgrover.cse,2020pcp5576,2019pis5479}@mnit.ac.in, abhiverma@iiitdmj.ac.in

[2] Swami Keshvanand Institute of Technology, Management and Gramothan, Jaipur 302017, India
girish@skit.ac.in

Abstract. Low power and lossy networks (LLN) are flourishing as an integral part of communication infrastructure, particularly for growing Internet of Things (IoT) applications. RPL-based LLNs are vulnerable and unprotected against Denial of Service (DOS) attacks. The attacks in the network intervene the communications due to the inherent routing protocol's physical protection, security requirements, and resource limitations. This paper presents the performance analysis of the Hatchetman attack on the RPL based 6LoWPAN networks. In a Hatchetman attack, an illegitimate node alters the received packet's header and sends invalid packets to legitimate nodes, i.e. with an incorrect route. The authorised nodes forcefully drop packets and then reply with many error messages to the root of DODAG from all other nodes. As a result, many packets are lost by authorised nodes, and an excessive volume of error messages exhausts node energy and communication bandwidth. Simulation results show the effect of Hatchetman attack on RPL based IoT networks using various performance metrics.

Keywords: Hatchetman attack · RPL protocol · IoT · LLN · IPv6

1 Introduction

The rapid expansion of the number of physical objects that can communicate over the Internet is helping people understand the concept of the IoT. In today's IoT technological world, where multi-scale devices and sensors are widely used, seamlessly blended, and communicate with one another. LLNs [9] are playing an essential role in the rapid growth of the IoT, as they help to build global communication and computing infrastructure. A Working Group of the Internet Engineering Task Force (IETF) has recommended RPL Routing Protocol for LLN Networks, as the basic solution for IoT networks [11].

IoT is helping in many domains of applications in the real world with the increase in communication technologies, cloud computing paradigms, sensor networks. We anticipate that Internet protocol and intelligent nodes that can be

V. E. Balas et al. (Eds.): ICETCE 2022, CCIS 1591, pp. 666–678, 2022.
https://doi.org/10.1007/978-3-031-07012-9_55

wirelessly connected under IoT can help to improve information availability and accessibility, as well as improve our lives [12].

However, due to a lack of resources and a shared medium, as well as security requirements and physical protection of network protocols, LLNs are unquestionably [6] vulnerable to attacks such as Denial of Service (DoS) [8]. As an example, consider a legitimate node that can be easily duplicated, altered, corrupted, overheard, or dropped an on-flying packet when compromised by an attacker or adversary. The RPL protocol standard includes the best security method for maintaining the integrity and confidentiality of control messages and the availability of routing detail. Existing RPL implementations, on the other hand, choose to suppress these safe operating modes owing to resource consumption, which has a significant influence on the performance of resource-constrained devices [4]. RPL protocol is prone to attacks because the threat analysis for securing RPL recognizes security issues with fundamental countermeasures only [10,13]. RPL allows illegitimate nodes to manipulate the content of a packet header to disrupt the routing protocol or interfere with current connections [17].

Hatchetman attack is a novel type of DoS-attack in RPL based IoT [3,16]. The malicious node modifies the source route header of incoming packets and sends a large number of erroneous packets to legal nodes with error routes in a hatchetman attack. Receiving nodes discard faulty packets with error routes because they are unable to forward packets with piggybacked error routes [5]. The receiving node will then send an error message to the DODAG root, notifying it of the issue with the source route header [14]. A significant number of incorrect packets with error routes causes the legitimate nodes to stop receiving packets and respond with an excessive amount of Error messages, resulting in a denial of service attack in RPL based LLNs [7,12]. We used a large number of simulations to look at the impact of this attack on RPL-based IoT performance in terms of packet delivery ratio, throughput, packet delivery delay, error messages issued, and power consumption. A mechanism to countermeasure this attack is also included in this paper.

The rest of this paper is organized as follows. In Sect. 2, we have presented the overview of RPL protocol. Functionality of Hatchetman attack is presented in Sect. 3. This section presents the design and implementation detail of Hatchetman attack. Section 4 presents the experimental setup and simulation results. A mechanism to countermeasure Hatchetman attack in RPL protocol is discussed in Sect. 5. Finally, conclusions of our work with future directions are presented in Sect. 6.

2 Overview of RPL

For LLN networks, RPL is a standard routing protocol. RPL Protocol has aided in the evolution of communication in the field of IoT or tiny embedded networking devices since its inception. RPL is a distance vector routing protocol, which means it uses a Distance oriented Acyclic Graph (DODAG) to route traffic Fig. 1.

RPL is defined as a proactive, distance-vector routing protocol for LLNs with 6LoWPAN as an adaptation layer. RPL works on the Network layer and also

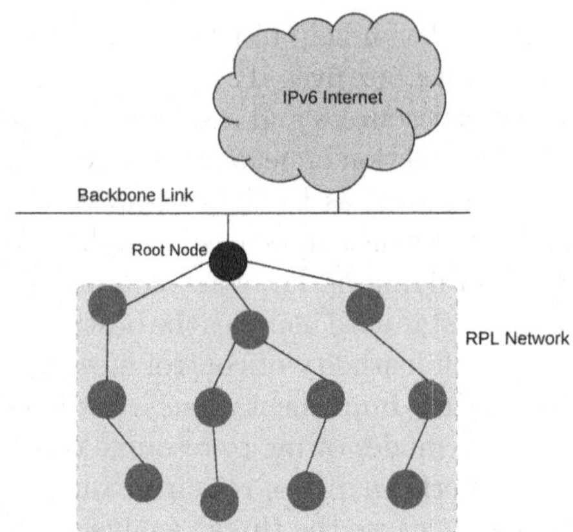

Fig. 1. Structure of RPL network

supports many link layer protocols, including 802.15.4 [18]. There are several constrained nodes or endpoints in LLN, and every node must pass the sensed information to the border router directly or via intermediate nodes.

Routing protocols in the network handle this whole procedure. RPL was accepted as a standard routing protocol by the IETF group in March 2012. RPL allows end-to-end bidirectional IPv6 communication on resource-constrained LLN devices, making it the best suitable option for IoT networks and devices. RPL mainly supports two modes of operation (MOP) specified as a MOP flag in DODAG: 1. Storing Mode 2. Non Storing Mode.

2.1 RPL Control Messages

Control messages are based on Internet Control Message Protocol version 6 (ICMPv6). Format of RPL control messages is described in Fig. 2. As per the RPL standard, there are four different forms of control messages.

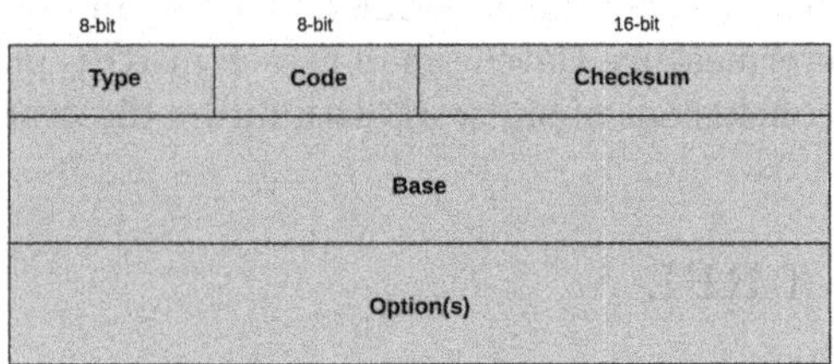

Fig. 2. Format of RPL control message

1. **DODAG Information Object (DIO):** In order to form DAG, DIO messages are transmitted by the root node (LBR) and then by subsequent nodes. It provides data such as routing limitations, metrics, rank [2], and the value of the Objective function that allows a node to find an RPL instance and learn how to configure parameters.
2. **DODAG Information Solicitation (DIS):** A node issues a DIS to receive DIO data from another node. The main application is to probe a neighbouring node, and it can also be used to request DIO data from a node.
3. **Destination Advertisement Object (DAO):** The path information from the child node is sent to the root through the preferred parent in this control message. In storing mode, it is generated in a unicast way and forwarded to the desired parent; in non-storing mode, it might be forwarded to the DODAG root or LBR node.
4. **Destination Advertisement Object Acknowledgement (DAO-ACK):** In response to a unicast DAO message, DAO-ACK is used. DAO-ACK, which is usually used in storing mode, is used by the DAO receiver to acknowledge the DAO.

2.2 Traffic Flow in RPL

RPL supports Point to point, multi-point to point and point to multi-point, the three different forms of traffic communication flows.

1. **Point to Point (P2P):** In point-to-point communication, the root (LBR) must route packets to the specified destination, which means that the root node must have routing information for all requested target nodes. The routing table can also be found on the intermediary node. A packet propagates towards the LBR until it reaches a node capable of providing the appropriate routing information. The packet will flow until it reaches the root node while intermediate nodes are in non-storing mode.
2. **Multipoint-to-point (MP2P):** The DODAG topology is built by traffic flow from any intermediate or leaf node to the gateway or root node in upward direction paths. The DODAG is built using DIO messages in MP2P traffic flow. This form of traffic flow is commonly utilised in implementation.
3. **Point-to-Multipoint (P2MP):** The P2MP paradigm is supported by the RPL Protocol, which allows traffic to flow from a gateway or LBR to other DODAG nodes. The root node in the P2MP paradigm sends packets to all DODAG leaves nodes. In order to construct the DODAG, the P2MP traffic flow uses the DAO and DAO-ACK control messages.

3 Design and Implementation of Hatchetman Attack in RPL

The primary goal of the hatchetman attack is to manipulate the source route header with the help of a malicious node in the network. Malicious node generates many invalid packets and sends a false node ID, causing legitimate nodes to try to find that false node in the network. If that node is unreachable, the node sends an error message back to the DODAG root. A significant number of error messages exhaust the node's communication information measure and energy, resulting in a DoS in RPL-based LLNs. In this paper, we consider that an adversary may capture and compromise legitimate nodes, collecting all knowledge and information, as well as public and personal keys, and rebuilding them to behave maliciously [1].

Whenever any node desires to send information to the other node, it sends this information to the root of the DODAG, i.e. Non-storing mode of RPL. In non-storing mode, nodes do not store any routing table to get information about different nodes. The nodes send information to the root of the DODAG, so the root of the DODAG can forward that information to the destination node. Malicious nodes play a key role here, retrieving the downstream information packet and piggybacking the cached supply route onto it. The authorized nodes can forward the error packet through the header. Structure of RPL source header is shown in Fig. 3.

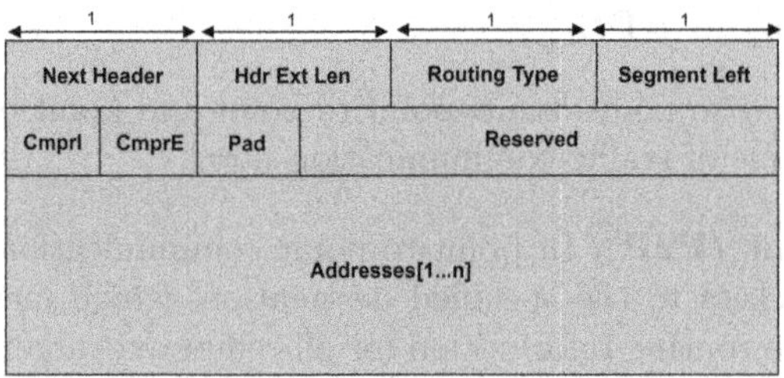

Fig. 3. RPL source route header

For example, consider that DODAG's root N_r sends a packet with the source route *([r,v,m,w,x,y,z])* to destination node N_z. When the attacker node N_m receives the downward packet, *pkt[r,v,m,w,x,y,z]*, the source of the route header is changed by the downward packet by replacing all the post hops (i.e., *[x,y,z])* of the authorized node (i.e., n_w) with a objective which is ctitious (i.e., n_f), and then transmits the non-valid packet with error-routes *([r, v, m, w, f])* to the next hop node, N_w. Here, *f* is the fictitious or unknown nodes of which it's node address that does not subsisted in the network. When N_b receives a packet , *pkt[r,v,m,w,f]*, it discards the packet received and then acknowledges with the error message to root of that DODAG. This happened because N_w, the packet to destination node cannot be forwarded to N_f which is specified in the header's address bit. Description of Hatchetman attack is explained in Fig. 4.

Suppose an attacker node generates several invalid packets in the network; it sends each packet to each post hop node along the forwarding route. The node rejects the packet because of an invalid path and sends an error message back to the DODAG's root. The malicious node N_m will generate and deliver several

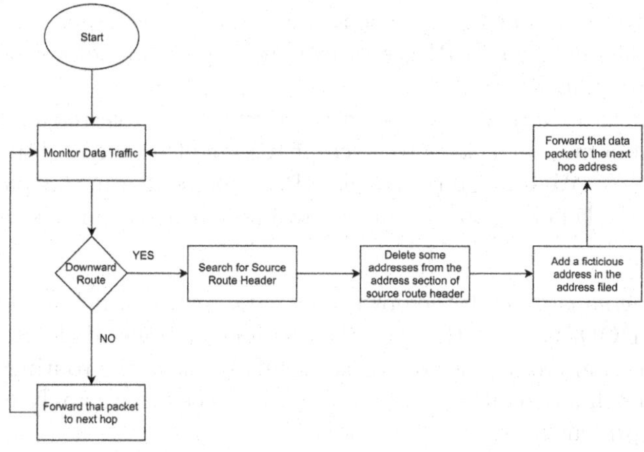

Fig. 4. Flowchart for implementation of Hatchetman attack

error packets with an error route to each post Hop Node, N_w, N_x, N_y, and N_z in turn. After receiving the packets, recipient nodes send an error message to the root, increasing the frequency of error messages in the network. As a result of the high frequency of error messages, node power consumption and bandwidth are disrupted, resulting in a DoS in RPL Protocol-based LLN networks [15]. Sequences of error messages propagated in Hatchetman attack are shown in Fig. 5.

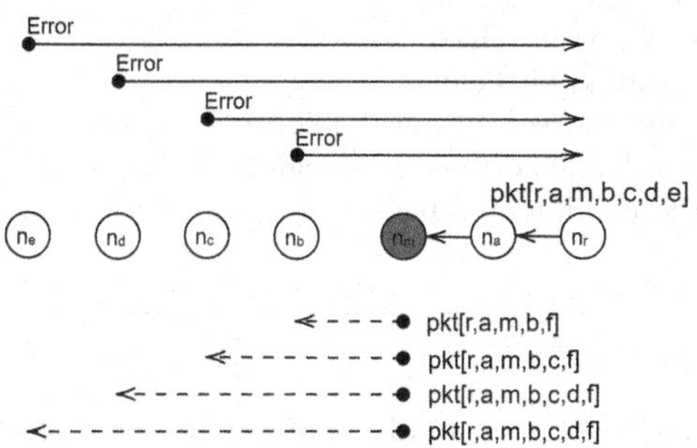

Fig. 5. Error messages sent in Hatchetman attack

We focused on the fundamental understanding of the working of RPL protocol and exploring various vulnerabilities in it. Some attacks are already being discussed in the referred research papers; the primary goal of most attacks is to try to exploit resources, topology or the traffic of the WSN. Our contribution in this paper is to implement an Hatchetman attack, which comes under the traffic exploitation category of the attacks in IoT. In this paper, we alter the

RPL source route header to implement this attack. This paper aims to primarily perceive the operating of RPL protocol and explore numerous vulnerabilities that could gift in it.

Some attacks are already being mentioned within the referred analysis papers. The first goal of most attacks is to attempt to exploit resources, topology, or the traffic of the IoT. We have sterilised the RPL route header in this paper to carry out this attack. Different steps required to implement the attack are discussed below.

1. Our main goal is to fetch the downward data packet, which might be coming from the DODAG root. Here we also assume that the mode of operation is Non-Storing, and only the root of the DODAG have the routing table.
2. After we fetch a downward route source route header, we analyse the address bit block present in it.
3. Add a fictitious address of a node N_f in the address bit.
4. After adding the fictitious node, forward that packet to the next-hop neighbour and repeat this process.

The Algorithm 1 shows the implementation technique for the Hatchetman attack. Here, we modify header of the packet to put false address so the node is not able to send it to the destination or other node.

Algorithm 1. Hatchetman Attack

```
High frequency error propagation
while MOP==downward do

Eavesdrop the source route header
    if Address bit is present then
        Fetch the address bit field
        Remove some addresses from the address sequence
        Add a unreachable address in the address field
        Forward the manipulated packet
    end if
end while
```

For the implementation, we made changes in the *rpl-icmp.c* file, which is responsible for receiving and sending all the control packets. The *rpl-icmp.c* is the part of *contiki* operating system which sends and receives all the DIO, DIS, DAO control packets.

4 Experimental Setup and Simulation Results

We performed a list of experiments in order to cover as many possible scenarios as we can cover in an RPL attack. All the experiments are done with and without the malicious nodes to compare the performance of the RPL protocol. The duration of each experiment is 600000 ms (10 min) in the Cooja simulator.

With the help of JavaScript, the packet delivery ratio, number of error messages and power-traces were calculated. The hatchetman attack impacts the network's PDR, latency, and power consumption, reducing overall network performance. The Table 1 shows our simulation setup for the experiments, and to validate the results, we use multiple scenarios having a different number of nodes. We simulated each experiment for 10 min in a $200 * 200$ m^2 environment using Z1 motes as the nodes.

Table 1. Experimental setup

Component	Name
OS	Ubuntu 20.04.2 LTS
OS type	64-bit
RAM	8 GB
Processor	Intel® Core^{TM} i7-4510U CPU @ 2.00 GHz
Simulation tool	Cooja from ContikiOS
Simulator used	Cooja 3.0
Node type	Z1
No. of nodes	10, 20, 30, 40, 50
No. of malicious nodes	10%, 20%, 30% of no. of legit nodes
Routing protocol	RPL protocol
Area	200 m * 200 m
Required simulation time	10 min
Transmission range	50 m
Interference range	100 m
Size of data packet	127 Bytes
Topology	Grid-center, Random

4.1 Comparing Packet Delivery Ratio (PDR)

The PDR drops when the number of malicious nodes and the total number of nodes increase due to the distance from the DODAG root. It can be illustrated in Fig. 6. As the distance from the DODAG root grows, so does the likelihood of packet drop. As a result, the longer the distance, the more likely packets will be dropped, lowering the PDR.

With reference to Table 2, we can see that if the number of malicious node increase in simulation, the PDR drops accordingly.

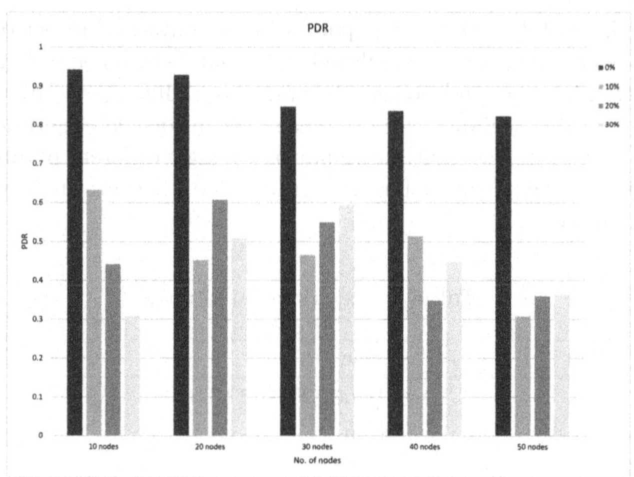

Fig. 6. Packet delivery ratio graph to show the relation between number of nodes and the evaluated PDR for the DODAG.

4.2 Comparing Number of Error Messages Propagated

The rate of error messages increases with increase of number of malicious nodes. With reference to Table 3, we can see the Error message count is increasing as we increase the number of malicious nodes in the simulation. It can illustrated from the Fig. 7.

4.3 Comparing Power Consumption

We are comparing the power consumption in each simulation and taking the average power per simulation. By this, we can see from Fig. 8 that more power is consumed in case of where more number of malicious nodes are present as compared to less number of malicious nodes. With reference to Table 4, we can see that if the number of malicious nodes increase, the average power consumed by DODAG also increase.

5 Countermeasure for Hatchetman Attack

We propose a simple yet effective countermeasure to reduce the effect of Hatchetman Attacks in RPL Protocol based LLN Networks. We can set a threshold value for the number of forwarded error messages in every node or every next node in order to decrease high frequency error message propagation in the RPL-based LLNs. The threshold can also be dynamically adjusted based on count of error packets travelling in the LLNs. On the basis of the count of error packets, threshold can be adjusted dynamically in LLN networks.

Table 2. PDR: without and with-malicious nodes

Sr no.	Number of legitimate nodes	Number of malicious node	PDRwithout-malicious	PDRwith-malicious
1	10	1	0.9418459887563478	0.6326530617645778
2		2		0.4420560747663551
3		3		0.3073005093378608
4	20	2	0.9285714285714286	0.4528099173553719
5		4		0.6073558648111332
6		6		0.5086538461538416
7	30	3	0.8475333333333333	0.4656389696452362
8		6		0.5497382198952879
9		9		0.594541910331384
10	40	4	0.8365793749936828	0.5139925373134329
11		8		0.3482293423271501
12		12		0.44815465729349735
13	50	5	0.8230769230769231	0.3073005093378608
14		10		0.3598553345388788
15		15		0.3637110016420361

Table 3. Error messages: without and with-malicious nodes

Sr no.	Number of legitimate nodes	Number of malicious node	Error without-malicious	Error with-malicious
1	10	1	6	225
2		2		270
3		3		272
4	20	2	28	362
5		4		268
6		6		306
7	30	3	56	417
8.		6		469
9		9		415
10	40	4	160	479
11		8		506
12		12		439
13	50	5	184	558
14		10		562
15		15		567

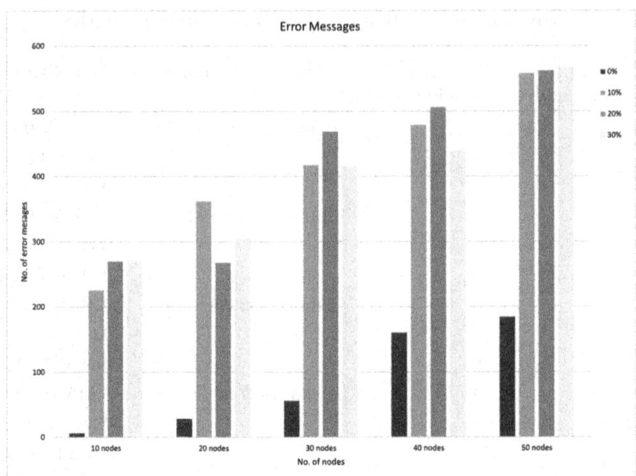

Fig. 7. Error messages propagation comparison to show the relation between number of nodes and number of error message generated by the DODAG.

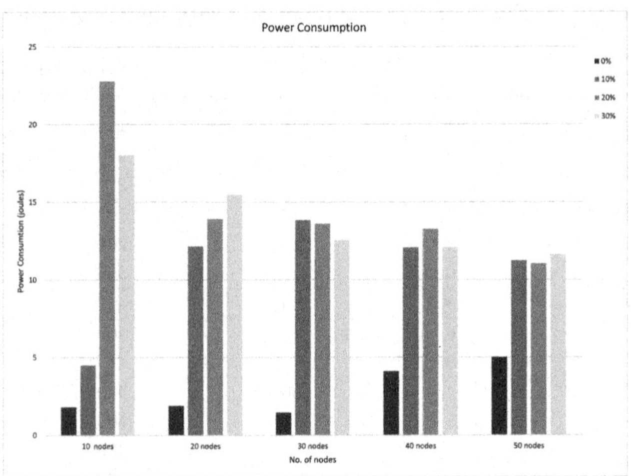

Fig. 8. Power consumption graph to show the relation between number of nodes and the power consumed by the DODAG.

Table 4. Average power consumption: without and with-malicious nodes

Sr no.	Number of legitimate nodes	Number of malicious node	Average power consumption without-malicious	Average power consumption with-Malicious
1	10	1	1.801918402	4.480618598
2		2		22.77752503
3		3		18.02295193
4	20	2	0.021419991	0.021419991
5		4		13.91407086
6		6		15.50007003
7	30	3	1.458142651	13.81565922
8		6		13.59225481
9		9		12.53360574
10	40	4	4.101293964	12.0585669
11		8		13.25455674
12		12		12.08830362
13	50	5	4.999591092	11.22578183
14		10		11.03130808
15		15		11.61657206

6 Conclusion

We have implemented and investigated the Hatchetman attack in RPL Protocol established in LLN Networks. On the basis of all these outcomes discussed above, we can conclude that it can be very severe if not taken care of in real time scenario. This attack generates a large amount of invalid data packets due to which frequency of error message propagation increases very rapidly. It is comparable to version number attack, hello-flooding attack, etc. Extensive simulation indicates that this attack can cause severe DoS attack by significantly decreasing the PDR and increasing the power consumption. As a part of future work, we want to design an efficient mechanism to countermeasure this attack. Each intermediary node in the forwarding network, for example, can define a threshold to limit the amount of Error messages transmitted in a certain period of time. All future Error messages will be refused if the number of forwarded Error messages exceeds the threshold.

References

1. Agiollo, A., Conti, M., Kaliyar, P., Lin, T., Pajola, L.: DETONAR: detection of routing attacks in RPL-based IoT. IEEE Trans. Netw. Serv. Manage. **18**, 1178–1190 (2021)
2. Almusaylim, Z.A., Alhumam, A., Mansoor, W., Chatterjee, P., Jhanjhi, N.Z.: Detection and mitigation of RPL rank and version number attacks in smart internet of things (2020)
3. Arış, A., Oktuğ, S.F.: Analysis of the RPL version number attack with multiple attackers. In: 2020 International Conference on Cyber Situational Awareness, Data Analytics and Assessment (CyberSA), pp. 1–8. IEEE (2020)

4. Butun, I., Österberg, P., Song, H.: Security of the internet of things: vulnerabilities, attacks, and countermeasures. IEEE Commun. Surv. Tutor. **22**(1), 616–644 (2019)
5. Ioulianou, P.P., Vassilakis, V.G., Logothetis, M.D.: Battery drain denial-of-service attacks and defenses in the internet of things. J. Telecommun. Inf. Technol. **2**, 37–45 (2019)
6. Kfoury, E., Saab, J., Younes, P., Achkar, R.: A self organizing map intrusion detection system for RPL protocol attacks. Int. J. Telecommun. Netw. (IJITN) **11**(1), 30–43 (2019)
7. Le, A., Loo, J., Luo, Y., Lasebae, A.: Specification-based ids for securing RPL from topology attacks. In: 2011 IFIP Wireless Days (WD), pp. 1–3. IEEE (2011)
8. Mayzaud, A., Badonnel, R., Chrisment, I.: A taxonomy of attacks in RPL-based internet of things (2016)
9. Mayzaud, A., Badonnel, R., Chrisment, I.: A distributed monitoring strategy for detecting version number attacks in RPL-based networks. IEEE Trans. Netw. Serv. Manage. **14**(2), 472–486 (2017)
10. Muzammal, S.M., Murugesan, R.K., Jhanjhi, N.: A comprehensive review on secure routing in internet of things: mitigation methods and trust-based approaches. IEEE Internet Things J. **8**, 4186–4210 (2020)
11. Napiah, M.N., Idris, M.Y.I.B., Ramli, R., Ahmedy, I.: Compression header analyzer intrusion detection system (CHA-IDS) for 6lowpan communication protocol. IEEE Access **6**, 16623–16638 (2018)
12. Pu, C., Hajjar, S.: Mitigating forwarding misbehaviors in RPL-based low power and lossy networks. In: 2018 15th IEEE Annual Consumer Communications & Networking Conference (CCNC), pp. 1–6. IEEE (2018)
13. Pu, C., Lim, S.: Spy vs. spy: camouflage-based active detection in energy harvesting motivated networks. In: MILCOM 2015–2015 IEEE Military Communications Conference, pp. 903–908. IEEE (2015)
14. Pu, C., Lim, S., Jung, B., Min, M. Mitigating stealthy collision attack in energy harvesting motivated networks. In: MILCOM 2017–2017 IEEE Military Communications Conference (MILCOM), pp. 539–544. IEEE (2017)
15. Pu, C., Song, T.: Hatchetman attack: a denial of service attack against routing in low power and lossy networks. In: 2018 5th IEEE International Conference on Cyber Security and Cloud Computing (CSCloud)/2018 4th IEEE International Conference on Edge Computing and Scalable Cloud (EdgeCom), pp. 12–17. IEEE (2018)
16. Pu, C., Zhou, X.: Suppression attack against multicast protocol in low power and lossy networks: analysis and defenses. Sensors **18**, 10 (2018)
17. Verma, A., Ranga, V.: Security of RPL based 6lowpan networks in the internet of things: a review. IEEE Sens. J. **20**(11), 5666–5690 (2020)
18. Winter, T., et al.: RPL: IPv6 routing protocol for low power and lossy networks. draft-ietf-roll-rpl-19 (2011)

Blockchain Framework for Agro Financing of Farmers in South India

Namrata Marium Chacko(✉), V. G. Narendra, Mamatha Balachandra, and Shoibolina Kaushik

Manipal Institute of Technology, Manipal Academy of Higher Education, Manipal 576104, Karnataka, India
namratamarium@gmail.com

Abstract. In India, small scale farmers face challenges acquiring crop loans and insurance. The government of India has many schemes for both. However, due to lapses in processing, these commercial bank loans are difficult to avail. Thus farmers approach Microfinance Institutions (MFI) or moneylenders, who charge high interest rates. To eliminate this difficulty, a novel blockchain based module is proposed in this paper, incorporating farmers' credibility, verifying insurance claims, and assured guaranteed repayment of loans to investors. The inherent property of blockchain ensures trust and transparency. The proposed blockchain module has four participating entities, the farmer, investors, customers, and the government agent, governed by the smart contract rules. The concept of trusted oracles is used to gather weather data, and the smart contract decides the payout and the subsidized loan repayment. This module is implemented and tested in the Ethereum live test network. Performance evaluation shows that the proposed module has an average latency of 30 s and minimal fees.

Keywords: Blockchain · Agriculture · Trusted oracles · Insurance · Loans

1 Introduction

Agriculture prevails to be the premier factor significantly influencing the Indian economy since time immemorial. For decades marginal and small farmers have contributed to the total gross cropped output, which is now almost 50% [1]. Typically small farmers are involved in growing local food, which is seasonal and beneficial to the environment. These farmers required financial assistance for the betterment of their livelihood. The government of India has various schemes to support farmers. Despite these reforms in the agricultural policies, there are still various issues still lurking [2]. Consequently, many farmers are reluctant to avail financial aid from commercial banks. Online registration has reduced the hassle of verification. However, various schemes proposed by the government often fail due to the local bureaucracy in addressing the financial constraints of the farmers

V. E. Balas et al. (Eds.): ICETCE 2022, CCIS 1591, pp. 679–690, 2022.
https://doi.org/10.1007/978-3-031-07012-9_56

[3]. Farmers still find the collateral imposed by the bank huge, and they worry about loan repayment in extreme weather conditions as the Indian agricultural practices are labor intensive and depend on the annual rainfall for irrigation [4]. Schemes like Weather Based Crop Insurance Scheme (WBCIS) and Pradhan Mantri Fasal Bima Yojana (PMFBY) aim to protect the farmers under unlikely weather conditions [5]. Nevertheless, these schemes have failed in timely claim settlements, procedural delays, lack of support from the state, unsustainable subsidy model, and skewed delay [6]. Consequently, farmers approach money lenders and Microfinance Institutions (MFIs), which impose substantial interest rates [7–10]. Centralized control does not guarantee fair results because many producers and intermediaries always assume the possibility of results falsification. Demand for organic products is rising, but small scale farmers face massive impediments in terms of funding [11].

Blockchain technology promises transparency, reliability, and trust by eliminating the need for a third party [12]. The use of blockchain in agricultural practices and funding is still nascent, with much potential to transform the sector [12]. Blockchain's smart contract can form agreements stating delivery and payment terms, which otherwise complicate the supply process. Investors will raise funds for agricultural loans on favorable terms with a payback guarantee through blockchain enabled funding [13]. This paper proposes a blockchain-based module that combines microfinancing and chit funds to provide a safe and transparent method of financing for small scale farmers. The major contributions of this paper are as follows:

- We propose a module for managing farmer credit based loan and insurance processing via smart contract, thus eliminating the need for a third party.
- We use concepts of oracles, which are trusted entities to connect to off chain data, to retrieve weather data used for an automatic insurance payout and reduction in the loan repayment.
- We present and discuss the system architecture. The sequence of the various interactions among all participating entities is also explored.
- We present algorithms and their key features along with various function interactions and evaluate them regarding time and cost.

2 Literature Review

Blockchain is one of the most promising technologies for securing financial transactions. It has the potential to revolutionize support for agricultural financial transactions. Blockchain allows increased transaction reliability, improved efficiency, and lower agricultural financing cost [14,15]. The benefits of microfinance in financially poorer sections of society are explored in [16–19]. They were emphasizing the need for microfinancing. Kaimathiri et al. [16] conducted a narrative inquiry into the daily challenges for the business growth of women's microenterprises in rural Kenya and concluded that besides the various barriers they face; microfinancing and crowdfunding could boost the productivity rate. Samineni et al. [17] carried out a regression model analysis to determine the performance of microfinance in India. Kshitij Naikadea et al. [18] evaluated the microfinance situation in India during the pandemic and its impact on the small and marginal entrepreneurs. Mohd. et al. [20] emphasized the importance of various microfinance institutions and the role they play in various socio-economic sectors. Rajarshi Ghosh [19], traced the evolution of microfinance institutions and studied their role in alleviating poverty and empowering women. Some papers propose a methodology to evaluate the risk of issuing loans to their clients [7,21–24]. Most of the papers focused on credit scoring models focusing on farmers and microfinancing institutions. To overcome the challenges faced with financing, many scholars have proposed various blockchain frameworks [8,25–29]. Notheisen et al. [25] proposed a novel method for mitigating transactional threats in the lemon markets using Blockchain. The authors Mukkamala et al. [26] have modeled a conceptual example of how blockchain technology can be used for microcredit. They have analyzed the useability of Blockchain in a business environment. The authors discovered that Blockchain technology could help to improve trust, transparency, and auditability in transactions. Cunha et al. [27] have proposed a novel framework that considers business, society, economy and finance, technology, and policy-related factors to understand better the opportunities Blockchain offers for development. Rakkini et al. [28], discussed the benefits of a model for decentralized autonomous organizations(DAO) with Blockchain. This model ensures that investors can issue loans based on assessment from an expert panel. The model also caters to crowdfunding. When borrowers repay the amount and become guarantors for other members, their credit will increase. It will provide more benefits and advantages than current microfinance, which is stuck in a vicious debt-loan cycle. García-Bañuelos et al. [29] have empirically evaluated a novel blockchain framework, which works as a threefold solution for executing business processes. The focus is on instantiating initial cost, costs of tasks execution, and increasing throughput at run time. The model for microloans is proposed in [8]. The paper highlighted the need for microfinance and how Blockchain can contribute to the farmers and investors for fair trading. From the literature, the gaps identified are:

- There is a need to eliminate the third party agents who charge high interest on loans while providing small scale farmers a trust assuring platform.
- There is a lack of accessibility to the Government based schemes.
- Investors need a fail proof method that guarantees returns.

In the next section, we propose a Blockchain based solution to address these gaps. Our proposed module offers a novel method to maintain the integrity of all the stakeholders involved.

3 Technologies and Proposed Framework

This section discusses the underlying technologies used for building the module, the proposed framework, and details of the system.

3.1 Technologies

- Blockchain: Blockchain provides a trustful environment. Researchers have found blockchain applications can go far beyond cryptocurrency [30–32]. Every transaction history is stored in a tamper proof ledger, which any member of the blockchain can access. These transactions are encrypted with private keys, therefore ensuring the privacy of the participants, and at the same time, making data authentic [33].
- Ethereum and smart contract: Ethereum is a blockchain, which allows developers to create contracts between different accounts within ethereum. These smart contracts ensure that the accounts abide by the rules laid down in them, thus working with minimal trust [34].
- Oracle: Smart contracts are not designed to access data from external sources. Oracles are trusted entities that fetch data from various external sources [35].

3.2 System Overview

This section presents a decentralized solution for loan processing via blockchain smart contracts. Figure 1 describes system architecture with key components and actors.

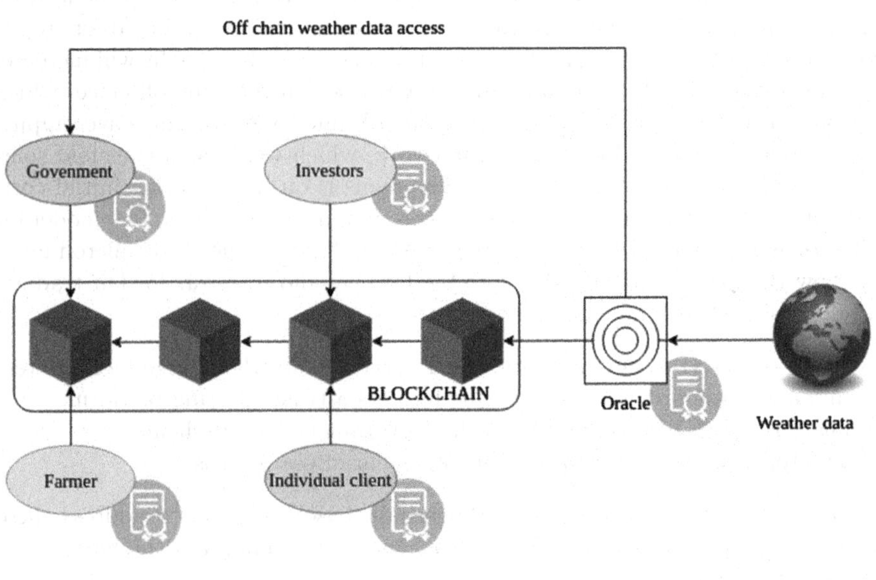

Fig. 1. System overview for loans using smart contracts

- Government: The role of the government entity is to regulate and manage the credits for loans, verify the customers and farmers, and issue the insurance. The government creates and owns the smart contract. The smart contract has a function to calculate the credit score of the farmer and triggers events for repayment after a set amount of time. It also communicates with the oracle to get weather information. This information is used to decide payout for the insurer in case of extreme weather. It also calculates a reduction in the loan repayment in unlikely weather.
- Farmer: Farmers apply for a loan by providing valid documentation like aadhar card or Voter ID. The farmer can create a listing for the crops he will be cultivating for the next season. The crop type, the season for harvesting, insurance details, expected harvest time are recorded.
- Individual Clients: Valid clients who have valid ID cards can register and create an account. Clients can view the farmer's listings. The contract terms will decide how the client will get the returns in terms of profit or harvested crops.
- Investors: These are Microfinance institutions, chit funds, or private banks verified by the government. The chit fund investors follow the chit fund methodology. The idea of this comes from the small chit funds group in rural villages in India.

3.3 Sequence Diagram

Figure 2 illustrates the flow between all the participants. Each entity has an account that is associated with an ethereum address. The participants interact with the smart contract by invoking various functions.

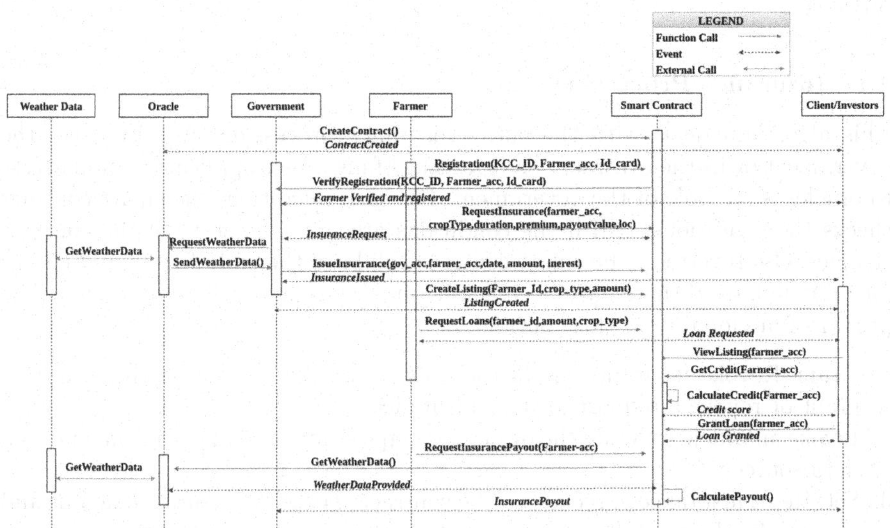

Fig. 2. Sequence diagram showing interaction between stakeholders

The execution of functions invokes events that update the participants regarding the change in the smart contract state. After the offline issue of some farmer ID cards, the farmers register on the blockchain. The government then executes the *VerifyRegistration()* function, which verifies the KCC ID and Farmer account and, on successful verification, triggers the event of *farmer verified and registered*, which invokes the government and farmer account to abide by the rules in the smart contract. The client and investors also register similarly. Then the farmer is eligible to request insurance and calls the *RequestInsurance()* function, with attributes farmer ID, crop type, duration, premium, payout value, and location. The event *InsuranceRequest* causes a call to the oracle to get rainfall data, and according to the Smart contract rules, new insurance is created and updated to all with event *InsuranceIssued.* Then farmers can create the listing of crops for the season by invoking the *CreateListing()* function. Once the event *ListingCreated* is triggered, all the clients and investors can view the listing. The investors and clients can request the credit score, which is calculated by the smart contract, based on an expert credit score model similar to the one proposed in [36]. After the specified season is over, the investor can be paid with tokens or crops according to the contract's conditions. Each time the farmer returns on time, they are awarded points to increase their credit score. In case of extreme weather conditions, the farmer can call the *RequestInsurancePayout()* function, which triggers the oracle to get the weather data and issue the payout amount.

4 Implementation Framework

This section outlines the two main algorithms: Insurance processing and Loans processing.

4.1 Insurance Processing

This algorithm deals with the entire insurance processing. As discussed, the government owns the contract. The granting of insurance applications and access to oracles is the role of the government. For each operation, the smart contract checks the conditions, and if the conditions are met, the state of the contract changes, else reverted. The smart contract validates the end of term and checks for rainfall data. If there have been heavy rains, the claim will be processed. The primary functionality of this contract is:

1. Grant\Revoke insurance application: Only the contract owner is eligible to grant or revoke a request for new insurance.
2. Invoke an oracle to: Only the owner can request for rainfall data from external data source.
3. Set\Reset insurance agreement: The contract checks the insurance details and also tracks the rainfall.
4. Process insurance claims: When an insurance claim in raised the owner checks for the validity and weather conditions before processing the claim.

Algorithm 1: Insurance Processing

```
Input: F list of FarmerAccount, crop type, duration,
       premium,payOutValue,location
Address of Farmer,
Address of Government
Data: Details for weather API
Output: Insurance agreement contact between farmer and Government
State of contract: Initialized
Farmer request insurance(All input variables) Government invokes
  GetNewPolicy(FarmerAccount)
if F = Account verified then
    Add to list of registered farmer RF
    emit event newPolicyCreated
    state of contract: NewPolicyAdded
else
    Ask to get registered
    state of contract: Reverted
Farmer Invokes
  RequestInsurancePayout(FarmerAccount,premium,PayoutValue)
Government InvokeGetweatherdata(path to API)
if RainfallCountDays=set days then
    Call CheckInsuranceValidity(farmerAcc,premium)
if FarmerAcc insurance still valid then
    issue Payout amount
    emit event insurance claimed
    state of contract: PayOutCompleted
else
    Insurance period completed
    state of contract: Reverted
else
    Weather conditions are favorable:claim not processed
    state of contract: Reverted
```

4.2 Crop Loan Processing

Algorithm 2 shows the process of calculating the credit score and sanctioning the crop loan. The farmer provides details like crop type, such as paddy or sugarcane, the season he intends to grow the crops Kharif or rabi if he intends to grow more than one crop per year, and the land size in acres. The weather data is used to provide a reduction in loan repayment. The primary functions in this algorithm are:

1. Calculate credit score: On request from the client, the credit score is calculated. The formula for calculating the credit score is similar to the one proposed in [36], except the transaction history variable is added.

$$CREDITSCORE = \sum_{v=1}^{n} f(a_1 X_1, ..., a_n X_n) \tag{1}$$

where X_n is the variables used for credit score calculations, and a_n is the weights assigned for each variable.

2. Loan approval: The government checks the various inputs and assigns loans accordingly.
3. Calculation of interest amount: The interest amount increases as the credit score reduces. Also, the weather data is used to determine any possible relaxation in terms of repayment of loans.

Algorithm 2: Loan Processing

```
Input: RF list of insured farmers, cropPerYear, cropType,landSize,
       pumpSetRepaired
Address of Framer, client and government,
Data: Details for weather API
Data: Details for calculating credit score
State of contract:Initialized
Contract invokes (calculateCredit)
if creditScore acceptable by client then
    initiate payment emit event  loan approved state of contract: Loan
     approved
else
    state of contract: Reverted
Contract invokes endOfTerm (FarmerAcc, ClientAcc,LoanAmt, ReturnAmt)
if  endOfterm then
    if Weather data shows rainfall then
        Loan Amount repayment reduction
    else
        Repay full loan amount
    emit LoanRepaid
else
    state of contract: Reverted
```

5 Results

In this section, first, a qualitative analysis of the proposed module is done concerning each stakeholder involved, and also a comparison with existing literature is carried out. Then a cost analysis for the amount of Gas used on the live test net for major algorithms is carried out. The latency is recorded and analyzed.

5.1 Qualitative Analysis

This research aims to ensure that farmers get the necessary financial support at minimal interest rates and protect the investors from loss. This module provides the following advantages:

- Farmer: The farmers will access and register onto the blockchain via mobile application. The registration process is straightforward, and the farmer can quickly get crops insured from the comfort of his home. The loan request can be quickly initiated with the ID issued by the government. The interest rate is automatically calculated, and in extreme weather, the smart contract triggers the insurance payout event.
- Clients: Clients can easily view the various products listing and verify the credit score of the farmer and then decide to invest. They can request repayment in the form of cryptocurrency or products of equivalent value.
- Investors: Various investors can pool in and provide funding, and the farmer with the highest need can receive the amount. Repayment will be triggered after a certain set period of time.

The Table 1 shows that the proposed solution additionally provides support for trusted oracles and policy management. Policy management defined here is calculating subsidized loan repayment and insurance payout amount.

Table 1. Comparison with existing literature

Features	[28]	[26]	[8]	Proposed solution
Decentralized transaction	Y	Y	Y	Y
Distributed data management	Y	Y	Y	Y
Credit score calculation	N	N	Y	Y
Trusted oracle	N	N	N	Y
Policy management	N	N	N	Y

5.2 Cost and Latency

Performance analysis of the module in terms of latency in seconds and transaction fee (Gas) in Gwei (it is a denomination of the cryptocurrency ether) used for various operations is given in Fig. 3. In the ethereum network, every action consumes a minimal amount of gas. The deployment of contracts, data storage, and all transactions executed on the chain consumes gas. The deployment cost of the smart contract and execution cost of various functions were calculated. The latency and transaction fees were tested by deploying the smart contract in the Ropsten test network. Ropsten is a live test network, which mimics the actual ethereum network. The average block time ranges between 10 and 30 s per block. To test the module, we ran the tests ten times. We followed the same series of events to ensure fairness and analyze how each function handles the requests-the chosen series of events combined with complete sequences were executed on the blockchain. The smart contract was executed via MetaMask. The series included deploying smart contracts, adding farmer accounts, requesting insurance, calculating a credit score, loan amount determination, and insurance payout in extreme weather conditions.

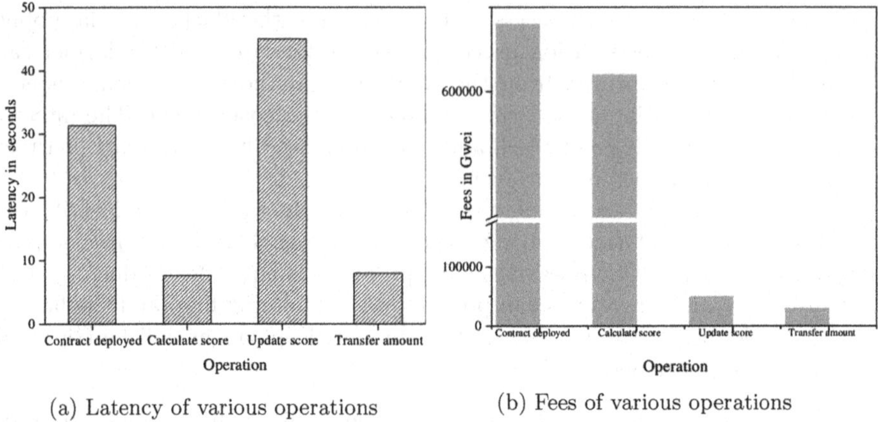

(a) Latency of various operations

(b) Fees of various operations

Fig. 3. Performance analysis in terms of latency in seconds and transaction fee (Gas) in Gwei

6 Conclusion and Future Scope

This paper has proposed a distributed alliance system between government, farmers, and clients for agro trading. The use of a trusted oracle ensures timely and non-reputable weather information. The smart contracts can regulate interest terms and conditions, easily verify insurance claims, effortless processing of credit calculation and loan repayment, thus providing cost effective and fair platform for all. The module also proved to have low latency in processing. Farmers can acquire loans without the hassle of traveling long distances. Blockchain opens new avenues for the farmers, and they can reach out to many investors and clients across the country.

The work is limited to short term loans and insurance for seasonal crops. The future scope of this work can be to implement a machine learning model for predicting the credit score of the farmers. This module can be integrated as a part of the food supply chain, using IoT to track and trace the farming practices of the farmers, thus ensuring the clients that the crops are grown organically and locally.

References

1. Department of agriculture ministry of agriculture and farmers welfare government of India: All India report on agriculture census 2015–16 (2020)
2. Mohan, R.: Agricultural credit in India: status, issues and future agenda. Econ. Polit. Wkly. 1013–1023 (2006)
3. World Bank Group: India Systematic Country Diagnostic: Realizing the Promise of Prosperity. World Bank (2018)
4. Singh, M.: Challenges and opportunities for sustainable viability of marginal and small farmers in India. Agric. Situat. India **3**, 133–142 (2012)

5. Department of agriculture ministry of agriculture and farmers welfare government of India. Operational guidelines: Restructured weather based crop insurance scheme (2015)
6. Tiwari, R., Chand, K., Anjum, B.: Crop insurance in India: a review of pradhan mantri fasal bima yojana (pmfby). FIIB Bus. Rev. **9**(4), 249–255 (2020)
7. Singh, A., Jain, N.: Mitigating agricultural lending risk: an advanced analytical approach. In: Laha, A.K. (ed.) Applied Advanced Analytics. SPBE, pp. 87–102. Springer, Singapore (2021). https://doi.org/10.1007/978-981-33-6656-5_8
8. Khara, R., Pomendkar, D., Gupta, R., Hingorani, I., Kalbande, D.: Micro loans for farmers. In: 2020 11th International Conference on Computing, Communication and Networking Technologies (ICCCNT). IEEE (July 2020)
9. Mahendra Dev, S.: Small farmers in India: challenges and opportunities, January 2012
10. Balkrisnan, D.K., Sahil, G., Sharma, S.: From license raj to unregulated chaos (a critical analysis of the shenanigans of the government). Int. J. Mod. Agric. **10**(2), 3997–4002 (2021)
11. Bhavsar, A., Diallo, C., Ülkü, M.A.: Towards sustainable development: optimal pricing and sales strategies for retailing fair trade products. J. Clean. Prod. **286**, 124990 (2021)
12. Rejeb, A., Keogh, J.G., Zailani, S., Treiblmaier, H., Rejeb, K.: Blockchain technology in the food industry: a review of potentials, challenges and future research directions. Logistics **4**(4), 27 (2020)
13. Bunchuk, N.A., Jallal, A.K., Dodonov, S.V., Dodonova, M.V., Diatel, V.N.: Perspectives of blockchain technology in the development of the agricultural sector. J. Complement. Med. Res. **11**(1), 415–421 (2020)
14. Tapscott, D., Tapscott, A.: How blockchain will change organizations. MIT Sloan Manag. Rev. (2), 10 (2017)
15. Swan, M.: Blockchain: Blueprint for a New Economy. O'Reilly Media, Inc., Sebastopol (2015)
16. Nyoroka, P.K.: A narrative inquiry into the daily challenges to the business growth of women's microenterprises in rural kenya (2021)
17. Samineni, S., Ramesh, K.: Measuring the impact of microfinance on economic enhancement of women: analysis with special reference to India. Glob. Bus. Rev. (2020). https://doi.org/10.1177/0972150920923108
18. Naikade, K., et al.: Critical analysis of relevance of microfinance in India in the post COVID era. Turkish J. Comput. Math. Educ. (TURCOMAT) **12**(10), 4264–4276 (2021)
19. Ghosh, R.: Microfinance in India: a critique (2005). Available at SSRN 735243
20. Mohd, S.: A study on the performance of microfinance institutions in India. Management **5**(4), 116–128 (2018)
21. Jha, S., Bawa, K.S.: The economic and environmental outcomes of microfinance projects: an Indian case study. Environ. Dev. Sustain. **9**(3), 229–239 (2007)
22. Hajji, T., Jamil, O.M.: Rating microfinance products consumers using artificial neural networks. In: Rocha, Á., Serrhini, M. (eds.) EMENA-ISTL 2018. SIST, vol. 111, pp. 460–470. Springer, Cham (2019). https://doi.org/10.1007/978-3-030-03577-8_51
23. Hajji, T., El Jasouli, S.Y., Mbarki, J., Jaara, E.M.: Microfinance risk analysis using the business intelligence. In: 2016 4th IEEE International Colloquium on Information Science and Technology (CiSt), pp. 675–680. IEEE (2016)

24. Tong, Z., Yang, S.: The research of agricultural SMEs credit risk assessment based on the supply chain finance. In: E3S Web of Conferences, vol. 275, p. 01061. EDP Sciences (2021)
25. Notheisen, B., Cholewa, J.B., Shanmugam, A.P.: Trading real-world assets on blockchain. Bus. Inf. Syst. Eng. **59**(6), 425–440 (2017)
26. Mukkamala, R.R., Vatrapu, R., Ray, P.K., Sengupta, G., Halder, S.: Converging blockchain and social business for socio-economic development. In: 2018 IEEE International Conference on Big Data (Big Data). IEEE (December 2018)
27. Cunha, P.R.D., Soja, P., Themistocleous, M.: Blockchain for development: a guiding framework (2021)
28. Jeyasheela Rakkini, M.J., Geetha, K.: Blockchain-enabled microfinance model with decentralized autonomous organizations. In: Smys, S., Palanisamy, R., Rocha, Á., Beligiannis, G.N. (eds.) Computer Networks and Inventive Communication Technologies. LNDECT, vol. 58, pp. 417–430. Springer, Singapore (2021). https://doi.org/10.1007/978-981-15-9647-6_32
29. García-Bañuelos, L., Ponomarev, A., Dumas, M., Weber, I.: Optimized execution of business processes on blockchain. In: Carmona, J., Engels, G., Kumar, A. (eds.) BPM 2017. LNCS, vol. 10445, pp. 130–146. Springer, Cham (2017). https://doi.org/10.1007/978-3-319-65000-5_8
30. Nakamoto, S.: A peer-to-peer electronic cash system. Bitcoin, April 2008. https://bitcoin.org/bitcoin.pdf
31. Ramamoorthi, S., Kumar, B.M., Sithik, M.M., Kumar, T.T., Ragaventhiran, J., Islabudeen, M.: Enhanced security in IOT environment using blockchain: a survey. Materials Today: Proceedings (2021)
32. Sun, J., Yan, J., Zhang, K.Z.K.: Blockchain-based sharing services: what blockchain technology can contribute to smart cities. Financ. Innov. **2**, 26 (2016). https://doi.org/10.1186/s40854-016-0040-y
33. Conoscenti, M., Vetro, A., De Martin, J.C.: Blockchain for the internet of things: a systematic literature review. In: 2016 IEEE/ACS 13th International Conference of Computer Systems and Applications (AICCSA), pp. 1–6. IEEE (2016)
34. Buterin, V.: A next generation smart contract & decentralized application platform (2014). http://www.ethereum.org/pdfs/EthereumWhitePaper.pdf
35. Adler, J., Berryhill, R., Veneris, A., Poulos, Z., Veira, N., Kastania, A.: Astraea: a decentralized blockchain oracle. In: 2018 IEEE International Conference on Internet of Things (IThings) and IEEE Green Computing and Communications (GreenCom) and IEEE Cyber, Physical and Social Computing (CPSCom) and IEEE Smart Data (SmartData), pp. 1145–1152. IEEE (2018)
36. Barry, P.J., Ellinger, P.N.: Credit scoring, loan pricing, and farm business performance. Western J. Agric. Econ. **14**(1), 45–55 (1989). http://www.jstor.org/stable/40988008. Accessed 13 May 2022

Image Encryption Using RABIN and Elliptic Curve Crypto Systems

Arpan Maity(✉), Srijita Guha Roy, Sucheta Bhattacharjee, Ritobina Ghosh, Adrita Ray Choudhury, and Chittaranjan Pradhan

Kalinga Institute of Industrial Technology (KIIT) Deemed to be University, Bhubaneswar, India
maityarpan797@gmail.com, gsrijita084@gmail.com, suchetabhatt8753@gmail.com, ghoshritobina@gmail.com, adritarchoudury@gmail.com, chitaprakash@gmail.com

Abstract. In this generation we know how much the internet matters to us. Every basic work can be done with the help of the internet. People have nowadays become very dependent on internet so it is very much needed to provide the users with enough security so that it becomes a reliable platform. This security can be provided by different cryptographic techniques. This makes it difficult for the hackers and the cryptanalyst from stealing crucial informations. This can be done efficiently with the help of asymmetric key and symmetric key cryptographic techniques. We can make a system more safe if we can combine both. So here we have we have used two layers of each encryption and decryption thus providing two layers of security. Here we have worked with different images so firstly, the image has been encrypted using Rabin Cryptography followed by ECC Cryptography. Similarly while decrypting, the decryption was first done using ECC and then Rabin Cryptography. We have tested this approach using different datasets and have received great accuracy. With this approach it can provide images with a double layer of security efficiently while sharing images.

Keywords: Cryptography · Decryption · Encryption · ECC · Key · Rabin · Security

1 Introduction

In today's modern world, the internet plays a very significant role, starting from money transfer to sending emails or purchasing things online or sending messages, images, audio, videos on different social media applications, data is transferred from one point to another at different times and in different forms. Even in the lives of all the people around the world, whether that person is a student or is making a living, everyone is reliant on the internet. Moreover, with the outbreak of the pandemic, people became even more dependent on the internet as the entire world started working in virtual mode. This digital transformation has completely changed the outlook of how businesses operate and compete today. In 2021, every minute, more than 2.5 quintillions of data is being generated and huge amounts of memory and time are invested to process the data and manage it. Even the computer machinery, data storage devices, software applications, and networks have become more complex which has also created an extended attack platform

V. E. Balas et al. (Eds.): ICETCE 2022, CCIS 1591, pp. 691–703, 2022.
https://doi.org/10.1007/978-3-031-07012-9_57

that has created more difficulties for the computer systems and the networks connecting it. Therefore, as the demand for data and modern technologies are increasing day by day, the need for its security, mainly data security, is also increasing simultaneously. But no matter how secure the network is, the chances of data breaches have escalated over the past few years.

In recent times, many cases of data breaches have shown up. On 31st March 2020, the hotel chain JW Marriott announced a security breach that affected 5.2 million hotel guests who used the loyalty application of their company [1]. Login credentials of two Marriott employees' accounts were also stolen and after that, the information was used to leak the data which contained critical data such as contact numbers, personal details, linked account data, etc. In addition to this, the Zoom application which provided us with the advantage of continuing our lives in the digital mode and gained a lot of popularity during the COVID-19 pandemic, also become exposed to many threats when in the first week of April, more than 500 passwords were stolen and were available for sale in the dark web [2]. On top of this, much more information such as the email address, login credentials, and personal details of the zoom participants was also stolen. In May 2020, EasyJet reported that they fell prey to a well-planned cyber attack wherein the email addresses and travel details of approximately nine million customers were accessed and the credit card details and CVV security codes of approximately 2000 customers were exposed. Usually, if there are traces of any kind of cyber attack then it has to be reported within 72 h but in this case, the reporting time lagged because of the sophisticated nature of the attack and the time taken to find out what and whose data has been taken away.

Hence, for the safety and security of valuable information, data security comes into the picture. Originally, data security was one of the important aspects of cryptography, which is one of the oldest forms of science that started thousands of years ago. Hence, data security refers to the protection of digital data from cyber attacks or data breaches so that the data remains safe from external manipulation. These data may refer to email address, personal details, login data, images, audio, video, etc. and one of the most common ways to protect data is through data encryption. Data encryption refers to the mixing up of data or substituting the values of the data with some other value in such a way that the data is secure from attacks from external sources. In data encryption, the original text, that is, the plain text is changed using various procedures so that a new text is formed. This new text is known as the ciphertext.

In this paper, two cryptography algorithms are used for image encryption and decryption, the first is the Rabin cryptography and the second is the ECC cryptography. Both of these encryption techniques are a part of asymmetric key cryptography. Rabin encryption algorithm is an advanced version of the RSA encryption algorithm and the advantage of using this particular algorithm is because of its complex integer factorization which is proven to be secure from any chosen plaintext attack if the attacker cannot factorize the integers efficiently. On the other hand, ECC makes use of elliptic curves in which the variables and coefficients are restricted to elements of a finite field. ECC performs computations using elliptic curve arithmetic instead of integer or polynomial arithmetic. This results in shorter key length, less computational complexity, low power requirement, and provides more security. So, our main contribution in this paper is the implementation of two major encoding methods to provide an effective way to protect the integrity of the actual image data.

The rest of the paper is structured as follows: In Sect. 2, we presented the existing related work. Section 3 presents the Proposed Work. Section 4 contains the details of the experimental setup, our working model, building that model. Section 5 contains information about the accuracy and performance. Section 6 concludes the paper.

2 Related Work

Xiaoqiang Zhang et al. used a digital image encryption technique that was based on the Elliptic Curve Cryptosystem (ECC) [3]. At first, in this technique the sender groups some pixel values of the actual image to generate large integer numbers. This was done to reduce the encryption time. Then, the sender encrypts the large integers using the ECC algorithm and chaotic maps. And finally, the encrypted image is produced from the large encrypted integers. The encryption or decryption process consists of the following steps which are: the generation of a chaotic sequence, generation of chaotic image, grouping the pixel values to form big integers, ECC encryption/decryption, and finally recovering the pixels.

Laiphrakpam Dolendro Singh et al. proposed a routing protocol known as the Cluster Tree-based Data Dissemination (CTDD) protocol on Wireless Sensor Networks which includes batteries, processor, sensor nodes, and a mobile sink [4]. The sink collects the data and sends it to the user over the Internet. The random mobility of the sink helps to create a more dynamic network with high energy efficiency and network lifetime. The CTDD protocol helps in reducing the latency delay and increases the network lifetime, it is also more energy-efficient and more reliable than any other tree or cluster-based routing protocols.

Aaisha Jabeen Siddiqui et al. focused on security protocols for Wireless Sensor Networks by providing strong security and efficient usage of the battery of sensor nodes [5]. In this, the location address of the nodes is considered as input for generating keys. Due to changes in distance and direction of the address of sensor nodes the key pairs for encryption will be different, thus resulting in more security and safety. To save battery power of WSNs, instead of using Elliptic Curves and RSA where large prime values are taken to generate keys resulting in more execution time, they used the Rabin algorithm which takes much lesser time for execution leading to the lesser power consumption of battery.

Hayder Raheem Hashim proposed the H-Rabin cryptosystem for better security [6]. Calculating the square roots modulo n is the fundamental of the decryption algorithm of Rabin where $n = p * q$ where p and q are prime numbers. But in this, the researcher modified $n = p * q * r$ where p, q and r are prime numbers respectively and are taken as private keys. Since the factorization is unknown, it is complex to compute the square root modulo composite. Dan Boneh researched Simplified OAEP for the RSA and Rabin Functions [7]. From this paper, we understand that OAEP is an approach for the conversion of RSA trapdoor permutation into a secure system with a selected ciphertext in a random oracle model. It can be simplified by applying RSA and Rabin functions.

3 Proposed Work

3.1 Elliptic Curve Cryptography (ECC)

ECC is a powerful cryptography approach. It generates security between key pairs for public-key encryption by using the mathematics of elliptic curves [8]. ECC has gradually been growing in popularity recently due to its smaller key size and ability to maintain high security [9]. This trend will probably continue as the demand for devices to remain secure increases due to the size of keys growing, drawing on scarce mobile resources. This is why it is so important to understand elliptic curve cryptography in context [10].

3.1.1 Encryption

A point G is considered, which lies in the elliptic curve, whose value is very large. Let n_a is the private key of user A and n_b is the private key of user B. The public keys can be easily calculated:

$$A's\,Public\,key;\ p_a = n_a * G \tag{1}$$

$$B's\,Public\,key;\ p_b = n_b * G \tag{2}$$

The secret key (K) can be calculated as:

$$K = n_a * p_b \tag{3}$$

$$\text{or } K = n_b * p_a \tag{4}$$

The message (M) is first encoded into P_m as a point on the elliptic curve. The cipher text (C_m) is generated as:

$$C_m = K * G, P_m + K * p_b \tag{5}$$

This cipher text will be sent to the receiver.

3.1.2 Decryption

The receiver will take the first coordinate of C_m and multiply with the private key:

$$T = (K * G) * n_b \tag{6}$$

Then the receiver subtracts T from second coordinate of C_m to get the message:

$$Message = P_m + K * p_b - T \tag{7}$$

Although ECC itself is a very strong algorithm, due to modern attackers with their power systems, ECC alone may be vulnerable. So, to make the encryption process more secure, we combined it with another strong asymmetric key cipher, known as the Rabin cryptosystem.

3.2 Rabin Cryptosystem

It uses asymmetric key encryption for establishing communication between two users and encrypting the message. The security of the Rabin cryptosystem is related to the difficulty of factorization. It has the advantage over the cryptographic algorithms that the problem on which it rests, has proved to be as complicated as integer factorization. It has a major disadvantage also, that each output of the Rabin function can be generated by any of four possible inputs. Therefore, the end-user has to choose what his desired plain text is.

In the Rabin cryptosystem, the public key is chosen as 'n_b' and private key as the combination of 'p' and 'q', where 'p' and 'q' are in the format of (4k + 3), where k is any integer. Generally, 'e' is chosen as 2.

3.2.1 Encryption

$$Public\ key;\ n_b = p * q \tag{8}$$

$$Cipher\ text\ (CT) = PT^e mod n_b \tag{9}$$

The cipher text (CT) is transferred to the receiver for decryption.

3.2.2 Decryption

The receiver first calculates a_1, a_2, b_1 and b_2 as:

$$a_1 = +\left(CT^{(p+1)/4}\right) mod\ p \tag{10}$$

$$a_2 = -\left(CT^{(p+1)/4}\right) mod\ p \tag{11}$$

$$b_1 = +\left(CT^{(q+1)/4}\right) mod\ q \tag{12}$$

$$b_2 = -\left(CT^{(q+1)/4}\right) mod\ q \tag{13}$$

Then, receiver finds the values of a and b as:

$$a = q\left(q^{-1} mod\ p\right) \tag{14}$$

$$b = p\left(p^{-1} mod\ q\right) \tag{15}$$

The receiver calculates four probable plain texts PT_1, PT_2, PT_3 and PT_4 as:

$$PT_1 = (a * a_1 + b * b_1) mod\ n_b \tag{16}$$

$$PT_2 = (a * a_1 + b * b_2) mod\ n_b \tag{17}$$

$$PT_3 = (a * a_2 + b * b_1) mod\ n_b \tag{18}$$

$$PT_4 = (a * a_2 + b * b_2) mod\ n_b \tag{19}$$

3.3 Proposed Model

Since the overall encryption speed of the Rabin algorithm is much higher than that of ECC, so at first, the Rabin algorithm is used to encrypt the original image. After that, a Rabin encrypted code is derived consisting of a large array of numbers. On that array, ECC encryption is applied, as ECC works with numbers, with very high efficiency. After that encryption, the final cipher text is achieved.

During the decryption, the reverse procedure is followed. First, the cipher text is put into the ECC decryption algorithm. As a result, an array of numbers comes as output which have been named as ECC decrypted cipher text. Each element of ECC decrypted cipher text must be equal to Rabin encrypted cipher text. After that with that ECC decrypted cipher text, Rabin decryption is applied, which gives the decrypted image as output. If all the above processes are executed flawlessly, then the decrypted image will be exactly equal to the original image.

The overall encryption and decryption of our proposed model is shown in Fig. 1.

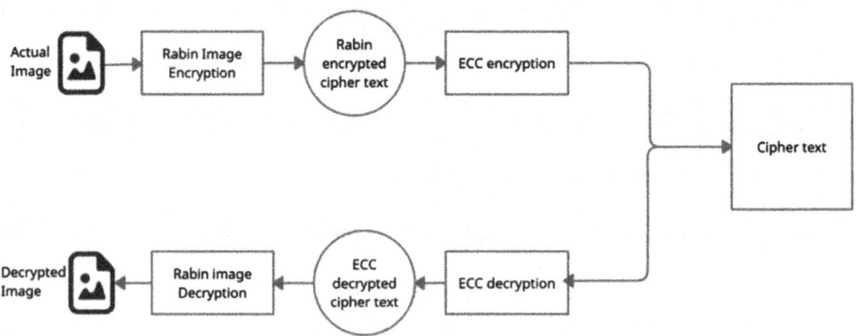

Fig. 1. Proposed model

4 Experimental Setup and Methodology

4.1 Image Processing and Device Specifications

Through this work, attempts were made to address the issues of the real world. So in the code also, we tried to mimic the real-life scenarios. In the real world, any image with any shape and size can be encountered (Fig. 2), very small to very large, color to black and white. So instead of imposing some limitations, work was done on every type of image. The users can give any kind of image as input, the code will handle it without any difficulty. That's why a negligible amount of image preprocessing is done and the image is fed to the code as it is so that the quality of the output image is not lost.

The whole code was done using an HP ProBook laptop with Intel(R) Core(TM) i5-8250U processor, which is a quad-core processor, CPU @ 1.60 GHz 1.80 GHz and 3.42 GHz, Max. RAM available is 16 GB and the hard drive is 512 GB SSD. So the processing time may vary a bit from device to device.

 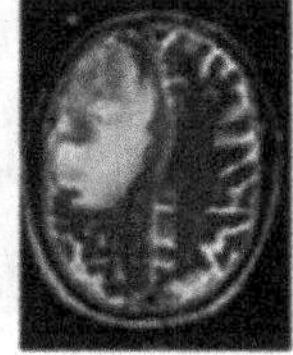 

Fig. 2. Test images

4.2 Implementation of Rabin Cryptosystem

Out of the two used algorithms, the implementation of the Rabin algorithm was much trickier. It is known that Rabin is normally used for digits. To use it to encrypt an image, some tweaks were needed in the implementation method. So, for the Rabin algorithm, the input is an image file of any size and the output is a 1D array of numbers. Step by step explanation of implementation of the Rabin algorithm is given below.

4.2.1 Rabin Encryption

- First the image is converted into a 3D array consisting or their pixel values using PIL and numpy libraries as shown in Fig. 3.
- After that for simplicity the 3D matrix is converted into 1D matrix.
- From the 1D matrix (or list), a random value, sval, was chosen, which was added with all the other values of the list, to increase the complexity a bit more.
- After that, the matrix can be simply put into the Rabin algorithm, for Rabin encryption, but instead of that, for making the code a bit more complex, every pixel value was converted to strings.
- After that the UTF values of each character were derived and then the Rabin encryption technique was applied to it as shown in Fig. 4.
- So, in this way the Rabin encrypted code was derived, and was sent to the ECC algorithm, for further encryption.

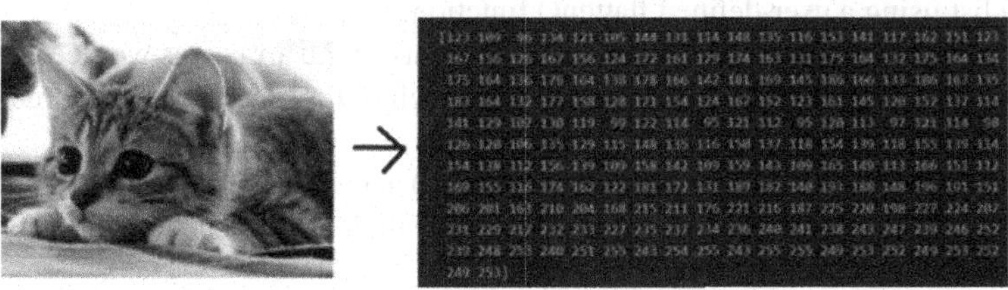

Fig. 3. Image to pixel values

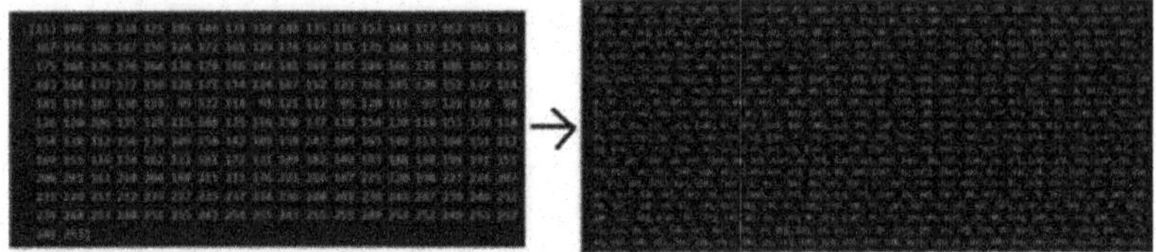

Fig. 4. Pixel values to Rabin encrypted cipher image

4.2.2 Rabin Decryption

- After ECC decryption, the reverse process was followed. The Rabin decryption algorithm was implemented to the array.
- After Rabin decryption, the UTF values were derived. So the UTF values were converted into characters and from characters, the string was derived.
- From that string, it was converted into an array of integers by typecasting.
- Then first the sval was subtracted from the array (which was added earlier), to get the original array K. That array of integers was nothing but the 1D representation of the image pixels. L.
- Then the 1D array was converted to a 3D matrix M. From that 3D matrix, we easily derived the image. That decrypted image was found to be similar to the original image N. Time complexity of implementing this was O(n3).

4.3 Implementation of ECC Cryptosystem

For the implementation of ECC algorithm, the code was not much complex. A matrix was derived consisting of [1 x 3] arrays of numbers as the output of Rabin encryption (Rabin encrypted cipher text). So, some numeric value was only needed to be dealt with in this case. This decreased the execution time to a large extent. Step by step explanation of implementation Rabin algorithm is given below.

4.3.1 ECC Encryption

- Working with a matrix of the array was difficult. So the matrix was first converted into a list using a user-defined flatten() function.
- Then the rest of the task was simple, the elements of the array was worked on, one by one and used the ECC encryption method to each of the elements, and stored them in a new array.
- To find the encrypted image (as shown in Fig. 5), a mod 256 operation was performed with the new array to find the new pixels, and then converted to image using user-defined reimg() function.

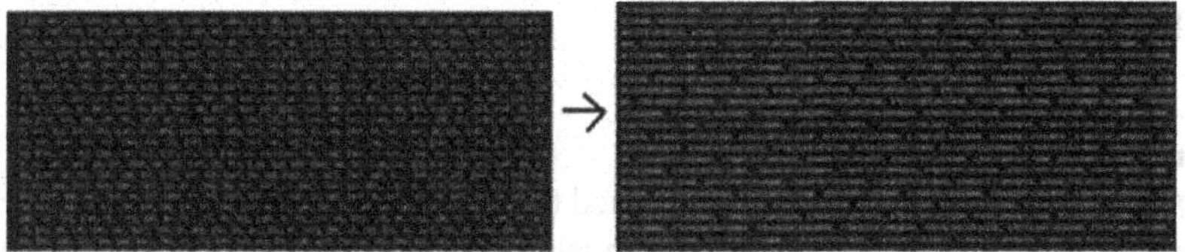

Fig. 5. Rabin encrypted cipher text to ECC encrypted cipher text

4.3.2 ECC Decryption

- During ECC decryption we followed the reverse pattern.
- Each element of the encrypted list was derived and it was applied with the ECC decryption algorithm there and stored the values one by one in a 1D array.
- Then the 1D array was converted into a matrix consisting of [1 X 3] arrays using a user-defined reframe() function and returned the matrix.
- This matrix will be further used by the Rabin Decryption algorithm to derive the original image. Time complexity of implementing this was O(n).

5 Result Analysis

No algorithm is called successful, if its performance and accuracy are not good enough. At every step, the code was tried to optimize, so that it takes the minimum time to execute, with maximum accuracy. After each encryption method, the numeric values were converted into pixel format, and then into array, to verify the effectiveness of our code. Few examples are given below (Fig. 6 and Fig. 7).

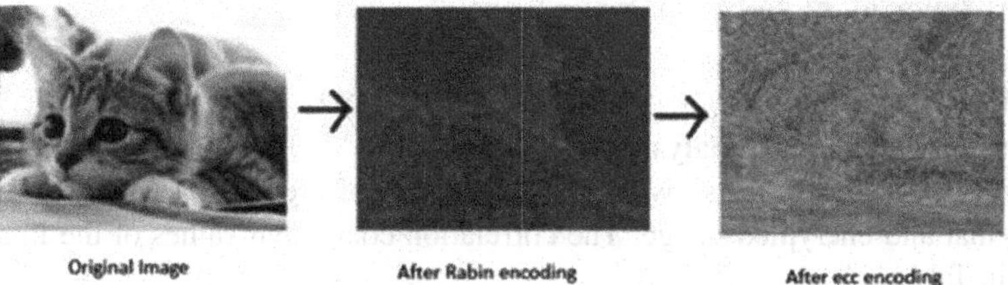

Fig. 6. Encryption of cat image

Fig. 7. Encryption of building image

5.1 Performance Analysis

Execution Time

The performance of the algorithm was tested thoroughly, by finding out their execution time and Correlation coefficients. Table 1 shows the Execution timings of different stages. We compared each execution time, by drawing a bar graph diagram, and observed how the execution time is changing, with change in size of the image as shown in Fig. 8.

From the results, it can confidently concluded that the encryption of the test images was successful and effective.

Table 1. Execution timings of different stages

Image size	Rabin encryption time (s)	ECC encryption time (s)	ECC decryption time (s)	Rabin decryption time (s)
400 × 300	2	0	0	9
1200 × 1600	30	8	5	113
758 × 433	8	2	1	26
300 × 228	1	0	0	5
294 × 380	2	0	0	8
1080 × 1080	26	7	4	92
600 × 400	8	2	1	30
640 × 1280	71	19	12	99
681 × 900	22	6	4	87
976 × 549	19	5	3	67

Correlation Coefficient Analysis

Correlation coefficient analysis is a statistical measure of strength of relationship between the original and encrypted image. The correlation coefficient values of the images are shown in Table 2.

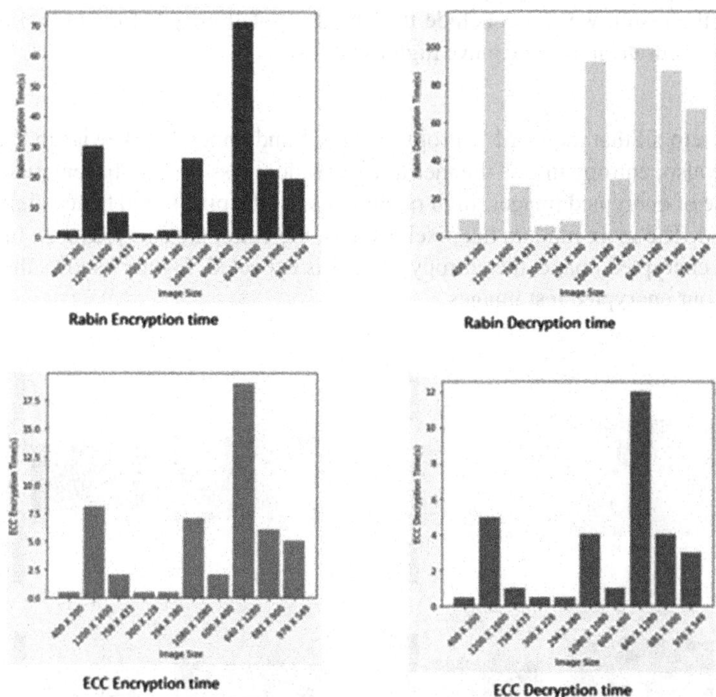

Fig. 8. Analysis of execution times of different stages

Table 2. Correlation coefficient of different encrypted images

Image size	Correlation coefficient (in %)
400 × 300	100
1200 × 1600	100
758 × 433	100
300 × 228	100
294 × 380	100
1080 × 1080	100
600 × 400	100
640 × 1280	100
681 × 900	100
976 × 549	100

From the result, we can conclude that the decrypted images formed by our model are highly accurate and hence, have high reliability.

Entropy

It is known to all that the word 'entropy' means "randomness". Likewise in the case of the image also, entropy means the measure of randomness of the adjacent pixel values. In the case of encrypted images, high randomness of the pixels, and hence high entropy is expected. The more random the pixel value is, the better the encryption is. In the case of a good encrypted image, the entropy of 4.5+ is expected. Figure 9 shows the entropy values of our encrypted test images.

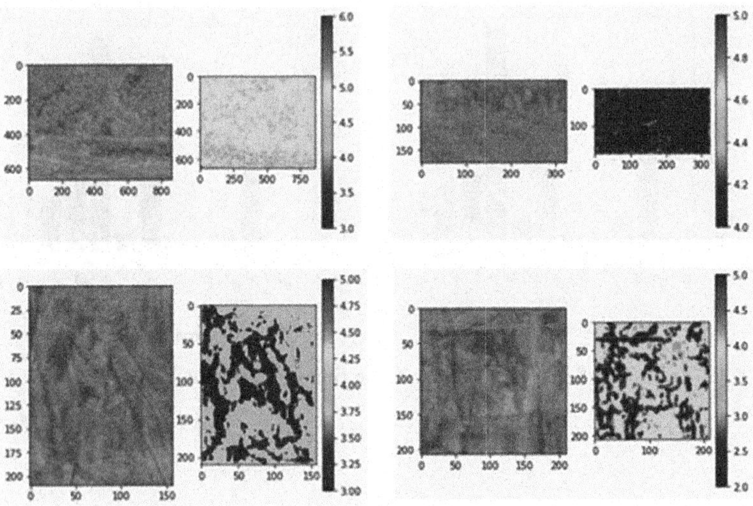

Fig. 9. Entropy analysis of different encrypted images

From the above result, it can be concluded that our working model is highly accurate, efficient and reliable.

6 Conclusion

Cryptography is the art of encryption and decryption of confidential and secret information and private messages. It must be implemented in the network security to prevent the involvement of other users. These can be done by any methods which come under these techniques so that a secure way of transaction is achieved. A few limitations that have been noticed in our work are that there is no such way in which the key which is generated can be sent, in a secured manner from the sender to the receiver. Though Diffie Hellman Key exchange can be used, there are chances of Man in the middle attack. It is a type of eavesdropping attack. The attackers get into the conversation and pose as one of the sender or receiver. Thus they get access to all the important data and information. The Diffie Hellman Key exchange method can be modified so that it can be used securely. We can also go further, to use block-chain methods, for secured key exchanges.

References

1. Marriott Data Breach 2020: 5.2 Million Guest Records Were Stolen. https://www.loginradius.com/blog/start-with-identity/marriott-data-breach-2020/
2. Why Zoom became so popular. https://www.theverge.com/2020/4/3/21207053/zoom-video-conferencing-security-privacy-risk-popularity
3. Zhang, X., Wang, X.: Digital image encryption algorithm based on elliptic curve public cryptosystem. IEEE Access **6**, 70025–70034 (2018)
4. Singh, L.D., Singh, K.M.: Image encryption using elliptic curve cryptography. In: International Multi-Conference on Information Processing, pp. 472–481. Elsevier (2015)
5. Siddiqui, A.S., Sen, P., Thakur, A.R., Gaur, A.L., Waghmare, P.C.: Use of improved Rabin algorithm for designing public key cryptosystem for wireless sensor networks. Int. J. Eng. Adv. Technol. **3**(3), 247–251 (2014)
6. Hashim, H.R.: H-RABIN CRYPTOSYSTEM. J. Math. Stat. **10**(3), 304–308 (2014)
7. Boneh, D.: Simplified OAEP for the RSA and Rabin Functions. In: Kilian, J. (eds.) Advances in Cryptology—CRYPTO 2001. CRYPTO 2001. LNCS, vol. 2139, pp. 275–291. Springer, Berlin, Heidelberg (2001). https://doi.org/10.1007/3-540-44647-8_17
8. Saha, B.J., Kabi, K.K., Pradhan, C.: A robust digital watermarking algorithm using DES and ECC in DCT domain for color images. In: International Conference on Circuits, Power and Computing Technologies, pp. 1378–1385. IEEE (2014)
9. Lauter, K.: The advantages of elliptic curve cryptography for wireless security. IEEE Wirel. Commun. **11**(1), 62–67 (2004)
10. Parida, P., Pradhan, C., Gao, X.Z., Roy, D., Barik, R.K.: Image encryption and authentication with elliptic curve cryptography and multidimensional chaotic maps. IEEE Access **9**, 76191–76204 (2021)

State-of-the-Art Multi-trait Based Biometric Systems: Advantages and Drawbacks

Swimpy Pahuja[1(✉)] and Navdeep Goel[2]

[1] Department of Computer Engineering, Punjabi University, Patiala 147002, Punjab, India
swimpy.pahuja@gmail.com

[2] Yadavindra Department of Engineering, Punjabi University Guru Kashi Campus, Damdama Sahib, Talwandi Sabo 151302, Punjab, India
navdeepgoel@pbi.ac.in

Abstract. Biometric authentication refers to the process of confirming the identity of an individual based on his physiological or behavioral features. This authentication method has outperformed the traditional authentication method of using passwords, one time pin etc. and now emerged as an overwhelming method of authentication. The biometric-based authentication is categorized into Unimodal and Multimodal authentication out of which Multimodal authentication provides better results in terms of accuracy, intrusion avoidance and user acceptance. This paper discusses some state-of-the-art Multimodal authentication systems along with their advantages as well as disadvantages. Also, some future research directions have been provided based on these disadvantages. Lastly, the utilization of deep learning methods in current Multimodal systems has been discussed followed by the analysis of some state-of-the-art Multimodal systems based on fingerprint and face traits.

Keywords: Unimodal biometrics · Multimodal biometrics · Authentication · Deep learning

1 Introduction

Digital image processing involves digital image or video frames as an input to a processing tool, performing the desired functions inside the tool and producing image or its characteristics as an output. This output can be used for decision making or further analysis. In real life applications such as surgical planning, diagnostic image analysis, medical field, object recognition, facial recognition etc., processing of digital images is so much prevalent that includes image enhancement, image restoration, image analysis, and image compression. As far as image enhancement is concerned, it is one of the most widely used image processing technique now-a-days as it enables applications to increase or enhance image quality for further information extraction. In image restoration, methods are used to restore the original image from distorted or degraded image. Image analysis includes various techniques to analyze the image and extract the required information attributes from it which can further be used for result declaration based on

V. E. Balas et al. (Eds.): ICETCE 2022, CCIS 1591, pp. 704–714, 2022.
https://doi.org/10.1007/978-3-031-07012-9_58

predefined methods. For example- features of a biometric [1, 2] image such as shape, size, color etc. can be extracted from it using various feature extraction methods and then these extracted features further can be compared with existing images present in the database to arrive at a solution. These features of biometric image are sometimes called biometric modalities, identifiers, characteristics, traits etc. Some of these biometric modalities have been shown in Fig. 1. Finally, image compression [3–6] is an image processing technique used to compress the image in order to transmit it via various communication channels with lesser bandwidth utilization.

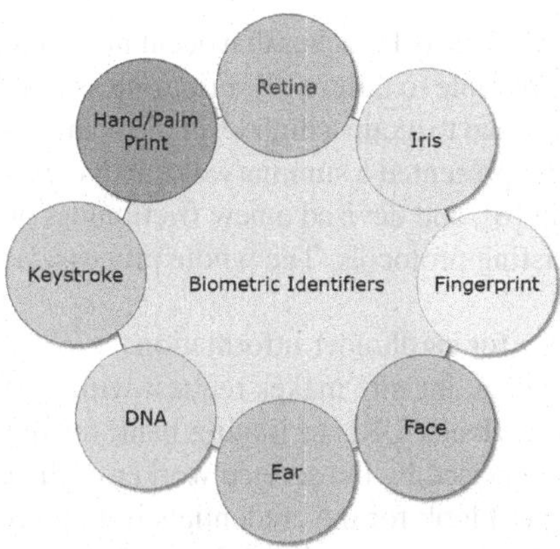

Fig. 1. Some of the Biometric Identifiers/Modalities/Characteristics/Traits

As mentioned above, biometric image can be processed in order to proceed with the decision making process of various authentication and authorization systems. As verifying a person is a necessary initial step in granting access to certain essential resources, the authentication technique should employ highly secure mechanisms. Authenticating a person incorporates a variety of ways, including knowledge-based, possession-based, and mixtures of the two. Password remembrance and hacking difficulties are raised by knowledge-based approaches, whereas lost or forgotten issues are raised by possession-based methods. In terms of enhanced security against hacking and no password-remembrance concerns, biometric-based authentication has shown to be one of the most promising solutions. This biometric-based authentication is further segregated into Unimodal and Multimodal based authentication, with the former employing a single biometric entity for authentication and the latter employing multiple biometric entities for the same reason.

Various researchers have undertaken extensive research on Unimodal-based systems, and it has been discovered that Unimodal systems have issues such as inter-class similarities, intra-class variability, relatively high error rates, noisy data, and high Failure to Enroll (FTE) rates. The utilization of several features in Multimodal systems solves the challenges associated with Unimodal systems. This paper discusses some of these Multimodal systems along with their advantages and disadvantages.

The rest of the paper is organized as follows: Sect. 2 provides a brief overview of multiple feature-based authentication systems; Sect. 3 gives an overview of current state-of-the-art Multimodal Biometric Systems in a tabular view; Sect. 4 presents application of machine learning and deep learning methods in the development of Multimodal Biometric Systems; Sect. 5 shows the comparative analysis of three fingerprint and face based Multimodal systems; Sect. 6 concludes the paper and discusses the future work.

2 Multiple Feature-Based Authentication

Due to the rising number of security breaches that occur nowadays, the traditional authentication mechanism is vulnerable to failure [7]. Frictionless multiple feature-based techniques are currently being used to attain a high level of security without requiring human participation. Corella et al. presented a summary of the two existing protocols for credit card based transactions in [8], and devised a new frictionless protocol for removing the shortcomings of both existing protocols. The whole process can be explained as under:

- Merchant's site requests for cardholder information.
- On reception of above data, the site makes request with the card information, transaction description and callback URL to issuing bank for the credit card credentials, but here the request is serviced by the service worker of the issuing bank.
- The service worker would look for the credentials in IndexedDB database using IIN (Issuer Identification Number) present in card information.
- These credentials are then sent to the cardholder's browser via confirmation page in the form of a script. Here, the confirmation page is the page of issuing bank not of merchant's site.
- Once the cardholder confirms, the script is sent to the merchant's site for signature verification.
- After verification, the payment is done and the merchant keeps the record for non-repudiation purposes.

The author of US Patent [9] designed a frictionless system which employs biometric technology for airline consumers. The whole process involves three main functional entities namely security station, backend server and airline computing system which can be explained as:

- Customer provides its biometric data to security station personnel at airport.
- The personnel look for identifier from backend server associated with the data provided.
- This identifier is used by security computing device to obtain the electronic boarding pass information from airline computing system.
- This pass information can further be used for security screening or for clearing the person.

In the above method, the backend server is vulnerable to be attacked. Similarly, in [10], authors used facial recognition techniques for completing the airport formalities.

The claimant has designed a new frictionless system in US 9,865,144 B2 [11]. Security control room, Verification and Tracking Centre (VTC), Positioning Unit (PU), Video Analysis System (VAS), and Door Controllers (DC) are some of the components used in the devised system. The whole functionality can be summarized as:

- PU receives user information and sends it to VTC
- VTC checks for match of user information in the stored database
- If matched, VTC queries for video information from VAS
- VAS sends the requested information to VTC
- VTC matches the information against user's data stored in the database
- If it matches, VTC signals the DC to give access to the person
- VTC records this event in its database

From the above discussion, it is clear that frictionless techniques will provide better results and high level of security when combined with other features.

3 Multimodal Biometric Systems

Biometric systems are categorized into two types namely Unimodal and Multimodal systems. The former systems include the use of single modality for the purpose of Identification/Verification while the latter use more than one modality to overcome the limitations of Unimodal biometric systems. In [12], face, ear, and iris modalities are used by Walia et al. to build a robust, secure, non-evident, revocable, and diversified multimodal. Some of the Multimodal systems have been provided in Table 1 along with the technique used, modalities used and other description.

Table 1. Some multimodal biometric systems with their shortcomings

Authors, Year	Traits used	Technique used	Dataset used	Fusion method used	Results	Advantages	Disadvantages
R.L. Telgad et al., 2014 [13]	Fingerprint and face	Euclidean distance matcher, Gabor filter, Principle Component Analysis	FVC 2004	Score level fusion	97.5% accuracy, FAR of 1.3% and FRR 4.01%	Increased recognition rate Decreased error rate	Occluded faces not considered, lighting conditions not considered

(continued)

Table 1. *(continued)*

Authors, Year	Traits used	Technique used	Dataset used	Fusion method used	Results	Advantages	Disadvantages
Amritha Varshini et al., 2021 [14]	ECG, face, and fingerprint	overlap extrema-based min–max (OVEBAMM) technique, confidence-based weighting (CBW) method and mean extrema-based confidence weighting (MEBCW) method	FVC2002/2004, Face94, and PhysioNet (MIT-BIH Arrythmia)	Hybrid fusion (Score level and feature level)	TPR of 0.99, FPR of 0.1, and EER rate of 0.5	Better recognition rates	Different input acquisition increases cost of system
Walia et al., 2019 [12]	face, ear and iris	graph-based random walk cross diffusion, cuckoo search optimisation and PCR-6 rules	ORL face, CASIA iris v1, Computer Vision Science Research Projects face database, MMU iris database, AMI ear database and IIT Delhi iris database	Feature level fusion	98.3% accuracy with 2.11s computational time	Robust, secure, revocable	Face images affected by surroundings, spoofing possible
Ming Jie Lee et al., 2021 [15]	Fingerprint and face	Feature Rescaling, IoM hashing and token less cancellable transform	FVC2002, FVC2004 for fingerprint and Labelled Faces in the Wild (LFW) for face	Feature level fusion	90% GAR and 0% FAR	Provides template protection, secure, irreversible	Occluded faces and varying lighting conditions not considered
Vijay et al., 2021 [16]	iris, ear, and finger vein	Multi- support vector neural network, deep belief neural network (DBN), chicken earthworm optimization algorithm (CEWA)	SDUMLAHMT, and the AMI ear database	Score level fusion	accuracy of 95.36%	Provides optimal authentication	Accuracy needs to be improved further
Al-Waisy Alaa et al., 2018 [17]	Left and right iris	Softmax classifier and the Convolutional Neural Network (CNN	SDUMLA-HMT, CASIA-IrisV3 Interval and IITD iris database	ranking-level	identification rate of 100%	real-time multimodal biometric system is made, reduced processing time	accuracy needs to be validated on larger databases
Jagadiswary et al., 2016 [18]	fingerprint, retina and finger vein	MDRSA(Modified Decrypted Rivest, Shamir and Adleman)	Not mentioned	Feature level fusion	GAR of 95.3% and FAR of 0.01%	Template protection	Datasets are not specified clearly

(continued)

Table 1. *(continued)*

Authors, Year	Traits used	Technique used	Dataset used	Fusion method used	Results	Advantages	Disadvantages
Xiong et al., 2021 [19]	Face and iris	Curvelet transform, 2D Log-Gabor, kernel extreme learning machine (KELM), chaotic binary sequence	CASIA multimodal iris and face dataset from Chinese Academy of Sciences	Feature level fusion	99.78% recognition rate	Reduced running time, improved recognition rate	Suffers from information loss
Peng et al., 2014 [20]	finger vein, fingerprint, finger shape and finger knuckle print	triangular norm	Hong Kong Polytechnic University Finger Image Database Version 1.0, m FVC2002 database Db1 set A, PolyU Finger-Knuckle-Print database,	score-level fusion	a larger distribution distance between genuine and imposter scores	No training required, less computational complexity	No template protection

Table 1 clearly shows that in case of Multimodal systems including face, the occluded faces, lighting conditions, clothing etc. are problematic and not considered in most of the systems developed. Also, the Multimodal systems lack security of templates which is another open research problem currently. To address these problems, researchers are now applying deep learning based algorithms in desire of better results in Multimodal systems as discussed in preceding section of this paper.

In advantages, they outperform the Unimodal based systems in terms of computational complexity, model accuracy, impostor access, error rates and running time.

4 Application of Machine Learning and Deep Learning in Multimodal Biometrics System

Lot of research has been done on the development of Multimodal systems and now the researchers are moving towards machine learning and deep learning based techniques for better results. Some current researches are discussed below.

IrisConvNet is a deep learning approach created by Alaa S. Al Waisy et al. [17], which blends the SoftMax classifier and the Convolutional Neural Network (CNN) for feature extraction of left and right iris images. This multi-biometric model leverages the training process to reduce overfitting and improve the neural network's classification accuracy. Similarly, Vijay et al. [16] makes use of Deep Belief Networks (DBN) which has been trained using Chicken Earthworm Optimization Algorithm (CEWA) for the purpose of optimal authentication of an individual.

Liu et al. suggested the FVR-DLRP (Finger Vein Recognition based on Deep Learning and Random Projection) verification technique, which is primarily aimed to preserve stored biometric data [21]. Individual finger veins are used as input to the acquisition device, which is then turned into a feature-based template for matching purposes.

Because the template is stored using transformation-based random projection and deep belief networks, the approach is resistant against diverse photos of the same person. However, the defined approach did not consider other recognition approaches with the ability of template protection.

Alotaibi et al. [22] deployed CNN for gait identification, reaching a 92% accuracy utilizing four convolutional layers and four subsampling layers. The technique was expanded by Alotaibi in [23], which used deep CNN architecture to account for cross-view variations and improved accuracy.

Alay et al. used CNN for feature extraction of face and iris for the purpose of successful recognition [24]. The recognition performance was first checked for each of the Unimodal systems and after that for the fused system. In order to enhance security further, they extended this work by integrating finger vein in the model [25] and used pretrained deep learning model for the development of secured and robust Multimodal Biometric system. The recognition accuracy is achieved using both feature level fusion and score level fusion and the results demonstrated that score level fusion methods provides better accuracy of 100%. Another work is reported by Yadav et al. [26] on pretrained VGG-32 CNN model leveraging iris, fingerprint and hand written signature.

Sengar et al. in [27] made use of Deep Neural Networks (DNN), Multilayer Perceptron (MLP) as well as Backpropagation Algorithm for the purpose of feature extraction of fingerprint and palmprint of an individual.

Tiong et al. in [28] utilized deep learning method for fusion of multiple features of face along with texture descriptors. They also considered the occluded faces and periocular region of faces thereby achieving an accuracy of 99.76%.

The above discussed papers illustrates that application of machine leaning and deep learning to Biometric Authentication will explore more avenues and lead towards more accurate and secured systems.

5 Comparative Analysis of Multimodal Biometric Systems

For the purpose of finding the best system in terms of accuracy, security and user acceptance, three Multimodal systems comprising of fingerprint and face modalities are taken into consideration. These systems are explained as under:

5.1 An Improved Biometric Fusion System (IBFS)

Tajinder et al. [29] proposed an Improved Biometric Fusion System (IBFS) which integrates face and fingerprint traits to achieve better recognition rates and minimize the fraudulent access to the system. Features of fingerprint are extracted using minutiae extraction and that of face are extracted using Maximally Stable Extremal Region (MSER) and then Whale Optimization Algorithm is applied in order to obtain the optimized features. These optimized features are then fused and trained using Pattern Net. Using this system, correct authentication attempts achieved were around 99.8% which means only 0.2% imposters can enter into the system. The system is resilient against spoof attacks and has enhanced security and accuracy.

5.2 Descriptor Generation System (DGS)

Atilla et al. proposed a method to enhance security and privacy of templates during communication [30]. Timestamp has been associated with the template generated for enhancing security and for privacy, a unique system identifier has been utilized. They used FaceNet model, Euclidean distance and Hyperbolic Tangent (TanH) activation function. The experiments are performed on FERET faces dataset and FVC2000 DB3 set B fingerprints dataset while the network training is done for 100 epochs.

5.3 An Optimized Multimodal Biometric Authentication System (OMBAS)

This work involves the facial features extraction using SIFT followed by optimization using PSO algorithm while the ridges and minutiae features are extracted in case of fingerprint. Implementation is done using Olivetti dataset for faces and the FVC2000 dataset for fingerprints [31].

Table 2 shows the different values of Accuracy, FAR and FRR of these three Multimodal systems.

5.4 Analysis and Discussions

From Table 2, numerical values clearly show that in comparison with other two systems, an Optimized Multimodal Biometric Authentication System (OMBAS) cannot be used in the applications which demand high grade of security due to its high value of False Acceptance Rate (FAR), also the system is not acceptable in terms of ease of access due to its high value of False Rejection Rate (FRR).

Out of multimodal systems IBFS and DGS, it is clearly visible from values in the Table 2 as well as in the Fig. 2 that IBFS system is most accurate but its FAR value is slightly high as compared to DGS system making it less secure. Therefore, DGS system will be preferred in terms of security while IBFS system will be preferred in terms of accuracy and ease of access.

Table 2. Results of multimodal systems

S. No	Multimodal system	Accuracy (in percentage)	FAR (in percentage)	FRR (in percentage)	EER (in percentage)
1	IBFS [29]	99.6	0.33	0.002	0.4
2	DGS [30]	99.41	0.21	0.97	0.59
3	OMBAS [31]	99.2	2	1.03	0.8

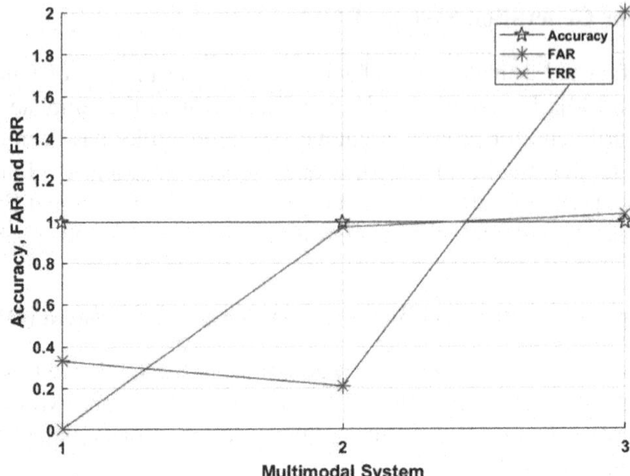

Fig. 2. Comparative analysis of mentioned multimodal systems

In the above figure, x-axis represents all the three Multimodal systems namely IBFS [29], DGS [30] and OMBAS [31] while y-axis represents Accuracy, FAR and FRR values for the respective systems.

6 Conclusion

This paper presented a roadmap from Multiple-feature based authentication to Deep learning based Multimodal Biometric authentication techniques covering the advantages and disadvantages of Multimodal systems. The paper clearly shows that there are still lot of open research issues that needs to be addressed by forthcoming researchers in the field of Multimodal biometric systems. Also, the application of deep learning algorithms can help solve these research issues. Lastly, three state-of-the-art Multimodal systems have been compared and analyzed in terms of security, ease of access and accuracy.

References

1. Jain, A., Ross, A., Prabhakar, S.: An introduction to biometric recognition. IEEE Trans. Circ. Syst. Video Technol. **14**(1), 4–20 (2004)
2. Alsaadi, I.M.: Physiological biometric authentication systems, advantages disadvantages and future development: a review. Int. J. Sci. Technol. Res. **12**, 285–289 (2015)
3. Subramanya, A.: Image compression technique. Potentials IEEE **20**(1), 19–23 (2001)
4. Zhang, H., Zhang, X., Cao, S.: Analysis and Evaluation of some image compression techniques. In: High Performance Computing in Asia- Pacific Region, 2000 Proceedings, 4th International Conference, vol. 2, pp. 799–803, 14–17 May 2000
5. Yang, M., Bourbakis, N.: An overview of lossless digital image compression techniques, circuits & systems. In: 48th Midwest Symposium IEEE, vol. 2, pp. 1099–1102, 7–10 August (2005)

6. Avcibas, I., Memon, N., Sankur, B., Sayood, K.: A progressive lossless/near lossless image compression algorithm. IEEE Signal Process. Lett. **9**(10), 312–314 (2002)
7. Simmonds, A., Sandilands, P., van Ekert, L.: An ontology for network security attacks. In: Manandhar, S., Austin, J., Desai, U., Oyanagi, Y., Talukder, A.K. (eds.) Applied Computing. AACC 2004. LNCS, vol. 3285, pp. 317–323. Springer, Berlin, Heidelberg (2004). https://doi.org/10.1007/978-3-540-30176-9_41
8. Corella, F., Lewison, K.P.: Frictionless web payments with cryptographic cardholder authentication. In: Stephanidis, C. (eds.) HCI International 2019 – Late Breaking Papers. HCII 2019. LNCS, vol. 11786. Springer, Cham (2019). https://doi.org/10.1007/978-3-030-30033-3_36
9. Physical Token-Less Security Screening Using Biometrics (US Patent US 9,870,459 B2 published on 16 Jan 2018)
10. SITA, 2018b. Société Internationale de Télécommunications Aéronautiques. Brisbane Airport leads with smart biometrics from check-in to boarding. https://www.sita.aero/pressroom/news-releases/brisbane-airport-leads-with-smartbiometrics-from-check-in-to-boarding/. Accessed 28 Mar 2018
11. Video recognition in frictionless acess control system (US Patent US 9,865,144 B2 published on 9 Jan 2018)
12. Walia, G.S., Rishi, S., Asthana, R., Kumar, A., Gupta, A.: Secure multimodal biometric system based on diffused graphs and optimal score fusion. IET Biom. **8**(4), 231–242 (2019)
13. Telgad, R.L., Deshmukh, P.D., Siddiqui, A.M.N.: Combination approach to score level fusion for Multimodal Biometric system by using face and fingerprint. In: International Conference on Recent Advances and Innovations in Engineering (ICRAIE-2014), 2014, pp. 1–8 (2014). https://doi.org/10.1109/ICRAIE.2014.6909320
14. Amritha Varshini, S., Aravinth, J.: Hybrid level fusion schemes for multimodal biometric authentication system based on matcher performance. In: Smys, S., Tavares, J.M.R.S., Bestak, R., Shi, F. (eds.) Computational Vision and Bio-Inspired Computing. AISC, vol. 1318, pp. 431–447. Springer, Singapore (2021). https://doi.org/10.1007/978-981-33-6862-0_35
15. Lee, M.J., Teoh, A.B.J., Uhl, A., Liang, S.N., Jin, Z.: A Tokenless cancellable scheme for multimodal biometric systems. Comput. Secur. **108**, 102350 (2021)
16. Vijay, M., Indumathi, G.: Deep belief network-based hybrid model for multimodal biometric system for futuristic security applications. J. Inf. Secur. Appl. **58**, 102707 (2021)
17. Al-Waisy, A.S., Qahwaji, R., Ipson, S., Al-Fahdawi, S., Nagem, T.A.: A multi-biometric iris recognition system based on a deep learning approach. Pattern Anal. Appl. **21**(3), 783–802 (2018)
18. Jagadiswary, D., Saraswady, D.: Biometric authentication using fused multimodal biometric. Procedia Comput. Sci. **85**, 109–116 (2016)
19. Xiong, Q., Zhang, X., Xu, X., He, S.: A modified chaotic binary particle swarm optimization scheme and its application in face-iris multimodal biometric identification. Electronics **10**(2), 217 (2021)
20. Peng, J., Abd El-Latif, A.A., Li, Q., Niu, X.: Multimodal biometric authentication based on score level fusion of finger biometrics. Optik **125**(23), 6891–6897 (2014)
21. Liu, Y., Ling, J., Liu, Z., Shen, J., Gao, C.: Finger vein secure biometric template generation based on deep learning. Soft. Comput. **22**(7), 2257–2265 (2017). https://doi.org/10.1007/s00500-017-2487-9
22. Alotaibi, M., Mahmood, A.: Improved gait recognition based on specialized deep convolutional neural networks. In: 2015 IEEE Applied Imagery Pattern Recognition Workshop (AIPR), pp. 1–7. IEEE Computer Society (October 2015)
23. Sharma, G., et al.: Reverse engineering for potential malware detection: android APK Smali to Java. J. Inf. Assur. Secur. **15**(1) (2020). ISSN 1554-1010

24. Alay, N., Al-Baity, H.H.: A Multimodal biometric system for personal verification based on Di_erent level fusion of iris and face traits. Biosci. Biotechnol. Res. Commun. **12**, 565–576 (2019)
25. Alay, N., Al-Baity, H.H.: Deep learning approach for multimodal biometric recognition system based on fusion of iris, face, and finger vein traits. Sensors **20**(19), 5523 (2020)
26. Yadav, A.K.: Deep learning approach for multimodal biometric recognition system based on fusion of iris, fingerprint and hand written signature traits. Turkish J. Comput. Math. Educ. (TURCOMAT) **12**(11), 1627–1640 (2021)
27. Sengar, S.S., Hariharan, U., Rajkumar, K.: Multimodal biometric authentication system using deep learning method. In: 2020 International Conference on Emerging Smart Computing and Informatics (ESCI), pp. 309–312. IEEE (March 2020)
28. Tiong, L.C.O., Kim, S.T., Ro, Y.M.: Implementation of multimodal biometric recognition via multi-feature deep learning networks and feature fusion. Multimed. Tools Appl. **78**, 22743–22772 (2019). https://doi.org/10.1007/s11042-0
29. Kumar, T., Bhushan, S., Jangra, S.: An improved biometric fusion system of fingerprint and face using whale optimization. Int. J. Adv. Comput. Sci. Appl. (IJACSA) **12**(1), 664–671 (2021)
30. Atilla, D.C., Alzuhairi, R.S.H., Aydin, C.: Producing secure multimodal biometric descriptors using artificial neural networks. IET Biom. **10**(2), 194–206 (2021)
31. Pawar, M.D., Kokate, R.D., Gosavi, V.R.: An optimize multimodal biometric authentication system for low classification error rates using face and fingerprint. In: Proceedings of the 2nd International Conference on IoT Based Control Networks and Intelligent Systems (ICICNIS 2021) (July 2021)

Study of Combating Technology Induced Fraud Assault (TIFA) and Possible Solutions: The Way Forward

Manish Dadhich[1(✉)], Kamal Kant Hiran[1], Shalendra Singh Rao[2], Renu Sharma[2], and Rajesh Meena[3]

[1] Sir Padampat Singhania University, Udaipur, India
manish.dadhich@spsu.ac.in
[2] Mohanlal Sukhadia University, Udaipur, India
[3] St. JKL College, Jaipur, India

Abstract. The study aims to identify modes of fraudulent payments and create awareness of such incidences to avoid decisive virtual activities. Disruptive developments such as contactless payments, mobile payments, and e-cloud payments witnessed a large-scale data breach. As a result, new fraud pathways have emerged and made traditional detection technologies less effective. Thus, there is a need to understand the contemporary fraud induced technology and corrective remedies—a relevant assessment of published approaches for detecting fraudulent card payment is critical to study. The researchers tried to identify fundamental issues using AI-ML-DL to detect fraud. An intellectual computing strategy is offered as a capable research avenue while boosting commercial data patronage. The paper discusses a standard method used by cybercriminals to mislead individuals and businesses. The study also focused on the methods utilized by cybercriminals and the economics of each confronted gadget. The paper talked about how systems can detect and block cybercriminals in three domains: card payment fraud, mobile payment fraud, and telephonic fraud.

Keywords: AI · ML · Cyber-crime · Fraud detection · Financial crime · Spear phishing

1 Introduction

Many facets of our lives are now connected to the Internet, including purchasing goods and services, business banking, booking, travel and vacation management, etc. [1, 5]. The number of crimes executed using the Internet has increased dramatically. Cybercriminals steal billions of dollars from service providers, regulators, and end-users every year due to targeted and organized misconduct [3]. According to [2, 4], criminals have used financial fraud techniques using advanced software through the Internet. The Internet-enabled cybercriminals conduct crimes in another jurisdiction while concealing their characteristics and evading the range of any prosecutor. In the words of [10], consumers, businesses, and governments worldwide will lose $608 billion due to cyber theft by

V. E. Balas et al. (Eds.): ICETCE 2022, CCIS 1591, pp. 715–723, 2022.
https://doi.org/10.1007/978-3-031-07012-9_59

2020. Mobile devices reported more than half of all financial transactions in the present scenario, providing channels for fraudulent transactions. However, few merchants are aware of the risks associated with mobile payments, and many feel that mobile gadgets are secure than desktops [6]. In the words of [8, 9], shoppers, brokers, insurance agents, healthcare professionals, and others can perpetrate insurance fraud at any phase in the insurance process. Crop, healthcare, and automotive insurance fraud are all included in this study. Usually, business fraud entails the activities viz. (i) fabrication of financial information, (ii) self-trading by corporate insiders, and (iii) stumbling block to hide any of the above-noted terms as criminal conduct [14, 15].

Virtual mass fraud is a common term for fraud in India using mass communication, telemarketing, and mass mailings by dummy identity. In the recent past, the RBI recorded roughly 7,400 bank fraud incidents across India in the financial year 2021. There was a reduction from the previous year and a reversal of the preceding decade's trend. The entire value of bank scams also reduced, from 1.85 trillion to 1.4 trillion Indian rupees [11, 23]. Figure 1 outlines recorded fraud cases in Indian banks between the years 2009–2021.

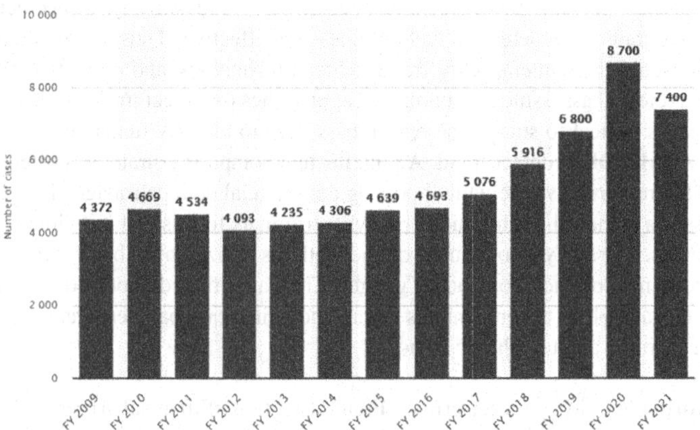

Fig. 1. Number of fraud cases in India (Source: www.statista.com)

Table 1 outlines recorded fraud cases in Indian banks between the years 2009–2021. In terms of both number and value, fraud has been most prevalent in the loan portfolio. Large-value scams were concentrated, with the top fifty credit-related frauds accounting for 76% of the total amount reported as fraud in 2019–20. Among the critical items on the RBI's agenda for FY22 is improving the fraud risk management system, particularly the efficacy of the early warning signal (EWS) framework. The regulator also aims to strengthen the mechanism for fraud governance and response. This involves enhancing data analysis for transaction monitoring, establishing a dedicated market intelligence (MI) unit for frauds, and implementing an automatic, system-generated number for each scam [12, 16].

Table 1. Number of fraud cases in banks

Bank Group/Institution	2018–19		2019–20		April–June 2019		April–June 2020	
	Number of frauds	Amount involved	Number of frauds	Amount involved	Number of frauds	Amount involved	Number of frauds	Amount involved
1	2	3	4	5	6	7	8	9
Public Sector Banks	3,568 *(52.5)*	63,283 *(88.5)*	4,413 *(50.7)*	1,48,400 *(79.9)*	1,133 *(56.0)*	31,894 *(75.5)*	745 *(47.8)*	19,958 *(69.2)*
Private Sector Banks	2,286 *(33.6)*	6,742 *(9.4)*	3,066 *(35.2)*	34,211 *(18.4)*	601 *(29.7)*	8,593 *(20.3)*	664 *(42.6)*	8,009 *(27.8)*
Foreign Banks	762 *(11.2)*	955 *(1.3)*	1026 *(11.8)*	972 *(0.5)*	250 *(12.4)*	429 *(1.0)*	127 *(8.2)*	328 *(1.1)*
Financial Institutions	28 *(0.4)*	553 *(0.8)*	15 *(0.2)*	2,048 *(1.1)*	4 *(0.2)*	1,311 *(3.1)*	3 *(0.2)*	546 *(1.9)*
Small Finance Banks	115 *(1.7)*	8 *(0.0)*	147 *(1.7)*	11 *(0.0)*	25 *(1.2)*	1 *(0.0)*	16 *(1.0)*	2 *(0.0)*
Payments Banks	39 *(0.6)*	2 *(0.0)*	38 *(0.4)*	2 *(0.0)*	10 *(0.5)*	0 *(0.0)*	3 *(0.2)*	0 *(0.0)*
Local Area Banks	1 *(0.0)*	0.02 *(0.0)*	2 *(0.0)*	0.43 *(0.0)*	1 *(0.0)*	0 *(0.0)*	0 *(0.0)*	0 *(0.0)*
Total	**6,799** ***(100.0)***	**71,543** ***(100.0)***	**8,707** ***(100.0)***	**1,85,644** ***(100.0)***	**2,024** ***(100.0)***	**42,228** ***(100.0)***	**1,558** ***(100.0)***	**28,843** ***(100.0)***

Source: www.rbi.org.in

Detection of financial fraud is critical for avoiding financial losses for individuals and organizations. It entails differentiating false financial statistics from authentic facts, revealing deceitful behavior or actions, and allowing decision-architects to implement suitable policies to reduce the impact of fraud. Data mining is a significant step in detecting monetary fraud because it is frequently used to extract and expose hidden truths from massive amounts of data.

According to [13, 16], data mining is the act of uncovering data patterns that may then be used to detect financial fraud. [12] define data mining as a process that extracts and identifies usable information from a vast database using statistical, mathematical, artificial intelligence, and machine-learning approaches. In the word of [17], data mining aims to discover usable, non-explicit information from data housed in vast repositories.

1.1 Frauds Involving IT Technology

Fraudsters utilize the Internet and decisive technology to conduct fraud in two steps: (a) transmitting risky information such as viruses that copy personal information from victims and (b) encouraging victims to divulge their private info using social number tactics. In the recent past, internet-based applications have evoked further opportunities for businesses and merchants and new ways for swindlers to leverage cutting-edge technology to defraud customers and enterprises [18].

1.2 Plastic Frauds

This section covers card payment systems relating to the overall system of payments. A card payment system is usually separated into two categories: active cards and inactive cards. The cardholder is fully present at the dealer store. Payment is made by swiping, inserting, or tapping (in the case of contactless) business accounts to the merchant, provided the terminal/reader has a point of sale (P-o-S) [19]. With card-present deals, the legality of the cardholder getting the transaction is verified either by requiring the cardholder's PIN or by requiring the cardholder's signature. When more protection mechanisms are added, present card transactions become more secure. The safety characteristics of a smart card with a mark, for example, allow you to store payment information safely. Payment security for present cards is enhanced by encryption techniques that link each transaction to a specific transaction code or cryptogram. Since its inception, the credit card has been associated with the phenomenon of fraud. Credit card fraud causes substantial financial losses, even though it only impacts a small percentage of all transactions [20].

1.3 Frauds Induced by Mobile

Over time, mobile technologies such as smartphones have begun to supplant desktop computers. Fraudsters have decided to take the risk and move their operations to this channel. The strategies used in mobile network fraud are highly like traditional online deception, making it easier to adapt cyber fraud to mobile environments. Furthermore, scammers have expressly explored new ways to profit from mobile applications [17, 18]. Technology adoption has been described as a process of complicated electro-techno transformations. Technical, financial, market, subjective, and intention elements can potentially stimulate the shift from conventional to AI-based surveillance systems. Due to rapid technological, legal, and financial innovation, the users are provided with a mode of easy virtual transactions, but such advanced conditions also create opportunities for fraud perpetrators [7, 13].

Thus, the rest of the paper is organized as follows. The following section examines the literature. The methodology of the empirical investigation of attacking methods in telecommunication frauds is then explained in the next section. The final portion explores the way forward to minimize frauds ramifications and, finally, makes certain recommendations.

2 Review of Literature

Several review articles related to financial frauds have appeared in recent publications; for example, [5, 14] studied statistical approaches for identifying fraud, such as plastic fraud, money laundering, and telecommunications fraud. Similarly, financial applications by the technique of data mining, such as stock market and bankruptcy projections, have been studied by [10].

According to [20], traditional means of payment are losing way to highly computer savvy criminals who live in an age of high-tech transmission and widespread usage of social media. As thieves begin to exploit AI and machine learning for attack purposes, more complicated and delicate fraud channels have emerged. [14, 18] used a stratiform k-folded cross-validation process to deliver results suggestive of success on a more generalized and unbiased dataset. Researchers have found that knowledge sharing in detecting fraud is substantially reduced due to defense and confidentiality concerns, particularly in the aftermath of public data cracks.

Some researchers in this study had to rely on fictional datasets that attempted to imitate real-world data [21]. Synthetic data may not be sufficient because legitimate and fraudulent conduct profiles change over time. There are various cryptic methods for preserving data linkages, but the data owner must ensure that the original data is not replicated [22].

Some rules, such as the EU General Data Protection Regulation, prevent such data from leaving national boundaries and ensure data security [23]. [25] presented a standard logistic regression in which a fraud detector trained on transactions, with additional derived fields, produce aggregated statistics across specified periods.

The review by [7, 12, 20] concentrated on current computational strategies for investigating various forms of anomalies in online social networks (OSNs). They divided the process of OSNs detecting anomalies into two phases: (i) selecting and calculating network characteristics and (ii) classifying annotations based on this feature space.

According to a literature assessment, most of the research focuses on the technical understanding of virtual fraud. In response to the highlighted research need, the paper seeks to combat technology-induced fraud assault (TIFA) and possible solutions by incorporating the various technical aspect collected from the literature review. Thus, the study is a novel attempt that consists of know-how associated with TIFA, business intelligence, monetary fraud, data mining and feature detection that may offer corrective remedies for a possible solution of detecting virtual fraud.

3 Objectives of the Study

The current research is based on theoretical modal, theories and secondary data sources gathered from reliable sources. Journals, publications, yearly reports, and books are among the sources. Having studied various articles, white papers, annual reports based on technology-induced fraud, the following objectives have been considered for the study:

- To explain and comprehend various types of fraud over Internet technologies.
- To find out the corrective remedies to minimize fraudulent activities on the virtual platforms.

4 Attack Methods in Telecommunication Frauds

4.1 Wangiri Phoney

It is a kind of hoax call that was initially published in 2002 in Japan. When an attack is carried out through Wangari, the attacker persuades receivers to pay a fee to call a domestic or international telephone line. The attacker made a short appeal of the victim's telephone and led curious callers to think that the authorized caller missed a call and redefined their numbers. They are then compensated for this call at the premium rate [26]. The customer is unaware that they have been duped until they get their regular bill from the service vendor.

4.2 Simbox or Bypass Fraud

The regulators and the telecommunications service vendor are the most expensive types of fraud affecting Bypass or SIM Box fraud. Bypass fraud is typical in areas or regional boundaries where international call prices are much higher than local fixed-line or mobile call fees. The cost of fraud is projected to be $4,3 billion each year. In such fraud, betrayers illegitimately set up or put a SIM Case on their premises. This case/link redirects all the international calls to the pre-set instrument. It is displayed as a call from a local fixed mobile or telephone exchange, avoiding the fees charged to the regulator on a long-distance international call [24].

4.3 International Revenue Sharing Fraud

Fraudsters commit fraud based on virtual revenue sharing illegally hacking or persuading a callee to dial a quality numeral. This results in financial losses not only for telecoms providers but also for end-users and small enterprises. There are two ways to defraud the IRSF; (i) A fraudster may use a stolen identity and insist them call back the assigned numbers. (ii) The fraudster will make calls for a premium number from the international supplier using the resources of the hacked telecoms operator or the company's private branch exchange [23].

4.4 Robo and Telemarketing Calls

A phone call employs a computerized auto dialer and a telecommunication medium to convey a post-telemarketing message to the call recipient [22]. Political contributions are received, holiday parcels are provided, political and religious beliefs are pushed, and legal and illicit things are marketed during these calls. These calls can be made any time of day and demand an immediate response, making it inconvenient to call receivers while at work, disrupting their family time [8].

4.5 Over the Top Bypass

Interaction applications that run over the Internet are known as over the top (OTT). Instant chat, Internet telephone, and streaming video are among the services they offer. People frequently utilize OTT applications to connect with pals and make long-distance, low-cost, or free phone calls. WhatsApp App is the most popular OTT app to communicate audio and instant messaging with more than 1.5 billion months of active users [4, 14].

5 Ways to Mitigate Frauds

Identity spoofing is common and can mimic legitimate entities such as banks, social safekeeping authorities, and tax branches. People are unaware of the work, credentials, and verification processes of the main public infrastructure and transport layer, which may be checked through social networks, the web, and email utilizing protocols on the main public infrastructure and transport layer [17, 25]. Even if the protocol activities are unknown, there will be an increase in injuries and social engineering casualties. Criminals take advantage of this medium to swindle users. The telephonic network also communicated sensitive mission and personal exchange data, such as two-factor authentication and exchanging one-time bank confirmation codes. Further, criminals exchange the codes for cracking the blockchain [16]. The telephony technique is secure to some extent, but it lacks an identity verification mechanism that allows handlers to see the other end.

5.1 Spear Phishing

Cybercriminals are astute, and they employ novel tactics to circumvent their victims' confidence. Spear phishing is one such approach used by cybercriminals, in which they pretend to be a member of the victim's organization, corporation, or social network group. The attacker uses spear-phishing to persuade victims to reveal personal information, which is then utilized for financial fraud. In spear phishing, the target is likely to recognize the email source, which leads him to believe the information in the email. Thus, users must be educated on the importance of not divulging personal information to an unidentified individual/organization while doing online transactions [5, 15].

5.2 Collaboration

The application of collaboration between service providers and handlers could dramatically augment cybercriminal protection. However, no structure exists that allows financial firms, government agencies, and ultimate users to collaborate efficiently to defend themselves against cyber criminals [18, 26]. Furthermore, businesses are still unwilling to comply due to privacy concerns and are competing. Designing a collaborative system is to ensure that the collaborators' privacy and integrity are protected. A cryptographic system and blockchain are methods to enable privacy-preserving collaboration [21].

5.3 Cyber Education and Training

Cybercriminals frequently use social engineering tactics and emotional appeals. Elderly individuals and techno-untouched fraternity are the easy targets of fraudsters in the present chaotic scenario. Financial firms and organizations must have an apparatus to train users when a consumer registers with any virtual transaction. Furthermore, financial institutions should use two-factor verification when transferring money from elderly customers [11, 14].

6 Conclusions and Discussion

Every year, clients who use Internet technology lose millions of dollars due to cyber criminals worldwide. Similarly, service vendors and authorities have invested substantial amounts in building a defense system to protect customers against fraud and hostile behavior of cybercriminals in the past few years. There is currently small evidence available regarding how crooks operate and how fraud occurs using ICT. This study investigates fraudsters' attack mechanisms and the safeguards put in place by service vendors to hedge the clients from fraudulent transactions. Because of their significant penetration and application in everyday life activities and interconnection, the researchers chose three technologies: credit card fraud, mobile payment, and e-fraud. To that end, the researcher attempted an inclusive conclusion on the tactics employed by fraudsters to deceive consumers using this machinery. The paper also discussed the economic impact of such fraud and rendered a protection mechanism to be followed by users and service providers (SP). Besides, the service providers should maintain an up-to-date defense system and educate their clients regularly. The regulator must create a collaborative stand for efficient attack detection where SP can communicate attack mechanisms and weaknesses of their existing model. According to a recent researcher viz, customers should also be educated and suggested not to disclose personal information to an unknown individual over the Internet. [10, 16, 19, 26], embracing collective protection and client training will effectively protect customers from cybercrime. Future research could investigate the possibility of inventing cleverly derived instruments to assist in better classifying transactions. The study developed attributes based on previous research, but a more extensive investigation of features best suited for fraud modelling, including the issue of transaction aggregation, which could be valuable in the future.

References

1. Gulati, R., Goswami, A., Kumar, S.: What drives credit risk in the Indian banking industry? An empirical investigation. Econ. Syst. **43**(1), 42–62 (2019). https://doi.org/10.1016/j.ecosys.2018.08.004
2. Yacob, P., Wong, L.S., Khor, S.C.: An empirical investigation of green initiatives and environmental sustainability for manufacturing SMEs. J. Manuf. Technol. Manag. **30**(1), 2–25 (2019). https://doi.org/10.1108/JMTM-08-2017-0153
3. West, J., Bhattacharya, M.: Intelligent financial fraud detection: a comprehensive review. Comput. Secur. **57**, 47–66 (2016). https://doi.org/10.1016/j.cose.2015.09.005
4. Fadaei Noghani, F., Moattar, M.: Ensemble classification and extended feature selection for credit card fraud detection. J. AI Data Min. **5**(2), 235–243 (2017)
5. Kamboj, M., Gupta, S.: Credit card fraud detection and false alarms reduction using support vector machines. Int. J. Adv. Res. Ideas Innov. Technol. **2**(4), 1–10 (2016)
6. Kumar, N., Dadhich, M.: Risk management for investors in stock market. Excel Int. J. Multidiscip. Manag. Stud. **4**(3), 103–108 (2014)
7. Subudhi, S., Panigrahi, S.: Use of fuzzy clustering and support vector machine for detecting fraud in mobile telecommunication networks. Int. J. Secur. Netw. **11**(1), 3–11 (2016). https://doi.org/10.1504/IJSN.2016.075069
8. Craja, P., Kim, A., Lessmann, S.: Deep learning for detecting financial statement fraud. Decis. Support Syst. **139**(September), 113421 (2020). https://doi.org/10.1016/j.dss.2020.113421

9. Nelson, L.A.: 2021 Nilson report (2020)
10. Leong, L., Hew, T., Tan, G.W., Ooi, K.: Predicting the determinants of the NFC-enabled mobile credit card acceptance: a neural networks approach. Expert Syst. Appl. **40**(14), 5604–5620 (2013). https://doi.org/10.1016/j.eswa.2013.04.018
11. Sharma, N., Dadhich, M.: Predictive business analytics: the way ahead. J. Commer. Manag. Thought **5**(4), 652 (2014). https://doi.org/10.5958/0976-478x.2014.00012.3
12. Chikkamannur, A., Kurien, K.L.: Detection and prediction of credit card fraud transaction using machine learning. Int. J. Eng. Sci. Res. Technol. **8**(3), 199–208 (2019). https://doi.org/10.5281/zenodo.2608242
13. Nagaraj, K.: Detection of phishing websites using a novel twofold ensemble model. J. Syst. Inf. Technol. **20**(3), 321–357 (2018). https://doi.org/10.1108/JSIT-09-2017-0074
14. Dadhich, M., Pahwa, M.S., Rao, S.S.: Factor influencing to users acceptance of digital payment system. Int. J. Comput. Sci. Eng. **06**(09), 46–50 (2018). https://doi.org/10.26438/ijcse/v6si9.4650
15. Sharma, S., et al.: Implementation of trust model on CloudSim based on service parametric model. In: Proceedings of 2015 IEEE International Conference on Research in Computational Intelligence and Communication Networks, ICRCICN 2015, pp. 351–356 (2015)
16. Dadhich, M., Chouhan, V., Adholiya, A.: Stochastic pattern of major indices of Bombay stock exchange. Int. J. Recent Technol. Eng. **8**(3), 6774–6779 (2019). https://doi.org/10.35940/ijrte.C6068.098319
17. Rao, S.S., Kumar, N., Dadhich, M.: Determinant of customers' perception towards RTGS and NEFT services. Asian J. Res. Bank. Financ. **4**(9), 253–260 (2014). https://doi.org/10.5958/2249-7323.2014.00960.2
18. Humpherys, S.L., Mof, K.C., Burns, M.B., Burgoon, J.K., Felix, W.F.: Identification of fraudulent financial statements using linguistic credibility analysis. Decis. Support Syst. **50**, 585–594 (2011). https://doi.org/10.1016/j.dss.2010.08.009
19. Merlini, D., Rossini, M.: Text categorization with WEKA: a survey. Mach. Learn. Appl. **4**(November), 100033 (2021). https://doi.org/10.1016/j.mlwa.2021.100033
20. Gianotti, E., Damião, E.: Strategic management of credit card fraud: stakeholder mapping of a card issuer. J. Financ. Crime **28**(1), 156–169 (2019). https://doi.org/10.1108/JFC-06-2020-0121
21. Kolli, C.S., Tatavarthi, U.D.: Fraud detection in bank transaction with wrapper model and Harris water optimization-based deep recurrent neural network. Kybernetes, Emerald Publishing Limited (2020). https://doi.org/10.1108/K-04-2020-0239
22. Piri, S., Delen, D., Liu, T.: A synthetic informative minority over-sampling (SIMO) algorithm leveraging support vector machine to enhance learning from imbalanced datasets. Decis. Support Syst. **106**, 15–29 (2018). https://doi.org/10.1016/j.dss.2017.11.006
23. Lokanan, M., Tran, V.: Detecting anomalies in financial statements using machine learning algorithm the case of Vietnamese listed firms. Asian J. Account. Res. **4**(2), 181–201 (2019). https://doi.org/10.1108/AJAR-09-2018-0032
24. Avortri, C., Agbanyo, R.: Determinants of management fraud in the banking sector of Ghana: the perspective of the diamond fraud theory. J. Financ. Crime **28**(1), 142–155 (2021). https://doi.org/10.1108/JFC-06-2020-0102
25. Mahrishi, M., Hiran, K.K., Meena, G., Sharma, P. (eds.): Machine Learning and Deep Learning in Real-Time Applications. IGI Global (2020). https://doi.org/10.4018/978-1-7998-3095-5
26. Fu, K., Cheng, D., Tu, Y., Zhang, L.: Credit card fraud detection using convolutional neural networks. In: Hirose, A., Ozawa, S., Doya, K., Ikeda, K., Lee, M., Liu, D. (eds.) ICONIP 2016. LNCS, vol. 9949, pp. 483–490. Springer, Cham (2016). https://doi.org/10.1007/978-3-319-46675-0_53

Blockchain Enable Smart Contract Based Telehealth and Medicines System

Apoorv Jain[1] and Arun Kumar Tripathi[2(✉)]

[1] Chegg India, New Delhi, India
[2] KIET Group of Institutions, Delhi-NCR, Ghaziabad, India
mailtoaruntripathi@gmail.com

Abstract. Healthcare system is one of the most prime concerns for mankind. The pandemic compelled to search new alternative solutions for present healthcare system and results in telehealth. The paper proposed a model for the "Telehealth and Medicine System" based on one of the most powerful transparent, decentralized, and secured technology i.e., "Blockchain". The basic aim of the proposed system is to provides the best remote healthcare facilities to people speedily and securely. The model also supports for managing the healthcare resources information for controlling the huge load of patients in hospitals. Present centralized system fails to maintain security, confidentiality, functional transparency, health-records immutableness, and traceability of medical apparatus and medicines. The present centralized system is also unable to detect scams correlated to patient's transaction prerogatives and doctor authorized prescription data. In this paper we explore the possible techniques and opportunities of health care system using blockchain technology in telehealth and medicine area. This paper provides the illustration of blockchain technology which gives essential data safety and confidentiality to this model. Moreover, the paper deals with secure functioning-transparency, health-records immutableness, and traceability of medical apparatus and medicines information. Smart Contract is one of the most popular and secure application of 'Blockchain'. It is based on public-blockchain framework. Blockchain technology can improve telehealth and medicine system by offering distant healthcare facilities on a decentralized platform along with tamper-proof transactions, transparency of data, flexible traceability of records, reliability to patients, and high security. This model identifies whole scenario of medicine and apparatus production from initial to final stage of delivery to the patient. This model also leads health specialists and doctors for treating patients correctly and share the prescriptions securely over the network. At the last, paper discuss numerous opportunities and techniques needs to be resolute the present system by acceptance of power decentralized technology 'Blockchain' in telehealth and health-care systems.

Keywords: Blockchain · Telehealth · Ethereum · Smart Contract · Peer-to-Peer

1 Introduction

The current pandemic of Coronavirus (COVID-19) teaches to review our current health system and requirements for establishing a consistent, strong, and full-bodied patientcare

V. E. Balas et al. (Eds.): ICETCE 2022, CCIS 1591, pp. 724–739, 2022.
https://doi.org/10.1007/978-3-031-07012-9_60

and health-service system. This coronavirus-pandemic lifts the interest of tele-health and distant-medicine tech. This system provides securely empower interaction with doctors and wellbeing experts over virtual networks for minimizing the blowout of infection. This system also able to provides better advise treatment of any issue as well as provide the full contact less medicine services. In current time many big corporations and organizations are working in this direct and provide the telehealth services. Companies and organizations like: Care-Health, Practo, and Doctor 24×7, etc. These have freshly saw a rapid surge in petition for tele-health and medicine-facilities for combating the feast of the coronavirus. Telehealth and distant sessions permit well-organized health care admittance and suggestion a healthier attention management and dealing-treatment results.

Telehealth and medicine services allows health-care specialists to virtually screen, analyze, and evaluate patients by providing cost-effective facilities, thus minimalizing patient admittance and workforce limitations, expanding technology capabilities, and mitigating the risk of exposure of doctors, nursing-staff, or patients in the pandemic time. Also, telehealth increases employments opportunities in the IT digital data platform and communication knowledges technologies. This is also help patients for managing their disease complete better-quality selfcare and admittance to education and care structures. The chief profits of the telehealth and medicine systems are that disclose computer-generated healthcare structures have the possible to positively alleviate.

In current time these telehealth services and system are based on 'Centralization Systems'. This current system existing in applications and systems is increases the danger of solo fact of letdown. As well as the information in existing applications are disposed to a diversity of outside and interior information cracks. This centralization system makes users for negotiating the dependability and obtainability of applications. In the centralized system there is high involvement of third parties and large risk of malicious attacks for stealing the patient and other transaction information. The decentralized-ledger technology 'Blockchain' can helps for addressing and improving these such critical difficulties.

The evolving decentralized-blockchain tech trails a distributed-manner for managing a public record of health-care data amongst various applicants. In blockchain tech, entirely every ledger facsimile is saved, proved and tamper-proof with each data-node. Lots of current challenges like trailing the places visited by patients, defending distant doctor-patient discussion information, tracking medicines and health-medical instruments & apparatus across the supply-chain management, authenticating the identifications of doctors, and verifying the source of productions health-medical instruments & apparatus, etc. are amongst various issues are discourse and improving thru the blockchain-technology [1].

1.1 Introduction to Blockchain

Blockchain is one of the most powerful distributed-ledger expertise which comprises entirely information in the form of ledger on a distributed network. These information's are kept in the node-blocks which links to an exclusive unique-hash value. Individually every node-block state to the previous data-block and composed in the form of chain manner. Therefore, the whole data-chain is formed which is called 'Blockchain'. Blockchain expertise is the speediest and most protected tech for sharing the information & material over the web. Respectively, every data-block might comprise a transaction-record or any additional information created on the system-application. These decentralized applications on Blockchain platform are known as DApps (Decentralized applications). Blockchain is a public distributed-ledger tech that allows user to save and transmission of information in a tamper-proof, provable, and unchallengeable mode. On a blockchain, the transaction-record of every data-node is existing on completely data-nodes, which are linked on a Peer-to-Peer (P2P) net which upsurges the accessibility of data rapidly [1]. Blockchain supports several features along with transparency and security (see Fig. 1).

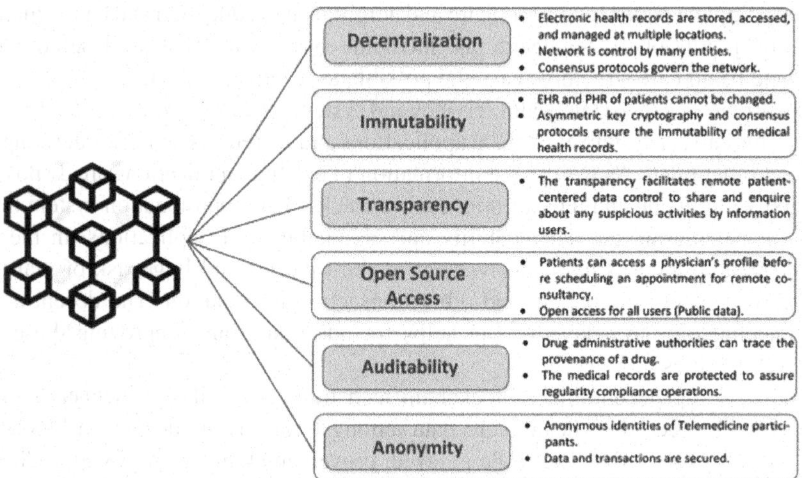

Fig. 1. Features of blockchain

Thus, the information is tamper-proof and transparent so, it can be simply proved by all the applicants. The perception of cryptography safeguards that all the information on Blockchain is unchallengeable i.e., Tamper-proof. Blockchain is divided into four categories, and each has unique features (see Fig. 2).

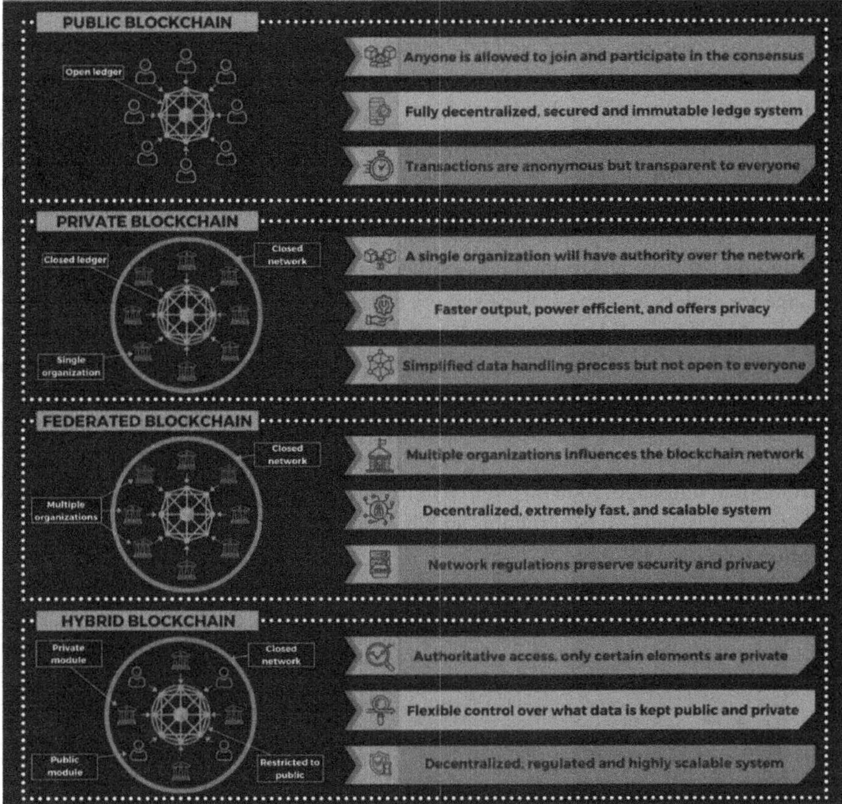

Fig. 2. Categorization of blockchain and features

1.2 Overview of Ethereum and Smart Contract

Ethereum is one of the greatest open-source public-blockchain framework. Ethereum holds one of the most powerful applications in smart-digitalized ledger-agreements called as Smart Contracts. Smart Contract comprises completely transaction information in encrypted-procedure on the blockchain-net. Many developers and creators comparable for developing and design the decentralized applications (DApps) on this blockchain framework. Ethereum is a permission-less & open-free system blockchain which previously uses Proof-of-work (PoW) consensus-procedure for proofing all data-block nodes. But now this framework uses the Proof-of-Stack (PoS) algorithm. This framework easily compatible with Python, Go, Node, JavaScript, and Solidity, etc. programming-languages for designing apps [2, 3].

Solidity is contract-oriented high-level contract design-language. Solidity is particularly planned for scripting and managing the hi-tech smart-distributed-ledgers called smart-contracts. This linguistic compatible easily with both blockchain that is: private as well as public. This open-free framework is extremely supporting this contract-language for designing and implementing the smart-contracts in the decentralized-applications (Dapps).

A smart-contract in the Solidity-linguistic is an assembly of program-code and it illustrates that exist this is present on definite-address on the Ethereum-blockchain framework. In this paper, we write four smart-contract of this model where we allocated the information such as Medicine data, Transaction-hash, addresses of producer, addresses of supplier, addresses of patient, addresses of doctor, disease-information, and prescription-information, etc.

For writing and deploying the contracts we use the most popular IDE called as Remix-IDE. The remix is a collection of gears to relate with the Ethereum-blockchain for debugging the full program-code and kept it in the Git-repository. GitHub is an open stage for coders and program-designers uses Git-repository. On this entirely every program-codes of apps and software-codes are kept at the cloud-stage. The Remix-IDE is an IDE for Solidity-DApp developers. The Remix-IDE is an open-online platform which is accessible by using the link: https://remix.ethereum.org. The remix is a web-browser built-compiler and IDE that allows workers for building Ethereum-Smart-Contracts thru Solidity-language also for debugging contacts. We have castoff this IDE for implementing our Solidity-language-code through 'extension.sol' which deploys entirely four contracts for transmission [2]. Smart contract has several benefits and use-cases in real life application (see Fig. 3).

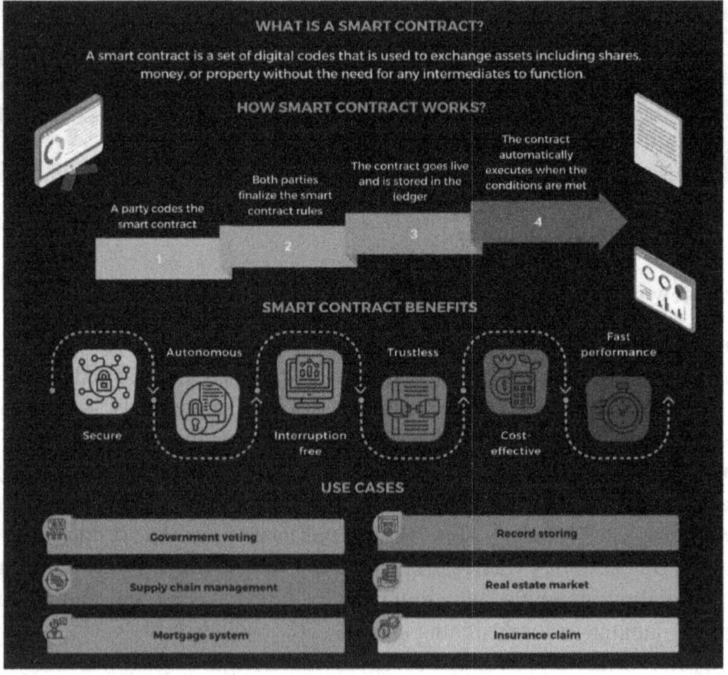

Fig. 3. Smart contract working, benefits and use cases

1.3 Consensus Algorithm

In the current time the Ethereum framework uses the Proof-of-Stake (PoS) consensus method which arises after Proof-of-Work (PoW). Equally PoW and PoS procedures have the similar effort for validating every data-block of the blockchain but PoS quite modest than PoW. This algorithm uses the staked procedure for securing each data-block of information on the net. PoS used fewer calculation hash-power than the PoW. PoS practices broadly for Ethereum System communications and transaction-information which direct from the sending person to the acknowledger on the blockchain net. PoS primary differs the verification and formerly authenticate the chunk of nodes of the chain. This shows proof of authenticating and confirming the PoS blockchain-consensus procedure [2–6].

This paper covers the possible prospects that blockchain-tech can powers the telehealth and medicine systems by improving their main drawbacks in terms of dependability, immutableness, transparency-of-data, faith, and safety. In this paper we design a model using smart contract on blockchain that illustrates case studies of providing peer-to-peer and most secure transmission connection in telehealth and medicine system.

The remainder of the paper is organized as follows. Section 2 presents potential opportunities for blockchain technology in telehealth and telemedicine Sect. 3 illustrates the role of smart in the designed model. Section 4 deliberates the conclusions and future enhancement.

2 Opportunities/Areas in Telehealth and Medicine System by Using Blockchain Technology

This section deliberates the vital areas and opportunities in the telehealth and medicine system by using the most powerful decentralized blockchain technology. Main opportunities and areas given below:

2.1 Role of Blockchain Application Smart Contracts in Patient Agreement Management

The efficacy of tele-care and health-nursing be contingent on the honesty of the electronic-health-records (eHRs). This eHRs contain whole affected people health antiquity, treatment information, and medicine details, etc. The eHRs contains extremely confidential and private information, which desires for transmitting the information in the P2P connection mode. The P2P (Peer-to-Peer) is securely setup helps for maintaining up to date the medical information between authorized parties like patient-to-hospitals, doctor-to-pharmacies, and patient-to-pharmacies, and between the high authorities, etc. Current traditional system is totally dependent on the intermediate service providers and third parties.

Blockchain-tech leads for enforcing faith as no mediators are intricate. Blockchain has a feature of computerized ledgers called smart contracts. Smart contracts provide the proof-of-work of eHRs records. Blockchain provides best managing power and establish the most secure P2P mode between the authorized users without involvement of third parties.

2.2 Tracking of Distant Treatment Records

The telehealth and medicine system needs an automated head-on meeting system between the patients and Doctors. This automated system used for providing the operative wellbeing examination to the distant patient. In current telehealth structures, health-administrations are incapable for managing the patient health information. All administrations are dependent on third party services which holds the data and records of patient. These parties also provide less storage for saving all patients information's. Blockchain-tech delivers a solitary and intelligible decentralization approach. Through this decentralization, all eHRs of patients are hold securely and in bulk quality without involvement of third parties on the network. This increases the faith, perceptibility, and transparency of patient-records. Smart-contracts empower the proof-of-work ledger for tracing the medical past-history of the remote-patients.

2.3 Tracking of Medicines, Medical-Apparatus, and Instruments

Current centralized telehealth and medical structures has deficiency of transparency, perceptibility, and temper-proof information. Current systems fail for tracking supply chain information about the medicines, medical-apparatus, and instruments. Blockchain-tech provides transparency and clearly tacking information of supply-chain. Blockchain linked the whole information of the supply-chain from the raw material supplier to final product and final product from shopkeeper to patient securely in the ledger. Smart-contract ledgers are design for recording and managing the whole information securely [6–8].

2.4 Fast and Secure Payment System

The existing telehealth structures regularly hire central third-party facilities for transmitting transactions between patients, hospitals, doctors, and other functions for using facilities. Still, the central transaction payment procedures are sluggish, possibly defenseless for hacking, and non-transparent. In central system, lots of transactions are occurs which are hard to detect and recover. Blockchain-tech stand provides crypto-currency tokens-based transaction system. This system is highly secure by using many cryptographic algorithms. Blockchain provides the straight transmission of transaction form wallet in P2P mode. This P2P provides a rapid, safe, transparent, and tamper-proof network structure that does not requirement a central third-party service for resolving the transmission of transactions [9, 10].

3 Role of Smart Contract

Smart Contract is the one of major player in this Telehealth and medicine service system which resolve all the entire issues of current centralized system. In this model four contracts are design which shows the transmitting of the information of medicine and medical apparatus from beginning stage to ending stage i.e., from Raw Material Supplier to final medicine information receiver (the Patient). In this smart contract whole data and information is present in encrypted form and only the authorized users seem the actual data. In the smart contract all the transmission information is immutable and tamper-proof on the blockchain network. This section deals with smart contract codes and their final representation.

3.1 Supplier to Producer Contract

This is the first smart contract which illustrate whole transaction information between the 'Raw Material Supplier' and the Production company of medicine and apparatus (Producer). This smart contract stores the information like Supplier Name, address of Supplier, address of Producer, Producer Name, Medicine salt information and apparatus, additional material information, quantity of the material and total transaction amount information. This contract provides the detail tracking and validating information transmitting between both users. In this the 'Supplier' has only authority to change the information in the contract. The code of smart contract from supplier to producer is depicted (see Fig. 4).

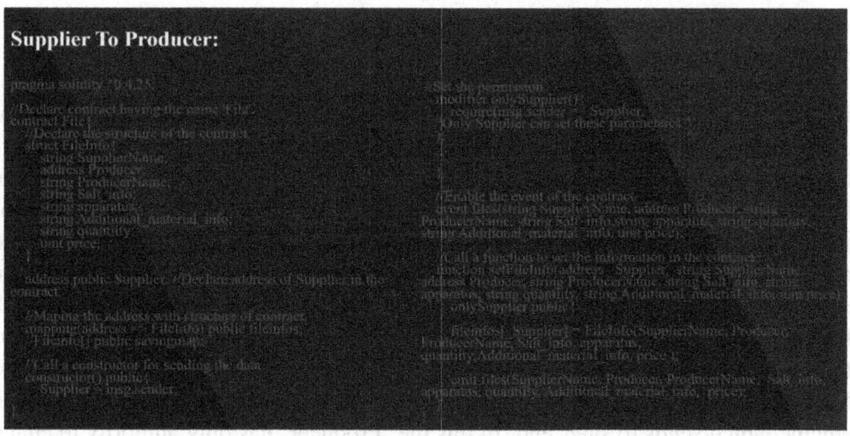

Fig. 4. Code for supplier to producer smart contract

3.2 Final Contract: Supplier to Producer

After the execution of the final smart contracts is generated, in which entirely whole information are existing in encrypted manner. The original information is only seen by the 'Supplier' and the 'Producer'. This contract is view by everyone, but the transmission input-data is only clear by the authorized-party. The final contract from supplier to producer is depicted (see Fig. 5).

Fig. 5. Final smart contract from supplier to producer

3.3 Producer to Wholesaler Contract

This is the second smart contract which illustrate whole transaction information between the 'Producer' and the 'Wholesaler'. This smart contract stores the information like Producer Name, address of Producer, address of Wholesaler, Wholesaler Name, Medicine API compound information, Medicine Exception information and expiry date information of medicine and kit, and price of medicine etc. Medicine API compound is the most important information which conveys the information of efficacy of the medicine as well as the salt composition of the medicine. This API compound also conveys the information the color of medicine. The second most important information is 'Exception information' which conveys the all the data of effect of medicine like allergies, and side effects etc. This contract provides the detail tracking and validating information of the medicine which sends to sales end. In this the 'Producer' has only authority to change the information in the contract. The code of smart contract from producer to wholesaler is depicted (see Fig. 6).

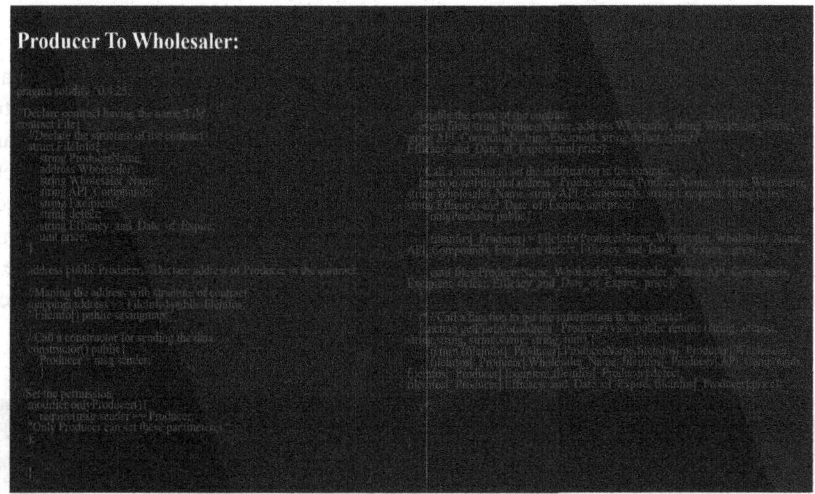

Fig. 6. Code for producer to wholesaler smart contract

3.4 Final Contract: Producer to Wholesaler

After the execution of the final smart contracts is generated, in which entirely whole information are existing in encrypted manner. The original information is only seen by the 'Producer' and the 'Wholesaler'. This contract is view by everyone, but the transmission input-data is only clear by the authorized-party. The final contract from producer to wholesaler is depicted (see Fig. 7).

Fig. 7. Final smart contract from producer to wholesaler

3.5 Wholesaler to Retailer Contract

In this smart contract illustrate whole transaction information between the 'Wholesaler' and the retail shopkeeper of medicine and apparatus (Retailer). This smart contract stores the information like Producer Name, address of Wholesaler, Wholesaler Name, address of Retailer, Retailer Name, Medicine information & apparatus, additional defect information of medicine, efficacy, and expiry date information as well as the price of medicine. This contract provides the detail tracking and validating information transmitting between both users. In this the 'Wholesaler' has only authority to change the information in the contract. The code of smart contract from wholesaler to retailer is depicted (see Fig. 8).

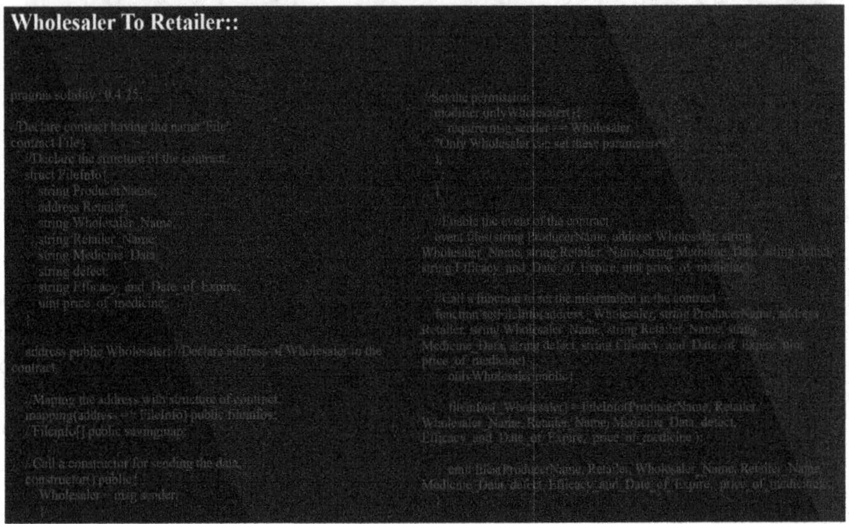

Fig. 8. Code for wholesaler to retailer smart contract

3.6 Final Contract: Wholesaler to Retailer

After the execution of the final smart contracts is generated, in which entirely whole information are existing in encrypted manner. The original information is only seen by the 'Retailer' and the 'Wholesaler'. This contract is view by everyone, but the transmission input data is only clear by the authorized party. The final contract from wholesaler to retailer is depicted (see Fig. 9).

Fig. 9. Final smart contract from wholesaler to retailer

3.7 Doctor to Patient Contract

This is the most important smart contract which illustrate whole discussion information between the 'Doctor' and the 'Patient'. This smart contract stores the information like Doctor Name, address of Doctor, address of Patient', Patient Name, Problem and disease information, Prescription information, and consultant fee information. This contract provides the detail tracking and validating information transmitting between both. In this the 'Doctor' has only authority to change the information in the contract. The code of smart contract from doctor to patient is depicted (see Fig. 10).

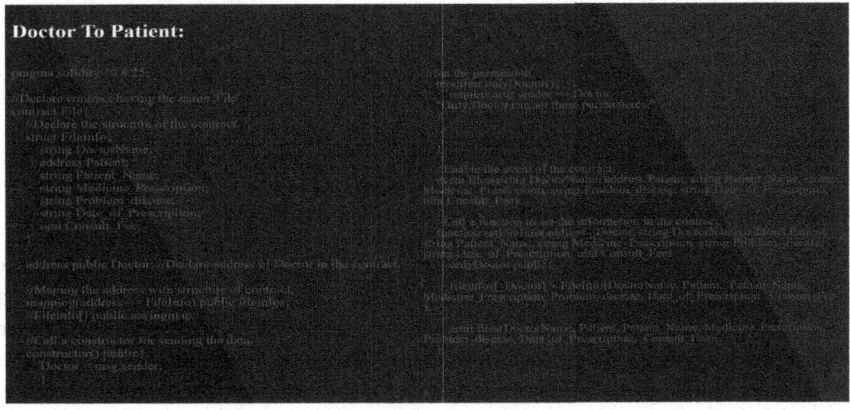

Fig. 10. Code for doctor to patient smart contract

3.8 Final Contract: Doctor to Patient

After the execution of the final smart contracts is generated, in which entirely whole information are existing in encrypted manner. The original information is only seen by the 'Doctor' and the 'Patient'. This contract is view by everyone, but the transmission input data is only clear by the authorized party. The final contract from doctor to patient is depicted (see Fig. 11).

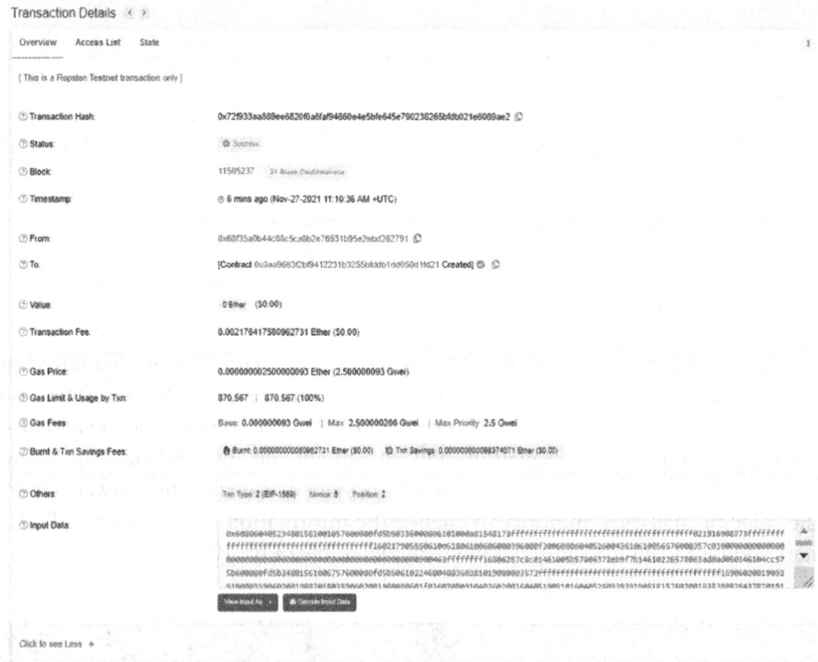

Fig. 11. Final smart contract from doctor to patient

3.9 Retailer to Patient Contract

This is the final smart contract which illustrate whole transaction information between the 'Retailer', the final seller of the medicine & apparatus and the Patient. This smart contract stores the information like Retailer Name, address of Retailer, address of Patient, Doctor Name, Medicine & apparatus billing information, date of bill, and amount information. This contract provides the detail tracking and validating information transmitting between both. In this the 'Retailer' has only authority to change the information in the contract. The code of smart contract from retailer to patient is depicted (see Fig. 12).

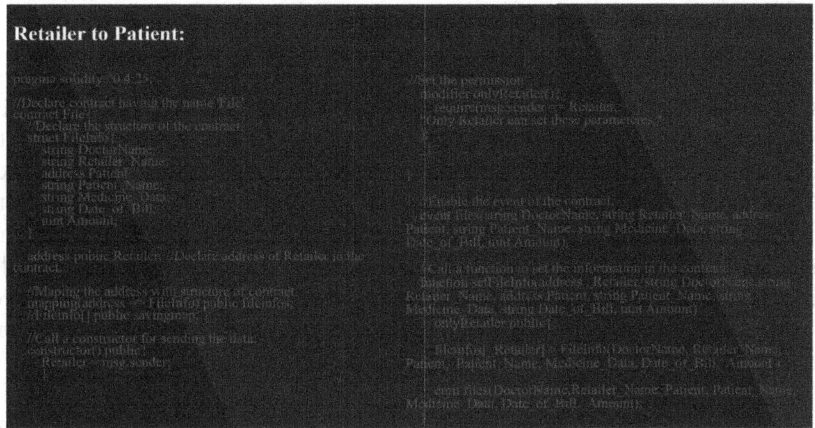

Fig. 12. Code for retailer to patient smart contract

3.10 Final Contract: Retailer to Patient

After the execution of the final smart contracts is generated, in which entirely whole information are existing in encrypted manner. The original information is only seen by the 'Retailer' and the 'Patient'. This contract is view by everyone, but the transmission input-data is only clear by the authorized-party. The final contract from retailer to patient is depicted (see Fig. 13).

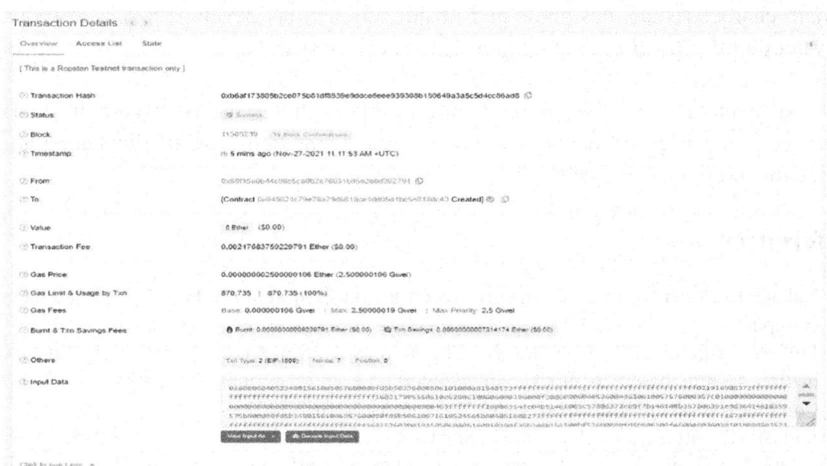

Fig. 13. Final smart contract from retailer to patient

4 Conclusion and Future Enhancement

This paper illustrates the overview of blockchain-tech in telehealth and medicine systems. Paper explains the important structures for providing distant healthcare facilities in easy and secure manner. Blockchain technology has an ability to provide this decentralization, tamper-proof, observable, absolute, and security to this system. In this, the detail exploration and discussion of the possible area & prospects presented by blockchain-tech is telehealth and medicine system are provided. In this, the detail discussion of the most popular application Smart Contract on Ethereum framework of blockchain-technology is given. The four smart contracts are deployed which show the speedy, tamper-proof, and most secure transmission of medicine data and telehealth record over the network. Main areas and opportunities trailed by closing explanations are given below:

- Blockchain-tech can provide an important part for effectively transmitting the health-information between the users. By implementing smart contracts, the data transmits securely, speedily, and tamper-proof manner.
- Second main area is to provides the faster and secure claim payment to distant patient over the network in the P2P connection. The smart contract leads to provide proper proof-of-work leader contract which speedily helps the patient. Thus, exciting revolution in the current blockchain-tech is used for minimizing the payment execution time and highly upsurge its correctness for the health-care area.
- The hereditary derivation characteristic of blockchain-tech can allow health specialists for tacking properly detect authentication details, medicines and medical-apparatus commonly used for distant treatment.
- Blockchain provide the very powerful information safety and confidentiality. By these main characteristics, designers and health originations develops a very appropriate system for digitization & computerization service system.

Finally, in the future we implement and demonstrates the supply chain management and procedures of tracking and monitoring the medicine and medical apparatus in telehealth and medicine system.

References

1. Nakamoto, S.: Bitcoin: a peer-to-peer electronic cash system (2008). https://bitcoin.org/bitcoin.pdf
2. Jain, A., Tripathi, A.K., Chandra, N., Chinnasamy, P.: Smart contract enabled online examination system based in blockchain network. In: International Conference on Computer Communication and Informatics (ICCCI), pp. 1–7. IEEE (2021)
3. Jain, A., Tripathi, A.K., Chandra, N., Shrivastava, A.K., Rajak, A.: Business service management using blockchain. In: International Conference on Issues and Challenges in Intelligent Computing Techniques (ICICT), pp. 1–6. IEEE (2019)
4. Campanella, P., et al.: The impact of electronic health records on healthcare quality: a systematic review and meta-analysis. Eur. J. Pub. Health **26**(1), 60–64 (2016)
5. Ben Fekih, R., Lahami, M.: Application of blockchain technology in healthcare: a comprehensive study. In: Jmaiel, M., Mokhtari, M., Abdulrazak, B., Aloulou, H., Kallel, S. (eds.) ICOST 2020. LNCS, vol. 12157, pp. 268–276. Springer, Cham (2020). https://doi.org/10.1007/978-3-030-51517-1_23

6. Cong, L.W., He, Z.: Blockchain disruption and smart contracts. Rev. Financ. Stud. 1 **32**(5), 1754–1797 (2019)
7. Krittanawong, C., et al.: Integrating blockchain technology with artificial intelligence for cardiovascular medicine. Nat. Rev. Cardiol. **17**(1), 1–3 (2020)
8. O'Donoghue, O., et al.: Design choices and trade-offs in health care blockchain implementations: systematic review. J. Med. Internet Res. **21**, e12426 (2019)
9. Franco, P.: Understanding Bitcoin: Cryptography, Engineering and Economics. Wiley, Chichester (2015)
10. Sharma, G., et al.: Reverse engineering for potential malware detection: Android APK Smali to Java. J. Inf. Ass. Sec. **15**(1), 26–34 (2020)

Author Index

Printed in the United States
by Baker & Taylor Publisher Services

Printed in the United States
by Baker & Taylor Publisher Services